Developments in Soil Classification,
Land Use Planning and Policy Implications

Shabbir A. Shahid • Faisal K. Taha
Mahmoud A. Abdelfattah
Editors

Developments in Soil Classification, Land Use Planning and Policy Implications

Innovative Thinking of Soil Inventory for Land Use Planning and Management of Land Resources

Springer

هيئة البيئة - أبوظبي
Environment Agency - ABU DHABI

Editors
Shabbir A. Shahid
International Center for Biosaline Agriculture
Dubai, United Arab Emirates

Faisal K. Taha
International Center for Biosaline Agriculture
Dubai, United Arab Emirates

Mahmoud A. Abdelfattah
Soil and Land Use Management
Environment Agency - Abu Dhabi
Abu Dhabi, United Arab Emirates

ISBN 978-94-007-5331-0 ISBN 978-94-007-5332-7 (eBook)
DOI 10.1007/978-94-007-5332-7
Springer Dordrecht Heidelberg New York London

Library of Congress Control Number: 2013931842

Printed on acid-free paper

Springer is part of Springer Science+Business Media (www.springer.com)

Foreword

Productive and fertile soils are a scarce resource in the arid and especially in the hyperarid desert environments. Using an internationally accepted soil classification system is a prerequisite for a common international and national understanding about soil typologies and their implications for sustainable land use planning.

Recognizing this need, the idea emerged to organize an international conference and invite top-level experts in soil science and related disciplines to discuss technological advancements in soil resources and land use planning for better agricultural and environmental outcomes in the future. Enhanced agriculture production and more sustainable outcomes are critical to improve food security.

The international conference, "*Soil Classification and Reclamation of Degraded Lands in Arid Environments*," was held under the patronage of H.H. Sheikh Hamdan Bin Zayed Al Nahyan (Ruler's Representative in the Western Region Abu Dhabi and Chairman of Environment Agency – Abu Dhabi (EAD)) in 17–19 May 2010. It proved to be the ideal forum to share and discuss with national and international audience the research findings and implications of these outcomes of the "Soil Survey of Abu Dhabi Emirate." The survey had been jointly completed by Environment Agency – Abu Dhabi and the Dubai-based International Center for Biosaline Agriculture through a contractor GRM International Pty Australia during 2006–2009. The survey findings published as the *Abu Dhabi Soil Survey Report* were officially launched at the conference.

During the conference, broad issues related to the advancement of soil classification systems, new trends in land degradation and desertification, and land use planning and policy directions were raised. It is these topics which are presented in this book to assist in a better understanding of the role of the soil in land use planning to ensure sustainable environmental and agricultural use.

I wish to thank H.E. Mohammed Al-Bowardi (Managing Director EAD), H.E. Majid Al Mansouri, then Secretary General EAD, and Her Excellency Ms Razan Al-Mubarak (Secretary General EAD) for their inspiration and respective endeavors to ensure the success of the conference and subsequent publication of its proceedings.

Director General, International Center for Biosaline Agriculture
Dubai, United Arab Emirates.

Shawki Barghouti

Preface

This book is the outcome of an International Conference on Soil Classification and Reclamation of Degraded Lands in Arid Environments held during 17–19 May 2010 in Abu Dhabi, United Arab Emirates. The Environment Agency – Abu Dhabi (EAD) and Dubai-based International Center for Biosaline Agriculture (ICBA) jointly organized the conference. The main objective of the conference was to bring renowned scientists, educators, and policy makers to share and discuss conference technical themes in broader perspectives and assist EAD in paving the way forward for future activities in soil and land management in the Emirate of Abu Dhabi.

We received overwhelming response to the call for papers, and over 250 abstracts were received from over 35 countries. The abstracts were reviewed, and those suitable were accepted for the submission of full manuscripts. The diversity of the conference themes made it necessary to publish these papers into two independent books. Prior to publication of the book, all preselected chapters went through rigorous technical review and through an iterative review process with authors before finalization for publication.

The chapters published in this book, *Developments in Soil Classification, Land Use Planning and Policy Implications: Innovative Thinking of Soil Inventory for Land Use Planning and Management of Land Resources*, represent part of the conference proceedings. The rest of the chapters are published in a separate book, *Developments in Soil Salinity Assessment and Reclamation: Innovative Thinking of Using Marginal Soil and Water Resources in Irrigated Agriculture*.

In this book, chapters pertaining to soil survey and classification, land use planning and policy implications, new trends in land degradation and desertification, modeling of soil and groundwater contamination, and innovations in research, development, education, and extension are presented in five parts divided into 50 chapters.

Part I deals with national soil inventories and advances in digital soil mapping, and examples from Abu Dhabi Emirate, Australia, Central Asia, Dubai Emirate, Egypt, Indonesia, Iran, Kuwait, Nigeria, Siberia, Sudan, Uzbekistan, and the USA are included. Part II deals with land use planning and policy implications, and topics such as land use planning and policy development, general land use framework, integrated agricultural land use, agricultural land conversion, common land resources, and participatory land evaluation have been included from countries such as Austria, Australia, Brazil, East Botswana, Egypt, Eritrea, India, Indonesia, Nigeria, South Africa, Southern Kazakhstan, and UAE. Part III presents chapters on new trends in land degradation and desertification from Germany, Italy, Kuwait, and Egypt, including topics such as controlling land degradation, methodological approaches for desertification measurement, managing the hazard of drought and shifting sands in drylands, and soil degradation assessment. Part IV includes chapters on modeling of soil and groundwater contamination from India and Oman. Part V shares innovative work in research, development, education, and extension, including information sharing, innovations in soil chemical analyses, mycorrhiza, environmental isotopes, QA standards in soil survey, and lombricompost use, from India, Jordan, Madagascar, Morocco, Thailand, UAE, and the USA.

We hope the book will be an excellent addition and contribution to the science and knowledge of soil survey and management of land resources in arid regions.

Shabbir A. Shahid
Faisal K. Taha
Mahmoud A. Abdelfattah

Acknowledgements

I would like to express my thanks and appreciation for the assistance received from the managements of Environment Agency – Abu Dhabi (EAD) and International Center for Biosaline Agriculture (ICBA) in organizing the conference and publication of this book. I would like to thank H.E. Majid Al Mansouri, then Secretary General (EAD) and Chairman of the conference, for his support, guidance, and encouragement to convene the conference. The staff from EAD and ICBA were instrumental in the planning and organizing of the conference and deserve recognition and special thanks. The contribution of members of both Steering and Technical Committees is greatly appreciated.

Special thanks go to Dr. Shawki Barghouti (ICBA Director General) for his kind support and approval of funds to publish the proceedings of the conference. Her Excellency Ms Razan Al-Mubarak (Secretary General EAD) endorsed this publication, and we extend our thanks to her. My sincere acknowledgements go to the editors of this book, as well as to Dr. Henda Mahmoudi (Visiting Scientist) and Ghazi Jawad Al-Jabri (Communications Coordinator) whose professionalism, dedication, and careful reviewing facilitated the printing. The Springer Publishing House deserves "thank you" for taking the challenging task of publishing this book and assisting with its distribution.

Last but not least, thanks to all competent scientists and partners for their invaluable contributions that made it possible to produce these proceedings.

Chairman
Technical Committee
International Conference on Soil Classification and
Reclamation of Degraded Lands in Arid Environments

Faisal K. Taha

About the Editors

Shabbir A. Shahid is salinity management scientist at Dubai-based International Center for Biosaline Agriculture (ICBA). He earned his Ph.D. degree from the University of Wales, Bangor, UK, in 1989 and B.Sc. Hons and M.Sc. Hons (soil science) from University of Agriculture, Faisalabad (UAF), Pakistan, in 1977 and 1980, respectively. He joined ICBA in 2004 and has over 30 years experience (Pakistan, the UK, Australia, Kuwait, and United Arab Emirates) in soil-related RD and E activities. He has held positions: associate professor soils (University of Agriculture Faisalabad, Pakistan); associate research scientist in Kuwait Institute for Scientific Research, Kuwait; and manager of soil resources department, Environment Agency – Abu Dhabi, UAE. He is a prolific author of over 150 publications in peer-reviewed refereed journals, proceedings, books, and manuals. He is life member and current councilor of the World Association of Soil and Water Conservation. Dr. Shahid is recipient of Sir William Roberts and David A. Jenkins Awards.

Mahmoud A. Abdelfattah an Egyptian national, earned M.Sc. degree in Soil Survey and Land Evaluation from ITC, the Netherlands (1998), and Ph.D. in soil science from Cairo University, Egypt (2002). He joined UAE University in 2000, and in 2002 he joined the Environment Agency–Abu Dhabi, as soil scientist. He was seconded to ICBA as deputy technical coordinator for Soil Survey of Abu Dhabi Project (2004–2009). Dr. Abdelfattah is currently a deputy manager, Soil and Land Use Management section, at EAD and managing Soil Survey of the Northern Emirates Project; he is also an associate professor at Fayoum University, Egypt. He has over 20 years experience in teaching and research. He participated in a number of international conferences, managed number of technical projects, and is author or co-author of over 50 published scientific papers.

Faisal K. Taha is director of technical programs at Dubai-based International Center for Biosaline Agriculture (ICBA). He earned Ph.D. degree from the University of Wyoming and has over 30 years of professional experience in research and development in the USA, Canada, Kuwait, and the United Arab Emirates. He has held key positions as project leader, program manager, department manager, chairman, professor, and technical program director at the Kuwait Institute for Scientific Research, Canada's Agriculture Development Fund, UAE University, and ICBA. Dr. Taha is an accomplished researcher and scientist with over 100 publications in refereed journals, proceedings, technical reports, and scientific books. He is also the winner of various regional and international awards in agricultural research and development.

Contents

Contributors

S.A. Abd El-Hafez Soil, Water and Environment Research Institute, ARC, Giza, Egypt

Tolba S. Abdel Aal Soil and Water Department, Faculty of Agriculture, El-Fayoum University, Al-Fayoum, Egypt

Mahmoud A. Abdelfattah Soil and Land Use Management, Environment Agency-Abu Dhabi, Abu Dhabi, UAE

Soils and Water Sciences Department, Faculty of Agriculture, Fayoum University, Fayoum, Egypt

Fawzy Hassan Abdel-Kader Remote Sensing and GIS Laboratory, Department of Soil and Water Sciences, Faculty of Agriculture, University of Alexandria, El-Shatby, Egypt

J. Abdullah Khan State Remote Sensing Applications Centre (APSRAC), Punjagutta, Hyderabad, India

Mohamed E. Abo-Waly Department of Soil Sciences, Faculty of Agriculture, University of Kafr El-Sheikh, Kafr El-Sheikh, Egypt

Fahmuddin Agus Centre for Agricultural Land Resources Research and Development, Bogor, Indonesia

Mushtaque Ahmed College of Agricultural and Marine Sciences, Sultan Qaboos University, Muscat, Oman

A.S. Ajayi Faculty of Agricultural Sciences, University of Ado-Ekiti, Ado-Ekiti, Ekiti, Nigeria

Jasem Al-Awadhi Earth and Environmental Science Department, Faculty of Science, Kuwait University, Kuwait, Kuwait

Mohamed S. Alhammadi Department of Aridland Agriculture, UAE University, Al-Ain, UAE

Rafaat A. Ali Department of Soil Sciences, Faculty of Agriculture, University of Kafr El-Sheikh, Kafr El-Sheikh, Egypt

Salem Al-Jabri College of Agricultural and Marine Sciences, Sultan Qaboos University, Muscat, Oman

Ghazi N. Al-Karaki Faculty of Agriculture, Jordan University of Science and Technology, Irbid, Jordan

Eman Al Khatib Dubai Municipality, Dubai, United Arab Emirates

Basma Nasser Al-Shidhani College of Agricultural and Marine Sciences, Sultan Qaboos University, Muscat, Oman

Khalil A. Ammar International Center for Biosaline Agriculture, Dubai, UAE

Mohamed Aoubouazza Service d'Ecologie, de Biodiversité et de la Conservation des Sols Centre de la Recherche Forestière, Agdal-Rabat, Morocco

J.O. Aruleba Faculty of Agricultural Sciences, University of Ado-Ekiti, Ado-Ekiti, Ekiti, Nigeria

Ivan Luiz Z. Bacic Centro de Informações de Recursos Ambientais e de Hidrometeorologia de Santa Catarina (CIRAM), Empresa de Pesquisa Agropecuária e Extensão Rural de Santa Catarina (EPAGRI), Florianópolis, SC, Brazil

Rupert Bäumler Institute of Geography, University of Erlangen-Nürnberg, Erlangen, Germany

Winfried E.H. Blum Institute of Soil Research, University of Natural Resources and Life Sciences (BOKU), Vienna, Austria

James Chakwizira School of Environmental Sciences, University of Venda, Thohoyandou, South Africa

Oagile Dikinya Department of Environmental Science, University of Botswana, Gaborone, Botswana

A.Z. El Bably Soil, Water and Environment Research Institute, ARC, Giza, Egypt

Abdelmagid Ali Elmobarak Land Evaluation Research Section, Land and Water Center, Agricultural Research Corporation (ARC), Wad Medani, Sudan

Razafindravoniarisoa Euphrasie Environmental Education and Cultural Center, Antsiranana, Madagascar

Abbas Farshad Faculty of Geo-Information Science and Earth Observation (ITC), University of Twente, Enschede, The Netherlands

R.W. Fitzpatrick CSIRO Land and Water, Adelaide, SA, Australia

School of Chemistry, Physics and Earth Sciences, Flinders University, Adelaide, SA, Australia

M.A. Fyzee National Remote Sensing Agency, Department of Space, Government of India, Balanagar, Hyderabad, India

Maria Gerasimova Faculty of Geography Moscow, Lomonosov University, Moscow, Russia

Mohamed S. Gheblawi Department of Agribusiness and Consumer Sciences, UAE University, Al-Ain, UAE

Dmitry L. Golovanov Department of Soil Geography and Landscape Geochemistry, M.V. Lomonosov Moscow State University, Russia

Gerard Grealish Acid Sulfate Soils Center, the University of Adelaide, Adelaide, SA, Australia

Davinder K. Grover Agro-Economic Research Centre, Department of Economics and Sociology, Punjab Agricultural University, Ludhiana, India

Hussein Harahsheh Global Scan Technologies, Dubai, UAE

A.M. Ibrahim Soils and Water Department, Faculty of Agriculture, El-Fayoum University, Al-Fayoum, Egypt

N. Jayaraju Department of Geology & Geoinformatics, Yogi Vemana University, Kadapa, India

Akmal Karimov International Water Management Institute (IWMI), Central Asia Regional Office, Tashkent, Uzbekistan

John A. Kelley United States Department of Agriculture-Natural Resources Conservation Service, Raleigh, NC, USA

Fikry I. Khalaf Earth and Environmental Science Department, Faculty of Science, Kuwait University, Kuwait, Kuwait

Anwar A. Khan Department of Drinking Water, Govt. of Uttarakhand, Deharadun, India

Peter King Department of State Development, East Perth, WA, Australia

Anil Kumar Environment Agency-Abu Dhabi, Abu Dhabi, UAE

Dilip Kumar Department of Geography, Shivaji College, University of Delhi, New Delhi, India

R. Kurbantoev The Soil Science Institute, The Academy of Science of Uzbekistan, Tashkent, Uzbekistan

Marina Lebedeva Laboratory of Soil Mineralogy and Micromorphology, V.V. Dokuchaev Soil Science Institute, Moscow, Russia

Bernhard Lucke Institute of Geography, University of Erlangen-Nürnberg, Erlangen, Germany

Ramez A. Mahjoory Michigan State University, East Lansing, MI, USA

Henda Mahmoudi International Center for Biosaline Agriculture, Dubai, UAE

Piotr Maloszewski GSF-Institute of Hydrology, Munich, Germany

Roberta P. Martins Conselho Nacional de Desenvolvimento Científico e Tecnológico (CNPq), Centro de Informações de Recursos Ambientais e de Hidrometeorologia de Santa Catarina (CIRAM), Empresa de Pesquisa Agropecuária e Extensão Rural de Santa Catarina (EPAGRI), Florianópolis, SC, Brazil

Yousef Marzouqi Dubai Municipality, Dubai, United Arab Emirates

Mohamed Mashroom Survey Department, Dubai Municipality, Dubai, United Arab Emirates

Heitor Matallo Junior UNCCD Secretariat Bonn, Bonn, Germany

Irina Mikheeva Institute of Soil Science and Agrochemistry, Siberian Branch of Russian Academy of Sciences, Novosibirsk, Russia

Raafat F. Misak Environment and Urban Development Division, Coastal and Air Pollution Department, Kuwait Institute for Scientific Research (KISR), Kuwait, Kuwait

Ruamporn Moonjun Faculty of Geo-Information Science and Earth Observation (ITC), University of Twente, Enschede, The Netherlands

Anny Mulyani Centre for Agricultural Land Resources Research and Development, Bogor, Indonesia

Abdul Munir Department of Geography, Faculty of Science, Aligarh Muslim University, Aligarh, Uttar Pradesh, India

Charles Nhemachena Human Sciences Research Council, Pretoria, South Africa

Iourii Nikolskii Colegio de Postgraduados, Texcoco, Mexico

A. Noble International Water Management Institute (IWMI), National Agriculture and Forestry Research Institute, Vientiane, Lao PDR

Abiodun E. Obayelu Department of Agricultural Economics and Farm Management, University of Agriculture Abeokuta (UNAAB), Abeokuta, Ogun State, Nigeria

Samira A.S. Omar Food Resources and Marine Sciences Division, Kuwait Institute for Scientific Research, Safat, Kuwait

Yasser Othman Environment Agency-Abu Dhabi, Abu Dhabi, UAE

John S. Pate School of Plant Biology, The University of Western Australia, Crawley, WA, Australia

Juniele R. Pivetta Conselho Nacional de Desenvolvimento Científico e Tecnológico (CNPq), Centro de Informações de Recursos Ambientais e de Hidrometeorologia de Santa Catarina (CIRAM), Empresa de Pesquisa Agropecuária e Extensão Rural de Santa Catarina (EPAGRI), Florianópolis, SC, Brazil

Adi Priyono Centre for Agricultural Land Resources Research and Development, Bogor, Indonesia

B.R.M. Rao National Remote Sensing Agency, Department of Space, Government of India, Balanagar, Hyderabad, India

Tewoldemedhin D. Rustu Department of Land Resources and Environment, Hamelmalo Agricultural College, Asmara, Eritrea

N. Salieva The Soil Science Institute, The Academy of Science of Uzbekistan, Tashkent, Uzbekistan

Fawzi Mohamed Salih Land and Water Research Center, Agricultural Research Corporation (ARC), Wad Medani, Sudan

Mohd Sadiq Salman Department of Geography, Faculty of Science, Aligarh Muslim University, Aligarh, Uttar Pradesh, India

Mohammad Hassan Sayyari-Zahan Department of Agronomy, Agriculture Faculty, Birjand University, Birjand, Iran

Noel Schoknecht Department of Agriculture and Food, 3 Baron-Hay Court, Western Australia, Australia

National (Australian) Committee on Soil and Terrain, South Perth, Australia

Sabry M. Shaheen Department of Soil Sciences, Faculty of Agriculture, University of Kafr El-Sheikh, Kafr El-Sheikh, Egypt

Shabbir A. Shahid International Center for Biosaline Agriculture, Dubai, UAE

Dhruba Pikha Shrestha Faculty of Geo-Information Science and Earth Observation (ITC), University of Twente, Enschede, The Netherlands

R.B. Singh Department of Geography, Delhi School of Economics, University of Delhi, New Delhi, India

Elena N. Smolentseva Institute of Soil Science and Agrochemistry, Russian Academy of Sciences, Novosibirsk, Russia

Samran Sombatpanit World Association of Soil and Water Conservation (WASWAC), Lat Phrao, Bangkok, Thailand

Willibad Stichler GSF-Institute of Hydrology, Munich, Germany

Faisal K. Taha International Center for Biosaline Agriculture, Dubai, UAE

Farid Talukder College of Agricultural and Marine Sciences, Sultan Qaboos University, Muscat, Oman

James E. Thomas United States Department of Agriculture-Natural Resources Conservation Service, National Soil Survey Center, Lincoln, NE, USA

Norair Toomanian Soil and Water Research Division, Agriculture and Natural Resources Research Center, Isfahan, Iran

William H. Verboom Department of Agriculture and Food, Narrogin, WA, Australia

Michael A. Wilson United States Department of Agriculture-Natural Resources Conservation Service, Lincoln, NE, USA

Irina A. Yamnova Soil Salinity, V.V. Dokuchaev Soil Science Institute, Moscow, Russia

Pandi Zdruli Land and Water Resources Management Department, International Centre for Advanced Mediterranean Agronomic Studies, Mediterranean Agronomic Institute of Bari, Valenzano, Bari, Italy

List of Figures

List of Tables

Part I
Soil Survey and Classification Strategies in Different Ecological Zones

Chapter 1
Innovative Thinking for Sustainable Use of Terrestrial Resources in Abu Dhabi Emirate Through Scientific Soil Inventory and Policy Development

Shabbir A. Shahid, Mahmoud A. Abdelfattah, Yasser Othman, Anil Kumar, Faisal K. Taha, John A. Kelley, and Michael A. Wilson

Abstract Abu Dhabi is the largest of the seven emirates that comprise the United Arab Emirates (UAE). The total area of the UAE is about 82,880 km^2. Abu Dhabi Emirate occupies more than 87% of mainland plus a string of coastal islands extending into the Arabian Gulf. The Emirate's leaders and population have a close affinity with the land and believe that careful agricultural development will be an important part of its future destiny and should be undertaken on a sustainable basis. With this aim, fourth-order extensive survey of Abu Dhabi Emirate was initiated in 2006 and completed in 2009. The field survey was completed through investigating 22,000 sites covering 5.5×10^6 ha, supplemented with typical profiles description,

S.A. Shahid (✉) • F.K. Taha
International Center for Biosaline Agriculture,
P.O. Box 14660, Dubai, UAE
e-mail: s.shahid@biosaline.org.ae; f.taha@biosaline.org.ae

M.A. Abdelfattah
Soil and Land Use Management, Environment Agency-Abu Dhabi,
P.O. Box 45553, Abu Dhabi, UAE

Soils and Water Sciences Department, Faculty of Agriculture,
Fayoum University, Fayoum, Egypt
e-mail: mabdelfattah@ead.ae; mahmoudalia@yahoo.com

Y. Othman • A. Kumar
Environment Agency-Abu Dhabi, Abu Dhabi, UAE
e-mail: yothman@ead.ae; akumar@ead.ae

J.A. Kelley
United States Department of Agriculture-Natural Resources Conservation Service,
Raleigh, NC, USA
e-mail: bettermark.john@gmail.com

M.A. Wilson
United States Department of Agriculture-Natural Resources Conservation Service,
Lincoln, NE, USA
e-mail: mike.wilson@lin.usda.gov

S.A. Shahid et al. (eds.), *Developments in Soil Classification, Land Use Planning and Policy Implications: Innovative Thinking of Soil Inventory for Land Use Planning and Management of Land Resources*, DOI 10.1007/978-94-007-5332-7_1,

laboratory analyses of soil samples, deep drilling to explore deep terrestrial resources, infiltration, permeability, and penetration resistance measurement. The survey was designed to take advantage of the latest technologies such as geographic information system (GIS), satellite image processing, and statistical analysis to produce *state-of-the-art* soil products. Sixty-two families and phases of soil families and 114 soil map units have been identified. The information were then used to publish soil map (1:100,000 and 1:500,000 scales) and 20 thematic maps at 1:500,000 scale. Using the extensive survey results, irrigated suitability map was prepared that led to delineate 1 million ha area, from which an area of 447,906 ha was surveyed at second-order level of USDA. The information collected will serve as a guide for future research and help to develop strategies that reduce the negative impact of the human activities on the natural surroundings and assist in the wise and sustainable use of its natural resources. In this chapter, methodologies used for extensive survey and results are presented and discussed for various uses. A brief introduction of the Abu Dhabi Soil Information System (ADSIS) developed to host all data for future retrieval, upgradation, and uses is also given, and policy issues are discussed.

Keywords Abu Dhabi • ADSIS • Extensive survey • Irrigated suitability • Soil map

1.1 Introduction

Soil inventory provides information used to facilitate a variety of natural resource planning, development, and implementation activities. Soils, by their nature, are a complex feature of the landscape with different soil properties changing at different rates. Soil surveys can be undertaken by using various survey orders at a variety of scales of final map production. These “orders” of soil survey are defined by Soil Survey Division Staff (1993) and include a range of site observation intensities applicable to the end-use requirements of the soil survey.

The purpose of extensive soil survey of Abu Dhabi Emirate was to have a comprehensive set of digital soil information to aid in broad land use planning and agricultural expansion in the Emirate. The technological advances in remote sensing, computers, terrain analysis, geostatistics, GIS data integration, and instrumentation were expected to make it possible to achieve unprecedented reliability and utility in the digital soil maps. Sound scientific and technical standards based on the United States Department of Agriculture-Natural Resources Conservation Service (USDA-NRCS) were used in the soil classification system, nomenclature, interpretation, and publications. These standards have been used in more than 75 countries of the world and in Gulf Cooperation Council countries such as the Kingdom of Saudi Arabia (MAW 1985), Sultanate of Oman (MAF 1990, 1993a, b, c), State of Kuwait (KISR 1999), and State of Qatar (Scheibert et al. 2005). A comparative soil classification using USDA and FAO systems for Abu Dhabi coastal area is discussed by Abdelfattah and Shahid (2007). A unique anhydrite soil has been discovered by Shahid et al. (2007) in the coastal area of Abu Dhabi Emirate and has now become the focus of change in US Soil Taxonomy at four different levels (EAD-ICBA-USDA 2011). In the past, little information was published on Abu Dhabi soils, and the pioneering efforts have recently been made to describe soils of Abu Dhabi Emirate (Shahid 2007; Shahid et al. 2007; Shahid and Abdelfattah 2008).

The survey was completed in two phases. Phase 1 was the extensive survey of the entire emirate (EAD 2009a), a fourth-order survey (1:100,000) based on the norms and standards of the USDA-NRCS (modified to fit Abu Dhabi conditions) intended to generate soil, thematic, suitability, salinity, and current land use maps (1:100,000; 1:250,000, and 1:500,000). From the interpretation of the phase 1 results, a soil suitability map for irrigated agriculture was prepared that led initially to identify 1 million ha with the highest potential for irrigated agriculture. Later, from this million ha, 447,906 ha was selected for phase 2 – an intensive, second-order survey at a scale of 1:25,000 on a grid basis; the results are presented elsewhere (EAD 2009b). A Soil Information System was included in the project to provide storage, processing, retrieval, and management of soil-related information for the use of a wide range of purposes. Rigorous quality assurance and quality control procedures were built into all aspects of the project with the Project Technical Support team closely monitoring the operations of the project team. A final quality control measure was the implementation of an independent review of the project in November 2008 by two highly experienced soil surveyors from USDA-NRCS. The rigor of the project was endorsed by their announcement that the project met and/or exceeded USDA standard field mapping procedures and data collection protocols.

The extensive survey is the first major soil survey to cover the mainland area of Abu Dhabi Emirate. A smaller soil survey was undertaken in 2004 that covered the coastal area (ERWDA 2004), and the results were incorporated with the new mapping. Preliminary work by the soil survey team began in May 2006 with analysis of remote-sensed satellite imagery (Landsat™) accompanied by targeted field work. The resulting overview of Abu Dhabi Emirate provided a broad categorization of the soils and landscapes and helped to guide the subsequent field survey operations and facilitated an understanding of soil-landscape relationships.

In this chapter, extensive survey results are discussed on a broader scale, including different landscape regions, generalized soil map and soil classification, current land use map and irrigated suitability map, and management issues.

1.1.1 Introduction to Survey Area

The UAE is in the southeast of the Arabian Peninsula and adjacent to the Arabian Gulf (Fig. 1.1). It borders Oman and Saudi Arabia. The total area of the country is about 82,880 km^2, of which Abu Dhabi is the largest of the seven emirates and accounts for 87% of the country's total landmass (National Media Council 2007). The Emirate lies between latitude 22° 29 N and 24° 53 N and longitude 56° 10 E and 51° 37 E. The extensive soil survey area consists of the entire Emirate of Abu Dhabi excluding the urban, industrial, and restricted areas and offshore islands. The total area covered in this survey was 5.5×10^6 ha excluding the coastal survey area, or 5.72×10^6 ha including it.

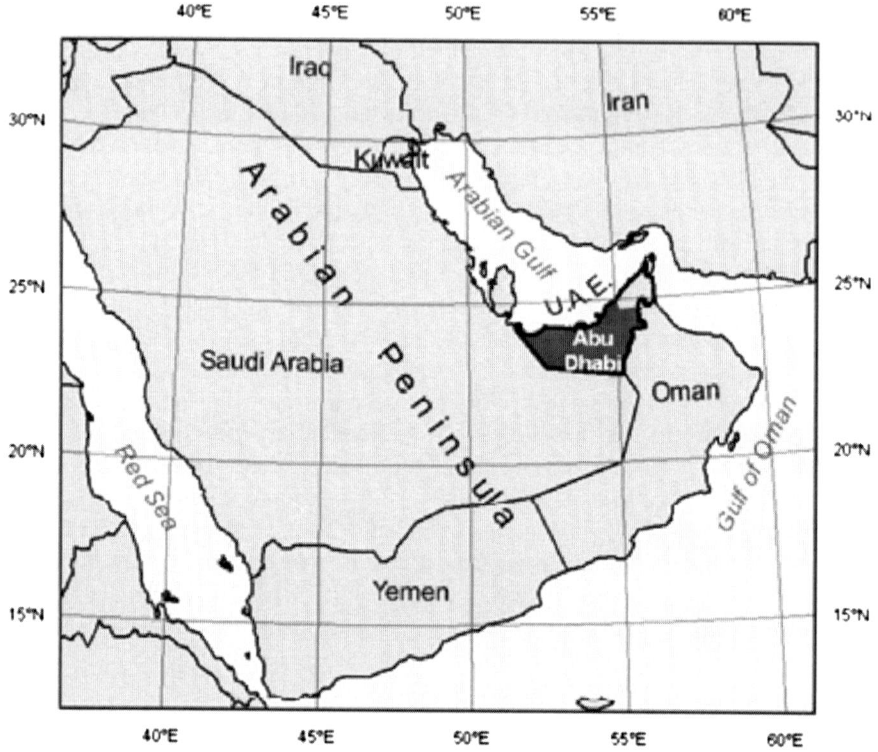

Fig. 1.1 Location of the Emirate of Abu Dhabi

Most of the land in the survey area is owned by the Emirate, with some farms near Madinat Zayed, Liwa, Al Qu', and Al Wijan being privately owned or controlled. There are extensive areas of land designated as oil field concession areas. Generally, only the active oil field areas have any infrastructure development, which is mostly restricted to oil wells, pipelines, gathering centers, and processing facilities. Rangeland and nomadic grazing of camels is the predominant land use, while a significant area is used for irrigated farming. Major concerns in managing the soils for irrigated agriculture are excessive drainage, poor nutrient-holding capacity, salinization, and encroachment of sand across the extensive survey landscape. Scanty protective vegetation has been excessively grazed by livestock, resulting in most of the soil surfaces in Abu Dhabi now being susceptible to wind erosion.

1.1.1.1 Geology

Geologically, Abu Dhabi shares the northeastern corner of the Arabian platform with Oman and has remained relatively stable for millions of years. The Emirate has remained in tropical and subtropical environments since the end of the Paleozoic

Fig. 1.2 Sand dunes and agriculture activities

though it has experienced a variety of climates significantly different to that of today. The surficial geology is dominated by sediments, predominantly eolian dunes, of Holocene and Pleistocene age. Outliers of the Hajar Mountains occur in the eastern parts of the Emirate, displaying the only true hard rock outcrops that occur in the Emirate. The oldest formations in the Emirate are represented by the salt diapirs of the Hormuz Formation seen at Jabal Az Zannah and a number of the offshore islands (e.g., Sir Bani Yas). These diapirs represent Cambrian materials that have been extruded through fissures in overlying materials as a result of the weight of overlying rock. Other minor occurrences of pre-Permian materials have been recorded on the eastern margins of the Emirate to the east of Jabal Hafit. Jabal Hafit comprises limestone and marls of Tertiary age that have been compressed and folded in the formation of the Hajar Mountains. The remainder of the Emirate is dominated by extensive sand and gravel plain with a thick blanket of eolian sand dunes (Fig. 1.2) formed by the prevailing wind.

1.1.1.2 Climate

The climate of Abu Dhabi Emirate is arid with extremely harsh and dry summers. The soil temperature regime of most of Abu Dhabi can be regarded as hyperthermic (Soil Survey Staff 2006). A prolonged dry summer of very high temperatures extends from April to November, followed by a mild to warm winter with a little rainfall between December and March. The wettest month is February, which receives an average of 24.4 mm (Table 1.1). Rainfall in the summer months is close

Table 1.1 Monthly climate characteristics of Abu Dhabi

Month	Mean daily temperature (°C)	Mean daily minimum temperature (°C)	Mean daily maximum temperature (°C)	Mean monthly rainfall (mm)	Monthly extreme rainfall (mm)	Mean monthly evaporation (mm)	Mean maximum monthly evaporation (mm)	Mean hourly wind speed (knots)
January	18.6	12.8	24.3	11.1	68.1	140	276	6.5
February	19.8	14.1	25.6	24.4	202.4	160	364	7.8
March	22.6	16.4	29.1	17.9	109.1	241	576	8.0
April	26.6	19.5	34.1	7.7	56.2	298	556	7.5
May	30.9	23.1	39.1	0.2	4.8	394	613	7.5
June	32.8	25.4	40.8	0.0	0.0	367	575	7.5
July	34.7	28.2	42.2	0.9	18.2	376	613	7.5
August	35.0	29.0	42.6	0.2	3.6	376	603	7.7
September	32.6	26.1	40.4	0.0	0.0	308	502	7.0
October	28.9	22.3	36.3	0.2	5.4	264	446	6.3
November	24.5	18.2	31.0	1.4	18.4	181	311	6.0
December	20.5	14.7	26.4	8.4	54.9	139	294	6.1
Average/total	27.3	20.8	34.3	(72.4)	(541.1)	(3,244)	(5,729)	7.1

Data adapted from Raafat (2006). Evaporation data from an unnamed site. All others represent 24-year data from Abu Dhabi International Airport to zero. Evaporation rate in all months exceeds rainfall many times over. Even during February, evaporation averages 160 mm – more than six times the likely rainfall. This ratio of evaporation to rainfall rises to nearly 400 to 1 from May to August. Around the rest of the Emirate, rainfall typically decreases to the south and west and increases toward the north and east.

1.1.1.3 Geomorphology and Landforms

The landscape of the Emirate is diverse, ranging from level coastal plains and sabkha to undulating desert sand plain; extensive areas of linear, transverse, and barchan sand dunes; and a single mountainous rocky outcrop, Jabal Hafit. This outcrop rises to 1,163 m in the eastern part of the Emirate. It is a mountainous outlier of the Hajar Mountains in neighboring Oman. In the south, the mega-barchans of the Liwa Crescent rise 250 m above the surrounding landscape interspersed with almost level deflation plains and flats. This area forms the northern tip of the Rub' al-Khali, or Empty Quarter of the Arabian Desert, that extends southward into Saudi Arabia. In the west, Sabkhat Matti represents an extensive sabkha formation.

The geomorphology and geology, combined with time and climate, are the main factors that influence the distribution of soils in Abu Dhabi Emirate. Understanding geomorphology is therefore useful in understanding soil patterns. While the current arid climate suggests that wind erosion is the dominant force shaping the geomorphology, this has not always been so. The extensive fold movements that created the Hajar Mountains in neighboring Oman, for example, have left outliers at Jabal Hafit that were eroded by water during the Pleistocene period. Such wetter phases also led to the creation of extensive sand and gravel alluvial fans, remnants of which are still found around the mountain.

Corresponding sea level changes have led to the development of sabkha (salt flats), coastal terraces, and minor scarps (including mesa-like features). The eolian processes of recent millennia have dominated the evolution of today's landscape. A number of dune-forming periods probably occurred in the last 20,000–30,000 years, and older dunes now contain cores of sandstone.

1.1.1.4 Vegetation

Until recently, studies of vegetation within the Emirate of Abu Dhabi were confined to those undertaken by amateur botanical or natural history groups. While the local Bedouin had an intimate knowledge and understanding of the flora, it was only in the later twentieth century that information on the local vegetative became more widely available. More than 667 species of plants are now known from the UAE and adjacent areas of Oman. However, many areas of the Emirate have a naturally sparse vegetative cover due to the harsh climate, and this sparsity of vegetation has been further degraded through overgrazing. Rapid changes in vegetation patterns can occur due to the erratic pattern of annual rainfall.

Table 1.2 Total water production in Abu Dhabi Emirate in 2001

Source	$Mm^3\ year^{-1}$	% of total
Falaj	12.0	0.38
Municipal wellfields	12.26	0.39
Forestry wellfields	362.38	11.60
Agriculture wellfields	1,741.43	55.72
Other wellfields	104.85	3.35
Desalination	742.41	23.76
Treated wastewater	149.89	4.80
Total	3,125.22	100%

Source: Dawoud (2008)

Vegetation mapping in the Emirate of Abu Dhabi (EAD 2009a) revealed 11 natural vegetation map units (*Acacietum* 0.3%, *Cyperetum-Haloxyletum-Zygophylletum* 22.7%, *Cyperetum-Tribuletum* 17.3%, *Cyperetum-Zygophylletum* 31.3%, *Halophyletum* 0.7%, *Haloxyletum-1* 9.7%, *Haloxyletum-2* 2.1%, *Panicetum* 1%, *Rhazyetum* 0.5%, *Zygophylletum-1* 0.7%, *Zygophylletum-2* 3.6%) and two modified communities (urban and disturbed area 2.2% and areas of plantations, farms, and oases 0.5% and areas devoid of vegetation 7.4%).

1.1.1.5 Agricultural Activities

Many forestry projects were established to shelter around cultivated areas and simply "green the desert." The main crops have been the traditional dates plus Rhodes grass and alfalfa (lucerne) as fodder supplements for livestock. Small areas of food crops such as vegetables and melons continue to be developed using water from underground sources.

1.1.1.6 Water Resources

Abu Dhabi has no rivers or lakes. Until the last two or three decades, water requirements were obtained from groundwater delivered from traditional hand excavated wells often via man-made channels. These wells now supply an insignificant amount of the total needs with the main water production coming from various wellfields using deep bores powered by mechanical pumps and supplemented by desalination and wastewater treatment plants (Table 1.2).

Groundwater resources are predominantly brackish or saline (Fig. 1.3). Due to intensive use of these groundwater resources, many aquifers are showing annual declines in groundwater levels of 1–2 m. At current abstraction rates, it is projected that fresh and brackish groundwater resources will be depleted in 50 years. Agriculture is the largest consumer of water resources (56% in 2006), followed by forestry and domestic use. Water use continues to increase.

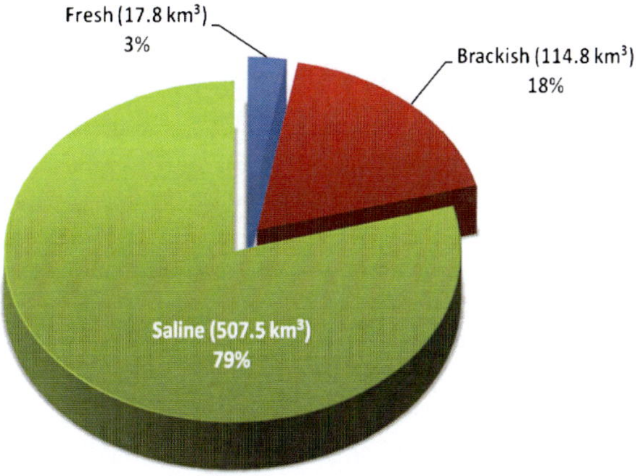

Fig. 1.3 Groundwater reserves in Abu Dhabi (Dawoud 2008)

1.2 Procedures for Extensive Soil Survey

Soils have been described using the "Field Book for Describing and Sampling Soils" (Schoeneberger et al. 2002), and soils have been classified according to the USDA system of Soil Taxonomy (Soil Survey Staff 1999) and the 10th edition of Keys to Soil Taxonomy (Soil Survey Staff 2006). Prior to the commencement of routine survey activities, a preliminary survey was undertaken of the entire survey area, in conjunction with Landsat image interpretation, to identify major landform patterns and soil types occurring within the Emirate. The resulting soil-landscape map was created from the results of ten transects across various parts of the Emirate to check soils and image interpretation.

The extensive survey has been completed using the latest norms and standards required of a fourth-order USDA soil survey (Soil Survey Division Staff 1993). There have, by necessity, been minor amendments to these standard procedures to meet the wider demands of the project. Considerable additional information has been collected. For example, a standard procedure includes full descriptions of soil profiles using soil augers to describe the diversity of soils; typical soil profiles (500) to characterize the soils and from which samples are collected for a broad range of physical, chemical, engineering, and mineralogical analyses; and check-sites that gather just sufficient information to enable a site to be allocated to a classification. For this soil survey, no sites were considered to be check-sites only; rather, all auger sites were fully described. Furthermore, soil information is normally only collected for the top 100 cm, but to ensure greatest potential use for the most stakeholders, soil information was recorded for the top 200 cm or until an impenetrable layer was struck. A sandspear developed in Western Australia was used to retrieve soil sample where it was not possible to explore depth to 200 cm. Fourth-order surveys for field mapping and publication from 1:63,360 to 1:250,000 may range in scale. A working scale of 1:50,000 was chosen for the present survey, and soil maps were published at scales of 1:100,000 and 1:500,000.

The extensive survey was completed by describing one site after every 250 ha. The survey area covers the entire mainland of Abu Dhabi excluding offshore islands, urban, industrial, military, and other restricted areas and totals about 5,500,000 ha and 22,000 observations sites. Further, field procedures permit plotting of soil boundaries by interpretation of remotely sensed data. The soils are identified by traversing representative areas to determine soil patterns and composition of map units. Transects are made in selected delineations for verification. According to USDA, most map units may be associations, some complexes, consociations, and undifferentiated groups. Components of map units are soil families and phases of families, and the minimum size of a map unit that can usefully be represented on a map is 40 ha.

A deep drilling (10 m) program was undertaken to provide information to assist developing thematic maps and an understanding of the landscape and soil development, determine occurrences of slowly permeable layers, and determine depth to water table to identify areas at risk of secondary salinization. The deep drilling program provided information on potential resources within the Emirate.

Statistical analyses allowed the final description of soil map units and the soil mapping process to meet the most rigorous standards possible. The statistical results included soil map unit purity, error rate of soil map units, partitioning of soil classes between soil map units, likelihood of property occurrence (water table, calcic, gypsic, petrocalcic, petrogypsic, salic, and hardpan layers), and comparison of soil map unit means for depths to diagnostic horizons, relative variance, and the associated relative variance between soil map units.

1.2.1 Satellite Imagery, Digital Elevation Model, and Projection Parameters

A more detailed visual interpretation of the satellite imagery was undertaken, supplemented by the results of image classification and a digital elevation model generated from the Shuttle Radar Topography Mission of 2006 data. This was used to delineate potential map units that were then captured on-screen using GIS. The map units so defined were overlain on 1:25,000 A3 size field maps using the Landsat imagery as the backdrop and incorporating a roads data layer provided by Environment Agency, Abu Dhabi. As a general rule, the complex areas were allocated 30–60 % more observations than areas considered to represent uniform sand plains or dunes. Field teams operated by conducting traverses over parts of the map sheets considered to be representative of wider areas and selecting observation sites according to the patterns and textures observable on the field map and in the field. Where appropriate, the soil surveyor also added map unit boundaries or adjusted those identified through pre-survey image interpretation.

The Landsat ETM data from the Global Land Cover Facility was geometrically transformed to the WGS84 datum and a user UTM projection. Precision checks were made using ground control points and differential GPS fixes to ensure the reliability of the transformations. Further adjustments to the selected images were performed to reduce the impacts of atmospheric scattering and to create a seamless image.

The projection parameters for this system are:

Projection	Transverse Mercator
Central meridian	54° 00 00″ E
Scale factor at the central meridian	0.99995
False easting of the origin	500,000 m
False northing of the origin	0 m
Reference spheroid	WGS84

1.2.2 General Site Description: Surface Landscape Features

The site features, where appropriate, were described (slope, slope class, slope morphological types, landscape, landform, relief model slope class, erosion type, runoff, land use, vegetation cover, surface condition, drainage and Ksat classes, moisture condition, surface coarse fragments, and moisture regime) using USDA procedures described in Field Book for Describing and Sampling Soils (Schoeneberger et al. 2002).

1.2.3 Morphological Description

The auger hole, sandspear, and typical profiles were described, where appropriate, for the following morphological characteristics for each soil layer identified: soil color, mottles, field texture, structure, consistence, segregations, coarse fragments, pores, roots, boundary condition, field EC, and pH (1:1 soil-water), using USDA procedures described in Schoeneberger et al. (2002).

1.2.4 Soil Map Units and Kinds of Map Units

Soil survey involves the simplification of a complex reality and the representation of that simplification through soil maps and reports. The success with which this representation is achieved reflects the usefulness of the final product for its intended purpose which in turn is dependent on the purity with which map units have been delineated and the accuracy with which they have been defined. The presence of inclusions within a map unit does not detract from the usefulness of the map so long as the map unit components have been appropriately defined and reflect the true variability of soils within the landscape. In an extensive survey, we recognize that map units are mostly associations with some complexes (Soil Survey Division Staff 1993) and consociations. Soil components are taxonomic families and phases of families.

1.2.4.1 Map Unit Legend Development

Depending on whether the map unit is a consociation, complex, or association, one or more soil taxonomic subgroup names are used to name the map unit together with a brief reference to the topography or landscape which dominates or is particularly characteristic of the map unit. In addition, each map unit is given a code. A total of 114 soil map units have been identified and described elsewhere (EAD 2009a). A typical map unit (TTP06) is described below.

1.2.5 TTP06 Typic Torripsamments, Consociation, and Rolling Dunes

This map unit consists of rolling dune fields that range in relative relief between 9 and 30 m. Some areas of deflation plains and closed depressions within the dune fields have calcareous sandstone fragments and pale brown sands over calcareous sandstone below 100 cm depth. The map unit occurs as small scattered polygons in the northeastern part of the Emirate in the nearby vicinity of Bida Hamama and to the east of Siwehan to Al Ain road. Polygons range in size from 93 to 28,464 ha. The land is used as low-density grazing. Vegetation in this map unit is dominated by *Cyperus conglomeratus* and *Zygophyllum* spp. The map unit forms part of the *Cyperetum-Haloxyletum-Zygophylletum* vegetation community. The soils of this map unit are dominated by Typic Torripsamments, mixed, hyperthermic (85%) in the rolling dune fields. Other components are Typic Haplocalcids, sandy, mixed, hyperthermic (5%); Typic Torripsamments, carbonatic, hyperthermic (5%); and Typic Torripsamments, mixed, hyperthermic, lithic phase (5%), all of which occupy the interdunal areas of the unit. Presence of high rolling dunes is the main limitation for irrigated agriculture in this unit. The unit covers an area of 118,698 ha (2.02% of the Emirate).

1.2.6 Map Series Published in Extensive Survey

Following maps (Table 1.3) have been published during the extensive survey. Details can be seen elsewhere (EAD 2009a); however, in this chapter, emphasis is given on the soil, irrigated suitability, current land use, and generalized soil-landscape region maps. The irrigated agriculture suitability map highlights potential areas where future agriculture can be expanded to bridge the gap between food import and local production.

Table 1.3 Maps published in the extensive soil survey

Title of map	Sheet size	Scale of publication	[a]No. of sheets
Soil map – satellite imagery	A1	1:100,000	29
Soil map – satellite imagery – subgroups	A0	1:500,000	1
Soil map – colored fill – subgroups	A0	1:500,000	1
Soil map – colored fill – great groups	A0	1:500,000	1
Generalized soil-landscape regions	A0	1:500,000	1
Current land use	A1	1:100,000	29
Current land use	A0	1:500,000	1
Vegetation map	A0	1:500,000	1
Suitability for irrigated agriculture	A0	1:500,000	1
Salinity in the 0–50-cm soil layer	A0	1:500,000	1
Salinity in the 50–100-cm soil layer	A0	1:500,000	1
Depth to water table	A0	1:500,000	1
Depth to hardpan	A0	1:500,000	1
Gypsum sources	A0	1:500,000	1
Gravel sources	A0	1:500,000	1
Carbonate sources	A0	1:500,000	1
Sand sources	A0	1:500,000	1
Sweet soil sources	A0	1:500,000	1
Anhydrite sources	A0	1:500,000	1
Land degradation	A0	1:500,000	1
Rangeland suitability	A0	1:500,000	1
Wildlife habitat suitability	A0	1:500,000	1
Forestry suitability	A0	1:500,000	1
Landfill (area) suitability	A0	1:500,000	1
1 million ha most suitable for irrigated agriculture	A0	1:250,000	4

[a]Number of sheets to cover study area in Abu Dhabi Emirate

1.2.7 Abu Dhabi Soil Information System (ADSIS)

To ensure that the information created in the soil survey is readily available to any potential user, a computer-based online information system was developed that links to the comprehensive soil database that resulted from the collection of survey information. The system was developed after broad consultation with a wide range of stakeholders who use soil information in Abu Dhabi and a review of the major existing soil information systems in other countries (United States of America, Australia, Canada, and Europe). Based on the needs identified, the system was created to support a wide range of information users and to provide tools for soil survey teams working on similar projects in the future. The Abu Dhabi Soil Information System (ADSIS) (Fig. 1.4) displays all the maps produced during the survey, with the powerful functionality of allowing a user to view particular locations of interest and extract data related to all the map units and observation sites described during

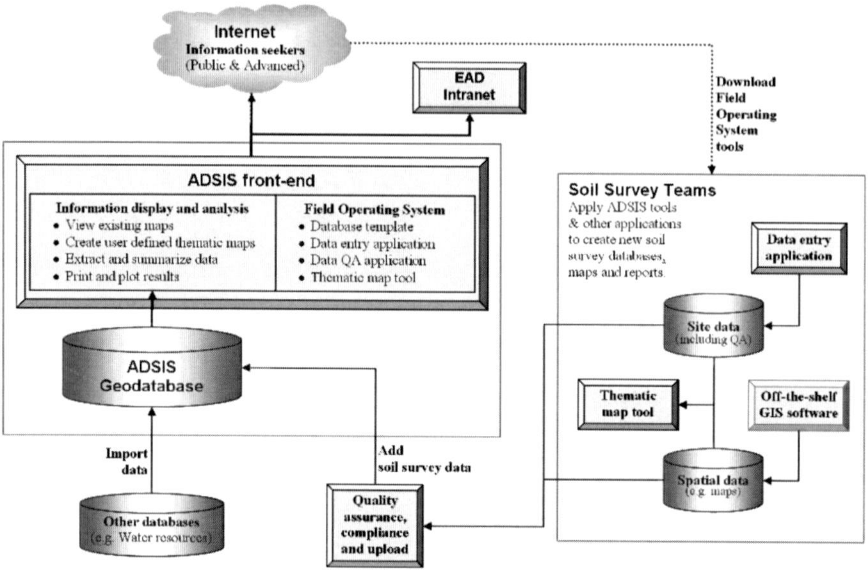

Fig. 1.4 Architecture of the Abu Dhabi Soil Information System

the survey. If registered as an advanced user, it is possible for a user to tailor the criteria for thematic maps based on the soil map properties or that of individual observation sites.

1.2.8 Laboratory Analytical Procedures

Laboratory analyses of soil samples are necessary to verify field observation data, to determine properties and characteristics that cannot be estimated accurately by field observations, and to help to characterize typical profiles. Standard USDA procedures (USDA 1954; USDA-NRCS 1995; Burt 2004) were used except where otherwise specified. Detailed particle-size distribution analyses were made using standard pipette method supplemented with wet sieving to quantify sand fractions, very coarse (1–2 mm), coarse (0.5–1 mm), medium (0.25–0.5 mm), fine (0.1–0.25 mm), and very fine (0.05–0.1 mm) sands. Using the proportion of sand, silt, and clay, the soil texture class was determined using USDA specifications (Soil Survey Division Staff 1993).

Soil moisture; total pretreatment loss; loss on acid treatment; carbonate equivalents; gypsum; extractable cations (Na, K, Ca, Mg); cation-exchange capacity (CEC); exchangeable sodium percentage (ESP); saturation percentage; soluble cations and anions; electrical conductivity of the soil saturation extract (ECe); pH of saturated soil paste; sodium adsorption ratio (SAR); osmotic potential (OP); engineering data (Unified Soil Classification System, American Association of State Highway and

Transportation Officials classification); Atterberg limits (plastic limit and liquid limit); water retention at 10, 33, and 1,500 kPa; organic carbon; whole soil and clay mineralogy (XRD analysis); elemental and oxides composition (XRF); and thin section study (soil micromorphology) in polarizing microscope were determined using standard USDA procedures (Burt 2004).

1.2.9 Field Tests: Infiltration Rate, Soil Permeability, and Penetration Resistance

Various determinations, such as saturated infiltration at the soil surface using a double ring infiltrometer (McKenzie et al. 2002), unsaturated permeability at the surface using a disc permeameter (McKenzie et al. 2002), and subsurface saturated permeability using a Guelph permeameter (Soil Moisture Corp 2005) were made beside the typical profile of each family identified. Penetration resistance of soil profiles was accomplished using a Bush penetrometer to a depth of 50 cm or to a hardpan. The information was collected at an interval of 2 cm.

1.3 Results and Discussion

In this section, extensive survey results are presented in broader perspectives to give overall view of Emirate landscapes, soils, current land uses, and areas having potential for irrigated agriculture. Restrictions to irrigated agriculture and management options as well as policy issues are discussed. More details can be seen elsewhere (EAD 2009a).

1.3.1 Generalized Soil-Landscape Regions

During the extensive survey, preliminary investigations of the landscapes recognized 27 generalized soil-landscape regions (Fig. 1.5). These regions have been identified on the basis of likely soils and characteristic landforms and are broadly consistent with major subdivisions identified by other researchers (Boer 1999):

1. *Sila* region presents what appear to be remnant alluvial fans. At its western margin, this unit is bound by a distinct scarp, and the ground surface typically has a gravely surface lag deposit.
2. *Sabkhat Matti,* a complex low-lying region, is characterized by saline flats and slightly better-drained gypsic rises. Groundwater levels are typically within 200 cm, and there is often a saline crust several centimeters thick on the soil surface. The Sabkhat Matti unit includes areas of a gilgai-type microrelief, often distinguishable on the satellite image by their darker appearance, plains

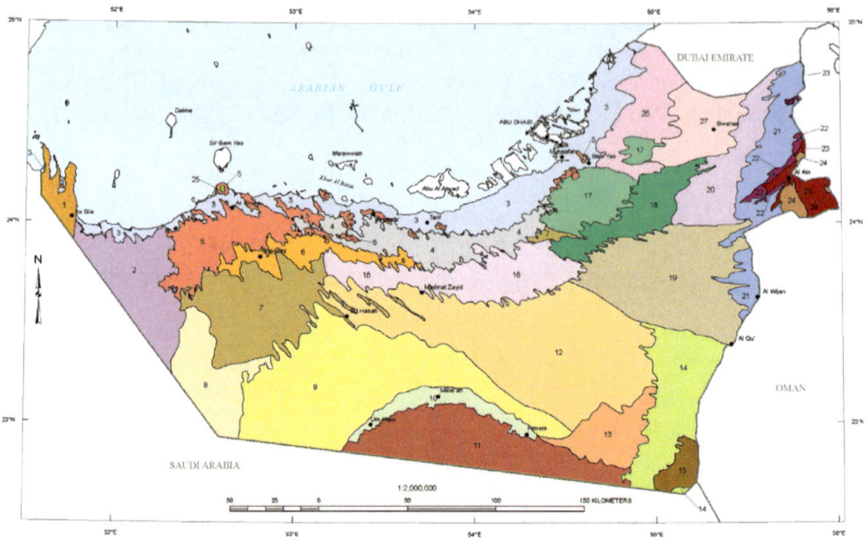

Fig. 1.5 Generalized soil-landscape regions. *Note: This map is not an authority on international or inter-emirate boundaries; it represents the extent of the extensive survey area

with lag gravels, and rock outcrop. Limited dune development has occurred in some areas resulting in low transverse dunes, known as Zibars, in an apparent "herringbone" pattern.

3. *Coastal flats* represent the low-lying coastal flats dominated by saline soils and slightly higher gypsic rises but including some areas of carbonatic sand sheets underlain by miliolite.
4. *Primary sand sheet* lies adjacent to the coastal plain in central parts of the Emirate. It represents carbonatic sand sheets with minor areas of saline soils. Some flat-topped mesas are capped by evaporate deposits from earlier soil-forming periods and left upstanding by erosion.
5. *Baynunah tertiary complex* lies in the north of the Emirate near Ghayathi. It comprises eroded Quaternary and Miocene sediments often with a deflation regolith of fine gravels.
6. *Baynunah red beds* lie to the south of Region 5. An area of undulating rises with small raised tablelands and associated scarps is capped by more resistant evaporite deposits.
7. *Baynunah barchanoid complex* occurs to the south of Ghayathi. It comprises extensive areas of barchanoid dunes occasionally partially overlain by paler carbonatic dunes. Interlayering of pale carbonatic sands with redder, coarser sands is common. Outcrops of miliolite are scattered and presumably underlie much of the dune systems.
8. *Al Maghrib* lies on the southern border with Saudi Arabia. It constitutes linear dune fields consisting of interlayered white carbonatic and red quartzite sands with minor exposure of Quaternary dunes and interdune formations.

9. *Liwa dunes* lie in the south of the Emirate and to the north of the Rub' al-Khali at Liwa. It comprises medium to high and rolling to steep linear and transverse dune systems that, in some areas, have been partially overlain by more recent barchanoid dunes. Scattered small deflation plains and sabkha flats are prominent in some areas.
10. *Liwa Oasis* lies in the south of the Emirate adjacent to the Rub' al-Khali. It comprises rolling to steep high mega valleys that have frequently been graded and developed for irrigated agriculture. Occasional interdunal depressions are deflated to the capillary fringe resulting in saline sabkha flats.
11. *Liwa mega-barchans* lies in the south of the Emirate and represents the northern extent of the Rub' al-Khali. It comprises high, often steep mega-barchans, frequently overlain by smaller barchan dune fields and associated intervening sabkha flats dominated by saline soils.
12. *Al Humrah* is an extensive region in central parts of the Emirate. It comprises older stable north linear dunes and intervening deflation flats. Occasional nested barchanoid dunes occur where more recent windblown sands have accumulated. The region includes broad, gently inclined sand ramps which may display a partial cover of fine lag gravels.
13. *Ramiat ar Rabbad* lies in the south and comprises sand sheets and relatively low relief transverse dunes comprised of red quartzite sands.
14. *Al Manadir rectangular dunes* occur in the south and comprise dune ridges set in a roughly rectangular pattern and comprising a mix of barchan, seif, and star dunes together with intervening deflation flats and inland sabkha. Surface lag gravels are common on many of the flats and often extend partway up the gently sloping windward face of the dune ridges.
15. *Al Manadir complex* occurs in the extreme south. It comprises rectangular dune ridges as for Region 14 but which overlie deflated lacustrine sediments including limestone. The region has a distinctive pale color on Landsat imagery.
16. *Table lands* occur in the center of the Emirate around Madinat Zayed. It comprises undulating to rolling terrain with frequent yardangs, mesas, and associated minor sabkha.
17. *Abu Dhabi complex* lies inland from Abu Dhabi City. It comprises lithified miliolite dunes and Quaternary to Miocene evaporites predominantly covered by eolian carbonatic sands that form low dunes and thin sand sheets. Deflation plains and sabkha flats are scattered throughout the region which has undergone significant human modification evidenced by extensive excavations and dumps.
18. *Abu Dhabi sand sheet complex* occurs to the east of Region 17 and has been identified as a separate unit on the basis of more even topography. It comprises sand sheets that overlie Quaternary and Miocene sediments and is generally less impacted by human activity.
19. *Bida Hamam* occurs in the east of the Emirate to the south of linear dune systems of moderate to high relief that partially overlie narrow interdunal flats and sabkhas.
20. *Al Khatam* occurs in the north Siwehan. It comprises linear dune systems of moderate relief with intervening broad plains and deflation flats.

21. *Al Ain region* occurs in the east of the Emirate, north and south of Al Ain. It comprises linear, transverse, and seif dune systems of moderate to high relief with associated interdunal flats.
22. *Outwash complex 1* occurs around Al Ain. It comprises gently inclined pediments and alluvial fans with a surface lag of fine gravels. In some areas, a partial covering by low to medium dunes is common.
23. *Outwash complex 2* occurs predominantly to the east of Jabal Hafit. It comprises almost level plains and pediments adjacent to the main mountain outcrops. Fan and gravely outwash deposits showing strong depositional stratification are common.
24. *Jabal Hafit Mountain* is in the east of the Emirate, near Al Ain. It comprises the rocky hills and mountains that represent outliers of the Hajar Mountains.
25. *Salt dome* is one small unit on the mainland at Jabal Az Zannah and comprises a salt dome against which eolian sediments have accumulated.
26. *Tawi Ghufar sand sheets* in the north of the Emirate comprise undulating carbonatic sand sheets that often overlie semi-lithified remnants of older dune systems with intervening sabkha flats and deflation plains.
27. *Tawi Ghufar desert plain* occurs around Siwehan in the northeast. It comprises branching linear dune systems of low to medium elevation with intervening broad deflation flats.

1.3.2 Current Land Use Map

Records of the current land use were also collected at all 22,000 observation sites and boundaries drawn on field map sheets. Existing digital land use information was also acquired from EAD. This information was combined with Landsat imagery and a reassessment of boundaries performed on-screen, leading to a draft map with a high level of confidence. Broad land use includes urban, farming, rangeland, oil fields, cemeteries, etc. Current land use maps were published at scale of 1:100,000 and 1:500,000 (Fig. 1.6). Table 1.4 shows area under each land use category. The information is a way forward for policy makers to make decisions to balance land uses in the Emirate.

1.3.3 Classification of Emirate Soils

Soil classification is used by pedologists to group and order components of a complex natural environment into categories that can be used for interpretation, evaluation, and planning purposes. One of the most widely accepted international systems is that of Soil Taxonomy: A Basic System of Soil Classification and Interpreting Soil Surveys (Soil Survey Staff 1999) and subsequent editions of the Keys to Soil Taxonomy (Soil Survey Staff 2006) that have been adopted for the extensive soil survey of Abu Dhabi Emirate. Soil Taxonomy is a classification system that uses

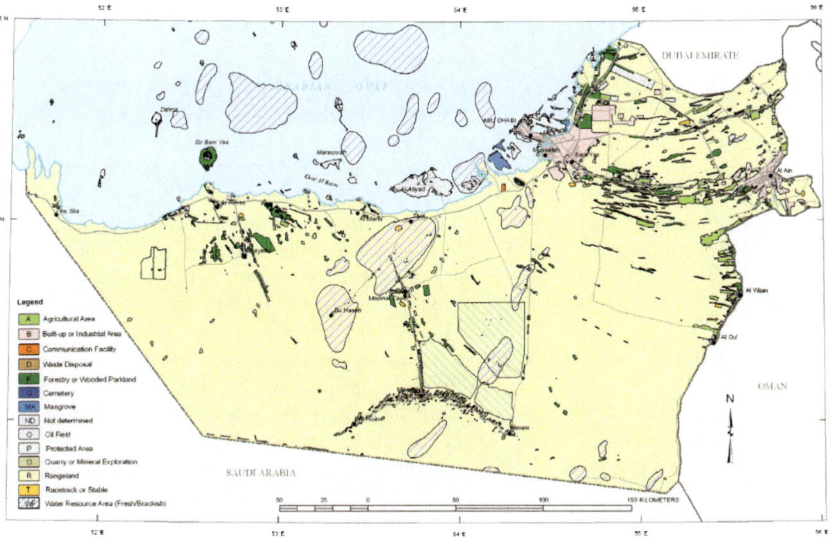

Fig. 1.6 Current land use map of Abu Dhabi Emirate. *Note: This map is not an authority on international or inter-emirate boundaries; it represents the extent of the extensive survey area

readily identifiable soil characteristics to define many of its classes at various levels of classification. By adopting the USDA Soil Taxonomy within Abu Dhabi, the Emirate is consistent with its neighbors (Kingdom of Saudi Arabia, Sultanate of Oman, State of Kuwait, and State of Qatar) who have utilized the same system, thereby allowing knowledge sharing, transfer of technology, and understanding throughout the region.

US Soil Taxonomy is a hierarchical system of soil classification that identifies six levels – order, suborder, great group, subgroup, family, and series. The soil order is the highest level of classification, and worldwide 12 different orders are recognized. Two orders have been recorded within Abu Dhabi during this survey – Aridisols and Entisols. The soil orders are subdivided into suborders according to properties that influence genesis or plant growth. Sixty-four potential suborders have been defined worldwide; only five have been identified in this survey; they are in order Aridisols (Calcids, Gypsids, and Salids) and Entisol (Orthents and Psamments). Soil great groups are defined on the basis of similarities in kind, arrangement and degree of profile development, soil moisture, and temperature status. The fourth level of classification is the subgroup.

Each great group has a typic subgroup that is considered to be the central concept of that great group. Other subgroups represent a range of other properties intergrading with other great groups. Soil families are established within a subgroup on the basis of physical or chemical properties that might affect management. For the soil survey of Abu Dhabi Emirate, they include particle-size class, mineralogy, soil temperature class, and a soil depth class. Soils have been classified at the family level of Soil Taxonomy according to the criteria defined in Soil Survey Staff (2006).

Table 1.4 Legend for 1:100,000 current land use map (as at 2007)

Code	Name	Description
Land uses other than oil fields and protected areas		
A	Agricultural areas 110,951 ha (1.87%)	Irrigated agriculture including open-field cultivation (fodder and horticulture, including date plantations), greenhouse cultivation, nurseries, and areas under animal farms
B	Built-up or industrial areas 126,935 ha (2.14%)	Populated places comprising residential, business and industrial centers, and associated infrastructure. These areas were excluded from the soil survey
C	Communication facility 845 ha (0.01%)	Only the larger and more important facilities are identified and indicated separately from the B unit (built-up or industrial area). The main facility is a base located 35 km southwest of Abu Dhabi City
D	Waste disposal 502 ha (0.01%)	Areas used for the disposal of waste. These sites vary from local disposal of waste to large facilities to receive the waste from the city of Abu Dhabi. Materials disposed of include household and industrial waste
F	Forestry or wooded parkland 134,948 ha (2.27%)	Includes the most extensive wooded park areas used for environmental objectives, recreation and camping, and zones of forest found on the outskirts of built-up areas or adjacent to major roadways
G	Cemeteries 65 ha (<0.01%)	Major cemeteries outside the built-up areas (B unit)
Q	Quarries or mineral exploration 9,111 ha (0.15%)	Ranges from open pit rock and gravel quarry and crusher operations to areas of borrow pits and associated tailings
R	Rangeland 5,469,636 ha (92.06%)	Represents the background land use unit. Small settlements, isolated buildings, etc., may be included in the unit which is used extensively for traditional animal grazing. Important secondary land uses include recreational activities and sporadic camping
T	Racetracks and stables 7,348 ha (0.12%)	Comprises camel and horse racing facilities located outside built-up areas
WF	Water resource areas 330 ha (0.01%)	Areal extent of unit delineations includes water wells, small reservoirs, pipelines, and power lines associated with the water fields
ND	Not determined 80,799 ha (1.36%)	The land use in some areas was not able to be determined using remote sensing and was either outside the scope of the ground-truthing component of this project (e.g., offshore islands) or was otherwise inaccessible (e.g., secure military areas)
Total area	5,941,441 ha (100.00%)	Note that as the oil fields and protected areas have coexisting land uses, their areas are not included in this total
Oil fields and protected areas		
O	Oil fields 292,608 ha (4.92%)	Represents the areal extent of oil fields comprising wells and associated infrastructure. Secondary land use periodically consists of animal grazing. The core oil fields were excluded from the soil survey. Oil fields were delineated on the basis of boundaries provided by the Abu Dhabi Company for Onshore Oil Operations (ADCO), where those areas coincided with land (mainland or islands)
PA	Protected areas 230,795 ha (3.88%)	Comprises designated protected areas. Land use includes recreation and sporadic camping. Some areas are intensively managed for the conservation of native wildlife of the Emirate

1.3.3.1 Soil Mineralogical, Particle Size, and Depth Family Classes

Soil mineralogy refers to the mineralogical composition of the soil. Gypsic, carbonatic, and mixed mineralogical classes were identified for Abu Dhabi Emirate soils:

- Gypsic – any particle-size class and more than 40% (by weight) carbonates (expressed as $CaCO_3$) plus gypsum with gypsum constituting more than 35% of the total weight of carbonates plus gypsum, either in the fine earth fraction or in the fraction less than 20 mm in size, whichever has a higher percentage of carbonates and gypsum
- Carbonatic – any particle-size class and more than 40% (by weight) carbonates (expressed as $CaCO_3$) plus gypsum and gypsum is less than 35% of the total weight of gypsum plus carbonates, either in the fine earth fraction or in the fraction less than 20 mm in size, whichever has a higher percentage of carbonates plus gypsum
- Mixed – other than gypsic or carbonatic class and also silica content is less than 90%

Sandy, sandy-skeletal, loamy, coarse loamy, and fine particle-size classes are identified in Abu Dhabi Emirate soils. The term "shallow" is added to the soil family name when a hardpan is encountered at or above 50 cm from the soil surface. This is applied in all soil subgroups except where the term is made redundant by the soil name as in Lithic subgroups.

1.3.3.2 Phases of Soil Families

Lithic, shallow, petrogypsic, anhydritic, calcic, and aquic phases are recognized:

- Lithic – when lithic contact occurs between 50 and 200 cm
- Petrogypsic – when petrogypsic horizon occurs between 100 and 200 cm
- Calcic – when a calcic horizon occurs between 100 and 200 cm
- Aquic – when water table occurs between 100 and 200 cm
- Anhydritic – when anhydrite was rich in soil profile in the upper 100 cm

A total of 62 soil families and phases of soil families have been identified in the extensive survey of Abu Dhabi Emirate (Fig. 1.7). A typical example of family description is shown below. Typical profile and associated landscape are shown in Fig. 1.8, and analytical results are presented in Table 1.5; details of other families and phases can be consulted elsewhere (EAD 2009a).

1.3.3.3 Soil Family: Typic Torripsamments, Mixed, Hyperthermic

Typic Torripsamments, mixed, hyperthermic are deep, sandy soils with mixed mineralogy. They occur on almost level plains to mega transverse and barchanoid dune fields and are widespread throughout the Abu Dhabi Emirate. They are typically

Order	Suborder	GreatGroup	Subgroup	Family/Phase of family
Aridisols		Haplocalcids	Typic Haplocalcids	sandy, carbonatic, hyperthermic
				sandy, carbonatic, hyperthermic, lithic phase
				sandy, mixed, hyperthermic
	Calcids			sandy, mixed, hyperthermic, lithic phase
				sandy -skeletal, mixed, hyperthermic
		Petrocalcids	Typic Petrocalcids	sandy, carbonatic, hyperthermic, shallow
		Calcigypsids	Typic Calcigypsids	sandy, mixed, hyperthermic
				sandy, mixed, hyperthermic, lithic phase
				sandy, mixed, hyperthermic, petrogypsic phase
		Haplogypsids	Leptic Haplogypsids	sandy, gypsic, hyperthermic
				sandy, gypsic, hyperthermic, lithic phase
				sandy, mixed, hyperthermic
	Gypsids			sandy, mixed, hyperthermic, lithic phase
				sandy, mixed, hyperthermic, petrogypsic phase
				sandy -skeletal, mixed, hyperthermic
			Lithic Haplogypsids	sandy, gypsic, hyperthermic
			Typic Haplogypsids	sandy, mixed, hyperthermic
				sandy, mixed, hyperthermic, petrogypsic phase
		Petrogypsids	Calcic Petrogypsids	sandy, mixed, hyperthermic
			Typic Petrogypsids	loamy, gypsic, hyperthermic, shallow
				sandy, gypsic, hyperthermic
				sandy, gypsic, hyperthermic, shallow
				sandy, mixed, hyperthermic
				sandy, mixed, hyperthermic, shallow
		Aquisalids	Gypsic Aquisalids	fine, gypsic, hyperthermic, anhydritic phase
				sandy, gypsic, hyperthermic
				sandy, mixed, hyperthermic
	Salids		Typic Aquisalids	sandy, carbonatic, hyperthermic
				sandy, mixed, hyperthermic
				sandy, mixed, hyperthermic, lithic phase
		Haplosalids	Gypsic Haplosalids	coarse -loamy, mixed, hyperthermic
				coarse -loamy, mixed, hyperthermic, lithic phase
				loamy, mixed, hyperthermic, shallow
				sandy, gypsic, hyperthermic, aquic phase
				sandy, mixed, hyperthermic
				sandy, mixed, hyperthermic, aquic phase
				sandy, mixed, hyperthermic, lithic phase
			Petrogypsic Haplosalids	coarse -loamy, mixed, hyperthermic
				loamy, gypsic, hyperthermic, shallow
				loamy, mixed, hyperthermic, shallow
				sandy, gypsic, hyperthermic
				sandy, gypsic, hyperthermic, shallow
				sandy, mixed, hyperthermic
				sandy, mixed, hyperthermic, shallow
			Typic Haplosalids	coarse -loamy, mixed, hyperthermic
				sandy, carbonatic, hyperthermic, aquic phase
				sandy, mixed, hyperthermic, aquic phase
				sandy, mixed, hyperthermic, lithic phase
				sandy, mixed, hyperthermic, shallow
Entisols	Orthents	Torriorthents	Lithic Torriorthents	sandy -skeletal, mixed, hyperthermic
			Typic Torriorthents	sandy -skeletal, mixed, hyperthermic
	Psamments	Torripsamments	Lithic Torripsamments	carbonatic, hyperthermic
				mixed, hyperthermic
			Typic Torripsamments	carbonatic, hyperthermic
				carbonatic, hyperthermic, calcic phase
				carbonatic, hyperthermic, lithic phase
				carbonatic, hyperthermic, shelly phase
				mixed, hyperthermic
				mixed, hyperthermic, calcic phase
				mixed, hyperthermic, lithic phase
				mixed, hyperthermic, petrocalcic phase
				mixed, hyperthermic, petrogypsic phase

Fig. 1.7 Soil Taxonomy hierarchy of extensive survey of Abu Dhabi

Fig. 1.8 Typical soil profile and associated landscape for Typic Torripsamments, mixed, hyperthermic

excessively drained or somewhat excessively drained and have rapid or very rapid permeability.

Site no.	: 271136		
Observation type	: Typical profile	Date	: Sep. 25, 2007
Described by	: Taj Muhammad		
Geographic coordinate in UTM	: 438347 E	2655622 N	
Physiography		Soil properties	
Slope	: 2%	Surface condition	: Loose
Slope class	: Nearly level	Microfeature	: Hummock
Slope morphological type	: Simple slope	Drainage class	: Excessively drained
Landscape	: Plains	Permeability class	: Very rapid
Landform	: Deflation plain	Ksat class	: Very high
Relief/modal slope class	: Undulating plain	Hydrological soil group	: A
Water table depth (cm)	: None	Root restriction depth (cm)	: Very deep
Erosion	: Wind erosion evident	Moisture condition	: 0–110 cm dry 110–200 cm moist
Runoff	: None	Surface coarse fragments	: 10% gravel (2–75 mm) mixed rock fragments
Land use/cover	: Rangeland (greater than 5% vegetation cover)	Moisture regime	: Torric

(continued)

(continued)

Morphological description		
Horizon	Depth (cm)	Description
Ak	0–20	Reddish yellow (7.5YR 6/6) dry, strong brown (7.5YR 5/6) moist, deflation crust; loamy sand; massive; soft, dry; very friable, moist; very weakly cemented by carbonates; nonsticky; nonplastic; low excavation difficulty; no roots; strongly effervescent; diffuse smooth boundary; EC (1:1) 0.256 dS m^{-1}; pH (1:1) 8.5
Ck1	20–70	Reddish yellow (7.5YR 6/6) dry, strong brown (7.5YR 5/6) moist, loamy sand; massive; soft, dry; very friable, moist; 6% fine (<2 mm) carbonate concretions; weakly cemented by carbonates; nonsticky; nonplastic; low excavation difficulty; common fine (1–2 mm) roots throughout; strongly effervescent; diffuse smooth boundary; EC (1:1) 0.367 dS m^{-1}; pH (1:1) 8.53
Ck2	70–110	Strong brown (7.5YR 5/6) moist, loamy sand; massive; very friable, moist; 5% fine (<2 mm) carbonate concretions; weakly cemented by carbonates; nonsticky; nonplastic; low excavation difficulty; common fine (1–2 mm) roots throughout; strongly effervescent; diffuse smooth boundary; EC (1:1) 0.25 dS m^{-1}; pH (1:1) 9.34
Ck3	110–160	Strong brown (7.5YR 5/6) moist, loamy sand; massive; very friable, moist; 8% fine (<2 mm) carbonate concretions; weakly cemented by carbonates; nonsticky; nonplastic; low excavation difficulty; common fine (1–2 mm) roots throughout; strongly effervescent; diffuse smooth boundary; EC (1:1) 0.493 dS m^{-1}; pH (1:1) 9.41
Ck4	160–200	Strong brown (7.5YR 5/6) moist, loamy sand; massive; very friable, moist; 7% fine (<2 mm) carbonate concretions; weakly cemented by carbonates; nonsticky; nonplastic; low excavation difficulty; no roots; strongly effervescent; EC (1:1) 0.555 dS m^{-1}; pH (1:1) 8.86

1.4 General Soil Map at Great Group Level

Nine soil great groups have been recorded as dominant map units (Fig. 1.9). Some great groups may also occur as subdominant soils in other map units. This generalized soil map, together with the following statements under each great group heading, is a summary of information that can be seen elsewhere (EAD 2009a). This section is useful for providing an overview of the soils of Abu Dhabi, and statements herein should be considered as broad generalizations. Reference to other soils refers to other great groups used to name the map units. Proportions of the map unit are given as part of the entire survey area including the coastal area (EAD 2009a).

1.4.1 Haplocalcids

This unit comprises 0.40% of the Emirate. It consists of deep to very deep well to somewhat excessively well-drained sandy to coarse loamy soils showing evidence of secondary calcium carbonate deposition. The unit occurs on flats and deflation plains across the Emirate. Other soils associated with this map unit include Torripsamments where windblown sands have accumulated, Calcigypsids in which accumulations of gypsum occur as well as carbonates, and Torriorthents that show little profile development but often contain significant amounts of gravel. This unit is used for rangeland grazing for livestock. It has moderate potential for irrigated agriculture being limited by the sandy nature of the soils.

1.4.2 Petrocalcids

This map unit has limited extent and comprises 0.04 % of the Emirate. It consists of shallow to moderately deep sandy or loamy soils with an indurated horizon cemented by calcium carbonate within the top 100 cm. It occurs in isolated deflation plains. Other soils associated with this unit include Haplosalids, usually occurring in lower landscape positions and containing high levels of accumulated salt. This unit is used for grazing of livestock. It has limited suitability for irrigated agriculture due to depth to hardpan.

1.4.3 Calcigypsids

This map unit has limited extent and comprises only 0.24 % of the Emirate. It consists of well to somewhat excessively drained sandy to coarse loamy soils that show accumulations of both calcium carbonate and gypsum. The unit occurs as deflation plains frequently partially obscured by windblown sands. Other soils associated with this map unit include Torripsamments where windblown sands have accumulated.

Table 1.5 Detailed characteristics (physical, chemical, mineralogical) and field test results of typical soil profile

Typic Torripsamments, mixed, hyperthermic

Site number – SSEAD 271136

Physical data

				Total			Silt		Sand	
Horizon	Depth cm	Texture	Field moisture	Clay <.002	Silt .002–.05	Sand .05–2	Fine .002–.02	Coarse .02–.05	VF .05–.10	F .1–.25
				% of <2 mm						
Ak	0–20	FS	–	1.1	1.2	97.5	0.1	1.3	11.0	71.5
Ck1	20–70	FS	–	1.2	0.7	98.1	–	0.7	8.3	77.0
Ck2	70–110	FS	–	1.0	–	99.0	–	–	5.6	63.2
Ck3	110–160	FS	–	0.7	1.4	97.9	0.1	1.3	9.2	61.2
Ck4	160–200	FS	–	0.8	1.2	98.0	–	1.2	9.5	62.9

Chemical data

		Extractable cations				CEC	ESP	Saturation extract amounts	
Horizon	Depth cm	Ca	Mg	Na	K			Ca	Mg
		cmolc kg^{-1}					%	meq L^{-1}	
Ak	0–20	18	0.9	0.4	0.2	0.1	1	10	3
Ck1	20–70	16	0.9	0.4	0.3	0.1	3	3	1
Ck2	70–110	15	0.7	0.5	0.2	0.1	4	5	3
Ck3	110–160	16	0.8	0.8	0.2	2.0	11	5	2
Ck4	160–200	16	0.7	0.8	0.3	3.0	9	4	2

Engineering Data

		Particles	Passing	Sieve	Number	Atterberg limits	Engineering class	
Horizon	Depth cm	4	10	40	200	Ip	USCS	AASHTO
		%						
Ak	0–20	100	98	96	2	np	SP	A-3
Ck1	20–70	100	100	99	9	np	SP-SM	A-3
Ck2	70–110	100	100	93	11	np	SP–SM	A-2-4
Ck3	110-160	100	100	98	2	np	SP	A-3
Ck4	160–200	100	99	96	2	np	SP	A-3

Whole soil mineralogical data

Horizon	Depth cm	Quartz	Plagioclase	Calcite	Palygorskite	Uncertain
Ck1	20–70	D	Mi	Tr	Mj	TrS
Ck2	70–110	D	Mi	Tr	–	–

Hydraulic conductivity

Surface saturated, 540 mm h^{-1}; surface unsaturated, 54 mm h^{-1}; subsurface (100 cm) saturated, 487 mm h^{-1}; soil strength, *FS* fine sand, *Mi* minor, *Mj* major, *WRD* water retention difference, *Tr* traces, *np* nonplastic, *ESP* exchangeable

								Coarse fraction				
M .25–.5	C .5–1	VC 1–2	TPL	LAT	$CaCO_3$	Gypsum		2–5	5–20	20–75	>75	>2
								% <75 mm				%of whole soil
13.2	0.7	1.1	21	19	19	0.3	–	1.9	–	–	–	1.9
12.6	0.2	–	5	5	5	0.2	–	–	–	–	–	–
28.2	2.0	–	4	3	4	0.2	–	0.2	–	–	–	0.2
26.9	0.5	0.1	4	4	5	0.2	–	0.4	–	–	–	0.4
25.2	0.4	–	4	4	5	0.1	–	1.2	–	–	–	1.2

Na	K	HCO_3	SO_4	Cl	NO_3	PO_4	SP	SAR unit	ECe	pHs	SAR	OP
							%	(mmoles $L^{-1})^{0.5}$	dS m^{-1}		(mmoles $L^{-1})^{0.5}$	atmos
4	0.7	1.5	16	3	0.4	–	28	–	1.5	8.23	2	0.5
5	0.6	1.4	4	3	0.7	–	28	–	0.9	8.42	3	0.3
8	0.5	3.0	5	10	0.2	–	27	–	1.0	8.94	4	0.4
16	0.7	1.0	10	11	0.1	–	28	–	2.0	8.30	9	0.7
13	0.6	1.7	6	12	–	–	27	–	1.9	8.40	7	0.7

Other data

Water content (<2 mm)			Bulk density	Particle density	Porosity	WRD	Organic carbon	Organic matter
1/10 bar	1/3 bar	15 bar						
%			g cm^{-3}	g cm^{-3}	%	cm cm^{-1}	%	%
3.9	1.9	0.9	1.61	2.69	40	0.016	0.09	0.15
3.2	1.8	1.3	1.62	2.66	39	0.008	0.03	0.05
2.5	1.6	1.1	1.63	2.65	39	0.009	0.01	0.03
3.0	1.6	1.1	1.62	2.67	40	0.009	0.03	0.05
3.3	1.5	0.9	1.62	2.66	39	0.009	0.01	0.03

depth at which soil penetration resistance reached 3.8 MPa, 12 cm

sodium percentage

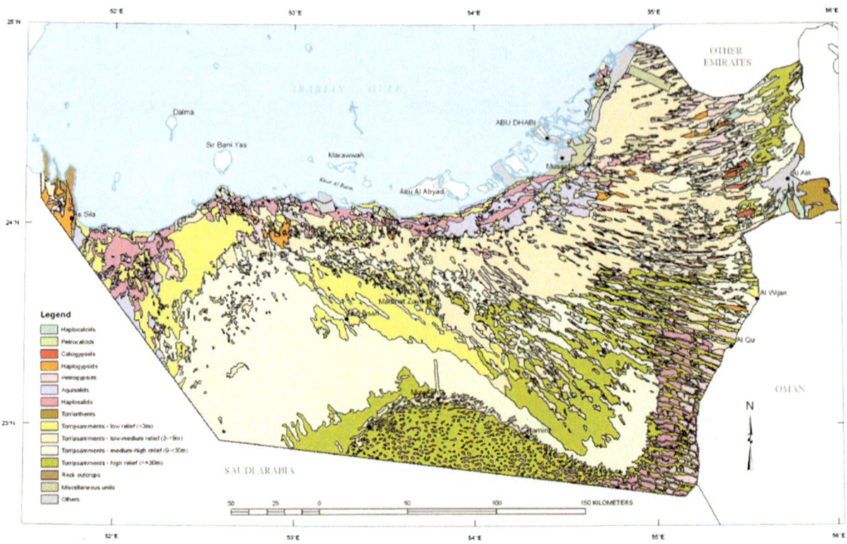

Fig. 1.9 Generalized great group soil map

This unit has some limited use for rangeland grazing. It has little suitability for irrigated agriculture due to high soil gypsum contents.

1.4.4 Haplogypsids

This map unit comprises 1.22% of the Emirate. It consists of deep to very deep well to somewhat excessively drained sandy to loamy soils. Accumulations of gypsum are common and can occur at any depth from the surface down. The unit occurs as almost level to gently undulating deflation plains occasionally with a partial covering of windblown sands. Other soils associated with this map unit include Torripsamments in areas of windblown sand and Petrogypsids where gypsum has accumulated and created an indurated layer. Areas of rock outcrop occur in some areas. This unit is mainly used for grazing of livestock. Some areas are used for forestry, and there are limited areas used for farming, predominantly fodder production.

1.4.5 Petrogypsids

This unit comprises 2.88% of the Emirate. It consists of shallow to moderately deep sandy or loamy soils with an indurated horizon cemented by gypsum within the top 100 cm. Other soils found in this map unit include Haplogypsids (no hardpan), Torripsamments where windblown sands have accumulated, and Haplosalids where soils have accumulated salt and many of which also have an indurated gypsic hardpan. This unit is used as rangeland grazing for livestock. It has little potential for irrigated agriculture due to limited depth of soil to hardpan and soil gypsum content.

1.4.6 Aquisalids

This map unit comprises 3.07% of the Emirate. It consists of poorly and somewhat poorly drained soils with water tables within 100 cm of the surface for much of the year. The soils are strongly saline, sandy to loamy, and frequently have a substantial surface salt crust. Often, soils contain an accumulation of gypsum. The unit occurs in coastal and inland sabkha flats. Other soils associated with this map unit include Haplosalids on slightly higher flats, some of which may include horizons strongly indurated by gypsum.

This unit is not used for any productive purpose and is considered permanently unsuitable for irrigated agriculture due to near-surface saline groundwater, high salt content, and gypsum content.

1.4.7 Haplosalids

This map unit comprises 6.53% of the Emirate. It consists of imperfectly to somewhat poorly drained, deep to very deep sandy to loamy soils. Water tables are occasionally within 100–200 cm of the surface. Many soils contain an accumulation of gypsum that is often cemented into a hard impenetrable layer. The unit occurs in coastal and inland sabkha flats across the Emirate. Other soils associated with this map unit include the Aquisalids with a saline water table within 100 cm, Torripsamments where sabkha are overlain by windblown sands, and Torriorthents that often contain high amounts of gravel. In some coastal flats, tertiary rock outcrop has been exposed through deflation. This map unit has little productive use. Small areas of forestry are occasionally developed on better soils found within the unit, but rock outcrop, salinity, and gypsum content all limit agriculture land uses.

1.4.8 Torriorthents

This map unit has limited extent and comprises only 0.59% of the Emirate. It consists of well-drained shallow to moderately deep sandy or loamy soils frequently containing large amounts of gravel or seashells. Some components have a lithic contact at less than 50 cm. They occur as deflation plains and flats. Other soils associated with this unit include Haplosalids, with or without gypsum, which typically occur in lower landscape positions, and Torripsamments that reflect accumulations of eolian sand. This unit is used for grazing of livestock and some limited farming and forestry activities. It has limited suitability for irrigated agriculture due to the gravel content of the soils.

1.4.9 Torripsamments

This is the most dominant and widespread unit in the survey area, comprising 81.09% of the Emirate. It consists of deep to very deep excessively well-drained sandy soils that occasionally overlie miliolite within 200 cm. They occur as extensive dune systems and sand sheets across the Emirate. Other soils associated with this map unit include most of the other great groups described above with the exception of Petrocalcids and Torriorthents. This map unit is used for rangeland grazing, farming, and forestry activities. It has a range of suitability for irrigated agriculture depending on the nature of the landform. Low dunes, sand plain, and sand sheet are considered moderately suitable, and suitability typically declines as the dune height and frequency increase. Because of the widespread nature of this unit, it has been subdivided in the generalized soil map on the basis of dominant relief within the map unit. In addition to the nine units described above, a number of other map units have been defined within which the dominant component is not a natural soil. These include rock outcrops, miscellaneous units, and others.

1.4.10 Rock Outcrops

This unit comprises 0.57% of the Emirate. It consists of outcrops of limestone, miliolite (lithified sand dunes), or tertiary sedimentary rocks. The unit occurs in coastal flats where it is often associated with coastal limestone and sedimentary rocks and as the massive rock outlier from the Hajar Mountains at Jabal Hafit. Other soils associated with this map unit include the Haplosalids and Aquisalids associated with coastal areas and the Haplogypsids in some deflation flats. This unit typically has little current use. It is generally unsuitable for irrigated agriculture due to the extensive areas of rock outcrop, steep gradients, and high salt levels of some soils.

1.4.11 Miscellaneous Units

A variety of units have been recorded and defined as miscellaneous units. They include refilled and leveled land, earthworks, rubbish tips, and quarries and amount to 0.97% of the Emirate. These units are not considered suitable for irrigated agriculture.

1.4.12 *Others*

Several other map unit types have been defined in areas that have not been surveyed during either this or the coastal survey project. They include areas identified specifically as "not surveyed areas," farms /forestry, urban areas, tidal flats, and highland. They amount to 2.40% of the Emirate. These areas are either already designated for an alternative land use or are unsuitable for irrigated agricultural use.

1.5 Uses and Management of the Soils

Evaluation of the land of Abu Dhabi for a variety of uses has been undertaken to provide a resource for land use planning. Land evaluation is used to match the requirements of each potential land use with the characteristics of each kind of land. Such information can help address soil-related problems and hence encourage better management. A major objective of the extensive soil survey was to develop irrigation suitability map. This assessment, plus others of generalized land uses including forestry, rangelands, wildlife habitats, sources of construction material, and sanitary landfill, has been undertaken. Land evaluation information is of value for many individuals and groups. The reader is referred to EAD (2009a), should he/she wishes to learn about the interpretation of survey results for other uses, to include all these results is beyond the scope of this chapter. However, as a guideline, the other map series published during extensive survey are presented as Table 1.3.

1.5.1 *Soil Suitability Classes for Irrigated Agriculture (FAO 1976)*

Soil interpretation for "irrigated agriculture" evaluated the soil's suitability for development of irrigated agriculture in the Emirate. The evaluation used follows the land suitability classification concepts developed by the FAO in its *Framework for Land Evaluation* (FAO 1976). The FAO system is generally considered as a benchmark for land evaluation. The framework does not in itself constitute an evaluation system but is a set of principles and concepts on which the basis of local, regional, and national evaluation systems can be constructed. The FAO system has been used as the basis for specific land evaluation applications such as rainfed agriculture (FAO 1983), extensive grazing (FAO 1984a), forestry (FAO 1984b), and irrigated agriculture (FAO 1985). Unlike the USDA land capability classification system (Klingbiel and Montgomery 1961), the FAO Framework does not contain predefined judgments about qualities of land in relation to specific land uses nor any proposed hierarchy of those land uses. An outline of the FAO land suitability classes is given below:

Suitability Class Definition

S1 *Highly suitable* land with no significant limitations to the specified use
S2 *Moderately suitable* land with moderate limitations to the specified use
S3 *Marginally suitable* land with severe limitations to the specified use
N1 *Currently unsuitable* land with severe limitations which cannot be corrected with existing knowledge and technology at currently acceptable costs
N2 *Permanently unsuitable* land with severe limitations which cannot be corrected

Important factors (Table 1.6) to consider in selecting land for irrigated agriculture in Abu Dhabi include the following: emphasis should be placed on profile depth and deep drainage because of the brackish nature of much of the irrigation water available. Areas having a limited capacity to dispose of excess irrigation water should be avoided. It is likely that the quality of water used will require a leaching fraction of between 20 and 25%. Therefore, suitable soils should be permeable in the surface and have deep, free draining subsoil material. Salinity values in the root zone should be below 4 dS m^{-1} ECe or have the potential, via leaching, to be reduced to this level. Sodicity is not considered critical as many soils contain sufficient amounts of gypsum or free calcium carbonate. Highly gypsic soils (>10% gypsum) should be avoided as, under irrigation, they may subside as the gypsum is dissolved from the soil. Sandy soils require careful water management because of low water-holding capacity. The sandy surface may be susceptible to wind erosion; therefore, wind breaks, mulches, and vegetative ground covers should be encouraged. Water-efficient irrigation systems should also be encouraged. These include drip irrigation, which is widely practiced in the Emirate, and subsurface irrigation may be a good option. Percent distribution of land suitability classes for irrigated agriculture in the Emirate is shown in Fig. 1.10.

In addition to the suitability class, subclasses indicating the most restrictive characteristics or qualities are noted.

These subclasses explain the type of restrictions for a specified land use and assist in identifying management options required to rectify the restriction (Table 1.7). Subclasses and their inherent limitations are originally defined in Part 620.04, National Soil Survey Handbook (USDA 2005). They are intended to assist users in identifying soil features important for use and provide some initial guidelines for their management. This information is intended as a guide to land users. The information presented, in its own right, does not restrict or control the development or use of land. It is intended as an input to the land use planning process and therefore can contribute to decisions on zoning and land use policy that lead to legislation on land use. Although the soil being assessed may have severe limitations or be poorly suited to a particular land use, this does not necessarily mean that it cannot successfully be put to that use. The restrictive features may be overcome by costly and difficult corrective measures such as engineering design or intensive maintenance. These interpretations

Table 1.6 Land suitability rating criteria for irrigated agriculture and summary of areas identified for each category in Abu Dhabi Emirate

Soil characteristic	Subclass code	Rating categories (see text for details)					Restrictive feature	Definition
		S1	S2	S3	N1	N2		
Hardpan or rock depth (cm)	m	>200	200 to >150	150 to >100	100 to >50	50 to 0	Restrictive layer	Impervious soil or rock layers inhibit movement of water or roots in soil
Water table depth (cm)	w	>200	200 to >150	150 to >100	100 to >50	50 to 0	Wetness	Soil is wet during the period of desired use
Salinity (ECe dS m^{-1}) weighted average for 0 to 50 cm	z	0 to 4	>4 to 8	>8 to 16	>16 to 40	>40	Excess salt	Excess water-soluble salts in the soil that restrict the growth of most plants
Salinity (ECe dS m^{-1}) weighted average for 50 to 100 cm	z	0 to 4	>4 to 8	>8 to 16	>16 to 40	>40	Excess salt	Excess water-soluble salts in the soil that restrict the growth of most plants
Gypsum-depth to upper boundary of gypsic diagnostic horizon (cm)	y	>200	200 to >100	100 to >50	50 to >20	20 to 0	Excess gypsum	Excess gypsum can result in soil subsidence after irrigation
Texture for surface 0 to 25 cm layer	t	LS, LFS, LVFS, FS	S				Too sandy	The soil is soft and loose, droughty, and low in fertility
Texture for surface 0 to 25 cm layer	t	SCL					Too clayey	The soil is slippery and sticky when wet and slow to dry
Slope gradient (%)	s	0 to 1	>1 to 3	>3 to 32	>32 to 56	>56	Too steep	Slope limits machinery use and exacerbates erosion risk
Relief-height above surrounding area (m)	r	0 to 1	>1 to 3	>3 to 9	>9 to 30	>30	Too high	The height restricts the ability to recontour the area
Area ('000 ha)		2	309	1,550	1,753	2,108		Total: 5,723

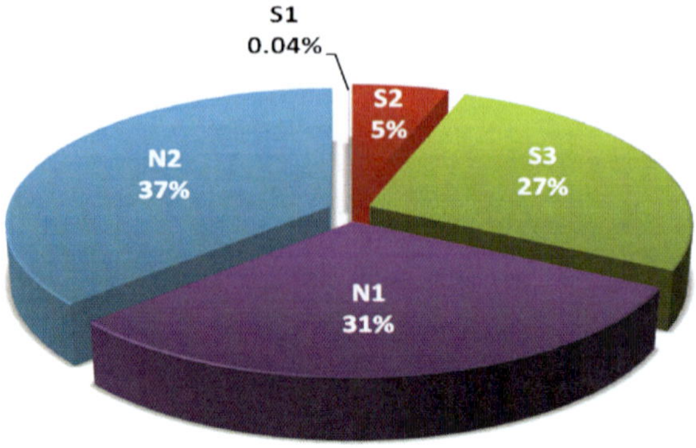

Fig. 1.10 Percent distribution of irrigated agriculture suitability classes in the Emirate

are useful for regional assessment and planning. They provide a regional perspective on restrictions or suitability for the particular land uses. They are by no means site-specific and do not eliminate the need for detailed onsite investigation.

The irrigated suitability map thus prepared for the Emirate is presented in Fig. 1.11; various soil suitability units in relation to general soil types and area occupied by suitability classes in the Emirate are briefly described below.

1.5.1.1 S1: Land That Is Highly Suitable for Irrigated Agriculture (2,000 ha = 0.04 %)

The soil is capable of producing sustained high yields for a wide variety of climatically adapted crops. The soils are nearly level and well drained. They are deep, fine sandy textured or finer, single grained thereby allowing for easy root penetration and retention of abundant air and water in the root zone. The soils have low soluble salts, sodicity, gypsum content, calcium carbonate content, and a neutral pH. Soils selected in this category in Abu Dhabi are lighter textured and contain more gravel and carbonate than might be considered highly suitable in other locations; however, these criteria are appropriate for the range of soils available in the Emirate.

1.5.1.2 S2: Land of Moderate Suitability for Irrigated Agriculture (309,000 ha)

The soils are inherently lower in productive capacity than soils ranked S1. Soil and land qualities may impose restrictions on irrigation; however, these restrictions may be relatively easily corrected or compensated for. S2 soils in Abu Dhabi have sandy texture and are single grain or massive. They are deep and somewhat excessively or well drained. The soils are typically very slightly saline, non-sodic,

Table 1.7 Limitations to land use for irrigated agriculture, management options, and corrective measures

Limitation class	Limitation	Management options/corrective measures
Depth to rock or hardpan	Efficient irrigation requires sufficient soil depth for plants to grow their roots without restriction as well as additional depth for leaching and drainage of salts. If a root-restricting horizon occurs close to the surface, there is only a limited volume of soil from which plants can extract water and nutrients. Low permeability of such a horizon may lead to periodic waterlogging and prevent the leaching of salts	Root-restricting layers such as hardpan can be broken down by deep ripping with a tined implement. However, the pans may reform. Consideration should be given to planting shallow rooting crops (e.g., onions) where root-restricting layer occurs within a meter of the surface. It will be necessary to determine if sufficient percolation to leach salts which accumulate in the soil as a result of irrigation can occur in these soils Irrigation scheduling and monitoring of soil-water to ensure sufficient water was continually available to plant roots while avoiding waterlogging due to ponding above the hardpan should also be undertaken. Appropriate drainage system to be installed at a suitable soil depth to remove drainage water The use of mulches and soil ameliorants may also be useful management tools Shallow soils require monitoring of soil salinity and water in conjunction with irrigation scheduling to ensure sufficient water is available to offset plant water requirements and to control salt buildup. To avoid ponding, these soils require suitable irrigation scheduling or artificial costly drainage systems
Water table depth	Below the water table, the soil is saturated and will restrict root growth. There may be not enough volume of soil from which plants can extract water and nutrients Efficient irrigation requires the leaching of salts and removal of excess water from the root zone. If the water table level is within this zone, then salt and water will build up	Water table levels can be lowered by building drains to remove excess water and salt. Careful irrigation scheduling will be required to ensure sufficient water is available to the plant roots and to avoid the water table rising so saturating the soil

(continued)

Table 1.7 (continued)

Limitation class	Limitation	Management options/corrective measures
Salinity (EC)	The potential and sustainability of irrigated agriculture are reduced by the presence of salts in the soil. High salinity adversely affects the growth of salt-sensitive plants. The high salinity restricts water intake by plants and cause physiological drought in plants	Flushing of saline soil with freshwater will leach salts from the root zone. However, care should be taken to monitor where the salts are flushed to, as a new salt problem may be generated. The risk of induced waterlogging must also be managed
	Excess salts also add impurities into other mineral resources such as gypsum and anhydrite and affect commercial value	Salt-tolerant crops/plants (biosaline agriculture) are the only options for exploiting highly saline conditions. On slightly saline soils, highly sensitive crops such as lettuce, onion, and pepper should be avoided
		Salts are to be removed to improve commercial value using appropriate technology
	Excess salts such as gypsum can corrode underground concrete and other metallic pipelines	Coating with suitable sealant around concrete pipes can control corrosion
Gypsum	Gypsum is partially soluble in water. Soils with >10 % gypsum may collapse if gypsum is removed by percolating water. This problem is exacerbated by irrigation practices. Small quantities of gypsum have positive effects in offsetting sodicity hazard of irrigation water. High gypsum also affects nutrient availability to plants	If subsidence is a problem, flushing of the soil with freshwater to dissolve and remove gypsum from the profile may be attempted prior to establishment of irrigation
		To be effective, this solution is likely to require considerable time and large volumes of freshwater. If gypsum is flushed from one area to another, the problem may merely be moved. Highly gypsic soils should be avoided
Relief	High areas, primarily dune height, will make cultivation and water management difficult for field crops as well as reducing the efficiency of farming operations	Leveling of dunes is used to overcome this problem. The practicality of leveling a dune becomes more difficult the higher the dune
Slope	Areas with steep slopes make cultivation and water management difficult for field crops as well as reducing the efficiency of farming operations	Slope is only prohibitive to flood irrigation practices. Leveling of dunes is used to overcome this problem. Sprinkler irrigation may be considered on sloping ground

Texture	Soil texture influences the ability of the soil to retain and release water and nutrients to the plant roots. The surface texture will also impact on how the soil is tilled for use. Seepage of water through the soil may be associated with sandy soils	For sandy soils minimize the tillage and maintain a surface vegetative cover or other covering to reduce erosion potential. Windbreaks are other options. Uses of organic matter and organic fertilizers can improve soil physical health and nutrient and water-holding capacity
	A sandy soil will be susceptible to erosion, and a clayey soil may be too sticky when wet; ideal textures are loamy soils	Clay, organic materials, and soil polymers can be used as soil ameliorants to improve the water- and nutrient-holding capacity of the sandy soils found in Abu Dhabi
		Careful nutrient management will be required to reduce nutrient losses through leaching as sandy soils have very high leaching capacity. Application of nitrogen fertilizers containing NO_3–N should be avoided. Fertilizer requirement can be split into small. Foliar application of micronutrients can be used for field crops
	If sandy soils are to be used as waste landfill sites (trench), the walls of the excavation can cave in or slough (cut banks cave), and for landfill (area) the excessive leaching of soluble materials can contaminate groundwater	Measures such as construction of compacted or hard layer, clay liners, or suitable geotextile layers can be used to avoid contamination at lower depths
		A trench can be excavated in "steps" or the sides of the trench angled out from the base near to the angle of repose of the soil
Flooding	Soil flooded by moving water from stream overflow, runoff, or high tides	Locate land use above high tide level and off active flood plains. Alternatively, for restricted areas, use protective structures such as levees and bunds to protect high-value infrastructure from flooding
Coarse fragments	These can range in size from gravel (2–75 mm) to rock fragments 75 mm or more across that adversely affect the specified use of the soil through a number of mechanisms	Where stones are present in the topsoil, crops requiring mechanical harvesting (e.g., potatoes) should be avoided as the stones may interfere with the harvesting machinery. Where localized patches of stones occur on the surface, they can be removed by raking or scraping with machinery. For forestry plantation (in pits), gravelly soil material is to be replaced with a suitable mixture of soil and organic material based on plant selection. For some tree crops, a surface gravel cover may prove an advantage by reducing soil evaporation.

(continued)

Table 1.7 (continued)

Limitation class	Limitation	Management options/corrective measures
	High quantities of gravels in potential resource deposits (e.g., gypsum, sand, clay) are a restriction to their use as resource material for soil improvement, cement factories, or construction material	Gravels can be removed through grinding and sieving at large scale to improve gypsum purity. Similarly, gravels can be separated through sieving to improve the quality of other materials such as sand, clay, sweet soil, and carbonates sources
Percs slowly	The slow movement of water through the soil adversely affects the specified use	Deep ripping with a tined instrument and application of gypsum and organic materials on selected soils can improve soil structure and water movement in the soil profile
Ponding	Standing water on soils in closed depressions that is removed only by percolation or evapotranspiration	Remove ponded water with shallow drains and other earthwork solutions. Alternatively plant vegetation suited to anaerobic environments in an effort to increase transpiration from the ponded area
Calcareousness	High levels of carbonates (most commonly calcium carbonate – $CaCO_3$) cause high pH and high buffering capacity and low nutrient availability to plants, particularly P, B, Fe, Cu, Mn, and Zn	Band placement of phosphorous fertilizers can be used to reduce reaction with soil material. Foliar sprays of micronutrients can be used
Mined material has low quality	The mined material has lower commercial value to be cost-effective, and the mining has impact on the environment	Prior to commercial mining of materials such as gypsum, gravels, anhydrite, and carbonates, a pilot-scale mining on a small area is recommended. Once it is proved that the material at a certain location has passed the criteria for commercial exploitation, it is only then large-scale mining is to be performed. In parallel to mining, impact of such practice on the environment is to be evaluated and suitable rehabilitation measures adopted

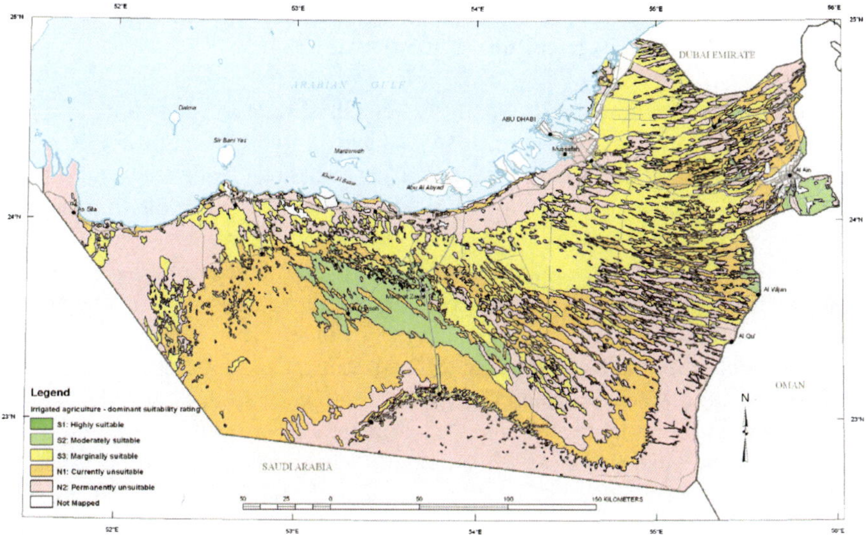

Fig. 1.11 Suitability for irrigated agriculture – map

have low gypsum content, and can have a hummocky microrelief. These moderate restrictions may be overcome with appropriate management strategies.

1.5.1.3 S3: Land of Marginal Suitability for Irrigated Agriculture (1,550,000 ha)

The productive capacity of S3 is less than S1 and S2 soils. The soils have severe limitations that may be corrected with appropriate management strategies. S3 soils in Abu Dhabi are moderately deep with a hardpan or water table occurring within 100–150 cm of the soil surface. They have sand to sandy loam textures and are single grained or massive. They are typically slightly saline and have moderate gypsum contents. These soils may have moderately steep gradient (up to 32 %) with moderately high relief (up to 9 m).

1.5.1.4 N1: Land That Is Currently Unsuitable for Irrigated Agriculture (1,753,000 ha)

Such soils in Abu Dhabi typically have shallow rooting depths with hardpans within 50–100 cm of the soil surface, high gypsum content close to the surface, or high relief (up to 30 m) and steep gradient (up to 56%).

1.5.1.5 N2: Land Considered to Be Permanently Unsuitable for Irrigated Agriculture (2,108,000 ha)

The N2 category includes soils which are very shallow, occur with rock outcrops, are on very steeply sloping land (over 56%), have a very high relief (over 30 m), are very poorly drained and strongly saline, or have shallow depth to gypsum. Under conventional economic conditions, these soils do not warrant further investigation for irrigation purposes.

The results of this evaluation indicated that well-drained Torripsamments and Haplocalcids with good deep drainage are the preferred soils for irrigated agriculture in Abu Dhabi. Other soils containing low quantities of gypsum and calcium carbonate may also be suitable, but again deep drainage must be adequate to remove excess irrigation water and facilitate the removal of excess salt through leaching.

In areas where impermeable layers underlie the eolian sands at a shallow depth, there is a danger that rising saline water tables may develop rapidly. The scale and frequency of sand dunes are a further constraint to the utilization of these otherwise suitable soils. High dunes generate long, moderately steep slopes, and frequent dunes create an irregular topography and are difficult to manage for agricultural purposes. However, perhaps with the exception of the bigger dunes, dune leveling is a feasible, though expensive, management option in these areas. Other soils are generally considered unsuitable for irrigated agriculture due to the shallow depth to hardpan (Lithic subgroups, Petrogypsids, and Petrocalcids), high salinity (Aquisalids and Haplosalids), or shallow depth to gypsum (Haplogypsids). The distribution of areas suitable for irrigated agriculture is shown in Figs. 1.10; 1.11. The most suitable areas for irrigated agriculture in the Emirate include a large elongated area of undulating sands running northwest to southeast, lying to the south of Madinat Zayed. A second area is indicated to the east of Jabal Hafit. Other much smaller areas lie near Al Ain and Al Wijan. Extensive areas of marginally suitable land occur inland from the coastal plains from the northeast of the Emirate, westward to Sabkhat Matti.

1.5.2 Limitations to Irrigated Agriculture and Management Options

A number of limitations and limitation classes to irrigated agriculture have been identified during the extensive survey. Management options to overcome these restrictions are addressed in Table 1.7.

1.5.3 Delineation of Area Having Potential for Irrigated Agriculture for Intensive Survey at 1:25,000 Scale

As stated earlier, one of the objectives of extensive survey was to delineate 400,000 ha area having potential for irrigated agriculture in the Emirate for fur-

Table 1.8 Extent of subareas for intensive survey

Subarea	Mapped area (ha)
Al Ain	198,596
Madinat Zayed	116,146
Ghayathi	105,355
As Sila'	27,809
Total	447,906

ther survey at 1:25,000 scale using norms and standards of second-order USDA protocols, modified to fit Abu Dhabi conditions. Using the extensive survey results and the irrigated suitability map (Fig. 1.11), 1 million ha area was delineated (Fig. 1.12), from which an area of 447,906 ha divided into four subareas was selected based on four broad criteria: (1) availability of suitable soil/landscape map units, (2) availability of suitable water resources, (3) access to infrastructure and workforce to support any development, and (4) strategic issues. The first two criteria were given in the highest weighting, while the third was used to corroborate the selection. The relative size of the four areas surveyed is shown in Table 1.8. The area mapped includes map units interpolated between or extrapolated away from survey sites.

The area (having the potential for irrigated agriculture) delineated was over 400,000 ha as described earlier, it was surveyed at intensive survey level, and results can be consulted elsewhere (EAD 2009b).

1.6 Future Opportunities to Use Emirates Terrestrial Resources and Policy Development

The extensive survey identified a number of issues facing the Emirate and further work that needs to be undertaken to build on this foundation. Such future work may include policy, research and development, and education and extension related to soil management.

1.6.1 Policy

The best management of the Emirate's soils will result from a planned, well-implemented approach based on policies that oversee research, development, education, regulation, and enforcement. Establishing and implementing the required policies must take account of these other objectives. This means that in addition to ensuring that adequate ongoing support is given to soil specialists with local knowledge, the activities of those specialists must be linked to the people and groups who are managing

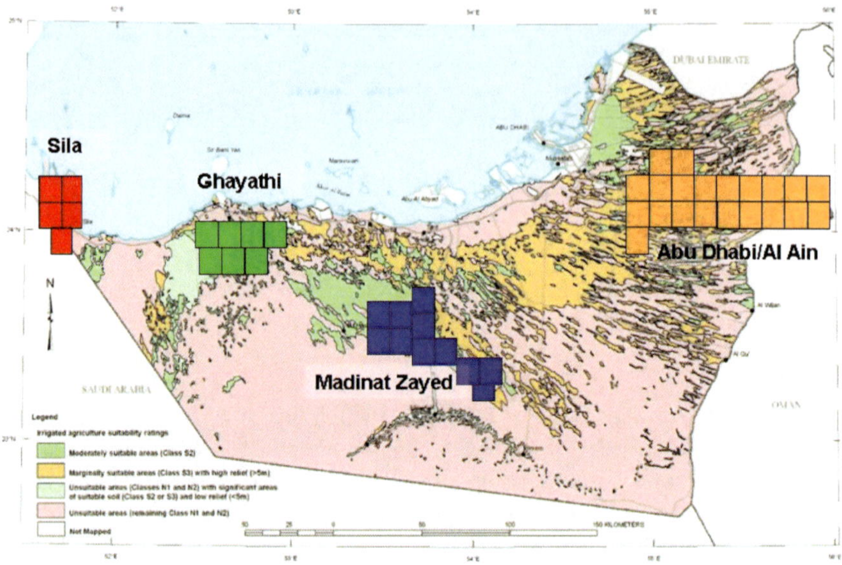

Fig. 1.12 Location of subareas in relation to 1 million ha of land identified in the extensive soil survey as the most suitable for irrigated agriculture (*green color* area)

the larger, integrated activities. Failing to provide the former (support for the specialists) will result in uninformed decisions being taken about soil management.

1.6.2 Establish a Soil Monitoring Program

The monitoring of soil quality is likely to be linked to monitoring of other environmental conditions (e.g., rangeland condition, wildlife status). In addition, soil monitoring by itself may not provide enough benefit in its own right to justify a single-focus monitoring program. However, it is an important part of an overall environmental monitoring program. This can be separated into at least two elements: monitoring of commercially used soil (primarily farmland), which has an impact on medium- to long-term economics, and monitoring of public lands (rangeland and nature reserves), which can impact on the ability of the land to support land uses of public benefit and avoid negative impacts on the community (e.g., excessive dust storms).

1.6.3 Establish a Soil Research, Development, Education, and Extension Program

In order to develop and implement an appropriate set of activities, it is recommended that the Emirate appoints a small group to oversee the establishment and management of a program of research, development, and education. This group should include a core of people with expertise in soil science, policy development, research implementation, public education, and contract management. Integration of soil data with other natural resource and cultural and political datasets allows it to be used for decision making, anticipating questions regarding land use that will arise, and using experienced soil people to correctly interpret the data to assist with providing answers. The group should work with other agencies and the private sector to maintain a technical knowledge base on the soils for the Emirate; recommend activities for partner organizations to undertake; implement research, development, and education activities, probably through subcontracts; and recommend policy settings for regulators to enforce in relation to activities that have significant impact on soils. Such a group will necessarily be part of a bigger organization because soil management objectives cannot be achieved in isolation from the real-world economy.

1.6.3.1 Integrated Soil Salinity Management

The huge amounts of salt in the soils and waters of the Emirate are one of the key features that impact on the management of land. The many factors that are impacted by the presence of salt mean that an integrated approach must be taken to its management.

1.6.3.2 Amelioration of Saline Soil

Options for saline soil management which could be explored as parts of an integrated management approach are physical (soil structure improvement and breaking hardpan), chemical (using soil ameliorants such as gypsum), hydraulic (understanding the leaching requirements of particular soils, drainage system), and biological (biosaline agriculture, where appropriate).

1.6.3.3 Irrigation Using Brackish Water

Water supplies most readily available for irrigation in the Emirate are brackish. This raises several issues that interact with the soil. There is a need to overwater in order to avoid accumulation of salts. Because of issues with shallow water tables and hardpans, it would be valuable to understand the interaction between irrigation strategy and soil properties in their impact on water use and productivity. Better use of saline and freshwater resources could be achieved by investigating when particular plant species can tolerate the existing groundwater resource and designing a farming system that utilizes freshwater during critical growth periods. This may need to be customized for each particular crop and saline water during "maintenance growth" periods. Part of this approach could involve using hydroponic systems for agriculture and using halophyte plants in all aspects of agricultural development (e.g., landscape, production farms, and agro forest).

1.6.3.4 Living with Shallow Saline Water Tables

Research should target this problem through (a) surveying the current extent of elevated water tables under plantings and relating any rise to soil type, longevity of the planting, irrigation methods, and landscape; (b) understanding the impact of rising water tables on productivity of the Emirate's plantings, initially by surveying existing plantings and examining the relationship between water table level, salinity, soil properties, and productivity and then, if required, undertaking controlled experiments to study this impact; and (c) forecasting the future development of problems by extrapolating the principles established in the research using modeling techniques with varying scenarios and so identifying the location, timing, and severity of future problems.

1.6.3.5 Selection of Salt-Tolerant Plants and Dewatering Options

The effect of salinity on plant growth may be magnified many times in the presence of waterlogging. Research should be undertaken into the particular interactions that apply in Abu Dhabi's environment, particularly accounting for shallow water tables and the potential overwatering through inefficient irrigation. If significant occurrence of induced shallow water tables is found, it may be worth considering evaluating efficient methods of drawing them down and ameliorating the salt bulges associated with them. An interesting alternative to plant production systems in addressing the issue of shallow water tables could be through development of artificial lakes for biodiversity or aquaculture (e.g., for brine shrimp or prawns or algae for beta-carotene and stock feed). It would be useful to understand the soil properties that would be best suited to the construction of such artificial lakes.

1.6.3.6 Understanding Shallow Hardpans

As with shallow water tables, Abu Dhabi Emirate has a high proportion of soils with shallow or very shallow hardpans. These vary in depth and in their nature. In some cases, they may be sufficiently fractured to allow water movement and root growth through them; however, hydraulic properties of the Emirate's hardpans and their effects on plant productivity are unknown. The management of areas with hardpans can include deep cultivation to disrupt the hard layer and so improve soil conditions. Whether this technique is effective on the local hardpans is unknown. Undertaking research to better understand these issues would allow a more rational approach to their management.

1.6.3.7 Gypsum and Soil Subsidence

In the extensive soil survey of Abu Dhabi Emirate, the presence of gypsum has been identified as a limitation for irrigated agriculture. Yet many existing farms have been established on soils with enough gypsum to be concerned that subsidence will occur as a result of irrigation water dissolving gypsum and creating voids in the soil profile. Research could focus on identifying the likely distribution of the problem and the high-risk areas, assessing the likely practical impacts and severity of the problem (e.g., by calculating the amount of subsidence if all the gypsum present was to dissolve), and evaluating the cost effectiveness of possible management options (e.g., importing additional topsoil to infill the subsided areas).

1.6.3.8 Managing Nutrients in a Desert Environment

Abu Dhabi's soils have little clay and very low organic matter levels, very high $CaCO_3$, and pH higher than optimum level. This results in soils' low nutrient-holding and availability capacity. Several inorganic nutrients (e.g., NO_3) leach readily in such conditions leading to high costs of fertilizer inputs and high risk of off-site pollution by the outflow of nutrients. Others are fixed in soil due to high $CaCO_3$ and pH (P, Fe, Cu, Mn, Zn). There is a significant body of knowledge in relation to soil-nutrient interactions which the Emirate can capitalize on. In addition to farmland, there is an opportunity to understand the cycling of nutrients in the natural, desert soils of the Emirate to contribute to an understanding of the management of the natural ecosystem and wildlife reserves.

1.6.3.9 Managing Fragile Soils in Grazing Systems

The best way to protect Abu Dhabi's fragile soils, particularly given its extremely harsh climate, is to have controlled grazing. The grazing impact on ecosystem can be minimized through adopting a conservative approach to grazing management. Commercial animal industries should be limited to as small an area as possible, consistent with productivity and animal welfare constraints, and those areas should be restricted to soils that are most resilient under animal grazing; the grazing should be managed carefully to minimize impact, and protective measures such as wind-break plantings should be implemented.

1.6.3.10 Fundamental Soil Research

While the topics suggested above as future opportunities in this section target issues that have clear, practical applications, it is also important to ensure the accuracy and rigor of the fundamental knowledge that underpins those immediately applicable activities. The following recommendations target specific research that is of particular relevance to Abu Dhabi and is not likely to be addressed elsewhere. These topics offer the opportunity for minimal investments that often suit being studied at universities or specialist research organizations, thereby also helping build the capacity of local scientists to work on the Emirate's soils in the future.

References

Abdelfattah MA, Shahid SA (2007) A comparative characterization and classification of soils in Abu Dhabi coastal area in relation to arid and semiarid conditions using USDA and FAO soil classification systems. Arid Land Res Manag 21:245–271

Boer BB (1999) Ecosystems, anthropogenic impacts and habitat management techniques in Abu Dhabi. University of Paderborn, Paderborn

Burt R (2004) Soil survey laboratory methods manual. United States Department of Agriculture, Natural Resources Conservation Service, Soil survey investigations report no 42, Version 4.0, Washington DC

Dawoud MA (2008) Strategic water reserves: new approach for old concept in GCC countries. Environment Agency, Abu Dhabi

EAD (2009a) Soil survey of Abu Dhabi Emirate, vol I, Extensive survey. Environment Agency, Abu Dhabi, p xx+506

EAD (2009b) Soil survey of Abu Dhabi Emirate, vol III, Intensive survey. Environment Agency, Abu Dhabi, p xviii+435

EAD-ICBA (2011) Proposal for adding anhydrite soils to USDA soil taxonomy. Proposal to amend soil taxonomy. Jointly submitted by Environment Agency – Abu Dhabi and International Center for Biosaline Agriculture. Proposal can be accessed at USDA website http://soils.usda.gov/technical/classification/taxonomy. Last Accessed 25 Oct 2011

ERWDA (2004) Soil survey for the coastline of Abu Dhabi. Environmental Research and Wildlife Development Agency, vol 1: Report, vol II: Soil maps

FAO (1976) A framework for land evaluation, FAO soils bulletin no 32. Food and Agricultural Organization, Rome

FAO (1983) Guidelines: land evaluation for rainfed agriculture, FAO soil bulletin 52. Food and Agricultural Organization, Rome

FAO (1984a) Guidelines: land evaluation for extensive grazing, FAO soil bulletin 58. Food and Agricultural Organization, Rome

FAO (1984b) Guidelines: land evaluation for forestry, FAO forestry paper 48. Food and Agricultural Organization, Rome

FAO (1985) Guidelines: land evaluation for irrigated agriculture, FAO soil bulletin 55. Food and Agricultural Organization, Rome

KISR (1999) Soil survey for the State of Kuwait, 5 vols. Kuwait Institute for Scientific Research, Kuwait

Klingbiel AA, Montgomery PH (1961) Land capability classification, Agricultural handbook. Soil Conservation Service, United States Department of Agriculture, Washington, DC

MAF (1990) General soil map of the Sultanate of Oman. Ministry of Agriculture and Fisheries, Directorate General of Agricultural Research, Muscat

MAF (1993a) South Batinah integrated study. Summary of conclusions and recommendations, vol 1, 1st edn. Directorate General of Agricultural Research, Ministry of Agriculture and Fisheries, Muscat, p 54

MAF (1993b) South Batinah integrated study, vol 2, 1st edn, Land resources. Directorate General of Agricultural Research, Ministry of Agriculture and Fisheries, Muscat, p 741

MAF (1993c) South Batinah integrated study. Part 1: Land use; Part 2: Irrigation report; Part 3: Crop water requirements report; Part 4: Farming systems report, vol 3, 1st edn. Directorate General of Agricultural Research, Ministry of Agriculture and Fisheries, Muscat, p 150

McKenzie NJ, Coughlan KJ, Cresswell HP (eds) (2002) Soil physical measurement and interpretation for land evaluation. CSIRO, Collingwood

Ministry of Agriculture and Water (1985) General soil map of the Kingdom of Saudi Arabia. Ministry of Agriculture and Water, Riyadh

National Media Council (2007) UAE at a glance, online at http://www.uaeinteract.com/uaeint_misc/glance//-ataglance.pdf

Raafat H (2006) Climate. In: Physical geography sector paper, AGEDI initiative. Environment Agency, Abu Dhabi, pp 72–92

Scheibert C, Stietiya MH, Sommer J, Abdalla OES, Schramm H, Memah Al M (2005) The atlas of soils for the State of Qatar. Ministry of Municipal Affairs and Agriculture, Doha

Schoeneberger PJ, Wysocki DA, Benham EC, Broderson WD (eds) (2002) Field book for describing and sampling soils, version 2.0. Natural Resources Conservation Service, National Soil Survey Center, Lincoln

Shahid SA (2007) Soils. In: Physical geography sector paper, AGEDI initiative. Environment Agency, Abu Dhabi, pp 6–41

Shahid SA, Abdelfattah MA (2008) Soils of Abu Dhabi Emirate. In: Perry RJ (ed) Terrestrial environment of Abu Dhabi Emirate. Environment Agency, Abu Dhabi, pp 71–91

Shahid SA, Abdelfattah MA, Wilson MA (2007) A unique anhydrite soil in the Coastal Sabkha of Abu Dhabi Emirate. Soil Surv Horiz 48:75–79

Soil Moisture Corp (2005) Operating Instructions April 2005, Eijkelkamp Agrisearch Equipment ZG Giesbeek The Netherlands

Soil Survey Division Staff (1993) Soil survey manual, USDA handbook no 18. U.S. Government Printing Office, Washington, DC

Soil Survey Staff (1999) Soil taxonomy – a basic system of soil classification for making and interpreting soil surveys, 2nd edn. United States Department of Agriculture, Natural Resources Conservation Service. Agriculture handbook number 436

Soil Survey Staff (2006) Keys to soil taxonomy, 10th edn. United States Department of Agriculture, Natural Resources Conservation Service, U.S. Government Printing Office, Washington, DC

USDA (1954) Diagnosis and improvement of saline and alkali soils, Agriculture hand book no 60. USDA, U.S. Government Printing Office, Washington, DC

USDA (2005) US Department of Agriculture, Natural Resources Conservation Service, 2005. National soil survey handbook, title 430-VI

USDA-NRCS (1995) Soil survey laboratory information manual, Soil survey investigations report no 45. National Soil Survey Center, Lincoln

Chapter 2
Demands on Soil Classification and Soil Survey Strategies: Special-Purpose Soil Classification Systems for Local Practical Use

R.W. Fitzpatrick

Abstract Classifying soils for a particular purpose involves the ordering of soils into groups with similar properties and for potential end uses. The classification of soil is a terrific conceptual and practical challenge, especially in arid environments. The challenge may spur on, or it may deter scientists or end users with an interest in soils. If a classification system proves to be relevant and user-friendly, it stimulates and encourages further work because it is recognised for its inherent capacity to create order and enhance the useful understanding and mapping of soils. General-purpose, internationally recognised soil classification systems such as Soil Taxonomy and the World Reference Base and other nationally recognised classification systems (e.g. Australian or South African) have proved to be tremendously useful for soil classification and advancing understanding of soils across the world. However, because the use of these general-purpose classifications requires considerable expertise and experience, there is a need for complementary special-purpose classification systems that are specifically tailored, for example, to particular environmental problems, land uses or local regions and that use plain language descriptions for soil types. General-purpose classification systems often lag in the incorporation of new terminologies, for example, classification of acid sulfate soils in the Murray-Darling Basin, Australia, has led to descriptions of soil types with subaqueous properties (submerged underwater), monosulfidic materials and hypersulfidic materials, to enable assessment of environmental risk and management options. In addition, new challenges face general-purpose soil classification systems, especially in

R.W. Fitzpatrick (✉)
CSIRO Land and Water, Glen Osmond, SA 5064, Australia

School of Chemistry, Physics and Earth Sciences,
Flinders University, Adelaide, SA, Australia
e-mail: Rob.fitzpatrick@csiro.au

S.A. Shahid et al. (eds.), *Developments in Soil Classification, Land Use Planning and Policy Implications: Innovative Thinking of Soil Inventory for Land Use Planning and Management of Land Resources*, DOI 10.1007/978-94-007-5332-7_2,

response to the following questions most frequently asked by soil users: (1) what soil properties are changing vertically and laterally in landscapes and with time, especially in acid sulfate soils? and (2) what are the most suitable approaches for characterising, monitoring, predicting and managing soil changes for environmental impact assessments, pollution incidents, waste management, product development and technology support? The purpose of this chapter is to address these challenges by presenting new ideas and concepts on how best to predict and solve practical problems by focussing on the development of special-purpose or more technical soil classification systems, which use plain language names for soil types. To demonstrate the critical importance of developing special-purpose technical soil classifications, the following five case studies are presented, which tackle difficult problems involving highly complex issues: (1 and 2) soil and water degradation in large aquatic environments from the River Murray and Lower Lakes region in South Australia (changing climatic and anthropogenic modified environments) and from the Mesopotamian marshlands in Iraq (anthropogenic modified arid environment); (3) acid sulfate soil as a new geochemical sampling medium for mineral exploration; (4) soil damage to the Australian telecommunication optic fibre cable network from shrink-swell soils and soil corrosion; and (5) soil landscape features to assist police in locating buried objects in complex terrain.

Keywords General-purpose • National • State • Regional • Special-purpose technical soil classification systems • Optical fibre cables • Acid sulfate soils • Mineral exploration • Forensic soil science

2.1 Introduction

Classifying soils for a particular purpose involves the ordering of soils into groups with similar properties and for potential uses. In general, soil classification systems currently used in most countries involve the use of the following three broad systems (Fitzpatrick 2004a): (1) general-purpose broad soil classifications, which communicate soil information at international scales (e.g. Soil Taxonomy (Soil Survey Staff 1999, 2010) and World Reference Base (2006)) and national scales (Australian (Isbell 1996)); (2) state, provincial or regional soil classifications, which are designed both to assist with 'user-friendly' communication of soil information and to account for the occurrence of soils that impact on existing and future industry development and prosperity (e.g. Western Australia (Schoknecht 2001) and South Australia (Hall et al. 2009)); and (3) special-purpose and more technical classification systems, which are used for local or single-purpose applications. These systems generally involve using detailed soil assessment criteria with recommendations for specific soil management practices for a range of specific industries (e.g. use of acid sulfate soils in mineral exploration (Skwarnecki and Fitzpatrick 2008)).

2.2 General-Purpose Soil Classification Systems: A Re-evaluation

The international (Soil Taxonomy: Soil Survey Staff 1999, 2010; the World Reference Base 2006) and national (Australian Soil Classification or ASC: Isbell 1996) soil classification systems have regularly provided soil scientists across the world with a valuable broad conceptual understanding of soils in terms of the morphological, chemical and physical differentiation of profiles, the relative development of various complex horizons, soil reaction trends, subsoil colour and mottle differences. However, a persistent problem with general-purpose soil classifications has been the limited appreciation of how the utility of a general-purpose soil classification is constrained by contrasting patterns of covariance in soil populations, especially at local, regional and national levels (Fitzpatrick 2004a; Fitzpatrick et al. 2003). For practical reasons, most general-purpose classification systems have depended heavily on morphological criteria. However, covariance between soil morphological properties and more practically relevant chemical, physical and mineralogical properties is complex and sometimes poor.

The substantial literature on spatial variability (Sanchez et al. 2009; Beckett and Webster 1971; Wilding and Drees 1983; Burrough 1993) demonstrates that soil properties have varying levels of covariance. As a consequence, we should have modest expectations of the capacity of general-purpose classification systems to discriminate soil types effectively when assessed against a broad range of criteria (e.g. soil fertility, hydraulic properties, engineering) across a wide range of soil conditions. In fact, several critics contend that general-purpose systems have inherent, and sometimes severe, limitations (e.g. Sanchez et al. 2009; Webster and Butler 1976; Webster 1977; Butler 1980; Dudal 1987; White 1993, Yaalon 1996), and Fitzpatrick (2004a) suggests the decline in the use of general-purpose soil classification systems can be ascribed to a range of factors, including (1) the specialised terminology used to name and classify soils in soil map legends and reports and the range of systems in use for classifying soils; (2) the need for adjustments in soil survey techniques and soil classification to meet the requirements of potential users; (3) insufficient attention given to presenting information in an accessible, purpose-orientated, user-friendly language and format; and (4) inadequate use of soil class criteria that are important to land use (i.e. physical properties such as porosity, infiltration rate and permeability of the surface layer of soil) and an overemphasis on taxonomic class criteria.

2.2.1 Needs of End Users

Some critics have also dismissed general-purpose broad soil classification systems as having limited value for soil fertility work (White 1996). However, classification systems do have considerable value for stratifying behaviour of soil groups in terms of nutrient dynamics. Such recognition (Fitzpatrick et al. 1999; NLWRA 2001) has led to more constructive dialogue between the soil testing and pedological communities in Australia. For example, according to Drohan et al. (2010): '… Some could effectively argue that in the last 20 years, the detail of Soil Taxonomy, and the information one needs to classify soil, has moved beyond the knowledge of non-soil scientists and even most soil scientists working outside of pedology (let's put aside the costs of characterizing a profile, too). This must change or Soil Taxonomy will certainly go the way of most extinct languages. How do we do this? Move Soil Taxonomy to a three-dimensional, digital, high-definition format, world-wide web interface that is landscape based in its presentation.'

There is also a recent general trend of soil and land resource survey programmes moving away from using conventional soil mapping procedures that use general-purpose soil classification systems to produce maps by means of a series of polygons, which limits resolution. According to Sanchez et al. (2009), these polygon-based maps do not adequately express the complexity of soils across a landscape in an easily understandable way. To address these many shortcomings, soil scientists have already used digital soil mapping techniques. A digital soil map is a spatial database of 'easy-to-measure soil properties', based on a statistical sample of landscapes (Sanchez et al. 2009). However, spatially inferred soil properties are used to predict the more difficult-to-measure soil properties, such as the potential for sulfides in acid sulfate soils to oxidise and form sulfuric acid. These inferred soil maps are produced using pedotransfer functions (Sanchez et al. 2009), which may include using (1) a special-purpose, technical classification system such as those used for classifying acid sulfate soils in a specific region and (2) soil-regolith toposequence conceptual models (see examples below).

2.3 Special-Purpose Soil Classification Systems

Special-purpose, technical classification systems have been devised and designed to cover a wide spectrum of practical issues and are required for finer scales of resolution. These include matching soils for viticulture and forestry (hardwoods and softwoods), engineering applications (defining best options for installing underground optical fibre cables), rehabilitation of disturbed mine sites, saline soils, coastal acid sulfate soils (with direct links to policy and jurisdiction), soil tillage (abrasive soils), topdressing soil for turf, urban planning for infrastructures and mineral exploration (Fitzpatrick 2004a). These special-purpose classification systems all involve using soil assessment criteria and also provide recommendations for improving soil management practices. These classification systems mainly rely on soil attributes but invariably also include relevant landscape features such as geology, terrain, vegetation, hydrology or soil chemical features. These together provide a more complete understanding of how soils and their properties vary and behave within landscapes and how this variability needs to be managed satisfactorily. Five contrasting case studies are presented in this chapter to illustrate how special-purpose soil classification systems are developed and utilised by a range of industries and institutions.

Special-purpose, technical classification systems are needed because general-purpose classification systems often do not yet incorporate new terminologies. For example, subaqueous soil features (submerged underwater), monosulfidic materials and hypersulfidic materials are critical soil properties required for classification of acid sulfate soils in the Murray-Darling Basin, Australia, to assess environmental risk and management options.

The advantages of using special-purpose, technical classification systems are that they are able to adequately address the following questions most frequently asked by soil users:

- What soil properties are changing, vertically and laterally in the landscape and with time?
- What are the most suitable approaches for characterising, monitoring, predicting and managing soil changes for environmental impact assessments, pollution incidents, waste management, product development and technology support?
- What soil measurements and user-friendly soil classifications are required to make suitable predictions about changing soil and landscape conditions and about sustainable land use?
- To what extent do soil processes and the management of soils influence engineering infrastructure and water quality?

These four inquiries can easily be addressed by using process-based technical classification systems that incorporate three or four dimensional soil landscape relationships (mechanistic toposequence models) to develop strategies to predict and manage spatial and temporal soil changes. This will also involve the development of special-purpose or more technical soil classification systems using plain language names for soil types for potential users with a range of practical needs. In turn, these technical classifications must be so constructed that they are able to contribute to modification or refinement of the general-purpose international and national classifications.

2.4 Objectives

The prime objective of this chapter is to present new ideas and concepts through a diverse range of five case studies to illustrate the importance of special-purpose technical soil classification systems to predict and solve practical problems among the following areas: (1) aquatic environments – special-purpose soil classification systems that describe temporal and spatial changes in soil and water characteristics of natural, drained and reflooded soils of the Mesopotamian marshlands in Iraq (anthropogenic modified environments) and River Murray and Lower Lakes region in South Australia (changing climate and anthropogenic modified environments); (2) mineral exploration – geochemical sampling medium for mineral exploration, based on a special-purpose soil classification system to identify specific types of acid sulfate soils overlying mineralised zones; (3) engineering – special-purpose soil classification to minimise soil damage from shrink-swell soils and soil corrosion to the Australian telecommunication optic fibre cable network; and (4) forensic soil science – special-purpose soil classification to assist police to locate buried objects in complex terrain.

This chapter outlines the benefits of providing practical soil class information for a broad range of land management issues required by land-holders, private enterprises, researchers and government agencies. There has been some excellent research leading to the development of numerical methods for soil classification (McBratney 1994) and digital mapping (Sanchez et al. 2009), but these are beyond the scope of this chapter.

2.5 Special-Purpose Soil Classification Systems: Five Case Studies

The following five case studies have been selected to illustrate some of the developments in devising special-purpose soil classification systems in Australia and Iraq and to solve a wide range of practical problems for end users. The case studies discussed are selective, but various other classification systems have also been developed and summarised in Fitzpatrick (2004a). These technical or special-purpose soil classification systems mainly rely on soil attributes but invariably also include other important environmental aspects such as geology, terrain, vegetation or hydrology, which may be relevant to a particular end user (Fitzpatrick 2004a, 2008). The combined use of soil and other information assists the understanding of how soils vary in landscapes so that strategies can be developed for managing both spatial and temporal changes within them.

2.5.1 Australia: River Murray, Lower Lakes and Tributaries

The current extreme drought in South-eastern Australia has had a major impact on the availability of water resources in the Murray-Darling Basin (MDB). Nowhere can this be more clearly seen than in the lower reaches of the River Murray, especially below Lock 1 (Blanchetown) where water levels are at an unprecedented low (below sea level at minus 0.80 m Australian Height Datum (AHD) in December 2009). The low water levels have caused a number of impacts related to inland acid sulfate soils (ASS) to be realised for the first time (Fitzpatrick et al. 2009a, c, 2010b). These include extreme soil acidification and more locally, water acidification and metal mobilisation. It is only over the past few years or so that the existence, extent and significance of these inland ASS has been fully appreciated. The nature, type and distribution of inland ASS, the environments in which they occur (lakes, rivers, wetlands) and the potential impacts on surrounding ecosystems make them more complex than their coastal equivalents.

Many of the ASS problems being experienced across much of the Lower Lakes region in South Australia are a consequence of wide-ranging and fundamental shifts in the 'environmental equilibrium' brought about by the impact of European settlement. These changes include building of locks and barrages to contain water flow and over-allocation of irrigation water as well as large-scale clearing of native vegetation. These changes have been exacerbated by extreme drought conditions since 2006, which have lowered water levels in rivers, lakes and wetlands. The effects of these changes have led to an accelerated accumulation, then drying and oxidation of inland ASS materials. The transformations of materials in inland ASS to acidic by-products arise from this disequilibrium.

Scientific studies have only recently begun to address the real impacts of these soils under future management and climate change scenarios, and much work still needs to be completed before future risks can be fully assessed (Fitzpatrick et al. 2009a, c, 2010a).

A representative case study in Lake Alexandrina and adjacent tributary (Finniss River) is summarised to illustrate the distribution, complex processes, environmental hazards and remediation options of inland ASS environments (Fitzpatrick et al. 2009a, c, 2010a). To aid in understanding the spatial heterogeneity of ASS properties and processes, soil landscape cross-sections in the form of conceptual soil-regolith toposequence models were constructed from field and laboratory data and surveyor knowledge (Fig. 2.1). Two categories of soil-regolith toposequence models have been found to be useful for ASS scenarios, namely, descriptive and predictive soil-regolith models. These models incorporate (1) photographs to illustrate the major soil ASS subtypes using the Australian ASS identification key (Table 2.1) and (2) soil-regolith-water processes involved and how to recognise specific types of ASS susceptible to land degradation. These figures provide the understanding needed to map ASS subtypes and underpin their best management.

2.5.1.1 Acid Sulfate Soil Technical Soil Classification System: Identification Key

Australia's current national soil classification (Isbell 1996) and other internationally recognised classification systems such as Soil Taxonomy (Soil Survey Staff 1999) require considerable expertise and experience to be used effectively. More importantly, these classification systems do not yet incorporate new acid sulfate soil terminologies such as (1) monosulfidic, hypersulfidic and hyposulfidic material (Sullivan et al. 2010) and (2) subaqueous soils, which are used in the nationally consistent legend of the 'Atlas of Australian Acid Sulfate Soils' (Fitzpatrick et al. 2008a; available on the Australian Soil Resource Information System: www.asris.gov.au).

To assist users to identify types and subtypes of soils, a special-purpose, technical classification system in the form of a user-friendly soil identification key (Tables 2.1 and 2.2) was developed to more readily define and identify the various types (Table 2.1) and subtypes (Table 2.2) of acid sulfate soil and nonacid sulfate soil (Fitzpatrick et al. 2008a, b, 2009a, c). The key is designed for people who are not experts in soil classification systems such as the Australian Soil Classification (Isbell 1996). Hence, it has been used to deliver soil-specific land development and

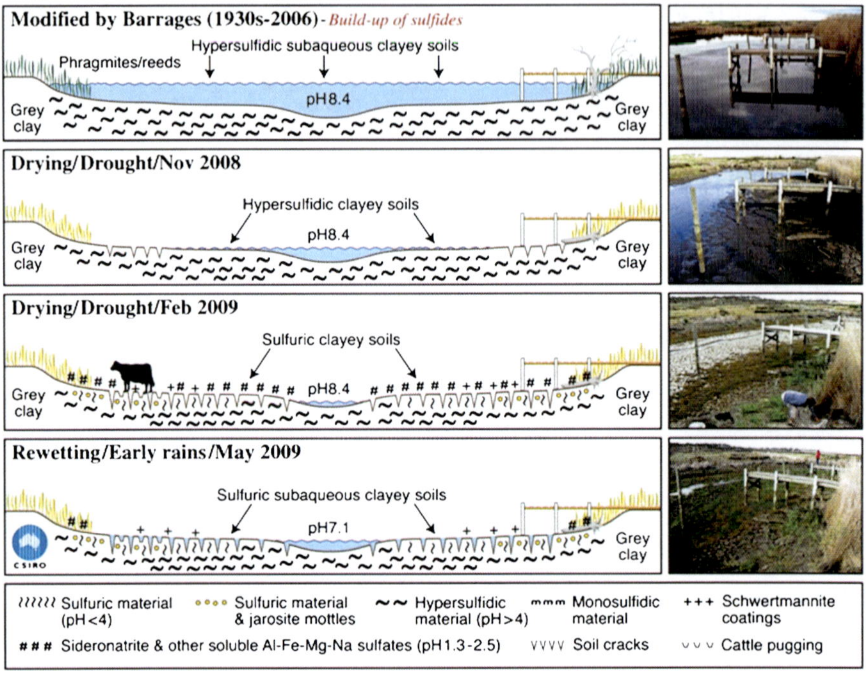

Fig. 2.1 Predictive soil-regolith toposequence models for the Finniss River at Wally's Landing illustrating modification of water levels by barrage installations causing the build-up of sulfides under continuous subaqueous ASS conditions from 1930s to 2006 followed by progressive drying (*middle* two panels) and finally a rewetting phase in May 2009 (*lower panel*) resulting in acidic waters in the cracks and cattle pugs and in running waters in adjacent wetlands (From Fitzpatrick et al. 2009a, c)

Table 2.1 Summary soil identification key for acid sulfate soil types

Diagnostic features for soil type	Soil type
Does the soil occur in shallow permanent flooded environments (typically not greater than 2.5 m)? No ↓ Yes →	Subaqueous soil 

(continued)

(continued)

Diagnostic features for soil type	Soil type
Does the upper 80 cm of soil consist of more than 40 cm of organic material (peat)? No ↓ Yes →	Organic soil 
Does the soil develop cracks at the surface, *Or* in a clay layer within 100 cm of the soil surface *Or* have slickensides (polished and grooved surfaces between soil aggregates)? *And* is the subsoil uniformly grey coloured (poorly drained or very poorly drained)? No ↓ Yes →	Cracking clay soil 

(continued)

(continued)

Diagnostic features for soil type	Soil type
Does a sulfuric layer (pH<4) occur within 150 cm of the soil surface? *And* is the subsoil uniformly grey coloured (poorly drained)? No ↓ Yes →	Sulfuric soil 
Does sulfidic material (pH>4 which changes on ageing to pH<4) occur within 100 cm of the soil surface? *And* is the subsoil uniformly grey coloured (poorly drained)? No ↓ Yes →	Hypersulfidic soil 
Other soils	Other soils

Modified from Fitzpatrick et al. (2009a)
After finding the soil type, use Table 2.2 to find the soil subtype

soil management packages to advisors, planners and engineers working in the Murray-Darling Basin.

The soil identification key uses non-technical terms to categorise acid sulfate soils and other soils in terms of attributes that can be assessed in the field by people with limited soil classification experience. Attributes include water inundation (subaqueous soils), soil cracks, structure, texture, colour, features indicating water logging and 'acid' status (soil already acidified, i.e. sulfuric material present, or soil

Table 2.2 Soil identification key for acid sulfate soil subtypes

Soil type	Diagnostic features for soil subtype		Soil subtype
[1]Cracking clay soil *Subaqueous soil*	Does hypersulfidic material occur within 100 cm of the soil surface? Does *hypersulfidic* material (pH > 4 which changes on ageing to pH < 4) occur within 100 cm of the soil surface?	Does a *monosulfidic black ooze (MBO)* material layer >10 cm thick occur within 50 cm of the soil surface?	Hypersulfidic subaqueous clayey soil with MBO
No ↓ Yes →	*And* does a clayey layer with slickensides occur within 100 cm of the soil surface? No ↓ Yes →	No ↓ Yes →	
	↓	→	Hypersulfidic subaqueous clayey soil
	↓	Does a *sandy or loamy* layer occur within 100 cm of the soil surface? No ↓ Yes →	Sulfidic subaqueous soil
	Does *sulfuric* material occur within 100 cm of the soil surface? No ↓ Yes →		Sulfuric subaqueous soil
			Subaqueous soil

(continued)

Table 2.2 (continued)

Soil type	Diagnostic features for soil subtype		Soil subtype
Organic soil	Does *hypersulfidic* material (pH >4 which changes on ageing to pH <4) occur within 100 cm of the soil surface?	Does a monosulfidic black ooze **(MBO)** material layer >10 cm thick occur within 50 cm of the soil surface?	Hypersulfidic organic clayey soil with MBO
No ↓ Yes →	*And*	No ↓ Yes →	
	Does a clayey layer with slicken-sides occur within 100 cm of the soil surface?		
	No ↓ Yes →		
			Hypersulfidic organic clayey soil
		Does a *sandy or loamy* layer occur within 100 cm of the soil surface?	Hypersulfidic organic soil
		No ↓ Yes →	
	Does *sulfuric* material occur within 100 cm of the soil surface?		Sulfuric organic clayey soil
	And		
	Does a clayey layer with slicken-sides occur within 100 cm of the soil surface?		
	No ↓ Yes →		
		Does a *sandy or loamy* layer occur within 100 cm of the soil surface?	Sulfuric organic soil
		No ↓ Yes →	

[a]*Cracking clay soil*	Does *hypersulfidic* material occur within 100 cm of the soil surface?	Does a monosulfidic black ooze (*MBO*) material layer >10 cm thick occur within 50 cm of the soil surface?	Hypersulfidic cracking clay soil with MBO
No↓ Yes →	*And*	No ↓ Yes →	
	Does a clayey layer with slickensides occur within 100 cm of the soil surface?		
	No ↓ Yes →	→	Hypersulfidic cracking clay soil
	Does *sulfuric* material occur within 100 cm of the soil surface?		Sulfuric cracking clay soil
	And		
	Does a clayey layer with slickensides occur within 100 cm of the soil surface?		
	No ↓ Yes →		Cracking clay soils
Sulfuric soil	Does *sulfuric* material occur within 100 cm of the soil surface?		Sulfuric soil
No↓ Yes →	No ↓ Yes →		
Hypersulfidic soil	Does *hypersulfidic* material *and* a sandy to loamy layer occur within 100 cm of the soil surface?	Does a monosulfidic black ooze (MBO) material layer >10 cm thick occur within 50 cm of the soil surface?	Hypersulfidic soil with MBO
No↓ Yes →		No ↓ Yes →	
		No ↓ Yes →	Hypersulfidic soil
Other soils			*Hydrosol – sandy or loamy*

[a]'Cracking clay soil' is equivalent to 'Vertosol' (Isbell 1996), for example, sulfuric cracking clay soil is similar to 'sulfuric Vertosol'. The latter terminology is used in the legend of the 'Atlas for Australian Acid Sulfate Soils' by Fitzpatrick et al. (2008a)

with the potential to acidify, i.e. sulfidic material present) and the depths at which they occur or change in the soil profile.

The key consists of a systematic arrangement of soils into five broad acid sulfate soil types, each of which can be divided into up to six soil subtypes. The key layout is bifurcating, being based on the presence or absence of particular soil profile features (i.e. using a series of questions set out in a key). A soil is allocated to the first type whose diagnostic features it matches, even though it may also match diagnostic features further down the key. The key uses a collection of plain language names for types and subtypes of ASS in accordance with the legend for the Atlas of Australian Acid Sulfate Soils (Fitzpatrick et al. 2008a, 2010a). It recognises the following five acid sulfate soil types: (1) subaqueous soils, (2) organic soils, (3) cracking clay soils, (4) sulfuric soils and (5) hypersulfidic soils (Table 2.1). These are further subdivided into 18 soil subtypes (Table 2.2) based on occurrence of sulfuric material, hypersulfidic material, clayey or sandy layers, monosulfidic material and firmness.

2.5.1.2 Application of Technical Soil Classification System

Predictive soil-regolith models: Simplified coloured cross-sectional diagrams and photographs have been used to illustrate the major soil-regolith-water processes involved and how specific subtypes of ASS susceptible to land degradation may be recognised (Fig. 2.1). These diagrams have also been used to help community groups to understand complicated scientific predictive processes and terminology and how this information can be used to underpin best management practices for ASS.

Management options: Water inundation inhibits further oxidation of sulfide and promotes anaerobic conditions for both sulfate and iron reductions – both of which are a source of alkalinity. Keeping ASS inundated will also prevent the potential odour problems that may result if certain subtypes of subaqueous ASS are exposed. Several hundred hectares of ASS with sulfuric material formed in Currency Creek (a tributary of the Lower River Murray) and wetlands of the Finniss River when they dried out over the summer of 2009 because of the low water levels in Lake Alexandrina (Fig. 2.1). Following rewetting in autumn and winter, acidic pools of water ($pH<4$) formed within the Currency Creek and Finniss River wetlands (Fig. 2.1). To counter the acidification risks, several management actions were implemented, including (a) managing water levels with structures such as temporary regulators built in the Goolwa Channel, Currency Creek and Lower Lakes and (b) neutralising acidity via bioremediation and dosing acidic water with ultrafine limestone.

2.5.2 *Iraq Marshlands*

The marshes of Mesopotamia, thought by biblical scholars to be the legendary site of the ‘Garden of Eden’, originally covering over 20,000 km^2 of interconnected lakes, mudflats and wetlands within modern-day Iraq and Iran, have now almost disappeared – and begun to transform to a desert, poisoned by the build-up of salt

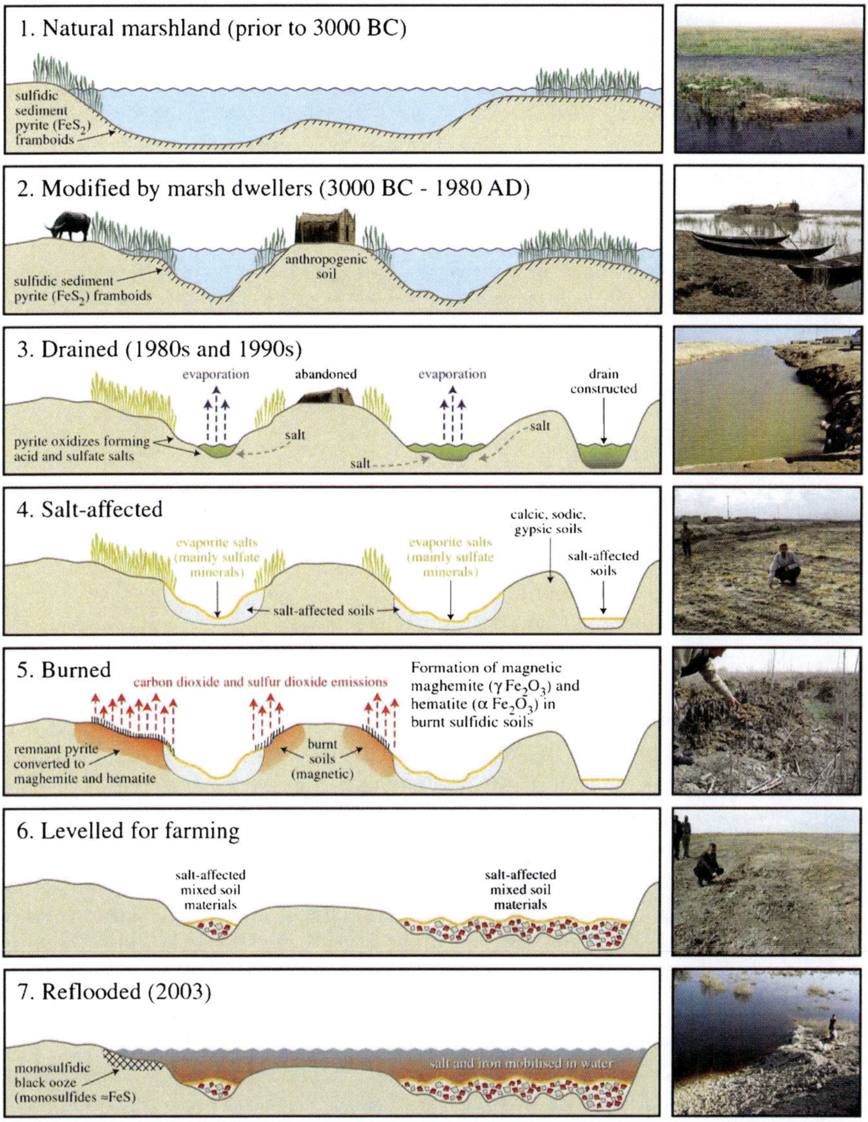

Fig. 2.2 Predictive soil-regolith toposequence model showing soil-water-landscape changes in the southern Mesopotamian marshlands from 3000 BCE to 2003 CE (present) (Modified from Fitzpatrick 2004b). *Notations BCE (before common era) = BC and CE (common era) = AD

and acid sulfate soils (Fitzpatrick 2004b). Historically this area between the Tigris and Euphrates rivers was a home to fish, millions of wading and migrating birds and the 5,000-year-old culture of the Marsh Arabs, living on artificial islands in houses made from tall reeds (Fig. 2.2). In what the United Nations has declared as 'one of the world's greatest environmental disasters', over 90% of the marshlands has changed by dewatering and the burning of reed vegetation through the combined actions of upstream damming and downstream drainage projects undertaken by the earlier regime of Iraq (Fig. 2.2).

Field and laboratory soil investigations undertaken across the marshlands in February 2004 (Fitzpatrick 2004b) gave rise to a comprehensive data set of soil properties, and soil-water-landscape models were also developed (Fig. 2.2). These investigations have advanced new concepts and practical information on the development of soils and minerals, forming after draining and reflooding. The salt crusts in the soils contain an assemblage of previously unrecorded types of sulfate-containing evaporite minerals, which are caused by the unique geochemistry of the region.

2.5.2.1 Special-Purpose Technical Soil Classification System for Mesopotamian Marshes

To assist users in the marshes of southern Mesopotamia to identify categories and subcategories of soils, a special-purpose, technical classification system in the form of a user-friendly soil identification key was developed (Table 2.3). The soil key was developed to enable easy identification of the various natural and degraded soils in the marshlands of southern Iraq by people who are not experts in soil classification. The soil identification key is an important tool for delivering soil-specific land development and soil management packages to advisors, planners and marsh dwellers in the marshlands.

The soil key provides the means to describe marshland soils in terms of attributes meaningful to land and water quality degradation. It also correlates these attributes with two international soil classification systems, namely, Soil Taxonomy (Soil Survey Staff 1999) and the World Reference Base (2006). The key essentially uses non-technical terms to categorise soils in terms of attributes that are important for characterising soil and water degradation (anthropogenic, wet, saline, gypsic, cracking clays, duplex, calcic, sodic, sulfidic). The soil features or indicators used in the key are easily recognised in the field by people with limited soil classification experience. The following important and mostly visual diagnostic features were used: depth to certain characteristic changes in wetness (waterlogging), consistency, colour, structure, salinity, gypsum occurrence, calcareousness, cracking and texture trends down profiles (e.g. texture contrast at A/B horizon boundary or duplex character).

The key layout is bifurcating, being based on the presence or absence of particular soil profile features. It consists of a systematic arrangement of soils into the following 7 broad soil categories and 15 soil subcategories (Table 2.3) and can be used as a practical vehicle for delivering soil-specific land development and soil management options to advisors, planners and marsh dwellers in the marshlands: (1) anthropogenic soils (dredgic, fusic/burned, garbic), (2) wet soils (sulfidic, redoxic), (3) saline soils (gypseous and calcareous), (4) gypsic soils (brown and grey), (5) cracking clays (brown and grey), (6) duplex soils (sodic/restrictive and non-restrictive) and (7) calcic soils (brown and grey).

Table 2.3 Key for identifying categories and subcategories of natural submerged, disturbed, drained, burned and reflooded marshland soils in the southern Mesopotamian marshlands in Iraq

Does the soil have one of the following diagnostic features?	Soil category	Subcategory
Has soil resulted from human activities and has a minimum depth of burial of 0.3 m YES? →→→→→→→ NO? ↓	1. Anthropogenic soil	Dredgic Formed by dredging mineral and organic materials from marshlands. Fusic (burned) Formed by high temperature burning of soils following the draining and desiccation of marshlands containing reeds to form abundant (>20%), irreversibly fused, particulate and discrete artefacts (often coarse fragment size) Garbic Formed from organic and mineral applications of domestic and industrial refuse
Has a water table within 50 cm of the surface for 3 months of the year or grey subsoil layers that may have yellow and/or reddish mottles (gleyed) YES? →→→→→→→→→→ NO? ↓	2. Wet Soil	Sulfidic Soils with sulfidic-like materials within the upper 1.5 m of the profile Redoxic Soils with the major part of the profile is mottled
Is bare, salt-encrusted, often with a soft fluffy surface, soil conductivity (ECse) is >16 dS m^{-1}, may or may not have halophytic plants and water table conductivity ranges from 2 to 50 dS m^{-1} YES? →→→→→→→→→→ NO? ↓	3. Saline soil thick white layer of crystals is visible when surface is dry (mostly sodium chloride) and thick cream salt crystals is visible when surface is moist (probably sulfate-rich salts gypsum and thenardite)	Gypseous Soils with more than 20% visible gypsum within the upper 0.5 m of the profile Calcareous Soils with more than 10% visible calcium carbonate within the upper 0.5 m of the profile.
Has more than 20% visible gypsum within the upper 0.5 m of the profile YES? →→→→→→→→ NO? ↓	4. Gypsic soil	Brown Dominant colour class is brown Grey Dominant colour class is grey
Is clayey to at least 50 cm, cracks on drying and has slickensides within 50 cm YES? →→→→→→→→ NO? ↓	5. Cracking Clay	Brown Dominant colour class is brown Grey Dominant colour class is grey

(continued)

Table 2.3 (continued)

Does the soil have one of the following diagnostic features?	Soil category	Subcategory
Has a sandy, loamy or clay loamy topsoil <80 cm thick abruptly (with sharp, abrupt or clear boundary) overlying a more clayey subsoil	6. Duplex soil	Sodic/restrictive Subsoil is hard and has a prismatic, columnar or coarse blocky structure and/or dull grey colours within 50 cm
YES? →→→→→→→→→→		
NO? ↓		Non-restrictive Subsoil does not have a prismatic, columnar or coarse blocky structure and/or dull grey colours within 50 cm
Is calcareous (>10% calcium carbonate) throughout or at least below 20 cm	7. Calcic soil	Brown Dominant colour class is brown
YES →→→→→→→→		Grey Dominant colour class is grey

From Fitzpatrick (2004b)

These soil categories and subcategories have also been correlated with Soil Taxonomy and the World Soil Reference Base by Fitzpatrick (2004b).

Finally, these soil categories and subcategories were used to develop a 'pictorial key' or 'soil identification key' for local advisors and marsh dwellers to easily identify those soil categories that are clearly suitable to be used to grow crops or pastures and those soil categories that should be avoided and fenced off (Fitzpatrick 2004b). The 'pictorial key' includes management options and the adoption of practices to reverse land and water degradation, thereby increasing agricultural productivity and sustainability.

2.5.2.2 Application of Technical Soil Classification System

Predictive soil-regolith models: A chronological evolution model comprising seven phases of the major soil-water-landscape changes in the southern Mesopotamian marshlands from 3000 BC to 2004 AD (present) was developed by Fitzpatrick (2004b). The model consists of a set of seven phases, which is illustrated in schematic cross-sections and photographs (Fig. 2.2). The model represents major mechanisms operating during successive evolutionary and environmental changes in the soil and water characteristics of the natural, disturbed, drained, burned and reflooded marshlands to form the following range of soils: sulfidic soils (phases 1 and 7), anthropogenic soils (phases 2 and 6), saline, sodic, gypsic and calcic soils (phases 3 and 4), burned soils (phase 5) and reflooded soils (phase 7).

The range of soil-forming processes operate in combinations at various rates from thousands of years (phases 1 and 2) to short periods such as weeks/months caused by drainage/flooding or even days caused by burning (e.g. phases 3–7). Within the floodplain alluvium, depositional facies are present but are not shown for the sake of clarity. Photographs depict at finer scales specific soil, water and topographic features (Fig. 2.2).

In summary, the model provides a powerful tool for communicating: (1) the unique formation of sulfidic materials in acid sulfate soils, salt storage, salt mobilisation and irreversible soil change knowledge for these complex landscapes affected by dredging, draining, burning and reflooding on a massive scale and (2) a framework for determining optimal patterns of regional land use and land management.

Management options: A pictorial key was developed for identification of soil indicators, land use options and best management practices (BMPs) for subcategories of marshland soils (Fitzpatrick 2004b). The 'pictorial key' and 'soil identification key' has been packaged as an easy-to-follow pictorial manual or brochure for local advisors and marsh dwellers to easily identify those soil categories that are clearly suitable to be used to grow crops or pastures and those categories to avoid and fence off. The manual can assist in the identification of the new unique categories of degraded marshland soils at any point of inspection and allocate a suitability assessment by (1) recognising soil morphological features such as soil colour and consistency; (2) using, where needed, simple tests for soil electrical conductivity (salinity), dispersion (sodicity) and pH (acidity); and (3) integration and adoption, where knowledge of soil and hydrological processes and production systems are brought together in recommendations for appropriate best management practices.

As a result, local landowners and advisors are able to decide on viable land use options and best management practices and procedures to ameliorate or reclaim identified categories of degraded marshland soils (farming based on soil type). This is more resource efficient than current 'trial-and-error' practices.

In some soil categories (e.g. calcic soils), soil and water degradation can be reversed; in others (e.g. anthropogenic fusic/burned soil or saline gypseous soil), the best thing to do is simply to fence off an area and leave it alone.

2.5.3 Mineral Exploration

This case study focuses on the Eastern Mt Lofty Ranges, South Australia, which have been the subject of a long programme of research by scientists in CSIRO (e.g. Fitzpatrick et al. 1996) and CRC LEME associated with mineral exploration research in the region (Skwarnecki and Fitzpatrick 2003, 2008). The landscape of the region is described as undulating low hills, with typical altitudes of 400–500 m. The underlying geology of the region is Cambrian metasediments of the Kanmantoo Group, which consist of interbedded, vertically dipping micaceous sandstones and schists (Fitzpatrick et al. 1996). There are numerous sulfide-rich lenses or bands (Fig. 2.3), some of which have been mined for lead and zinc. The hydrology has been described by several workers (e.g. Fitzpatrick et al. 1996; Salama et al. 1999). There are two water tables in all landscape positions upslope of the seepage and marsh areas

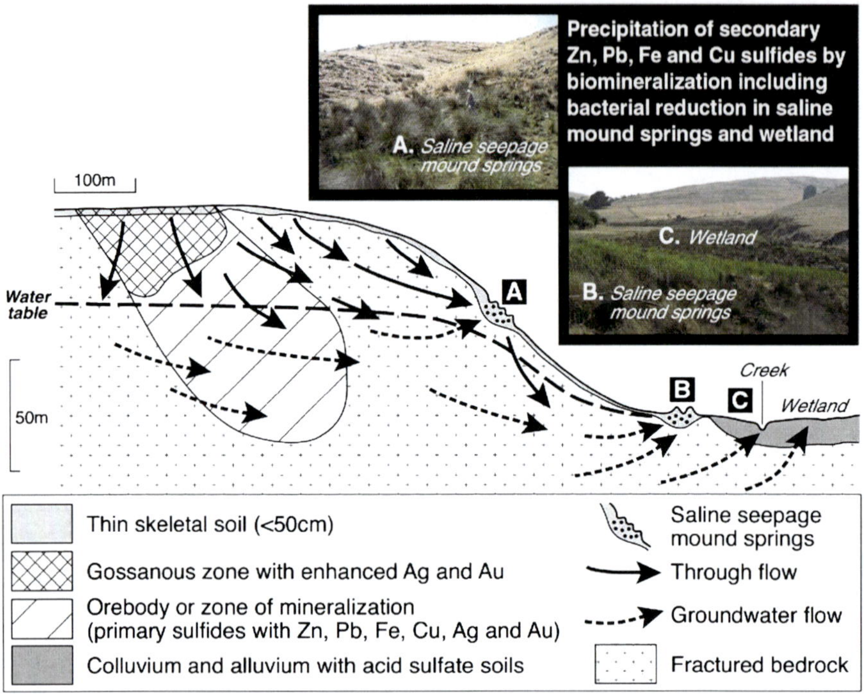

Fig. 2.3 Descriptive soil-regolith toposequence model showing geochemical dispersion from mineralised zones in sulfidic/sulfuric materials from seeps, springs and wetlands, eastern Mount Lofty (After Skwarnecki and Fitzpatrick 2003)

(Fig. 2.3). The perched, usually fresh water table fluctuates within the upper soil layers and is very responsive to rainfall events. The deeper aquifer water table level fluctuates more slowly and contains saline-sulfidic water due to contact with the deep sulfidic mineral lenses. The descriptive soil-regolith model in Fig. 2.3 illustrates groundwater and surface water movement, which is connected to the seepage and marsh areas, where various types and subtypes of the inland ASS are found. Hence, the ASS overlying mineralised zones in the Mount Lofty Ranges differs from the ASS in the Murray River flood plains system (see Case Study 1) because the source of sulfur is from sulfatic-rich groundwaters derived from sulfide mineralisation (ore bodies) of the Kanmantoo rocks.

Regional sampling by Skwarnecki and Fitzpatrick (2003, 2008) showed that a range of materials (sulfidic and sulfuric materials, salt efflorescences and Fe- and Al-rich precipitates) are typically anomalously enriched in elements such as As, Bi, Cd, Cu, Pb, Ti and Zn, especially where they are spatially related to sulfide mineralisation (cf. elements likely to be present in gossans). Thus, the sulfidic/sulfuric material carry clear indications of the presence of blind or concealed ore deposits, making these sediments a potential sampling medium for mineral exploration (Fig. 2.3; Table 2.4).

Table 2.4 Summary of special-purpose soil classification system used as a mineral exploration soil sampling medium

Landscape attributes	Sampling medium attributes: soil features, surface precipitates and salt efflorescences
Saline seeps, springs and wetlands (see Fig. 2.3)	Sulfidic, hypersulfidic and hyposulfidic materials in ASS subtypes (see Table 2.2)
	Monosulfidic material in ASS subtypes (see Table 2.2)
	Iron-rich precipitates (reddish and orange) and aluminium-rich precipitates (white) in ASS subtypes
	Mottles and redox concentrations in sulfuric material in ASS subtypes (see Table 2.2)
	Salt efflorescences in ASS subtypes

These refined models were used to help explain the complex pedological, hydrological and biogeochemical interactions that occur in the regolith environment. This increased understanding of environmental degradation processes has also aided in mineral exploration in the Eastern Mt Lofty Ranges.

2.5.4 *Optical Fibre Communication Cable*

Managing Soil Damage to the Australian Telecommunication Optical Fibre Cable Network

This case study summarises the development of a suitable special-purpose soil classification system to minimise soil damage to the Australian telecommunication optic fibre cable network (Fitzpatrick et al. 1995, 2001). Some types of optical fibre cables can develop transmission faults due to (1) soil movements caused by soil shrink-swell properties (Fig. 2.4a, b) and (2) corrosion from saline soil solutions. Such faults are very costly to repair and, if avoided, can save millions of dollars.

2.5.4.1 Special-Purpose Technical Soil Classification System for Managing Soil

Damage to the Australian Telecommunication Optical Fibre Cable Network

Field and laboratory investigations on a representative range of soils known to cause faults in optical fibre cables were undertaken (Fitzpatrick et al. 1995). Close liaison between soil scientists and engineers ensured that research investigations led to the development of a practical special-purpose soil classification system comprising a user-friendly 1–10 rating of soil shrink-swell risk. This can be derived logically by using a series of questions and answers set out in a manual entitled 'Soil Assessment Manual: A Practical Guide for Recognition of Soils and Climatic Features with Potential to Cause Faults in Optical Fibre Cables'. The manual describes practical,

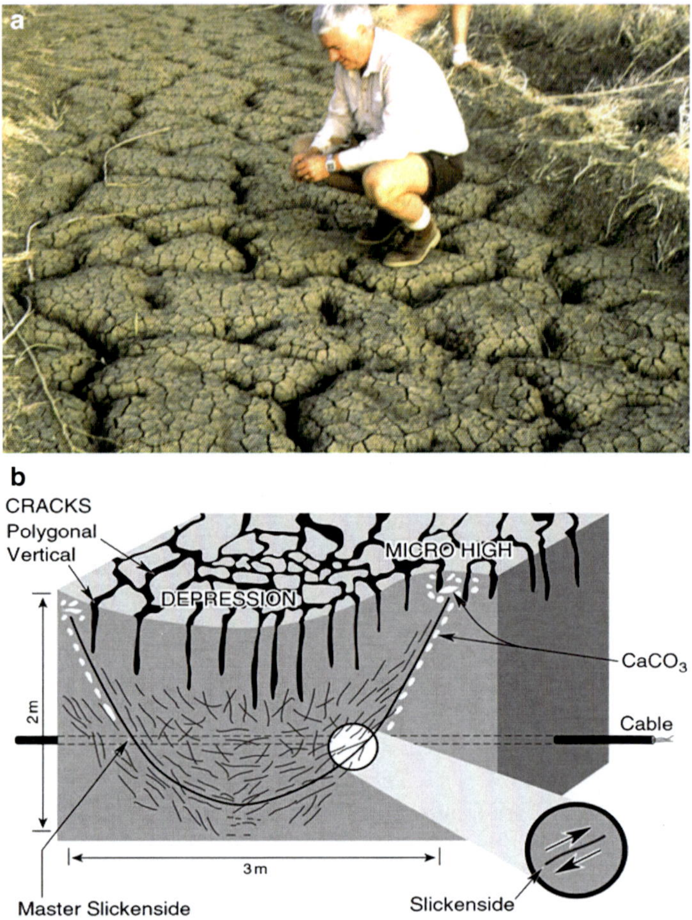

Fig. 2.4 (**a**) Cracks greater then 40 mm wide to a depth of 1 m in a deep clay soil, which shrinks and swells during seasonal wetting and drying cycles (Hughenden, north Queensland), (**b**) Schematic section through a swelling clay soil or Vertosol (Fig. 2.4a) showing micro relief (gilgai), cracks zones with slickensides (*shearing zone*), where cable distortion occurs due to soil movement (*shearing action*) (From Fitzpatrick et al. 1995)

surrogate methods to assist engineers to estimate soil shrink-swell indices by using either published soil maps in office assessments (Atlas of Australian Soils, Northcote et al. 1960–68) or by undertaking simple visual observations and chemical measurements of soil properties in the field (Table 2.5). This information is incorporated in a planning operations and procedure manual for engineers.

Guided by the manual, telecommunication engineers have learnt how to use pedological, climatic and soil chemical information to (1) avoid the shrink-swell and corrosive soils along optical fibre cable routes; (2) use the appropriate type of cable for a particular soil, thereby utilising more expensive, heavy duty cables in a

Table 2.5 Soil shrink-swell risk rating groups (from Fitzpatrick et al. 1995) for soil orders within the Australian Soil Classification

Shrink-swell risk rating groups	Index rating	Australian Soil Classification soil order[a]	Index rating	Shrink-swell risk
High to very high	8–10	Vertosols, *clayey* Sodosols	10	Very high
			9	High
			8	High
Moderate	6–7	Tenosols, Kurosols, Chromosols, Sodosols, *clayey* Calcarosols, Hydrosols, Dermosols, Ferrosols	7 6 5	Moderate Moderate Little
No or little	2–5	Rudosols, Podosols, Calcarosols, Kandosols, *sandy* Tenosols, *sandy* Hydrosols, Organosols	4 3	Little None
None	1	Rocks	2	None
			1	Hard rock

[a]The Australian Soil Classification (Isbell 1996) is extremely useful in the development of a labelling system for soil landscape mapping units

cost-effective manner whilst still ensuring the reliability of the optical link; and (3) correct problems affecting cables previously installed in troublesome soil types and rectify problems of reinstatement of fragile soils following cable burial.

Telstra (Australian National Telecommunication) engineers have used this methodology to install their telecommunications network. It has saved millions of dollars by predicting and overcoming practical problems.

The Telstra network relies on thousands of kilometres of buried optical fibre cable. Optical fibre cable technology, one of the most important innovations in decades, has catapulted the world into a new era of networking. Telephones, internet, facsimiles, cable television and electronic services rely upon these miniature 'tunnels of light'. Burying the cable protects it from all sorts of damage – but some soils move in a way that can damage vital and expensive telecommunication links. The pressure exerted by these shrink-swell soils on optical cables through soil movement (shearing action) can cause significant damage and at times even stop transmission (Fig. 2.5). In fact, the pressure exerted by these soils can be great enough to crack large concrete foundations beneath buildings and damage pipes. Investigations in the field revealed that soil induced optical fibre cable faults were confined to sections of routes through soils that have shrink-swell properties. These soils are often referred to as Cracking Clays or Vertosols (Fig. 2.5). In these soils, optical fibre cables can be stretched enough to cause the optical signals, being transmitted though them, to become distorted.

In the 'Soil Assessment Manual' referred to above, each of the original 3063 soil landscape (map) unit descriptions of the Atlas of Australian Soils (Northcote et al.

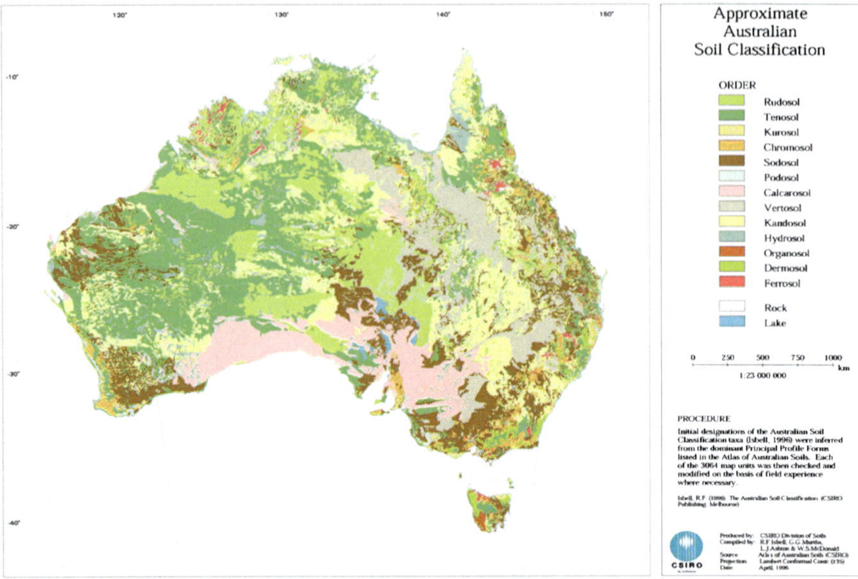

Fig. 2.5 Generalised soil map of Australia from the Atlas of Australian Soils (Northcote et al. 1960–68) showing the approximate distribution of the various soil orders according to the Australian Soil Classification (Isbell 1996). See Tables 2.5 and 2.6 for shrink-swell risk rating group approximately corresponding to each soil order

1960–68) were individually considered and classified into one of the four following risk rating groups by Fitzpatrick et al. (1995):

1	Hard rock shallower than cable depth.
2–5	No, or little, shrink-swell risk, sand to loam soils.
6–7	Moderate shrink-swell risk, clay soils and/or duplex soils.
8–10	High to very high shrink-swell risk, clay soils (Vertosols or cracking clays)

The 'Soil Assessment Manual' describes practical methods for engineers to estimate soil shrink-swell indices and assists them to categorise soil landscapes across Australia that have (1) high to very high shrink-swell risk ratings of 8–10 (e.g. Vertosols or grey clays and black earths), (2) moderate shrink-swell risk ratings of 6–7 (e.g. Sodosols and Chromosols or red and yellow duplex clays) and (3) low shrink-swell risk ratings of 2–5 (e.g. Rudosols, sandy Tenosols, Podosols or sandy and loamy soils (Tables 2.5 and 2.6)).

2.5.4.2 Application of Technical Soil Classification System

Guided by the Soil Assessment Manual, telecommunication engineers have learnt how to use soil map, field, climatic and soil chemical information to, firstly, avoid

Table 2.6 Soil orders within the Australian Soil Classification[a] and their approximate soil shrink-swell risk rating groups

Soil order[a]	Shrink-swell risk rating groups	Index rating
Rudosols	No or little	2–5
Tenosols	Moderate	6–7
Sandy Tenosols	No or little	2–5
Kurosols	Moderate	6–7
Chromosols	Moderate	6–7
Sodosols	Moderate	6–7
Clayey Sodosols	High to very high	8–10
Podosols	No or little	2–5
Calcarosols	No or little	2–5
*Clayey*Calcarosols	Moderate	6–7
Vertosols	High to very high	8–10
Kandosols	No or little	2–5
Hydrosols	Moderate	6–7
Sandy Hydrosols	No or little	2–5
Organosols	No or little	2–5
Dermosols	Moderate	6–7
Ferrosols	Moderate	6–7
Rocks	None	1

From Fitzpatrick et al. (1995)

[a]The Australian Soil Classification (Isbell 1996) is extremely useful in the development of a labelling system for soil landscape mapping units

the shrink-swell and corrosive soils along optical fibre cable routes. Secondly, it enables them to use the appropriate type of cable for a particular soil, thereby utilising more expensive, heavy duty cables in a cost-effective manner whilst still ensuring the reliability of the optical link. Finally, correct problems affecting cables previously installed in troublesome soil types and rectify problems of reinstatement of fragile soils following cable burial (e.g. by application of water, gypsum or agricultural lime).

Soil types can vary a number of times along a cable route (Fig. 2.6). Soil types have been broadly mapped for the entire Australian continent and have been published as soil maps (Northcote et al. 1960–68; Fig. 2.5). Based on the soil maps, troublesome shrink-swell soils along any proposed optical fibre cable route should be avoided if possible and the most appropriate cable type selected (Fig. 2.6). In the top half of Fig. 2.6, between towns A and B, the small area with a high shrink-swell risk (8–10) can easily be avoided so as to allow the cheaper standard cable to be ploughed into the low shrink-swell risk soil area (2–5). Wherever possible, problem clay soils will be avoided when cable is being laid. If the soils cannot be avoided, tougher and more expensive cable will be used in that section of the route.

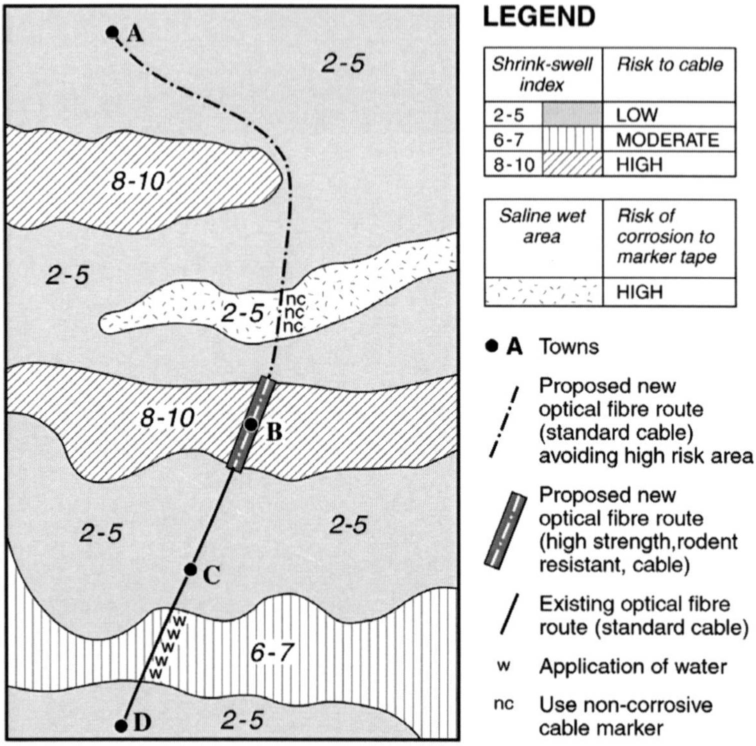

Fig. 2.6 A schematic example, showing the relationships between soil shrink-swell risk rating groups and a suitable optical fibre installation linking towns *A*, *B*, *C* and *D*. Selection of new routes, cable types and marker types are optimised according to local environmental factors. The need for temporary amelioration is indicated (e.g. application of water using water tankers) (From Fitzpatrick et al. 2001)

2.5.5 Forensic Soil Science

These days, 'forensic soil science' as a newly developed discipline of soil science has matured to the extent that well-defined questions and successful crime scene investigations can be addressed in increasingly refined ways (e.g. Fitzpatrick et al. 2009b; Fitzpatrick 2009). In this case study, consideration of a specific example highlights the kind of investigation that has been carried out on highly complex soil materials obtained from shovels, vehicles and crime scenes. The significance and relevance of established concepts and standard terminologies used in soil science but especially in pedology with practical relevance to forensic science are discussed.

2.5.5.1 Approaches, Methods and Models for Making Comparisons Between Soil Samples

Forensic soil examination can be complex because of the diversity and heterogeneity of soil samples that can be encountered. However, such diversity, heterogeneity and complexity enable forensic soil examiners to distinguish between soils, which may appear to be similar (e.g. Fitzpatrick et al. 2009b; Fitzpatrick 2009; Murray and Tedrow 1992). Forensic soil scientists must first determine if uncommon and unusual particles, or unusual combinations of particles, occur in the soil samples and must then compare them with similar soil in a known location. Methods for characterising soils for a forensic comparison broadly involves subdividing methods into four stages each comprising several steps and involving a combination of techniques (e.g. various descriptive or morphological and analytical methods outlined by Fitzpatrick (2009)). The progression of a soil examination through each of the four stages will depend on a number of factors such as the amount of sample available and the results from the early stages of the examination. Consequently, soil characterisation requires a multidisciplinary approach, which combines descriptive, analytical and spatial information (e.g. mapping) steps in the following four stages:

Stage 1 – Morphological characterisation of composite soil particles in whole soil or bulk samples for screening of samples (Fitzpatrick et al. 2009b; Fitzpatrick 2009). Initial screening (i.e. morphological comparison examination) of whole soil samples is to visually compare samples (i.e. hand-held samples/specimens, soil profiles and samples/specimens under a low magnification light microscope). To do this properly, the soil must first be systematically described and characterised using standard Australian (McDonald and Isbell 2009) and international (Schoeneberger et al. 2002) soil morphological methods, to deduce whether a soil sample can be used as evidence.
Stage 2 – Identification, characterisation and semi-quantification of minerals and organic matter in composite and individual soil particles following sample selection and size fractionation (usually <50 μm).
Stage 3 – Detailed characterisation and quantification of minerals and organic matter in composite and individual soil particles following sample selection; size or magnetic or heavy mineral fractionation using advanced analytical methods.
Stage 4 – Landform and soil mapping

- Step 4.1: Soil, geological and vegetation maps, terrain analysis, remote sensing and geophysics.
- Step 4.2: Construction of soil-regolith conceptual models (e.g. Fig. 2.7). The integrated and extrapolated soil information (Step 4.1) is used to ensure (1) better-informed sampling when pedometrically testing, for example, how 'similar' soil on a suspect's shoe is to soil from a scene of crime (e.g. Junger 1996) and (2) construction of a coherent model of soil information from microscopic observations to the landscape scale (e.g. physical, chemical or biological mechanistic process models). Decision-making in forensic soil science is sometimes guided by mechanistic process models describing processes

with physical, chemical or biological mechanisms (e.g. Ruffell and McKinley 2005, 2008; Murray and Tedrow 1992; Fitzpatrick et al. 2009b). Some models use multiple data layers as spatial input. The data required for the analysis are frequently found in and extracted from soil and environmental databases such as geographic information systems (GISs).

The systematic forensic soil examination approach applied in all investigations is to use soil morphology (e.g. colour, consistency, texture and structure), mineralogy (X-ray powder diffraction) and chemistry. Forensic soil characterisation usually combines the descriptive and analytical steps for (1) rapid characterisation of whole soil samples for screening (Stage 1) and (2) detailed characterisation and quantification of composite and individual soil particles after sample selection, size fractionation and detailed mineralogical and organic matter analyses using advanced analytical methods (Stage 2). X-ray powder diffraction methods are arguably the most significant for both qualitative and quantitative analyses of solid materials in forensic soil science.

2.5.5.2 Application of Approach to Case Study

Soil layers, regolith layers, whole soil profiles, soil-regolith toposequence or cross-sectional models and soil/geological maps are commonly being used by forensic soil scientists in developing models to predict where sites of particular soil materials are located. An example of this relationship is from the staff in the Centre for Australian Forensic Soil Science (CAFSS) using soil maps, conducting field soil survey investigations and developing soil-regolith toposequence conceptual models to solve a double murder case (Fitzpatrick and Raven 2012; Zala 2007). Morphological, chemical, physical and mineralogical properties were also used to identify similarities between soil found on a shovel taken from the suspect's vehicle and soil subsequently located in a quarry. Samples were indistinguishable or strongly matched in terms of all comparison criteria used, thus revealing the location of two buried bodies.

Another example is from an investigation where shovels, mattocks and a rake retrieved from the boot (trunk) of the suspect's vehicle contained soils that were identified to be unique to a specific soil-regolith province where forensic investigators and police had circumstantial evidence that the suspects had buried weapons. To help forensic investigators and police in the field to determine if samples from this soil-regolith province 'do compare or do not compare' with soils previously identified and characterised on implements (i.e. shovels, mattocks and rakes retrieved from the boot of a suspect's vehicle), a soil-regolith field guide was compiled, which incorporates a typical sequence of soil subtypes and vegetation found on a slope in a densely or thickly vegetated landscape (Fig. 2.7). The soil-regolith toposequence model (Fig. 2.7) uses a plain language special-purpose technical soil classification system, which places strong emphasis on soil colour (e.g. grey swampy wet soil, yellow-brown soil and yellow-gravelly soil) so as to rapidly closely match these various soil subtypes in this complex/rugged terrain to the soil on the digging implements retrieved from a suspect's vehicle.

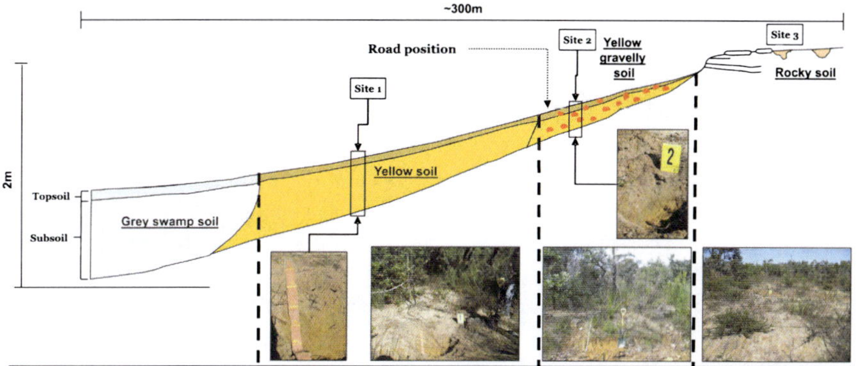

Soil type	Grey swamp soil	Yellow soil	Yellow gravelly soil	Rocky soil
Brief description	Grey swampy wet soils in low lying areas	Yellow-brown soils	Yellow-brown soils with ironstone gravels	Sandstone outcrops and pavements with pockets of shallow soils and ironstone gravels
Soil of interest?	NO	YES	NO	NO
Vegetation type	Swampy upland open, often treeless areas, dominated by dense vegetation and sedges	Predominantly uncleared open heath, closed-heath and scrub, with patches of eucalypt low-woodland	Predominantly uncleared open heath, closed-heath and scrub, with patches of eucalypt low-woodland	Predominantly uncleared open heath, closed-heath and scrub, with patches of eucalypt low-woodland
Surface soil features	White/grey sand	Yellow-brown with some patches of grey sand accumulated by surface water wash. Charcoal and burnt stumps may be present	As with "**Yellow Soil**", plus ironstone local accumulations of ironstone gravels	Flat and steep areas of sandstone with pockets of shallow yellow gravelly soil
Topsoil	White/grey sand	Yellow-brown sandy loam	As with "**Yellow Soil**", plus ironstone accumulations of ironstone gravels	Pockets of shallow yellow gravelly soil
Subsoil	White/grey sand	Yellow-brown sandy clay loam	As with "**Yellow Soil**", plus ironstone accumulations of ironstone gravels	Pockets of shallow yellow gravelly soil

Fig. 2.7 Typical sequence of soil types and vegetation found on a slope in a densely or thickly vegetated landscape

2.6 Conclusions and Future Directions

The special-purpose or technical soil classification systems or keys were specifically developed to assist users who are not soil experts to more readily use and interpret soil information and soil survey data to make practical decisions. They can also be scaled to classify the level of detail required. These technical classification systems have enabled transfer of soil information to land managers, regional and urban planners, engineers, water scientists, researchers and police in a way that is better tailored to suit client needs (end users) and is less 'pedocentric' (Fitzpatrick 2004a). This has involved improvements in the communication of critical soil information, especially to predict and solve practical problems among the following disciplines: (1) water science – involving changes in soil and water characteristics of natural, drained and reflooded soils of the Mesopotamian marshlands in Iraq and the River Murray and associated Lower Lakes in South Australia, (2) mineral exploration – discovery of new mineral ore deposits, (3) engineering – avoiding damage to telecommunication optic fibre cable networks from shrink-swell soils and corrosion and (4) forensic science – assisting police to locate buried objects in complex terrain. These soil classification systems will continue to be developed and become more dominant within Australia and overseas (e.g. Brunei – Grealish et al. 2007, 2013).

The soil-regolith toposequence models, which incorporate soil types or subtypes defined using technical soil classification systems can be scaled to show the level of detail required. They illustrate soil properties that are changing with time and space. As has been shown by these five case studies, the soil-regolith toposequence models do not replace soil survey reports and maps but bridge the gap between collection of technical soil data and the provision of information in a form that supports decision-making. Thus, the models allow soil scientists to translate their findings and to present them in an accessible and useable form to people who are not soil experts. Finally, there is a need to develop an integrating system or approach for describing soils in terms of landscape dynamics (e.g. with aid of predictive soil-regolith toposequence models), and this is being actively researched. There is also a need to ensure that useful soil characteristics developed in special-purpose soil classifications are incorporated into the general-purpose classifications used by soil scientists.

Acknowledgements I would especially like to thank the acid sulfate soil, forensic soil and Telstra optical fibre cable soil teams in CSIRO Land and Water (Richard Merry, Paul Shand, Mark Raven, Steve Marvanek, Warren Hicks, Stuart McClure, Brett Thomas, Mark Thomas, Nathan Creeper, Andrew Baker, Stuart Simpson, Peter Self, Gerard Grealish, Nilmini Jayalath, Sonia Grocke, Sean Forrester, Malcolm Wright, Phil Slade and Paul Peter) for assistance over several decades and Greg Rinder, CSIRO, for artwork.

References

Beckett PHT, Webster R (1971) Soil variability: a review. Soils Fertil 34:1–15

Burrough PA (1993) Soil variability: a late 20th century view. Soils Fertil 56:529–562

Butler BE (1980) Soil classification for soil survey. Oxford University Press, Oxford

Drohan PJ, Havlin JL, Megonigal JP, Cheng HH (2010) The "Dig It!" Smithsonian soils exhibition: lessons learned and goals for the future. Soil Sci Soc Am J 74:697–705

Dudal R (1987) The role of pedology in meeting the increasing demands on soils. Soil Surv Land Eval 7:101–110

Fitzpatrick RW (2004a) Classification systems: Australian. In: Daniel Hillel (Editor-In-Chief) Encyclopedia of soils in the environment. Elsevier, Oxford, pp 211–216. © 2005. ISBN (Set): 0-12-348530-4

Fitzpatrick RW (2004b) Changes in soil and water characteristics of natural, drained and re-flooded soils in the Mesopotamian marshlands: implications for land management planning. CSIRO Land and Water Client report, p 181. http://www.clw.csiro.au/publications/consultancy/2004/Mesopotamian-marshlands-soil.pdf

Fitzpatrick RW (2008) Soils and natural resource management (Chapter 12). In: Scott KM, Pain CF (eds) Regolith science. CSIRO Publishing/Springer, Collingwood, pp 307–339 (colour plates: pp 172–174)

Fitzpatrick RW (2009) Soil: forensic analysis. In: Jamieson A, Moenssens A (Editors-In-Chief) Wiley encyclopaedia of forensic science. Wiley, Chichester, pp 2377–2388. http://mrw.interscience.wiley.com/emrw/9780470061589/home/

Fitzpatrick RW, Wright MJ, Slade PG, Hollingsworth ID, Peter P (1995) Soil assessment manual: a practical guide for recognition of soils and climatic features with potential to cause faults in optical fibre cables. Confidential report to Telstra, p 52

Fitzpatrick RW, Fritsch E, Self PG (1996) Interpretation of soil features produced by ancient and modern processes in degraded landscapes: V Development of saline sulfidic features in non-tidal seepage areas. Geoderma 69:1–29

Fitzpatrick RW, McKenzie NJ, Maschmedt D (1999) Soil morphological indicators and their importance to soil fertility. In: Peverell K, Sparrow LA, Reuter DJ (eds) In soil analysis: an interpretation manual. CSIRO Publishing, Melbourne, pp 55–69

Fitzpatrick RW, Slade PM, Hazelton P (2001) Soil-related engineering problems: identification and remedial measures (Chapter 3). In: Gostin VA (ed) Gondwana to greenhouse: Australian environmental geoscience, Geological Society of Australia special publication 21., pp 27–36

Fitzpatrick RW, Powell B, McKenzie NJ, Maschmedt JD, Schoknecht N, Jacquier DW (2003) Demands on soil classification in Australia. In: Eswaran H, Rice T, Ahrens R, Stewart BA (eds) Soil classification: a global desk reference. CRC Press LLC, Boca Raton, pp 77–100

Fitzpatrick RW, Raven MD (2012) How Pedology and mineralogy helped solve a double murder case: Using forensics to inspire future generations of soil scientists. Soil Horizons. 53 (5) doi: 10.2136/sh12-05-0016

Fitzpatrick RW, Powell B, Marvanek S (2008a) Atlas of Australian acid sulfate soils. In: Fitzpatrick RW, Shand P (eds) Inland acid sulfate soil systems across Australia, CRC LEME open file report no 249. (Thematic volume). CRC LEME, Perth, pp 75–89

Fitzpatrick RW, Shand P, Marvanek S, Merry RH, Thomas M, Simpson SL, Raven MD, McClure S (2008b) Acid sulfate soils in subaqueous, waterlogged and drained soil environments in Lake Albert, Lake Alexandrina and River Murray below Blanchetown (Lock 1): properties, distribution, genesis, risks and management. Prepared for Department of Environment and Heritage, SA. CSIRO land and water science report 46/08. CSIRO, Adelaide, p 167. http://www.clw.csiro.au/publications/science/2008/sr46-08.pdf

Fitzpatrick RW, Grealish G, Shand P, Simpson SL, Merry RH, Raven MD (2009a) Acid sulfate soil assessment in Finniss River, Currency Creek, Black Swamp and Goolwa Channel, South Australia. Prepared for the Murray Darling Basin Authority. CSIRO land and water science report 26/09. CSIRO, Adelaide, p 213. http://www.clw.csiro.au/publications/science/2009/sr26-09.pdf

Fitzpatrick RW, Raven MD, Forrester ST (2009b) A systematic approach to soil forensics: criminal case studies involving transference from crime scene to forensic evidence. In: Ritz K, Dawson L, Miller D (eds) Criminal and environmental soil forensics. Springer Science+Business Media BV, Dordrecht, Netherlands pp 105–127

Fitzpatrick RW, Shand P, Merry RH (2009c) Acid sulfate soils. In: Jennings JT (ed) 'Natural history of the Riverland and Murraylands'. Royal Society of South Australia (Inc), Adelaide, pp 65–111

Fitzpatrick RW, Marvanek S, Powell B, Grealish G (2010a) Atlas of Australian Acid Sulfate soils: recent developments and future priorities. In: Gilkes RJ, Prakongkep N (eds) Proceedings of the 19th world congress of soil science; soil solutions for a changing world; ISBN 987-0-646-53783-2; Published on DVD; http://www.iuss.org; Symposium WG 3.1 processes in acid sulfate soil materials; 1–6 Aug 2010, Brisbane; IUSS; 2010, pp 24–27

Fitzpatrick RW, Shand P, Hicks W (2010b) Technical guidelines for assessment and management of inland freshwater areas impacted by acid sulfate soils. CSIRO client report: Water for a Healthy Country National Research Flagship. CSIRO, Adelaide, p 155 http://www.clw.csiro.au/publications/science/2010.pdf

Grealish G, Fitzpatrick RW, Ringrose-Voase AJ (2007) Soil fertility evaluation/advisory service in Negara Brunei Darussalam report P1-2 – soil properties and soil identification key for major soil types. Science report 76/07, CSIRO land and water, Australia

Grealish G, Fitzpatrick RW, King P, Shahid SA (2013) Conceptual soil-regolith toposequence models to support soil survey and land evaluation (Chapter 7). In: Shahid SA, Taha FK, Abdelfattah MA (eds) Developments in soil classification, land use planning and policy implications-innovative thinking of resource inventory for sustainable use and management of land resources., pp 165–174

Hall JAS, Maschmedt DJ, Billing NB (2009) The soils of Southern South Australia. The South Australian land and soil book series, vol 1. Geological survey of South Australia; Bulletin 56, Vol 1. Department of Water, Land and Biodiversity Conservation, Government of South Australia, Adelaide

Isbell RF (1996) The Australian soil classification. CSIRO Publishing, Melbourne

Junger EP (1996) Assessing the unique characteristics of close-proximity soil samples: just how useful is soil evidence. J Forensic Sci 41:27–34

McBratney AB (1994) Allocation of new individuals to continuous soil classes. Aust J Soil Res 32:623–633

McDonald RC, Isbell RF (2009) Soil profile. National Committee on soil and terrain (2009) Australian soil and land survey field handbook, 3rd edn. CSIRO Publishing, Melbourne

Murray RC, Tedrow JCF (1992) Forensic geology. Prentice Hall, Englewood Cliffs

NLWRA (2001) Australian agricultural assessment. National Land and Water Resources Audit, Canberra

Northcote KH, Beckmann GG, Bettenay E, Churchward HM, Van Dijk DC, Dimmock GM, Hubble GD, Isbell RF, McArthur WM, Murtha GG, Nicholls KD, Paton TR, Thompson CH, Webb AA, Wright MJ (1960–68) Atlas of Australian soils. Sheets 1–10 with explanatory booklets. CSIRO Publishing, Melbourne. Soil maps may be purchased at AUSLIG Sales, Belconnen

Ruffell A, McKinley J (2005) Forensic geoscience: applications of geology, geomorphology and geophysics to criminal investigations. Earth Sci Rev 69:235–247

Ruffell A, McKinley J (2008) Geoforensics. Wiley, Chichester

Salama RB, Otto CJ, Fitzpatrick RW (1999) Contributions of groundwater conditions to soil and water salinisation. Hydrogeol J 7:46–64

Sanchez PA, Ahamed S, Carré F, Hartemink AE, Hempel J, Huising J, Lagacherie P, McBratney AB, McKenzie NJ, de Lourdes Mendonça-Santos M, Minasny B, Montanarella L, Okoth P, Palm PCA, Sachs JD, Shepherd KD, Vågen TG, Vanlauwe B, Walsh MG, Winowiecki LA, Zhang GL (2009) Digital soil map of the world. Science 325(5941):680–681. doi:10.1126/science.1175084

Schoeneberger PJ, Wysocki DA, Benham EC, Broderson WD (eds) (2002) Field book for describing and sampling soils, version 2.0. Natural Resources Conservation Service, National Soil Survey Center, Lincoln

Schoknecht NR (ed) (2001) Soil groups of Western Australia. Resource management technical report 193. Agriculture Western Australia, Perth

Skwarnecki MS, Fitzpatrick RW (2003) Regional geochemical dispersion in materials associated with acid sulfate soils in relation to base-metal mineralisation of the Kanmantoo Group, Mt Torrens-Strathalbyn region, eastern Mt Lofty Ranges, South Australia. Cooperative Research Centre for Landscape Environments and Mineral Exploration. CRC LEME restricted report no 185R, 193 pp; Reissued in 2008 as: CRC LEME open file report no 205, p 193

Skwarnecki MS, Fitzpatrick RW (2008) Geochemical dispersion in acid sulfate soils: implications for mineral exploration in the Mount Torrens-Strathalbyn area, South Australia. In: Fitzpatrick RW, Shand P (eds) Inland acid sulfate soil systems across Australia, CRC LEME open file report no 249. (Thematic volume). CRC LEME, Perth, pp 98–128

Soil Survey Staff (1999) Soil taxonomy: a basic system of soil classification for making and interpreting soil surveys, 2nd edn. USDA-Natural Resources Conservation Service, Agriculture handbook number 436

Soil Survey Staff (2010) Keys to soil taxonomy, 11th edn. USDA-Natural Resources Conservation Service, Washington, DC

Sullivan LA, Fitzpatrick RW, Bush RT, Burton ED, Shand P, Ward NJ (2010) The classification of acid sulfate soil materials: further modifications, Southern cross GeoScience technical report no 310. Southern Cross University, Lismore, p 12

Webster R (1977) Quantitative and numerical methods in soil classification and survey. Oxford University Press, Oxford

Webster R, Butler BE (1976) Soil survey and classification studies at Ginninderra. Aust J Soil Res 14:1–24

White RE (1993) The role of soil science in shaping policies for sustainable land management. Soil News 93:1–4

White RE (1996) Soil science – raising the profile. In: Uren N (ed) Proceedings of the Australian and New Zealand soils conference 1996, vol 2, pp 1–11

Wilding LP, Drees LR (1983) Spatial variability and pedology. In: Wilding LP, Smeck NE, Hall GF (eds) Pedogenesis and soil taxonomy. I. Concepts and interactions, Developments in soil science 11A. Elsevier, Amsterdam

World Reference Base (2006) World reference base for soil resources: a framework for international classification, correlation and communication. IUSS working group WRB. World soil resources reports no 103. ftp://ftp.fao.org/agl/agll/docs/wsrr103e.pdf

Yaalon DH (1996) Soil science in transition: soil awareness and soil care research strategies. Soil Sci 161:3–8

Zala K (2007) Dirty science: soil forensics digs into new techniques. Science 318:386–387

Chapter 3
Reconnaissance Soil Survey for the State of Kuwait

Samira A.S. Omar and Shabbir A. Shahid

Abstract In 1999, the government of Kuwait in collaboration with international consultant (AACM) completed a soil survey project at two levels: reconnaissance at scale 1:100,000 and a semi-detailed at scale 1:25,000. The survey followed the latest USDA-NRCS norms and standards for the fourth-order extensive soil survey, called as reconnaissance survey. Field mapping was completed by describing one point per 200 ha making a total of 8,400 observation points in the entire state, covering an area of 16,800 km^2. To support field mapping, 105 typical soil profiles representing different soil taxa were described, and 570 horizons were sampled and analyzed for their physical, chemical, engineering, and mineralogical characteristics. Eight diagnostic horizons and eight soil great groups (Haplocalcids, Petrocalcids, Haplogypsids, Calcigypsids, Petrogypsids, Aquisalids, Torriorthents, and Torripsamments) were mapped. Of 12 soil orders distributed worldwide, only two Aridisols and Entisols were mapped. Aridisols occupy 70.8% and Entisols 29.2% of the surveyed area. A total of 23 soil taxa at the family level of USDA soil taxonomy hierarchy were mapped and included as major and minor components of 71 soil map units. The survey results were interpreted for several uses and translated to a number of thematic maps such as sand and gravel sources and uses for shallow excavations, septic tanks, sewage lagoons, sanitary landfills (area and trench), seedling mortality, and herbaceous desert plants. The major outcome was the delineation of 207,309 ha area with highest potential for irrigated agriculture, this area was surveyed at the second-order (semi-detailed) level of USDA-NRCS standards, and suitability map

S.A.S. Omar (✉)
Food Resources and Marine Sciences Division, Kuwait Institute for Scientific Research, P.O. Box 24885, Safat 13109, Kuwait
e-mail: somar@kisr.edu.kw

S.A. Shahid
International Center for Biosaline Agriculture, P.O. Box 14660, Dubai, UAE
e-mail: s.shahid@biosaline.org.ae

S.A. Shahid et al. (eds.), *Developments in Soil Classification, Land Use Planning and Policy Implications: Innovative Thinking of Soil Inventory for Land Use Planning and Management of Land Resources*, DOI 10.1007/978-94-007-5332-7_3,

for irrigated agriculture was prepared. The reconnaissance survey results are valuable source to base future land-use planning and will serve as a guide for decision makers and land-use planners. The reader is referred to (KISR, Soil survey for the State of Kuwait – vol II Reconnaissance survey. AACM International, Adelaide, 1999) for details of these interpretations.

Keywords Reconnaissance survey • Semi-detailed soil survey • Soil analysis • Soil characteristics • Soil mapping • GIS

3.1 Introduction

National soil surveys are essential to understand distribution and extent of soils of varying land-use capabilities particularly for designating potential agricultural areas. In any country, good soils and sufficient water are taken as granted for successful agriculture and food security. However, due to hyperarid condition and general water scarcity in Kuwait, both resources are not enough to meet national requirement. It is, therefore, necessary to find ways for efficient utilization of these resources; this is only possible if soils are characterized by using internationally recognized standards. Acknowledging these needs, the Public Authority for Agriculture and Fish Resources (PAAFR) and Kuwait Institute for Scientific Research (KISR) jointly implemented "Soil Survey for the State of Kuwait" through an International Australian Contractor AACM International. The project was completed in 1999 (KISR 1999).

Prior to this comprehensive survey, several soil surveys had previously been completed in Kuwait at scales ranging from reconnaissance (1:500,000) to detailed (1:5,000) levels (Ergun 1969). Soil classification was based on the FAO system as outlined in USDA (1938) and modified in 1949 (Ergun 1969), which classified the soils into great groups. Ergun (1969) reported two soil great groups, 10 soil associations, and 18 detailed soil series descriptions. Mapping units comprised of soil associations and soil series. In 1970, the Societe Centrale pour I'Equipement du Territoire Cooperation (SCET) was contracted to map the soils of 264,000 ha that were delineated by Ergun (1969) as having the highest potential for horticulture. Mapping was based on a grid survey for both scales (1:40,000) and (1:5,000). However, details of map production and spatial accuracy were not reported in SCET (1970).

Apart from the above-mentioned surveys, no other formal soil surveys have been carried out at the national scale. However, some soil-related investigations in Kuwait were conducted, such as survey of area for wildlife habitats (Omar et al. 1986), area vulnerable to sand encroachment (Omar et al. 1988), first-order soil survey of demonstration sites and proposed management (Shahid and Omar 1999), and soil survey for farm planning in northern Kuwait (Shahid et al. 2004).

To maintain and upgrade soil information in Kuwait, the present soil survey was carried out between 1996 and 1999, using the sixth edition of the Keys to

Soil Taxonomy (USDA-NRCS 1994) available at the commencement for soil classification in Kuwait. This survey updates the previous surveys (Ergun 1969; SCET 1970) by providing additional information, a wider coverage of the state, and a large database of spatial, site, and soil information, as well as an opportunity for their international correlation (Abdelfattah and Shahid 2007) and its use in the vegetation mapping in Kuwait (Omar et al. 2001). The rigor in using the USDA-NRCS specifications is that the USDA soil taxonomy has been used by over 75 countries in the world including Kingdom of Saudi Arabia (MAW 1985), Sultanate of Oman (MAF 1990), State of Qatar (MMAA 2005), and recently in Abu Dhabi Emirate (ERWDA 2004; EAD 2009).

The main objective of the soil survey was to provide basic information for broad land-use planning and irrigated agriculture development in Kuwait. The surveys conducted at different scales serve as guideline for sustainable farming (SFRI 1995), as agriculture is the most related and the widest field of soil survey data use. Among many deliverables, soil suitability map for irrigated agriculture for the State of Kuwait was the major outcome for future informed planning for irrigated agriculture.

3.1.1 Description of the Study Area

The State of Kuwait is situated at the northwestern corner of the Arabian Gulf between latitudes 28°30 and 30°05 N and longitudes 46°33 and 48°35 E. The total land area is about 17,818 km^2. The study area includes the entire State of Kuwait, excluding all military areas; oil areas in the vicinity of infrastructure; urban areas; industrial areas; minefields; and areas not cleared of ordinance. Collectively, the nonsurvey area is considered as miscellaneous unit.

Kuwait experiences extremely high temperatures during summer, short mild winters, strong sunshine, low humidity, and generally dry conditions. The daily maximum temperature averages 45°C in July, but temperatures as high as 51°C are not uncommon at this time and 18°C in January. The daily minimum temperature fluctuates between 29°C in July and 8°C in January. Rainfall is light, averaging 110 mm per annum and mainly falling between November and April. Annual rainfalls vary widely and have ranged from 20 mm in 1964 to 242 mm in 1976. The average evaporation rate ranges from 21 mm per day for July to 3 mm per day for January. The prevailing winds blow from the northwest and the southeast. In the summer months between June and September, northwesterlies are most common.

Most surface and near-surface formations of Kuwait belong to the Miocene and Pliocene Epochs and Quaternary Period. They consist of clastic deposits with subordinate carbonates and evaporites, locally known as the Kuwait Group. The dominant rock types are calcareous sandstone, fine-grained limestone, and muddy sand with minor quantities of granules and scattered pebbles. Calcrete is widely spread in the thick sand sequence of the undifferentiated Ghar and Fars Formations in the south. Gypcrete is mostly confined to the upper member of the Dibdibah Formation in the north (Al-Sulaimi and El-Rabaa 1994; Al-Sulaimi and Pitty 1995).

Most land outside the metropolitan area is owned by the state. Rangeland grazing by livestock is the predominant land use, covering almost three-quarters of the country (KISR 1999). Some 24,000 ha of privately owned or controlled farms are used for agriculture (of which 5,000 ha is cultivated). Kuwait City, surrounding towns, and industrial areas occupy about 60,000 ha. Oil wells, pipelines, and collection and processing areas also occupy localized areas.

3.2 Materials and Methods

3.2.1 Field Soil Mapping

The reconnaissance field survey was conducted at 8,400 sites. Each site was investigated to a depth of 100 cm using shovel and hand augers. Sites were located accurately with global positioning system (GPS) equipment attached to each vehicle and marked on field maps and aerial photographs. During the survey, pedologists recorded the key characteristics of each investigated site. At the sites, auger holes were dug, and the sequence of natural layers or horizons is described (Soil Survey Division Staff 1993). The descriptions included color, texture, structure, rock fragments, segregation, and other features as appropriate that enabled identification, classification, and mapping of the soils. To support field mapping, 105 typical soil profiles representing different soil taxa were described, and 570 horizons were sampled and analyzed for their physical, chemical, engineering, and mineralogical characteristics. The soils were described at the family and phases of soil family level using fourth-order level of soil survey.

In addition to routine soil mapping, 12 transacts were established throughout the country to test the previous studies to provide an overview of the soil variation. Soil descriptions were recorded on field cards and information transferred into the Soil Information System (SIS).

Soil taxonomy is based on soil properties observed or inferred during the field survey and confirmed by laboratory assessments of soil samples. Soil taxonomy takes into account the soil's morphology, physical and chemical characteristics, soil temperature, and soil moisture status.

3.2.2 Soil Information System and Development of Soil Map

Information from preliminary mapping and routine soil mapping was used to compile the final soil map and reports of the reconnaissance survey. Compilation of the map line work was then transcribed onto a base map provided by 1:100,000 scale Satellite Pour l' Observation de la Terre (SPOT) and Landsat Thematic Mapper (TM) satellite imagery rectified to remove distortions. These were used because they allowed map line work to be placed accurately in relation to observed features, then efficiently digitized and captured in the GIS. Map boundaries were defined using field data, photo interpretation, and existing information. This enabled preparation of a map legend, definition of map units, and characterization of the main soils and production of a soil map.

Data collected was stored and managed in a Soil Information System (SIS) that integrates a textural database and GIS. Soil maps were prepared at scale of 1:250,000, nine A1 size map sheets at scale 1:100,000, and an atlas (A3 size) at scale 1:100,000. Although, the soil survey was carried out according to the protocols of the fourth order (Soil Survey Division Staff 1993) at the family level, in this chapter, the soil distribution at soil great groups levels are generalized. The reader is referred to KISR (1999) for further information.

3.2.3 Laboratory Soil Analyses and Sampling Archive

Soil samples were collected from shovel/auger holes and typical soil profiles, air-dried and stored in the soil sampling archive at Kuwait Institute for Scientific Research (KISR). A total of 105 profiles (570 horizons) were analyzed to confirm the soil classification made during the field survey. The soil samples were analyzed at the State Chemistry Laboratory of Victoria, Australia, having ISO 9000 accreditation and familiar with USDA methods and standards (USDA-NRCS 1996). Soil samples were analyzed for their physical, chemical, engineering, and mineralogical characteristics. Soil-saturated paste was analyzed for pH, soil saturation extract for EC, and soluble Ca, Mg, Na, and K using atomic absorption spectroscopy, HCO_3 by acid titration, chlorides by chloride analyzer, and SO_4 by ICP (USDA-NRCS 1996). CO_3 was undetectable in all samples. Sodium adsorption ratio (SAR) was calculated using standard calculation procedure (Richards 1954). Calcium carbonate equivalents were measured by a standard calcimeter and gypsum by loss on hydration (Page et al. 1982) and acetone precipitation procedures (USDA-NRCS 1996). Particle size distribution analyses were made by plummet balance method (Ross 1996) after the removal of soluble salts and gypsum (USDA-NRCS 1996).

3.2.4 Land-Use Assessment for Irrigated Agriculture

Several land evaluations of both agricultural and nonagricultural land uses have been undertaken for the 1:100,000 reconnaissance scale map data. One major outcome was the delineation of 207,309 ha area having the highest potential for irrigated agriculture (Fig. 3.1, Table 3.1), later examined at semi-detailed scale 1:25,000 level. The results can be found in Volume IV (KISR 1999). The FAO land evaluation system of (FAO 1976) and important soil characteristics were considered in selecting land for irrigated agriculture. Criterion was developed, which considered rooting depth, site drainage, profile coarse fragments, slope, microrelief, surface stone, salinity, sodicity, soil reaction, gypsum, and $CaCO_3$ percentage.

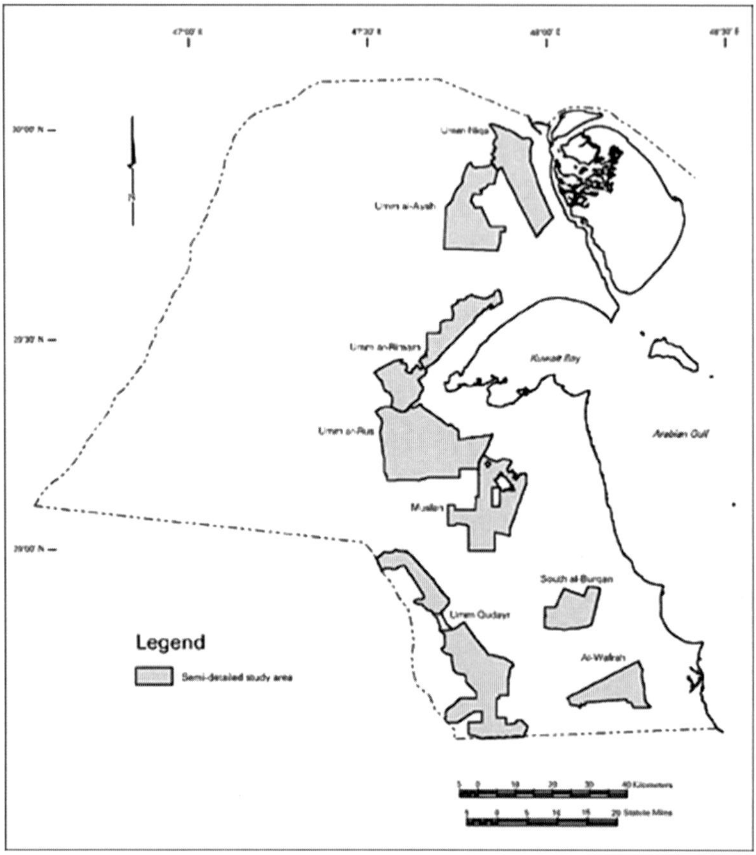

Fig. 3.1 Delineation of area having highest potential for irrigated agriculture

Table 3.1 Extent of areas studied in the semi-detailed survey

Subarea	Area (ha)
Umm al-Aish	23,121
Umm Niqa	21,430
Umm ar-Rimam	24,894
Umm ar-Rus	41,808
Muslan	25,571
South al-Burqan	11,838
Al-Wafrah	13,538
Umm Qudayr	45,109
Total area	207,309

3.3 Results and Discussion

The soils of Kuwait are generalized into great group map units using the USDA-NRCS specifications (USDA-NRCS 1994; Soil Survey Division Staff 1993).

3.3.1 Diagnostic Horizons Identified in Kuwait

Eight diagnostic soil horizons (1 surface and 7 subsurface) were recognized in Kuwait. The surface horizon (ochric) is light colored with little organic matter. Subsurface horizons are argillic, with accumulation of clay; calcic, rich in calcium carbonate; cambic, with loam texture and/or structural development; gypsic, rich in gypsum; petrocalcic, characterized by cemented calcium carbonate; petrogypsic, characterized by cemented gypsum; and salic, that is extremely salty. The presence/absence or combination of these diagnostic horizons in the pedon identifies the soil great group in Kuwait.

3.4 Soil Taxonomy Hierarchy (Reconnaissance Survey)

Soil taxonomy hierarchy from reconnaissance soil survey of Kuwait is presented as Fig. 3.2 and is briefly described below.

3.4.1 Aridisols

Aridisols are dry and do not have moisture available for plants for long periods. They have one or more of the subsurface diagnostic horizons. The Calcids and Gypsids are the major suborders identified in Kuwait; however, in the coastal area, Salids and Orthents in minor quantities are also mapped. All soils in Kuwait have an aridic soil moisture regime except for those in low-lying areas such as sabkhas where shallow groundwater sits within the soil profile. These soils are classified as the great group Aquisalids.

Gypsids cover 39.6% area of Kuwait. In many investigation sites, Gypsids have a calcic horizon above the gypsic or petrogypsic horizon. Gypsids are recognized as Haplogypsids, Petrogypsids, and Calcigypsids. Haplogypsids are minor and are included in Calcigypsids mapping unit. The Typic Haplogypsids (Fig. 3.3a) are soils with illuvial accumulations of gypsum (gypsic horizon), and the illuvial gypsic layer is not cemented sufficiently to give a petrogypsic horizon. When the gypsic horizon extends to within 18 cm of the surface, the soils are a Leptic Haplogypsid (Fig. 3.3d). If there was a calcic horizon, then the soil is Typic Calcigypsid (Fig. 3.3g).

Order	Suborder	Great Group	Subgroup	Family
Aridisols	Salids	Aquisalids	Gypsic Aquisalids	sandy, mixed, hyperthermic
			Typic Aquisalids	sandy, mixed, hyperthermic
	Gypsids	Petrogypsids	Petrocalcic Petrogypsids	sandy, mixed, hyperthermic
				sandy, mixed, hyperthermic, shallow
				coarse-loamy, mixed, hyperthermic
			Calcic Petrogypsids	sandy, mixed, hyperthermic
			Typic Petrogypsids	sandy, gypsic, hyperthermic
				sandy, gypsic, hyperthermic, shallow
		Calcigypsids	Typic Calcigypsids	sandy, mixed, hyperthermic
				sandy, gypsic, hyperthermic
				coarse-loamy, mixed, hyperthermic
		Haplogypsids	Leptic Haplogypsids	sandy, gypsic, hyperthermic
			Typic Haplogypsids	sandy, mixed, hyperthermic
				sandy, gypsic, hyperthermic
				coarse-loamy, mixed, hyperthermic
	Calcids	Petrocalcids	Calcic Petrocalcids	sandy, mixed, hyperthermic
				sandy, mixed, hyperthermic, shallow
				coarse-loamy, mixed, hyperthermic
			Typic Petrocalcids	sandy, mixed, hyperthermic
				sandy, mixed, hyperthermic, shallow
		Haplocalcids	Typic Haplocalcids	sandy, mixed, hyperthermic
				coarse-loamy, mixed, hyperthermic
Entisols	Psamments	Torripsamments	Typic Torripsamments	mixed, hyperthermic
	Orthents	Torriorthents	Typic Torriorthents	sandy-skeletal, carbonatic, hyperthermic

Fig. 3.2 Soil taxonomy hierarchy from reconnaissance soil survey of Kuwait

In gypsic horizon, the gypsum translocates from overlying horizons so that in gypsiferous material, the absence or relatively lower gypsum in A horizon and its presence and enrichment in B horizon are strong evidences of a gypsic horizon, particularly in Calcigypsids identified in Kuwait. It is visualized that gypsic horizon due to looser gypsum crystals remained soft, except where they were identified as Petrogypsid (Fig. 3.3f). Van Alphen and Romero (1971) noted that gypsum is redistributed in the soil and frequently forms cemented and indurated layers. It is believed that in Petrogypsids, the gypsic horizon like argillic and calcic horizons and duripans form by feedback process (Torrent and Nettleton 1978). Petrogypsic horizon may have some amount of gypsum as in a normal gypsic horizon; however, the gypsum is indurated rather than being soft. They become hard due to continuous interlocking crystals of gypsum. Dissolution of gypsum can cause subsidence, particularly in irrigated farming, where leveling becomes necessary. Cracked building foundations and uneven roadbeds can be other victims of subsidence. Gypsum is also present in some soil parent materials, particularly those of the Dibdibah Formation in Kuwait. The soil depth ranges from very shallow, when a restrictive layer is present, to very deep. Textures are sandy or loamy, rarely sandy-skeletal. They are extensive throughout the west and north of Kuwait.

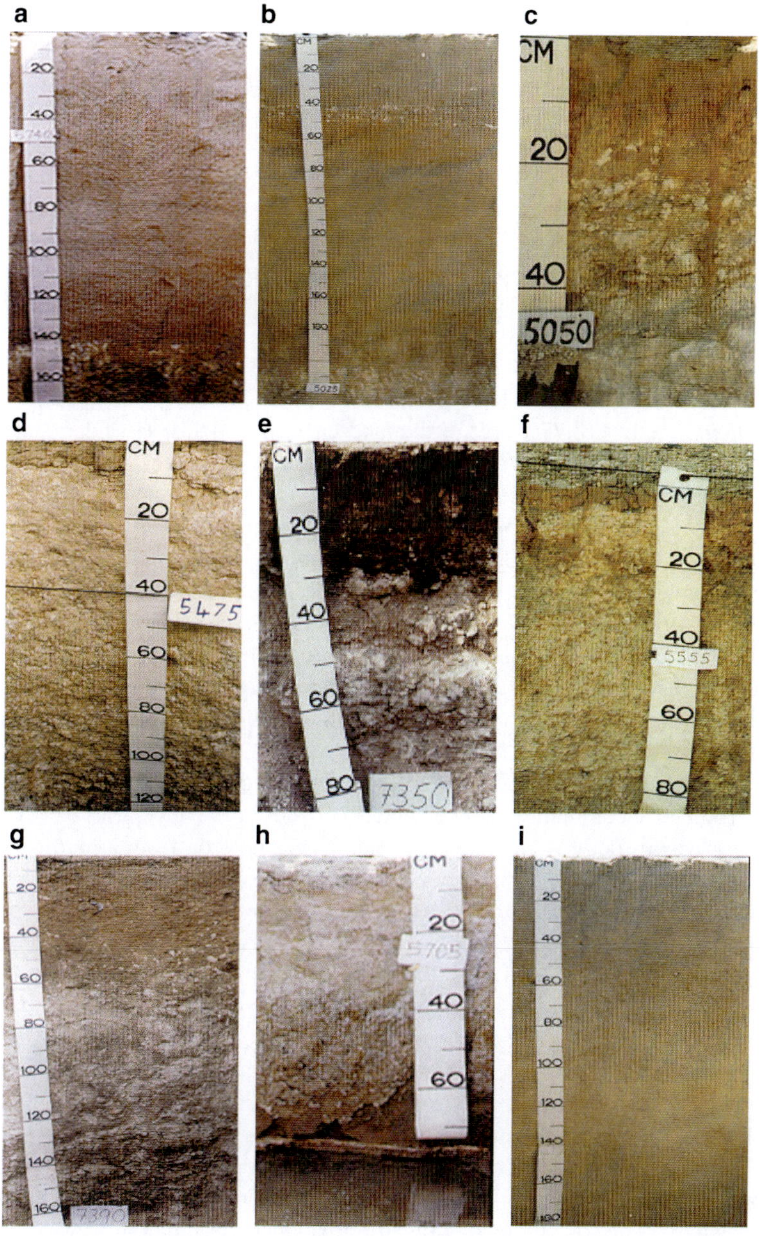

Fig. 3.3 Typical soil profiles of major soil families identified in Kuwait. (**a**) Typic Haplogypsids, (**b**) Typic Haplocalcids, (**c**) Calcic Petrogypsids Hardpan 40 cm, (**d**) Leptic Haplogypsids, (**e**) Petrocalcic Petrogypsids, (**f**) Typic Petrogypsids, (**g**) Typic Calcigypsids, (**h**) Gypsic Aquisalids, (**i**) Typic Torripsamments

Petrocalcic Petrogypsid (Fig. 3.3e) soils contain a petrogypsic horizon and in addition have a carbonate-cemented horizon (petrocalcic horizon) that does not overlie the gypsic or petrogypsic horizon. Sometimes, the petrogypsic and petrocalcic horizons are separated within the profile, but more commonly, they overlap or coincide to give a very hard layer that is cemented by both carbonate and gypsum. Such a layer has a slow permeability and presents a physical barrier to root penetration. If carbonate cementation is not a feature, but a calcic horizon is still present, the soils are identified as Calcic Petrogypsids (Fig. 3.3c). When the carbonates were even less evident, then the soils are identified as Typic Petrogypsids (Fig. 3.3f).

Argids were encountered during routine field survey, but they are of minor extent and therefore are not described in this survey as a separate suborder. They occurred on the sand plains in the south of Kuwait where they were found as a minor soil in association with the Typic Haplocalcids.

Calcids occupy 18.77% area of the landscape. In Calcids, secondary carbonates have accumulated to form a calcic or petrocalcic diagnostic horizon that has its upper boundary within 100 cm of the soil surface. Only two great groups, Petrocalcids and Haplocalcids, were mapped; they occupy 11.05 and 7.72% of the Kuwaiti landscape, respectively. It is visualized that in Kuwait the rainfall has not been sufficient to leach carbonates from the upper part of the soil. As a result, these soils are usually calcareous from the surface down to, and below, the calcic horizon. Calcids have sandy and loamy textures and range from very shallow, when a restricting layer is present, to very deep. They occur throughout Kuwait but are more common on the sand sheets in the south and center of the country, and in other areas, they may occur in depressions.

Petrocalcids and Haplocalcids are the recognized great groups. In the soils where the carbonate accumulation and cementation occur in the upper part of a hard substrate, it was difficult to distinguish a petrocalcic horizon from a hard calcareous substrate (a petrocalcic horizon rather than a lithic contact). However, if there is clear evidence of carbonate illuviation in the soil above the cemented layer, then it is highly likely that it will be a petrocalcic horizon. A one-centimeter laminar capping on the hard substrate material is all that is required for defining a petrocalcic horizon. If the level of carbonate accumulation above the petrocalcic horizon is too low for a calcic horizon, the soil was identified as Typic Petrocalcids.

Typic Haplocalcids (Fig. 3.3b) are soils that have accumulated sufficient carbonates to form a calcic horizon. It is recognized that in Kuwait many of the Typic Haplocalcids occurring in the south of the country are very similar to the Typic Torripsamments (Fig. 3.3i). There is only a small difference in the quantity of carbonate in the top 100 cm of the profile between calcareous Typic Torripsamments and Typic Haplocalcids. The soil survey of Kuwait project visualized that in Kuwait probably the Typic Haplocalcids soils are the best for agricultural purposes (USDA 1998).

Salids occur in landscape positions that receive saline or brackish groundwater through lateral flow or through seawater intrusion in the coastal area. In these areas, the water evaporates from the soil surface leaving behind the soluble salts. The water extract (1:1 soil:water) from a salic horizon is more saltier than seawater.

Soils are strongly saline, sometimes wet, occurring in basins, such as playas and coastal flats above the high tide level locally named "sabkha." In most places, the surface layers when dry has a salt crust of varying thicknesses. They are deep soils with textures that range from sand to clay. The Aquisalid is the only great group recognized in Kuwait. At the subgroup level, they are identified as Typic Aquisalids and Gypsic Aquisalids (Fig. 3.3h). When the water table remains below 100 cm, the soils are Haplosalids. The Haplosalids were encountered during fieldwork, but they are of relatively minor extent and are not mapped separately. The Salids are mainly confined to coastal and associated depression areas (sabkhas) and show significant salinity development (Shahid et al. 1998).

3.4.2 Entisols

Entisols occupy 29.2% of the surveyed area; it comprises soils that show little or no evidence of soil development. They are loose calcareous sandy material. They may have an ochric epipedon but have none of the other diagnostic horizons. The Entisols are found in recent eolian sand deposits on level to gently undulating plains in the south of Kuwait, as well as on steep actively eroding slopes, and in drainage depressions that have been filled with sand (Khalaf and Al Ajmi 1993). The Psamments (dominant) and Orthents (minor) are the suborders recognized in Kuwait.

Psamments (suborder) are formed in poorly graded (well-sorted) sands on dunes and in other sandy deposits on plains and wadi floors. They occur throughout Kuwait but are more extensive in the south of the country on the sand sheets. The soils are deep and have a texture of sand or loamy sand throughout the top 100 cm of the profile or to a lithic contact at a depth of 50–100 cm.

Typic Torripsamments are usually deep and have a texture of sand or loamy sand throughout the top 100 cm of the profile. They are mostly nonsaline and calcareous, but carbonates have not been accumulated enough to form a calcic horizon (Table 3.2). Areas of shallow Torripsamments have also been recognized in places where there is a thin sand cover (<50 cm) over bedrock. These soils are Lithic Torripsamments; these are very minor soils and are not described in this survey as a separate map unit.

Orthents (suborder) are soils that have a rock fragment content exceeding 35% (by volume) or a texture of loamy fine sand or finer (and there is no cambic horizon) in any horizon within the top 100 cm of the soil surface. Torriorthents is the only recognized great group in Kuwait. This is an Orthent that has a torric moisture regime (dry and does not have moisture available for plants for long periods). Typic Torriorthents have formed mostly in deposits on coastal dunes and beach ridges or in colluvium on actively eroding slopes. Some have formed in alluvium on wadi floors. They occur throughout Kuwait but are more extensive on the coast, islands, Jal Az Zur scrap, and in wadis in the northeast. They range from nonsaline to strongly saline.

Table 3.2 Physical, chemical, and morphological characteristics of typical soil taxa from great groups identified in Kuwait

Horizon	Depth (cm)	Munsell color Dry	Munsell color Moist	Sand %	Silt	Clay	Texture	Gravel %	Structure	Consistence	Stickiness	Effervescence	Boundary	$CaCO_3$ % eq.	ECe dS/m	pHs	Gypsum %
Haplocalcids																	
Ak	0–40	2.5Y 6/4	10YR 5/4	92.1	2.9	5.0	S	6.0	SG	LM	–	VE	CS	6.9	0.31	8.3	<0.1
Bk1	40–55	2.5Y 7/3	10YR 6/4	92.0	0.7	7.3	S	29.0	SG	LM	–	VE	CW	10.1	0.87	8.1	<0.1
Bk2	55–190	2.5Y7/3	10YR 6/4	89.9	5.8	4.4	S	2.0	M	FM	–	VE	GS	10.3	1.75	8.2	<0.1
Petrocalcids																	
Ak	0–15	10YR 7/4	10YR 5/3	95.0	3.3	1.7	S	1.0	SG	LD	–	SE	AS	8.6	2.20	7.9	–
Bk1	15–40	10YR 7/4	10YR 6/3	97.3	1.7	1.0	S	35.0	SG	LD	–	SIE	AS	3.4	2.50	7.9	–
Bk2	40–60	10YR 5/4	10YR 5/4	95.7	1.7	2.6	S	48.0	M	LD	–	SE	AS	3.0	1.20	7.9	–
Bkm	60–90	10YR 8/2	10YR 5/6	–	–	–			–	–	–	SE	–	21.5	–	–	–
Calcigypsids																	
Ak	0–10	10YR 6/4	10YR 4/3	78.1	12.1	9.8	SL	26	PS	MHD	–	SE	CS	15.2	0.99	7.9	1.3
Bk1	10–45	10YR 7/3	10YR 5/3	77.8	7.9	14.3	SL	52	M	MHD	–	VE	GW	15.2	0.46	8.1	3.8
Bk2	45–70	10YR 7/3	10YR 5/4	79.1	6.4	14.5	SL	31	SG	LD	–	SE	GW	8.0	2.60	7.8	2.5
By1	70–110	10YR 7/3	10YR 6/4	86.8	4.3	8.9	LS	37	M	HD	–	NE	GW	2.0	4.30	7.9	12.0
By2	110–150	10YR 7/3	10YR 6/3	93.9	1.1	5.0	S	33	M	MHD	–	NE	GW	1.5	4.20	7.8	8.1
By3	150–200	10YR 7/3	10YR 6/3	–	–		–	44	SG	LM	–	NE	–	0.4	–	–	3.5
Haplogypsids																	
Ak	0–10	2.5Y 6/4	10YR 5/4	94.9	1.4	3.7	S	4.0	P	SD	–	SIE	CW	3.9	2.75	8.3	1.1
Bk1	10–30	2.5Y 7/4	2.5Y 6/3	87.9	2.1	10.0	LS	1.0	M	MHD	–	SE	GS	2.5	1.50	7.8	1.3
Bk2	30–70	2.5Y 7/4	2.5YR 6/3	84.7	2.6	12.8	LS	0.0	M	MHD	–	SIE	GS	1.0	0.31	7.9	2.4
By1	70–110	5Y 7/3	5Y 7/3	88.1	1.1	10.8	LS	0.0	M	MHD	–	VSIE	GS	0.1	2.60	7.8	5.8
By2	110–160	5Y 7/3	5Y 7/2	88.7	1.6	9.6	LS	0.0	M	MHD	–	NE	–	0.1	2.50	7.8	5.5
Petrogypsids																	
Ak	0–6	2.5Y 6/3	10YR 7/3	93.1	2.1	4.8	S	7.0	SG	LM	–	VE	CS	13.2	2.60	7.8	1.5
Bky	6–20	10YR 6/3	10YR 5/3	77.6	9.9	12.5	SL	20.0	AB	FM	–	SE	CS	11.2	3.10	7.8	10.7

By	20–40	10YR 7/3	10YR 6/4	86.8	8.0	5.2	LS	24.0	M	FM	–	SE	CI	2.8	3.90	7.9	29.5
Bym	40–140	10YR 8/3	10YR 6/4	88.2	5.9	5.9	S	–	–	–	–	NE	–	0.3	6.00	7.7	49.4
Aquisalids																	
Akz	0–2	2.5Y 6/4	2.5Y 5/3	96.2	2.5	1.3	S	0	M	VFrM	NS	SE	AS	3.5	358.4	7.4	9.5
Bkz	2–30	2.5Y 6/4	2.5Y5/3	95.7	3.8	0.5	S	0	M	VFrM	SS	SE	CS	5.6	64.3	8.0	3.1
Bkyz1	30–60	2.5Y 6/4	2.5Y5/3	94.3	3.1	2.6	S	0	M	VFrM	SS	SE	CW	4.7	46.4	7.9	14.0
Bkyz2	60–90	2.5Y 7/4	2.5Y5/3	92.1	5.9	2.0	S	0	M	VfrM	SS	SE	CW	6.9	30.9	7.9	14.0
Bkyz3	90–120	2.5Y 7/4	10YR 5/4	95.0	3.8	1.2	S	0	M	FM	SS	SE	–	9.0	44.2	7.8	5.0
Torripsamments																	
Ak1	0–10	2.5Y 6/4	10YR 5/3	93.7	2.5	3.8	S	1.0	M	LM	NS	VE	–	8.4	0.31	8.2	0
Ak2	10–60	2.5Y 6/4	10YR 5/3	92.8	2.2	5.0	S	1.0	M	LM	NS	VE	–	7.3	0.29	8.3	0
Ak3	60–120	2.5Y 6/4	10YR 5/3	93.4	2.2	4.4	S	2.0	M	LM	NS	VE	GW	7.5	0.31	8.3	0
Bk1	120–135	2.5Y 6/4	10YR 6/4	91.6	1.0	7.4	S	11.0	SG	FrM	NS	VE	AS	3.8	0.41	8.3	0
Bk2	135–145	2.5Y 6/4	10YR 7/2	83.3	5.9	10.8	LS	65.0	M	VHD	NS	SE	AS	21.6	–	–	0
C	145–190	2.5Y 7/3	10YR 7/3	93.4	0.2	6.4	S	8.0	SG	VFM	NS	VSE	AW	3.6	–	–	0
2Btb	190–215	5YR 4/4	2.5Y 7/3	65.6	9.0	25.4	SCL	0.0	AB	VHD	SS	NE	–	4.1	–	–	0
Torriorthents																	
A	0–20	2.5Y 7/3	2.5Y 5/3	42.1	33.0	24.9	L	11	P	SlHD	–	SE	GW	29.7	8.27	8.2	0
C1	20–40	10YR 5/2	10YR6/2	37.2	42.1	20.7	L	41	M	MHD	–	VE	DW	52.5	29.2	7.4	0
C2	40–70	10YR 6/2	10YR 6/2	73.4	11.8	14.8	SL	52	M	MHD	–	VE	DW	54.7	28.3	7.6	0
C3	70–100	10YR 6/2	10YR 6/2	72.8	15.4	11.8	SL	64	M	MHD	–	VSlE	–	55.9	26.6	7.6	0

S sand, SL sandy loam, LS loamy sand, SCL sandy clay loam, L loam, SG single grain, M massive, P platy, AB angular blocky, LM loose moist, FM firm moist, LD loose dry, MHD moderately hard dry, HD hard dry, VFrM very friable moist, FM firm moist, FrM friable moist, SlHD slightly hard dry, NS nonsticky, SS slightly sticky, VE violently effervescent, SE strongly effervescent, VSE very strongly effervescent, SlE slightly effervescent, VSlE very slightly effervescent, NE noneffervescent, CS clear smooth, CW clear wavy, GS gradual smooth, AS abrupt smooth, GW gradual wavy, CI clear irregular, DW diffuse wavy, ECe electrical conductivity of soil saturation extract, pHs pH of saturated soil paste, dS/m deci Siemens per meter, %eq percent equivalents of $CaCO_3$

3.5 Physical and Chemical Characteristics

Some important physical and chemical characteristics of representative soil profiles from each great group are presented in Table 3.2. It is apparent that all the pedons are calcareous to varying degrees of $CaCO_3$ accumulation, the highest being in the Torriorthents due to broken shells in the soil material. Minor to high quantities of gypsum were identified in the pedons based on the soil characteristics (Table 3.2). It is apparent from soil characteristics that the soils in the great groups Haplocalcids, Calcigypsids, and Torripsamments can be exploited for irrigated agricultural activities; however, the variable quantities of $CaCO_3$ may affect the nutrient availability to plants such as P, Mo, Fe, Zn, and Mn. The deficiencies of which are quite common in plants growing in calcareous soils (Tisdale et al. 1993). Calcium carbonate may induce buffering capacity and resist a change in soil pH while managing calcareous soils. Minor quantities of gypsum are related to its presence in the fallen dust through wind erosion and the highest (Petrogypsids) to pedogenically formed gypsic horizon. The primary soil particles are dominant in sand, which give sand a sandy loam and loamy sand texture to different horizon depth; clay is the highest at 0–40 cm in Torriorthent and gives loamy texture.

The Ca and SO_4 ions dominate the solution chemistry of all pedons. In salids and orhents, Na and Cl are dominant ions, this is due to the major influence of seawater intrusion in the coastal areas where these were identified. In these pedons, higher sodium content developed high exchangeable sodium percentage (ESP) compared to other pedons. The native soils of Kuwait are generally nonsaline (Torripsamments); however, where there is influence from seawater in the coastal areas, high levels of salinity are developed (Salids and Orthents); in others, variable quantities of gypsum and calcium carbonate add some electrolytes in the soil solution. The pH is buffered between slightly to moderately alkaline (7.4–8.3) range (Soil Survey Division Staff 1993), which is above the optimum levels (6.7–7.3) where most of the nutrients become available to plants (Tisdale et al. 1993).

3.6 Great Group Level Generalized Soil Map of Kuwait

Of 12 soil orders reported (USDA-NRCS 1998, 1999), only two have been identified in the State of Kuwait; they are Aridisols and Entisols. Excluding the nonsurveyed restricted areas (6.64%), the surveyed area constitutes 70.8% Aridisols and 29.2% Entisols. The area and extent of soil great groups and their component soil subgroups are shown in Table 3.3.

The general distribution of soils in Kuwait (at the great group level) is shown in Fig. 3.4. Seven major (Torripsamments, Haplocalcids, Aquisalids, Calcigypsids, Petrocalcids, Petrogypsids, and Torriorthents) and a minor (Haplogypsids, 0.05%) soil great group are mapped characterizing the soils of Kuwait. Miscellaneous map unit comprises 6.64% of the state area and includes quarries and dumps (2.94%) and urban and industrial areas (3.7%). The distribution of great groups into units is associations and complexes and therefore may contain other great groups in minor quantities.

Table 3.3 Area and extent of soil great groups and their component soil subgroups

Soil great group	Soil subgroup	Hectares	Percent
Haplocalcids		**133,929**	**7.72**
	Typic Haplocalcids	133,929	7.72
Petrocalcids		**191,764**	**11.05**
	Calcic Petrocalcids	166,118	9.58
	Typic Petrocalcids	25,646	1.47
Calcigypsids		**107,255**	**6.18**
	Typic Calcigypsids	107,255	6.18
Haplogypsids		**812**	**0.05**
	Leptic Haplogypsids	812	0.05
Petrogypsids		**578,289**	**33.39**
	Calcic Petrogypsids	81,755	4.72
	Petrocalcic Petrogypsids	202,654	11.70
	Typic Petrogypsids	293,880	16.97
Aquisalids		**122,630**	**7.08**
	Gypsic Aquisalids	104,537	6.03
	Typic Aquisalids	18,093	1.05
Torripsamments		**473,627**	**27.27**
	Typic Torripsamments	473,627	27.27
Torriorthents		**10,769**	**0.62**
	Typic Torriorthents	10,769	0.62
Miscellaneous unit		**115,337**	**6.64**
	Dump and quarries	51,134	2.94
	Urban and industrial	64,203	3.70
Total		**1,734,412**	**100.00**

The soil map units in great group soil map cover broad areas that have a distinctive pattern of soils, relief, and drainage. Each map unit is a unique natural landscape and consists of a number of soil families (component soils) or miscellaneous areas. The generalized soil map can be used to determine the suitability of large areas for broadly determined land uses and provides an overview of the general soil distribution in Kuwait. Because of its small scale, the map is not suitable for detailed on-site planning, and the statements made in the map unit description are generalizations; however, this can be used as a guideline for more detailed investigation for specific purposes. The general soil map units are described below.

3.6.1 Haplocalcids

This map unit consists of well-drained, deep or very deep, sandy to loamy soils, which have a layer of carbonate masses and nodules (calcic horizon) in the profile. The soils occur on level to gently undulating plains and in depressions or basins. This unit makes up 8% of the survey area. Other soils found in this unit are Torripsamments, Calcigypsids, and minor areas of Petrocalcids. The unit is used for rangeland grazing,

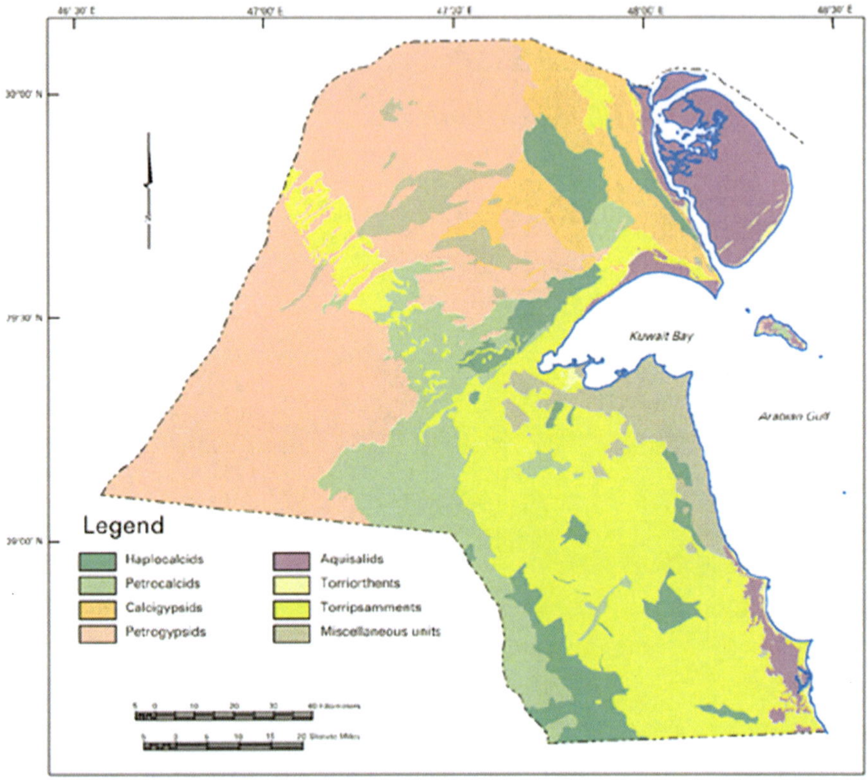

Fig. 3.4 Great group level generalized soil map of Kuwait

and in some areas, there are oil fields. It is moderately suited for irrigated agriculture; the main limitations are areas of other soils that have a shallow rooting depth, impeded drainage, and medium gypsum content.

3.6.2 Petrocalcids

The map unit consists of well-drained or moderately well-drained, shallow or moderately deep, sandy to loamy soils overlying a calcic hardpan (petrocalcic horizon). Within the soil profile, there is usually a layer of carbonate masses and nodules (calcic horizon). The soils occur on elevated plains or along the rim of escarpments. This unit makes up 11% of the survey area. Other soils found in this unit are Torripsamments, Calcigypsids, and Haplocalcids, which tend to occur in shallow depressions, in the lee and on slopes of rises, and as low sand ridges. The unit is used for rangeland grazing, and in some areas, there are oil fields. Most of this unit is marginally suited for irrigated agriculture, but where the soils are shallow, it is currently unsuitable; the main limitations are a shallow rooting depth and impeded drainage.

3.6.3 Haplogypsids

The map unit consists of well-drained, deep or very deep, sandy to loamy soils, which have a layer of gypsum crystals (gypsic horizon) in the profile. The soils generally occur on slopes and ridges. This unit makes up of less than 0.5% of the survey area and therefore not shown in Fig. 3.4. This small area has been included with the Calcigypsids. Other soils found in this unit are Calcigypsids, Petrogypsids, and Torripsamments. The unit is used for rangeland grazing. It is unsuitable for irrigated agriculture; the main limitation is high gypsum content.

3.6.4 Calcigypsids

The map unit consists of well-drained, deep or very deep, sandy to loamy soils containing a layer of carbonate masses and nodules (calcic horizon) and a layer of gypsum crystals (gypsic horizon) within the profile. The soils occur on slopes and ridges within level to gently undulating plains. This unit makes up 6% of the survey area. Other soils found in this unit are Torripsamments and Haplocalcids that occur in drainage depressions or where sand has accumulated and Haplogypsids and Petrogypsids that occur on the crests of rises. The unit is used for rangeland grazing. It is currently unsuitable for irrigated agriculture; the main limitation is a medium to high gypsum content.

3.6.5 Petrogypsids

The map unit consists of moderately well-drained, shallow or moderately deep, sandy to loamy soils overlying a gypsic hardpan (petrogypsic horizon) or a gypsic horizon underlain by a calcic hardpan (petrocalcic horizon). The hardpan is commonly cemented by both gypsum and carbonate. The soils generally occur on a level to gently undulating plain formed on the sand and gravel deposits of the Dibdibah Formation. Within the plain, low rises and shallow drainage depressions are common, and in the central and western areas, there are a number of recognizable gravel ridges and intervening valleys. This unit makes up 33% of the survey area. Other soils found in this unit are Haplocalcids and Petrocalcids which occur in the shallow depressions. The unit is used for rangeland grazing. It is currently unsuitable for irrigated agriculture; the main limitations are a shallow rooting depth, impeded drainage, and high gypsum content.

3.6.6 Aquisalids

The map unit consists of poorly or somewhat poorly drained, deep or very deep, sandy to clayey soils. Within the soil, there is a layer of salt accumulation (salic horizon) that usually occurs near the surface, and often there is a layer of gypsum crystals and water table within upper 1 m. The soils occur on coastal flats and in inland playas. This unit makes up 7% of the survey area. Other soils found in this unit are Torripsamments on sand dunes and Torriorthents. The unit is not used for any form of agricultural production or rangeland grazing. It is permanently unsuitable for irrigated agriculture; the main limitations are the high salt content, poor drainage, and high gypsum content.

3.6.7 Torriorthents

The map unit consists of excessively drained to well-drained, moderately deep or very deep, sandy soils. Within the soil profile, there is a high content of shell fragments and some salt and gypsum accumulations. The soils occur on coastal dunes or beach ridges. This unit makes up 1% of the survey area. Other soils found in this unit are Torripsamments on sand dunes. The unit is not used for any form of agricultural production or rangeland grazing. It is permanently unsuitable for irrigated agriculture; the main limitations are excessive drainage, high salt content, and low water storage capacity (due to the high shell content).

3.6.8 Torripsamments

The map unit consists of well to somewhat excessively drained, deep or very deep, sandy soils. The soils generally occur on an extensive sand sheet in central and southern areas that is aligned in a northwest to southeast direction. They also occur at the base of escarpments where fall dunes have formed, along the coast on dunes, and in the northwest on isolated barchan dunes and sand ridges. This unit makes up 27% of the survey area. Other soils in this unit are Haplocalcids and Calcigypsids that occur in low-lying and deflated areas and Petrocalcids that occur on rises. The unit is used for rangeland grazing, and in some areas, there are oil fields with their associated infrastructure. It is moderately suitable for irrigated agriculture; the main limitations are the rapid permeability and low water storage capacity (somewhat excessive drainage).

3.6.9 Miscellaneous Units

The map unit consists of areas that are not soils, such as areas where the soils have been removed or severely disturbed or covered by urban and industrial activities.

These areas include, in the north, mined areas comprising gravel quarries and overburden and tailing dumps and, in the center of Kuwait, sand quarries, dumps of building rubble, rubbish dumps, and the urban and industrial areas of Kuwait City and associated satellite towns, including vacant land reserved for urban or industrial use. This unit makes up 7% of the survey area. Soils found in this unit include Haplocalcids, Torripsamments, Haplogypsids, and Petrogypsids. The unit is used for mining, refuse disposal, and urban and industrial activities. It is permanently unsuitable for irrigated agriculture. Some areas such as overburden and tailing dumps could be rehabilitated, but the cost would be high.

3.7 Spatial Distribution of Gatch-Like Deposits

A common feature of many Kuwaiti soils is a calcic and/or gypsic pan, colloquially named gatch. As discussed above and defined in the soil classification section, a pan formed from precipitated and cemented carbonate or gypsum is called a petrocalcic or petrogypsic horizon where the pan is continuous and provides a barrier to plant

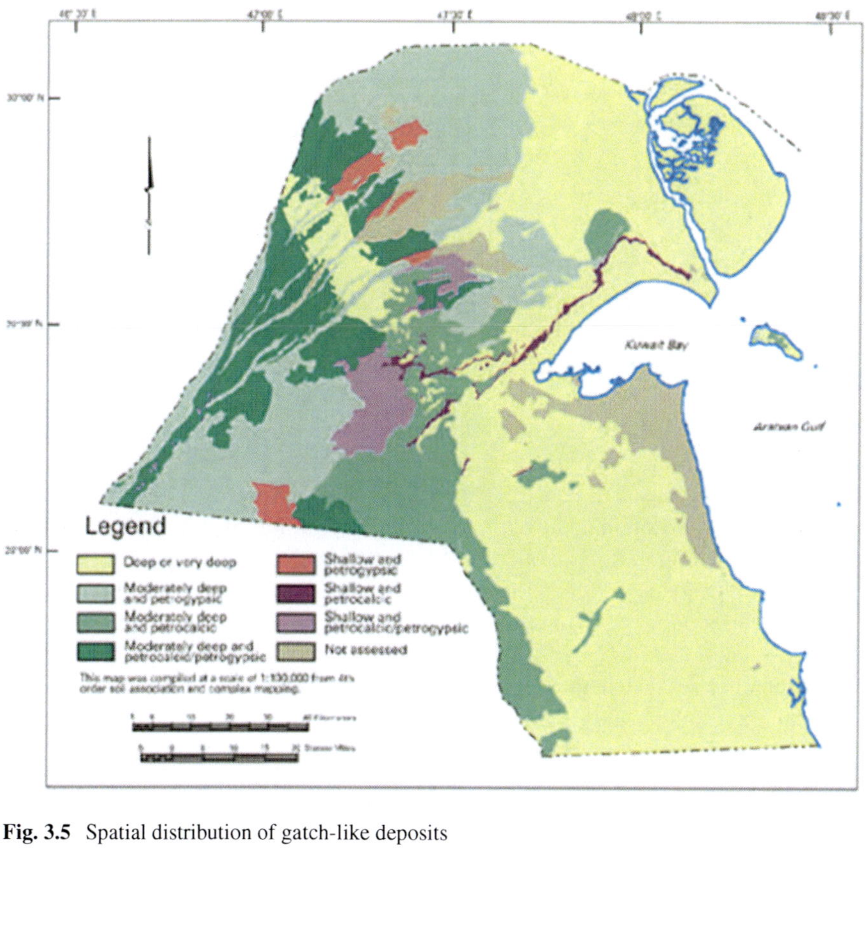

Fig. 3.5 Spatial distribution of gatch-like deposits

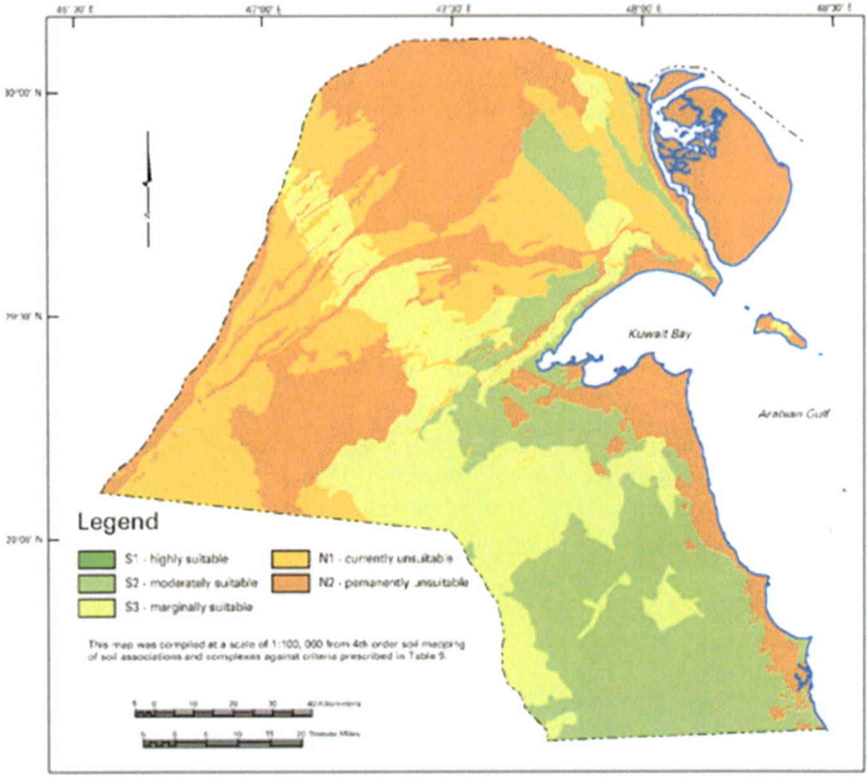

Fig. 3.6 Irrigated suitability map of Kuwait

roots. Figure 3.5 shows the spatial distribution of gatch-like deposits in Kuwait. The gatch layer is more common in the center and northwest of Kuwait. The map identifies where the gatch layer becomes an impediment to land use, namely, within 50 cm of the soil surface (shallow), and where it is within 50–100 cm of the soil surface (moderately deep). The eastern and southern parts of Kuwait comprise deeper soils with the gatch layer below 100 cm if present.

Gatch layers have the effect of hindering water movement in soils, decreasing the soil air proportion in soil pores, and ultimately limiting the growth of plants. If soils with shallow and moderately deep gatch layers are developed for irrigated agriculture, the gatch will increase the likelihood of perched water tables and development of secondary salinization and make irrigation management extremely difficult. Therefore, irrigation development on such soils should be avoided. If development does take place, it may be necessary to deep plow or rip the gatch layer to break the gatch up and install a subsurface drainage system. In soils where gatch occurs, shallow-rooted crops can be grown and proper irrigation scheduling is practiced to minimize the amount of excess water in the soil profile.

3.8 Land-Use Assessment for Irrigated Agriculture

The outcome of evaluating the soils of Kuwait against thc criteria developed in the soil survey project (KISR 1999) is presented in Fig. 3.6, which presents the suitability of soils for irrigated agriculture. The results of the evaluation indicated that well-drained Torripsamments and Haplocalcids with good drainage are the preferred soils for irrigated agriculture in Kuwait.

3.9 Conclusions

A number of uses and resources were identified to be part of the soil survey project activities. These were achieved by developing criteria of their assessment and uses through using Soil Information System. The survey results were interpreted for several uses and translated to a number of thematic maps such as sand and gravel sources and uses for shallow excavations, septic tanks, sewage lagoons, sanitary landfills (area and trench), seedling mortality, and herbaceous desert plants. The major outcome of the soil survey for the State of Kuwait was the delineation of more than 207,309 ha area with highest potential for irrigated agriculture; in Kuwait, this area was surveyed at the second-order (semi-detailed) level of USDA-NRCS standards, and suitability map for irrigated agriculture was prepared.

Acknowledgements The authors express their thanks and gratitude to Mr. Jasem Habib, the Director General of the Public Authority for Agriculture and Fish Resources (PAAF) for approving the publication of the manuscript and to the PAAF managers and staff members for their contribution in terms of project initiation, funding, supervision, participation, and technical advising to the "Soil Survey and Associated Activities for the State of Kuwait" project (FA006). Thanks are further extended to Kuwait Institute for Scientific Research (KISR) upper management for their support and encouragement to publish the manuscript and providing all necessary information. Thanks are also addressed to Mr. John Farger, Peter King, and Gerard Grealish from AACM International for their contribution to the Soil Survey for the State of Kuwait.

References

Abdelfattah MA, Shahid SA (2007) A comparative characterization and classification of soils in Abu Dhabi Emirate coastal area in relation to arid and semiarid conditions using USDA and FAO soil classification systems. Arid Land Res Manag 21:245–271

Al-Sulaimi JS, El-Rabaa SM (1994) Morphological and morphostructural features of Kuwait. Geomorphology 11:151–167

Al-Sulaimi JS, Pitty AF (1995) Origin and depositional model of Wadi Al-Batin and its associated alluvial fan, Saudi Arabia and Kuwait. Sediment Geol 97:203–229

EAD (2009) Soil survey of Abu Dhabi Emirate. Environment Agency, Abu Dhabi (5 vols)

Ergun HN (1969) Reconnaissance soil survey of Kuwait, FAO project FAO/KU/TF-17. Food and Agriculture Organization, Rome

ERWDA (2004) Soil survey for the coastline of Abu Dhabi Emirate, 2 vols (vol I: Reconnaissance survey and vol II: Soil map). Environmental Research and Wildlife Development Agency (ERWDA), Abu Dhabi, UAE

FAO (1976) A framework for land evaluation, FAO soils bulletin 32. Food and Agriculture Organization, Rome

Khalf FI, Al-Ajmi D (1993) Aeolian process and sand encroachment problems in Kuwait. Geomorphology 6:111–134

KISR (1999) Soil survey for the State of Kuwait – vol II Reconnaissance survey. AACM International, Adelaide

MAF (1990) General soil map of the Sultanate of Oman. Ministry of Agriculture and Fisheries, Muscat

MAW (1985) General soil map of Kingdom of Saudi Arabia. Ministry of Agriculture and Water, Riyadh, Saudi Arabia

MMAA (2005) Soil classification and land use specifications for the State of Qatar. Phase IV: The atlas of soils for the State of Qatar. Ministry of Municipal Affairs and Agriculture, Qatar

Omar SAS, Al-Sdirawi F, Agrawal V, Hamdan A, Al-Bakri D, Al-Shuaibi F (1986) Criteria for establishment of Kuwait's first national park/nature reserve, vol 1. Inventory and zoning. Kuwait Institute for Scientific Research. AG51. Final report, Kuwait

Omar SAS, El-Bagouri I, Anwar M, Khalaf MF, Hashash M, Nassef A (1988) Measures to control mobile sand in Kuwait. Kuwait Institute for Scientific Research, Report EES78, vol 1. KISR8790. Kuwait

Omar SAS, Misak R, King P, Shahid SA, Abo-Rizq H, Grealish G, Roy W (2001) Mapping the vegetation of Kuwait through reconnaissance soil survey. J Arid Environ 48:341–355

Page AL, Miller RH, Keeney DR (1982) Methods of soil analysis, part 2. Madison, Wisconsin

Richards LA (ed) (1954) Diagnosis and improvement of saline and alkaline soils. Agriculture handbook 60. USDA. US Government Printing Office, Washington, DC

Ross D (1996) Method number 30018. SCL method manual, vol 4. State Chemistry Laboratories, Werribee

SCET (1970) Report to the Ministry of Public Works, Department of Agriculture, Kuwait, vol 1. Economic and agricultural study; vol 2. Soil survey; and vol 3. Pilot farm. Societe Centrale pour l'Equipement du Territoire Cooperation, France

SFRI (1995) From soil survey to sustainable farming. In: Proceedings of the conference to 35th anniversary of the Soil Fertility Research Institute Bratislava. Organized within the ENCY activities. 3–5 Oct 1995, Higa Tatras Lesna-2 vols, p 461

Shahid SA, Omar SAS (1999) Order 1 soil survey of the demonstration farm sites with proposed management. Kuwait Institute for Scientific Research, KISR 5463, p viii+144

Shahid SA, Omar SAS, Grealish G, King P, El-Gawad MA, Al-Mesabahi K (1998) Salinization as an early warning of land degradation in Kuwait. Probl Desert Dev 5:8–12

Shahid SA, Omar SAS, Jamal ME, Shihab A, Hanna A-R (2004) Soil survey for farm planning in northern Kuwait. Kuwait J Sci Eng 31(1):43–57

Soil Survey Division Staff (1993) Soil survey manual, Agriculture handbook 18. USDA-NRCS. US Government Printing Office, Washington, DC

Tisdale SL, Nelson WL, Beation JD, Harlin JL (1993) Soil fertility and fertilizers, 3rd edn. Macmillan, New York

Torrent J, Nettleton WD (1978) Feedback processes in soil genesis. Geoderma 20:281–287

USDA (1938) Soil and man, Yearbook of agriculture. USDA. US Government Printing Office, Washington, DC

USDA (1998) Quality assurance issues for phase 1 of the soil survey of Kuwait. Report submitted to KISR by USDA-NRCS Staff

USDA-NRCS (1994) Keys to soil taxonomy, 6th edn. USDA-NRCS. US Government Printing Office, Washington, DC

USDA-NRCS (1996) Soil survey laboratory methods manual, Soil survey investigation report 42. USDA-NRCS. US Government Printing Office, Washington, DC

USDA-NRCS (1998) Keys to soil taxonomy, 8th edn. USDA-NRCS. US Government Printing Office, Washington, DC

USDA-NRCS (1999) Soil taxonomy. A basic system of soil classification for making and interpreting soil surveys, USDA agriculture handbook 436. USDA-NRCS. US Government Printing Office, Washington, DC

Van Alphen JG, de Los Rios Romero F (1971) Gypsiferous soils. International Institute for Land Reclamation and Improvement, Wageningen Netherlands, Bulletin 12

Chapter 4
Do the Emerging Methods of Digital Soil Mapping Have Anything to Learn from the Geopedologic Approach to Soil Mapping and Vice Versa?

Abbas Farshad, Dhruba Pikha Shrestha, and Ruamporn Moonjun

Abstract The use of soil maps and the feasibility of the existing soil survey procedure are often questioned by both surveyors and users. Thanks to the advances in the fields of remote sensing (RS) and geographic information system (GIS), a new trend – digital soil mapping – is emerging which might have answers to some of the questions. With a glance to some of the definitions and concepts, such as 'what is a soil?' and 'what is the content of a soil map?' we intend to highlight the complexity of the soil and its mapping. At the same time, we apply some of the geopedologic-oriented techniques of the digital terrain modelling to soil mapping in order to show the role of geomorphology in the mapping. The exercise was carried out as case studies in several areas in Thailand. Various soils at subgroup levels (Fluventic, Arenic, Aquic, Aeric, Ultic, Ustic, Vertic) belonging to the soil orders Entisols, Mollisols, Inceptisols, Alfisols, and Ultisols occur in different geomorphic surfaces, following well the physiographic setup of the landscapes. The case studies demonstrate the conventional predictive mapping (the ITC approach) and the geopedologic approach to soil survey, based on parameterisation of the soil-forming factors and their integration: in one case through applying decision trees, followed up by a statistical validation, and in another case by means of Artificial Neural Network (ANN). It is hoped to open up a discussion, which should lead to (1) clarifying the term 'digital soil mapping' and (2) finding out whether or not the shortcomings of the conventional approach of soil mapping can be recovered using the new trend and does the new trend suggest changes in the current definitions and concepts.

Keywords Digital soil mapping • Predictive soil mapping • Geopedology • Soil survey

---

A. Farshad (✉) • D.P. Shrestha • R. Moonjun
Faculty of Geo-Information Science and Earth Observation (ITC), University of Twente, P.O. Box 6, 7500 AA Enschede, The Netherlands
e-mail: farshad@itc.nl

S.A. Shahid et al. (eds.), *Developments in Soil Classification, Land Use Planning and Policy Implications: Innovative Thinking of Soil Inventory for Land Use Planning and Management of Land Resources*, DOI 10.1007/978-94-007-5332-7_4,

4.1 Introduction

Since the release of the Soil Survey Manual (Soil Survey Division Staff 1951) aerial photo interpretation-based techniques for mapping soils have been, and still are, widely used in many countries. The stereoscopic analysis of aerial photographs was on the basis of identification and classification of some relevant photographic and morphologic elements such as elevation (slope/relief), shape, size, texture, pattern, shadow and (grey) tone. Several approaches, some synthetic, some analytic, and others holistic, were developed, meant to partition landscape into meaningful compartments (units), in terms of soils content.

Buringh (1960) came up with his (synthetic) approach of 'element analysis', very much resembling the map overlaying techniques in geographic information system (GIS). He also introduced 'physiognomic analysis', roughly resembling 'pattern analysis', and some of the ideas used in 'Definiens' (http://www.definiens-imaging.com/ecognition/). This was further developed, not only in the Netherlands but also in Australia and in the United States of America (USA), becoming more and more physiography based. For several years, many surveyors were trained at ITC (the Netherlands) in 'physiographic analysis to aerial photo interpretation' (Goosen 1967; Bennema and Gelens 1969), very much based on establishing mental models of soil-landform relationship. Identifying geomorphic processes, responsible for the formation of landforms, and their appearance on the aerial photograph underlines the delineation of the physiographic units which in turn provide a good basis for the distribution pattern of soil mapping units. In the late 1980s, the 'geopedologic approach to soil survey' (Zinck 1988/1989), obviously benefitted from the preceding ones, rejuvenated the 'clorpt' principles and the use of geomorphology in soil survey. Through a rather strict application of the geopedologic rules, 'soilscape (soil [land] scape)' is mapped more efficiently. Geopedology can be defined as a science and, at the same time, an art of modelling the occurrence of soil(s) in the landscape, the basis of which is mental integration of extracted knowledge, for the study area, on climate, geology, geomorphology, sedimentology, hydrology, vegetation, and pedology. A geopedologic map is a soil map, which includes not only soil information but also some facts of landscape ecology-based aspects. A well-prepared large-scale geopedologic map, which is associated with a database, can be transformed into several digital maps of various soil attributes (as layers) in a GIS environment (Kumar 2008).

Since the tendency turned to go digital, that is, quantification began to overcome qualification, conventional soil survey was often questioned. The most frequently asked question is the following: Do the soil maps provide what is needed, considering the advances in mathematical modelling? The newly developing technology in the field of data acquisition and management is gradually taking over the conventional approaches. Here too, relevant questions are asked, such as is soil property mapping (e.g. topsoil texture) considered a 'soil survey'? In this chapter, we touched upon this and several other related questions, with the intention to follow up the topic in other papers.

4.2 Geopedologic Approach to Soil Mapping

Geopedology (a term that consists of 'geo' [geology, geomorphology] and 'pedology' [soil science]) is used to connote application of geomorphology (in its broad sense) in soil science. In this chapter, geopedologic approach to soil survey is meant to include all the conventional techniques of mapping and interpreting soils, which are based on soil-landform relationship, the latest of which is the one developed at ITC (Zinck 1988/1989).

4.2.1 What Does Soil Mean to Us and How 'A Soil' Is Defined

Soil means different to different people. To soil scientists, soil is a natural body (of inorganic, organic, water, and air), with a unique morphology resulted from a unique interaction of forming factors (climate, living matter, relief, parent material, and time, abbreviated as cl, o, r, p, t), as stated by Jenny (1941, 1980).

A soil: defining a minimum limit to the area of 'a soil' seemed inevitable. The concept 'pedon' was introduced to provide a clear basis for soil description and for the selection of representative soil samples. A pedon is the smallest volume of the soil that represents the nature and arrangement of its horizons and the variability in the properties that are preserved in samples. The minimal horizontal area of a pedon (USDA 1975) is arbitrarily set at 1–10 m^2, depending on the degree of soil variability. Australians (McKenzie et al. 2008), as an example of those who have their own standards, speak of 'soil individual' and propose a surface area of 25 m^2.

4.2.2 Soil Survey Procedure

A conventional soil surveyor describes soils and, after classifying them, plots their boundaries on a map and makes predictions about their behaviour versus a selected number of management practices (Soil Survey Division Staff 1993). Initially, the only procedure was the 'grid survey'. Surveying instruments (the ancestors of theodolite) were used to plot observation points, which could later be clustered to form map units. This time-demanding procedure was abandoned when aerial photo interpretation technique was introduced in soil survey. It was in the 1950s when the link between soil delineations and geomorphic units was realised, and so the application of geomorphology was promoted. Practically, 1960, when the Manual of Photographic Interpretation was published, can be considered as the beginning of a new era, although aerial photography was known earlier.

Classically, soil survey is done in several levels of detail (survey intensity), namely, exploratory, reconnaissance, semi-detailed, and detailed depending on the purpose of the survey. The appropriate scale, also for aerial photography, is determined according to the chosen level. Customarily, the corresponding scales for different levels are 1:250,000 to 1:100,000, 1:100,000 to 1:50,000, 1:50,000 to 1:25,000, and 1:25,000 to 1:5,000, respectively. Generally, aerial photograph does not play an important role in too detailed surveys. In the USA, 1:12,000 and 1:24,000 are the standard mapping and publication scales (Zhu et al. 2004). Geopedologic modelling, on the other hand, is important in all levels, although its application is more straightforward in the smaller scales.

4.2.3 What Is the Content of a Soil Map?

All definitions and concepts in soil science, such as soil delineation and taxonomic and map unit, are given in the 'guidelines for using soil taxonomy in the names of soil map units' (van Wambeke and Forbes 1986). Depending on the user's perspective, the map scale is decided upon, which is determinant for the content of map units. The larger the publication scale, the higher the number of observations and, as a result, the more discovery about the soil occurrence. Although soil delineation can be considered as the smallest polygon on the map in terms of soil content, it may still be too large to contain an individual soil body (consociation). Even in very detailed soil survey, the mapped soil can be one or another phase of 'a soil class'. The reason to this is that soils are often intergrading, and that is not easy to plot on a map. The fact is that soils seldom occur as discrete bodies. Since most of soil properties are hidden below the surface, it is difficult to follow their actual boundary on the ground surface. Soil surveyors have to rely on surface indicators such as topography, vegetation, surface soil colour, drainage ways, and drainage condition to draw lines showing the location of different soils.

While soil survey is still being done, several questions are repeatedly asked by many users, simply because soil survey has become too expensive. The most common questions are the following: Do soil maps provide what is needed? Is the soil map a good communication tool (Hudson 1992)? Soils are often mapped by the soil surveyors who are trained to document their mental models in maps and reports. The fact is that the quality of the mental models and, consequently, of the soil map is very much experience dependent.

4.3 The Emerging Methods in Soil Mapping

Although the shortcoming of the Boolean logic in soil mapping (polygon-based), that is, the transitional border between soil bodies, and the issue of soil variability are known to soil surveyors, there is not much to be done about it. In small-scale surveys, the transitional units are either shown on the map by different series name, which also carry a different classification name (e.g. one soil class intergrading towards another soil), or are described under the flag of 'range of characteristics', in soil survey report, howsoever, not quite satisfactory for some users, in particular in the field of modelling.

The speed with which the advances in the scientific world have been taking place is striking. The introduction of space technology and advances in information and communication technology (ICT), coupled with the derived facilities in automation, urged many scientific fields to go for adaptation. The changes were not of an equal speed in all disciplines. Some could cope with the adaptations easily, and some could not. In some cases, existing concepts and methodologies disappeared before the new ones were properly settled. Soil science may be considered as one of these disciplines. A quick review shows that (1) the Jenny's equation (Jenny 1941) has been improved and used in different contexts (Jenny 1994; Naveh and Liebermann 1983), (2) Soil Survey Manual (Soil Survey Staff 1951) was revised after almost half a century (Soil Survey Division Staff 1993), and (3) Soil Taxonomy (USDA 1975) has regularly been revised; its eleventh edition is issued in 2010 (Soil Survey Staff 2010). At the same time, similar concepts were used to introduce a more applicable soil classification system, the World Reference Base for Soil Resources (IUSS [in Montpelier] 1998) and a World Soil Resources Reports No. 103 (FAO 1998).

With the improvement in hardware computational power, interpolation techniques using geostatistical application became a common practice in soil mapping. Above all, the introduction of the digital terrain surface modelling (Zhilin et al. 2005) is a great achievement.

4.3.1 What Is Digital Soil Mapping?

Digital soil mapping (DSM), or 'predictive soil mapping' (Scull et al. 2003), is the computer-assisted production of digital maps of soil type and soil properties (Wikipedia on digital soil mapping). The International Union of Soil Sciences (IUSS) Working Group on Digital Soil Mapping (WG-DSM) has defined it as 'the creation and the population of a geographically referenced soil databases, generated at a given resolution by using field and laboratory observation methods coupled with environmental data through quantitative relationships'. The phrase 'using field and laboratory observation' in this definition does not only cover the collection of soil samples to determine the spatial variability of a given soil property (Robinson and Metternicht 2003) but also implies the field of image spectrometry (applying remote sensing), where spectral differences are plotted and interpreted to distinguish between different soils, for instance, on the basis of the clay mineralogy differences (Ben-Dor 2002; Kariuki 2003).

Despite the definitions that are given by some highly positioned workers, the term 'digital soil mapping' is used sometimes to distinguish between the conventional manual soil mapping (also called polygon-based) and the computer-based (raster data model) soil mapping, sometimes (as in the Wikipedia) to mean 'predictive mapping', and sometimes it even refers to the digitised copy of an existing soil map (Hengl and Reuter 2009). Moreover, the term digital soil mapping is frequently used to map a certain soil property of interest (Krol 2008; McBratney et al. 2003; Mora-Vallejo et al. 2008; Peng et al. 2003) and not necessarily the whole soil body, as envisaged in the original concept of soil mapping. Pedometrics, on the other hand, is the field that covers the quantitative approaches in soil science (Bouma 1997).

To remove confusion, clear distinction ought to be made between predictive soil mapping, digital predictive soil mapping, digital soil mapping, and digital soil property mapping.

4.3.2 *Advances in Inventory Tools and Data Management*

4.3.2.1 Use of Digital Terrain Model (DTM)

A digital terrain model is a mathematical (or digital) model of the terrain surface (Zhilin et al. 2005). The mathematics takes care of the interpolation process, which has been advanced with increasingly efficient and cheap computation power and storage, availability of digital contour, stream, and orthophotographic data (www.ffp.csiro.au/nfm/mdp/softdem.htm). Zhilin et al. (2005) classify the surface modelling approaches as point-based modelling, triangle-based modelling, grid-based modelling, and a hybrid approach combining any pair out of the three approaches. The required input data for DTM generation may come either from field survey (e.g. use of conventional surveying instrument or GPS), from stereo pairs of aerial (or space) images using photogrammetric techniques, or from digitisation of the existing topographic maps. The latter source is the most commonly used technique, although more and more use is made of the freely available DEMs, downloadable from SRTM (Shuttle Radar Topography Mission) at http://srtm.usgs.gov/. However, this product (with 90 m resolution, except for the USA, with 30 m resolution) will not satisfy those who need high-resolution data. For this, LIDAR (Light Detection and Ranging) will be an option. This is an optical remote-sensing technology that helps measure differential height leading to register the terrain surface topography. Almost all well-known commercial GIS packages are equipped with a sub-module taking care of generating DEM. The ARCINFO, for instance, is equipped with ANUDEM, a programme developed in the Centre of Resource and Environmental Sciences of the Australian National University in Canberra, which supports production of grid-based DEMs using contour line map. Or in the ENVI software, the sub-module 'topography' supports generating DTM using ASTER images. The GRASS GIS software is also equipped with a number of terrain analysis procedures, especially for hydrological modelling and erosion mapping. ILWIS, which became recently open-source GIS package, has also extensive DEM hydro-processing possibilities. There are also a few freely available packages, such as TARDEM and TauDEM developed at the Utah Water Research Laboratory (http://www.itc.nl/library/Papers_2003/misca/hengl_digital.pdf).

4.3.2.2 Some Other Tools and Techniques Available to Soil Surveyors

Process of inventory/mapping includes collecting data (usually from points) and storing them with their geographic (spatial) properties, which should be put in a map to show soil distribution in a given area (map presentation). The advances in remote sensing, although not fully benefitted from, and in geographic positioning system (affordable GPS receivers) facilitate the process of data collection/fieldwork (Hengl and Reuter 2009; McKenzie et al. 2008). In the same way, the advances in database management system (DBMS), both spatial and nonspatial, in a GIS environment, and the newly developed technology in the fields of expert system, decision support system, etc. (Bui 2004; Bui and Moran 2001, 2003; Hengl and Reuter 2009; Hengl et al. 2002; Lagacherie et al. 2007; Moran and Bui 2002) are all at the service of the mapping. Furthermore, the possibility that the field of mathematical modelling offers seems promising.

4.4 Materials and Methods

The once well-formulated set of soil survey procedures, which made soil survey looks as if it were a routine activity, was often questioned. Introduction of the 'geopedologic approach to soil survey' (Zinck 1988/1989) made surveyors asking themselves whether it was just to roughly delineate polygons on aerial photographs, which should be described for their soils content. Should the interpretation of aerial photographs and the survey not be based on understanding of soil-landform relations? This is where 'soil geomorphology' or 'geopedology' comes in, that is, when geomorphology, in its broad sense (covering lithology, hydrology, sedimentology, etc.), forms the backbone of the survey (the knowledge system). To predict the occurrence and distribution of soils, it is essential to understand the geomorphic processes responsible in the formation of landforms and the reasoning on the landform-soil relationship.

Based on a strong integration of geomorphology and pedology, the former can be used as a tool to improve and speed up the survey. The discipline of geomorphology as used in this approach operates through a taxonomic system of six levels (Zinck 1988/1989). In regional surveys, four of the six levels are differentiated: (1) landscape as the first entry in the legend, (2) relief/moulding, (3) lithology/facies, and (4) landform. On the first level, seven landscape units (including the mapping symbols) are distinguished: mountain (Mo), hill land (Hi), plateau (Pu), piedmont (Pi), peneplain (Pe), plain (Pl), and valley (Va).

Each landscape unit can be composed of a number of relief types (e.g. hill, mesa, glacis), which may occur on different materials (lithology/facies). Any relief type may then comprise of more than one landform. An obvious advantage of the geopedologic analysis is in the stepwise approach in image interpretation, according to the hierarchical levels of the system. Contrary to other approaches in visual image interpretation, the interpreter applying the principles of geopedologic analysis builds up a legend as he/she proceeds, by going through the following steps (Farshad and Rossiter 2008):

1. Photo reading: stereoscopically studying sample image (photo pairs, or more recently, DEM overlain by satellite imageries).
2. Identifying master lines: Lines separating major units are traced on an overlay or on-screen. The major unit of stratification is the geoforms. It depends on the mapping scale and degree of detail, and at any detail, it represents morphology plus geology.
3. Sketching: A number of cross sections are selected to cross the geoforms.
4. Pattern recognition: A number of units are recognised along each of the cross sections.
5. Delineation: The recognised units are delimited; at the same time, the legend columns (from the hierarchical classification) are filled in.
6. Composition of legend: A coherent overall legend is constructed from the legend columns, and the original legend columns are adjusted to match the overall legend.
7. Photo interpretation of the complete area to be mapped, using the developed legend: Here, a thorough knowledge of geomorphology is required to delineate the map units.
8. Addition of columns to the legend specifying the soil mapping unit type and component(s) according to standard USDA nomenclature (Soil Survey Manual and Soil Taxonomy).
9. Optional addition of columns to the legend specifying other soil information beyond the taxonomic name, for example, depth and details of parent material; much of this would be in a soil series description in detailed soil survey.

The geopedologic approach to soil survey demands understanding of relationships between soil on one side and the landscape constituents on the other side. The approach has been applied and taught to international students following soil survey procedures at ITC, the Netherlands, for almost a decade. In the past few years, some attempts were made to automate soil mapping using the concepts of geopedology. In the following sections, three case studies are presented. The case studies are of different areas in Thailand, and the geopedologic approach to soil survey forms the backbone of all the cases.

4.5 Results and Discussion

4.5.1 Case Study 1 (Upper Pa Sak Valley, Thailand)

The case study area is located in Lom Kao and Lom Sak districts, Phetchabun Province, Thailand. It is located between 101°30′–101°45′ E and 16°45′–17°15′ N, with an approximate surface area of 750 km^2 (Hansakdi 1998). Under the influence of past climate fluctuations, tectonic movements, and lithologic-structural controls, the Pa Sak valley has been formed, banked by a few terraces and a series of hills and glacis surfaces, of which the middle level appears in scattered patches. The area falls in the 'central highlands', a complex physiographic unit, mainly composed of hills, strongly incised plateaux and piedmont, varying in altitude as from 1.200 to about 300 m above sea level (masl). The complex mountainous area in the north serves as the catchment basin of several important rivers such as the Pa Sak and the Nam Phung, originating from the heights of the Dan Sai district, Loei Province. The Pa Sak river flows north to south along a steep cliff, where the river changes into a streamlet which is overflown in rainy seasons. The Nam Phung is another waterway which contributes to the total water resource used for cultivation and drinking purposes.

The geopedologic-based photo interpretation map was used to locate sample areas. These are to economise on the fieldwork costs. Sample areas should represent all map units (to the level of landform) obtained from the image interpretation. Sample areas are selected in the easily accessible parts of the area under study, and each must be a sample of a larger area. As a rule of thumb, the total surface of sample areas should be 1/10th of the total surface area under study. Obviously, sample area is not applicable to large- and very-small-scale surveys, as in the former full check is applied, and in the latter in a short visit to field, a quick check is carried out. Sample areas are also not required when the petitioner requests for a full check.

An ideal survey procedure started with collecting and studying the existing materials and publications, followed up by a short visit to the area (if not known to the surveyor and not too far from office). This is very useful when doing photo interpretation of small-scale aerial photograph coverage of the area under study. Sample areas are then selected on the base map. In an ideal situation, aerial photograph coverage of the sample areas is of larger scale and is interpreted in a more detailed and precise way, followed by a detailed field survey in the sample areas. The survey in the sample areas is quite intensive as the selected sites are studied applying catena/toposequence sampling. The advantage of this sampling technique guarantees full understanding of soil-landform relationships. It is in this stage – studying the sample areas – that the surveyor can check and continuously improve the hypothesis (model) that is originally established for the soil (pattern) distribution in the area. To economise on time and cost, observations are in mini-pits, that is, reduced sized pits. A mini-pit is $100 \times 100 \times >50$ cm pit. The depth (>50 cm) is varying depending on when the subsurface diagnostic horizon appears. The next phase is reinterpreting the aerial photographs coverage of the whole area, where all what is learnt about relationships soil-landform is applied. This is the extrapolation phase. This phase is then followed by a random checking outside the sample areas. A satisfied surveyor would be someone whose predictions on all check sites outside the sample areas are fulfilled. This means the mental model developed by the surveyor is correct. The resulting geopedologic map and legends (shortened) are given in Fig. 4.1 and Tables 4.1 and 4.2.

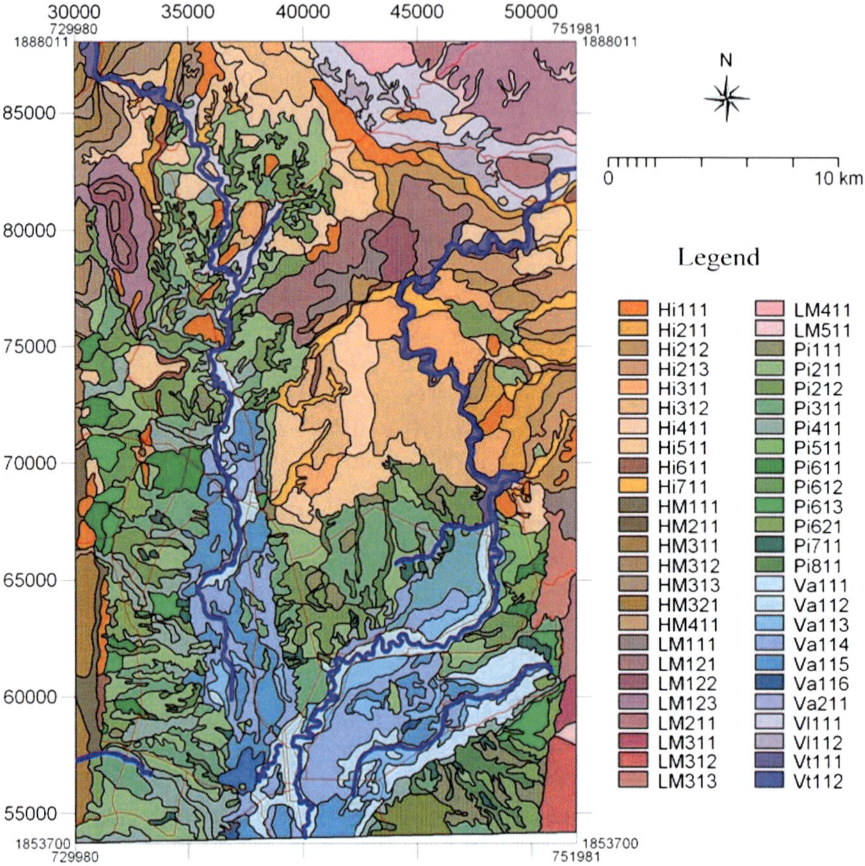

Fig. 4.1 Geopedologic map of Upper Pa Sak valley, Phetchabun, Thailand

4.5.2 Case Study 2: Doi Ang Khang, Thailand

The study area is located in Fang district, Chiang Mai Province, in northern Thailand (Udomsri 2006). The area lies between 19°56′ 00″ N and 19°51′ 02″ N and between 99°01′ 27″ E and 99°04′ 25″ E, with about 27 km^2 of surface area. The altitude varies from 1,040 to 1,920 masl. The area has humid subtropical climate with average temperature of 18 °C with a mean maximum value of 24°C (April) and mean minimum value of 12 °C (January). The mean annual rainfall is about 2,055 mm, most of which falls during May to October with the highest peak in August.

The area comprises of mountains and hills which are strongly dissected with high relief. The sedimentary rock with predominantly shale (including also phyllite and sandstone) occurs in the eastern part and with predominantly limestone in the western part. Between the two mountain ranges lies the piedmont which is narrow

Table 4.1 Geopedological legend for the 'high mountain' landscape

Landscape	Relief type	Lithology	Landform	Map unit	Map unit description	Dominant taxa and inclusion	
						Name	%
High mountains	Hills	Predominantly sandstone, siltstone, and shale	Slope facet complex	HM111	Convex-concave complex, steep to very steep, stony and rocky surface	Inferred:	
						Dystropepts	
						Hapludults	
						Rocks	
	Ridge	Sandstone and shale	Slope facet complex	HM211	Straight, elongated, steep to very steep dissected slopes	Inferred:	
						Lithic Ustorthents	
						Rocks	
	Hogback	Sandstone and shale	Reversal	HM311	Straight, steep to very steep slopes	Lithic Ustropepts	~40
						Ultic Haplustalfs	~40
						Arenic Paleustults	~10
						Fluventic Dystropepts	~10
			Front	HM312	Straight, steep to very steep slopes	Lithic Dystropepts	~40
						Typic Ustropepts	~40
						Typic Argiustolls	~20
			Talus	HM313	Concave-convex complex	Typic Hapludults	~50
						Lithic Ustorthents	~40
						Rocks	~10
		Predominantly limestone, siltstone, and chert	Slope facet complex	HM321	Concave-convex complex, steep to very steep slopes	Inferred:	
						Haplustalfs	
						Argiustolls	
						Rocks	
	Vale	Alluvium-colluvium		HM411	Concave, level to nearly level bottom, colluvial deposition, and agricultural terraces	Fluventic Ustropepts	>75

Table 4.2 Geopedological legend for the 'valley' landscape

Landscape	Relief type	Lithology	Landform	Map unit	Map unit description	Dominant taxa and inclusion	
						Name	%
Valley	Terrace	Alluvium	Tread-riser complex	Va111	Nearly level, wide agricultural alluvial terraces, incised by streams	Aeric Tropaquepts	~40
						Fluventic Ustropepts	~30
						Vertic Ustropepts	~10
						Oxyaquic Dystropepts	~10
						Fluventic Eutropepts	~10
			Levee	Va112	Nearly level, dominated by villages and home gardens	Fluventic Ustropepts	~60
						Typic Ustropepts	~40
			Levee/overflow mantle complex	Va113	Nearly level	Aeric Tropaquepts	~40
						Fluventic Ustropepts	~30
						Typic Ustropepts	~20
						Aquic Eutropepts	~10
			Overflow mantle	Va114	Nearly level	Fluvaquentic Eutropepts	~30
						Typic Ustropepts	~30
						Fluventic Ustropepts	~30
						Typic Tropaquepts	~10
			Overflow basin	Va115	Slightly depression	Fluvaquentic Eutropepts	~30
						Aquic Eutropepts	~30
						Typic Tropaquepts	~20
						Typic Ustropepts	~10
						Fluventic Dystropepts	
			Overflow mantle/basin	Va114/115	Nearly level	Fluvaquentic Eutropepts	
						Typic Ustropepts	
						Fluventic Ustropepts	
						Aquic Eutropepts	
						Typic Tropaquepts	
						Fluventic Dystropepts	

			Decantation basin	Va116	Depression, poorly drain, no outlets	Aquic Eutropepts	~40
						Aeric Tropaquepts	~30
						Fluventic Eutropepts	~20
						Typic Tropaquepts	
			Overflow/decantation basin	Va115/116	Depression	Fluvaquentic Eutropepts	
						Aquic Eutropepts	
						Typic Tropaquepts	
						Aquic Eutropepts	
						Aeric Tropaquepts	
						Fluventic Eutropepts	
	Flood plain	Alluvium	Levee/basin complex	Va211	Narrow river banks	Fluventic Ustropepts	~30
						Aquic Ustropepts	~30
						Fluvaquentic Eutropepts	~30
						Typic Eutropepts	~10
Lateral valley	Terrace complex	Colluvium-alluvium	Bottom/side complex	Vl111	Undulating to rolling	Aeric Tropaqualfs	~40
						Typic Haplustalfs	~40
						Aquic Eutropepts	~10
						Typic Ustifluvents	~10
	Depression		Bottom/side complex	Vl112	Nearly level	Aeric Tropaquepts	~60
						Aeric Tropaqualfs	~40
Trench valley	Terrace complex	Alluvium/residual	Bottom/side complex	Vt111	Nearly level	Typic Ustifluvents	~50
						Typic Haplustalfs	~30
						Ultic Paleustalfs	~20
	High terrace		Bottom/side complex	Vt112	Nearly level to undulating	Ultic Paleustalfs	~50
						Ultic Haplustalfs	~50

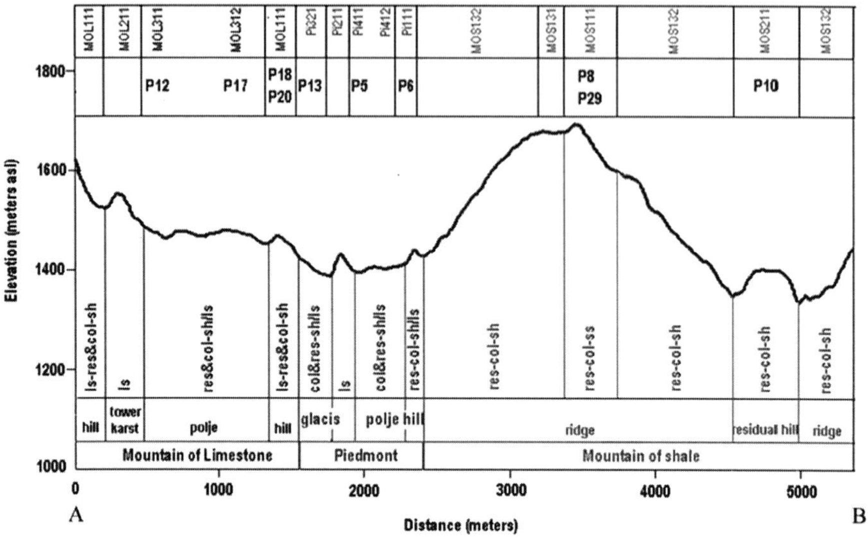

Fig. 4.2 An example of a transect crossing different geomorphic units. Location of the transect line *AB* is given in Fig. 4.4

and elongated and which also serves as the basin of the whole area (Fig. 4.2). Soil data was collected along a few transects, and observations were made in mini-pits and auger holes.

The Jenny equation (Jenny 1941) was parameterised in a GIS environment to map the soils of sloping and inaccessible areas. According to the equation, soil is the function of five factors [S = f (cl, o, r, p, t), where cl = climate, o = organism, r = relief, p = parent material, and t = time]. The equation is classically used to discuss soil genesis, but in the time of automation, the essential questions are how to parameterise the factors in order to generate the different GIS map layers and in what way the layers can best be integrated to obtain the soil map (representing soil bodies). The soil-forming factor maps were generated as described below:

Climate (cl) map

For climatic map, soil moisture regime was used as the indicator although it is only partially a function of the climate. The moisture regime of a soil is an important property of the soil as well as a determinant of processes that can occur in the soil (USDA 1975).

Organism (o) map

For organism map, it is difficult to state what kind of vegetation grew in a given (the study) area in particular when soils were forming. Using correlation between lithology and the vegetation type (Santisuk 2006), two vegetation types were selected for the organism map as follows:

1. Lower montane oak-pine forest: occurring in the eastern part of the study area, where shale, phyllite, and sandstone are dominant

2. Lower montane oak forest: occurring in the western part of the study area, where limestone is dominant

Relief (r) map
For the relief map, slope position and slope steepness classes were used. They were generated using digital elevation data, in combination with photo interpretation. Slope positions are classified as summit-shoulder complex, footslope, backslope-footslope complex, sinkhole, toeslope, and slope facet complex.

Parent material (p) map
For generating parent material map, available geology map, aerial photo interpretation, the data from site characterisation, and field observation in transects were used.

Time (t), (soil profile development) map
Time is considered in terms of maturity or stage of soil development (Van Reeuwijk 1997). To estimate time for soil profile development, soil morphology as related to lithology and landforms and taxonomic classification at soil order level (Ultisols as being very old to Entisols as being young) were used.

To integrate different soil-forming factors (in five GIS map layers), a decision tree was set up with soil classes as output. Decision trees are useful tools, because they perform classification through a sequence of simple, easy-to-understand statements, whose semantics are intuitively clear to domain experts (Murthy et al. 1994), and because of their simplicity, they have been used for various classification purposes (Mather 2004; Rossiter 1990; Shrestha et al. 2004). An example of the decision tree used for integrating different soil-forming factors to map soil of the study area is given in Fig. 4.3. The resulting soil map is shown in Fig. 4.4 (Udomsri 2006).

4.5.3 Case Study 3: Hoi Num Rin, Chiang Mai, Thailand

The study area is located in Mae Ka Chan district, in Chiang Rai Province, Thailand, and lies between 99°02′ and 99°07′ east and between 19°25′ and 19°27′ north (Moonjun 2007). The area is mountainous, and the altitude varies from 700 m to 1,720 masl. Annual rainfall is about 1,973 mm, and average annual temperature is 24°C. The rainy season starts at the beginning of May which lasts until late October. The area had dense forest cover in the past with evergreen forest and deciduous dipterocarp forest types, which was deforested and followed by shifting cultivation. Deforestation has stopped now, and because of strict government policy, dense forests exist in some areas, while it is degraded elsewhere.

The area is characterised by having three main geomorphologic zones: mountains, hills, and valley. The area is mostly hilly and mountainous with slope ranging between 12 and 75%. Flat to gently sloping area is confined to only a few parts of the vale and valley.

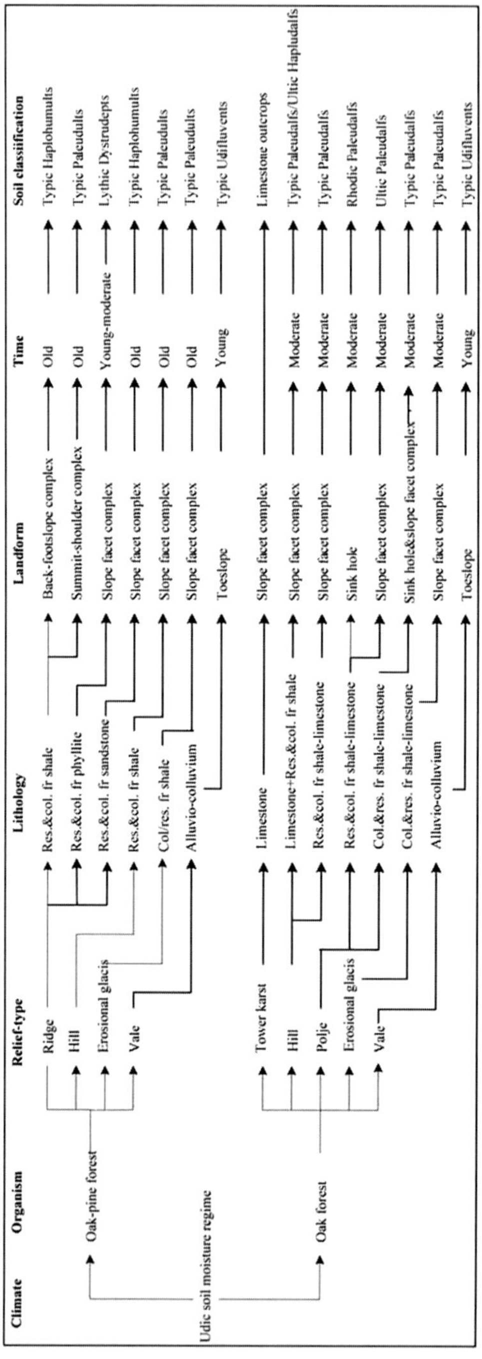

Fig. 4.3 Decision tree for soil classification

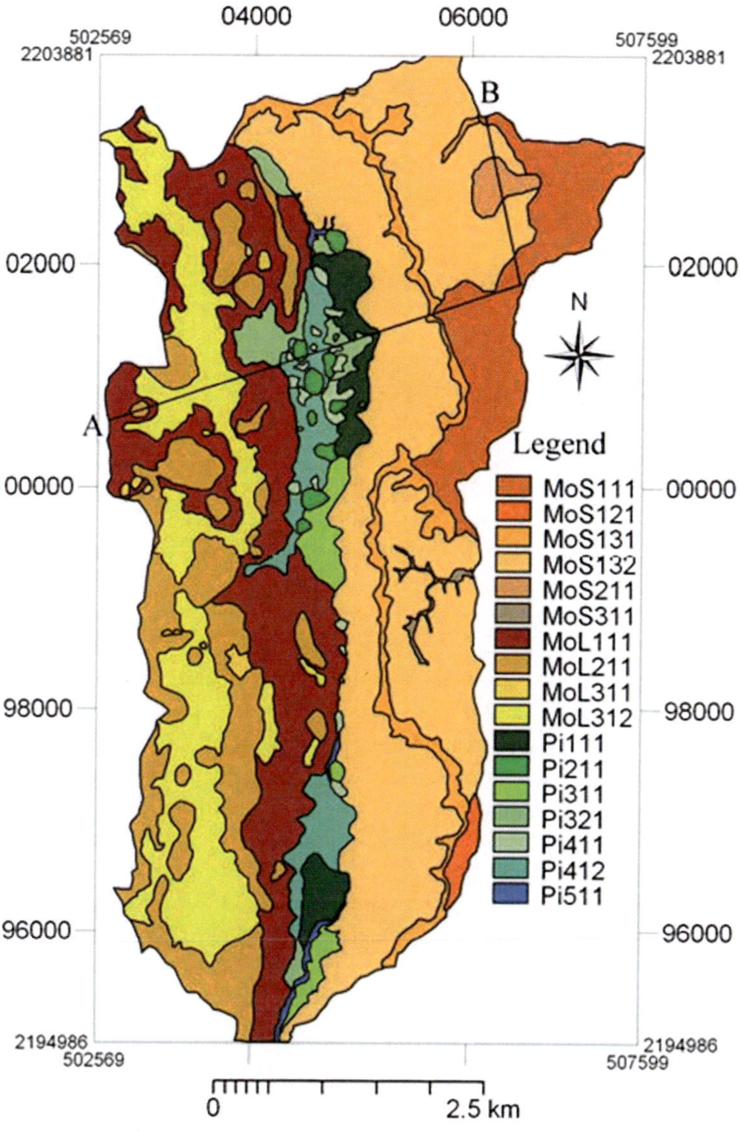

Fig. 4.4 Soil map of Doi Ang Khang, Chiang Mai, Thailand; the line *AB* crosses various geomorphic units as shown in Fig. 4.2

For mapping soil, the soil-forming factors as referred to in the SCORPAN method (McBratney et al. 2003) were integrated using Artificial Neural Network (ANN). Artificial Neural Network is the mathematical model of a neural network consisting of a set of simple functions linked together by weights. The application of neural networks as 'pedotransfer functions' for predicting soil hydraulic properties is well known (McBratney et al. 2003). Here, the map layers on climate, organism, relief, parent material, and time were derived from an image interpretation map at scale 1:25,000 (geoform map). This geoform map was used to select the sample areas where soil observations were made (Fig. 4.5). The soils were classified according to USDA Soil Taxonomy (USDA 1975). Soils outside the sample areas were classified using ANN. The resulting map is shown in Fig. 4.6.

4.6 Discussion and Conclusions

Soil mapping using aerial photo interpretation has been carried out since a long time. Geopedologic approach to image interpretation for soil survey helps increase mapping efficiency. Sample area-based survey helps better understand landform-soil relationship, and consequently, while increasing mapping efficiency and quality, it also reduces costs. The case study 1 is a typical example of a detailed reconnaissance geopedologic mapping. The published soil map by Land Development Department of Thailand (LDD publication) shows the occurrence of soil series (as used in Thailand) without giving much information on the landforms where they occur. Main entry in the legend is the series name. Obviously, the shortage in the geomorphic information limits the applicability of such a map, for instance, in land-use planning. No doubt that proper land-use plans minimise land degradation problems.

In the case study 2, the landform-soil relationship is coupled with the 'soil-forming factors' and put into decision trees (rule-induction algorithms). For mapping soils of the inaccessible mountainous areas, decision tree model proved to be a very useful tool. Comparison between the output of this study and the conventionally prepared soil map (LDD product) shows promising results: 95% match at soil order level, 84% match at suborder level, and 43% at subgroup level, considering the fact that to make distinction, for instance, between 'Typic' and 'Ultic' subgroups is not an easy job in the field, particularly for a beginner (Farshad et al. 2005).

In the case study 3, the revised Jenny equation (SCORPAN-like approach) was incorporated in understanding landform-soil relationship. Soil mapping was carried out using the same guiding principles as in the case 2. To integrate the parameterised factors, use was made of the Artificial Neural Network (ANN). Here too, results were satisfactory (Moonjun et al. 2008).

Soil mapping using conventional technique (as used at Land Development Department) is quite time demanding and hence expensive. This is even more the case in poorly accessible mountainous areas. To increase efficiency in mapping soils, various tools and techniques within the fields of remote sensing, GIS, and modelling have been tried out in the past years. The guiding principles of geopedology were tried out to map soil classes, and, to a lesser extent, some given soil properties such as soil organic carbon content and salinity (EC) occurrence and its distribution in topsoil. To animate what is going on in the mind of an experienced soil surveyor, various indicators (terrain parameters) such as normalised difference vegetation index (NDVI); topographic wetness index (TWI) or compound topographic index

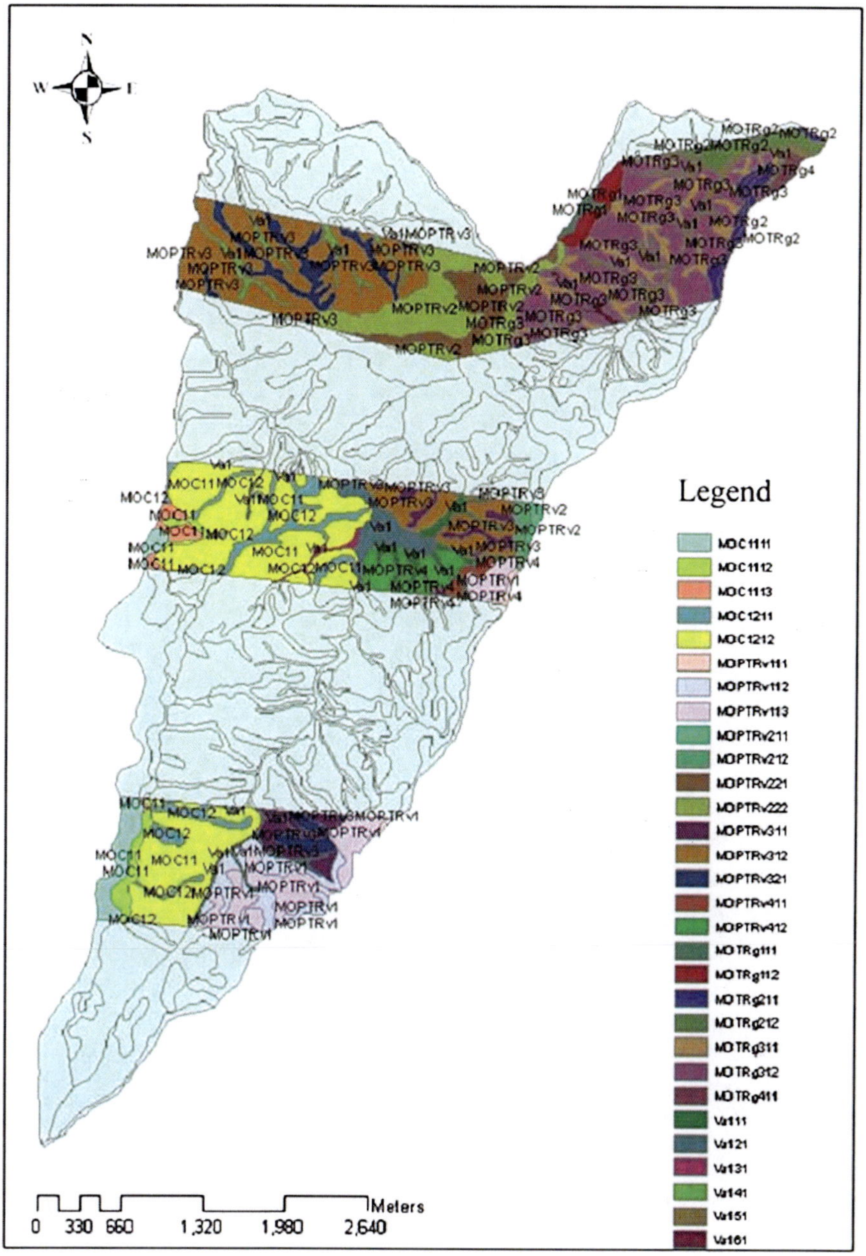

Fig. 4.5 Selected sample areas in Hoi Num Rin, Chiang Rai, Thailand

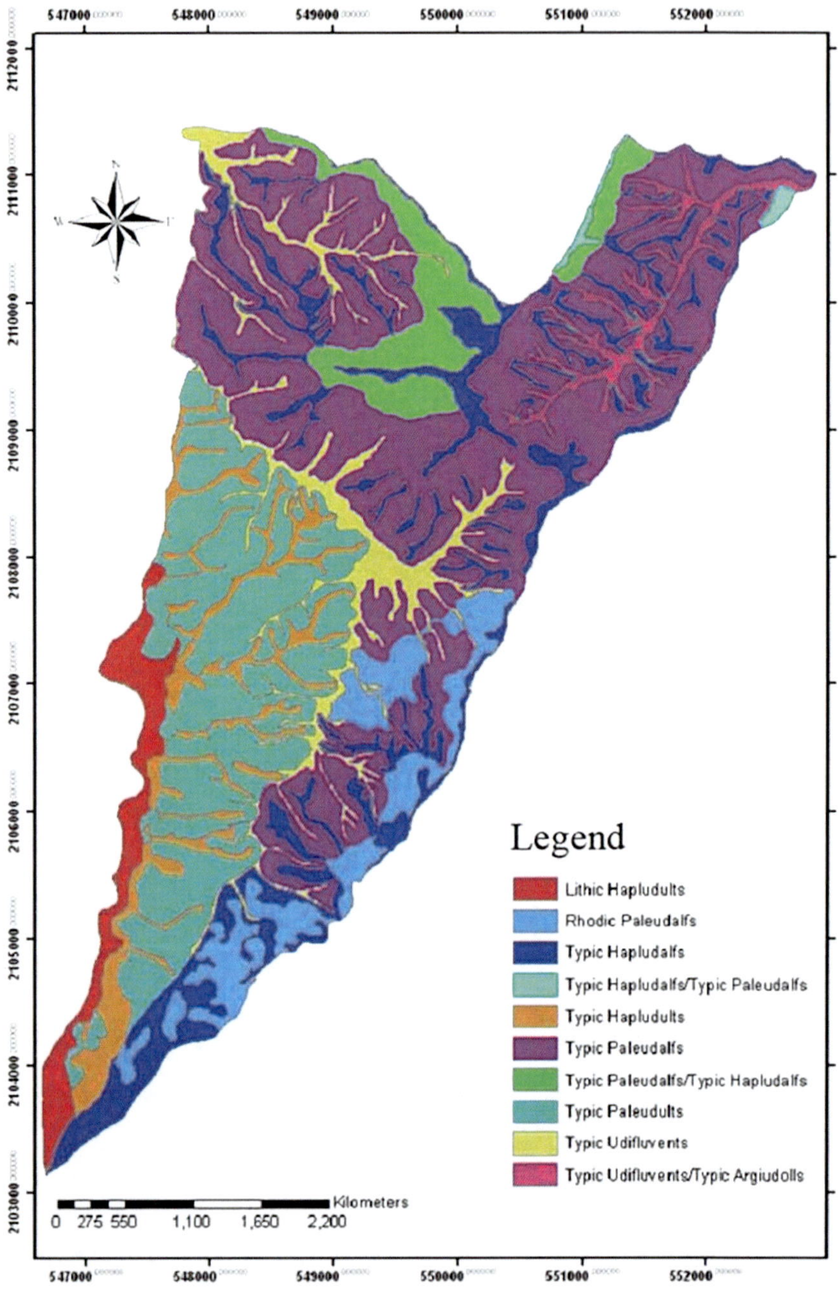

Fig. 4.6 Soil map of Hoi Num Rin, Chiang Rai, Thailand

(CTI), some derived from image interpretation and some from generated DEM; and/or various types of kriging (geostatistics) for interpolation have been used by many researchers (Corbane et al. 2008; McBratney et al. 2003; Mora-Vallejo et al. 2008). Many of the techniques used to map certain soil properties, rather than mapping the soil body, are producible and have been applied, not necessarily by soil scientists. Incorporating thorough understanding of landform-soil relationship in the (mental) model considerably helps automating (computer-based) the survey procedure. This, in combination with an appropriate soil database, will increase the map quality and the possibility of extraction of some attribute maps (e.g. pH of topsoil). Obviously, such information (subject of soil property mapping) can only be extracted from a geopedologic map when scale allows and if soil series (USDA-based) are mapped.

The current situation in soil survey is not quite clear. Most of 'soil science divisions' either in university or in ministry of agriculture and in many countries have been either completely shut down or placed in or under one or another newly created department, such as 'land-use planning', 'physical land resources', and 'earth system analysis'. Seeing the situation, soil surveyors have been trying to demonstrate the importance of soil in environmental studies (Hartemink 2006) by producing joint studies with other specialists such as hydrologists and archaeologists. The results appear in some scientific papers in such fields as 'hydropedology' and 'archaeopedology'.

Despite all these, there is little or no discussion around some fundamental issues, such as the following: (1) What is a soil? (2) What is soil survey? (3) Is soil mapping the same as soil survey? (4) Do we do soil mapping (survey) when a soil property is the study subject? (5) Is soil map still needed, or we should map a given soil property, at a time, what is needed for a given purpose? (6) Who should do soil mapping (survey), if soil maps are still needed? (7) Where do we train soil surveyors? These questions and anything in this line must be thoroughly debated and published, for instance, as an IUSS publication, considering that IUSS is the home to many soil scientists.

Acknowledgements Materials used in the case studies were derived from a joint research project of ITC, the Netherlands, with the Land Development Department (LDD) and Ministry of Agriculture and Cooperatives, Bangkok, Thailand. Contribution of our course participants, especially Anukul Suchinai, Ekanit Hansakdi, and Satira Udomsri, is duly acknowledged.

References

Ben-Dor E (2002) Quantitative remote sensing of soil properties. Adv Agron 75:173–243, Academic Press

Bennema J, Gelens HF (1969) Aerial photo-interpretation for soil surveys. International Institute for Aerial Survey and Earth Sciences, Delft, p 87

Bouma J (1997) The role of quantitative approaches in soil science when interacting with stakeholders. Geoderma 78:1–12

Bui EN (2004) Soil survey as a knowledge system. Geoderma 120(1–2):17–26
Bui EN, Moran CJ (2001) Disaggregation of polygons of surficial geology and soil maps using spatial modelling and legacy data. Geoderma 103(1–2):79–94
Bui EN, Moran CJ (2003) A strategy to fill gaps in soil survey over large spatial extents: an example from the Murray-Darling basin of Australia. Geoderma 111(1–2):21–44
Buringh P (1960) The application of aerial photographs in soil surveys, Manual of photographic interpretation. American Society of Photogrammery, Washington, DC
Corbane C, Raclot D, Jacob F, Albergel J, Andrieux P (2008) Remote sensing of soil surface characteristics from a multiscale classification approach. Catena 75(3):308–318
FAO (1998) World reference base to soil resources. FAO, ISRIC and ISSS, Rome
Farshad A, Rossiter DG (2008) Computer-assisted geopedology for predictive soil mapping. In: 3rd Global workshop on digital soil mapping, Utah State University, Logan
Farshad A, Shrestha DP, Moonjun R, Suchinai A (2005) An attempt to apply digital terrain modeling to soil mapping, with special attention to sloping areas; advances and limitations. Wrap up seminar of the LDD-ITC research project, Hua Hin
Goosen D (1967) Aerial photo interpretation in soil survey. Food and Agriculture Organization of the United Nations, Rome, p 108
Hansakdi E (1998) Geopedologic approach to soil survey in LomSak/LomKao area in Thailand. ITC, Enschede
Hartemink A (ed) (2006) The future of soil science. International Union of Soil Sciences, Wageningen, p 165
Hengl T, Reuter HI (eds) (2009) Geomorphology; concepts, software, application, vol 33, Develop soil science. Elsevier, Amsterdam/Oxford, p 765
Hengl T, Rossiter DG, Husnjak S (2002) Mapping soil properties from an existing national soil data set using freely available ancillary data. In: 17th WCSS, Bangkok, Thailand
Hudson BD (1992) The soil survey as paradigm-based science. Soil Sci Soc Am J 56(3):836–841
Jenny H (1941) Factors of soil formation. McGraw-Hill, New York, p 281
Jenny H (1980) The soil resource; origin and behavior, vol 37, Ecological studies. Springer, New York
Jenny H (1994) Factors of soil formation: a system of quantitative pedology. Dover Publications, New York, p 281
Kariuki PC (2003) Spectroscopy and swelling soils, an integrated approach. ITC dissertation no 109, Ph.D. thesis, Technical University of Delft, Enschede, p 158
Krol BGCM (2008) Towards a data quality management framework for digital soil mapping with limited data. In: Hartemink AE, McBratney A, Mendonça Santos ML (eds) Digital soil mapping with limited data. Springer, Dordrecht, pp 137–149
Kumar N (2008) Investigating the potentiality of regression kriging in the estimation of soil organic carbon versus the extracted result from the existing soil map. Unpublished MSc thesis, ITC, Enschede
Lagacherie P, McBratney AB, Voltz M (eds) (2007) Digital soil mapping; an introductory perspective, vol 31, Develop soil science. Elsevier, Amsterdam/Boston, p 600
Mather PM (2004) Computer processing of remotely – sensed images: an introduction. Wiley, Chichester, p 324
McBratney AB, Mendonça Santos ML, Minasny B (2003) On digital soil mapping. Geoderma 117(1–2):3–52
McKenzie N, Webster R, Ryan P (2008) Sampling using statistical methods. In: McKenzie N, Grundy M, Webster R, Ringrose-Voase A (eds) Guidelines for surveying soil and land resources. CSIRO Publishing, Melbourne
Moonjun R (2007) Application of artificial neural network and decision tree in a GIS – based predictive soil mapping for landslide vulnerability study: a case study of Hoi Num Rin sub – Watershed, Thailand. ITC, Enschede, p 104
Moonjun R, Farshad A, Shrestha DP, Vaiphasa C (2008) Application of artificial neural network and decision tree in a GIS-based predictive soil mapping in sloping areas; a case study of Hoi

Num Rin sub-watershed, Thailand. In: 3rd global workshop on digital soil mapping. Utah State University, Logan

Mora-Vallejo A, Claessens L, Stoorvogel J, Heuvelink GBM (2008) Small scale digital soil mapping in Southeastern Kenya. Catena 76(1):44–53

Moran CJ, Bui EN (2002) Spatial data mining for enhanced soil map modelling. Int J Geogr Info Sci 16(6):533–549

Murthy SK, Kasif S, Salzberg S (1994) A system for induction of oblique decision trees. J Artif Intell Res 2:1–32

Naveh Z, Liebermann SA (1983) Landscape ecology, theory and application. Springer, New York/Berlin/Heidelberg/Tokyo

Peng W, Wheeler DB, Bell JC, Krusemark MG (2003) Delineating patterns of soil drainage class on bare soils using remote sensing analyses. Geoderma 115(3–4):261–279

Robinson TP, Metternicht G (2003) A comparison of inverse distance weighting and ordinary kriging for characterizing within-paddock spatial variability of soil properties in Western Australia. Cartography 32(1):11–24

Rossiter DG (1990) ALES: a framework for land evaluation using a microcomputer. Soil Use Manag 6(1):7–20

Santisuk T (2006) Vegetation types of Thailand. Wildlife and Plant Conservation Department, National Park

Scull P, Franklin J, Chadwick OA, McArthur D (2003) Predictive soil mapping: a review. Prog Phys Geogr 27(2):171–197

Shrestha DP, Zinck JA, Van Ranst E (2004) Modelling land degradation in the Nepalese Himalaya. Catena 57(2):135–156

Soil Survey Division Staff (1993) Soil survey manual, Handbook no 18. United States Department of Agriculture, Washington, DC, p 437

Soil Survey Staff (1951) Soil survey manual, Handbook no 18. United States Department of Agriculture, Washington, DC, p 503

Soil Survey Staff (2010) Keys to soil taxonomy, 11th edn. USDA-Natural Resources Conservation Service, Washington, DC

Udomsri S (2006) Application of computer assisted geopedologic to predictive soil mapping and its use in assessing soil erosion prone areas: a case study of Doi Ang Khan, Ang Khang Royal agricultural station, Thailand. ITC, Enschede, p 119

USDA (1975) Soil taxonomy: handbook no 436. U.S. Government Printing Office of Agriculture, Washington, DC, p 754

Van Reeuwijk LP (1997) Introduction to physico-chemical aspects of soil formation. ITC, Enschede, p 83

van Wambeke A, Forbes TR (1986) Guidelines for using soil taxonomy in the names of soil map units, SMSS technical monograph, 10. Soil Conservation Service, USDA, Washington, DC

Zhilin L, Qing Z, Gold C (2005) Digital terrain modeling; principles and methodology. CRC Press, Boca Raton/London/New York/Washington, DC

Zhu AX, Hudson B, Burt J, Lubich K, Simonson D (2004) Soil mapping using GIS, expert knowledge, and fuzzy logic. Department of Geography, University of Wisconsin, Madison

Zinck JA (1988/1989) Physiography and soils. International Institute for Aerospace Survey and Earth Sciences, Enschede, p 156

Chapter 5
Soil Thematic Map and Land Capability Classification of Dubai Emirate

Hussein Harahsheh, Mohamed Mashroom, Yousef Marzouqi, Eman Al Khatib, B.R.M. Rao, and M.A. Fyzee

Abstract The soils of Dubai were mapped using remote-sensing satellite data (IRS-P6 LISS-IV) at 1:25,000 scale and classified to soil series level and their associations as per the USDA-NRCS Keys to Soil Taxonomy. A total of 26 soil series have been identified in Dubai Emirate, of which 13 were identified in Hatta area. The soils in general are coarse textured, sandy, highly calcareous and least developed. In the coastal and low-lying areas and depressions, the soils are highly saline; in the inland, soils are either saline or sodic. The Hatta area is characterized by mountains with steep side slopes, which are devoid of vegetation. Almost all the hills are barren (80–90%) and rocky without any soil cover. The soils in the hilly area are shallow to very shallow, skeletal in nature and highly calcareous. The soils have been assessed for land capability classes, and 17% of the total study area falls under land capability class IV, with the major limitations of climate and soil characteristics. These soils are suitable for marginal agriculture with the condition that sufficient water is available to offset water requirements of crops. The dominant land capability class identified in Dubai Emirate is class VI covering 65% of the study area. The dominant land capability class in Hatta is class VIII. Thus, the soil has major limitations of climate and soil which can be improved by adopting various soil conservation measures like sand dune levelling and stabilization, shelter belts and afforestation.

H. Harahsheh (✉)
Global Scan Technologies, Dubai, P.O. Box 1286, United Arab Emirates
e-mail: hussein@belhasa.ae

M. Mashroom
Survey Department, Dubai Municipality, Dubai, United Arab Emirates

Y. Marzouqi • E. Al Khatib
Dubai Municipality, P.O. Box 67, Dubai, United Arab Emirates

B.R.M. Rao • M.A. Fyzee
National Remote Sensing Agency, Department of Space, Government of India, Balanagar, Hyderabad 500 037, India

S.A. Shahid et al. (eds.), *Developments in Soil Classification, Land Use Planning and Policy Implications: Innovative Thinking of Soil Inventory for Land Use Planning and Management of Land Resources*, DOI 10.1007/978-94-007-5332-7_5,

Keywords Soil thematic map • Land degradation • Land capability classification • Soil survey • Dubai Emirate

5.1 Introduction

The socio-economic development of any country is mainly based on land and water resources. Due to general increase of population worldwide and food demand, these precious resources are overexploited often leading to resource depletion. It is apparent from literature that majority of the world's land resources are under severe threat of degradation at an unacceptable rate (Kharin et al. 1999; Harahsheh 2001). A recent study on Middle East desertification shows the gravity of land degradation problems in all levels. It is apparent that the Middle East is subject to degradation, mainly by vegetation degradation, where 40% of the study area is severely and very severely affected by vegetation degradation, followed by soil degradation process (27% severe and very severe wind erosion). Undoubtedly, these results show the gravity of land degradation problem in the study area (Harahsheh 2001). It is therefore necessary to understand these resources for better management and sustainable uses.

Dubai Emirate prevails harsh hyperarid conditions. It is highly sensitive to a number of critical environmental issues. Soil is the one which is a nonrenewable natural resource. Soil supports all terrestrial life and is vital to human existence. Soil information with regard to their nature, extent and spatial distribution along with their potential and limitations is essential for a variety of uses, such as agricultural development, engineering, sanitary, recreation and aesthetics. In addition, such information is required for modelling and environmental impact assessment analysis. It is imperative that the soils are managed and conserved judiciously to meet the growing need for food, fodder, fibre and fuel. To achieve this, it is prerequisite to generate scientific-based soil information including soil classification, physical, chemical, engineering properties, limitations to different uses, their extent and distribution and presented in the form of soil and other thematic maps.

In the modern soil classification systems, remote-sensing imagery and GIS techniques are effectively used to generate preliminary information to aid in field soil survey and developing database for sustainable development. The RS-GIS is a powerful tool, enabling study resources in spatial domain in time and in the cost-effective manner. The literature reveals that satellite data such as Landsat, SPOT, IRS, IKONOS, and WORLDVIEW1 are useful for soil mapping studies.

Owing to the need to manage soil resources and for land use planning, the Dubai Municipality has taken the initiative to conduct emirate-wide soil mapping using latest technologies to produce soil and other thematic maps for optimal utilization of various land and water resources and other applications such as demarcation of potential ground water zones and to select potential sites for various uses. Dubai Municipality implemented soil-mapping project through Global Scan Technologies (GST) in partnership with National Remote Sensing Agency (NRSA), Department of Space, Government of India.

5.2 Remote-Sensing Application in Soil Resources Study

Use of RS imagery plays key role in the field survey and in the production of preliminary and final soil and thematic maps. It helps in speeding up field soil mapping, produces better quality soil maps and improves understanding about the interaction of soil, weather and crop growth. It is to be remembered that remote sensing is a tool to identify and map soils and monitor degradation; however, it is not a substitute for field data collection (Evans 1990). Most remote-sensing techniques use radiation, which shows only a shallow penetration upon interaction with soil, rock and plants. By using these techniques, it is only possible to obtain direct information about the soil surface or vegetation cover. Remote sensing also allows deeper penetration (microwaves) and provides data (with thermal waves) which are the result of soil physical structure that is not limited to the soil surface alone. It is inappropriate to interpretate remote-sensing data alone as decisive for soil distribution without the ground truthing (Mulders 1987).

Until the late 1920s, soil surveys had been carried out through conventional methods, which are tedious, time-consuming, cost-prohibitive and impractical in difficult terrain. Remote sensing has augmented the efficiency of soil survey programmes by providing synoptic coverage of the Earth's surface at regular intervals.

Initially, aerial photographs were used as a topographic base and then for deriving information on soils, which has helped improving the efficiency of soil survey programmes. During the period from the 1980s to mid-1990s, the second generation of remote-sensing satellites, viz. Landsat TM, SPOT and IRS satellites, was launched with higher spatial and spectral resolutions which had enabled to map soils at 1:50,000 scale at the level of soil association including soil series as component soils. The information on soil resources was generated by understanding of the spectral response pattern of soils (Westin and Frazee 1976; Dwivedi 1985).

The use of Landsat TM, SPOT and IRS satellites had set the trends of rapid development and wider acceptability of remote-sensing application in soil resources study (Frazier and Cheng 1989; NRSA and AIS & LUS 1986). Similar studies have been conducted using SPOT HRV (Agbu 1991; NRSA 1995) and Indian Remote Sensing Satellite (IRS-IA, IRS 1B, IRS IC/ID), Linear Imaging Self-Scanning Sensor (LISS-I, LISS-II-, LISS-III) data (Rao et al. 1998, 2001). Using these basic foundations, many national soil-mapping projects were completed (NRSA 1995, 1996, 2001).

5.3 Objective

The objective of the present study is to generate soil thematic map and land capability classification on soil resources of Dubai Emirate.

The soil map of Dubai and Hatta was prepared at scale 1:25,000 using the Indian Remote Sensing Satellite (IRS-ID), Linear Imaging Self-Scanning Sensor (LISS IV) data. The soils of the study area were classified as per USDA, NRCS (2003), up to soil series and their association level.

5.4 Study Area

The project area covers the Dubai Emirate (UAE), and an area of 4,000 km^2 (Fig. 5.1) was surveyed. The climate of the Emirate is subtropical, warm and arid. Air temperatures range between 35 and 50°C from May to October during the middle of the day and between 20 and 35°C at midday during the winter months. The average annual rainfall of the Emirate which falls mostly during winter months is about 100 mm. The rainfall, however, is very erratic and varies extremely both from year to year and place to place. Some moisture also condenses in the form of fog and dew, especially in the coastal belts. Strong winds

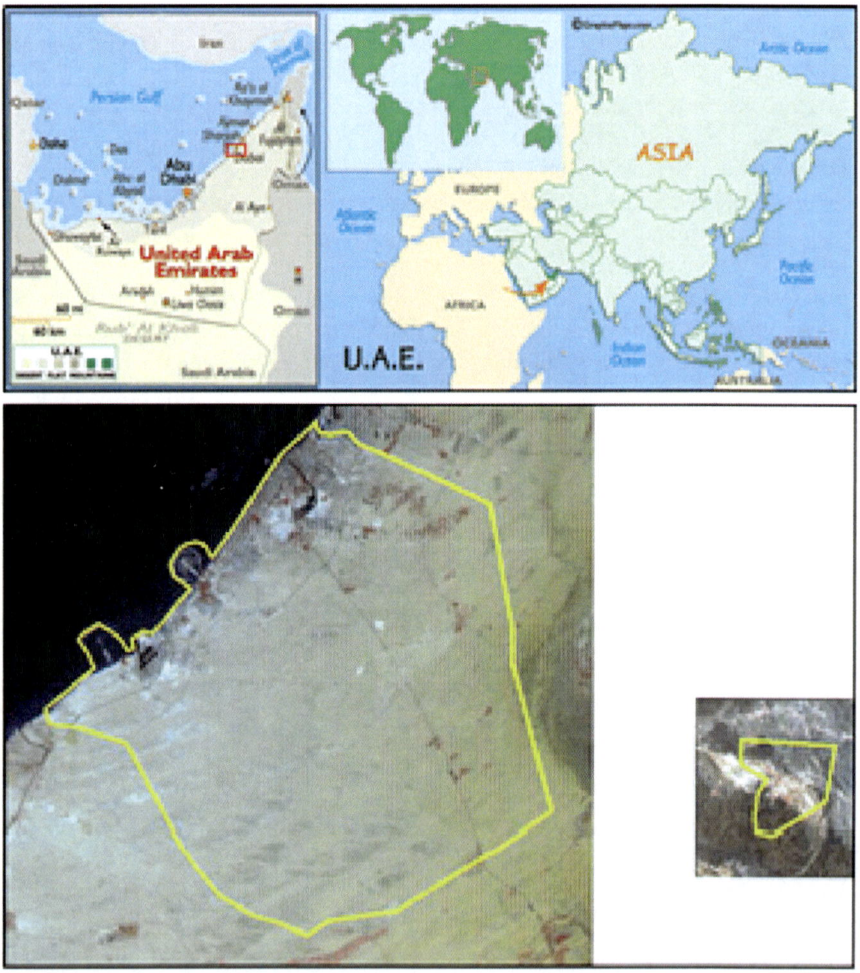

Fig. 5.1 Map showing study area

and sand storms are common throughout the Emirate and are more frequent and severe during summer months. Sand dunes are the dominant feature of the landscape over most of the Emirate.

5.5 Input Data

To generate baseline information on soil resources of Dubai and Hatta area, the Indian Remote Sensing Satellite (IRS-ID), Linear Imaging Self-Scanning Sensor (LISS-III & LISS IV) data have been used. The orthorectified IKONOS data is provided by Dubai Municipality for geo-referencing of IRS-P6, LISS IV data. Besides the satellite data, the base maps were prepared by the project team. Existing soil and topographic maps and climatic data also collected from different sources and used as collateral data.

5.6 Methodology

Soil survey involves extensive fieldwork for ground truthing and soil mapping. It is the systematic examination, description, classification and mapping of soils of an area, and it comprises of a group of interlinked operations including but not limited to preliminary visual interpretation of satellite data; fieldwork to study important characteristics of soils and associated land characteristics such as landform, natural vegetation and slope; laboratory analysis of soil samples from key soils mapped to support field descriptions; correlation and classification of soils into defined taxonomic units; and mapping of soils –establishing and drawing soil boundaries of different kinds of soils on standard geographical base map. The overall methodology for soil mapping is shown in Fig. 5.2.

5.6.1 Preliminary Visual Interpretation

The steps involved in the pre-field interpretation are monoscopic visual interpretation of Indian Remote Sensing Satellite (IRS) ID LISS-III and IRS P6 LISS IV data at scale of 1:25,000 based on the standard remote-sensing techniques using image characteristics such as tone, texture, pattern, shape, size and association in conjunction with the collateral information available in the form of published maps and reports. A tentative interpretation key in terms of lithology, physiography, land use-land cover, erosion-salinity- alkalinity hazards and image elements was developed.

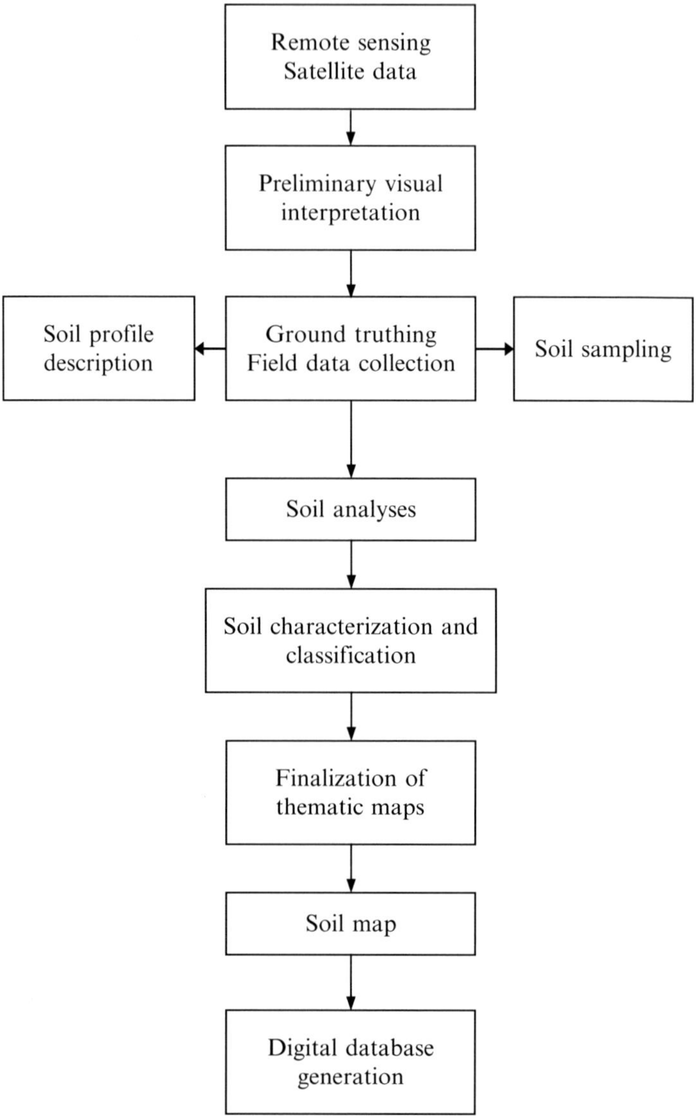

Fig. 5.2 Schematic diagram of soil mapping

5.6.2 Field Surveying

A field visit was undertaken in Dubai and Hatta study areas to overview important soil characteristics and associated land features. The first step in soil survey begins with general fact-finding exercise through a reconnaissance visual observations of the area to observe landform, geology, climate and vegetation conditions. The field teams traversed the study area with a base map and satellite imagery during the first 3 days; they recorded variations in landforms and associated soils and correlated the ground and image features.

5.6.3 *Description of Soil Profiles and Soil Sampling*

The detailed soil-site description was made in each soil-mapping unit by general traversing and by collecting surface soil, minipit and soil profile observations at intervals depending on soil variability. The soil profile is a vertical cut from the surface down to the hard rock, where several successive layers can be identified. A profile pit with exposed vertical face of approximately 1 m across to an appropriate depth (nearly 1 m) was satisfactory for soil classification. Each layer (horizons) was studied for various morphological features such as colour, texture, structure and consistency. The description of soil profiles and subsequent soil sampling enabled determination of the depth of various horizons and to identify soil-forming processes such as accumulation in horizons (illuviation) and losses (eluviation). Visual observations and field testing (finger test) procedures were used to determine soil texture, which is an important soil characteristic controlling nutrient and water-holding capacity of soils.

Detailed description of each profile was recorded on the standard field sheets, which include list of parameters essential for soil classification purposes. The GPS coordinates of each investigation site (latitude and longitude), physiographic unit, parent material, geology, slope (%), soil characteristics, soil depth, soil texture and consistency were recorded. The effervescence test was performed using 10% HCl on the soil to determine the free carbonates.

In Dubai and Hatta area, 129 soil profiles were excavated and large number of observations recorded as well as 333 soil samples were collected for laboratory analyses.

5.6.4 *Laboratory Analysis of Soil Samples*

5.6.4.1 Physical and Chemical Characteristic

Soil texture (particle-size distribution analysis – PSDA) was determined by pipette method using sodium hexametaphosphate as a dispersing agent. The textural class was determined using the USDA textural triangle (Soil Survey Division Staff 1993). The soil samples were analysed for important chemical characteristics such as soil reaction (pH), electrical conductivity (EC), organic carbon (OC), calcium carbonate equivalents ($CaCO_3$) and cation exchange capacity (CEC) using standard USDA-NRCS procedures (Burt 2004).

5.6.5 Post-field Interpretation

Preliminary drawn soil boundaries from IRS-P6 were modified using field results, and final thematic details were transferred on to the base map. The soils were classified in the light of soil morphological features and physical and chemical characteristics (USDA-NRCS 2003). The landscape map was converted into soil map in terms of soil series and associations. Based on the variations in the soil and site characteristics, 26 soil series have been identified in Dubai area and 13 soil series in the Hatta area.

5.7 Description of Soils of Dubai Emirate

The soils are generally coarse, sandy, highly calcareous and least developed. They are deficient in organic matter. The soils in the coastal belt and low-lying areas and depressions are highly saline, whereas the soils in the interior of the desert are generally normal (desert sand) and occasionally either saline or sodic where salts were accumulated in salt flats.

The major landscapes identified in the study area are coastal plain, lower aeolian plain and upper aeolian plain. These major landscape units were further subdivided into different physiographic units such as beach, tidal flats/mudflats, salt flats (young and old) and dunes over the coastal plain. The lower aeolian plain has low sand dunes, longitudinal dunes, interdunal flat areas (sandy, saline and sodic), dunal complex areas and residual hills and linear ridge. The upper aeolian plain has dunal complex, interdunal flats (sandy and sodic) and low sand dunes. The soil map of Dubai is shown in Fig. 5.3, and Table 5.1 describes the soil series indentified in Dubai Emirate.

In Hatta area, the major physiography units identified are structural valley region, piedmont area, residual hills, denudational hills (peridotite/dunite/gabbro), denudational hills (limestone/dolomite/marble) and structural hills (peridotite/dunite/gabbro interbedded).

The soil temperature regime of Dubai and Hatta is hyperthermic, which can be inferred from the climatic data of Dubai. The soil moisture regime is aridic/toric. In general, all the soils of Dubai and Hatta areas are calcareous. As the study area falls under arid region, the soils occurring in these areas will normally have an aridic (toric) moisture regime. The other features of the soils occurring in the study area are discussed separately.

5.8 Land Capability Classification

Soil resources map of Dubai Emirate provides information on location, spatial extent and physical and chemical characteristics of soils.

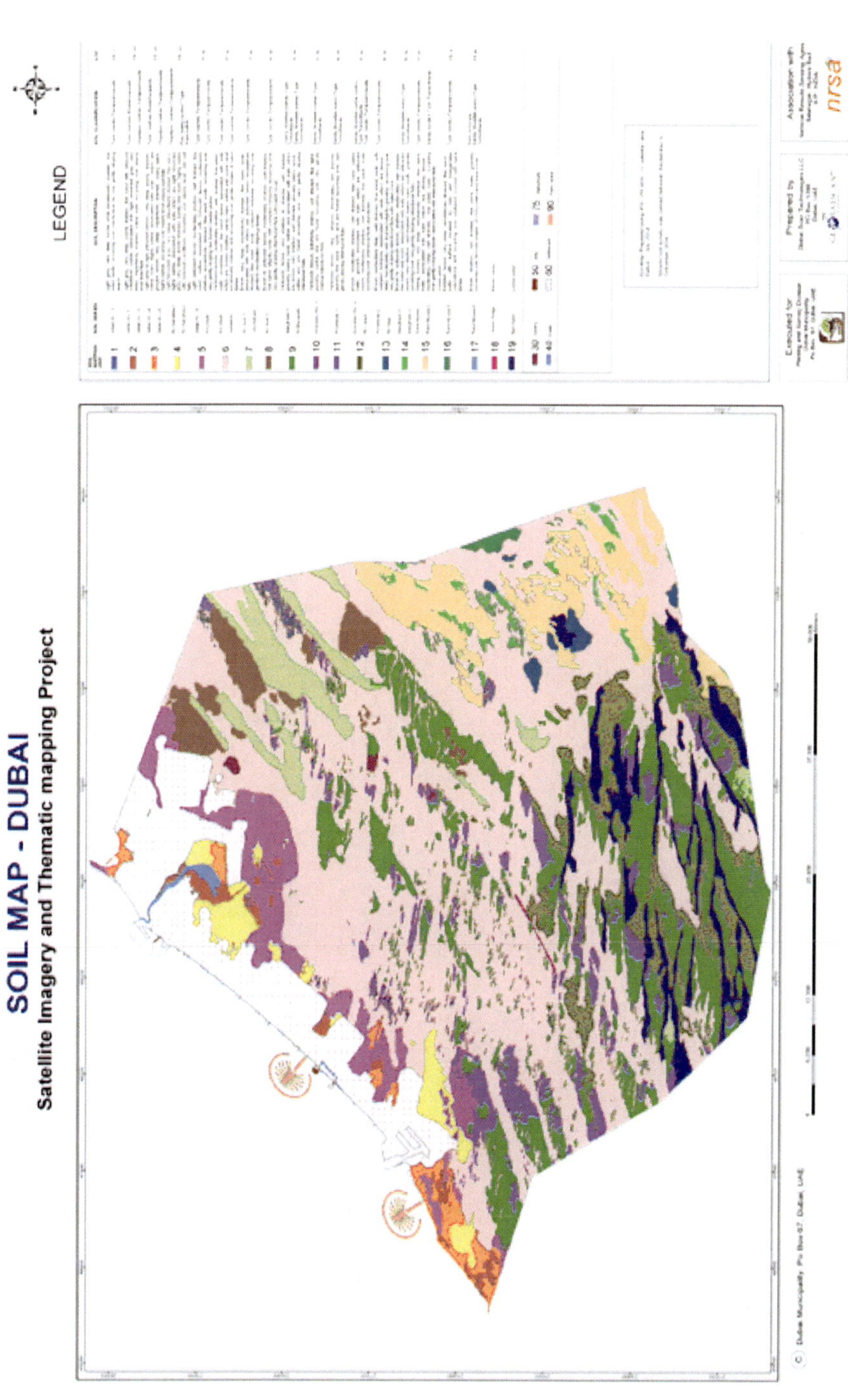

Fig. 5.3 Soil map of Dubai Emirate

Table 5.1 Description, classification and land capability of Dubai soils

S. No.	Soil series	Soil description	Soil classification	Land capability class	Area (ha)
1	Jebel Ali 1	Light grey, very deep, somewhat excessively drained, fine sand, sodic, occurring over nearly level to very gently sloping beach	Typic (sodic) Torripsamments	VIII s	155
2	Jebel Ali 2 Jebel Ali 3	Light grey, very deep, poorly drained, fine sand with reduced conditions, saline; associated with light brownish grey, very deep, imperfectly drained, saline soils occurring over nearly level tidal flats	Typic (saline) Psammaquents Oxyaquic (saline) Torripsamments	VIII sw	3,261
3	Jebel Ali 4 Jebel Ali 5	Light grey to light yellowish brown, very deep, poorly drained, sandy loam, highly saline associated with soils which are greyish brown, very deep, imperfectly drained, loamy sand, highly saline, occurring over nearly level young salt flats	Typic (saline) Psammaquents Oxyaquic (saline) Torripsamments	VIII sw	6,120
4	Al Karama Al Karama 2	Light brownish grey, very deep, imperfectly drained, fine sand, highly saline; associated with soils which are light brownish grey, very deep, poorly drained, sandy clay loam, highly saline with reduced conditions; occurring over nearly level old salt flats.	Oxyaquic (saline) Torripsamments Fine-loamy (saline) Typic Aquisalids	VIII sw	7,695
5	Jebel Ali 6 Al Lisaili	Light yellowish brown, moderately shallow, well drained, fine sand, saline associated with yellowish brown, moderately shallow, excessively drained, fine sand, sodic, occurring over very gently sloping low dunal coastal plains	Typic (saline) Torripsamments Typic (sodic) Torripsamments	VI es	16,653
6	Al Lisaili Jumeirah	Yellowish brown, moderately shallow, well drained, fine sand, sodic, occurring over low sand dunes associated with soils which are brown, moderately shallow, sandy loam, saline, and somewhat excessively drained and occurring over gentle slopes of sand dunes	Typic (sodic) Torripsamments Typic (saline) Torripsamments	VI es	169,721

7	Al-Awir 1 Umm Nahad	Brown, very deep, excessively drained, fine sand, sodic, associated with soils which are yellowish brown, excessively drained, fine sand, sodic, compact horizons occurring over gentle to moderately sloping dunes	Typic (sodic) Torripsamments	VI es	13,722
8	Al-Awir 2	Brown to yellowish brown, moderately shallow, well drained, fine sand, slightly sodic with compact horizons occurring over very gently sloping interdunal flats with sand cover	Typic (sodic) Torripsamments	VI es	9,108
9	Margham 1 Al Murqab	Brown, very shallow, well drained, gravelly loamy sand, gravelly, saline associated with soils which are yellowish brown, very shallow to shallow, well drained, gravelly, loamy sand, saline occurring over very gently sloping interdunal flats	Sandy-skeletal (saline) Typic Torriorthents Loamy-skeletal (saline) Typic Torriorthents	VI es	51,456
10	Emirates Rd 1	Yellowish brown, extremely shallow, well drained, fine sand, gravelly, saline occurring over very gently sloping interdunal flats	Sandy-skeletal (saline) Typic Torriorthents	VI es	16,516
11	Al Lahbab 1	Yellowish brown, very shallow, moderately well drained, gravelly fine sand, gravelly, sodic occurring over very gently sloping interdunal flats	Sandy-skeletal (sodic) Typic Torriorthents	VI es	1,731
12	Emirates Rd 2 Al Lisaili	Brown, moderately shallow, well drained, fine sand associated with yellowish brown, very shallow, well drained, loamy sand, gravelly, saline occurring over interdunal flats and dunes	Sandy-skeletal (saline) Typic Torriorthents Typic (sodic) Torripsamments	VI es	20,639
13	Al Lahbab 2 Al-faqa	Brown, moderately deep, well drained, fine sand, sodic, with compact horizons, associated with soils which are brown, deep, moderately well drained slightly gravelly; occurring over very gently sloping interdunal flats with sand cover	Typic (sodic) Torripsamments	VI es	2,387

(continued)

Table 5.1 (continued)

S. No.	Soil series	Soil description	Soil classification	Land capability class	Area (ha)
14	Margham 2 Margham 3	Brown to yellowish brown, shallow, moderately well drained, fine sand and sodic; associated with soils which are yellowish brown, very shallow, well drained, loamy sand, sodic, gravelly occurring over very gently sloping interdunal flats	Sandy-skeletal (saline) Typic Torriorthents	VI es	3,907
15	Tawi Nizwa 1 Tawi Nizwa 2	Strong brown, very deep, excessively drained, fine sand, sodic; associated with soils which are strong brown, moderately deep, well drained, fine sand, sodic, occurring over gently sloping dunes associated with interdunal flats	Typic (sodic) Torripsamments Sandy (sodic) Typic Torriorthents	VI es	19,946
16	Tawi Nizwa 3	Loose sand, reddish brown, very deep, excessively drained, fine sand, sodic, with surface covered by iron and manganese concretions and occurring over piedmont covered with sand dunes	Typic (sodic) Torripsamments	VIII e	1,513
17	Tawi Nizwa 4	Brown, shallow, well drained, fine sand, sodic, gravelly occurring over hillside slopes (8–15%) with sand dune cover	Rock outcrops Sandy-skeletal (sodic) Typic Torriorthents	VII es	36
18	Linear ridge				205
19	Barchans				17,411
20	Quarry/mine				509
21	Creek				931
22	Jetty				20
23	Settlement				

Land capability classification is an interpretive grouping of soils mainly based on (1) the inherent soil characteristics, (2) external land features and (3) environmental factors that limit the uses of land. Scientific soil surveys provide information on the first two aspects. Effective soil depth, soil texture, permeability of subsoil and substratum, available moisture capacity, reaction, inherent fertility, organic matter content, salinity and sodicity are some of the important inherent soil characteristics. Natural surface drainage, slope, erosion, wetness and gravels are also important land features. Besides the aforesaid factors, climate does play a very significant role in deciding the potential of a given piece of land for sustainable development.

In the land capability classification, there are eight classes. Classes I, II and III include the land suited for regular cultivation. Class IV land is fairly good for cultivation, but its safe use for cropping is very limited by natural features such as slope, erosion, unfavourable soil characteristics and adverse climate. Classes V, VI and VII are not suited for any cultivation but may be used for grazing or forestry, according to adaptability. Class VIII land is suited only for wildlife or recreation. The results of land capability classification and associated limitations are presented in Table 5.1, which clearly illustrates common land capability classes as IV, VI, VII and VIII.

References

Agbu PA (1991) Comparisons between spectral mapping units derived from SPOT image texture and field soil map units. Photogram Eng Rem Sens 57(4):397–405

Burt R (ed) (2004) Soil survey laboratory methods manual, Soil survey investigations report 42, version 4.0. United States Department of Agriculture, Natural Resources Conservation Service, National Soil Survey Center, Washington, DC

Dwivedi RS (1985) A multistage approach to mapping soil resources from remotely sensed data. Soil Surv Land Eval 5(1):13–18

Evans R (1990) Discrimination and monitoring of soils. In: Steven MD, Clark JA (eds) Applications of remote sensing in agriculture. Butterworths, London

Frazier BE, Cheng Y (1989) Remote sensing of soils in the Eastern Palouse region with Landsat thematic mapper. Rem Sens Environ 28:317–325

Harahsheh H (2001) Development of environmental GIS database and its application to desertification study in Middle East, Chiba University, Chiba, p 155

Kharin N, Tateishi R, Harahsheh H (1999) Degradation of the drylands of Asia, Chiba University, Chiba, p 80

Mulders MA (1987) Remote sensing in soil science. Elsevier, Oxford

NRSA (1995) Soil resource mapping for part of Kurnool District, Andhra Pradesh, India. IMSD project report

NRSA (1996) Soil survey and land evaluation for agricultural land use planning in tribal areas of Andhra Pradesh, India

NRSA (2001) Management of salt-affected soils and rational land use at village level using remote sensing and GIS techniques in part of south coastal region of Andhra Pradesh. Progress report of National Remote Sensing Agency Kadapa, India

NRSA and AIS & LUS (1986) Utility and relative efficiency of various remote sensing techniques in soil and land use data abstraction, Chitradurga (RS) project, Karnataka, India, Project report

Rao BRM, Sreenivas K, Fyzee MA, Sankar TR (1998) Evaluation of IRS-IC PAN data for mapping soil resources. NNRMS Bull 22:68–71

Rao BRM, Fyzee MA, Thammappa SS, Ramana KV (2001) Utility of space borne multispectral data for soil and land irrigability assessment-A case study from Southern part of India. Geocarto Int J 16(2):31–36

Soil Survey Division Staff (1993) Soil Survey Manual. USDA handbook no 18, Washington, DC

USDA-NRCS (2003) Keys to soil taxonomy, 10th edn. United States Department of Agriculture, Natural Resources Conservation, Washington, DC

Westin FC, Frazee CJ (1976) Landsat data; its use in soil survey programme. Soil Sci Soc Am J 40:81–89

Chapter 6
Land Evaluation Interpretations and Decision Support Systems: Soil Survey of Abu Dhabi Emirate

Peter King, Gerard Grealish, Shabbir A. Shahid, and Mahmoud A. Abdelfattah

Abstract The soil survey of Abu Dhabi Emirate was completed in two stages, the extensive (4th-order level) and intensive (2nd-order) levels of USDA-NRCS classification system. Both surveys have generated an enormous amount of primary soils data that is now available to land use planners and decision-makers in the Emirate. The soil information provides farmers, land managers, planners and the like with baseline information upon which they can base future land use and environmental management decisions and policies. As such, the information can be regarded as a great asset for future generations of the UAE. The information is stored in the Abu Dhabi Soil Information System (ADSIS) database that has been designed to provide ready online access to users. In its raw form, the majority of the soil data is only usable by specialist soil scientists and geoscientists. Land evaluation methods provide a mechanism for the soil information to be synthesised, simplified, interpreted and presented to a far wider audience. Several land evaluations were conducted on the extensive and intensive data sets generated by the soil survey of Abu Dhabi

P. King (✉)
Department of Agriculture, South Perth, WA, Australia
e-mail: peter.king@dpa.wa.gov.au

G. Grealish
Acid Sulfate Soils Center, the University of Adelaide, Adelaide, SA, Australia
e-mail: Gerard_grealish@adelaide.edu.au

S.A. Shahid
International Center for Biosaline Agriculture, P.O. Box 14660, Dubai, UAE
e-mail: s.shahid@biosaline.org.ae

M.A. Abdelfattah
Soil and Land Use Management, Environment Agency-Abu Dhabi,
P.O. Box 45553, Abu Dhabi, UAE

Soils and Water Sciences Department, Faculty of Agriculture,
Fayoum University, Fayoum, Egypt
e-mail: mabdelfattah@ead.ae; mahmoudalia@yahoo.com

S.A. Shahid et al. (eds.), *Developments in Soil Classification, Land Use Planning and Policy Implications: Innovative Thinking of Soil Inventory for Land Use Planning and Management of Land Resources*, DOI 10.1007/978-94-007-5332-7_6,

Emirate. They included assessments to identify soils suitable for the generalised land use irrigated agriculture, which was subsequently used to delineate areas for more detailed examination in the intensive survey. Assessments of other land uses including afforestation, range management, recreation, urban development, construction material and sanitary landfill were also conducted. These assessments can be used to identify the potentials and limitations of soils for the different land uses. However, a more detailed analytical and modelling approach is required to extract the full worth of the data set and solve complex management issues such as sustainable irrigation practices for intensive agricultural development in the Emirate.

Keywords Abu Dhabi Emirate • Land evaluation • Soil survey • Decision support systems • ADSIS

6.1 Introduction

Land evaluation as a discipline is originated from the realisation that mapping of natural resources alone does not provide sufficient information on land use and its consequences. Hence, the voluminous amount of primary soil information collected by the soil survey can only tell part of the story. Further, much of the data is only relevant to specialist soil scientists and geoscientists, therefore restricting the broader application of this valuable wealth of data.

Land evaluation provides a mechanism for the soil information to be synthesised, simplified, interpreted and presented to a far wider audience and, in so doing, expands the utility of the soil data. Land evaluation provides an added dimension to basic resource studies by relating the characteristics of soil, plants and climate to the requirements of different kinds of land use.

This chapter presents a history of the development of land evaluation methodologies internationally and how they can be applied in the Abu Dhabi Emirate context. It will then highlight the land evaluation interpretations undertaken by the soil survey of Abu Dhabi Emirate and discuss their practical application to land use decision-makers in the Emirate. Finally, suggestions for future analysis of the comprehensive data set will be discussed.

6.2 Methods

6.2.1 Soil Survey

The extensive soil survey of Abu Dhabi Emirate was conducted at the fourth-order level at a scale of 1:100,000 according to the standards of the United States Department of Agriculture-Natural Resources Conservation Service (USDA-NRCS). The Abu Dhabi soil survey results are reported in final publications (EAD 2009a). The extensive soil survey identified the soil families and phases of families and their distribution in the Emirate. The survey objectives included providing a scientific inventory of the Emirate's soil resources, interpreting those soil resources for their suitability for different uses, identifying land with the greatest potential for irrigated agriculture and developing a soil database in a GIS environment. The extensive soil survey area consists of 5.74×10^6 ha covering most of the Emirate but excluding area already used and non-accessible areas.

The subsequent intensive soil survey was conducted in order to develop a more detailed soil map and associated interpretations for 447,906 ha identified from the extensive survey. The survey was undertaken as a second-order survey at a scale of 1:25,000 (EAD 2009b). Four subareas were surveyed including an area extending from Al Ain in the eastern region to near the city of Abu Dhabi (198,596 ha) and three areas in Abu Dhabi's western region, namely, Madinat Zayed (116,146 ha), Ghayathi (105,355 ha) and As Sila' (27,809 ha).

6.2.2 *Soil Information System*

The information generated by the soil survey is stored in the Abu Dhabi Soil Information System (ADSIS). The ADSIS is a web-based system that facilitates the storage, retrieval, interpretation, analysis and presentation of soils information. The system was developed for the project after consultation with local users and a comprehensive review of existing soil information systems elsewhere in the world.

The ADSIS provides access to all the soil data and interpretations generated during the survey, plus additional functionality to allow users to examine and extract information to suit their individual needs. Users are able to view both the raw soils data and the extensive range of maps that were compiled during the soil survey. This includes being able to select a particular map generated (e.g. one of the soil maps or the map showing suitability for irrigated agriculture), zoom in to select a geographic area of interest and then view the detail for the map units in that area. Maps can be generated based on the data from the 22,000 sites across the Emirate examined during the extensive survey and 33,000 sites in the four subareas surveyed (Fig. 6.1) during the intensive survey. Figure 6.1 also shows the distribution of survey sites across the Emirate (extensive survey).

The ADSIS has a number of analytical tools including thematic map generation, symbolisation, statistical analysis and data extraction. The ADSIS provides output as maps or text and can be delivered to a printer, plotter or external digital file for use in other applications.

6.2.3 *Techniques for Assessing Complex Map Units*

The soil map units defined for the extensive and intensive surveys generally consisted of several components. In order to generate a thematic map or conduct any sort of land evaluation, it was necessary that each soil map unit delineation has a single rating index. Therefore, criteria to process a map unit with multiple components were required.

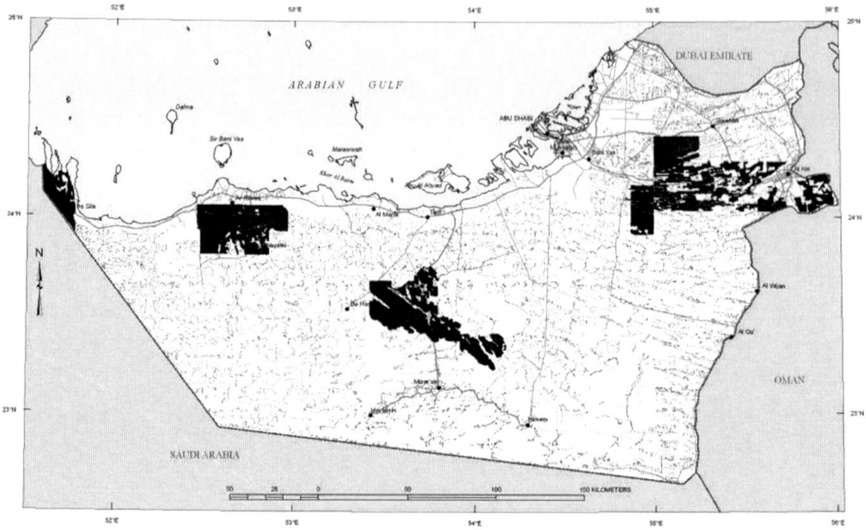

Fig. 6.1 Distribution of 22,000 observation sites for extensive survey and location of four subareas for intensive survey (Abu Dhabi Emirate)

Methods used for this included:

- Dominant rating by percent
- Weighted average of major components
- Presence/absence of a soil property

The dominant rating by percent is allocated to map units as follows:

Step 1: A map unit is selected, and the component soils are listed with their rating and the proportion they make up of the map unit.
Step 2: Starting with the lowest rating (e.g. 1), the soil components with that rating are identified, and the percentages of those are summed.
Step 3: Step 2 is repeated for all ratings.
Step 4: The rating associated with the highest of the summed percentages calculated in steps 2 and 3 is allocated to the map unit as the dominant rating by percent.

Steps 1–4 are repeated for all other map units.

The weighted average method allocates ratings to map units as follows:

Step 1: A map unit is selected, and the component soils are listed with the numeric value of the property under consideration and the proportion they make up of the map unit.
Step 2: For each component, the product of the numeric value of the property under consideration and the percentage of that component are found.
Step 3: The products calculated in step 2 are summed and then divided by 100 to give a weighted average.

Step 4: The weighted average is related to a predefined set of ranges (e.g. EC 0–1.99 is rated as 1 and EC 2–3.99 is rated as 2) and the appropriate rating allocated to the map unit.
Steps 1–4 are repeated for all other map units.

The presence/absence method allocates ratings to the map units as follows:

Step 1: A map unit is selected, and the component soils are listed with the property under consideration being rated as present or absent (as calculated using the rules and criteria) and the proportion they make up of the map unit.
Step 2: Soil components that are shown to have the property as present are identified, and the percentages of those are summed.
Step 3: The calculated percentage area is related to a predefined set of ranges (e.g. 0–9.9% is rated as 1 and 10–19.9% is rated as 2) and the appropriate rating allocated to the map unit.
Step 4: Steps 1–3 are repeated for all other map units.

These techniques allow land evaluations to be conducted on a single rating value per map unit and have been used in all the interpretations (EAD 2009a, b).

6.3 Discussion

The process of estimating the potential of land for one or more kinds of land use is known as land evaluation. Land evaluation as a discipline originated from the realisation that mapping of natural resources alone does not provide sufficient information on land use and its consequences. Land evaluation allows the characteristics of soil, plants and climate to be related to the requirements of different kinds of land use, that is, the requirements of land use are compared with the qualities of the land for a given land use.

Several land evaluations of both agricultural and nonagricultural land uses were undertaken on the 1:100,000 extensive and 1:25,000 intensive scale map data compiled during the soil survey of Abu Dhabi Emirate. These evaluations portray various land management scenarios in the Emirate and serve as base information for land use planning and land management decision-making. These evaluations are only as good as their level of abstraction and do not negate the need for detailed on-site investigation that is a prerequisite for any detailed land use planning.

The land evaluation interpretations conducted on the soil survey data are physical land evaluations. Physical land evaluation assesses the performance of specific land uses in terms of constraints imposed by land but does not consider economic and sociological factors or the influence of existing infrastructure. Performance is rated using a suitability index. Land suitability can be defined as the fitness of a given type of land for one clearly defined homogenous activity. This may be either a major use such as irrigated agriculture or a more specific land utilisation type such as a cropping system in a specific biophysical, technical and socio-economic setting.

Table 6.1 FAO land suitability classes (FAO 1983)

Suitability class	Definition
S1	*Highly suitable* land with no significant limitations to the specified use
S2	*Moderately suitable* land with moderate limitations to the specified use
S3	*Marginally suitable* land with severe limitations to the specified use
n1	*Currently unsuitable* land with severe limitations which cannot be corrected with existing knowledge and technology
n2	*Permanently unsuitable* land with severe limitations which cannot be corrected

The suitability may be determined with respect to the current condition or may suggest potential suitability if components of the system were changed.

In general, land suitability classification schemes currently in use internationally have applied the concepts developed by the Food and Agriculture Organization in its framework for land evaluation (FAO 1976). The FAO system is generally considered a benchmark for land evaluation. The framework does not in itself constitute an evaluation system but is a set of principles and concepts on which the basis of local, regional and national evaluation systems can be constructed. The FAO system has been used as the basis for specific land evaluation applications such as rainfed agriculture (FAO 1983), extensive grazing (FAO 1984a), forestry (FAO 1984b) and irrigated agriculture (FAO 1985). Unlike the USDA land capability classification system (Klingebiel and Montgomery 1961), the FAO framework does not contain preconceived judgements about qualities of land in relation to specific land uses nor any proposed hierarchy of those land uses.

The FAO land suitability classes are listed in Table 6.1. In addition to the suitability class, subclasses indicating the most restrictive characteristics or qualities are noted. These subclasses explain the type of restrictions for a specified land use and assist in identifying management options required to rectify the restriction.

6.3.1 Assessment for Irrigated Agriculture

One of the primary outputs of the extensive soil survey was the identification of areas having the highest percentage of potentially arable and irrigable soils in the Emirate, in order to delineate areas suitable for development of irrigated agriculture. This interpretation of the extensive survey identified the most suitable 1,000,000 ha of land in the Emirate. Of this 1,000,000 ha, 447,906 ha was selected for further investigation at the intensive 1:25,000 scale.

Factors taken into consideration in selecting land for irrigated agriculture in Abu Dhabi Emirate include:

- Profile depth and deep drainage because of the brackish nature of much of the irrigation water available.
- Avoidance of poorly drained areas having a limited capacity to dispose of excess irrigation water.
- Soils need to be permeable in the surface and have deep, free draining subsoil material capable of sustaining a desired leaching fraction.
- Salinity values in the root zone should be below 4 dS m^{-1} ECe or have the potential, via leaching, to be reduced to this level.
- Highly gypsic soils (>10% gypsum) should be avoided due to the risk of subsidence under irrigation as the gypsum is dissolved from the soil.
- Sandy soils require careful water management because of their low water-holding capacity. Sandy surfaces may be susceptible to wind erosion; therefore, wind breaks, mulches and vegetative ground covers are encouraged.

These factors were embedded in the criteria used to assess areas suitable for irrigated agriculture in Abu Dhabi. The criteria (EAD 2009a, b) are presented in Table 6.2. They represent a refinement on criteria previously developed for other surveys in the region including soil surveys of the Kingdom of Saudi Arabia (Ministry of Agriculture and Water 1985), Sultanate of Oman (Ministry of Agriculture and Fisheries 1990) and the State of Kuwait (KISR 1999).

The evaluation results (EAD 2009a, b) reveal that only 2,000 ha or 0.04% of Abu Dhabi has soils rated highly suitable for irrigated agriculture and capable of producing sustained high yields for a wide variety of climatically adapted crops. The landscape needs to be nearly level, with well-drained soils that are characteristically deep, fine sandy-textured or finer and single-grained structure allowing easy root penetration and retention of abundant air and water in the root zone. The soils have low soluble salts, sodicity, gypsum content, calcium carbonate content and a neutral pH.

309,000 ha or 5.40% of Abu Dhabi is moderately suitable for irrigated agriculture. These soils have sandy texture and are single grain or massive structure. They are deep and somewhat excessively or well drained. They are typically very slightly saline, non-sodic, have low gypsum content and can have a hummocky microrelief.

1,550,000 ha or 27.09% of the Emirate is marginally suitable. The soils are moderately deep with a hardpan or water table occurring within 100–150 cm of the soil surface. They have sand to sandy loam textures and are single grained or massive. They are typically slightly saline and have moderate gypsum contents. These soils occur on moderately steep gradient (up to 32%) with moderately high relief (up to 9 m).

$1,753 \times 10^3$ ha or 30.63% of soils in Abu Dhabi are currently unsuitable for irrigated agriculture. These soils typically have shallow rooting depths with hardpans within 50–100 cm of the soil surface, high gypsum content close to the surface or high relief (up to 30 m) and steep gradient (up to 56%). These soils have severe limitations that may be corrected with appropriate management strategies.

Table 6.2 Land suitability rating criteria for irrigated agriculture in Abu Dhabi (EAD 2009a)

Soil characteristic	Sub-class mode	Rating categories (see text for details) S1	S2	S3	N1	N2	Restrictive feature	Definition
Hard pan or rock depth (cm)	m	>200	200 to >150	150 to >100	100 to >50	50 to 0	Restrictive layer	Impervious soil or rock layers inhibit movement of water or roots in soil
Water table depth (cm)	w	>200	200 to >150	150 to >100	100 to >50	50 to 0	Wetness	Soil is wet during the period of desired use
Salinity (ECe dS/m) weighted average for 0 to 50 cm	z	0–4	>4 to 8	>8 to 16	>16 to 40	>40	Excess salt	Excess water-soluble salts in the soil that restrict the growth of most plants
Salinity (ECe dS/m) weighted average for 50–100 cm	z	0 to 4	>4 to 8	>8 to 16	>16 to 40	>40	Excess salt	Excess water-soluble salts in the soil that restrict the growth of most plants
Gypsum – depth to upper boundary of gypsic diagnostic horizon (cm)	y	>200	200 to >100	100 to >50	50 to >20	20 to 0	Excess gypsum	Excess gypsum can result in soil subsidence after irrigation
Texture for surface 0 to 25 cm	t	LS, LFS, LVFS, FS	S				Too sandy	The soil is soft and loose, droughty and low in fertility
Texture for surface 0 to 25 cm	t	SCL					Too clayey	The soil is slippery and sticky when wet and slow to dry
Slope gradient (%)	s	0 to 1	>1 to 3	>3 to 32	>32 to 56	>56	Too steep	Slope limits machinery use and exacerbates erosion risk
Relief – height above surrounding area (m)	r	0 to 1	>1 to 3	>3 to 9	>9 to 30	>30	Too high	The height restricts the ability to re-contour the area

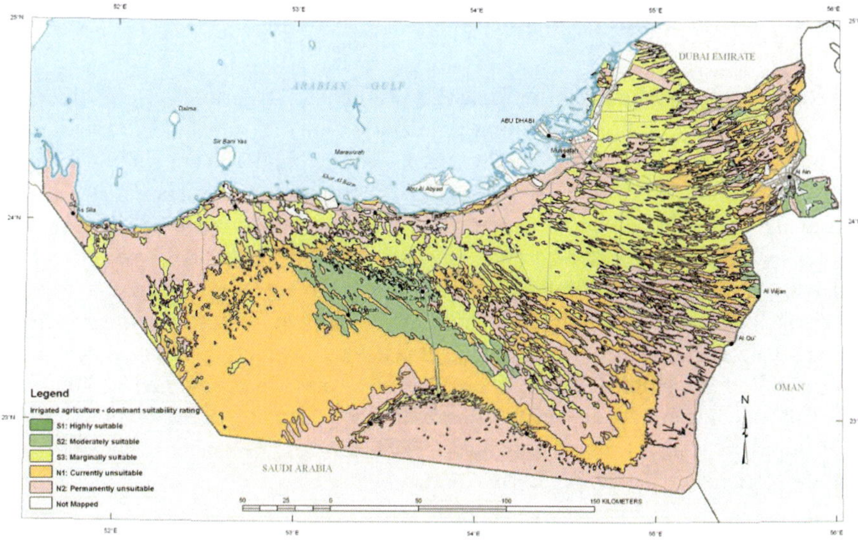

Fig. 6.2 Suitability for irrigated agriculture (EAD 2009a)

Finally, $2{,}109 \times 10^3$ ha or 36.84% of soils in Abu Dhabi are permanently unsuitable for irrigated agriculture according the prescribed criteria. These soils are generally very shallow; occur with rock outcrops, on very steeply sloping land (over 56%); having very high relief (over 30 m); are very poorly drained and strongly saline; or have shallow depth to gypsum. These soils do not warrant further investigation for irrigation purposes.

Well-drained Torripsamments and Haplocalcids with good deep drainage are the preferred soils for irrigated agriculture in Abu Dhabi. Other soils containing low quantities of gypsum and calcium carbonate may also be suitable; however, deep drainage must be adequate to remove excess irrigation water and facilitate the removal of excess salt through leaching. In areas where impermeable layers underlie the aeolian sands at a shallow depth, there is a risk that rising saline water tables may develop. The morphology of the sand dunes is a further constraint to the utilisation of these otherwise suitable soils. Dune levelling is a feasible, though expensive, management option in these areas.

The remainder of the soils in Abu Dhabi is considered unsuitable for irrigated agriculture due to the shallow depth to hardpan (lithic subgroups, Petrogypsids and Petrocalcids), high salinity (Aquisalids and Haplosalids) or shallow depth to gypsum (Haplogypsids).

The distribution of areas suitable for irrigated agriculture is shown in Fig. 6.2. The most suitable areas for irrigated agriculture (green-shaded areas in Fig. 6.2) in the Emirate include a large elongated area of undulating sands running northwest to southeast, lying to the south of Madinat Zayed. A second area is located east of Jebel Hafeet. Other smaller areas lie near Al Ain and Al Wijan. Extensive areas of marginally suitable land (yellow-shaded areas) occur inland from the coastal plains from the northeast of the Emirate, westwards to Sabkhat Matti.

6.3.2 Other Land Evaluations

The soil survey of Abu Dhabi Emirate (EAD 2009a, b) undertook several other land evaluation interpretations including assessments of soil and land properties that impact on land use, including salinity in two soil layers (0–50 cm and 50–100 cm), shallow water tables (shallower than 200 cm from the soil surface) and shallow hardpans (or rock). Potential construction and other material resources that were evaluated included gypsum, gravel, clay, calcium carbonate, sand, sweet soil and anhydrite. Assessments of land degradation were made based on criteria developed for Abu Dhabi Emirate's desert conditions. Other land use suitabilities included broad-scale assessments of rangeland, wildlife habitat and forestry. Finally, assessments were made for landfill disposal. These interpretations are not exclusive. They were targeted at the immediate needs of Emirate land managers. Future pressures for development in the Emirate will necessitate new and enhanced evaluations and interpretations of the comprehensive data sets.

6.3.2.1 Soil Salinity

Evaluations of soil salinity were made for the 0- to 50-cm and 50- to 100-cm layer depths to provide information throughout the rooting zone of most shallow-rooted crops. The assessments are based on electrical conductivity of the soil saturation extract (ECe). High concentrations of neutral salts, such as sodium chloride and sodium sulphate, may interfere with the absorption of water by plants because the osmotic pressure in the soil solution is nearly as high as or higher than that in the plant cells. Salts may also interfere with the uptake of nutrient ions, thereby causing nutritional deficiencies in plants.

The EC values measured in the field (EC 1:1) were used to calculate an average EC value (weighted for horizon thickness) for the first two 50-cm layers of soil at each site. EC values (1:1) were converted to ECe by multiplying with a factor of 3. These values were then used to calculate the median ECe value for the 0- to 50-cm layer and 50- to 100-cm layer for each investigation site. The weighted average method was then used to calculate an ECe value for each map unit, which was then categorised according to the ranges specified in Table 6.3.

The results of these assessments on the extensive survey data set for 0- to 50-cm and 50- to 100-cm layer are shown in Figs. 6.3 and 6.4, respectively. The maps show that highly saline soils (purple-shaded areas) are confined to the coastal plain and areas of deflation plain and inland sabkha where groundwater levels approach the surface, creating large areas of Aquisalids. These soils are unsuitable for all plants except salt-tolerant halophytes or most other land uses. The coastal flats extend several kilometres inland in a band from Abu Dhabi City westwards to Sabkhat Matti. Other saline areas (red- and pink-shaded areas) include the low-lying area of Sabkhat Matti, the sabkha flats amongst the mega barchan dunes around Liwa and many of the deflation plains in the east and southeast of the Emirate. Surface salt

Table 6.3 Salinity categories (EAD 2009a)

Rating categories	Electrical conductivity (ECe dS/m)	Yield restriction
Nonsaline	0 to <2	Salinity effects mostly negligible
Very slightly saline	2 to <4	Yields of very sensitive crops may be restricted
Slightly saline	4 to <8	Yield of many crops restricted
Moderately saline	8 to <16	Only tolerant crops yield satisfactory
Strongly saline	16 to <40	Only a few very tolerant crops yield satisfactory
Very strongly saline	40	Halophytes are the only option
Total		

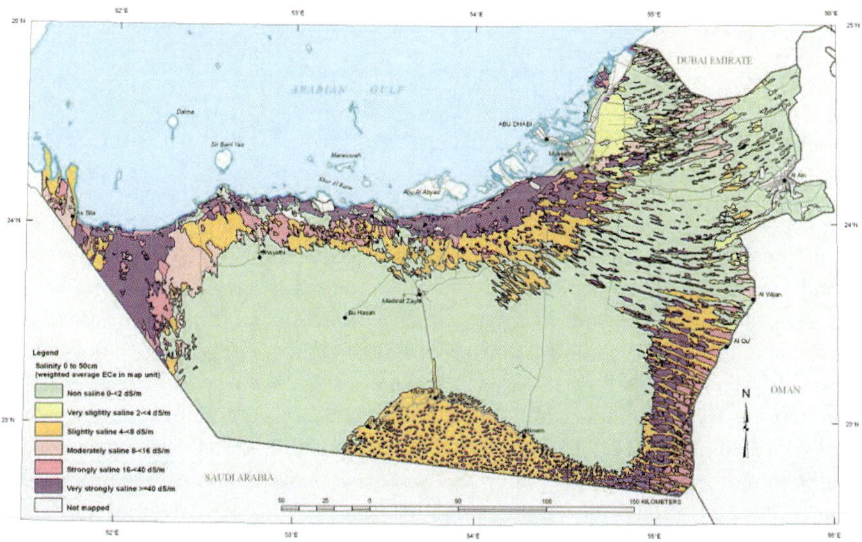

Fig. 6.3 Salinity in the top 50 cm of the soil profile (EAD 2009a)

crusts were observed on most sabkha surfaces but were thicker and more widespread in coastal locations and in Sabkhat Matti.

6.3.2.2 Construction Materials

Soils are a potential source of material for a wide range of applications in the construction industry. However, different applications such as road base, bricks and cement have their own unique requirements, and relatively few soils have profile characteristics that meet defined criteria and performance standards for specific purposes.

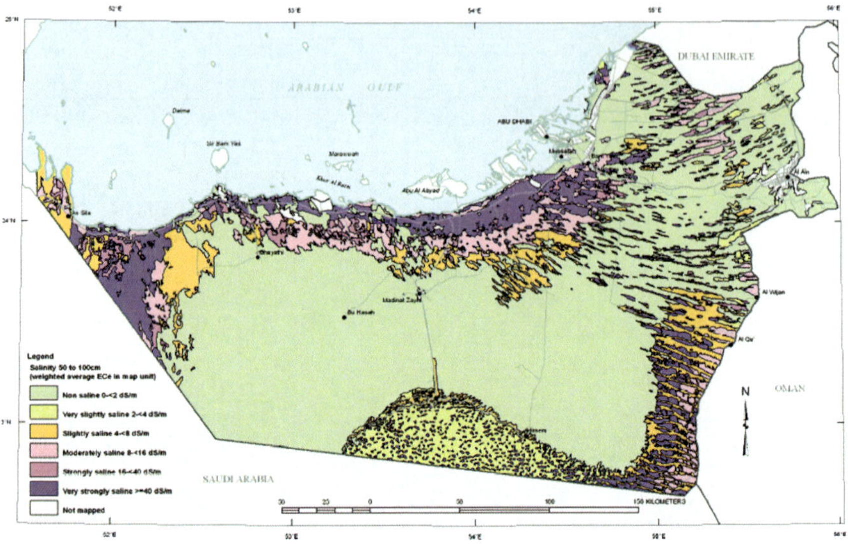

Fig. 6.4 Salinity in the 50 to 100-cm soil layer (EAD 2009a)

The soil survey results provide a useful guide for selecting potential sources of construction materials for further evaluation. Each soil was rated in its existing condition. Assessments were made against a number of criteria including soils as sources for sand, sweet soil (suitable for agriculture and landscape purposes), clay, carbonate, gravel, road-fill and topsoil. Suitability ratings of good, fair or poor and restrictive features were given for soils used as a source of sweet soil, clay, road-fill and topsoil. Ratings of probability based on presence or absence of suitable material in soils rated as a source of sand, carbonate and gravel. Such assessments are useful when looking for stockpiling materials for borrow pits and land reclamation; source material used to rehabilitate areas of soil disturbance; and cover material for parking areas, roads, tracks and other uses.

These assessments help to minimise the need for random exploratory investigation by pinpointing potential areas. Final site evaluation and selection requires field inspections of the areas to quantify the suitability of the materials for the intended purpose. Ultimately, individual soils or groups of soils may be selected as potential source materials because their source is close at hand, is the only source available or meets some or all of the physical or chemical properties/composition required for the intended application.

The assessment of gravel sources serves to demonstrate the utility of the information. For the purposes of the evaluation (EAD 2009a), gravel as a construction material is defined as particles ranging in diameter from 4.76 mm (sieve no. 4) to 75 mm. Gravel is used in great quantities in many kinds of construction, and the specifications for each purpose vary widely. The intent of the assessment undertaken in Abu Dhabi was to show the probability of finding material in any given map unit. The suitability of the gravel for specific purposes is beyond the scope of the data set and requires follow-up targeted site assessment.

Table 6.4 Frequency of occurrence of potential gravel sources (EAD 2009a)

Rating categories: % of map unit area with gravel 'present'
None or rare (0–9.9%)
Low probability (10–29.9%)
Moderate probability (30–49.9%)
Common (50–100%)
Total

The assessment for gravel source was based on the presence/absence method. It assessed the likely occurrence of coarse fragments in the appropriate size range, or the presence of rock that may be crushed to produce gravel. Thus, gravel was evaluated as being present, if:

1. Family particle size class is skeletal.
2. Suitable rock is assessed present using the following criteria:
   (a) Component is rock outcrop or the subgroup is lithic.
   (b) Bedrock kind is conglomerate, calcareous sandstone, claystone, dolomite or limestone.

For each map unit, the percentages of the components with gravel present were summed to obtain the percentage presence for the map unit. Table 6.4 describes the classes to which each percentage presence of gravel was allocated.

Figure 6.5 depicts the distribution of gravel sources in Abu Dhabi against the categories specified in Table 6.4. The majority of soils found across the Emirate are Typic Torripsamments developed in windblown sands, a process not conducive to the accumulation of gravels. In some areas, deflated sands have led to a surface accumulation of fine gravels that then protect the underlying sand from further erosion. This surface lag probably represents an accumulation from a considerable thickness of sand and rarely is it representative of a significant gravel source deeper within the soil profile. The only areas of significant gravel accumulation (green-shaded areas) in the Emirate occur in the far west near Sila and in the east and southeast. At Jabal Hafit, the surrounding piedmont plains contain large stones and boulders nearer to the mountain with more gravel-sized material further from the mountain. These limestone gravels are often intermixed with darker-coloured gravels washed down from the Oman Mountains. Smaller quantities of fine gravels were also recorded on the various deflation plains between Sweihan and Al Ain and south around Al Wijan and Al Qu'. However, deep drilling data indicated that these occurrences are relatively thin.

6.3.3 Modelling and Decision Support Systems

Whilst the various land evaluations reported in the final report (EAD 2009a, b) and highlighted above serve to enhance the soil survey data, they do not go into the next realm of predictive interpretation or modelling, specifically crop modelling and

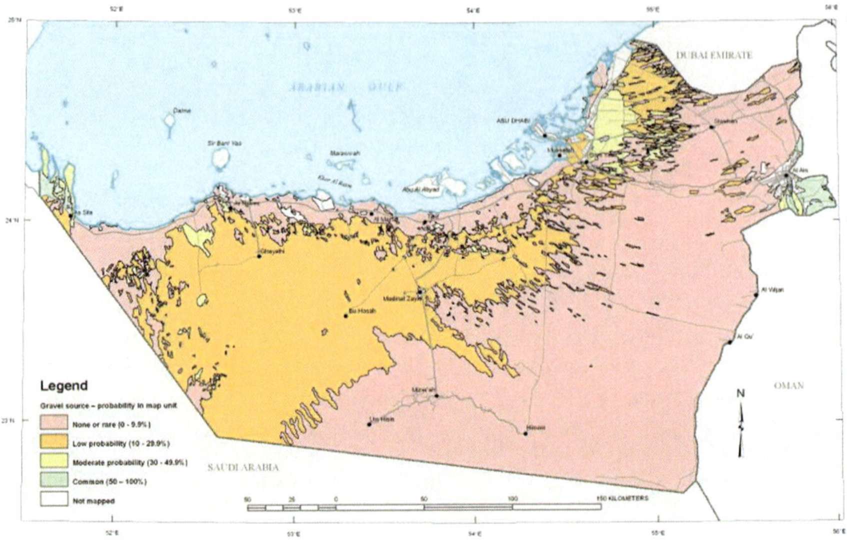

Fig. 6.5 Gravel sources (EAD 2009a)

irrigation management for agricultural planning purposes. This leap can be achieved by combining the survey's comprehensive soil data sets with climatic information and crop characteristics/requirements to generate (1) crop potential modelling, resulting in a quantifiable assessment of crop suitability and yield determination; (2) crop water requirement (which combined with crop areas determines irrigation water requirements); (3) economic assessments/scenario planning; and (4) ultimately some form of decision support system (DSS) designed to better manage agricultural and irrigation development against multiple objectives including maximisation of profit, optimisation of water use and/or deriving environmental benefits. The author of this chapter developed and applied such an approach in Uttar Pradesh (UP), India, in the development of a DSS for the Uttar Pradesh Irrigation Department (UPID) to better plan and manage the ageing state irrigation infrastructure (SMEC International Pty. Ltd. 2005a, b). The DSS developed for the UPID allows decision-makers to review alternative planning scenarios or options and make informed decisions and develop solutions to complex and competing land uses and the allocation of scarce water resources.

The approach used in UP calculates an achievable yield which then in turn provided base line data for an economic appraisal of land use and future land use patterns, which in turn drove water demand. The PlantGro™ model was commissioned to undertake this task. It used soil and climatic information, matched with crop requirements to assess crop suitability and compute achievable yields for selected crops.

The assessment of crop water requirements is required for carrying out water balance computations and quantifying irrigation requirements for various cropping patterns and crop rotations. Crop water requirements for selected crop types in the Ghaghra-Gomti Basin in Uttar Pradesh were determined from the FAO CropWat model and ET-DSS. CropWat and ET-DSS are essentially decision support systems in their own right. CropWat was developed by the Land and Water Development Division of FAO for planning and management of irrigation (FAO 1998, 1999a and 1999b). It was designed as a practical tool to carry out standard calculations for reference evapotranspiration, crop water requirements and crop irrigation requirements and the design and management of irrigation schemes. It allows the development of recommendations for improved irrigation practices, the planning of irrigation schedules under varying water supply conditions and the assessment of production under rainfed conditions or deficit irrigation.

Calculations of crop water requirements and irrigation requirements in CropWat are carried out with inputs of climatic, crop and soil data. Procedures for calculation of the crop water requirements and irrigation requirements are based on methodologies presented in FAO (1977, 1979). CropWat uses the Penman-Montieth (FAO 1993) method for calculating reference crop evapotranspiration. CropWat can also be used for developing irrigation schedules and irrigation practices using various options for water supply and irrigation management. As such, it could be used for detailed planning of irrigated agriculture in Abu Dhabi.

The DSS-ET is a software package for estimating reference crop evapotranspiration (Swarnakar 1999). The DSS-ET allows assessment of time-series data sets against some 20 ET methodologies. It also appraises the suitability of input data and ranks the performance of various ET assessments against the data, before calculating ETm for all suitable methods.

Assessment of potential crop production is integral to the development of a DSS, because potential crop production or yield is the driver for irrigated water use. It is also linked to environmental issues such as fertiliser leaching and water body eutrophication, pesticide/herbicide leaching and subsequent water body contamination. The assessment of potential crop production provides (1) a measure of crop production which is unlikely to be exceeded however good management may be; (2) a benchmark against which to measure current production levels and estimate the scope for improvement; (3) a uniform assessment that can be used to compare the relative performances of parts of a catchment or study area; (4) a way of defining the suitability of the present crop production systems and identifying presently unrecognised opportunities for crop production; and (5) a basis for determining cropping patterns and crop water use.

Considerable effort has been devoted around the world to developing methods for land evaluation for crop production. They have resulted in a number of systems that estimate plant productivity by taking land characteristics and matching them with the requirements of crops to estimate suitability for plant production. In FAO's agro-ecological zones (FAO 1978), the methodology uses a two-step process involving (1) estimating potential (constraint free) productivity and (2) estimating the effects of constraints in reducing productivity below its defined potential. The work in UP followed the FAO methodology. It combined estimates of potential biomass and yield with a crop response function methodology for estimating limitations to crop production imposed by land characteristics. The framework is implemented as a computer model called PlantGro™ (Topoclimate Services 2005).

PlantGro™ attempts to quantify crop response to climatic and edaphic factors using the simple plant response function approach developed by Hackett (1991). It takes plant, soil and climate information and produces (1) limitation ratings for plant productivity based on the known (or estimated) plant responses to the soil and environmental factors believed to influence productivity and (2) an estimate of achievable yield obtained by reducing the potential yield in proportion to the greatest limitation rating during the growing season.

PlantGro™ provided a level of sophistication in crop modelling appropriate for the DSS of the Ghaghra-Gomti Basin. It does not have the considerable data requirements of more detailed crop simulation modelling. The model, whilst simple enough to be readily used, makes use of the very large store of existing information about crop requirements. It estimated achievable yield for a specific crop and calculated the climatic and soil limitations associated with a particular soil or land unit. Output from the model provided input to the DSS and other project assessments such as agricultural economics.

PlantGro™ was a suitable tool for the work in UP. It may also have application in Abu Dhabi given that the methodology is universal, that is, it matches measured environmental and soil properties with crop response, to the best resolution of the data. However, at some level of assessment, and for some very detailed applications, the approximations inherent in the biomass calculation and in the semi-quantitative limitation rating may mean that more mechanistic simulation models, such as the CERES set of crop simulation models formatted by IBSNAT into the DSSAT package, are required to give more realistic results. For these models to be used, their demanding data requirements need to be met, and the ability to calibrate mechanistic simulation models against experimental data will need to be attained. Such a level of sophistication may or may not be warranted in the Abu Dhabi context.

6.4 Conclusions

There is considerable potential to significantly add value to the results of the soil survey of Abu Dhabi Emirate with land evaluation interpretations. Land evaluation has the ability to integrate disparate disciplines, all the time building on the primary data sets, enhancing the utility of the information and widening the potential audience for the information. The land evaluation process is explicit, using rules and criteria to make a decision. These criteria can be modified as new information becomes available or to investigate alterative scenarios. Land evaluation has the ability to produce land use planning results and scenarios in its own right or to provide input to more thorough analysis such as the development of detailed DSS capable of conducting multiple objective evaluations to maximise profits, optimise water use and/or derive some form of environmental benefit for a biological or cropping system.

In order to analyse the complex and voluminous amount of data captured by the soil survey of Abu Dhabi Emirate in an integrated land evaluation exercise, it is imperative to have the necessary tools and techniques. The techniques and interpretations conducted on the Abu Dhabi soil data sets generate useful information for regional assessment and planning. They enhance the raw soils data and provide a perspective on restrictions or suitability for the particular land uses. However, they are by no means site-specific and do not eliminate the need for detailed on-site investigation. Further, there are many land evaluation tools available that can be applied to elevate the analysis of the soil survey results to a predictive level and, in so doing, aid in complex land management decision-making and planning in the Emirate.

References

EAD (2009a) Soil survey of Abu Dhabi Emirate-Extensive survey, vol I. Environment Agency, Abu Dhabi, pp xx + 506

EAD (2009b) Soil survey of Abu Dhabi Emirate-Intensive survey, vol III. Environment Agency, Abu Dhabi, pp xviii + 435

FAO (1976) A framework for land evaluation, Soils bulletin 32. Food and Agriculture Organisation, Rome

FAO (1977) Crop water requirements. FAO Irrigation and Drainage papers no 24. Food and Agriculture Organisation, Rome

FAO (1978) Report on the agro-ecological zones project, vol 1. Methodology and results for Africa. World soil resources report 48/1. Food and Agriculture Organization, Rome

FAO (1979) Yield response to water. FAO irrigation and drainage no 33. Doorenbos J, Kassam AH (eds). Food and Agriculture Organization, Rome

FAO (1983) Guidelines: land evaluation for rainfed agriculture. FAO soils bulletin no 52. Food and Agriculture Organization, Rome

FAO (1984a) Guidelines: land evaluation for extensive grazing. FAO soils bulletin. Food and Agriculture Organization, Rome

FAO (1984b) Land evaluation for forestry. FAO forestry paper 48. Food and Agriculture Organization, Rome

FAO (1985) Guidelines: land evaluation for irrigated agriculture. FAO soils bulletin no 55. Food and Agriculture Organization, Rome

FAO (1993) ClimWat for CropWat: a climatic database for irrigation planning and management (by Smith M. IIDS, NWRC). Food and Agriculture Organization, Rome. ISBN 92-5-103416-8/9251034168

FAO (1998) Crop evapotranspiration (Guidelines for computing crop water requirements). Irrigation and drainage paper No 56. by Allen RG, Pereira LS, Raes D, Smith M, 1998. Food and Agriculture Organization, Rome, p 290. ISBN 92-5-104219-5

FAO (1999a) ECOCROP. Plant production and protection information system (PPPIS) at http://pppis.fao.org. Food and Agriculture Organization, Rome

FAO (1999b) CropWat for Windows, version 4.3. (C) 1996–1999, IIDS, NWRC. Food and Agriculture Organization, Rome

Hackett C (1991) PLANTGRO: a software package for coarse prediction of plant growth. CSIRO, East Melbourne

KISR (1999) Soil survey for the State of Kuwait, 5 vols. Kuwait Institute for Scientific Research, Kuwait

Klingebiel AA, Montgomery PH (1961) Land capability classification. USDA agriculture handbook 210. Soil Conservation Service, Washington, DC

Ministry of Agriculture and Fisheries (1990) General soil map of the Sultanate of Oman. Ministry of Agriculture and Fisheries, Muscat

Ministry of Agriculture and Water (1985) General soil map of the Kingdom of Saudi Arabia. Ministry of Agriculture and Water, Riyadh

SMEC International Pty Ltd (2005a) Ghaghra-Gomti Basin plans and development of a decision support system. Interim report for Jaunpur & Imamganj branch sub-basins. Unpublished report prepared for PACT, Water Sector Reconstruction Project, Lucknow, India. Uttar Pradesh Irrigation Department and World Bank

SMEC International Pty Ltd (2005b) Ghaghra-Gomti Basin plans and development of a decision support system. Inception report for Ghaghra-Gomti Basin. Unpublished report prepared for PACT, Water Sector Reconstruction Project. Lucknow, India. Uttar Pradesh Irrigation Department and World Bank

Swarnakar RK (1999) Decision support system for the estimation of evapotranspiration. Version 3.1. Department of Agricultural and Food Engineering. IIT Kharagpur, Kharagpur

Topoclimate Services (2005) PlantGro™-matching plants to soils and climates. User manual. Version 4.0 for Windows. International Global Change Institute, Hamilton, Gary Hutchinson, Topoclimate Services, Coffs Harbour

Chapter 7
Conceptual Soil-Regolith Toposequence Models to Support Soil Survey and Land Evaluation

Gerard Grealish, R.W. Fitzpatrick, Peter King, and Shabbir A. Shahid

Abstract Soil maps and the accompanying soil survey report are used to portray the spatial variation of soils in landscapes by indicating what soils, their proportion and their soil properties are likely to occur at a particular location or within a soil map unit. Soil surveyors intuitively understand this soil variation and how it may occur by reading the landscape. However, soil maps and soil survey reports are often too technical and not easily understood by land managers and decision-makers who are not specialist soil scientists. This chapter demonstrates how conceptual soil-regolith toposequence models can be used to describe (supporting soil survey map data and reports), explain (providing an understanding of the processes) and predict (supporting land evaluation) soil spatial variability in a range of complex landscapes. Case studies from Australia and Brunei are provided to illustrate how soil toposequence models are critical to explain, predict and solve practical land use problems, especially in complex soil-landscape environments. These conceptual

G. Grealish (✉)
Acid Sulfate Soils Centre, The University of Adelaide, Adelaide, SA, Australia
e-mail: gerard.grealish@adelaide.edu.au

R.W. Fitzpatrick
CSIRO Land and Water, Adelaide, SA 5064, Australia

Acid Sulfate Soils Centre, The University of Adelaide, Adelaide, SA, Australia
e-mail: rob.fitzpatrick@csiro.au

P. King
Department of Agriculture, South Perth, WA, Australia
e-mail: peter.king@dpa.wa.gov.au

S.A. Shahid
International Center for Biosaline Agriculture,
P.O. Box 14660, Dubai, UAE
e-mail: s.shahid@biosaline.org.ae

S.A. Shahid et al. (eds.), *Developments in Soil Classification, Land Use Planning and Policy Implications: Innovative Thinking of Soil Inventory for Land Use Planning and Management of Land Resources*, DOI 10.1007/978-94-007-5332-7_7,

models provide the following critical data to support land evaluation and management decisions by illustrating soil properties that are changing in time and space, which is especially important in salt-affected and acid sulphate soils (e.g. seasonal and climatic changes in occurrences of salt efflorescences), and the most suitable approaches for characterising, monitoring, predicting, managing and displaying soil changes for environmental impact assessments, pollution incidents, waste management and technology support.

Keywords Soil regolith • Toposequence • Land evaluation • Brunei Darussalam • Haplohumults

7.1 Introduction

Soils provide a valuable service to humans and the environment, including shelter for seeds and physical support for plant roots, water purification and water storage, retaining and delivery of nutrients to plants, decomposition of dead organic matter and wastes and regulating the earth's major element cycles (carbon, nitrogen and sulphur) (Daily et al. 1997; Costanza et al. 1997). However, soils are at risk to physical degradation from wind and water erosion and chemical degradation such as salinity, acidification, oxidation (acid sulphate soils) and waste contamination (heavy metals, pesticides, oil, etc.). Remediation of degraded soils can take hundreds to thousands of years; therefore, managing soils in a sustainable manner to minimise degradation and improve soil resources is essential to maintain services provided by soils.

Sustainable management of soils requires robust decisions to be made by land managers (e.g. farmers and farm advisories) and land use planners (e.g. landowners, policymakers, state and federal government). Hence, decisions must take into account soil variability and behaviour with time.

The first step in providing soil information to assist with managing soil sustainability is to conduct a soil survey to provide essential baseline soil data and to interpret that data through land evaluation processes, such as the work completed with the Soil Survey of Abu Dhabi Emirate (EAD 2009). Soil maps and their accompanying soil survey reports are used to portray the spatial variation of soils in the landscape by indicating what soils, their proportion and soil properties are likely to occur at a particular location or area. The problem is that the users of soil information who make the land use planning decisions and determine on-ground actions are often not specialist soil scientists, and the information provided in soil maps and soil survey reports is often too technical and not easily understood by them and therefore not used to support a decision.

To enhance the utility of the highly technical soil information, conceptual soil-regolith toposequence models are used. A soil toposequence shows a catena (or cross section) of soils and their relationship in the landscape where soils differ from one place to the next, and there is a succession of soils down a slope (Milne 1935a, b).

Understanding of soil catena is a key part of the soil surveyor developing a mental model of soil patterns in the landscape and is described by Conacher and Darylmple (1977). Conceptual soil-regolith process models enable workers to develop and present a mechanistic understanding of complex spatial and temporal soil-regolith environments (e.g. Fritsch and Fitzpatrick 1994; Fitzpatrick and Merry 2002). The regolith is the unconsolidated earth material present above bedrock and includes the upper soil layers. Soil-regolith models have been chosen to help visualise the results of several acid sulphate soil investigations performed at typical sites with complex surface and subsurface acid sulphate soil features, including several regolith layers and shallow surface water interface systems (e.g. Fitzpatrick et al. 2009). Two categories of soil-regolith toposequence models have been found to be useful:

- Descriptive soil-regolith models
- Predictive soil-regolith models: generalised and specific (Fitzpatrick et al. 2010a)

Descriptive soil-regolith process models essentially depict typical scenarios of a soil toposequence or cross section. These models characterise the lateral and vertical spatial variability of current soil-regolith layers, horizons, materials and features (salt efflorescences). Such models can represent current and past water levels and help to develop practical frameworks and solutions for managing soils. These models are in turn used to help develop predictive soil-regolith models (4D = space × time) for a particular environment (generalised models) or transect (specific models).

Predictive soil-regolith models (generalised and specific) are constructed to illustrate either generalised or specific transects and scenarios. Predictive soil-regolith models (4D) are subsequently constructed using a collage of figures, which illustrate several stages of soil-regolith condition in response to natural or human-induced (e.g. management) changes over time. These conceptual models can be used to predict processes and potential consequences but not the timing of events, which will depend on weather, changes in water level and land management (Fitzpatrick et al. 2010a).

This chapter presents two case studies from Brunei and South Australia where conceptual soil-regolith toposequence models have been successfully used to provide understanding of soil variation to support land use decisions and the benefits of this approach are described.

7.2 Methods

The soil fertility and evaluation study of Negara Brunei Darussalam (Grealish et al. 2007a, b, 2008) was a soil survey conducted to assist the country to achieve the goal of increasing agricultural production and reducing dependence on imported food products, by increasing the area of land used for agriculture and by increasing productivity on existing agricultural land. The soil survey provided supporting information by assessing the soil characteristics and their suitability for a variety of crops and provided recommendations for profitable and sustainable land management. To assist the Bruneian users, this project presented information and assessments in a variety of formats from detailed technical reports to simple common language field manuals that could readily be used to guide decision-making.

The soil spatial heterogeneity of acid sulphate soils in the Lower Lakes, South Australia (Fitzpatrick et al. 2010b), was a soil survey conducted to assist with determining the impact of dropping lake water levels due to the severe drought reducing water inflows. Unprecedented drought during the past decade has recently led to significantly lowered water levels, previously from about +0.75 m AHD down to −0.5 m AHD (Australian Height Datum), exposing subaqueous soils. There is concern that these soils will be a hazard to ecosystem function and water quality through acidification, release of toxic metals and deoxygenation of water and environmental degradation of landscapes due to (1) acidic soil, (2) airborne dust, (3) transport of acidic materials and metals once water levels rise and (4) acidic pulses during and following rainfall events. The survey identified acid sulphate soil materials and their occurrence throughout the area. Predictions of change resulting from lowering and rising of the water levels were then made.

Details of the soil survey approaches are described in the referenced reports. This chapter presents and discusses the ability of conceptual toposequence models and how they were used to convey information to decision-makers.

7.3 Results and Discussion

7.3.1 Case Study 1: Negara Brunei Darussalam

The Brunei soil survey project recognised that farmers and agricultural advisors identified and understood soil distribution on their farms by relating soil type to positions in the landscape and that they were more familiar with this relationship than interpreting a soil map or reading a text in a soil survey report. To assist them, simple soil toposequences were drafted as presented in Fig. 7.1, which shows the expected distribution of the soil in the landscape and the soil distribution in the landscape including a key to identification. Accompanying the toposequence is a separate page that provides a summary description of the soils in landscape position and crop suitability. A field manual using plain language provides further information to allow users to quickly identify important soil features and the main limitations and determine which crops are suitable. An example is shown in Fig. 7.2.

7.3.2 Case Study 2: Lower Lakes, South Australia

For the Lower Lakes soil survey, a more complex conceptual toposequence model was prepared to illustrate the spatial variability (changing location) of sulphuric and sulphidic materials in acid sulphate soils with time. The landscape that these soil

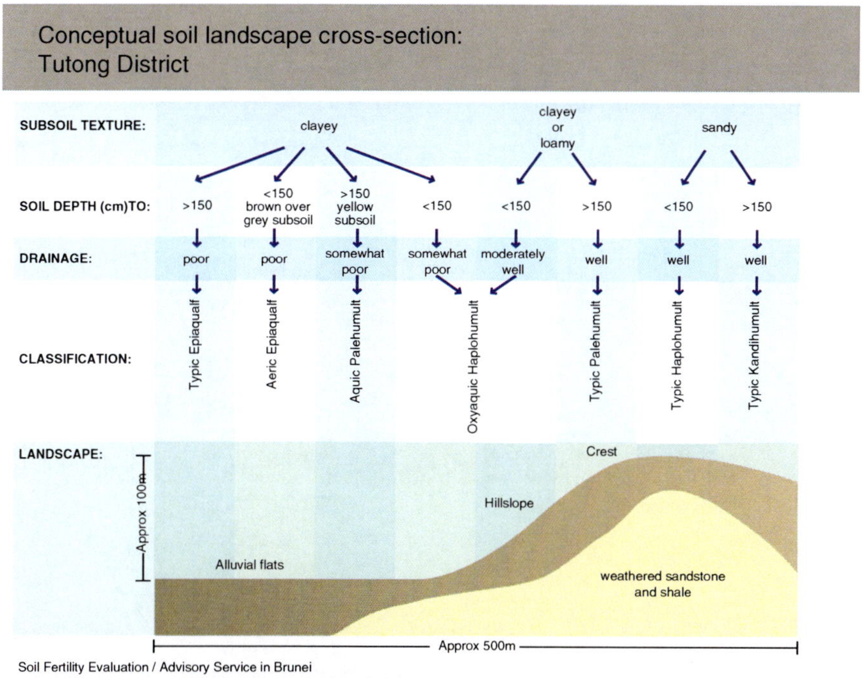

Crop suitability and soil landscape position:
Tutong District
(Kupang, Maraburong, Padnunok/Sg Burong, Batang Mitus (Buah), Batang Mitus (Halaman), Birau, (P.P.Muda), Birau (Penyelidikan) and Sg Tajau (Brunei-Muara)

Landscape position	Alluvial valley flats and lower slopes	Mid slopes, upper slopes and crests
Soil type	Brown over grey soils and Very deep yellow soils	Yellow soils and Very deep yellow soils
Suitable Crops	Rice Grass species Fodder legume species adapted to wet areas	*Where slope <55%* Grass species Fodder legume species adapted to wet areas
Moderately suitable crops	Leafy, fruit and root vegetables Groundnuts Soya and mungbean Maize Ginger and turmeric Cassava and sweet potato All fruit crops assessed (except Durian, Langsat-duku, Papaya, Citrus) Fodder legume species adapted to well drained conditions	Fodder legume species adapted to well drained conditions *Where slope >55%* Grass species Fodder legume species adapted to wet areas *Where slope <65%* All fruit crops assessed *Where slope <55%* Cassava and sweet potato *Where slope <35%* Leafy, fruit and root Vegetables Groundnuts Soya and mungbean Maize Ginger and tumeric

Soil Fertility Evaluation / Advisory Service in Brunei

Fig. 7.1 Descriptive soil-regolith toposequence model with a soil identification key indicating the soil positions, and related crop suitability (From Grealish et al. 2007a)

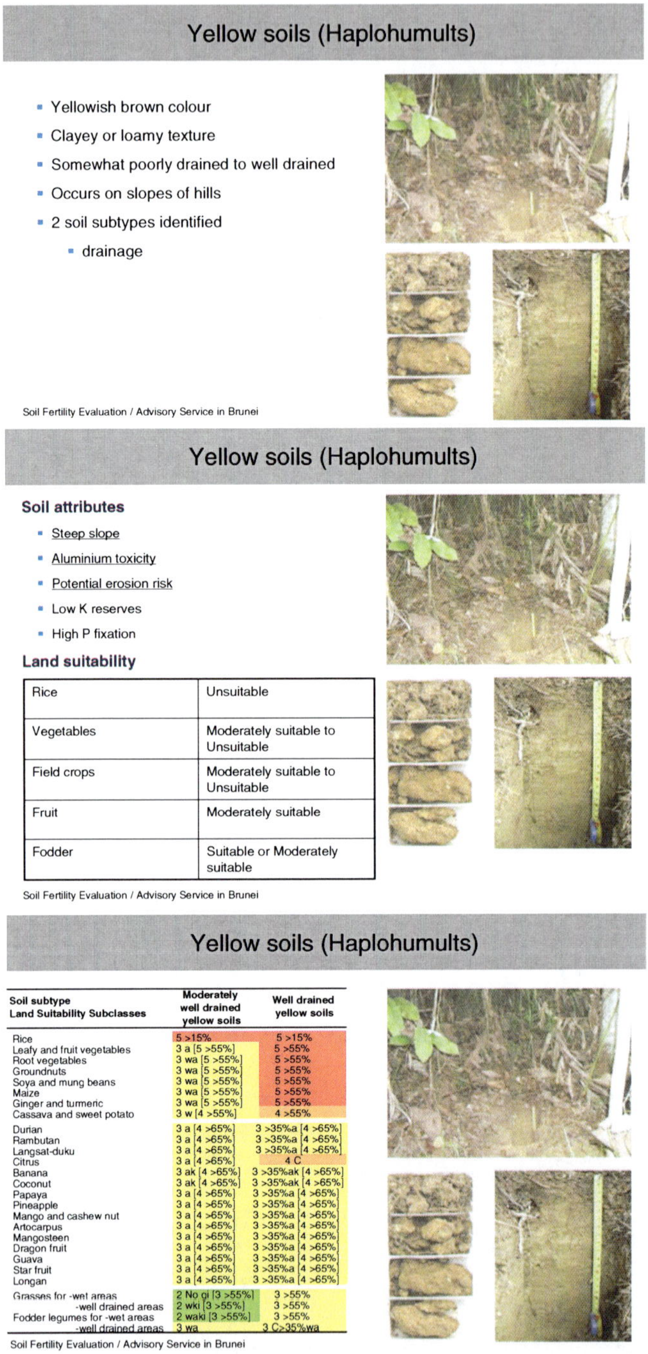

Rice	Unsuitable
Vegetables	Moderately suitable to Unsuitable
Field crops	Moderately suitable to Unsuitable
Fruit	Moderately suitable
Fodder	Suitable or Moderately suitable

Soil subtype Land Suitability Subclasses	Moderately well drained yellow soils	Well drained yellow soils
Rice	5 >15%	5 >15%
Leafy and fruit vegetables	3 a [5 >55%]	5 >55%
Root vegetables	3 wa [5 >55%]	5 >55%
Groundnuts	3 wa [5 >55%]	5 >55%
Soya and mung beans	3 wa [5 >55%]	5 >55%
Maize	3 wa [5 >55%]	5 >55%
Ginger and turmeric	3 wa [5 >55%]	5 >55%
Cassava and sweet potato	3 w [4 >55%]	4 >55%
Durian	3 a [4 >65%]	3 >35%a [4 >65%]
Rambutan	3 a [4 >65%]	3 >35%a [4 >65%]
Langsat-duku	3 a [4 >65%]	3 >35%a [4 >65%]
Citrus	3 a [4 >65%]	4 C
Banana	3 ak [4 >65%]	3 >35%ak [4 >65%]
Coconut	3 ak [4 >65%]	3 >35%ak [4 >65%]
Papaya	3 a [4 >65%]	3 >35%a [4 >65%]
Pineapple	3 a [4 >65%]	3 >35%a [4 >65%]
Mango and cashew nut	3 a [4 >65%]	3 >35%a [4 >65%]
Artocarpus	3 a [4 >65%]	3 >35%a [4 >65%]
Mangosteen	3 a [4 >65%]	3 >35%a [4 >65%]
Dragon fruit	3 a [4 >65%]	3 >35%a [4 >65%]
Guava	3 a [4 >65%]	3 >35%a [4 >65%]
Star fruit	3 a [4 >65%]	3 >35%a [4 >65%]
Longan	3 a [4 >65%]	3 >35%a [4 >65%]
Grasses for -wet areas	2 No gi [3 >55%]	3 >55%
-well drained areas	2 wki [3 >55%]	3 >55%
Fodder legumes for -wet areas	2 waki [3 >55%]	3 >55%
-well drained areas	3 wa	3 C>35%wa

Fig. 7.2 Soil type description, capability classification for crops and key soil attributes presented in field manuals for users who are not soil specialists (From Grealish et al. 2007a)

Fig. 7.3 Descriptive soil-regolith model for Loveday Bay in Lake Alexandrina showing various acid sulphate soil materials formed due to fluctuating water conditions (From Fitzpatrick et al. 2010; 2011)

materials occur in had limited relief (or topography) to assist with placement of the different soil types; therefore, the use of surface features such as location of soil cracks, surface salts and photographs was used to assist with locating the soil materials in the landscape, as shown in Fig. 7.3, which shows the general pattern and layering of the soil layers and how these vary across the landscape.

Acid sulphate soil materials can change characteristics relatively quickly with time depending on the water conditions. The users of this soil survey information wanted to understand how these hazardous soil materials formed, their relationship in the landscape and how they would change with fluctuating water conditions. To convey this information, a time-series toposequence conceptual model was prepared showing three different events (see Fig. 7.4). The first of these toposequences (a) shows the pre-drought conditions where the acid sulphate soil materials are sulphidic and not of concern because they are submerged (subaqueous) in an anaerobic state. The second toposequence (b) shows the same area after water levels have declined; sulphidic soil material near the soil surface has oxidised to form sulphuric

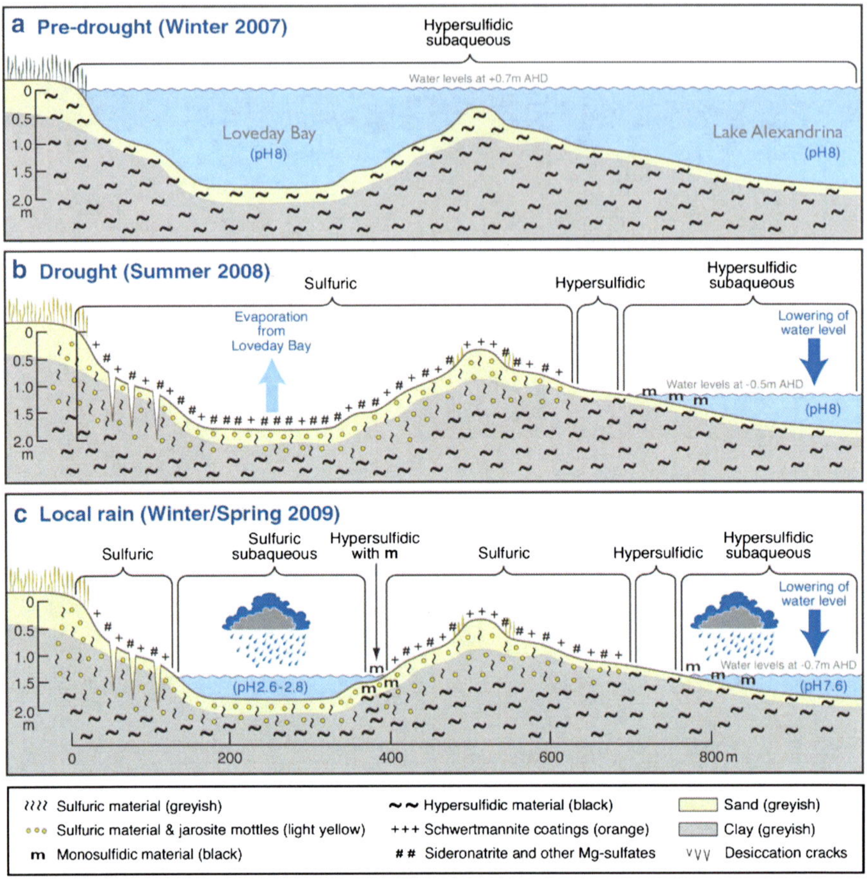

Fig. 7.4 Predictive specific conceptual model for Loveday Bay in Lake Alexandrina showing changes of various acid sulphate soil materials with time due to fluctuating water conditions (From Fitzpatrick et al. 2010; 2011)

soil material, and many types of salts related to acid sulphate soil conditions occur on the surface. Sulphidic soil material occurs at depth in the soil profile. Three soil types are identified, subaqueous hypersulphidic, hypersulphidic and sulphuric, and their general locations in the landscape are shown in Fig. 7.3. The third toposequence (c) shows the consequences of allowing this hazardous soil material to be exposed; the sulphuric soil material when in contact with water will acidify the water, and as shown in Fig. 7.3, for this location the water pH was dropped to a very low value of 2.5. Additionally, Fig. 7.3 shows how the soil types have changed in terms of soil materials and classification, with a reduction in the area of sulphuric soils and the inclusion of subaqueous sulphuric soils where the area has been inundated.

The primary users of this soil survey information were policy and environmental land use planners. They did not have a technical understanding of soil survey maps and data. Presenting the toposequence results allowed them to understand the types of soil materials that were occurring and where they are formed in the landscape (Fig. 7.3) and provided them with a predictive indication that explains how they form and what impacts they have on the water quality (n).

7.4 Conclusions

This chapter demonstrates how soil toposequence conceptual models can be used to (1) describe (supporting soil survey map data and reports), (2) explain (providing an understanding of the processes) and (3) predict (supporting land evaluation) soil spatial variability in a range of complex landscapes. The toposequence models were constructed and used to assist users who are not soil experts to more readily use the soil survey data to make land use decisions. The toposequence models can be scaled to show the level of detail required. They illustrate soil properties that are changing with time and space. As has been shown by these case studies, the toposequence models do not replace soil survey reports and maps but bridge the gap between technical soil data and providing information in a form to support decision-making, allowing soil scientists to translate their findings and therefore present them to people who are not soil experts.

Acknowledgements The soil fertility evaluation soil survey project was commissioned and funded by the Department of Agriculture, Negara Brunei Darussalam, and the assessment of acid sulphate soils was commissioned and funded by the Department of Environment and Heritage, South Australia. We would like to acknowledge the input of many staff from CSIRO and the funding organisations that assisted with this project.

References

Conacher AJ, Darylmple JB (1977) The nine unit landsurface model and pedogeomorphic research. Geoderma 18:127–144

Costanza R, d'Arge R, de Groot R, Farber S, Grasso M (1997) The value of world's service and natural capital. Nature 387:253–260

Daily GC, Alexander S, Ehrlich PR, Goulder L, Lubchenco J, Matson PA, Mooney HA, Postel S, Schneider SH, Tilman D, Woodwell GM (1997) Ecosystem services: benefits supplied to human societies by natural ecosystems. Issues Ecol 2:1–16

EAD (2009) Soil survey of Abu Dhabi Emirate-Extensive Survey, vol I. Environment Agency-Abu Dhabi, Abu Dhabi, pp xx, 506

Fitzpatrick RW, Merry RH (2002) Soil-regolith models of soil-water landscape degradation: development and application. In: McVicar TR, Rui L, Fitzpatrick RW, Changming L (eds) Regional water and soil assessment for managing sustainable agriculture in China and Australia, vol 84. Australian Centre for International Agricultural Research, Canberra, pp 130–138. http://www.eoc.csiro.au/aciar/book/PDF/Monograph_84_Chapter_09.pdf

Fitzpatrick RW, Grealish G, Shand P, Simpson SL, Merry RH, Raven MD (2009) Acid sulfate soil assessment in Finniss River, Currency Creek, Black Swamp and Goolwa Channel, South Australia. Prepared for the Murray Darling Basin Authority. CSIRO Land and Water Science report 26/09. CSIRO, Adelaide, p 213. http://www.clw.csiro.au/publications/science/2009/sr26-09.pdf

Fitzpatrick RW, Grealish G, Chappell A, Marvanek S, Shand P (2010) Spatial variability of subaqueous and terrestrial acid sulfate soils and their properties, for the Lower Lakes, South Australia. Report prepared for the SA Department of Environment and Natural Resources (DENR). Client report R-689-1-15: CSIRO Sustainable Agriculture National Research Flagship, p 122. http://www.clw.csiro.au/publications/science/2010/SAF-Lower-Lakes-SA-sulfate-soils.pdf

Fitzpatrick RW, Shand P, Hicks W (2011) Technical guidelines for assessment and management of inland freshwater areas impacted by acid sulfate soils. CSIRO Land and Water Science Report, 05/11. p 160. http://www.clw.csiro.au/publications/science/2011/sr05-11.pdf

Fritsch E, Fitzpatrick RW (1994) Interpretation of soil features produced by ancient and modern processes in degraded landscapes. I. A new method for constructing conceptual soil-water-landscape models. Aust J Soil Res 32:880–885

Grealish GJ, Fitzpatrick RW, Ringrose-Voase AJ (2007a) Soil fertility evaluation/advisory service in Negara Brunei Darussalam. Report P1-2-Soil properties and soil identification key for major soil types. Science report 76/07, CSIRO Land and Water, Australia

Grealish GJ, Ringrose-Voase AJ, Fitzpatrick RW (2007b) Soil fertility evaluation/advisory service in Negara Brunei Darussalam report P1-1.2-Soil Maps. Science report 75/07, CSIRO Land and Water, Australia

Grealish GJ, Fitzpatrick RW, Ringrose-Voase AJ, Hicks W (2008) Brunei: summary of acid sulfate soils. In: Fitzpatrick RW, Shand P (eds) Inland acid sulfate soil systems across Australia. CRC LEME open file report no. 249 (Thematic volume) CRC LEME, Perth, Australia, pp 301–309

Milne G (1935a) Composite units for the mapping of complex soil associations. Trans 3rd Int Congr Soil Sci Oxf 1:345–347

Milne G (1935b) Some suggested units of classification and mapping particularly for East African soils. Soil Res 4:183–198

Chapter 8
Anhydrite Formation on the Coastal Sabkha of Abu Dhabi, United Arab Emirates

Michael A. Wilson, Shabbir A. Shahid, Mahmoud A. Abdelfattah, John A. Kelley, and James E. Thomas

Abstract A fluvial marine sabkha along the coastal area of Abu Dhabi, United Arab Emirates, is hypersaline from evaporative losses of groundwater originating from rain, seawater intrusion, lagoons that border the sabkha, and inland groundwater sources. Anhydrite ($CaSO_4$) is present in these soils and is regarded to be both a neoformed mineral and a product of gypsum transformation. Six pedons (designated 1–6) were described, sampled, and characterized from a 13-km transect across the sabkha in order to better understand the distribution of anhydrite across the sabkha, determine suitable laboratory methods for detection and quantification of this mineral, and evaluate soil genesis and mineral formation. Soils were highly saline

M.A. Wilson (✉)
United States Department of Agriculture-Natural Resources Conservation Service, National Soil Survey Center, 100 Centennial Mall North, Rm 152, Lincoln, NE 68508–3866, USA
e-mail: mike.wilson@lin.usda.gov

S.A. Shahid
International Center for Biosaline Agriculture, P.O. Box 14660, Dubai, UAE
e-mail: s.shahid@biosaline.org.ae

M.A. Abdelfattah
Soil and Land Use Management, Environment Agency-Abu Dhabi, P.O. Box 45553, Abu Dhabi, UAE

Soils and Water Sciences Department, Faculty of Agriculture, Fayoum University, Fayoum, Egypt
e-mail: mabdelfattah@ead.ae; mahmoudalia@yahoo.com

J.A. Kelley
United States Department of Agriculture-Natural Resources Conservation Service, Raleigh, NC, USA
e-mail: bettmark.john@gmail.com

J.E. Thomas
United States Department of Agriculture-Natural Resources Conservation Service, National Soil Survey Center, 100 Centennial Mall North, Rm 152, Lincoln, NE 68508-3866, USA

S.A. Shahid et al. (eds.), *Developments in Soil Classification, Land Use Planning and Policy Implications: Innovative Thinking of Soil Inventory for Land Use Planning and Management of Land Resources*, DOI 10.1007/978-94-007-5332-7_8,

with electrical conductivity (EC1:2) ranging from 11 to 167 dS m^{-1}. Evaporative minerals identified by x-ray diffraction include calcite, gypsum, halite, aragonite, and anhydrite. Together, salts, gypsum, and anhydrite composed 5–100% of the mineral matter of sabkha soils. Quantification of anhydrite was achieved by the difference in the acetone method (gypsum + anhydrite quantification) and low-temperature weight loss (for gypsum quantification). Both a thermal gravimetric analyzer (TGA) and an oven were tested for the latter procedure. The TGA method was found to provide the most reliable data, while the oven method yielded inconsistent results. Anhydrite was identified in the two sites (pedons 5 and 6) most distant from the coast, ranging up to 43% of the <2-mm fraction and occurring in thicknesses of 70 and 55 cm, respectively.

Keywords Fluvial marine • Anhydrite formation • Coastal sabkha • Hypersaline • Evaporative • Abu Dhabi • Thermal analysis

8.1 Introduction

Sabkhas (salt flats) are widely distributed in the coastal areas around the world. Detrital sediments, plus soluble elements from the marine environment, create a suite of carbonate, chloride, and sulfate minerals. The climate of the Arabian Peninsula creates unique conditions that influence the resulting mineralogical suite in sabkhas of this area. Anhydrite ($CaSO_4$) is a mineral that has been documented in this area, both a neoformed mineral and as a product of gypsum ($CaSO_4$•$2H_2O$) transformation, due to the combined effects of high salinity and temperature. Anhydrite develops in these soils and remains stable in the vadose zone above the water tables (Aref et al. 1997; Sanford and Wood 2001; El-Tabakh et al. 2004; Shahid et al. 2007). This process is well documented to occur during carbonate deposition where anhydrite is a common by-product (along with gypsum) associated with seawater evaporation (El-Tabakh et al. 2004; Melim and Scholle 2002). Often this process occurs under elevated burial temperatures or due to hot fluid-rock interactions (Kasprzyk and Orti 1998).

The coastal area of Abu Dhabi, United Arab Emirates (UAE), is hypersaline from evaporative losses of groundwater originating from rain, seawater intrusion, lagoons that border the sabkha, and inland groundwater sources. Sanford and Wood (2001) documented that the water within the sabkha soils of Abu Dhabi is principally derived from rainfall, with the major source of solutes from the slowly permeable Tertiary formations of marl and gypsum that underlie the sabkha and create episaturated conditions within soils. Their results show that the lateral intrusion of seawater is very low due to the slow groundwater flux due principally to the low gradient of the sabkha (typically 2×10^{-4} m m^{-1}) (Sanford and Wood 2001).

The recent soil survey of the Abu Dhabi Emirate (EAD 2009) has documented the presence of mapping units principally composed of soils dominated by anhydrite within the coastal sabkha. This discovery resulted in a formal proposal for the modification of Soil Taxonomy (Soil Survey Staff 2010). This proposal included the formation of an anhydritic family mineralogy class to separate these soils dominated by anhydrite from gypsum-rich soils (Abdelfattah and Shahid 2007; Shahid et al. 2007). The inclusion of this family mineralogy class in Soil Taxonomy requires the quantification of anhydrite. This study was created to help determine the distribution of this mineral in the broad, level mapping units of Abu Dhabi sabkhas and to investigate laboratory methods that would assist in the identification and quantification of anhydrite. Thus, the objectives of this study are to investigate methods to identify and/or quantify anhydrite in soils, determine the distribution of anhydrite in selected soils across a coastal sabkha, and discuss factors affecting the genesis and distribution of this mineral. Knowledge gained in this study should have applicability across the region, but also globally in other arid, hot coastal and inland regions (e.g., playas) that are locations of salt accumulations.

8.2 Materials and Methods

8.2.1 Soil-Forming Factors

Abu Dhabi is located on the north coast of the Arabian Peninsula within the United Arab Emirates (Fig. 8.1). The coastal area of Abu Dhabi is composed of three geomorphic areas: islands, shoal and channel areas of the lagoons, and the coastal sabkha (Butler 1970). The fluvial marine sabkha (broad coastal plain) is over 400 km in length and up to 16 km in width (de Matos 1989). The sabkha slopes toward the sea at about 0.4 m km^{-1} (Evans et al. 1969). Tertiary-aged geologic materials underlie and border the sabkha on the southeastern perimeter, partially buried at the surface by Holocene-aged dunes and gravels. The sabkha is composed of Quaternary (Pleistocene and Holocene-aged) sediments derived from aeolian and fluvial materials, with overlying lagoon and tidal deposits, and other evaporative and chemically precipitated minerals developed in this current environment (Evans et al. 1969; Butler 1970). The water table forms by water perching on the Tertiary-aged strata that underlie the sands (Sanford and Wood 2001).

The aeolian sediments were principally deposited during a stage when the Arabian Gulf was dry and winds transported the exposed sands from the gulf bottom onto the land area along the coast (Alsharhan and Kendall 2003). The sand is currently about 10 m deep (Sanford and Wood 2001). Following the last glacial maximum (18,000 years before present-ybp), water transgressed and reached the present shoreline by 6,000 ybp (Lambeck 1996). The maximum transgression of the sea inundated the area for a short time following this period, then regressed to its present level when the current shoreline was established (Evans et al. 1969). The area has been relatively stable and undergoing pedogenic development for about 4,000–5,000 years (Evans et al. 1969; Lambeck 1996).

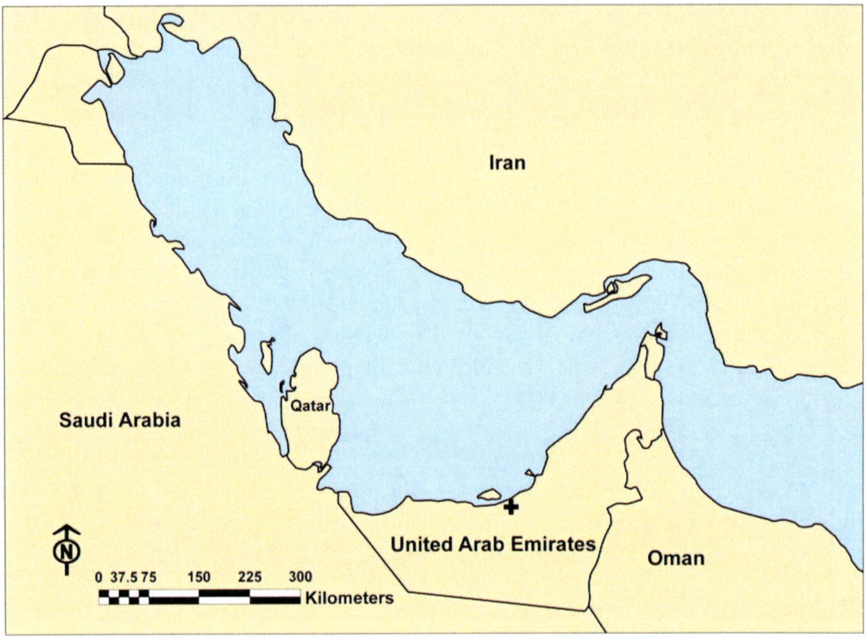

Fig. 8.1 Map showing the location of the United Arab Emirates. The "x" denotes to general area of the study

8.2.1.1 Climate

Summer temperatures are extremely hot (mean summer temperature of 32.8°C) with maximum temperatures between June and September, ranging up to and exceeding 50 °C. Winters (December to February) are characterized as short and mild, with a mean temperature of 19.0 °C (minimum temperature as low as 3°C). Mean annual rainfall is 29 mm (NCMS 2009). Vegetation is largely absent in the sabkha but includes blue-green algal mats in intertidal zones and sparsely populated halophytes.

8.3 Sampling and Laboratory Analysis

A 13-km-long transect was established across the sabkha starting at the shoreline (Fig. 8.2). Six pedons were sampled by horizon and macromorphology described (Schoeneberger et al. 2002). Bulk samples were air dried and sieved to <2-mm. Laboratory analyses followed by codes were performed according to Burt (2004). For particle size distribution analysis (PSDA) (method 3A1), the fine-earth fraction was dispersed following removal of organic matter and soluble salts, and the sand fraction was separated by wet sieving. Silt and clay fractions were measured by pipette method. Total C, N, and S was determined by dry combustion (6A2f) and

Fig. 8.2 Study area showing the six sites. Distance from site 1 to site 6 is 13 km

calcium carbonate equivalent (CCE) by use of an electronic manometer to quantify gas evolution following acid contact in a closed vessel (6E1g). Total analysis of the <2-mm fraction was determined by microwave digestion in concentrated HF, HNO_3, and HCl, with determination of elements by ICP spectroscopy (4H1b). Electrical conductivity (EC1:2) was measured on a 1:2 (soil to solution) mixture (4F1a1a1). The liquid from a saturated paste was extracted and electrical conductivity (ECsat), cations, and anions determined (4F2). The equivalent gypsum content (EGC) was measured using the method of Elrashidi et al. (2007). Intact pieces of soil fabric were collected and dipped in an elastic saran mixture, with bulk density (4A1d) and water retention (4B1c) quantified at 33 kPa. Water retention at 1,500 kPa was measured on <2-mm air-dried soil (3C2a1a).

Anhydrite was quantified by the difference in two measurements: (a) water dissolution/acetone precipitation (4E2a) that measures both gypsum and anhydrite and (b) gypsum by weight loss. Measurement of gypsum by quantifying the weight of the crystalline waters of hydration was determined by using both thermal gravimetric analysis (TGA) and the method of Artieda et al. (2006). The TGA method measured the weight loss by heating the sample from 20 to 200 °C at a rate of 2 °C min^{-1}. The weight of water loss between 75 and 115°C was used to quantify the gypsum based on a theoretical weigh loss of 20.9% (Karathanasis and Harris 1994).

The Artieda method measured the weight loss of a sample between the temperatures of 70 and 90°C, and gypsum was calculated by

$$\begin{aligned}\text{Percent Gypsum} &= \{[(ws - wf)/(ws - wt)]\,100\,[(100/14.95)]\} \\ &= [(ws - wf)/(ws - wt)]\,669\end{aligned} \tag{8.1}$$

with

ws = weight of sample dried at 70°C plus weight of sampling dish
wf = weight of sample dried at 90°C plus sampling dish
wt = weight of sampling dish
14.95 = the recovery factor of gypsum between 70 and 90 °C

A Pyrex crystallizing dish was used for drying the sample. Weights were collected for several days at multiple temperatures. Based on agreement with both a gypsum mineral standard plus laboratory soil standard with gypsum quantified by the acetone method, reported data was recorded at both 70 and 90 °C after 3 days of equilibration in the oven at the respective temperature.

Mineralogy of the <2-mm fraction was performed by crushing the soil material to less than 0.02 mm, placing on a deep-well slide holder, and analyzing by Cu-K alpha radiation from 2 to 60°2Θ. The mineral composition of the fine sand (0.1–0.25 mm) or very fine sand (0.05–0.1 mm) fraction (following particle size analysis procedure that removes salts and gypsum) was determined by mounting the grains on a glass slide with an epoxy cement (refractive index = 1.54) and counting 300 grains using a petrographic microscope with plane- and cross polarized light (7B1a2). Thin sections of intact soil material were prepared by epoxy impregnation (refractive index = 1.54). The soil fabric was examined with a polarizing petrographic microscope and features described.

8.4 Results

8.4.1 Field Morphology

All pedons were weakly developed and commonly light yellowish or pale brown at the surface (10YR 6/4 or 6/3) to white or light gray (10YR 8/1 or 7/2) in the subsoil (Table 8.1). The soils are generally sandy textured. White or light gray carbonate or gypsic cemented layers were present in the subsoil of most pedons. Sanford and Wood (2001) described similar features as algal and dolomitic crusts. Small seashells were found in the subsoil (generally at depths between 50 and 100 cm) of pedons 3, 4, and 5, suggesting that these layers were developed when the area was submerged in water. The depth to the water table at time of sampling ranged from 105 to 130 cm in all soils with the exception of the tidally influenced pedon nearest the seashore. Thin, cemented surface crusts were present at most sites.

Table 8.1 Selected physical properties of the soils

Horizon	Depth (cm)	Field texture	Moist color[a]	Clay (%)	Silt (%)	Sand (%)	Bulk density 33 kPa ($g\ cm^{-3}$)	Water retention 33 kPa (%)	Water retention 1,500 kPa (%)
Sabkha site 1									
Cz1	0–3	Fine sand	2.5Y 6/3	5.2	14.6	80.2	1.40	11.6	3.8
Cz2	3–7	Coarse sand	2.5Y 7/3	5.6	9.6	84.8	1.66	11.6	3.9
Cz3	7–16	Sand	2.5Y 6/3	1.9	5.6	92.5	1.65	5.1	2.7
Cz4	16–42	Sand	2.5Y 6/3	1.6	0.1	98.3	1.65	3.8	2.0
R	42–46	Cemented		2.8	11.8	85.4			1.4
Sabkha site 2									
Salt crust	–	–	–	4.6	11.4	84.0	–	–	3.3
Az1	0–10	Sand	10YR 7/4	5.2	19.2	75.6	–	–	4.4
Az2	10–20	Sand	10YR 8/3	8.7	12.6	78.7	1.24	17.8	7.3
Bkz1	20–42	Sand	10YR 8/1	7.2	7.3	85.5	1.25	16.1	7.2
Bkz2	42–68	Sand	2.5Y 8/2	6.4	7.4	86.2	1.60	16.6	5.2
Bkzm	68–92	Cemented	2.5Y 5/3	3.4	13.2	83.4	1.78	6.0	3.3
Byz	92–107	Sand	5Y 4/2	1.6	6.2	92.2	1.65	6.8	2.2
Cyzg	107–150	Sand	5Y 5/1	1.9	4.3	93.8	–	–	1.5
Sabkha site 3									
Salt crust	–	–	–	5.4	12.3	82.3	–	–	2.9
Az	0–8	Sand	2.5Y 7/2	9.3	21.9	68.8	1.37	23.8	4.1
Bkyz	8–22	Loam	5Y 8/1	15.9	15.7	68.4	1.18	34.1	7.5
Bkyz	22–38	Sand	2.5Y 7/3	5.2	4.5	90.3	1.29	16.4	3.5
Bkyzm	38–56	Cemented	2.5Y 7/3	2.6	3.2	94.2	1.69	15.6	2.5
B'kyz	56–59	Coarse sand	2.5Y 7/3	2.1	2.8	95.1	1.48	13.5	2.6
B'kyzm	59–70	Cemented	2.5Y 7/3	1.4	4.6	94.0	1.45	10.6	2.1
BCkyz	70–105	Sand	2.5Y 6/2	0.9	1.8	97.3	1.70	6.3	2.3
Ckyzg	105–150	Sand	5Y 5/1	1.8	2.1	96.1	1.62	8.8	2.5

(continued)

Table 8.1 (continued)

Horizon	Depth (cm)	Field texture	Moist color[a]	Clay (%)	Silt (%)	Sand (%)	Bulk density 33 kPa (g cm^{-3})	Water retention 33 kPa (%)	Water retention 1,500 kPa (%)
Sabkha site 4									
Salt crust	–	–	–	3.0	11.9	85.1	–	–	3.4
Az	0–7	Sand	2.5Y 6/3	3.2	14.7	82.1	1.51	13.7	5.0
Bkyz	7–20	Sand	2.5Y 6/3	0.7	4.5	94.8	1.50	13.9	4.5
Bkyzm	20–23	Cemented	2.5Y 6/2	2.0	5.2	92.8	1.62	15.0	3.5
B'kyz1	23–60	Sand	2.5Y 6/2	3.1	6.8	90.1	1.46	11.7	4.9
B'kyz2	60–90	Fine sandy loam	5Y 8/1	9.8	8.5	81.7	1.43	19.3	6.2
B'kyz3	90–115	Loam	5Y 8/2	14.2	11.5	74.3	1.29	24.8	9.0
BCkyz	115–123	Loamy sand	2.5Y 6/2	5.7	4.9	89.4	1.31	15.5	4.3
Cyzg	123–150	Sand	5Y 6/1	8.4	15.3	76.3	1.19	20.0	5.5
Sabkha site 5									
Akyz1	0–10	–	10YR 6/4	3.9	10.7	85.4	1.53	16.8	4.1
Bkyz	10–27	Sand	2.5Y 5/3	4.8	16.5	78.7	1.44	15.7	7.5
Bz	27–35	Clay	10YR 8/1	10.0	52.2	37.8	0.98	56.2	5.4
Bkyz1	35–50	Sand	10YR 7/2	5.3	21.1	73.6	1.23	27.5	14.4
Bkyz2	50–70	Sand	Gley 1 8/10Y	16.6	34.2	49.2	1.14	28.1	12.0
2Bk1z1	70–95	Sand	Gley 1 7/10Y	10.2	15.8	74.0	1.10	31.7	7.6
2Bk2z2	95–110	Sand	Gley 1 7/10Y	5.0	11.7	83.3			4.1
Sabkha site 6									
Akz	0–5	Sand	10YR 6/3	7.2	23.1	69.7	1.31	23.0	3.1
Bz	5–40	Clay	10YR 8/1	8.4	81.4	10.2	0.96	49.3	4.1
Bkyz	40–50	–	Gley 1 7/N	5.1	62.7	32.2	1.12	42.0	24.4
Byzm	50–55	Cemented	10YR 8/1	3.0	40.0	57.0	1.65	13.5	1.7
Ck1z1	55–100	Sand	2.5Y 5/4	1.8	1.6	96.6	1.60	4.2	1.4
Ck2z2	100–140		2.5Y 5/3	2.7	1.2	96.1	1.62	4.4	1.6

[a] Munsell color notation

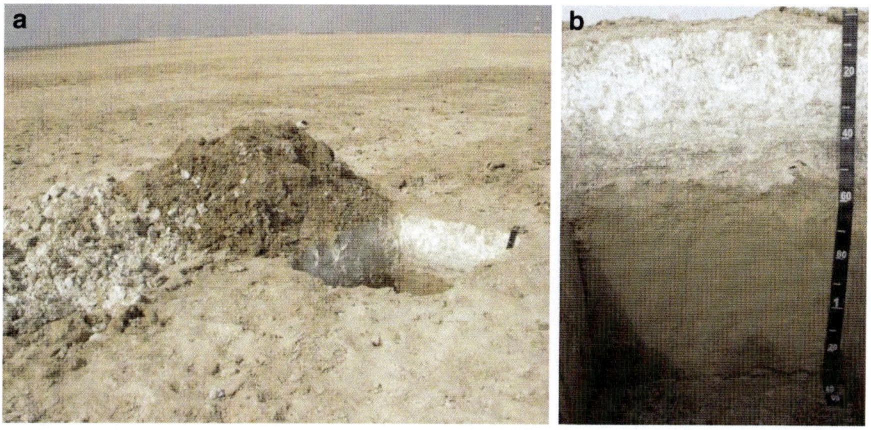

Fig. 8.3 Landscape (**a**) and profile of pedon 6 (**b**). Anhydrite is the *white zone* in the *upper part* of the profile

Specific morphological indicators were regarded by the authors as unique to mineral concentrations of anhydrite in this environment. They were a combination of white color (5Y or 10YR 8/1 or 8/2), clayey/silty textures, and/or thixotropic nature. Based on these field criteria, the Bkyz horizon (8–22 cm) of pedon 3, B'kyz2 (60–90 cm) and B'kyz3 horizons (90–115 cm) of pedon 4, Bz horizon (27–35 cm) of pedon 5 (Fig. 8.3a, b), and Bz (5–40 cm) and Bkyz (40–50 cm) of pedon 6 were regarded as possible layers with accumulations of anhydrite.

8.4.2 Carbonates, Equivalent Gypsum, and Salts

Pedon 1, the site closest to the shoreline, has the lowest salt concentrations of all sampled pedons with EC1:2 ranging from 10.7 to 23.7 dS m^{-1} (Table 8.2). The site is tidally influenced (component of the intertidal zone) and any salts that precipitate by evaporation would be transient with diurnal fluctuation of the water. The pH is highest in this pedon (8.5–8.9) relative to the other sites in the study.

The other pedons (2–6) have salt accumulation in surface horizons, with EC1:2 values (Table 8.2) ranging from 84 to 162 dS m^{-1}. The ECsat on some surface horizons exceeded the detection limits of the instrument (>200 dS m^{-1}). The EC1:2 in subsoil horizons varies from 29 to 95 dS m^{-1} (ECsat = 90–190 dS m^{-1}). Saturated paste EC values are higher than EC1:2 due to lower dilution ratios and provide a reasonable measure of soluble salt concentrations present in solution at field capacity, though all salts are not dissolved at this water content. The EC1:2 is advantageous as this value represents a fixed soil to water ratio, resulting in more comparable salinity values between horizons.

Equivalent gypsum content is a good measure of total water soluble minerals (gypsum, anhydrite, and minerals more soluble) of the <2-mm soil material.

Table 8.2 Selected chemical properties of the soils

	Depth	EC (1:2)	EC (sat)	Equiv. gypsum	$CaCO_3$	ΣEqGy+$CaCO_3$	
Horizon	(cm)	(dS m^{-1})		%			pH
Sabkha site 1							
Cz1	0–3	16.5	86.2	29	47	76	8.7
Cz2	3–7	10.7	53.4	33	48	81	8.5
Cz3	7–16	13.5	74.0	6	61	67	8.5
Cz4	16–42	23.7	113.7	7	67	74	8.7
R	42–46	19.5			84		8.9
Sabkha site 2							
Salt crust		129.9	>200	67	44	111	7.6
Az1	0–10	140.1	>200	66	42	108	7.6
Az2	10–20	55.3	186.4	23	69	92	8.1
Bkz1	20–42	48.1	151.6	18	82	100	8.5
Bkz2	42–68	48.4	167.1	15	80	95	8.3
Bkzm	68–92	36.1	123.6		74		8.1
Byz	92–107	46.7	141.1	16	63	79	8.0
Cyzg	107–150	40.0	148.5	11	61	72	7.9
Sabkha site 3							
Salt crust		167.4	>200	76	44	120	7.5
Az	0–8	130.1	>200	62	52	114	7.7
Bkyz	8–22	65.6	175.4	23	77	100	7.9
Bkyz	22–38	38.5	119.4	15	87	102	8.3
Bkyzm	38–56	38.1	110.1	16	87	103	8.1
B'kyz	56–59	29.6	89.6	12	89	101	8.2
B'kyzm	59–70	41.8	119.7	16	82	98	8.1
BCkyz	70–105	45.4	141.2	17	81	98	8.2
Ckyzg	105–150	55.0	176.1	19	73	92	7.9
Sabkha site 4							
Salt crust		147.6	>200	68	42	110	7.2
Az	0–7	83.7	>200	34	45	79	7.7
Bkyz	7–20	29.4	103.6	10	55	65	7.9
Bkyzm	20–23	31.5	97.2	31	47	78	7.9
B'kyz1	23–60	38.2	115.3	18	57	75	7.6
B'kyz2	60–90	48.7	146.8	16	77	93	7.7
B'kyz3	90–115	65.1	146.1	24	81	105	7.8
BCkyz	115–123	50.3	151.0	18	84	102	7.9
Cyzg	123–150	67.8	151.6	22	73	95	7.6
Sabkha site 5							
Akyz1	0–10	135.1	>200	75	39	114	7.4
Bkyz	10–27	30.8	147.0	38	44	82	7.8
Bz	27–35	91.4	190.3	95	14		7.6
Bkyz1	35–50	47.5	175.0	76	30	106	7.9
Bkyz2	50–70	49.3	164.1	56	48	104	7.9
2Bk1z1	70–95	74.4	181.3	25	72	97	8.0

(continued)

Table 8.2 (continued)

Horizon	Depth (cm)	EC (1:2) (dS m^{-1})	EC (sat)	Equiv. gypsum %	$CaCO_3$	ΣEqGy+$CaCO_3$	pH
2Bk2z2	95–110	57.1	162.7	26	80	106	8.1
Sabkha site 6							
Akz	0–5	161.9	>200	98	26	124	7.5
Bz	5–40	95.9	129.9	108	4	112	7.7
Bkyz	40–50	52.9	104.5	71	35	106	8.1
Byzm	50–55	47.3	171.0		6		7.9
Ck1z1	55–100	29.0	155.1	9	34	43	8.3
Ck2z2	100–140	31.7	140.5	11	28	39	8.2

These EGC values (Table 8.2) ranged from 34 to 98% in pedons 2–6. Equivalent gypsum generally decreased in the subsoil and was <25% in the subsoil of all pedons. Carbonates were generally the lowest in the surface where they ranged from 26 to 52% and up to 89% in the subsoil. Together, these two mineral groups (as measured by EGC and CCE) composed from 70 to 100% of the mineral matter in most horizons. The salt crusts sampled for pedons 2, 3, and 4 were composed of about 40% carbonates and 60% equivalent gypsum materials. For two of the three pedons, the equivalent gypsum is only slightly higher in the salt crust than in the A horizon (which included the salt crust). These results suggest that the "salt crust" is similar in composition to the surface soil horizon in these sites.

Elemental analysis of saturated paste extracts (Table 8.3) shows that Na >>Mg> Ca>K. Sodium had the highest concentration due to the greater solubility of Na salts in the saturated paste compared to other minerals. This was reflected in sodium adsorption ratio (SAR) values exceeding 300 in certain horizons. Magnesium is between 2 and 3 times higher in concentration in saturated paste extracts relative to Ca. Suarez (2005) indicates a ratio of Mg/Ca in seawater of 5, similar to these results. Mg is commonly higher due to the lower solubility of Ca minerals such as calcite, aragonite, and gypsum, thus creating a sink for the Ca ions in solution. Total analysis of the <2-mm fraction (Table 8.3) indicated a much greater abundance of Ca overall, with a Ca to Na ratio ranging from 1:1 to 28:1 and an average of 10:1. This ratio is similar for Mg, with an average Ca to Mg ratio of 11:1. These results illustrate an overall greater abundance of Ca in the soil matrix.

Chloride was the dominant anion in saturated paste extracts, generally 30–60 times higher than SO_4. Carbonate anions were not present; bicarbonate was <1 mmol L^{-1}. This ratio decreased in horizons associated with anhydrite formation ranging from 10 to 35%, reflecting an increase in SO_4 content of these horizons. This increase in S is also reflected in the total S, ranging up to 16%S.

8.4.3 Measurements for Gypsum and Anhydrite

The acetone measurement cannot differentiate between gypsum and anhydrite and extracts both minerals together. The results (Table 8.4) show that minerals in this

Table 8.3 Elemental data from saturated paste extracts and total analysis

		Saturated paste							Total elemental analysis						
	Depth	Ca	Mg	Na	K	Cl	SO_4		Si	Al	Ca	Mg	Mn	Na	S
Horizon	(cm)	(mmol L^{-1})						SAR	(mg kg^{-1})						(%)
Sabkha site 1															
Cz1	0–3	51.7	232.0	777.6	20.6	1021.6	176.9	65	92,953	14,775	210,547	33,805	263	11,812	3.06
Cz2	3–7	47.8	122.5	433.6	12.6	536.2	120.4	47	70,897	11,063	227,988	35,762	215	8,134	4.33
Cz3	7–16	57.6	129.0	660.4	17.6	825.9	114.9	68	99,565	15,251	240,036	20,514	227	15,733	0.18
Cz4	16–42	54.9	270.0	1245.0	32.4	1513.6	137.9	98	114,458	17,822	226,107	25,199	284	13,213	0.07
R	42–46								43,963	5,707	320,107	9,155	105	10,518	0.07
Sabkha site 2															
Salt crust		158.6	304.0	5030.0	83.8	5418.5	47.3	331	114,838	12,552	176,108	42,766	307	46,712	1.65
Az1	0–10	186.8	578.0	4468.0	124.0	5402.1	36.4	228	55,586	8,777	184,178	18,593	185	104,935	0.95
Az2	10–20	131.9	309.0	2733.0	63.0	3259.0	75.4	184	40,820	6,748	279,683	20,719	124	28,896	0.76
Bkz1	20–42	87.0	264.0	1882.5	45.5	2281.7	42.2	142	71,512	10,157	134,704	22,950	199	115,744	0.14
Bkz2	42–68	100.2	322.0	2204.5	64.5	2736.4	49.0	152	30,322	3,837	311,350	14,352	77	24,053	0.16
Bkzm	68–92	120.9	193.0	1609.0	40.6	1849.2	98.5		41,361	4,629	307,514	14,098	81	29,462	0.70
Byz	92–107	107.3	253.0	1913.0	52.2	2286.0	78.5	143	74,250	8,616	258,980	20,206	125	32,336	0.27
Cyzg	107–150	108.5	274.0	2004.0	59.8	2509.2	76.8	145	61,601	8,376	170,842	15,524	185	127,318	0.22
Sabkha site 3															
Salt crust		144.0	303.0	5184.0	43.6	5699.1	45.2	347	79,850	12,341	216,292	32,677	176	16,236	1.34
Az	0–8	145.4	337.0	4801.0	62.4	5436.3	49.4	309	117,363	14,594	183,164	37,060	268	8,622	1.45
Bkyz	8–22	127.4	239.0	2936.0	71.8	3289.3	56.2	217	66,988	9,436	183,766	23,728	172	5,889	0.16
Bkyz	22–38	77.4	167.0	1495.0	37.0	1791.2	39.6	135	88,428	9,425	198,334	25,335	161	7,241	0.22
Bkyzm	38–56	114.4	200.0	1298.0	35.0	1594.7	86.1	104	54,256	7,777	291,112	16,258	130	24,052	0.42
B'kyz	56–59	105.3	275.0	955.4	27.6	1200.1	81.2	69	28,983	2,996	246,131	10,558	74	24,599	0.33
B'kyzm	59–70	93.7	178.5	1507.5	37.7	1817.3	53.7	129	37,243	4,280	331,251	12,630	82	22,157	0.19
BCkyz	70–105	89.7	282.0	2153.0	56.4	2334.7	37.3	158	68,200	7,626	261,557	18,145	113	32,774	0.12
Ckyzg	105–150	124.2	356.0	2895.0	79.6	3322.6	45.2	187	70,386	10,802	152,708	21,684	162	109,002	0.14

Sabkha site 4															
Salt crust		222.6	447	4,371	97.0	5393.3	29.3	239	68,910	10,903	159,374	23,119	196	117,432	1.32
Az	0–7	123.0	171	4,545	49.2	4972.4	69.1	375	32,093	5,308	205,284	33,941	82	50,490	0.95
Bkyz	7–20	121.8	157	1,198	28.0	1463.0	72.7	101	33,740	5,807	208,800	30,564	100	22,631	0.28
Bkyzm	20–23	126.2	157	1,072	36.0	1334.6	76.3	90	34,360	6,590	227,854	28,953	76	34,249	2.83
B'kyz1	23–60	132.3	191	1,389	45.8	1725.5	77.9	109	38,587	7,600	263,596	21,245	82	43,602	1.48
B'kyz2	60–90	135.2	263	2,032	55.6	2417.4	60.4	144	59,277	10,288	144,541	21,036	164	128,800	0.10
B'kyz3	90–115	108.3	241	2,056	51.8	2391.5	39.7	156	45,156	7,956	489,204	47,209	125	91,522	0.17
BCkyz	115–123	104.6	249	2,167	54.0	2542.3	37.2	163	26,765	4,832	181,216	37,736	79	57,385	0.18
Cyzg	123–150	111.0	270	2,162	59.6	2540.6	37.3	157	30,387	5,059	152,839	73,866	115	29,573	0.21
Sabkha site 5															
Akyz1	0–10	240.6	808	4,342	159.0	6030.0	83.7	190	61,186	10,421	165,026	20,779	152	99,918	2.33
Bkyz	10–27	145.6	369	1,793	68.4	2485.5	108.6	112	79,243	12,557	213,787	32,645	173	16,381	3.99
Bz	27–35	153.6	416	3,076	97.8	4172.0	119.1	182	25,542	4,638	186,512	29,988	74	46,320	10.21
Bkyz1	35–50	128.6	303	2,643	71.8	3448.0	112.1	180	33,766	5,667	211,478	31,892	101	23,129	8.79
Bkyz2	50–70	121.2	277	2341.5	60.2	3003.6	120.2	166	32,414	6,174	224,128	27,915	76	34,173	5.12
2Bk1z1	70–95	104.2	340	2,824	14.8	3696.9	96.7	189	38,781	7,470	262,301	21,777	83	43,856	0.28
2Bk2z2	95–110	120.8	279	2,355	59.6	3016.4	132.6	167	17,089	2,676	308,656	9,089	43	32,864	0.68
Sabkha site 6															
Akz	0–5	152.6	598.0	4621.0	96.8	6085.8	106.0	239	51,115	9,666	143,060	20,322	157	121,752	4.20
Bz	5–40	129.7	144.5	1684.5	24.6	1886.5	143.4	144	24,786	4,624	166,060	35,720	74	53,986	11.82
Bkyz	40–50	103.0	91.5	1220.5	19.1	1341.4	122.8	124	30,419	5,085	155,998	71,344	118	30,493	8.34
Byzm	50–55	117.8	194.0	2634.0	35.6	3082.6	130.9		23,415	2,241	239,798	11,442	73	19,981	16.18
Ck1z1	55–100	68.1	167.5	2232.5	29.4	2573.8	77.1	206	135,389	17,266	231,816	11,736	254	21,381	0.18
Ck2z2	100–140	60.5	161.0	1904.0	27.8	2187.2	70.8	181	144,362	18,687	193,455	12,268	237	25,094	0.08

Table 8.4 Selected mineralogical properties of the soils

Horizon	Depth (cm)	Gypsum + anhydrite (acetone)[a]	Gypsum, 70°C/90°C (oven wt)[a]	Anhydrite (by difference with oven wt)	Gypsum (thermal)	Anhydrite (by difference with thermal wt)	XRD of the < 2-mm[b] (Relative quantities)	Mineral composition of fine (0.1–0.25 mm) or very fine sand (0.05–0.10 mm) fraction[b] (% of the fraction)
		(%)						
Sabkha site 1								
Cz1	0–3	18.6	17.9	1	–	–	–	
Cz2	3–7	23.6	22.5	1	–	–	–	CB49 CA32 FK7 QZ7 AR1 CD1 FP1 OP1
Cz3	7–16	0.2	1.1	−1	–	–	CA2 HA1 GY1	
Cz4	16–42	0.0	1.1	−1	–	–	–	CB42 CA26 QZ20 FK8 CD2 FP1 PR1
R	42–46	0.0	0.7	−1	–	–	–	
Sabkha site 2					–	–	–	
Salt crust		9.4	9.7	0	–	–	–	
Az1	0–10	13.1	13.8	−1	–	–	–	CB57 QZ15 CA13 FK12 CD2 AR1 FF1
Az2	10–20	4.2	6.9	−3	–	–	–	
Bkz1	20–42	0.0	2.2	−2	–	–	–	CB90 CA5 QZ4 FK2
Bkz2	42–68	0.0	1.6	−2	–	–	AR2 HA1 QZ1 CA1 GY1	
Bkzm	68–92	4.1	5.1	−1	–	–	–	CB79 QZ16 CA7 FK5 CD1
Byz	92–107	0.1	1.9	−2	–	–	–	
Cyzg	107–150	0.0	1.6	−2	–	–	–	

Sabkha site 3					–	–	–	
Salt crust		6.6	5.8	–1	–	–	–	
Az	0–8	8.0	7.2	–2	–	–	–	CB50 CA26 QZ14 FK4 CD2 AR1 FP1 GN1 HN1
Bkyz	8–22	0.1	2.3	–2	–	–	AR3 HA2 CA1	
Bkyz	22–38	0.0	1.3	–1	–	–	–	CB99 CA1 QZtr
Bkyzm	38–56	2.0	3.5	–2	–	–	–	
B'kyz	56–59	0.5	2.4	–2	–	–	–	CB98 CA1 QZ1
B'kyzm	59–70	0.1	1.8	–2	–	–	–	
BCkyz	70–105	0.1	1.7	–2	–	–	–	CB77 QZ13 CA5 FK4 CD2
Ckyzg	105–150	0.1	1.9	–2	–	–	–	
Sabkha site 4	–	–			–	–		
Salt crust		6.3	9.1	–3	–	–	GY2 HA2 CA1 AR1 QZ1	
Az	0–7	4.0	7.8	–4	–	–	CA3 HA3 GY2 AY1	
Bkyz	7–20	0.8	1.9	–1	–	–	CA3 QZ3 GY2 HA1	CB48 CA26 FK16 QZ8 CD2 FF1
Bkyzm	20–23	20.1	22.1	–2	–	–	GY2 CA2 QZ1 HA1	
B'kyz1	23–60	5.7	7.2	–1	–	–	HA3 CA2 QZ2 GY1 AR1	CA46 CB33 QZ13 FK6 CD1 FP1
B'kyz2	60–90	0.1	1.9	–2	–	–	CA2 HA2 QZ1 GY1 AR1	CB68 CA20 QZ6 FK3 CD1
B'kyz3	90–115	0.1	2.1	–2	–	–	HA2 AR2 GY1 QZ1 CA1	CB90 CA5 QZ3 CD1 FK1
BCkyz	115–123	0.0	1.7	–2	–	–	HA2 AR2 CA1 QZ1 GY1	

(continued)

Table 8.4 (continued)

Horizon	Depth (cm)	Gypsum + anhydrite (acetone)[a]	Gypsum, 70°C/90°C (oven wt)[a]	Anhydrite (by difference with oven wt)	Gypsum (thermal)	Anhydrite (by difference with thermal wt)	XRD of the < 2-mm[b] (Relative quantities)	Mineral composition of fine (0.1–0.25 mm) or very fine sand (0.05–0.10 mm) fraction[b] (% of the fraction)
		(%)						
Cyzg	123–150	0.0	2.6	–3	–	–	–	
Sabkha site 5								
Akyz1	0–10	13.0 (±0.0)	12.7 (±1.1)	0	5.8	7	HA3 AY2 CA1 GY1 FK1	
Bkyz	10–27	23.6 (±0.0)	31.1 (±5.1)	–7	7.9	16	HA3 AY2 GY2 CA1 QZ1 FK1	CB52 CA22 QZ13 FK11 FF1
Bz	27–35	41.0 (±0.0)	5.1 (±0.5)	36	2.0	39	AY3 HA2 CA1	CB46 AY21 CA18 QZ13 FK2
Bkyz1	35–50	42.6 (±3.6)	65.1 (±7.4)	–23	11.5	31	CA3 AY1 HA1 GY1 AR1 QZ1	CB64 CA28 FK6 CD1 QZ1
Bkyz2	50–70	28.0 (±0.0)	61.9 (±22.4)	–34	8.0	20	HA2 CA2 GY1 AR1 QZ1	CB71 CA20 FK5 QZ4 CD1
2Bk1z1	70–95	0.2 (±0.0)	3.0 (±0.2)	–3	2.3	0	HA3 AR2 QZ2 CA1 GY1	CB84 CA7 QZ4 FK2 CD1 FF1
2Bk2z2	95–110	2.1 (±0.0)	4.4 (±0.1)	–2	3.8	0	AR2 HA2 QZ1 CA1 GY1	
Sabkha site 6								
Akz	0–5	16.5 (±0.0)	4.0 (±1.0)	12	2.6	14	AY3 CA3 HA3 QZ2 GY2 AR1 FK1	CB40 CA34 QZ12 FK9 CD3 AYtr

Bz	5–40	47.0 (±0.0)	3.3 (±0.6)	44	2.0	45	AY3 HA2 CA1 QZ1 GY1	AY48 CB23 QZ10 CA8 FK7 FD3 HN1
Bkyz	40–50	40.0 (±0.0)	52.4 (±4.7)	−12	6.6	33	HA2 AY2 CA2 GY1	CB10 CA24 QZ12 FK9 AY6 GY6 CD1
Byzm	50–55	30.0 (±0.0)	2.5 (±0.1)	27	0.8	29	AY3 HA1 CA1	AY92 CA5 CB1 FK1
Ck1z1	55–100	0.1 (±0.0)	0.7 (±0.0)	−1	0.5	0	CA3 QZ3 HA1 AY1	CB34 CA29 QZ20 FK11 CD2 FP2 FF1
Ck2z2	100–140	0.0 (±0.0)	0.8 (±0.1)	−1	0.5	0	CA3 QZ1 AY1 GY1 HA1 AR1	

AR aragonite, *FK* orthoclase feldspar, *AY* anhydrite, *CD* chalcedony, *FP* plagioclase feldspar, *OP* opaques, *PR* pyroxene, *GN* garnet, *HN* hornblende

[a] Analyses for pedon 5 and 6 were duplicated. Number in parentheses represents the standard deviation

[b] X-ray data represent relative quantities ranging from 1 to 3 based on peak height. *CA* calcite, *CB* carbonate aggregates, *QZ* quartz, *HA* halite, *GY* gypsum

"group" (gypsum + anhydrite) were concentrated in upper horizons (upper 50 cm) of pedons 1–4, ranging from 0 to 24%. Surface salt crust in pedons 2, 3, and 4 had gypsum + anhydrite ranging from 5 to 9%. Gypsum + anhydrite was concentrated in the upper 70 and 55 cm for pedons 5 and 6, ranging from 13 to 47%. Concentrations were <2% below these depths for the latter two pedons.

Analysis for gypsum by weight loss (Artieda et al. 2006) found reasonably close agreement with the acetone method for pedons 1–4 (Table 8.4). This agreement suggests that no anhydrite was present in the first four pedons. Anhydrite was suspected during the field examination in the Bkyz horizon of pedon 3 and in the B'kyz2 (60–90 cm) and B'kyz3 horizons (90–115 cm) of pedon 4. But very minor amounts (≤0.1%) of gypsum + anhydrite were found in these horizons.

The presence of anhydrite (as indicated by the positive difference between the acetone and weight loss methods; i.e., Eq. 8.1) was found in pedons 5 and 6. Anhydrite was determined by this approach to be in the Bz horizon (27–35 cm) of pedon 5 and the Akz, Bz, and Byzm horizons of pedon 6 in amounts ranging from 12 to 44%.

Negative values for calculated anhydrite using Eq. 8.1, ranging from −12 to −34% (Bkyz1 and Bkyz2 of pedon 5 and Bkyz of pedon 6), reflect greater gypsum measured by weight loss than gypsum + anhydrite by the acetone method. Small negative differences were common for other samples as the difference between methods for pedons 1–4 ranged from 1.0 to −3.8%. These similar values were regarded as generally equivalent, representing methodological differences.

Large negative values for anhydrite using the oven weight loss method (Eq. 8.1) suggest a possible interference or methodological problem. To better understand the problem, both the acetone and weight loss procedures were duplicated for pedons 5 and 6. Low standard deviations for the acetone method suggest good reproducibility. Standard deviations of the weight loss method range from 5 to 22%, illustrating that this method has limited reproducibility. Possible interferences in these samples contributing to the excessive weight loss between 70 and 90 °C may be other minerals (hydrated salts) or clay minerals such as smectite (Lebron et al. 2009). These soils have low clay percentages, and analyses show that no aluminosilicate minerals such as smectite were present in the clay fraction to contribute as a source of errors.

A possible positive interference in the weight loss method by hydrated salts is not apparent. The ECsat for these three horizons in question ranged from 104 to 175 dS m^{-1}, very similar to ECsat from other horizons in these two soils where no negative calculated anhydrite values occurred. Also, other horizons with similar ECsat show good agreement between the two methods. For example, horizons Az1 in pedon 2 and Bkyzm in pedon 3 both have ECsat >97 dS m^{-1} and have gypsum + anhydrite by the acetone methods of 20 and 13%. The corresponding values for the weight loss method are 22 and 14%, respectively.

Thermal gravimetric analysis (TGA) (Fig. 8.4) was performed on all horizons from pedons 5 and 6 as a second method to quantify gypsum by weight loss. While positive interferences are possible with the thermal analysis method, overall, gypsum measured by TGA (Table 8.4) was lower than quantities measured by the oven

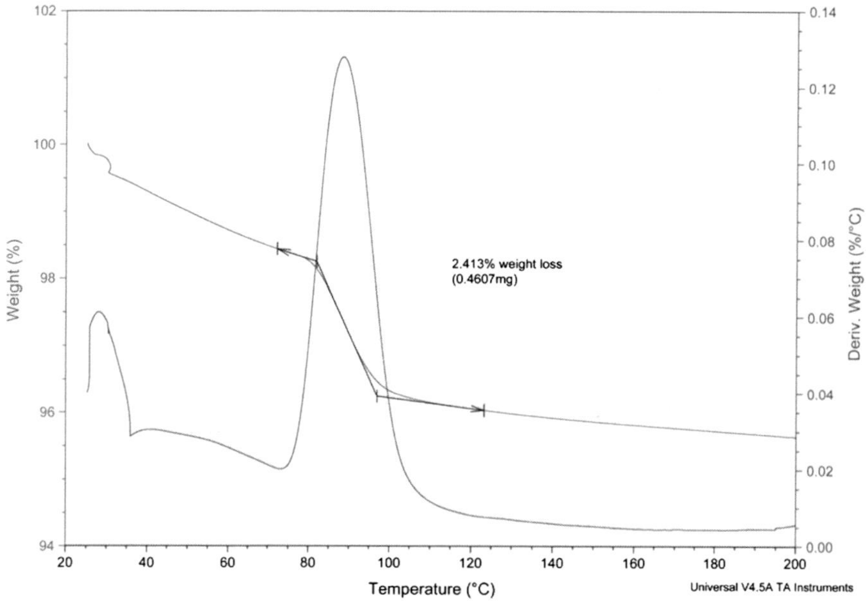

Fig. 8.4 Thermal gravimetric analysis (TGA) pattern illustrating loss of weight from heating the <2-mm soil from the Bkyz1 horizon of pedon 5

method. Good agreement between the two weight loss gypsum methods was achieved with some samples, while large differences in measured gypsum between the two methods also occurred. Large differences in results were those horizons with large negative anhydrite values when using data from the oven method. For example, the Bkyz2 horizon of pedon 5 had 62% gypsum by the oven weight loss method versus 8% gypsum by TGA. Since both methods measure the weight of water loss from gypsum, these TGA data suggest that the oven method has problems for gypsum quantification that are not necessarily interferences of water loss from other minerals.

The TGA results suggest that anhydrite was present in the upper 70 and 55 cm of pedons 5 and 6, respectively, with anhydrite quantities ranging from 7 to 45%. These concentrations are somewhat lower than found in other studies of the sabkha that reported percentages ranging up to 98% (Shahid et al. 2009). The ratio of anhydrite to gypsum in these horizons ranged from 1:1 to 36:1. Largest ratios were present in Bz horizons of both pedons.

8.4.4 X-Ray Diffraction and Optical Analysis

Evaporative minerals in the <2-mm fraction identified by x-ray diffraction (Table 8.4) include calcite, gypsum, halite, aragonite, and anhydrite. No conclusive evidence for bassanite ($CaSO_4 \cdot ½\ H_2O$), mineral that would be considered the intermediary between gypsum and anhydrite, was found in these samples. Anhydrite was detected by this method in the upper 4 horizons of pedon 5 and all horizons of pedon 6. Detection of this mineral is based primarily on a peak at 0.350 nm, with possible secondary peaks at 0.285, 0.233, and 0.387 nm, and other minor peaks. Relative amounts, based on peak intensity of the 0.35 nm peak, vary. The greatest quantities were in the Bz horizons of pedon 5 and the Akz, Bz, and Byzm horizons of pedon 6. These results are in reasonable good agreement with the anhydrite detected by the TGA method. The only major difference is the Bkyz2 horizon (50–70 cm) of pedon 5, with 20% anhydrite measured by the acetone/TGA method, while no anhydrite was detected in the <2-mm fraction by x-ray diffraction. This may suggest that the acetone/TGA method overestimates anhydrite for some samples, but also the lack of the major 0.350 nm anhydrite XRD peak in the sample does not rule out the presence of the mineral. There were several minor peaks present in the XRD analysis of that horizon that could be assigned to anhydrite.

Anhydrite occurs in a variety of forms and sizes. For example, it forms microcrystalline to coarse crystalline fillings as well as lath- or fiber-like forms (El-Tabakh et al. 2004). Sand-sized anhydrite under the microscope is distinct from other minerals, especially gypsum. It has a refractive index >1.54 and a high birefringence under cross polarized light. Cleavage is generally at right angles or parallel to the mineral axis resulting in elongated, angular mineral forms. Gypsum can be distinguished from anhydrite due to a refractive index <1.54 and low birefringence. Calcite has a variable (double) refractive index (can be both below and above 1.54) depending on direction of orientation, as well as extreme birefringence and rhombohedral cleavage in the macrocrystalline form (Kerr 1977). Often, calcite is microcrystalline in soils and tends to segregate from other minerals (not adsorbed to other mineral surface), forming homogeneous deposits in voids (Chadwick et al. 1987).

Optical analysis (Table 8.4) of the fine and very fine sand fractions shows that carbonate aggregates and calcite were the most common mineral forms identified. Quartz and orthoclase feldspar were also identified in amounts up to 20%. The sand fractions analyzed are derived from the particle size analysis procedure. Washing pretreatments in this procedure will remove all salts and possibly much of the gypsum and anhydrite. Gypsum was found only in the Bkyz horizon of pedon 6, while anhydrite was optically identified in several horizons of pedon 5 and 6, composing 21% in the sand fraction of Bz horizon of pedon 5 and 92% of the sand fraction in the Byzm horizon of pedon 6. The grains of anhydrite are principally elongated and highly birefringent under cross polarized light (Fig. 8.5a, b). These results suggest that gypsum is much more soluble than anhydrite (relative to the PSDA washing procedure).

In thin sections of natural fabric, anhydrite was present as randomly oriented grains in aggregates. This nodular form agrees with both Evans et al. (1969), who identified anhydrite as soft plastic masses that range from having sharp to diffuse boundaries with the surrounding soil matrix, and El-Tabakh et al. (2004), who identified anhydrite beds with this mineral in a nodular form. The largest and most well-developed anhydrite grains were in Byzm horizon of pedon 6. These grains (Fig. 8.6a) ranged up to 0.5 mm in length and were highly birefringent. Figure 8.6b, c

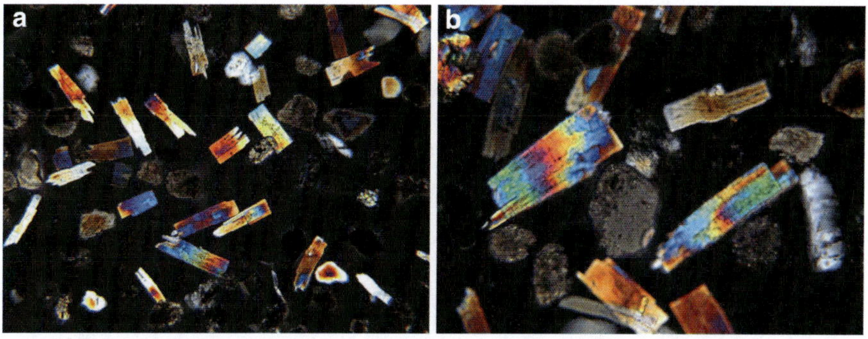

Fig. 8.5 Photomicrograph of anhydrite grains from the very fine sand (0.05–0.1 mm) fraction pedon 6 (Bz horizon). Cross polarized light

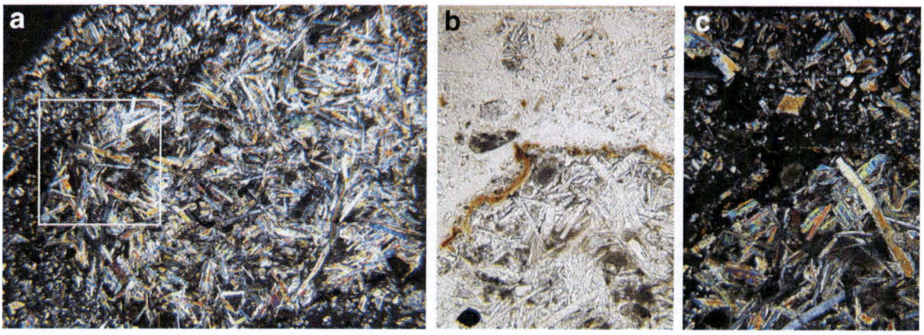

Fig. 8.6 Photomicrograph of the soil fabric of the Bz horizon from pedon 6 illustrating the nodular form of anhydrite aggregates. Anhydrite grains composing this horizon are in the very fine sand size range. Cross polarized light. Photo width = 2.5 mm (**a**). Higher magnification images of the area enclosed by the box in (**a**) both in (**b**) plain and (**c**) cross polarized light. These images illustrate the variety of elongated grains as well as the Fe deposit along the boundary of the nodule. Photo width = 1.1 mm

illustrates frequent occurrence of Fe oxide deposition around the perimeter of the nodules, indicating a hydrologic discontinuity between the nodules of anhydrite and the rest of the soil matrix. The grains in the Bz horizons of pedons 5 and 6 were smaller in size. This is especially true for pedon 5 where anhydrite grains appear to be principally clay and fine silt (Fig. 8.7). These results illustrate that anhydrite occurs in a range of particle sizes in these soils. This difference in the dominant size fraction of anhydrite between horizons suggests a difference in mode of deposition, conditions of formation, or age of the deposit. This difference in size of grains likely accounts for the better agreement between the x-ray diffraction data on the <2-mm fraction and the optical grain count data on a specific sand fraction (Table 8.4) for certain horizons. For example, the Bkyz horizon of pedon 5 has no anhydrite identified in the sand fraction by optical analysis, while anhydrite was present in the x-ray diffraction data for the <2-mm fraction, suggesting anhydrite was present in

Fig. 8.7 Photomicrograph of the soil fabric of the Bz horizon from pedon 5 illustrating the nodular form of anhydrite aggregates. Much of the anhydrite in this horizon appears to be in the clay and fine silt size range. Cross polarized light. Photo width = 0.53 mm

the smaller particle size fractions only. Also, the Akz horizon of pedon 6 had an XRD anhydrite peak with the relative size of 3, yet had only trace amounts of anhydrite identified in the sand fraction. As mentioned previously, these differences could also be attributed to the dissolution of anhydrite from the sand fraction during PSDA pretreatments used to derive the sand fraction for optical analysis.

8.4.5 Physical Properties

These soils (Table 8.1) are sandy textured, with >75% sand in the <2-mm fraction. Due to pretreatments, the particle size data reflects the loss of soluble salts and much of the gypsum and anhydrite from these samples, but still reflect similar textures as determined in the field. Carbonates and carbonate-coated minerals (e.g., carbonate-coated gypsum) likely remain following pretreatments.

Horizons with large anhydrite concentrations had greater amounts of silt than comparable horizons. For example, the Bz horizon of pedon 5 and the Bz and Bkyz horizons of pedon 6 (all with >30% anhydrite) had silt contents ranging from 52 to 81%. This particle size is generally accompanied by a greater 33 kPa water retention and lower bulk density than other horizons. The bulk density of the three horizons cited above ranged from 0.96 to 1.12 g cm^{-3}, and 33 kPa water retention ranged from 42 to 56%. In contrast, the Byzm horizon in pedon 6 was determined to have 92% anhydrite by optical analysis, but a higher sand content and bulk density than other anhydrite-rich horizon. The difference in this horizon was that it is cemented, unlike the other more friable, anhydrite-rich horizons.

8.5 Discussion

Mineral formation in soils is controlled by the elements present, their concentration and form, as well as chemical and environmental conditions of the pedoenvironment. In the level, coastal region of the Abu Dhabi Emirate, elements in soil solution derived from soluble mineral components of Tertiary deposits underlying the sabkha, as well as from sea and lagoon water intrusion or overland flow, were concentrated via evaporation in the soil vadose zone. The most stable minerals, such as calcium carbonates, were initially deposited. As alkaline anions (principally HCO_3^{-1}) were depleted via calcite deposition, then sulfate minerals, either gypsum or anhydrite, precipitated.

Anhydrite was identified in pedons 5 and 6 by the chemical/weight loss technique, x-ray diffraction, thin section, and optical analysis. The oven method used to measure gypsum and then to calculate anhydrite exhibited poor reproducibility, though appeared to work reasonably well in some samples. Measurement of gypsum by thermal analysis (TGA), followed by anhydrite quantification, did provide more consistent results when compared to other methods (x-ray diffraction, optical analysis). Thermal analysis appears to have promise as a method of choice to quantify gypsum in samples that also contain anhydrite.

X-ray diffraction is an excellent tool to identify the presence of anhydrite in soils, but it only provides qualitative or possibly semiquantitative information. Optical analysis is a reliable method to quantify anhydrite in the particle size range of sand and coarse silt. This mineral has distinct characteristics under the microscope. Anhydrite, unlike gypsum, appears to survive the washing treatment used for particle size analysis, but dry sieving the soil is likely the best option for separating a sample for analysis. The limitation of optical analysis would be that only a limited size range can be easily identified under the microscope (0.02–2 mm) and anhydrite exists in smaller particle size classes as well.

There was no anhydrite detected in pedons 1–4. These four pedons lack the conditions to promote the formation or stability of anhydrite. Pedon 1 is located in the intertidal zone, an area subjected to daily fluctuations of the tide. Pedons 2, 3, and 4 are part of the lower supratidal zone (Shinn 1973), a zone of gypsum formation. Reoccurring seasonal flooding likely limits the formation and stability of anhydrite in these soils.

Pedons 5 and 6, on the highest and most stable part of the sabkha, have anhydrite. This mineral was detected in the upper horizons (0–50 cm) of pedon 5 and all horizons in pedon 6. The Bz horizon of pedon 5 and Bz and Byzm horizons in pedon 6 were the zones of maximum anhydrite concentration in these soils. In Bz horizons, sand, bulk density, and carbonates are at minimum values within the profiles, while silt, water retention, and total S and SO_4 values are at a maximum. While these trends are slightly different for the cemented Byzm horizon, it had the highest total S of all horizons (16.2%). These trends in the data reflect the impact of anhydrite on soil properties. Overall, based on the acetone/TGA method, anhydrite composed 22 and 39% of the mineral matter in the upper 50 cm of pedons 5 and 6, respectively, and about 16% of both pedons between 25 and 100 cm depth, the mineralogy control section of Soil Taxonomy (Soil Survey Staff 2010). Based on optical analysis data, the weighted average percent of anhydrite in the control section for these two soils are 2 and 17%, respectively.

It appears that anhydrite formation occurs principally by neoformation under extreme conditions of salinity and temperature in these soils. These conditions commonly take place in the capillary zone above the water table, where salts are concentrated by a process referred to as "evaporative pumping" (Hsü and Siegenthaler 1969). Al-Youssef et al. (2006) indicated that necessary conditions for anhydrite formation were Na >0.13 μg L^{-1}, Cl>0.18 μg L^{-1}, total dissolved solids >0.32 μg L^{-1}, and temperatures >25 °C. These concentrations are well below levels of Na and Cl in the saturated paste extracts. Anhydrite appears to initially form nodules that later coalesce into layers or a mosaic of anhydrite. The Bz horizons of pedons 5 and 6 have high amounts of anhydrite and are also the zone of maximum subsurface salinity (91 and 96 dS m^{-1}). In these two Bz horizons, EC1:2 values are nearly double the horizon below and are only exceeded in the pedon by the surface salt concentrations. They are the zones of minimum calcite concentration within the pedons (14 and 4%, respectively), and gypsum is about 5% of the total amount of $CaSO_4$ minerals (anhydrite is 95% of the total). Anhydrite is also present in horizons below and above this zone of maximum concentration, suggesting that anhydrite formation fluctuates somewhat within the profile, likely due to changes of water table depth or temperature. The Byzm horizon in pedon 6 has larger anhydrite crystals than the two Bz horizons. This accounts for the high amount of anhydrite as determined in the sand fraction by optical analysis. It also suggests that this horizon may be the site of older deposits of anhydrite with larger, better-developed crystals.

There is little physical evidence in these soils of alteration of gypsum to anhydrite as suggested by other researchers. Butler (1970) cited that anhydrite is converted from gypsum by either dissolution-precipitation or conversion by dehydration with or without bassanite as an intermediary. He identified solution cavities in gypsum crystals with a rind of coarsely crystalline bassanite and traces of anhydrite. Aref et al. (1997) cited the formation of anhydrite laths enclosing relics of corroded gypsum. El-Tabakh et al. (2004) cited the presence of anhydrite varying between vertically elongated and equidimensional shapes. They regarded the former as an indication of gypsum as a precursor of anhydrite. In this study, some grains of anhydrite in the surface horizons of pedons 5 and 6 were found to have a lenticular appearance similar to gypsum. Figure 8.8a illustrates some grains from the Akyz1 horizon of pedon 5 with the high birefringence characteristic of anhydrite, but with a common shape that is typical of gypsum, such as in the Bkyz horizon of pedon 6 (Fig. 8.8b).

While soil and environmental conditions of this sabkha promote the formation of anhydrite, it is commonly regarded that anhydrite readily converts to gypsum upon exposure and common weathering conditions found in soils (Holliday 1970). Conditions that promote this conversion do not likely exist in this sabkha, as well as other landscapes in the Emirate. Evidence of the stability of anhydrite is present

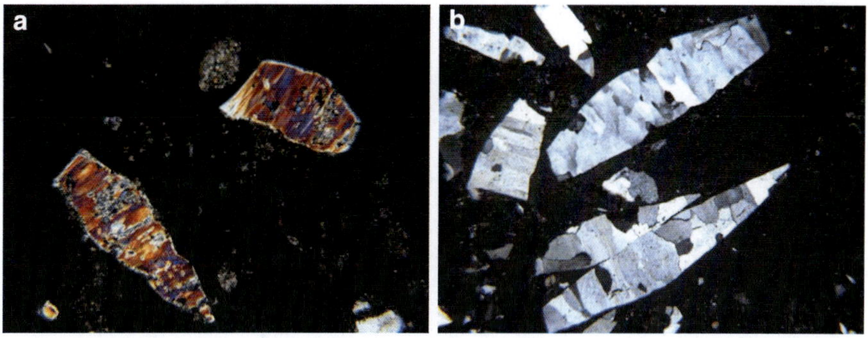

Fig. 8.8 (**a**) Weathered grains of anhydrite from the Akyz1 horizon of pedon 5. Cross polarized light. Photo width = 1.1 mm. (**b**) Gypsum crystal with lenticular morphology similar to Fig. 8.5a from the Bkyz horizon of pedon 6. Cross polarized light. Photo width = 2.5 mm

in anhydrite-rich outcrops scattered throughout the Emirate in upland landscape positions.

Based on regional models of anhydrite formation (Shinn 1973), greater concentrations of anhydrite should be present further inland on the sabkha. Additional evaluation of these landscapes should be conducted to determine the areas of greatest anhydrite content. With further information, the soil maps can be better refined relative to this mineral.

8.6 Conclusion

The formation of anhydrite is a regional phenomenon in hot desert climates of the Middle East, commonly associated with saline coastal areas and a component of carbonate deposition. Six pedons across a sabkha of Abu Dhabi, United Arab Emirates, were sampled for analysis and characterization. Anhydrite was identified and quantified in two of six pedons in this study. Anhydrite quantification by the difference in the acetone dissolution method (gypsum + anhydrite) and weight loss by thermal analysis (gypsum) appears reasonable as a methodology for quantification. This method is in general agreement with other data (XRD, optical grain count of sand) produced in the study. This method will prove useful for the quantification of anhydrite for the purpose of classification of these soils in Soil Taxonomy or other classification systems. Anhydrite quantification was one of the important requirements for inclusion of an anhydritic soil mineralogy family in Soil Taxonomy.

Overall, the presence of anhydrite is documented in the two pedons that are located at the greatest distance from the coast. Zones of maximum anhydrite concentration are occurring in subsurface horizons above the water table, suggesting that the capillary rise and evaporation of groundwater result in the concentration of salts and formation of anhydrite under these environmental conditions.

References

Abdelfattah MA, Shahid SA (2007) A comparative characterization and classification of soils in Abu Dhabi coastal area in relation to arid and semiarid conditions using USDA and FAO soil classification systems. Arid Land Res Manag 21:245–271

Alsharhan AS, Kendall CGSC (2003) Holocene coastal carbonates and evaporates of the southern Arabian Gulf and their ancient analogues. Earth Sci Rev 61:191–243

Al-Youssef M, Stow DAV, West IM (2006) Salt Lake area, Northeastern part of Dukhan Sabkha, Qatar (Chapter 13). In: Khan MA, Boer B, Kurst GS, Barth HJ (eds) Sabkha ecosystems. West and Central Asia, vol II. Springer, Dordrecht, pp 163–181

Aref MAM, Attia OEA, Wali AMA (1997) Facies and depositional environment of the Holocene evaporates in the Ras Shukeir area, Gulf of Suez, Egypt. Sediment Geol 110:123–145

Artieda O, Herrero J, Drohan PJ (2006) Refinement of the differential water loss method for gypsum determination in soils. Soil Sci Soc Am J 70:1932–1935

Burt R (2004) Soil survey laboratory methods manual. SSIR No 42, Ver 4.0. US Government Printing Office, Washington, DC

Butler GP (1970) Holocene gypsum and anhydrite of the Abu Dhabi sabkha, Trucial Coast: an alternative explanation of origin. In: Rau JL, Dellwing LF (eds) Third symposium on salt. Northern Ohio Geological Society, Cleveland, pp 120–152

Chadwick OA, Hendricks DM, Nettleton WD (1987) Silica in duric soils: I. A depositional model. Soil Sci Soc Am J 51:975–982

de Matos JE (1989) The coastal sabkha of Abu Dhabi. Bull 37. Emirates Natural History Group. Available at http://www.enhg.org/bulletin/b37/37_16.htm. Accessed 5 Jan 2010, verified 5 Jan 2010

EAD (2009) Soil survey of Abu Dhabi Emirate. Extensive survey report. Environment Agency, Abu Dhabi, UAE

Elrashidi MA, Hammer D, Seybold CA, Engel RJ, Burt R, Jones P (2007) Application of equivalent gypsum content to estimate potential subsidence of gypsiferous soils. Soil Sci 172(3):209–224

El-Tabakh M, Mory A, Schreiber BC, Yasin R (2004) Anhydrite cements after dolomitization of shallow marine Silurian carbonates of the Gascoyne Platform, Southern Carnarvon Basin, Western Australia. Sediment Geol 164:75–87

Evans G, Schmidt V, Bush P, Nelson H (1969) Stratigraphy and geologic history of the sabkha, Abu Dhabi, Persian Gulf. Sedimentology 12:145–159

Holliday DW (1970) The petrology of secondary gypsum rocks: a review. J Sediment Pet 40(2):734–744

Hsü KJ, Siegenthaler C (1969) Preliminary experiments on hydrodynamic movement induced by evaporation and their bearing on dolomite problem. Sedimentology 12:11–25

Karathanasis AD, Harris WG (1994) Quantitative thermal analysis of soil minerals. In: Amonette JE, Zelazny LW (eds) Quantitative methods in soil mineralogy. SSSA Miscellaneous Publication, Soil Science Society of America, Madison, pp 360–411

Kasprzyk A, Orti F (1998) Paleogeographic and burial controls on anhydrite genesis: the Badenian basin in the Carpathian Foredeep (Southern Poland, Western Ukraine). Sedimentology 45:889–907

Kerr PF (1977) Optical mineralogy, 4th edn. McGraw-Hill, New York

Lambeck K (1996) Shoreline reconstructions for the Persian Gulf since the last glacial maximum. Earth Planet Sci Lett 142:43–57

Lebron I, Herrero J, Robinson DA (2009) Determination of gypsum content in dryland soils exploiting the gypsum-bassanite phase change. Soil Sci Soc Am J 73(2):403–411

Melim LA, Scholle PA (2002) Dolomitization of the capitan formation for reef facies (Permian, West Texas and New Mexico): seepage reflux revisited. Sedimentology 49:1207–1227

NCMS (2009) United Arab Emirates National Center of Meteorology and Seismology. On-line climatic data. Available at: http://www.ncms.ae/english/. Accessed 29 Dec 2009

Sanford WE, Wood WW (2001) Hydrology of the coastal sabkhas of Abu Dhabi, United Arab Emirates. Hydrogeol J 9:358–366

Schoeneberger PJ, Wysocki DA, Benham EC, Broderson WD (eds) (2002) Fieldbook for describing and sampling soils, Ver 2. USDA-NRCS, National Soil Survey Center, Lincoln

Shahid SA, Abdelfattah MA, Wilson MA (2007) A unique anhydrite soil in the coastal sabkha of Abu Dhabi emirate. Soil Surv Horiz 48:75–79

Shahid SA, Abdelfattah MA, El-Saiy AK, Mufti KA (2009) Commercial value assessment of newly discovered anhydrite soil in the coastal lands of Abu Dhabi Emirate. Eur J Sci Res 29(1):36–46

Shinn EA (1973) Sedimentary accretion along the leeward, SE coast of Qatar Peninsula, Persian Gulf. In: Purser BH (ed) The Persian Gulf: Holocene carbonate sedimentation and diagenesis in a shallow epicontinental sea. Springer, Berlin/Heidelberg/New York, pp 199–209

Soil Survey Staff (2010) Keys to soil taxonomy, 11th edn. US Government Printing Office, Washington, DC

Suarez DL (2005) Chemistry of salt-affected soils. In: Tabatabai MA, Sparks DL (eds) Chemical processes in soils. SSSA book series, no 8. Soil Science Society of America, Madison, pp 689–705

Chapter 9
Fundamental Steps for Regional and Country Level Soil Surveys

Norair Toomanian

Abstract The spatial and temporal data and information are essential for decision and policy making within each governing system as well as in conservation and sustainable management programs through the execution of soil surveys. The soil survey data are used to establish national and regional level databases. A unique method of soil survey was executed to map some soil attributes in 300,000 ha of Zayandeh-rud Valley, Isfahan, Central Iran. To establish a powerful database, it is important that soil surveys address the environmental impacts. To do so, the following steps were considered: (1) fundamental factors and processes for landscape formation, (2) evolution pathways of geomorphic surfaces, and (3) mapping of pedologic properties and visualization of collected information. Execution of mentioned steps highlighted some historical facts in study area. It has observed that some geomorphic surfaces have developed before Pleistocene period; the Zayandeh-rud River had three different pathways in Quaternary period; the pedodiversity indices are directly related to soil evolution and time; and the soil evolution pathways in this valley does not follow the convergence pathway of the Jenny's theory. Results also indicate that the digitally extracted continuous maps have the ability to accurately show the spatial distribution of pedologic properties.

Keywords Spatial data • Soil-surveying steps • Pedodiversity • Landscape evolution • DSM

N. Toomanian (✉)
Soil and Water Research Division, Agriculture and Natural Resources Research Center, Amir Hamzeh Town, P.O. Box 81785-199, Isfahan, Iran
e-mail: norairtoomanian@gmail.com

S.A. Shahid et al. (eds.), *Developments in Soil Classification, Land Use Planning and Policy Implications: Innovative Thinking of Soil Inventory for Land Use Planning and Management of Land Resources*, DOI 10.1007/978-94-007-5332-7_9,

9.1 Introduction

It is critical to have proper laws for the use, management, and protection of the natural resources. Such legislation is related to the fact that the resources belong to all generations of a nation or country and protects resources against degradation and physical extinction. Soil survey execution, generation of scientifically based soil information, and development of comprehensive database are aimed to address the public interests for sustainable land uses (Capelin 2008). To establish a powerful earth-related database, it is prerequisite that soil surveys must be executed considering performance of all environmental attributes. Soil survey process distinguishes different soil populations and their extent and distribution in the study area and offers fundamental quantitative and qualitative data and information for policy and decision-making processes. Soil data can be pedologic or edaphologic aspects of needed information serving as knowledge bases for different disciplines. Execution of precise soil survey depends on the ability of surveyor to distinguish the processes by which soils are formed, developed, and distributed during the elapsed time and on the precision of their delineation.

Recognition of spatial soil diversity is a fundamental notion of soil mapping at various scales (Yaalon 1998). Owing to the effect of different soil-forming factors and their interaction and interplay of different geomorphic, hydrologic soil-forming processes, the soil cover patterns are basically complex and chaotic nature (Yaalon 1998). In other words, the morphology of the earth surface results from the interaction of the endogenic and exogenic processes.

The processes are scale-dependent, hierarchically structured complex systems which are composed of a large number of heterogeneous components interacting nonlinearly (Hay et al. 2002). The necessity to understand the importance of landscape dynamics, heterogeneity, and environmental changes is the main responsibility of geomorphologic and pedologic studies (Toomanian et al. 2006). There exist different school of contextual approaches for defining the causes and forms of soil distribution in pedology. The emphasis is now on elucidating the cause factors and processes responsible for the observed patterns. In this chapter, different cycles of activities are proposed for execution in traditional (polygon base) or digital soil mapping approaches. Each proposed cycle iteratively recorrects the process of soil population's determination and distinguishes the extent and distribution of soil patches and lastly visualize them in study area. In the following section, basic cycles and associated steps are discussed.

9.1.1 A: Cycle of Landscape Differentiation

It distinguishes the processes which define the soil populations' distribution.

9.1.1.1 Landscape Stratigraphy

The stratigraphic studies start with Butler (1959) and Beattie (1972) works which place emphasis on the soil mantel rather than individual profiles. The soil mantels are formed by sedimentologic (glacial, alluvial, or eolian) processes, or directly on parent rocks, in spatiotemporal positions of landscapes. Soil stratigraphy investigates the distribution and extent of soils having polygenetic development and evolution. Each polygenic soil has been developed in different depositional parent materials and different period of past times (Mackenzie and Austin 1993). In stratigraphic studies, the paleosols (Brewer 1964) are now central context of discussion. Paleosols in different buried, exhumed, and relict positions are formed in the past on any landscape (Beckmann 1984) without any relation with present climatic condition. A paleosoil represents a past period in which it has been developed. Paleosoils are any kind of soil layers formed on different depositional sediments in previous environmental conditions (Mackenzie and Austin 1993). These are mappable unit of unconsolidated sediments (pedologically evolved or not) either on land surface or partially or wholly buried by other layers.

The sedimentologic and pedologic differences of the layers permit the consistent recognition and mapping of these layers. Dating and defining the processes resulting these paleosols and recognizing the relationships between them provide the evidences for deducing soil history (vertical development) and soil evolution (temporal development). Due to above-mentioned causes, soil properties vary spatially depending on the complexity of the way in which soils have been formed. The main tasks of fieldwork in soil stratigraphic studies are to (1) establish that specific soil layers are independent of each other, (2) identify buried soils, and (3) locate the edges or boundaries of pedoderms (Mackenzie and Austin 1993). An understanding of the nature of these surface deposits is required for the management of environmental issues like sand encroachment, land-use planning, and soil and groundwater pollution. Pedostratigraphy is the study of stratigraphic relationships and implications of soils (including buried soils) and paleosols. Soil horizons are morphologically distinct, laterally traceable, and a time marker. The soil stratigraphic relationships are important for determining the geomorphic history of an area (Douglas et al. 2005).

9.1.1.2 Landscape Evolution

The debate on landscape evolution starts with the theory of Davis on geographical cycle. In the William Morris Davis's geographical cycle, the landscapes are seen to evolve through youth, maturity, and old stages. Landscapes are subject of evolution by the influence of different processes; in other words, they are directionally changing, developing, and evolving temporally progressively and/or regressively (Phillips 1999) by different processes; therefore, different ancient landforms (relict, exhumed, and stagnant) are able to form during changing environment of each region.

Process geomorphology studies the processes responsible for landform formation and development. The geomorphic processes have different rates of progression and infer short-term and long-term changes during landforms formation.

Phillips (1999) has offered 11 principles that have immense relevance to geomorphology and may help to reconcile the split between the process and historical aspects of the subject. To have a complete induction from landscape, the understanding of landforms must be based on knowledge of both history and process. Without an understanding of process, history is undecipherable, and without knowledge of history, process lacks a meaningful context. Process and history together lead to better appreciation of forms, behavior, and the earth surface evolution (Huggett 2005). All landscapes are affected by environmental changes. Fluvial system response to environmental change is usually complex. Phillip's principles of earth surface systems promise to help bridge the gap between process geomorphology and historical geomorphology (Huggett 2005).

Landscape evolution is governed by a balance of forces: on the one hand, vertical tectonic movements resulting from the interaction between lithospheric plates and, on the other, erosion and deposition controlled by a range of processes whose relative importance depends on local climatic conditions, vegetation, and rock type (Kaufmann 2003). Different processes during geologic time, such as tectonic movements and erosion deposition sequences, affect the nature of landscape evolution. Local climate condition controls the vegetation and the rate of erosion and deposition. Climate changes immensely affect on nature of geomorphic and hydrologic processes to form specific landforms. In the other hand, evolutionary sequences of landscape formation are responses to changes of hydrologic, geomorphologic, and pedogenic processes. Consequently, tectonic activities, hydrologic and geomorphic processes, and climatic changes govern the landscape evolution and the soil development. All of these changes are reflected and recorded in soil profiles during its developing periods. The proxy records of historical events and environmental changes are kept in soil layers. Study of the relationship between these proxy data (pedogenic and geogenic) and their formative processes helps us to reconstruct the sequence of historical events that has formed the landscape and confirms its evolutionary pathways. As a good source of information, the soil paleo-environmental proxy records are used in different scientific disciplines for reconstructing the historic sequences of concerned studies. Paleosoils can provide much more information concerning the evolution of landscapes in every region.

9.1.1.3 Landscape Stratification

During soil development phase, soil horizons are produced that are superimposed onto landforms and parent materials. Sequential and temporal erosion and deposition (eolian and fluvial) can remove, truncate, and/or bury soil horizons. Spatial and/or temporal landscape stability permits initial and subsequent soil development. Thus, most landscapes are a mosaic of various-aged landforms, parent materials, soils, eroded deposited segments, and geomorphic surfaces. Surficial soils, buried soils, and paleosols (globally polygenic), therefore, are a unique portion of the geomorphic and stratigraphic record in a region (Douglas et al. 2005). The defining and delineating of these patches of landscapes which have different origin and characterization is called stratification. This step is being done via various methods within soil-surveying approaches. Traditionally, land system and physiographic definitions were mostly used to stratify, which newly has changed to geomorphic bases (Zinck 1988). But because earth crust is formed by scaled and hierarchical processes, to stratify the land surface, there is a great need of a geomorphologic taxonomic hierarchy to define the way of their formation (the history of formation and responsible processes). Additionally, there is a great need to establish universally accepted procedures to delineate the defined geomorphic taxa. In general, a solid structure for defining and stratifying the geoforms is lacking. Obviously, some authors have tried to follow a sort of structure, but none of them were successful (Farshad 2006). The problem is that geomorphology is quite a controversial subject and that a real taxonomic classification is lacking, whereas some other disciplines such as botany and soils have succeeded to establish one (Farshad 2006). A systematic hierarchy has been proposed (Zinck 1988) in concordant with USDA Soil Taxonomy to define different geoforms in any region. This structure and its fundamental contexts are defined in Farshad (2006). A three level geomorphic taxonomy has also been proposed by USDA-NRCS which is now being used within the soil-surveying steps (Schoeneberger and Wysocki 2008; Schoeneberger et al. 2002).

The classic cartographic method of delineating the landscape patches is the air photo interpretation (API), and if this method is combined with a proper geomorphic taxonomy, it is still considered as one of the best landscape-stratifying methods. The digital terrain modeling (DTM) is a mathematical (or digital) approach to delineate the terrain surface (Li et al. 2005). Recently, soil scientists and hydrologists have tried to digitally stratify the landscape (Drãguþ and Blaschke 2006). They have used the methods described by Hengl and Reuter (2009) which are geomorphometric approaches. The required data for the digital terrain modeling is either obtained from field survey (use of conventional surveying instrument or GPS), from stereo pairs of aerial (or space) images using photogrammetric techniques, or from the digitization of the existing topographic maps (Farshad et al. 2005).

9.1.2 B: Cycle of Soilscapes Differentiation

It distinguishes the diversity of soil types and soil evolution pathways in each geomorphic unit. A geomorphic surface, smallest division of geoforms, is a geomorphic unit which is formed by a unique process within a defined span of time (Ruhe 1975). As a consequence, the soils formed by one process but in different time spans are considered as different geomorphic surfaces. It is assumed that the soils developed under one geomorphic process should have uniform distribution in defined geoform; however, the influences of minor differences in soil-forming factors (parent material, spatially different orientation and distribution of soil particles, minor spatially differences of environmental conditions) and minor spatially differences in the operation of soil-forming processes and execution of different history on minute part of soil profiles lead to the formation of somehow different soils (Phillips 2001). To conclude, to what extent the sampling scheme has defined the soil populations and how accurate soil types can be delineated, and also to find out that soils have evolved convergent or divergent evolution pathways, the pedodiversity analyses should be carried out in each geomorphic surface. In case of more diversity, complimentary samples should be taken.

9.1.2.1 Pedodiversity Analysis

Pedodiversity is a way of showing soil variation in an area or category (McBratney 1992) usually using soil taxa. Different diversity indices are presented in literature (Ibanez et al. 1995). Most of them used entropy bases to simulate the biodiversity essentials in pedology. From a methodological point of view, the different ways of measuring diversity may be grouped into three classes (Magurran 1988): (1) indices of richness (number of objects in the site, i.e., a count of the number of biological species or soil types known to occur in a defined sampling unit), (2) object abundance models (these models describe the distribution of objects abundance), and (3) indices based on proportional abundance of objects (in this case, diversity is defined in terms of a function of the number of different objects and their relative abundance or cover) (Ibanez et al. 1995). The definition of pedodiversity, indices, and its usage in pedology are defined by various workers (Ibanez et al. 1995, 1998; Martin and Rey 2000; Guo et al. 2003; Phillips and Marion 2005; Toomanian et al. 2006).

The application of diversity analysis to geomorphology and soil has recently attracted attention (Ibanez et al. 1998, 2005; Guo et al. 2003; Saldana and Ibanez 2004; Phillips and Marion 2005). Soil Taxonomy is hierarchically based on breaking the soil continuum into discrete ranges of soil types and bodies (soilscapes) which, in the lowest level, could be considered as individual soil species that are produced through natural development processes (Ibanez and De Alba 1999; Guo et al. 2003). Relationships between species richness and area have long been used in biogeography and biodiversity studies. This approach has also been adapted to soils (Ibanez et al. 1995, 1998; Phillips 2001, 2005; Guo et al. 2003). Using this diversity approach, soil-type complexity in a category or defined area or polygon can be shown. Using the diversity indices, it is possible to show the increasing or decreasing rate of entropy in soil pedogenesis among different levels of geomorphic hierarchical levels or area (Phillips 2001; Toomanian et al. 2006).

9.1.3 C: Cycle of Predicting Soil Patterns in Study Area

Soil surveys are conducted to distinguish and map the distribution of soil types using various sampling scheme suitable to the study area and objective of survey. The mapping is accomplished by predicting soil types in the area of interest through traditional or pedometrics methods. Traditional soil-surveying paradigm was based on interpretative ability of experts to mentally relate the soil–landscape relationship concept with the initial soil-forming processes and environmental factors to extract soil distribution in a completely subjective manner; however, this method is now subject to some criticisms (Zhu et al. 2001).

During the last decades, quantitative methods of describing, classifying, and studying the spatial distribution pattern of soils in a more objective manner have been developed to address the criticism using pedometrics. This is the application of mathematical and statistical methods for the quantitative modeling of soils, with the purpose of analyzing its distribution, properties, and behavior. The definition covers quantitative mathematical and statistical measurements and predictions of soil-related modules and roughly pedology. "In this sense, pedometrics deals with uncertainty in soil-related problems due to deterministic or stochastic variation, vagueness and lack of knowledge of soil properties and processes."

9.2 Materials and Methods

The study area is located in the central basin of Iranian plateau (Fig. 9.1). This area includes 0.3×10^6 ha of Zayandeh-rud Valley. The geologic infrastructure of the area is mainly Cretaceous limestone, Mesozoic shale, and sandstone. Erosion and deposition processes, especially in the late Tertiary and early Quaternary, have been the main geo-formation processes in the area. After uplifting of Zagros Mountains (Alpine Orogeny), the Zayandeh-rud River downcut its bed, forming the terraces along its path to the Gavkhuni marsh in the eastern part of the study area. The salinity of soils simultaneous with aridity increases eastwardly while altitude decreases in that direction. Ascending water table, extreme drought, increase in salinity, and some human activities resulted in high rate of wind erosion. The eolian deposition is now covered all eastern part of the study area. The methods considered in various steps of this study are used in the following sequence.

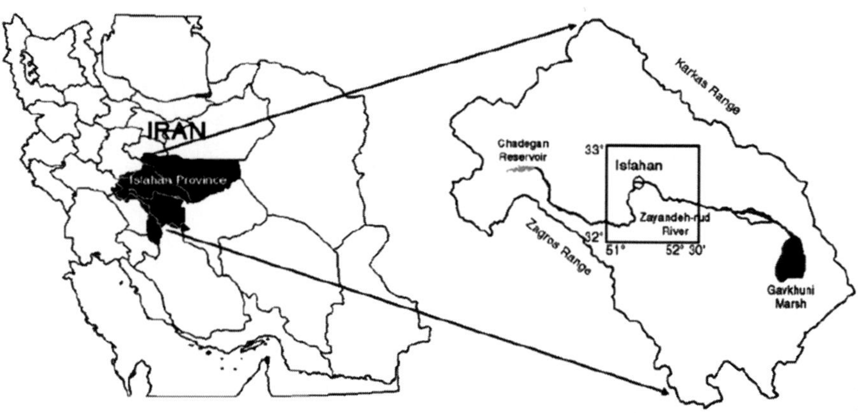

Fig. 9.1 The study area in central Iran

9.2.1 Landscape Stratification

Having geomorphic knowledge in mind and considering the topography, geology, land use, and vegetation cover, the study area is stratified by API method. In this context, the geomorphic taxonomy is one of the most needed subjects. The API differentiated geomorphic patterns (nested geomorphic hierarchy) based on their formation processes, general structure, and morphometry. The terminology of interpreting legend was based on Zinck (1988). A four-level geomorphic hierarchy is used to break down the complexity of different landscapes in the study area. In the lower level of the hierarchy, the geomorphic surfaces, which were formed by a unique process during a specific geologic time, were defined. The principle of this technique is to search for edges, or discontinuities, by breaking down the neighborhood of adjacent areas into subgroups that are internally as homogeneous as possible. In API, the hierarchy of field geographical organization was delineated on air photos (1:55,000). The cartographic scale of these geomorphic surfaces was of a reconnaissance soil survey (1:100,000–1:250,000). Stereoscopically interpreted air photos of the study area were imported into a geographic information system (GIS) environment, and after ortho-photo geo-referencing, geomorphic surfaces were mapped and glued via on-screen digitization.

9.2.2 Field Check and Soil Sampling

The delineated geomorphic surface map overlaid on a registered color composite image was used in the field to check the boundaries and to allocate sampling points within delineations.

9.2.2.1 Sampling

A purposive soil sampling method based on the extent of geomorphic surfaces and direction of changing gradients like slopes was used for proportional sampling throughout the study area excluding mountains and rocky hills. A total of 191 soil profiles were excavated and classified using Keys to Soil Taxonomy (Soil Survey Staff 2003), and their genetic horizons were sampled. The samples were analyzed for important physical and chemical characteristics using standard USDA procedures (Soil Survey Staff 1996). Already existing data from 66 profiles in the study area was also used for digital mapping of soil properties.

9.2.3 Landscape Stratigraphy and Evolution

The geologic (tectonic, fault, etc.) and paleoclimatic history of the area is considered to differentiate the soil layers. The stratigraphic layers are distinguished by landform unconformities in margins of geoforms formed by unique geomorphic processes. Landscape evolution is also defined by relating the soil layers to find the precedence or recency of layers which were formed by different deposition processes. This enables us to find whether a geoform has passed different historic evolutionary sequences compared with other points within each geomorphic surface. This was critical when eolian deposits were covering the surface of the most part of Segzi playa.

9.2.4 *Pedodiversity Analysis*

In this study, taxonomic diversity at the family level and genetic diversity at horizon level is investigated; both are important soil individual entities for diversity analysis. Pedodiversity indices including Shannon K-entropy, richness, and evenness for each geomorphic category were calculated in each category (Guo et al. 2003). To calculate the diversity indices in each landscape, the number of profiles belonging to a given landscape and the total number of profiles in the study area were considered. The number of different objects or entities including soil families and horizon sequences in a certain ecosystem or predefined territory and geomorphic categories was considered as richness of species. The diversity indices are measured by relative abundance of soil families to total sampled points in geomorphic units (Ibanez et al. 1995; Phillips 2001). The proportional abundance of objects is the most frequently used method to estimate the diversity. Evenness is another index which refers to the relative abundance of each object in a defined area. Logically, when the evenness of objects is equally probable, the diversity is highest when the richness of comparing units is the same (Ibanez et al. 1995).

9.2.5 *C: Cycle of Predicting Soil Patterns in Study Area*

Execution of survey, i.e., the interpolation of soil classes and attributes in unsampled points of study area, could be executed by using methods described in *Soil Survey Manual* (Soil Survey Division Staff 1993) or through mapping the distribution of soil types or attributes using quantitative pedometric approaches. In this study, digital soil mapping of some soil attributes was accomplished.

9.3 Results and Discussion

In soil-surveying methods, regardless of the approach used, three fundamental steps must be accurately undertaken. The first is the landscape evolution which should be defined to show how the geoforms are developed and what evolutionary history they had. Second, the complexity of soil types in each geoform and the cause of diversity

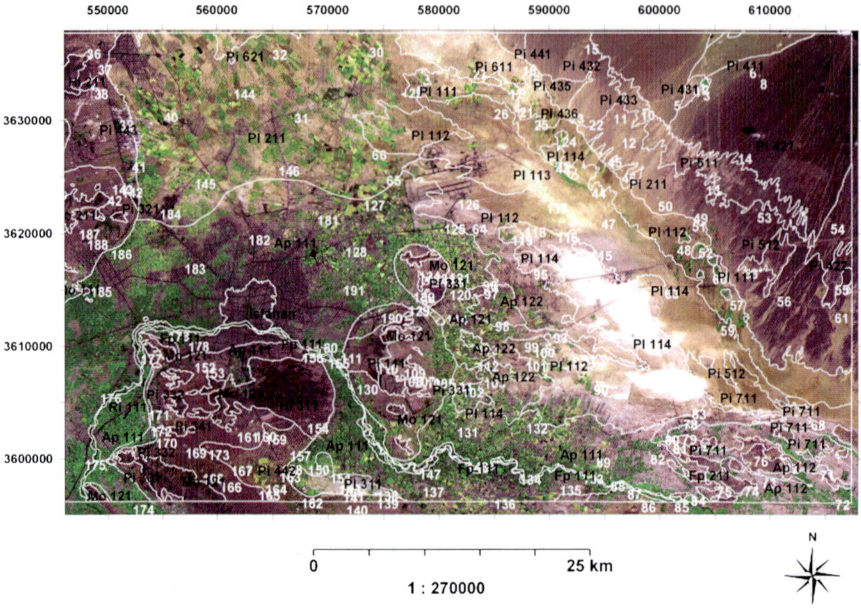

Fig. 9.2 Delineated geomorphic units in study area with its hierarchic code (*black*) and studied points (*white*)

should be distinguished. Third, iterate the stratification process and sampling design with the incorporation of landscape evolution results and pedodiversity analysis. This enables the surveyor to be sure that all soil-type populations have been accounted in soil mapping process.

9.3.1 Air Photo Interpretation

Through photo interpretation, the knowledge of soil–landscape relationships, geology, topography, and geomorphologic factors affecting the soil formation of the study area was considered. In the first attempt, the API has distinguished seven kinds of landscapes and 46 geomorphic surfaces in the study area. The delineated surfaces resulted from photo interpretation, and the legend of interpretations is shown in Fig. 9.2 and Table 9.1, respectively. The drawn delineations for all geomorphic surfaces in the field were checked, and corrections were inserted in the GIS map of the landforms. Upon the processes ruling the formation and development of all soil types, the sampling scheme was designed and the soil profiles described. The soil samples were analyzed and amount of all the soil attributes measured.

Table 9.1 Legend of delineated geomorphic surfaces before inserting the landscape evolution results

No	Landscape	Landform	Lithology[a]	Geomorphic surfaces	Code
1	Mountain	Dissected ridge	Marly limestone (K4, K2) and shale	Structured surface	Mo 111
2	Mountain		Marly limestone (K4, K2)	Structured surface	Mo 121
3	Mountain	Rock pediment	Eroded calcareous and dark shale	Scarp slope	Mo 211
4	Hill land	Dissected ridge	Basal conglomerate (OMC)	Slope facet complex	Hi 111
5		Eroded ridge	Dissected dark-gray shale	Structured surface with braded stream	Hi 211
6		Rocky high hill	Remnant of shale (J)	Slope facet complex	Hi 311
7	Piedmont	Pediment	Remnant of shale (J)	Slope facet complex	Pi 111
8		Flash flood fan delta	Quaternary alluvium	Outwash sediment finer and white	Pi 211
9				Outwash sediment (coarser)	Pi 212
10		Alluvial fan	Alluvium of OM, OMC	Apical part	Pi 311
11			Alluvium of dark-gray shale	Apical part, slope facet complex	Pi 321
12			Alluvium of marly limestone	Slope facet complex	Pi 331
13				Slope facet complex, cultivated	Pi 332
14			Alluvium of marly limestone	Active fan	Pi 341
15		Bahada	Alluvium of OM, OMC, Ev	Middle part	Pi 411
16			Alluvium of andesite, granodiorite	Apical part	Pi 421
17				Apical part (extremely braded drainage)	Pi 422
18			Alluvium of foraminifera limestone	Apical part, with dense drainage network	Pi 431
19				Middle part	Pi 432
20				Distal part, with dense drainage network	Pi 433
21				Distal part with dense drainage network, finer	Pi 434
22				Distal part calcareous	Pi 435
23				Distal part, salt crusted, gypsiferous	Pi 436
24			Alluvium of K4, K2, Tn	Middle part with parallel drainage pattern	Pi 441
25				Middle part with less drainage	Pi 442
26				Distal part with dense drainage network	Pi 443
27		Dissected old bahada Alluvium of foraminifera limestone	Paleoterrace, undulating plateau	Pi 511	

(continued)

Table 9.1 (continued)

No	Landscape	Landform	Lithology[a]	Geomorphic surfaces	Code
28				Paleoterrace, with braded intense network	Pi 512
29		Old bahada	Fine marly gypsiferous sediments	Piedmontal terrace, flat, salty fine alluviums	Pi 611
30			Fine marly alluvium	Piedmontal terrace, distal part, fine alluviums	Pi 621
31		Rolling old bahada	Coarse, gypsiferous alluvium	Paleoterrace, gypsic plateau	Pi 711
32	Alluvial plain	Alluvial flat, river terraces	Zayandeh-rud river alluviums	Cultivated terraces	Ap 111
33				Playa/river terraces, cultivated, salty	Ap 112
34			Old river sediments	Meandering complex facet	Ap 121
35				Cultivated old river terrace	Ap 122
36	Flood plain	Lowest river terrace	Recent alluviums	Channel margin alluvium, cultivated	Fp 111
37				Channel margin alluvium, cultivated, salty	Fp 112
38	Seasonal drain	Recent alluviums systems	Salty gleyed fine alluviums	Fp 211	
39	River	River sediments	Recent gravelly alluviums	Channel sediments	Ri 111
40	Playa	Segzi basin	Alluvio-lagoonary fine sediments	Wet zone, flat, salty, cultivated	Pl 111
41				Wet zone, flat, very salty	Pl 112
42				Soft clay flat, with drained groundwater	Pl 113
43				Soft clay flat, gypsiferous, extremely salty	Pl 114
44		Borkhar basin	Alluvial fine sediments, slightly salty	Soft clay flat, cultivated	Pl 211
45		Margh basin	Alluvio-lagoonary fine sediments	Puffy ground, lagoonary, gypsiferous	Pl 311
46		Jarghuye basin	Alluvial fine sediments, salty	Soft clay flat, cultivated	Pl 411

[a]K2 – Red conglomerate and sandstone with a yellow sandy dolomite bed at the top and intercalations of dolomite beds locally
K4 – Gray limestone containing orbitolinas and ammonites
Tn – Dark-gray shale with intercalations of lenticular limestone (containing corals and Heterstridium) and sandstone (Naiband formation)
OMC – Basal conglomerate
J – Shale containing ammonites, with intercalations of conglomerates, sandstone, radiolarite sandstone, and volcanics
OM – Foraminiferal limestone (Qom formation)
Ev – Tuff breccia and andesitic volcanics

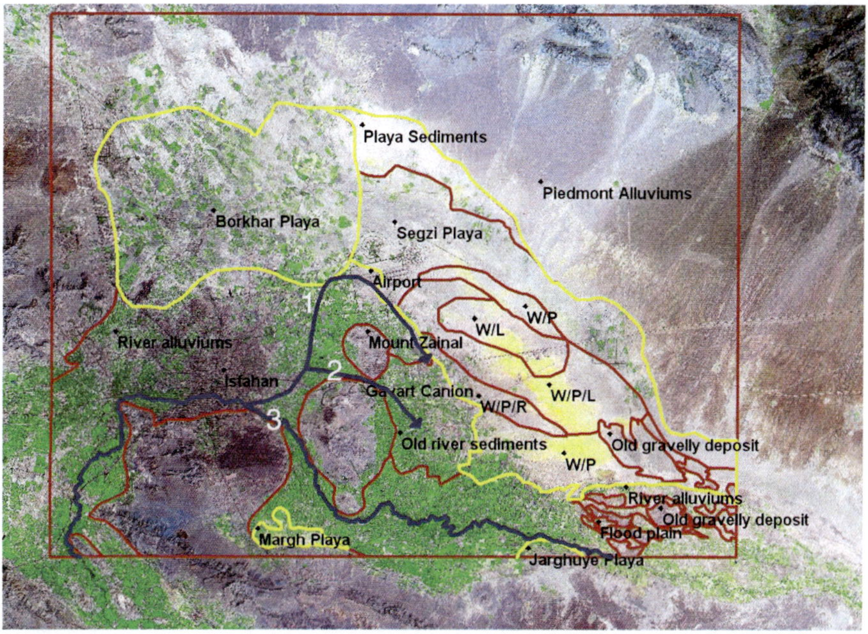

Fig. 9.3 Different zones of parent materials delineated upon landscape evolution. The sequential river pathway changes are shown with *1*, *2*, *3*

9.3.2 Landscape Evolution

Soil profile descriptions proved that the delineations are more complex in Segzi playa than those determined by API method (underneath differences were not distinguished by API). In this playa, different hydrologic and geomorphic processes have created some heterogeneous parent materials. These heterogeneous parent materials were formed by intermittent sedimentation of eolian, lagoonal, and river alluvial layers which have been laid during the Tertiary and Quaternary. These geomorphic variations were not detected by API method because the surface of these map units has been covered by uniform eolian materials. The distribution map of these parent materials was resulted from upscaling of the profile data (Fig. 9.3). Overlaying the parent materials mapped on air photo interpreted delineations inferred more details in some geomorphic surfaces (API map and legend), which are presented in Table 9.2.

The evidences recorded in geologic and geomorphic units and soil profiles of the study area have shown different critical evolutionary steps in Zayandeh-rud Valley formation which are sequentially shown in Fig. 9.4. The inherited proxy records of past environmental changes are used to reconstruct the past evolutionary history during late Tertiary and Quaternary. These evidences and proxies are used to find the sequential steps of landscape evolution in study area.

Table 9.2 Legend of delineated geomorphic surfaces which was changed after inserting the landscape evolution results

Landscape	Landform	Lithology	Geomorphic surfaces	Code
Piedmont	Bahada	Alluvium of foraminiferal limestone	Distal part, with dense drainage network (Pi 433)	Pi 433
			Fine	Pi 4331
			Coarse	Pi 4332
Alluvial plain	River terraces	Old river sediments	Meandering complex facet (Ap 121)	Ap 121
			Fine/coarse	Ap 1211
			Fine	Ap 1212
Playa	Segzi basin	Alluvio-lagoonary fine sediments	Wet zone, flat, very salty (Pl 112)	Pl 112
			Playa	Pl 1121
			Windy/playa	Pl 1122
			Windy/playa/old river	Pl 1123
			Windy/playa/lagoon	Pl 1124
			Soft clay flat, with drained groundwater (Pl 113)	Pl 113
			Playa	Pl 1131
			Playa/lagoon	Pl 1132
			Windy/playa	Pl 1133
			Windy/playa/old river	Pl 1134
			Windy/playa/lagoon	Pl 1135
			Soft clay flat, gypsiferous, extremely salty (Pl 114)	Pl 114
			Playa	Pl 1141
			Playa/lagoon	Pl 1142
			Windy/playa	Pl 1143
			Windy/playa/lagoon	Pl 1144
	Margh basin	Alluvio-lagoonary fine sediments	Puffy ground, lagoonary (Pl 311)	Pl 311
			Windy/playa/lagoon	Pl 3111
			Playa/lagoon	Pl 3112

9.3.3 Pedodiversity Analysis

The diversity indices for each geomorphic surface were calculated using the total number of profiles studied in the area of unit and the number of profiles belonging to each soil family within that unit. Table 9.3 shows the pedodiversity indices in some geomorphic surfaces which have more diversity. This seems to be due to simultaneous increasing of the richness and evenness through this hierarchical downscaling method. Although using the result of landscape evolution in study area decreases the diversity indices (Table 9.4) and subdivided some geomorphic units, but there remain some high diversity indices in these geomorphic

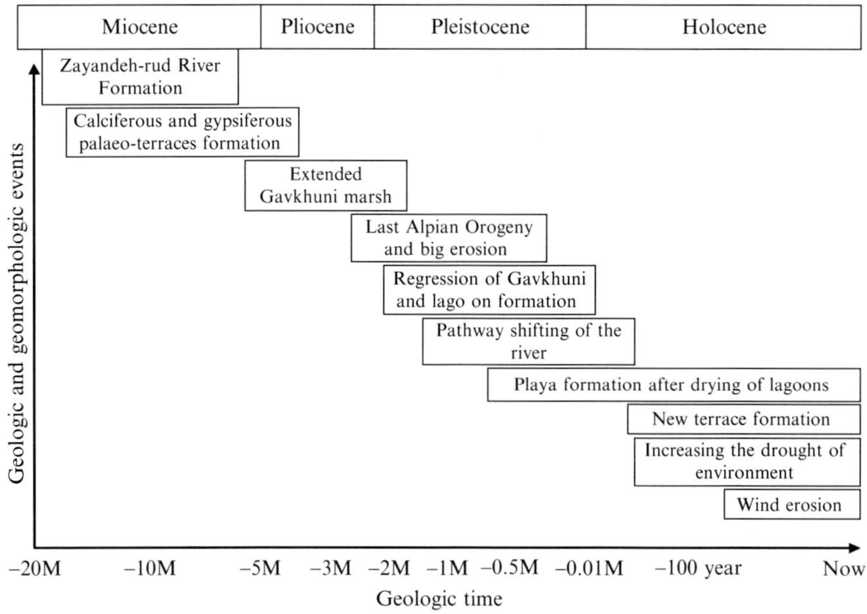

Fig. 9.4 Sequences of landscapes evolved during time

units. This increasing rate of diversity through geomorphic and taxonomic hierarchies confirms the existence of divergent soil evolutionary pathway in this area.

In Ap 111 and Ap 121 units (Table 9.3), the general trend of increasing the diversity indices from order to subgroup is not the same as the others. In most part of the Ap 111 mapping unit, where intensive cultivation has been carried out, the argilification (clay illuviation) was the only soil-forming process leading to local convergence soil development. The only factor responsible for differentiating the undeveloped soils in Ap 121 mapping unit is irregular sedimentation of fine and coarse materials by meandering older Zayandeh-rud channel. Ibanez et al. (1990, 1994) have used this method to show how the evolution of fluvial systems induces an increase in pedo-geomorphological landscape complexity. Ibanez et al. (1995) studied the diversity of soil mapping units of Spain and reached to the same conclusion. Saldana and Ibanez (2004) stated that an increase in soil heterogeneity through a geomorphic hierarchy verifies the existence of divergent soil evolution. However, in the units in which the amounts of diversity indices were high, the complementary sampling was conducted to reduce the probable inaccuracies.

The relationship between K-entropy and time or soil evolution was studied in Zayandeh-rud River terraces. This river has three different terraces along its path to Gavkhuni marsh. The entropy measured in these terraces (lower, middle, and upper) significantly increases from lower to upper terraces (Fig. 9.5) indicating the positive relationship between the K-entropy and time or soil evolution in the absence of any variation in climate, parent material, and topography. In the lower terrace, Torriorthent was the only developed soil great group. In the middle terrace,

Table 9.3 Pedodiversity indices measured for diverse units before inserting the landscape evolution results

Code	Total	Family classification (Soil Taxonomy 2003)	%	*S*	*H′*	H_{max}	*E*
Pi 321	6	Loamy-skeletal, mixed, thermic, typic haplocalcids (143-142-186)	50				
		Sandy-skeletal, mixed, thermic, typic torriorthents (187)	16.66	4	1.24	1.386	0.895
		Loamy-skeletal, mixed, thermic, typic torriorthents (42)	16.66				
		Loamy-skeletal, mixed, thermic, typic haplogypsids (188)	16.66				
Pi 331	23	Fine-loamy, mixed, thermic, typic calciargids (166-165)	8.7				
		Loamy-skeletal, mixed, thermic, calcic argigypsids (169-159-153-122)	17.5				
		Fine-loamy, mixed, thermic, typic torriorthents (156)	4.3				
		Loamy-skeletal, mixed, thermic, typic haplogypsids (104, 168)	8.7				
		Loamy-skeletal, mixed, thermic, typic haplocalcids (106-107-108-109)	17.5				
		Fine-loamy, mixed, thermic, typic calcigypsids (189-130)	8.7	11	2.27	2.4	0.95
		Loamy-skeletal, mixed, thermic, typic calciargids (160-152- 105)	13				
		Loamy-skeletal, mixed, thermic, typic torriorthents (161)	4.3				
		Fine-silty, mixed, thermic, typic torriorthents (121)	4.3				
		Fine-silty, mixed, thermic, typic calcigypsids (190)	4.3				
		Loamy-skeletal, mixed, thermic, typic calcigypsids (123-124)	8.7				
Pi 442	8	Coarse-loamy, gypsic, thermic, typic haplogypsids (20)	12.5				
		Loamy-skeletal, mixed, thermic, typic torriorthents (167-173)	25				
		Loamy-skeletal, mixed, thermic, typic haplogypsids (157)	12.5				
		Loamy-skeletal, mixed, thermic, typic calcigypsids (158-163)	25	6	1.73	1.79	0.97
		Fine-loamy, mixed, thermic, typic calciargids (164)	12.5				
		Loamy-skeletal, mixed, thermic, typic calciargids (162)	12.5				

(continued)

Table 9.3 (continued)

Code	Total	Family classification (Soil Taxonomy 2003)	%	*S*	*H'*	H_{max}	*E*
Pi 512	8	Coarse-loamy, gypsic, thermic, typic haplogypsids (11, 61)	25				
		Fine-loamy, gypsic, thermic, leptic haplogypsids (12)	12.5				
		Loamy-skeletal, gypsic, thermic, typic haplogypsids (50, 53)	25	6	1.73	1.79	0.97
		Fine-silty, mixed, thermic, leptic haplogypsids (13)	12.5				
		Coarse-loamy, mixed, thermic, typic haplosalids (49)	12.5				
		Fine-loamy, mixed, thermic, typic haplogypsids (56)	12.5				
Ap 111	27	Fine, mixed, thermic, typic haploargids (128-129-131-133 88-175-125-150-176-182-136-137-174-180-181-183-185-191)	66.7				
		Fine-silty, mixed, thermic, typic haplocambids (82-84-87-132)	14.8				
		Loamy-skeletal, mixed, thermic, typic haplocalcids (171)	3.7	6	1.11	1.79	0.62
		Fine, mixed, thermic, typic haplocambids (89-135)	7.4				
		Fine, mixed, thermic, typic torriorthents (111)	3.7				
		Fine-loamy, mixed, thermic, typic torriorthents (178)	3.7				
Ap 121	7	Coarse-loamy, mixed, thermic, typic torriorthents (112)	14.286				
		Fine-silty over sandy, mixed, thermic, typic torriorthents (103)	14.286				
		Fine-silty, mixed, thermic, typic torriorthents (99)	14.286				
		Loamy-skeletal, mixed, thermic, typic torriorthents (101)	14.286	7	1.946	1.946	1
		Fine, mixed, thermic, typic torriorthents (114)	14.286				
		Coarse-silty, mixed, thermic, typic torriorthents (120)	14.286				
		Fine-silty over sandy, mixed, thermic, typic haplocambids (97)	14.286				
Fp 111	6	Coarse-silty, mixed, thermic, typic torriorthents (134- 147- 148-179)	66.67	2	0.636	0.69	0.92
		Loamy-skeletal, mixed, thermic, typic torriorthents (177-155)	33.33				

(continued)

Table 9.3 (continued)

Code	Total	Family classification (Soil Taxonomy 2003)	%	S	H'	H_{max}	E
Pl 112	13	Fine, mixed, thermic, gypsic haplosalids (43, 66, 69, 48)	30.77				
		Fine, mixed, thermic, typic haplosalids (60-51)	15.4				
		Fine-silty, mixed, thermic, gypsic haplosalids (25, 92, 93)	23	6	1.67	1.79	0.93
		Fine, mixed, thermic, calcic haplosalids (64-116)	15.4				
		Loamy-skeletal, mixed, thermic, gypsic haplosalids (91)	7.75				
		Fine, mixed, thermic, gypsic haplosalids (126)	7.75				
Pl 113	11	Coarse-silty, gypsic, thermic, gypsic haplosalids (26)	9.1				
		Fine, mixed, thermic, gypsic haplosalids (70, 78, 59, 80, 118, 47, 68, 117)	72.7	4	0.885	1.39	0.64
		Fine-silty, mixed, thermic, typic haplosalids (95)	9.1				
		Fine-silty over sandy, mixed, thermic, gypsic haplosalids (83)	9.1				
Pl 114	7	Fine, mixed, thermic, gypsic haplosalids (44, 62, 63, 94, 119)	71.43				
		Fine, mixed, thermic, calcic haplosalids (115)	14.285	3	0.796	1.1	0.72
		Fine-silty, mixed, thermic, gypsic haplosalids (90)	14.285				
Pl 211	12	Fine-silty, mixed, thermic, typic haplocambids (30)	8.3				
		Fine, mixed, thermic, typic torriorthents (40-127)	16.74				
		Fine-silty, mixed, thermic, typic haploargids (31)	8.3				
		Fine-silty, mixed, thermic, typic calciargids (184)	8.3	8	1.9	2.08	0.91
		Fine, mixed, thermic, typic haploargids (146)	8.3				
		Fine-silty, mixed, thermic, gypsic haplosalids (29)	8.3				
		Fine, mixed, thermic, typic haplosalids (65)	8.3				
		Fine, mixed, thermic, typic calciargids (145-144-34-32)	33.46				

Table 9.4 Pedodiversity indices[a] measured for diverse units after inserting the landscape evolution results

Unit	*N*	Family classification (Soil Taxonomy 2003)	%	*S*	*H′*	H_{max}	*E*
Ap 1211	6	Coarse-loamy, mixed, thermic, typic torriorthents (112)	16.666				
		Fine-silty over sandy, mixed, thermic, typic torriorthents (103)	16.666				
		Fine-silty, mixed, thermic, typic torriorthents (99)	16.666				
		Loamy-skeletal, mixed, thermic, typic torriorthents (101)	16.666	6	1.79	1.79	1.00
		Fine, mixed, thermic, typic torriorthents (114)	16.666				
		Coarse-silty, mixed, thermic, typic torriorthents (120)	16.666				
Pl 1121	7	Fine, mixed, thermic, gypsic haplosalids (43, 66, 69, 48)	57.000				
		Fine, mixed, thermic, typic haplosalids (60-51)	28.600	3	0.96	1.10	0.87
		Fine-silty, mixed, thermic, gypsic haplosalids (25)	14.400				
Pl 1132	3	Fine-silty over sandy, mixed, thermic, gypsic haplosalids (83)	33.333	2	0.64	0.69	0.93
		Fine, mixed, thermic, gypsic haplosalids (78, 59)	66.666				
Pl 1142	3	Fine, mixed, thermic, calcic haplosalids (115)	33.333	2	0.64	0.69	0.93
		Fine, mixed, thermic, gypsic haplosalids (62, 63)	66.666				

[a] *S* (richness) – number of soil types in the reference area; *H′* – negative entropy or diversity Index of the population; *E* (evenness) – relative abundance of each soil type among the others; H_{max} – richness when all objects in reference area are equiprobable; *N* (total number of soil types)

Haplocambids have formed, whereas in the upper terrace, Haplosalids and Haplargids have been developed. These findings are in agreement with those of Saldana and Ibanez (2004), and Phillips (2001) on river terraces. The increase in K-entropy from younger to older soil cover in such condition is expected in a chaotic system (Phillips 2001). In this case, dynamic instabilities and chaos in pedogenesis result in the magnification of initial differences and effect of perturbations to produce an increasingly diverse soil cover (Phillips 1999, 2001).

Another testable hypothesis is examining the regional geomorphic evolution. It is accomplished by plotting the calculated K-entropies within geomorphic surfaces versus the ranked age of these units. It has been found that the soils are more developed on older geomorphic surfaces compared with those on younger ones. The relationship between K-entropy and richness of soil types versus the relative age of geomorphic surfaces are presented in Fig. 9.6. This, in turn,

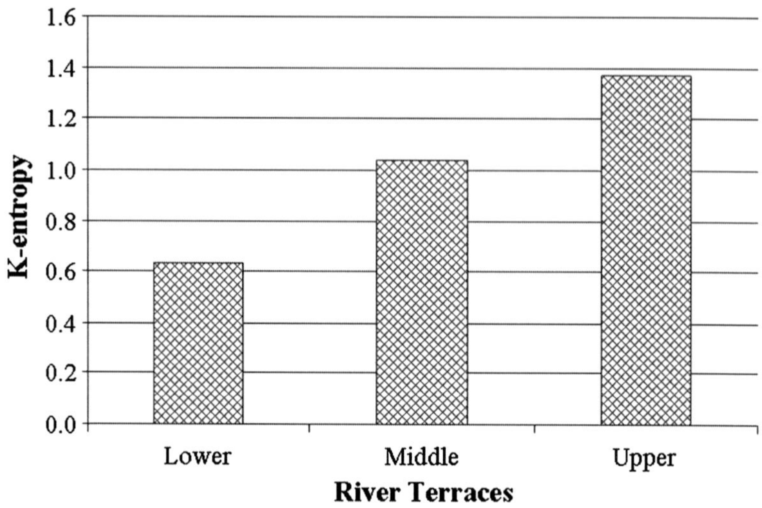

Fig. 9.5 Comparison of the diversity in river terraces

indicates that, in the study area, the soil pedogenesis follows a chaotic and divergent pathway, similar to what has been reported earlier (Ibanez et al. 1990; Phillips 2001).

9.3.4 *Predicting Soil Patterns*

In this study, attempts have been made to map the thickness of A and B horizons (cm) and clay amount in subsurface layer. The target variables are produced and assessed. Figure 9.7a presents the thickness and its estimating error maps of A horizon. The estimated standard error map shows the quality of calculated thicknesses of "A" horizon in the study area (Fig. 9.7b). The standard error in sparsely sampled area is 7.5–15 cm and in intensely sampled areas is 6–6.7 cm. It means that the most of the studied area has been predicted with error around 7 cm thicknesses for "A" horizon. The unsampled area (mountains and some rocky hills) are masked (white areas in the map) (Fig. 9.8) shows the relation of predicted depths of "A" horizon by undertaken digital soil mapping method with landforms stratified by manual API approach. As it shows, the model could differentiate the landforms and predict the proper depth for "A" horizon in study area. The predicted and standard error maps of other variables are not shown. The estimated error for the thickness of B horizons is mostly between 42 and 55 cm, which are considerable. The estimation error of some points in this map is 160 cm; this is due to low sampling intensity, weak relationships of this variable with predictors, and weak relationships of this horizon formation with current environmental condition (the most of B

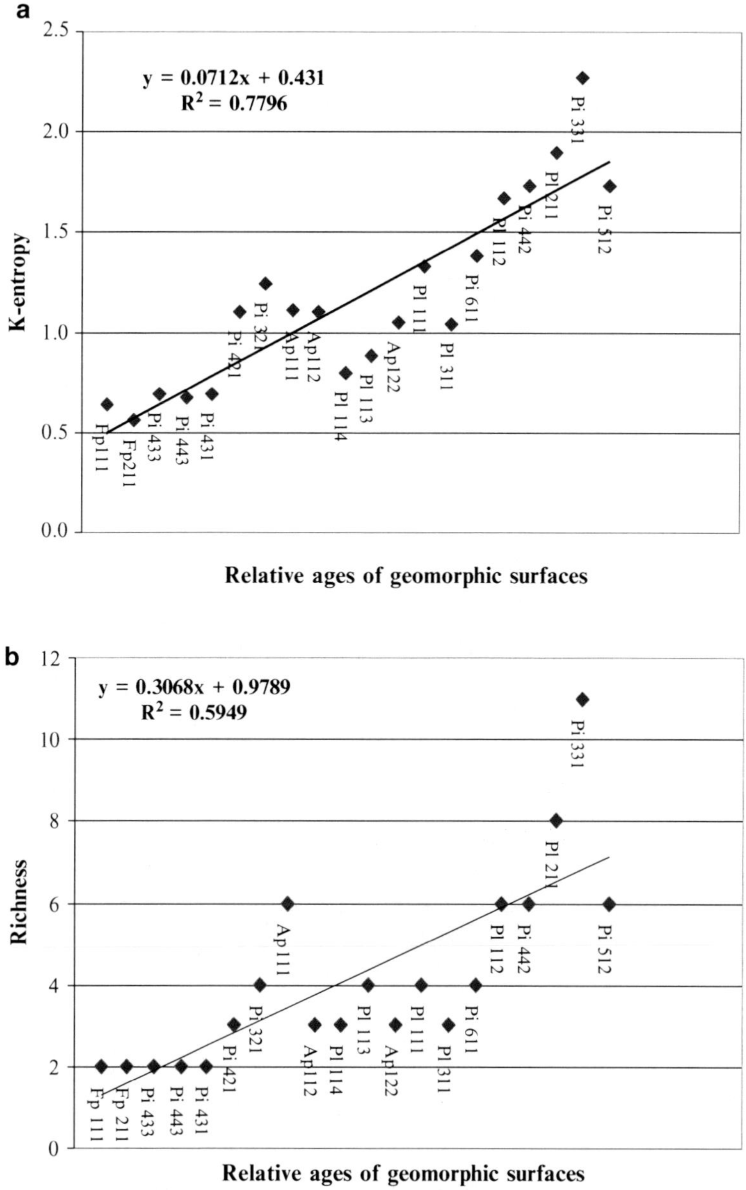

Fig. 9.6 Relation between the diversity (**a**) and richness (**b**) of geomorphic units with age

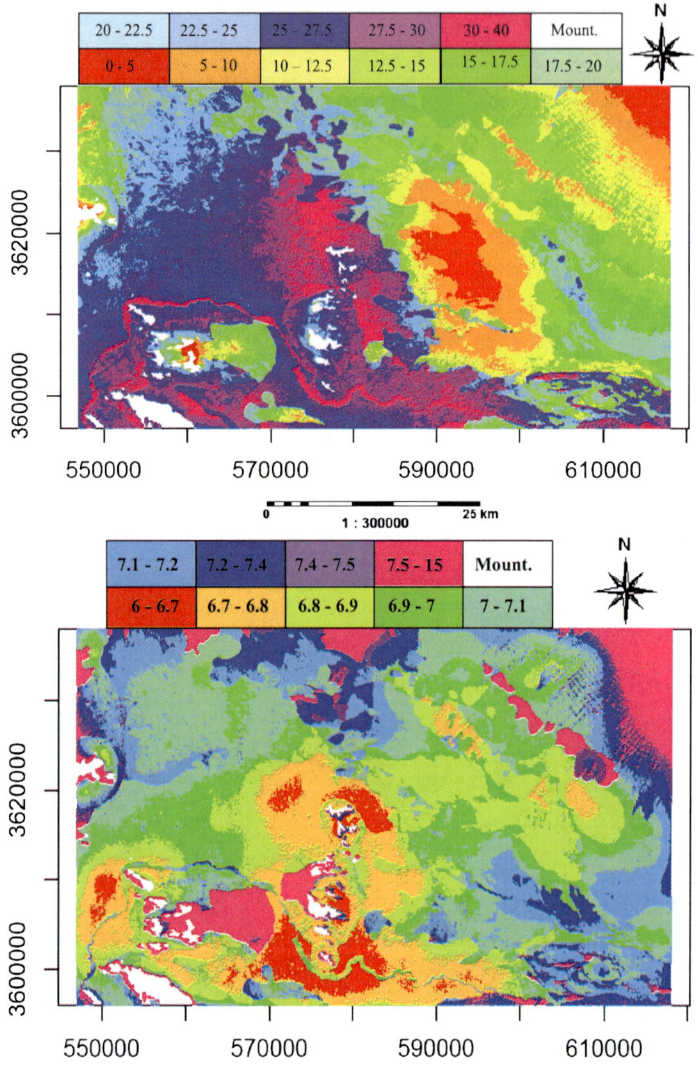

Fig. 9.7 Extracted map of "A" horizon depth (**a**) and the error of prediction in each pixel of study area (**b**)

horizons are developed under different paleoclimatic conditions). Interpolating the clay amount (%) in second layer was carried out on transformed asin $y^{1/2}$ (arc sin of square root of target variable y) data. Therefore, due to nonlinearity of back transformation of kriging variance, it was hard to calculate the standard error image in this case, but instead the lower and upper confidence interval boundaries of predicted variance was calculated for a 0.975 probability, $\hat{Y}_{UK} \pm 1.96\,\mathrm{sqrt}(\mathrm{Var}_{UK})$, to present the boundary interval maps.

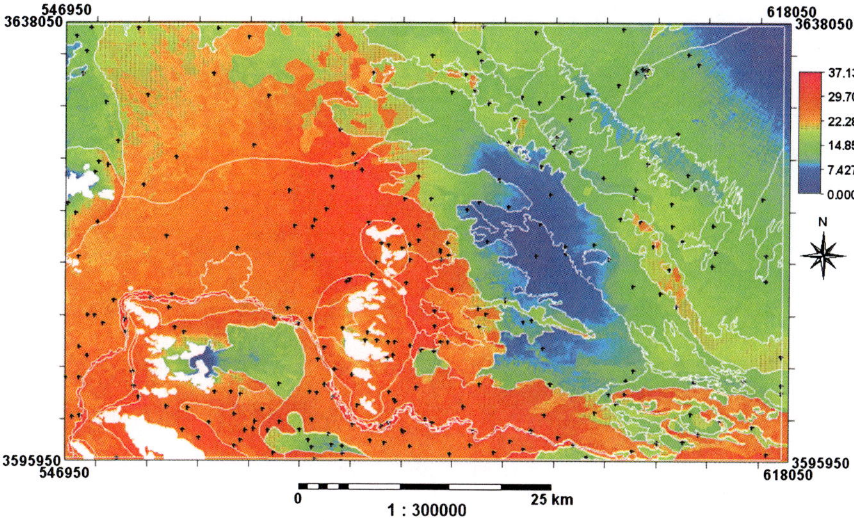

Fig. 9.8 Predicted depths (cm) of "A" horizon compared with delineation of geomorphic surfaces

9.4 Conclusions

It is concluded that incorporating the upper mentioned steps in any soil survey sequences upgrades the quality of survey and increases the accuracy and precision of extracted maps. It also enables to highlight the localities which need more sampling points to account the distribution of minute soil types.

References

Beattie JA (1972) Groundsurfaces of the Wagga Wagga region, NSW. Soil publication no 28. CSIRO Australia, Canberra

Beckmann GG (1984) Paleosols, pedoderms and problems in presenting pedological data. Aust Geogr 16:15–21

Brewer R (1964) Fabric and mineral analysis of soils. Wiley, New York, p 470

Butler BE (1959) Periodic phenomena in landscape as a basis for soil studies. CSIRO Aust Soil Publ No 14. CSIRO Australia, Canberra

Capelin M (2008) Legal and planning framework. In: McKenzie NJ, Grundy MJ, Webster R, Ringrose-Voase AJ (eds) Guidelines for surveying soil and land resources, 2nd edn. CSIRO, Canberra

Douglas AW, Schoenebergera PJ, LaGarry HE (2005) Soil surveys: a window to the subsurface. Geoderma 126:167–180

Drăguþ L, Blaschke T (2006) Automated classification of landform elements using object-based image analysis. Geomorphology 77:1–15

Farshad A (2006) A syllabus on soil geomorphology (geopedology), in the framework of landscape ecology. ITC, Enschede

Farshad A, Udomsri S, Yadav RD, Shrestha DP, Sukchan S (2005) Understanding geopedologic setting is a clue for improving the management of salt-affected soils in Non Suang district, Nakhon Ratchasima, Thailand. Wrap up seminar of the LDD-ITC funded research project, Hua Hin, Thailand

Guo Y, Gong P, Amundson R (2003) Pedodiversity in the United States of America. Geoderma 117:99–115

Hay GJ, Marceau DJ, Bouchard A (2002) Modeling multi-scale landscape structure within a hierarchical scale-space framework. In: Symposium on geospatial theory, processing and applications, Ottawa, Canada

Hengl T, Reuter HI (2009) Geomorphometry concepts, software, applications, vol 33, Developments in soil science. Elsevier, Amsterdam/Oxford

Huggett RJ (2005) Fundamental of geomorphology. Routledge fundamentals of physical geography. Taylor & Francis Publication, p 401. New York

Ibanez JJ, De Alba S (1999) On the concept of pedodiversity and its measurement. A reply. Geoderma 93:339–344

Ibanez JJ, Jimenez-Ballesta R, Garcia-Alvarez A (1990) Soil landscapes and drainage basins in Mediterranean mountain area. Catena 17:573–586

Ibanez JJ, Perez A, Jimenez-Ballesta R, Saldana A, Gallardo J (1994) Evolution of fluvial dissection landscapes in Mediterranean environments. Quantitative estimates and geomorphological, pedological and phytocenotic repercussions. Z für Geomorphol NF 38:105–119

Ibanez JJ, De Alba S, Bermudes FF, Garcia-Alvarez A (1995) Pedodiversity: concepts and measurements. Catena 24:215–232

Ibanez JJ, De Alba S, Lobo A, Zucarello V (1998) Pedodiversity and global soil patterns at coarse scales. Geoderma 83:171–214

Ibanez JJ, Caniego J, San Jose F, Carrera C (2005) Pedodiversity-area relationships for islands. Ecol Model 182:257–269

Kaufmann G (2003) Karst landscape evolution. Peleogenesis and evolution of Karst aquifers. Virtual Sci J 1(3):22–31

Li Z, Qing Z, Christopher G (2005) Digital terrain modeling: principles and methodology. CRC Press, Boca Raton/London/New York/Washington, DC

MacKenzie NJ, Austin MP (1993) A quantitative Australian approach to medium and small-scale surveys based on soil stratigraphy and environmental correlation. Geoderma 57:329–355

Magurran AE (1988) Ecological diversity and its measurement. Croom Helm, London, p 179

Martin MA, Rey JM (2000) On the role of Shannons entropy as a measure of heterogeneity. Geoderma 98:1–3

McBratney AB (1992) On variation, uncertainty and informatics in environmental soil management. Aust J Soil Res 30:913–935

Phillips JD (1999) Earth surface systems: complexity, order and scale. Blackwell, Oxford

Phillips JD (2001) Divergent evolution and the spatial structure of soil landscape variability. Catena 43:101–113

Phillips JD (2005) Weathering instability and landscape evolution. Geomorphology 67:255–272

Phillips JD, Marion D (2005) Biomechanical effects, lithological variations, and local pedodiversity in some forest soils of Arkansas. Geoderma 124:73–89

Ruhe RV (1975) Geomorphology. Houghton Mifflin, Boston

Saldana A, Ibanez JJ (2004) Pedodiversity analysis at large scales: an example of three fluvial terrain of the Henares River (central Spain). Geoderma 62:123–138

Schoeneberger PJ, Wysocki DA (eds) (2008) Geomorphic description system, version 4.1. Natural Resources Conservation Service, National Soil Survey Center, Lincoln

Schoeneberger PJ, Wysocki DA, Benham EC, Broderson WD (eds) (2002) Field book for describing and sampling soils, version 2.0. Natural Resources Conservation Service, National Soil Survey Center, Lincoln

Soil Survey Division Staff (1993) Soil survey manual, Agricultural handbook 18. USDA-NRCS. US Government Printing Office, Washington, DC

Soil Survey Staff (1996) Soil survey laboratory methods manual, Report no 42. USDA-NRCS

Soil Survey Staff (2003) Keys to soil taxonomy, 9th edn. USDA-NRCS. US Government Printing Office, Washington, DC

Toomanian N, Jalalian A, Khademi H, Eghbal MK, Papritz A (2006) Pedodiversity and pedogenesis in Zayandeh-rud Valley, Central Iran Geomorphology. Elsevier, Amsterdam

Yaalon DH (1998) Pedodiversity and global soil patterns at coarse scales. Geoderma 83:193–196

Zhu AX, Hudson B, Burt J, Lubich K, Simonson D (2001) Soil mapping using GIS, expert knowledge, and fuzzy logic. Soil Sci Soc Am J 65:1463–1472

Zinck JA (1988) Physiography and soils, ITC lecture note SOL.4.1. International Institute for Geoinformation Science and Earth Observation (ITC), Enschede, p 156

Chapter 10
Assessment of Soil Diversity in Western Siberia Using WRB 2006

Elena N. Smolentseva

Abstract Soil classification has practical importance. It is a base for the mapping of soil resources, data registration, and development of land-use and soil management. In this study, during the period from 2000 to 2008, new data on soil cover, properties of its components, their genesis, and ecology have been collected in the semiarid and dry subhumid regions of Western Siberia. The study enables to upgrade information on soil variability, properties and to a certain extent filled the gap in fundamental knowledge of the area. New data includes soil genesis, soil properties, landscape, and environmental conditions of formation and the classification diversity. Basic types of soil cover disturbances and main reasons of its limited use have been discussed. In our opinion, the international soil classification system (IUSS, Working group WRB. World reference base for soil resources, 2nd edn. World Soil Resources Reports No 103. FAO, Rome, 2006; IUSS, Working group WRB. World reference base for soil resources, first update. World Soil Resources Reports No 103. FAO, Rome, 2007) addresses soil formation characteristics in semiarid and dry subhumid regions of Western Siberia fairly adequately. Present study has shown that great soil diversity exists, presenting 12 of 32 reference soil groups in the study area.

Keywords Soil classification • Soil diversity • Solonetz • Solonchaks • Western Siberia

10.1 Introduction

Soil classification is one of the fundamental branches of soil science discipline. The main purpose of soil classification is to create a system that reflects basic laws of pedogenesis as well as systematic scientific-based soil information for more

E.N. Smolentseva (✉)
Institute of Soil Science and Agrochemistry, Russian Academy of Sciences, Siberian Branch, Novosibirsk, Russia
e-mail: parabraunerde@rambler.ru

S.A. Shahid et al. (eds.), *Developments in Soil Classification, Land Use Planning and Policy Implications: Innovative Thinking of Soil Inventory for Land Use Planning and Management of Land Resources*, DOI 10.1007/978-94-007-5332-7_10,

convenient uses of soil resources as well as unification in international communication. Correlations between soil objects are ascertained within the scope of soil classification, through grouping soils based on their similarities in characteristics and occurrences. Soil classification generates new knowledge and is of great practical importance. It is a base for the mapping of soil resources, their registration, and development of land-use plan and soil management.

Considerable success has been achieved in creating a unified international soil classification system for wider global use. Second edition (World Reference Base for Soil Resources: IUSS 2006) is the result of many years' efforts in soil classification work. It has been used worldwide in various regions. The system allows revision, addition of new soil names (soil nomenclature), and information base according to diagnostic soil characteristics. However, information about Siberian soils is presented poorly in the global information. Siberia covers large area; it is a complex soil-forming territory and therefore presents great soil diversity. The international soil scientists are poorly acquainted with the soil cover of semiarid areas of Siberia due to the reasons that the soil information of cold northern territories is dominated in the literature.

The use of new international soil classification system provides additional opportunities to Siberian soil scientists to diagnose regional soils and improve their predictions about soil-forming processes and correlate soils and the environment. To fill the gap in soil information, the present study was initiated with the objective to characterize the component diversity of the soil cover using the second edition of the international soil classification of World Reference Base for Soil Resources (WRB: IUSS 2006, 2007) as well as to estimate adequacy of the nomenclature and proposed diagnostic criteria for soil units of southern regions of Western Siberia.

10.2 Materials and Methods

10.2.1 Characteristic of the Study Area

Western Siberia is a large region situated in the Asian part of Russia. It extends eastward from the Ural Mountains to the valley of river Yenisei and southward from the Arctic Ocean to the hills of north central Kazakhstan and the national borders of both Mongolia and China. The total area of Western Siberia is 243×10^6 ha. This region includes geomorphologic units such as West Siberian plain and mountains of southern Siberia (Altai, Kuznezki Alatau, Salair). Ninety percent of the West Siberian territory is a plain, whereas the mountain systems occupy only 10% area. The West Siberian plain is one of the largest plains of the world. Its extension from the north to the south is 2,500 km, and from the west to the east is 1,900 km.

The climate of Western Siberia is continental with cold winter and relatively hot summer. Most part of the territory is humid; however, there is a region in the south where potential evaporation is higher than the rainfall. Droughts are common in summer. The average annual rainfall varies from 230 to 350 mm on the region

Fig. 10.1 Location of the studied area and its boundary (1)

territory. The ratio of the annual rainfall to potential evapotranspiration is within the range of 0.35–0.65, whereas during the vegetation period, it varies from 0.20 to 0.45. The UNCCD (1992) categorizes such territories as dry subhumid and semiarid areas (Fig. 10.1). Average annual temperature is 0.6°C in dry subhumid areas and 2.9°C in semiarid ones. In the latitude 52°, the average long-term temperature of January achieves −18°C; for July, it is +20°C. The snow cover lies longer than 5 months a year.

Dry subhumid and semiarid areas of Western Siberia are characterized by the following landscape conditions. Here low plains (Baraba and Kulunda) are prevailing. Their surface is flat or slightly wavy. Positive relief elements are presented by low ridges. They have the absolute height above sea level 120–150 m and the relative elevation above the local lows 5–12 m. Such relatively high-relief positions are characterized by deep groundwater level.

The river network is either poorly developed or it is absent completely. There is great number of salty lakes. They are situated in lows between ridges. This superfluous moistening of soils and grounds is a result of weakened drainage of surface and groundwaters because of the flat relief.

Soil parent material is mostly the Quaternary deposits, aeolian and rewashed aeolian (loess), and lacustrine-alluvial sediments having different granulometric composition. Loesses overlay positive relief elements (ridges), whereas lacustrine-alluvial sediments embed lows between ridges.

The low accumulative plains are the areas where the processes of continental salt accumulation are developing. Therefore, the surface lake waters and groundwaters contain salts, and therefore, saline soils are formed. Specific bioclimatic conditions affect the formation of the soil cover characterized by great diversity of soil units.

Table 10.1 Important characteristics of the soils

Depth of soil sample, cm	pH-H_2O	Organic C (%)	Clay (<2 μm) (%)	$CaCO_3$ (%)	Gypsum (%)	Exchangeable bases, cmolc kg^{-1} Ca^{2+}	Mg^{2+}	Na^{+}	Σ cations
Calcic Chernozem (Molliglossic, Siltic)									
0–10	6.8	4.1	27.1	–	–	31.8	4.1	0.3	36.2
10–20	7.3	2.3	30.5	–	–	27.8	3.8	0.2	31.8
23–34	7.9	1.8	30.8	0.8	–	24.4	3.3	0.5	28.2
50–60	8.3	0.8	40.8	6.2	–	19.6	3.0	0.4	23.0
90–100	8.4	0.1	37.9	10.0	–	16.2	3.4	0.2	19.8
Calcic Gypsic Kashtanozem (Siltic)									
0–10	7.1	2.0	20.8	–	–	18.9	1.7	0.4	21.0
20–30	7.3	1.7	16.8	–	–	17.1	1.7	0.1	18.9
40–50	7.8	0.6	20.0	0.7	–	14.8	4.2	0.6	19.6
70–80	8.4	0.2	18.5	13.4	5.2	12.4	3.8	0.6	16.8
Calcic Solonetz (Magnesic, Siltic)									
0–12	6.6	2.0	18.2	0.2	–	9.3	1.8	0.6	11.7
12–18	9.8	0.5	25.5	2.5	–	2.5	1.7	2.2	6.4
20–30	10.0	0.3	27.9	2.1	–	1.0	2.2	3.0	6.2
40–50	10.2	0.1	28.5	10.8	–	0.1	1.6	2.7	4.4
70–80	10.3	0.1	25.5	9.1	–	0.1	1.5	2.0	3.6

10.2.2 Analytical Procedures

Standard classification rules of the World Reference Base for Soil Resources (IUSS 2006) have been used. To diagnose horizons and establish soil properties, the following analyses were carried out: particle-size distribution analyses, pH-H_2O, organic carbon, carbonates equivalents, gypsum, and exchangeable cations (Ca^{2+}, Mg^{2+}, Na^{+}, K^{2+}). The soil saturation extract was analyzed for electrical conductivity (ECe, soil salinity), pH, soluble cations (Ca^{2+}, Mg^{2+}, Na^{+}, K^{+}), and anions (CO_3^{2-}, HCO_3^{-}, Cl^{-}, SO_4^{2-}). In saline soils, exchangeable bases (Ca^{2+}, Mg^{2+}, Na^{+}, K^{+}) have been determined by the Pfeffer (1956) method with modification (Molodtsov and Ignatova 1975).

Analytical methods have been applied considering the WRB (IUSS 2006) recommendations. The details of standard analytical procedures can be found in Van Reeuwijk (2006) and Burt (2004). Results of analytical procedures are presented in Tables 10.1 and 10.2.

10.3 Results and Discussion

The occurrence of soils on the investigated territory depends on their position on the relief. Firstly, we discuss those soils which are developed on the high-relief positions-ridges. The important factors in such formations are deep groundwaters and lack of their influence on soil formation. The soils developed have the mollic horizon, and the soils are Chernozems, Kastanozems, and Phaeozems. The mollic horizon is well

Table 10.2 Composition of soil saturation extract

Depth of soil sample (cm)	pH	ECe, dS m^{-1}	cmol$_c$ kg^{-1}						
			CO_3^{2-}	HCO_3^-	Cl^-	SO_4^{2-}	Ca^{2+}	Mg^{2+}	Na^+
Endosalic Chernozems (Molliglossic, Siltic)									
0–15	6.8	0.1	–	0.36	0.16	0.06	0.12	0.11	0.25
22–32	7.2	0.2	–	0.20	0.75	0.06	0.06	0.05	0.90
60–70	8.5	4.2	–	0.40	3.20	7.93	5.00	1.90	4.63
100–110	8.2	0.6	–	0.60	2.00	0.44	0.25	0.30	2.49
Gleyic Mollic Solonchak (Sulphatic, Siltic)									
0–10	8.3	9.8	0.04	0.64	16.40	39.04	12.30	12.00	31.82
20–30	8.9	7.2	–	0.40	11.60	29.15	13.50	8.30	19.35
50–60	9.2	6.8	0.16	0.34	10.80	28.32	14.63	6.69	18.48
Calcic Salic Solonetz (Abruptic, Magnesic)									
0–7	7.9	0.1	OTC	0.44	0.17	0.14	0.11	0.11	0.53
13–19	9.8	1.0	1.40	3.40	1.00	0.12	0.35	0.40	5.17
19–27	10.1	1.5	1.40	4.00	1.27	1.85	0.50	0.35	7.67
60–70	9.2	0.5	0.16	0.88	0.50	0.87	0.17	0.21	2.03
Gleyic Salic Fluvisols (Hypereutric, Siltic)									
0–10	9.4	0.8	0.48	1.84	0.40	1.31	0.18	0.06	3.79
10–20	10.0	4.6	1.90	2.65	0.47	2.62	0.20	0.17	7.27
30–40	10.3	5.8	3.56	4.84	0.83	5.04	0.24	0.21	13.82

structured and dark colored and has sufficient thickness, high base saturation, and moderate to high organic matter contents. Table 10.3 lists the RSGs and second-level units of the WRB (IUSS 2006) in the study area.

10.3.1 Chernozems

The mollic horizon of Western Siberia soils has some characteristics differences from the analogous European one, such as morphological characteristics (relatively small thickness and tonguing of the mollic horizon into an underlying layer (Fig. 10.2a). Siberian soil scientists (Gorshenin 1955; Kovalev et al. 1966) described the tonguing as a result of sharp oscillations in the soil mass volume as well as its cracking under the influence of deep frost penetration in winter and intense shrinkage in summer. These processes cause formation of deep cracks in soils and intrusion of organic matter in lower depths. Some tongues can reach to the lower boundary of the cambic horizon. Therefore, the Chernozems can be distinguished with molliglossic suffix qualifier, which characterize exactly the regional feature of the Chernozems in dry conditions of Western Siberia.

Thickness of the mollic horizon under natural vegetation ranges between 35 and 43 cm. In the more arid conditions (rainfall <350 mm per annum), the thickness reduces to 25 cm. The Chernozems of the arable land have the mollic horizon thickness less than 25 cm that makes the criteria unrepresentative for the standard mollic horizon. Soil structure, color, and base saturation are the typical diagnostic characteristics for

Table 10.3 List of the RSGs and second-level units of the WRB (IUSS 2006) in the semiarid and dry subhumid regions of Western Siberia

Reference soil group	Prefix qualifiers	Suffix qualifiers
1. Chernozems	Endo-Hyposalic, Gleyic, Gypsic, Calcic, Luvic, Haplic	Molliglossic, Siltic, Clayic
2. Kastanozems	Endofluvic, Endo-Hyposalic, Gypsic, Calcic, Luvic, Haplic	Glossic, Arenic, Siltic, Novic
3. Phaeozems	Calcic, Haplic	Glossic, Pachic, Arenic, Siltic
4. Luvisols	Lamellic, Cutanic, Stagnic, Haplic	Humic, Arenic, Siltic
5. Planosols	Solodic, Endosalic, Endogleyic, Mollic, Calcic, Luvic, Haplic	Albic, Manganiferric, Ferric, Eutric, Greyic, Siltic, Clayic
6. Stagnosols	Solodic, Endosalic, Endogleyic, Mollic, Calcic, Luvic, Haplic	Albic, Manganiferric, Ferric, Eutric, Greyic, Siltic, Clayic
7. Gleysols	Histic, Endosalic, Mollic, Gypsic, Calcic, Haplic	Abruptic, Colluvic, Humic, Eutric, Siltic, Clayic, Novic
8. Solonetz	Gleyic, Salic, Mollic, Gypsic, Calcic, Haplic	Glossalbic, Albic, Abruptic, Colluvic, Magnesic, Humic, Siltic, Clayic, Transportic, Novic
9. Solonchaks	Hypersalic, Puffic, Gleyic, Mollic, Gypsic, Calcic, Haplic	Chloridic, Sulphatic, Carbonatic, Arenic, Siltic
10. Fluvisols	Salic, Gleyic, Calcic, Haplic	Humic, Eutric (Hypereutric), Siltic, Clayic
11. Arenosols	Lamellic, Protic, Endosalic, Endogleyic, Haplic	Calcaric, Hyposalic, Eutric, Placic, Novic
12. Regosols	Colluvic, Stagnic, Haplic	Humic, Hyposalic, Sodic, Eutric, Arenic, Siltic

Fig. 10.2 Calcic Chernozem (Molliglossic, Siltic): tonguing of the mollic horizon into an underlying layers (**a**) and columnar element of soil structure with rounded top in the natric horizon (**b**)

the mollic horizon. The soil structure is strong and well developed in natural soils; often it loses its properties in arable soils. The typical Munsell color of the mollic horizon is with a chroma of 3 or less when moist, a value of 3 or less when moist, and 5 or less when dry on broken samples. The mollic horizon has a moderate content of organic matter; more often the organic carbon content ranges between 2.3 and 4.1% (Table 10.1). In the mollic horizon and below it, the Chernozems have the base saturation percentage (by 1 M NH_4OAc) equal or more than 80.

In the Chernozems of Western Siberia, mostly the secondary carbonates are usually present at 30–50 cm depth from the lower depth of the mollic horizon. The calcic horizon is present in soils located in the semiarid conditions. The secondary carbonates are present in soil profile as coatings or filaments (pseudomycelia). The quantity of calcium carbonates equivalents ranges between 6 and 12%. Gypsum ($CaSO_4$•$2H_2O$) occurs only in Chernozems where they are developed on clayey soil parent material and generally observed at a depth of 70–100 cm from the soil surface. These gypsic soils are characterized as Gypsic Chernozems (Clayic). They account for 5–10% (Chernozems Clayic) in the Western Siberia. They have silt, silt loam, and silty clay loam soil texture, and therefore 70–80% of the Chernozems are siltic.

The Chernozems those are developed on the coarser substrates (sandy loam) have deeper occurrence of secondary carbonates; the accumulation of secondary carbonates begins at 50 cm below the lower depth of the mollic horizon. Therefore, such soils are described as Phaeozems, whereas Chernozems (Arenic) do not present such features.

10.3.2 Phaeozems

The Phaeozems are considered as soils of the more humid conditions; however, in the Western Siberia, they do occur in the semiarid climate regions. If the soils are developed on the coarser substrate (sandy loam) with low quantity of primary carbonates (3–6% calcium carbonate equivalents), the secondary carbonates are accumulated at 80–120 cm depth from the soil surface. Because of the arid conditions, the thickness of the mollic horizon is 25–30 cm, and the accumulation of secondary carbonates begins at 50 cm below the lower depth of the mollic horizon. It is now clear from the above discussion that in semiarid and dry subhumid regions of Western Siberia, Phaeozems are developed on the coarser sediments and can be distinguished from Chernozems by deeper occurrence of secondary carbonates. Mostly for Phaeozems, the typical prefix qualifiers are Calcic and Haplic, and the suffix qualifiers are Glossic, Pachic, Arenic, and Siltic.

10.3.3 Solonchaks

The Solonchaks are the intrazonal soils of arid and semiarid regions that contain large quantities of soluble salts in the soil solum. Solonchaks can be developed naturally or anthropogenically. Naturally they are formed from saline parent materials and salt accumulation from groundwaters through evaporation (West Siberian Solonchaks); in the latter case, saline soil parent materials are absent. So, a salic horizon is formed. In the Western Siberia, the Solonchaks are most often found where surface accumulation of salts is the highest (external Solonchaks); under such conditions, a salt crust formation is a very typical phenomenon. Solonchaks formation does not occur under the influence of irrigation water because irrigation does not take place on this territory in the post-soviet time.

In the West Siberian region, soluble salts are deposited from the groundwater to earlier formed soils; therefore, the characteristics of the salic horizon are often combined with characteristics of other diagnostic horizons, properties, or materials. Thus, many Solonchaks have a gleyic color pattern at some depth. The solums of the Solonchaks often have diagnostic horizons as Mollic, Gypsic, and Calcic. In the West Siberian Solonchaks group, the prefix qualifiers, e.g., Hypersalic, Puffic, Gleyic, Mollic, Gypsic, Calcic, and Haplic are distinguished.

The Solonchaks of the West Siberian region have diverse chemical composition of soluble salts. The analyses of the soil saturation extract revealed the presence of cations (Ca^{2+}, Mg^{2+}, Na^{+}) and anions (CO_3^{2-}, HCO_3^{-}, SO_4^{2-} and Cl^{-}) of the dissolved salts in variable quantities. Based on analyses, the suffix qualifiers, e.g., Chloridic, Sulfatic, or Carbonatic, are used. The suffix qualifiers characterizing the soil texture are Siltic and Clayic. The Mollic Gleyic Solonchak (Carbonatic, Siltic) commonly occurs, as well as the Gypsic Gleyic Solonchak (Sulfatic, Siltic). Generally the Solonchaks occur as small spots together with Solonetz and other saline soils where salt concentration is less than in normal Solonchaks.

10.3.4 Fluvisols

In the southern part of the Western Siberia, besides Solonchaks, there are other strongly salt-affected soils belonging to Fluvisols. These soils occur in depressions on a place of salty shallow lakes. In the southern part of Western Siberia, there are many closed depressions where lakes are formed and salts accumulate. In the past three decades, the aridity intensification resulted in the drying of many lakes, and therefore, saline soils are formed. These soils are developed on the stratified lacustrine sediments with obvious accumulation of soluble salts on the soil surface up to the depth of 50 cm. This upper layer qualifies as salic horizon where the ECe is 20–30 dS m^{-1} or more in the upper 30 cm depth. Also, these lacustrine sediments have a gleyic color pattern at some depth or even beginning from the surface. In the Fluvisols group, the prefix qualifiers Salic and Gleyic are diagnosed at the second-level unit. The base saturation (by 1 M NH_4OAc) is more than 80% in the solum. This property is characterized by the suffix qualifier Hypereutric.

Since these lacustrine sediments have a texture of silt, silt loam, loam, silty clay, or clay (usually in the upper part where soil is formed), suffix qualifiers as Siltic and Clayic are used. The most widespread saline soils formed as a result of the drying of shallow lakes are Gleyic Salic Fluvisols (Hypereutric, Siltic). Usually these soils occupy areas of 2–3 km^2.

Other soils of the Fluvisols group are poorly extended in the studied territory. Where found these soils are formed on alluvial deposits. Alluvial deposits do not have soluble salts but have coarser texture, and they are frequently more stratified compared with lacustrine sediments. Therefore, the soils developed from alluvial deposits differ significantly by their properties from those developed on lacustrine sediments. Fluvisols formed from alluvial deposits have mostly the characteristic Calcic Gleyic Fluvisols (Humic, Siltic) or Haplic Fluvisols (Eutric, Arenic) characteristics.

10.3.5 Solonetz

Landscape conditions of the southern part of Western Siberia are more favorable for the formation of Solonetz soils. Solonetz soils are associated with flatlands in a climate with hot, dry summer and an annual precipitation of 250–400 mm. Solonetz occurs predominantly in areas with a shallow groundwater table and with impeded vertical and lateral drainage. The basic morphological diagnostic criterion of Solonetz is a very dense, strongly structured horizon with clayey subsurface. The soil structure is coarse columnar or prismatic (Fig. 10.2b), sometimes blocky or massive. Rounded and often whitish tops of the structural elements are characteristics. The horizon is high in exchangeable Na and/or Mg cations and qualifies as natric horizon. Many Solonetz have a field pH of about 8.5, indicative of the presence of free sodium carbonate (Na_2CO_3). On the investigated territory, Solonetz is often strongly alkaline in subsurface horizons with pH (H_2O) 9.0–10.5 (Table 10.1).

Solonetz is widely distributed in the Western Siberia. Many of them are well developed and have indicators of other soil-forming processes along with the natric horizon. Correspondingly, we have diagnosed many second-level units of the WRB classification. The shallow groundwater table considerably affects the development of Solonetz soils and their morphological features. The developed soils have reducing conditions and a gleyic color pattern within the upper 100 cm. The brackish groundwater accumulates salts in the middle part of the soil profile or sometimes on the soil surface after evaporation; therefore, very often salic gleyic Solonetz soil occurs, where the ECe is low and ranges between 0.1 and 1.5 dS m^{-1} (Table 10.2). A calcic and/or gypsic horizon may be present in the solum below the natric horizon, usually at a depth from 30 to 60 cm. Therefore, we diagnose Gypsic and Calcic Solonetz soils. Some Solonetz present mollic horizon with thickness less than 30 cm.

Properties characterized by suffix qualifiers represented in Solonetz are diverse. Well-developed Solonetz can have an albic eluvial horizon lying directly over the natric horizon. Often the albic horizon is showing tonguing into a natric horizon called Glossalbic. In Western Siberia, Solonetz has an average exchangeable Na content within the limits of 20–40%, total with Mg up to 90% (n). Rarely the exchangeable Mg content is higher than the exchangeable Na. In this case, Solonetz is characterized as Magnesic. Many Solonetz have abrupt textural change: a very sharp increase in clay content of the natric horizon in comparison with overlaying horizons.

Altogether in Solonetz, we have diagnosed six prefix qualifiers (Gleyic, Salic, Mollic, Gypsic, Calcic, Haplic) as well as ten suffix qualifiers (Glossalbic, Albic, Abruptic, Colluvic, Magnesic, Humic, Siltic, Clayic, Transportic, and Novic).

10.3.6 Kastanozems

Kastanozems are zonal soils of the short-grass steppe belt. They are formed in the driest conditions of the investigated territory. Kastanozems have profile similar to that of Chernozems with a humus-rich surface horizon. The main diagnostic criteria are the following: the mollic horizon and secondary carbonates or a calcic horizon in the subsoil. The mollic horizon is thinner and not as dark as that of the Chernozems. Generally the mollic horizon is dark brown or brown with the thickness from 25 to 40 cm. The content of organic carbon is 1.2–4% in the upper part of the mollic horizon. In average, this is lower by 20–30% than that in Chernozems.

The accumulation secondary carbonates or a calcic horizon occurs at a depth of 30–45 cm and rarely at 60–70 cm. The secondary carbonates are present as concretions or spheroidal aggregates (white spots) or as soft coatings on structural faces. The carbonates content in the solum is 9–12% and 3–4% in the soil parent material.

In some cases, Kastanozems soils have secondary gypsum in the subsoil (120–140 cm). Sometimes in the lower part of the solum (50–60 cm), soluble salts also occur; however, their quantity is low. The ECe is >4 dS m^{-1} in some layers within upper 100 cm. This corresponds to hyposalic characteristic not recommended for the given soil group. Probably, it is necessary to add this prefix qualifier to Kastanozems. Some Kastanozems soils present argic horizon with cation exchange capacity (CEC) (by 1 M NHOAc) of 24–28 cmolc kg^{-1} in the upper 100 cm. Prefix qualifiers of Kastanozems are Hyposalic, Calcic, Luvic, and Haplic. Suffix qualifiers are Glossic, Arenic, Siltic, and Novic. The most widely distributed soils are Haplic Kastanozems (Siltic, Novic).

10.3.7 Planosols and Stagnosols

Planosols and Stagnosols soils are formed under the influence of stagnating water. These soils are located in the closed drainless depressions occupying small areas. Groundwater table varies from 1 to 10 m. The depressions are mostly occupied by forest vegetation. The moistening regime of soils is characterized by clear alternation of wet and dry seasons. During winter, snow is accumulated in these depressions. The snow thawing in the spring (from week to a month) leads to water stagnation. In the upper part of soil solum, the reduction conditions are created, and in summer, oxidative conditions are prevalent. Planosols and Stagnosols are formed at the study territory under similar soil-forming conditions and have the same features. However, these soils present reducing conditions for some time of the year in half or more of the soil volume and a stagnic color pattern and/or an albic horizon.

Planosols are soils with a surface horizon that shows signs of periodic water stagnation and abruptly overlies a dense slowly permeable subsoil with significantly more clay than the surface horizon. An abrupt textural change is a principal diagnostic criterion that differentiates Planosols from Stagnosols. Planosols of the studied region are characterized by the following second-level units: prefix qualifiers are Solodic, Endosalic, Endogleyic, Mollic, Calcic, Luvic, and Haplic, and suffix qualifiers are Albic, Manganiferric, Ferric, Eutric, Greyic, Siltic, and Clayic. The most part of qualifiers is typically associated with those qualifiers (Mollic, Calcic, Haplic) which are present in all the diagnosed RSGs.

Planosols have qualifiers Solodic, Albic, Manganiferric, and Ferric, those we have not considered. Prefix qualifier Solodic describes the soil with the columnar or prismatic structure of the natric horizon but lacking its sodium saturation requirements. These soils have along with structure of natric horizon well-developed mollic horizon. Earlier they have been called in Russia as deep Solonetz. The diagnostic albic horizon often meets at Planosols of Western Siberia too. Planosols have been found which had diagnostic criteria for Manganiferric and Ferric. There was a ferric horizon with segregation of Fe or Fe and manganese (Mn). They were discrete reddish to blackish nodules with a diameter of 2 mm or more, with the exteriors of the nodules being at least weakly cemented or indurated. We faced difficulties to choose suffix qualifiers, when soil horizons had some diagnostic criteria. For example, albic horizon with segregation of Fe as discrete reddish to blackish nodules or ferric horizon with stagnic color pattern.

Stagnosols are soils with a perched water table showing redoximorphic features caused by surface water. Stagnosols are periodically wet and mottled in the topsoil and subsoil, with or without concretions and/or bleaching, and have structural or moderate textural discontinuity. As described above, in Western Siberia conditions for the formation of Stagnosols are similar to those at Planosols. Therefore, both have identical prefix and suffix qualifiers.

10.3.8 Gleysols

Gleysols are wetland soils that, unless drained, are saturated with groundwater for long periods to develop a characteristic gleyic color pattern. These soils occupy depression areas and low landscape positions with shallow groundwater. The solum has evidence of reduction processes with segregation of Fe compounds within 50 cm of the soil surface. Groundwater affected soils of the region have sometimes histic horizon, which has thickness of 10–15 cm. Calcic horizon occurs at 60–80 cm depth and accumulated in the form of hard nodules. Some soils have the mollic horizon of 25–45-cm thickness. Gleysols with distinguishable surface soluble salts (ECe of 4–6 dS m^{-1}) were also found. The second-level units of Gleysols are the following: prefix qualifiers as Histic, Endosalic, Mollic, Gypsic, Calcic, and Haplic. Suffix qualifiers are Abruptic, Colluvic, Humic, Eutric, Siltic, Clayic, and Novic.

10.3.9 Arenosols

Arenosols are the sandy soils that have a weighted average texture of loamy sand or coarser. They are often recorded but do not occupy large areas. The sandy substrates are ancient (from the middle Pleistocene) lacustrine and alluvial sediments which were rewashed and redeposited subsequently in the Holocene. As a result, redeposited lacustrine and alluvial sediments became homogenous and do not have criteria of fluvic material. Prefix qualifiers are lamellic, protic, endosalic, endogleyic, and haplic. Suffix qualifiers are Calcaric, Hyposalic, Eutric, Placic, and Novic. Arenosols (Novic) are formed by the accumulation of the aeolian sediments on the soil surface, and above the solum, there is a layer of recent sediments of 12–25-cm thickness.

10.3.10 Luvisols

Luvisols are extended locally in more damp parts of investigated territory. They have an argic horizon starting within 100 cm of the soil surface. Prefix qualifiers are Lamellic, Cutanic, Stagnic, and Haplic. Suffix qualifiers are Humic, Arenic, and Siltic.

10.3.11 Regosols

The Regosols form a group of all soils that could not be accommodated in any of the other reference soil groups (RSGs). Regosols are weakly developed soils in unconsolidated material and have no diagnostic horizons. Profile development is minimal as a consequence of young age and/or slow soil formation. This group concerns Chernozems which are strongly degraded by deflation (wind erosion) and Solonetz where natric horizon is destroyed mechanically or chemically. Prefix qualifiers are Colluvic, Stagnic, Haplic, and suffix qualifiers are Humic, Hyposalic, Sodic, Eutric, Arenic, and Siltic.

10.4 Conclusions

The present investigation revealed 12 of 32 reference soil groups (RSGs) in the studied territory. This does not include soils with strong human influence (Anthrosols, Technosols). Thirty-eight units of the 182 second-level units of WRB classification have been diagnosed in the explored region. It evidences high representativeness of the diagnostic criteria. Mostly often Mollic, Calcic, Gypsic, Salic, Gleyic, Haplic, Eutric, and Siltic are the qualifiers for regional soils. These reflect pedogenesis features as well as secondary soil-forming processes. The study is conditioned by the fact that the use of new diagnostic criteria requires an experience as well as supporting analytical data. The Russian analytical procedures differ to some extent for soils diagnostics within the scopes of WRB (IUSS 2006, 2007). Large analytical results were carried out in accordance with the WRB (IUSS 2006) requirements. These data are based for soils diagnostics for the present study. We strongly believe the second edition of the WRB (IUSS 2006, 2007) addresses soil formation characteristics in semiarid and dry subhumid regions of Western Siberia adequately; however, some differences occur in characterizing diagnostic horizons.

References

Burt R (ed) (2004) Soil survey laboratory methods manual. Soil Survey Investigations Report No 42, Version 4.0. Natural Resources Conservation Service, Washington DC

Gorshenin KP (1955) Pochvi uzhnoi chasti Zapadnoi Sibiri (Moskva). AN SSSR, Moscow

IUSS (2006) Working group WRB. World reference base for soil resources, 2nd edn. World Soil Resources Reports No 103. FAO, Rome

IUSS (2007) Working Group WRB. World reference base for soil resources, first update. World Soil Resources Reports No 103. FAO, Rome

Kovalev RV, Selakov SN, Trofimov SS, Panin PS (1966) Pochvi Novosibirskoi oblasti. Nauka SO AN SSSR, Novosibirsk

Molodtsov WA, Ignatova WP (1975) Testing of exchangeable bases composition in the saline soils. Euras Soil Sci 6:123–127

Pfeffer P (1956) Zur Bestimmung der austauschbaren Basen und des S-Wertes von Böden. Pflanzenernährung, Düngung, Bodenkunde 75 H1, pp 17–32

UNCCD (1992) Definition of desertification. http://www.unccd.entico.com/cnglish/glossary.htm. Accessed 1 July 2007

Van Reeuwijk LP (2006) Procedures for soil analysis, 7th edn. Technical Report 9. Wageningen, Netherlands, ISRIC-World Soil Information. Financial support for this research was provided by the Russian Academy of Sciences (Project 16.14)

Chapter 11
Classification of the Topsoil Fabrics in Arid Soils of Central Asia

Marina Lebedeva, Maria Gerasimova, and Dmitry Golovanov

Abstract Existing data on the soil micromorphology of arid regions are sparse; in the present study, micromorphological features of a wide spectrum of arid soils in Russia, Uzbekistan, Kazakhstan, and Mongolia are studied to develop diagnostic criteria for the new substantive-genetic soil classification systems. The diversity, functioning, and resilience of arid soils are determined by the properties of their topsoil, which reflects the recent environment, whereas subsoil reflects the paleoenvironment. Each of three upper horizons (light-humus, solonetzic eluvial, and xerohumus) as recognized in the new Russian system of soil classification (2004) and which can be found in arid soils presents similar micromorphological features in different soils. However, present study reveals that, in a sequence of soils, there are some specific micromorphological features indicating the increasing trends of aridity. In a soil sequence with increasing aridity, the diagnostic horizons and properties are combined in a regular way corresponding to the changes in environmental conditions and soil-forming processes; at the same time, the sequence is in good agreement with diagnostic elements of substantive Russian and WRB classification systems. Thus, the arid soils present two groups: one with a distinct light-humus

M. Lebedeva (✉)
Laboratory of Soil Mineralogy and Micromorphology, V.V. Dokuchaev Soil Science Institute, Pyzhevskii per., 7, 119017 Moscow, Russia
e-mail: m_verba@mail.ru

M. Gerasimova
Faculty of Geography Moscow, Lomonosov University, Leninskie gory, Moscow 119992, Russia

D. Golovanov
Department of Soil Geography and Landscape Geochemistry, M.V. Lomonosov, Moscow State University, Moscow 119992, Russia

S.A. Shahid et al. (eds.), *Developments in Soil Classification, Land Use Planning and Policy Implications: Innovative Thinking of Soil Inventory for Land Use Planning and Management of Land Resources*, DOI 10.1007/978-94-007-5332-7_11,

horizon, and the other with a xero-humus horizon composed of crusty and subcrusty subhorizons. These groups correspond to two different types of pedogenesis. The micromorphological features of the topsoil make it possible to identify the mechanisms of some phenomena, for example, aeolian deposition, structural rearrangement, dynamics of secondary carbonates, and cryptosolonetzic manifestations.

Keywords Topsoil • Former USSR • Soil fabrics • Micromorphology • Central Asia

11.1 Introduction

In the Union of Soviet Socialist Republic (USSR), Feofarova was the first scientist to apply micromorphological methods to aridic soils, where the structural and mineralogical modifications are the most conspicuous (Feofarova 1950, 1956, 1958). Further micromorphological studies of genetic horizons in aridic soils were made by Minashina (1966, 1973). Other studies (Yarilova 1963, 1966; Tursina et al. 1983, 1984; Romashkevich and Gerasimova 1977, 1982; Gubin 1984; Golovanov et al. 2005; Yamnova 2006; Gerasimova et al. 1992; Gerasimova 2003; Gerasimova and Lebedeva 2008; Lebedeva-Verba and Gabchenko 2006; Lebedeva-Verba et al. 2008) were focused on micromorphological diagnostics of soil types in arid regions with the objective to clarify some pedogenetic aspects. These studies were conducted on chestnut, solonetzes, serozems, brown semidesert, gray-brown desert, extra-aridic, takyr soils in the USSR classification system (Classification and Diagnostics of Soils of the USSR 1977).

The results of these studies helped to understand the details of some pedogenic processes and revealed the major fabric elements in different types of saline soils. First soil micromorphological information on aridic soils was completed by Romashkevich and Gerasimova (1977). The subdivision of pedogenic and lithogenic fabric elements, diagnostics, and origin of textural, carbonate, gypsum, and salt pedofeatures were discussed in many foreign publications (Fedoroff and Courty 1987; Khademi and Mermut 2003; Miedema and Slager 1972; Netlleton et al. 1969; Pagliai and LaMarca 1979; Poch et al. 1998; Stoops and Ilaiwi 1981; Stoops and Poch 1994; Shahid and Jenkins 1994; Shahid and Mufti 1994; Mills et al. 2008), and recently in Russia, work has also been published (Khokhlova 2008; Kovda 2008; Kovda et al. 2009; Lebedeva-Verba and Gerasimova 2009; Lebedeva-Verba et al. 2009).

The application of micromorphology to diagnostic horizons and properties of Aridisols was attempted by Allen (1985) for the special issue of soil micromorphology and soil classification published by the Soil Science Society of America (SSSA). Although much have been published worldwide on soil micromorphology, the micromorphological data on aridic soils of Russia are very sparse, this is mainly due to small areas of arid lands in Russia, difficulties of sampling the fragile aridic soils, and preparation of thin (25–30-μm thickness) sections (especially from the topsoils and saline soils). When investigating soils, a problem was faced for non-differentiated sampling "from the upper horizon as a whole," that is, from 10- to 15-cm-thick layer, whereas the most informative soil depth in terms of current pedogenesis is the upper 2–5-cm layer, as well as the microprofiles or subhorizons.

At present, the classification of zonal soils in arid lands attracted more interest because of the changing classification paradigm in Russia. The significance of soil-forming agents as taxonomic criteria is mitigated in favor of soil properties. Hence, the micromorphological method acquires significant importance. The aridic soils are commonly perceived as paleosols with imprints of pre-Holocene processes, as soils dominated by parent rock elements and phenomena related to salts. However, in authors' opinion this is true for the subsoils, whereas topsoils are recent dynamic formations and they are responsible for the diversity and functioning of aridic soils; they obviously reveal their dependence on the existing soil-forming agents.

The objective of this chapter is to depict the micromorphological background for the diagnostic elements of genetic-substantive classification systems. These systems are Classification and Diagnostic System of Russian Soils 2004; Field Guide on Soil Identification 2008; World Reference Base for Soil Resources (WRB) 1998, 2006.

11.2 Materials and Methods

In this study, micromorphological features of a wide spectrum of arid soils in Russia, Uzbekistan, Kazakhstan, and Mongolia have been examined to develop the diagnostic criteria for the new substantive-genetic soil classification systems (Classification and Diagnostic System of Russian Soils (2004) and Field Guide on Soil Identification (2008)).

The diagnostic elements in the new Russian classification system were studied in soils of conventionally virgin lands arranged in a sequence with increasing aridity. All soils are formed on slightly saline parent materials. The most arid soils in Russia, for which the diagnostic elements are discussed, the brown aridic soils (Gypsic Calcisols) of the Caspian semidesert; Episalic Solonetzes; and chestnut (Gypsic Kastanozems) soils are referred to as soils of lower aridity. More arid soils are siltic serozems and gray-brown soils (Yermic Calcisols) of Uzbekistan deserts, and the next in this sequence, are the desert soils of Mongolia with the extra-aridic (Yermic Regosols) as final members.

Diagnostic horizons and diagnostic (genetic) properties have similar functions in the new Russian classification and the WRB systems. In different versions of both systems, their taxonomic significance slightly changes, whereas the definitions remain the same.

Diagnostic horizons in Field Guide on Soil Identification (2008) comprise the following: light-humus (AJ) horizon, which is characteristic of brown aridic (Gypsic Calcisols), part of solonetzes (Episalic Solonetz), and chestnut soils (Gypsic Calcisols) in Russia, its central image was derived from serozems (siltic serozems of Central Asia).

In the former version of the Russian classification system (Classification and Diagnostic System of Russian Soils 2004), the xero-humus or crusty-subcrusty horizon (AKL) was introduced. It was defined as a paragenetic association of shallow subhorizons: porous crust and platy "subcrust." In Russia, it was identified only in the brown aridic soils, where it was discontinuous, rather weakly expressed, or its thickness was insufficient to fit criteria for a diagnostic horizon. That is why, its diagnostic significance was reduced in the version of 2008, and it was qualified for a diagnostic property or microprofile-[akl]. Nevertheless, it preserves its morphology and genetic essence as an association of thin subhorizons composing a microprofile, 3–7 cm deep.

The crusty subhorizon is rather compact although porous (up to spongy), brownish light gray, effervescent, and its depth does not exceed 2–4 cm. The subcrusty subhorizon is loose, light in color, and has a lenticular-platy structure; both subhorizons are salt-free.

Irrespectively of the taxonomic significance, the association of the subhorizons discussed has much in common with the yermic horizon in the WRB system. By definition, the yermic horizon has the same components—the porous crust with stony inclusions of desert pavement, underlain by a fragile platy layer. The mutual position of components was indicated as the above said in the definition of yermic horizon in WRB-1998, while in WRB-2006, this sequence was indicated as not obligatory. There are also aeolian phenomena and aridic properties in both definitions of the yermic horizon (WRB 1998, 2006). The latter diagnostic property (recognized by low organic carbon content, aeolian features, achromatic color, high base saturation) is included in the definition of one more horizon of aridic soils—the takyric one.

In the Russian system (2004), these elements of aridic environments and soils are recorded at the level of diagnostic properties, which is justified by the geographic reasons. The following diagnostic properties are specified: takyric (kt), saline (s), pendant-carbonate (ic), gypsic (cs), aeolian-accumulative (ael), and water-accumulative (aq).

Sampling from soil pits was made at small intervals to characterize the subhorizons. Thin sections were prepared by M.A. Lebedev in the Laboratory of Dokuchaev Soil Institute; synthetic resins were used for the impregnation procedure under vacuum, which provides the preservation of crystalline pedofeatures. The resin-impregnated soil blocks were then converted to soil thin sections (25–30-μm thickness) using diamond saw and subsequently the lapping and polishing machines and Al_2O_3 grit. The final lapping and polishing was achieved using diamond pastes of different diameter. The desired thickness (25–30-μm) was achieved by observing the quartz grain under the polarizing microscope to gray birefringent color.

11.3 Results and Discussion

Table 11.1 illustrates light-humus AJ horizon characterized in the serozems (siltic serozems), chestnut (Gypsic Kastanozems), and brown aridic soils (Gypsic Calcisols). The topsoil of a typical serozem (Uzbekistan, piedmont plain of the Turkestanskii

Table 11.1 Diagnostic elements of aridic topsoils identified at a micromorphological level

Soil (Nogina et al. 1977; USSR 1977)	Soil (WRB-2006)	Diagnostic horizons[a]				Genetic features[b]					
		AJ	AKL		Desert pavement						
			K	L		[akl]	ael	ic	dc	s	cs
Serozem	Siltic Calcisol	+[c]									
Chesnut soil	Gypsic Kastanozem	+									
Brown aridic soil	Gypsic Calcisol	(+)	+		(+)	+	(+)				(+)
Crusty solonetz	Episalic Solonetz (Yermic)		+	+		+	(+)				
Pale-brown soil	Aridic Calcisol		+	+	(+)		+	+	(+)		(+)
Gray-brown soil	Yermic Calcisol		+	+	+		+		(+)		(+)
Extra-aridic soil	Yermic Regosol		+	+	+		+		+	+	(+)

[a]Symbols for horizons: AJ – light-humus; AKL – xero-humus, K – crusty and L –subcrusty subhorizons

[b]Genetic features: microprofile - [akl]; aeolian-accumulative – ael; dispersed-carbonate – dc; pendant-carbonate – ic; saline – s; gypsum-containing – cs

[c]Conventions: + diagnostic horizon or genetic feature is present; (+) – horizon or feature is optionally manifested; empty column – horizon or feature has no manifestation

Ridge) is light brown and weakly compacted and has moderate crumb structure and many fine roots and faunal chambers. It presents a set of micromorphological features (Fig. 11.1a) that are high pedality and porosity, carbonate-clay plasma; the majority of the aggregates are rounded (0.1–0.2 mm in diameter), the largest are the faunal castings; and loessic microaggregates (Minashina 1966) of 0.02–0.05 mm in size are recognized in some parts of thin sections. Plasma is weakly anisotropic and b-fabric is crystallitic. Packing voids with clear walls are predominant among pores. Fine organic residues including root remains in voids are few, as the fine-dispersed organic matter. The skeleton is mostly composed of silt-size particles of quartz, feldspars, mica, calcite, amphiboles, and epidote (Verba 1990).

The enumerated micromorphological characteristics of the upper horizon of serozems (siltic serozems) are typical for aridic soils on loess (Minashina 1966), and they may be regarded as components of a “central image,” or micromorphotype of the light-humus horizon (Gerasimova and Lebedeva 2008). The transformation of parent rock (loess) fabric into that of the AJ horizon is due to the activity of soil fauna, structuring by roots, and microbial transformation of organic residues during the period of spring rains.

The light-humus horizon of chestnut soil (Gypsic Kastanozems) has the darkest color among the AJ horizons of other soils, the strongest crumb structure with a higher proportion of biogenic aggregates, and the most abundant root residues; it may be even partially free of carbonates. The light-humus horizon was studied in soils of Dzhanybek experimental station of Russian Academy of Sciences in the Caspian Lowland. It was recently found that properties of this horizon are not stable

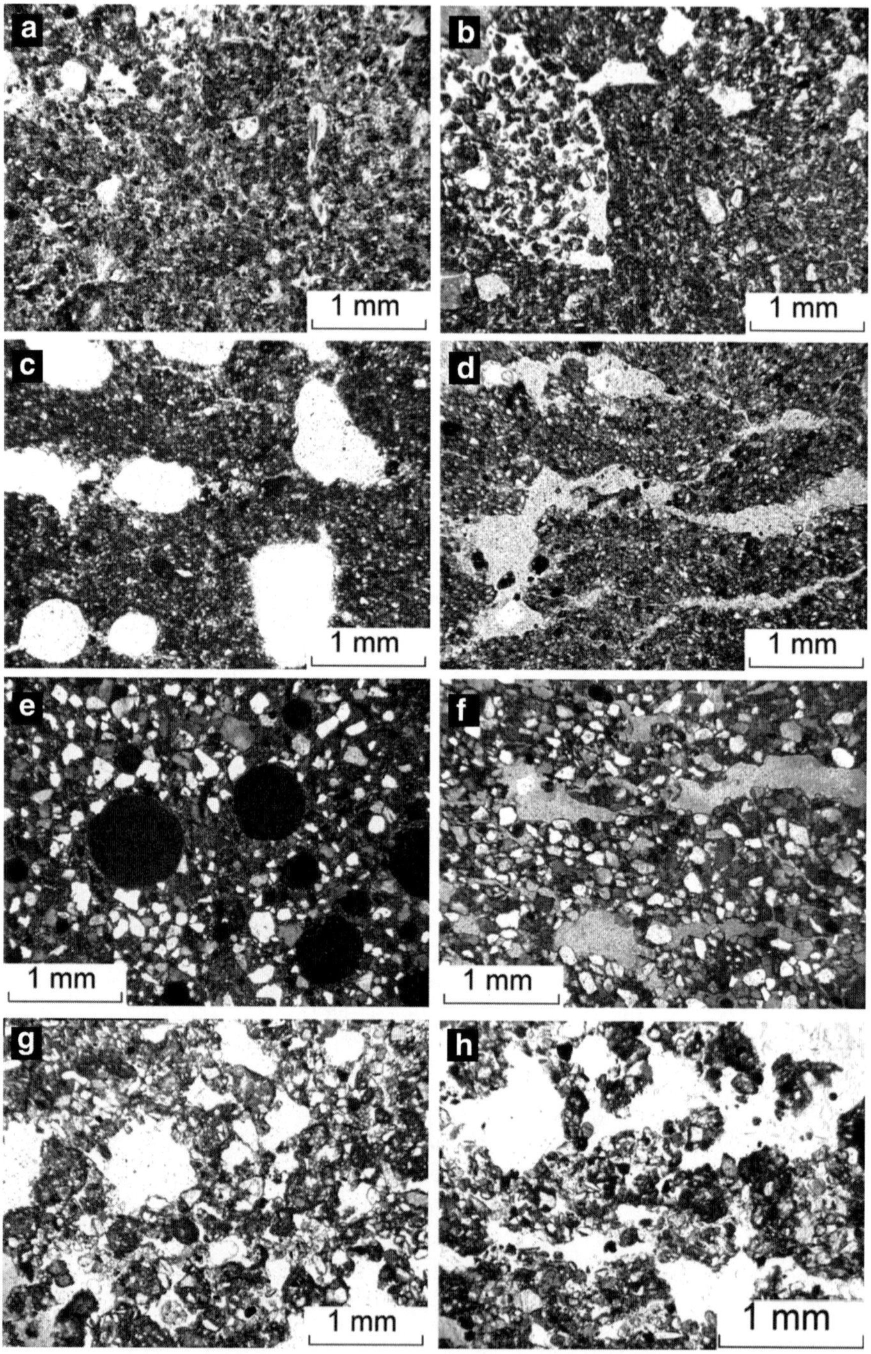

Fig. 11.1 Topsoil of typical serozem showing high pedality (**a**), topsoil in chestnut soil (**b**), micromorphology of the crusty (K) and subcrusty (L) horizons (**c**, **d**), plasmic-silty groundmass and platy microstructure (**e**, **f**), irregular pattern of sand particles (**g**) and strong platy microstructure, and plasmic-silty-sandy groundmass (**h**)

because of the changes in microrelief (Bol'shakov and Borovskii 1937; Rode and Pol'skii 1958) and activity of burrowing mammals removing the carbonates and salt-enriched material to the surface (Abaturov 1985). Therefore, the topsoil properties range from light-colored and highly calcareous variants to rather dark, structured, and leached variants.

In terms of micromorphology, the AJ horizon in chestnut soils has the following divergences from the "central image": clay plasma, brown disperse humus, predominance of moderately decomposed plant residues, and absence of "loessic microaggregation" (Fig. 11.1b). Detailed studies in Dzhanybek microcatena revealed specific microstructures of AJ horizon: rounded aggregates arranged in fine plates (Lebedeva-Verba and Gerasimova 2009). Admixtures of apedal micrite-enriched material were also found; they testify to the translocation of subsoil substrate into the AJ horizon and thus maintaining its properties.

Crusty solonetzes (Episalic Solonetz Yermic) were also studied in microcatenas on the territory of Dzhanybek experimental station in the Caspian Lowland. Morphologically, they differ from other (dominant) solonetzes by shallow topsoil above the solonetzic horizon. Crusty solonetzes (Episalic Solonetz) are confined to the driest microhighs in microcatenas, they mostly lack vegetation, and their surface is dissected by fissures into small polygons. Unlike other solonetzes in this area, they have crusty and subcrusty subhorizons, each 2–4 cm thick. The lower part of the porous crust is gradually acquiring a horizontal stratification and merges in the subcrusty subhorizon. In this case, the stratification may be maintained by freezing/thawing cycles.

In thin sections, the crusty K subhorizon is homogeneous and compact, has silty-plasmic elementary fabric, isolated isometric pores, including the rounded ones, mostly vesicular (Fig. 11.1c). The term "elementary fabric" borrowed from Kubiena's system is traditional in Russian micromorphology and mostly characterizes texture; in some cases, the authors use another similar but more precise term "c/f related distribution" recommended in the guidelines by Stoops (2003). The subcrusty L subhorizon is of the same texture but has elongated or horizontal voids, partially fine planes (Fig. 11.1d).

A comparative analysis of pedofeatures in topsoil of crusty solonetzes developed in the dry-steppe zone and the xero-humus AKL horizon of desert soils showed that they are represented by a brittle spongy crust of any thickness combined with the platy-cellular subcrust. It means that the morphological structural peculiarities of topsoil in desert soils (from brown aridic soils, gray-brown to extremely arid soils) and in crusty solonetzes seem to be practically similar.

Judging by publications (Budina and Medvedev 1966; Lobova 1965), the brown aridic soils (Gypsic Calcisols) have their AJ horizons basically corresponding to its central image, and with inclusions of morphons of the crusty-subcrusty microprofile (akl). The latter may be either complete, composed of both layers, or represented by one of its components. Our observations in the Caspian Lowland have shown that the proportion between the AJ and [akl] morphons depends on texture: the lighter the texture, the lower is the share of the latter. In other words, the [akl] morphons are rather confined to loamy plots with sparse vegetation, while under the bunch of grasses, the AJ morphons are prominent.

Micromorphology of Gypsic Calcisols in the southern Caspian Lowland was studied in the near-coastal hummocky area. The loamy sandy light-humus horizon has a coarse to fine (c/f) related distribution close to gefuric: sand-size grains (coarser particles) are linked by clay (finer material) braces or coatings, and coarser particles are not in contact with each other and thus have no skeletal function; however, packing voids prevail, plasma is carbonate-clay with clear micrite concentrations, very few plant residues are weakly decomposed, and black tissue residues occur as well, and faunal casts are very few. Vesicular pores were recorded in an AJ horizon of heavier texture, sometimes in K subhorizon (Fig. 11.1e). Grain coatings are laminated: the inner ones are clayey, while the outer layers are carbonate-clayey. There were no evidences of platy microstructures (in the subcrusty subhorizon), which may be explained by light texture (Fig. 11.1f).

The xero-humus AKL horizon or [akl] genetic property is inherent to desert soils. In the sequence of soils with increasing aridity up to the extra-aridic Mongolian soils, the [akl] genetic property is gradually replaced by the AKL horizon, as the latter is described in Classification and Diagnostic System of Russian Soils (2004). Moreover, the depth of xero-humus horizon increases, and its differentiation into crusty and subcrusty subhorizons becomes more distinct (Lobova 1965; Pankova 1992; Nogina et al. 1977). Special sampling for the topsoils enabled us to give a micromorphological characterization of the AKL horizon and reveal some mechanisms of its formation.

Micromorphology of the gray-brown desert soil (Calcisols) of Uzbekistan is affected by current aeolian accumulation of loessic weakly calcareous material with inclusions of rounded (pedogenic) aggregates. The crusty ingredient of the AKL horizon abounds in voids-vesicular and packing voids among weakly separated aggregates; the void system is basically open (Fig. 11.1g). The clay plasma is impregnated by carbonates, although there are few fine and discontinuous grain coatings. The skeleton grains are weakly sorted and sand- and silt-sized, few of them are weakly rounded, and they produce a porphyric c/f ratio. Similar properties were described in Yermic Calcisols of Ust'-Urt Plateau (Gubin 1984). The subcrusty ingredient of the AKL horizon has a horizontal stratification—fine platy microstructure; in some soils, weak rounded aggregates included in the plates were recognized (Fig. 11.1h). The vesicular pores are substituted for horizontal packing voids, the impregnation by carbonates is low in some parts, and no evidence of clay plasma mobility was found; however, few fresh castings of woodlice were observed.

Aeolian accumulation is an additional mechanism contributing to the formation of xero-humus horizon or that overlaying it. Aeolian aggregates are easily identified in the in-blown silt by their rounded shape, distinct boundaries, higher clay content, and c/f ratio different of enclosing material; besides, the boundary between the wind-blown deposit and the surface of the crust is clear. Such rounded aeolian aggregates participate in the fabric of the AKL horizon, and they are better seen in the subcrusty soil material, where they form primary peds in platy microstructures. In the upper part, the rounded aeolian aggregates coalesce; voids between them may be mitigated by the infillings responsible for the c/f-related distribution heterogeneity.

Thus, major elements of the AKL horizon being well expressed macro- and micromorphologically in the gray-brown desert soil (Yermic Calcisols) of Uzbekistan are complemented by aeolian phenomena, which in the new Classification and Diagnostic System of Russian Soils (2004) correspond to the "ael" diagnostic property (*n*) (Plate 11.1g,h).

The next group of soils in the sequence of aridic soils under consideration is composed of Mongolian soils of Trans-Altai Gobi, developed from the Quaternary slightly saline fanalluvial deposits. They also have crusts and subcrusty layers (AKL horizon) and desert stony or gravelly pavement with desert varnish.

Micromorphology of the AKL horizon of the pale-brown soil (subtype of gray-brown soils) according to Pankova (1992) or Aridic Calcisols (WRB 2006) is specific and depends on the content of clay particles and their mineralogy in the K and L subhorizons.

The fabric of crusty subhorizon is heterogeneous: the more compact plasmic-silty zones have vesicular microstructure, and those of lighter texture are spongy and have an open-pore system. The mineralogical composition of the skeleton grains is diverse and comprises grains of individual minerals and rock fragments. A particular feature of crust in these soils is the great abundance of various carbonate pedofeatures, namely, fine nodules in the groundmass, micritic concretions, rather thick micritic coatings on sand grains, and micrite "intrusions" into cracks in the fragments of sandstone. All these pedofeatures are interpreted as the recent ones formed owing to the quick evaporative precipitation of carbonate-saturated solutions at any interface. On the other hand, in both crusty and subcrusty subhorizons, thick carbonate-clay coatings on the coarse skeleton grains occur; they are heterogeneous and have a specific wavelike pattern. Presumably, these coatings are relic, since the present-day environment could hardly provide such an active mobilization of both clay and carbonates that are required for their development.

The subcrusty material (its lower boundary is at the depth of 10 cm) is more porous, less compact, and lighter in texture. Sand grains are partially surrounded by clay coatings which form bridges between the grain-bridged-grain microstructure (Stoops 2003). Few fine micritic nodules seen adjacent to the voids seemed to be recent formations produced by short but intensive migrations of carbonate-rich solutions. It can be explained that in the past, clay migration in this subhorizon occurred under moister climate; later, clay coatings might be impregnated by carbonates through calcification process.

The difference in the mineralogical composition of skeleton grains between crust and subcrusty subhorizons may be attributed to aeolian mechanisms. The gray-brown soil of Mongolia (Yermic Calcisols) is confined to drier conditions than the Uzbek soil having the same name; it is also more calcareous in both components of its AKL horizon. The crusty subhorizon is compact, and its plasma is impregnated by carbonates and encloses sand grains, mostly of quartz and also apatite, while coarse grains are rock fragments. The porosity is high and dominated by large rounded pores with clay coatings (illuviation) dissected by fine cracks on pore walls. The layers in coatings differ in birefringence intensity and microgranularity. Presumably, coatings are subject to the effect of carbonate solutions.

The subcrusty subhorizon lacks any horizontal stratification. Coarse sand grains have fine carbonate-clay coatings forming bridges and developing pellicular grain microstructure, where the sand grains are surrounded by fine material which bridges and welds the grains together. It is worth emphasizing that any manifestations of faunal activity or organic matter accumulation are negligible in both parts of the AKL horizon.

Very specific extra-aridic soils (Yermic Regosols) were first described in Mongolia by Efstifeev (1977), some of their properties may serve as keys to disclose mechanisms of aridic pedogenesis. The Yermic Regosols have the desert pavement on the surface, maintaining the AKL horizon stability. Nogina et al. (1977) described that if the size of rock fragments on the soil surface reaches 10–15 cm, the topsoil is better developed and preserved, while between fragments, it is destroyed or outblown by wind.

The crusty material of the AKL horizon is compact and very porous, with closed (isolated) rounded pores. It is peculiar by the rusty hue of plasma and low concentration of micrite in it. The color of plasma is in good agreement with the occurrence of iron oxides pedofeatures: mottles, segregations, and fine threads of iron-oxidating bacterial colonies on the fragments of mafic rocks. There are very few fine silty clay and homogenous layered clay coatings and few fine light-colored residual skeletons in some vesicular pores formed by iron and clay removal. These pedofeatures indicating the mobilization of iron and clay were also described in Yermic Regosols on saline parent material (Golovanov et al. 2005).

Strong enrichment in carbonates is recorded in the subcrusty subhorizon. Clay papules assimilated in the groundmass and iron oxides pedofeatures are also present. The microstructure is platy, and in some plates, vesicular pores are visible. Presumably, the subhorizon was once a crust strongly impregnated by carbonates; later on, it was buried by the wind-blown material serving as a substratum for the present-day crust. The difference in clay mineralogy between the AKL components confirms the contribution of aeolian input of salts together with the fine earth (Chizhikova et al. 1988). Additionally, clay papules may be regarded as witnesses of the solonetzic process.

Summarizing the interpretations of micromorphological characteristics of the Yermic Regosols, it is revealed that traces of solonetzic process and mobilization of iron were not recorded in soils of lower aridity. Former investigations revealed the role of microorganisms in the latter process (Golovanov et al. 2005), and their existence is maintained (during an extremely short period) by moisture stored in the topsoil owing to the closed-pore system. Probably, the evaporation of moisture that is formed due to condensation is limited by the closed shape of vesicular pores; water permeability is the lowest in these soils as compared to other soils of Mongolian deserts (Gunin 1990). The probability of transient solonetzic phenomena is not excluded; Glazovskaya and Gorbunova (2002, 2004) described a similar process as a cryptosolonetzic.

11.4 Conclusions

In a soil sequence with increasing aridity, the diagnostic horizons and properties are combined in a regular way corresponding to the changes in soil formation conditions and soil-forming processes; in the same time, the sequence is in good agreement with diagnostic elements of substantive classification systems. Thus, the soils studied form two groups: with a distinct light-humus AJ horizon and with a xero-humus AKL horizon composed of K and L subhorizons; these groups correspond to two types of pedogenesis. The former is the dry-steppe humus-accumulative type (minimal variant); the latter is the desertic "crust formation" manifested in the structural rearrangement of the solid phase. With this approach, Gypsic Calcisols may be regarded as intergrades between steppe and desert pedogenesis types; Episalic Solonetzes, despite their occurrence in the dry steppe, seem to be closer to the desert type unlike deep and shallow light-humus solonetzes. The principles of the new classification system defined as substantive-genetic strongly facilitate the application of micromorphology for substantiating properties of the diagnostic horizons and interpreting the pedogenetic mechanisms. Since aridic soils are scarce in Russia and occur in the periphery of the aridic world, their diagnostic elements have been micromorphologically tested on a broader range of aridic soils in other countries. The main diagnostic elements (diagnostic horizons and genetic properties) in the new Russian system basically coincide with those in the WRB system, although their taxonomic significance may be different. Thus, the best coincidence exists between the xero-humus horizon (AKL), or genetic property (akl), and the yermic horizon; the aeolian-accumulative genetic property (ael) has a narrower meaning and definition than the aridic property in WRB; takyric is either horizon or diagnostic (genetic) property in different versions of both system, and this does not affect their good correlation in terms of characteristics and origin. The light-humus horizon had a partial analogue in WRB-1998, namely, ochric horizon that was cancelled in WRB-2006.

Nonetheless, the light-humus horizon AJ has a quite definite set of properties, part of which is inherent to some soils of the Caspian semidesert. Being at its northern climatic limit, this horizon is an unstable formation and follows some environmental changes in its characteristics; they are clearly recorded in thin sections and concern the biotic aspects of pedogenesis. Most typical for aridic environments is the xero-humus AKL horizon. It does not reach the status of a diagnostic horizon in brown aridic soils of Russia, where it is only a diagnostic (genetic) property. However, it comprises two thin (below 10 cm in total) subhorizons—crusty and subcrusty—whose thickness and development increase with growing aridity. Each subhorizon has its own set of micromorphological properties. The interpretation of features of the AKL horizon permits to identify processes induced by the dynamics of carbonates (HCO_3-CO_3 equilibrium) as related to hydrothermal regimes, dispersion, and mobilization of clay plasma during short rainy periods and aeolian processes. Modifications of AKL horizon properties are in good agreement with the parent rock features.

Acknowledgments This work was supported by the Russian Foundation for Basic Research (project no. 12-04-00990). We are grateful to Prof. Yevgenia Pankova for providing samples from Mongolia and to Michael Lebedev for the preparation of soil thin sections.

References

Abaturov BD (1985) The development of microtopography and soil complexes. The Northern Caspian semi-deserts as the result of living activity of susliks. In: Mammals in terrestrial ecosystems. Izd-vo Akad. Nauk SSSR, Vsesoyu. Teriologich. Obshch., Moscow, pp 224–249 (in Russian)

Allen BL (1985) Micromorphology of aridisols: soil micromorphology and soil classification. Soil Science Society of America, Madison, pp 197–217

Bol'shakov AF, Borovskii VM (1937) Soils and microrelief of the Caspian lowland (on the basis of materials obtained at the Dzhanybek experimental station of the Soil Science Institute of the USSR Academy of Sciences). Izd-vo VASKhNIL, Moscow, VII, pp 134–169 (in Russian)

Budina LP, Medvedev VP (1966) Brown semidesert soils. Genesis and classification of semidesert soils. Nauka, Moscow, pp 23–58 (in Russian)

Chizhikova NP, Evstefeev Yu G, Pankova EI (1988) Mineralogical composition of the clay fraction in desert soils of Mongolia. Pochvovedenie 8:44–54

Classification and Diagnostic System of Russian Soils (2004) Oikumena, Smolensk, p 342 (in Russian)

Classification and Diagnostics of Soils of the USSR (1977) Kolos, Moscow, p 65 (in Russian)

Efstifeev YG (1977) Soils of extremely arid territories of Mongolia. In: Abstr Rep at the 5th delegate —Conf. of the All-Union Soil Sci Soc, Minsk 6, pp 172–175 (in Russian)

Fedoroff N, Courty MA (1987) Morphology and distribution of textural features in arid and semiarid regions. In: Micromorphologie des Sols – soil micromorphology. AFES, Paris, pp 213–219

Feofarova II (1950) Pseudomorphic substitution of calcite for gypsum. Tr Pochv Inst im VV Dokuchaeva Akad Nauk SSSR 34:202–206 (in Russian)

Feofarova II (1956) Micromorphological characterization of Takyrs. Takyrs of Western Turkmenia and the ways of their agricultural development. Akad. Nauk SSSR, Moscow, pp 351–381 (in Russian)

Feofarova II (1958) Determination of carbonates in salt-affected soils by the microscopic method. Tr Pochv Inst im VV Dokuchaeva Akad Nauk SSSR 53:75–89 (in Russian)

Field Guide on Soil Correlation (2008) Izd-vo RASKhN, Moscow, p 182 (in Russian)

Gerasimova M (2003) Higher levels of description: approaches to the micromorphological characterization of Russian soils. Catena 54(3):319–337

Gerasimova MI, Lebedeva M (2008) Contribution of micromorphology to classification of aridic soils. In: Kapur S, Stoops G (eds) New trends in soil micromorphology. Springer, Berlin, pp 151–162

Gerasimova MI, Gubin SV, Shoba SA (1992) Micromorphological features of the USSR zonal soils. Izd-vo ONTI, Pushchino, p 176 (in Russian)

Glazovskaya MA, Gorbunova IA (2002) Genetic analysis and classification of brown arid soils. Euras Soil Sci 35(11):1139–1148

Glazovskaya MA, Gorbunova IA (2004) Biogenic alkalization of soils as a factor of their textural differentiation. In: Glazovskii NF (ed) Soils, biogeochemical cycles, and the biosphere: the Development of idea advanced by V.A. Kovda. Tovvo Nauchn. Izd. KMK, Moscow, pp 343–357 (in Russian)

Golovanov DL, Lebedeva-Verba MP, Dorokhova MF, Slobodkin AI (2005) Micromorphological and microbiological characterization of elementary soil-forming processes in desert soils of Mongolia. Euras Soil Sci 38(12):1290–1300

Gubin SV (1984) Micromorphological diagnostics of brown and gray-brown soils of the Northern Ustyurt Plateau. In: Nature, soils, and problems of development of the Ustyurt Desert. Pushchino Scientific Centre RAS, pp 127–134 (in Russian)

Gunin PD (1990) Ecology of desertification processes in arid ecosystems. Izd-vo VASKhNIL, Moscow, p 354 (in Russian)

Khademi H, Mermut A (2003) Micromorphology and classification of argids and associated gypsiferous aridisols from central Iran. Catena 54:439–455

Khokhlova OS (2008) Pedogenic carbonates as carriers of memory of soil-forming factors (Case study in steppe areas of the Russian plain). In: Targulian VO, Goryachkin SV (eds) Soil as a memory of biosphere-geosphere-anthroposphere interactions. Izdvo URSS, Moscow, pp 406–437 (in Russian)

Kovda IV (2008) Information on former pedogenic properties and agents recorded in secondary carbonates. In: Targulian VO, Goryachkin SV (eds) Soil as a memory of biosphere-geosphere-anthropo-sphere inter-actions. Izd-vo URSS, Moscow, pp 352–405 (in Russian)

Kovda I, Sycheva S, Lebedeva M, Inozemtzev S (2009) Variability of carbonate pedofeatures in a loess-paleosol sequence and their use for paleoreconstructions. J Mountain Sci 6(2):155–161 (Science Press, co-published with Springer)

Lebedeva-Verba MP, Gabchenko MV (2006) Micromorphological analysis of recent processes in soils of solonetzic complexes of the Northern Caspian region. Pedogenetic Processes. Pochv Inst Im VV Dokuchaeva, Moscow, pp 236–256 (in Russian)

Lebedeva-Verba MP, Gerasimova MI (2009) Macro-and, micromorphological features of genetic horizons in a solonetzic soil complex at the Dzhanybek Research Station. Euras Soil Sci 42(3):237–250

Lebedeva-Verba M, Gerasimova M, Konyushkova M (2008) Micromorphological features of solonetzic horizons as related to environmental events in the Caspian Low-land. In: Proceedings of the 13th international conference on soil micromorphology, Chengdu, China, pp 64–65

Lebedeva-Verba MP, Golovanov DL, Inozemtsev SA (2009) Microfabrics of desert soils of Mongolia. Euras Soil Sci 42(11):1204–1217

Lobova EV (1965) Classification of desert soils of the sub-boreal zone: genesis and classification of Asian soils, Part 1. Akad. Nauk SSSR, Moscow, pp 11–38 (in Russian)

Miedema R, Slager S (1972) Micromorphological investigation of clay illuviation. J Soil Sci 23:309–314

Mills A, Fox S, Poch RM (2008) Micromorphology of surface crusts in the Knevrsvlakte, South Africa. In: Soil micromorphology. Micro-investigation on the Earth's critical zone. Conference Abstracts, China, Chengdu, p 23

Minashina NG (1966) Micromorphology of loess, sierozems, and heilutu soils and some problems of their paleogenesis. In: Zonn SV (ed) Micromorphological method in the study of soil genesis. Nauka, Moscow, pp 76–92 (in Russian)

Minashina NG (1973) Specific features of the microfabric of desert takyric soils in Central Asia. In: Dobrovol'skiy GV (ed) Micromorphology of soils and loose deposits. Nauka, Moscow, pp 42–52 (in Russian)

Netlleton W, Flach D, Brasher G (1969) Argillic horizons without clay skins. Soil Sci Soc Am J 33(1):121–125

Nogina NA, Evstifeev Yu G, Ufimtseva KA (1977) Soils of low-mountain and plain steppes and deserts of Mongolia (Systematization and diagnostics). In: Arid soils: genesis, geochemistry and use. Nauka, Moscow, pp 165–195 (in Russian)

Pagliai M, LaMarca M (1979) Micromorphological study of soil crust. Agrochimica 23:16–25

Pankova EI (1992) Genesis of salinization of desert soils. Izd-vo VASKhNIL, Moscow, p 132 (in Russian)

Poch RM, De Cosle W, Stoops G (1998) Pore space characteristics as indicators of soil behaviour in gypsiferous soils. Geoderma 87:87–109

Rode AA, Pol'skii MN (1958) Soils of the northwestern Caspian semidesert and their reclamation. Tr Pochv Inst Im VV Dokuchaeva Akad Nauk SSSR Moscow 56:3–214

Romashkevich AM, Gerasimova MI (1977) Major aspects of the micromorphology of arid soils of the Soviet Union. In: Kovda VA, Lobova EV (eds) Arid soils: genesis, geochemistry and use. Nauka, Moscow, pp 239–253 (in Russian)

Romashkevich AI, Gerasimova MI (1982) Micromorphology and diagnostics of soil formation. Nauka, Moscow, p 125 (in Russian)
Shahid SA, Mufti KA (1994) A modern submicroscopic method for in-situ micromorphological and microchemical analysis of soil components. I. Salt-crust. Mod Agric 5(4,5):32–39
Shahid SA, Jenkins DA et al. (1994) Mineralogy and micromorphology of salt crusts from the Punjab, Pakistan. In: Ringrose-Voase AJ, Humphreys GS (eds) Soil Micromorphology: studies in management and genesis, vol 22, Development in soil science. Elsevier, Amsterdam, pp 799–810
Stoops G, Ilaiwi M (1981) Gypsum in arid soils: morphology and genesis. In: Proceedings of the 3rd international soil classification workshop, Damascus, Syria, pp 75–185
Stoops G, Poch RM (1994) Micromorphological classification of gypsiferous soil materials. In: Ringrose-Voase AJ, Humphreys GS (eds) Soil micromorphology: studies in management and genesis, vol 22, Developments in soil science. Elsevier, Amsterdam, pp 327–332
Stoops G (2003) Guidelines for analysis and description of soil and regolith thin sections. Soil Science Society of America, Madison, 184 pp
Tursina TV, Pankova EI, Yamnova IA (1983) Micromorphology of readily and difficultly soluble salts and microfabrics of salt-affected soils. In: Targulian VO (ed) Micromorphological diagnostics of soils and soil-forming processes. Nauka, Moscow, pp 89–109 (in Russian)
Tursina TV, Verba MP, Nikol'skii AG (1984) Specificity of the microfabric of soils buried in loess deposits of Central Asia (by the example of the Tashkent and Nanaisk Paleosol-Loess Sequences). Pochvovedenie 4:98–109
Verba MP (1990) Macro- and microfabrics of soils of the Syr Darya experimental station. In: Soil-forming conditions and properties of difficultly reclaimed soils of the Dzhizak steppe. Izdvo VASKhNIL, Moscow, pp 48–55 (in Russian)
World Reference Base for Soil Resources (1998) FAO/ISRIC/ISSS, World Soil Resources Rep 84. FAO, Rome, pp 46–51
World Reference Base for Soil Resources (2006) FAO/ISRIC/ISSS, World Soil Resources Rep 103, FAO, Rome, pp 36–43
Yamnova IA (2006) Recent soil-forming processes in low-sodium solonelzes and in chestnut soils. Pedogenetic processes. Izdvo RASKhN, Moscow, pp 281–294 (in Russian)
Yarilova EA (1963) Micromorphology of soils of the Ergeni ipland. Pochvovedenie 2:33–39
Yarilova EA (1966) Specific micromorphological features of solonetzes in chernozemic and chestnut soil zones. In: Zonn SV (ed) Micromorphological method in the study of soil genesis. Nauka, Moscow, pp 58–75 (in Russian)

Chapter 12
Digital Mapping of Gypsic Horizon Morphotypes and Soil Salinity in an Old Alluvial Piedmont Plain of Uzbekistan

Dmitry L. Golovanov and Irina A. Yamnova

Abstract The digital maps of soil salinity, gypsum, and gypsum pedofeatures in soils (gypsic calcisols, gypsisols, and solonchaks) of the Dzhizak experimental station (Uzbekistan) are compiled from the results of soil survey conducted during 1980 and 2008, using digital elevation model (DEM) and remote sensing data. The study area represents a part of the Golodnaya Steppe piedmont plain to the north of Turkestan ridge. The macro- and micromorphological descriptions of gypsic horizons made it possible to distinguish three different morphotypes. The morphological features of gypsic horizons were classified, and their distribution map was prepared using GIS tool. The spatial distribution of different morphotypes of gypsic horizons showed distinct correlations with the soil salinity, the groundwater level, and the character of soil water regime. It was established that the presence of different gypsic horizons should be taken into account in the new substantive-genetic classification of Russian soils at high taxonomic levels. Studies completed after 20 years, when the groundwater level dropped by about 1 m, confirmed that the gypsum content decreased in soils; however, the major morphotypes were preserved. The micromorphological investigations demonstrated that certain changes took place in the microfabric of gypsum pedofeatures. Thus, fine dispersed crystals of gypsum disappeared from the soil profiles, and the number of pseudomorphic substitutions of calcite for gypsum crystals increased significantly attesting to the progressive calcification of soil profiles.

D.L. Golovanov
Senior Lecturer, Department of Soil Geography and Landscape Geochemistry, M.V. Lomonosov Moscow State University, Moscow 119991, Russia

Faculty of Geography Moscow, Lomonosov University, Moscow, Russia
e-mail: dm_golovanov@mail.ru; golovanov.dmitry.leonid@gmail.com

I.A. Yamnova (✉)
Soil Salinity, V.V. Dokuchaev Soil Science Institute, Pyzhevsky per. 7, Moscow 119017, Russia
e-mail: irinayamnova@mail.ru

S.A. Shahid et al. (eds.), *Developments in Soil Classification, Land Use Planning and Policy Implications: Innovative Thinking of Soil Inventory for Land Use Planning and Management of Land Resources*, DOI 10.1007/978-94-007-5332-7_12,

Keywords Digital mapping • Gypsic • Micromorphology • Soil salinity • Uzbekistan

12.1 Introduction

Soluble salts and gypsum ($CaSO_4$•$2H_2O$) are common soil constituents of arid region. Gypsiferous soils contain more than 1% gypsum in the horizon. The qualitative indices of soil gypsum and the shape and size of gypsum pedofeatures are important characteristics of gypsiferous soils that should be taken into account along with quantitative gypsum data as well as the depth and thickness of gypsum-bearing horizons (Kerzum 1974; Minashina and Shishov 2002; Stoops and Ilaiwi 1981). The morphological characteristics of gypsum in soil are important not only from the pedogenetic viewpoint but also from the viewpoint of ameliorative practices (Pankova and Yamnova 1987). In this study, the morphology of gypsum pedofeatures, their role in the microfabric of gypsiferous soils, their pedogenetic interpretation, the regularities of spatial distribution of different morphotypes of gypsiferous horizons, and the means of their cartographic representation are investigated.

12.2 Materials and Methods

Soils of the Dzhizak experimental station were studied. The soils represent part of the Golodnaya Steppe piedmont plain to the north of the Turkestan ridge, composed of the sediments of merging alluvial fans. The study plot (100.67 ha) exists to the northwest of the settlement of Buston, 17 km to the east of Dzhizak city (Uzbekistan). According to the scheme of natural zoning (Pankova and Muradova 1990), 21 natural districts are distinguished within the Dzhizak Steppe area. With respect to the soil, lithological, and geomorphic conditions, the upper and lower zones of the piedmont plain can be separated. Automorphic sierozemic soils predominate in the upper zone, whereas semihydromorphic and hydromorphic meadow-sierozemic and solonchak soils with different contents of soluble salts and gypsum predominate in the lower zone (Fig. 12.1a). The territory of the station occupies a local plateau with nonsaline and non-gypsiferous soils (within the upper 1 m) and the valley of a temporary stream (sai) with hydromorphic strongly saline and gypsiferous soils. Meadow-sierozemic soils occupy an intermediate position (Kochubey 1990).

The study area is characterized by the continental subtropical climate; the frostless period lasts for 207–217 days. The mean annual precipitation is 300–400 mm, of which 80% falls in the winter and spring seasons. The relative humidity of the air is very low, as the territory lies close to the vast Kyzylkum Desert. Vegetation conditions are differentiated: the valleys of temporary streams with salt-affected soils are occupied by halophytic plants (*Halocnemum strobilaceum*, *Salicornia rerbacla*, *Halostachis caspica*, *Kalidium caspicum*, and numerous species belonging to the *Salsola genus*). Elevated interfluves with typical sierozemic soils are occupied by

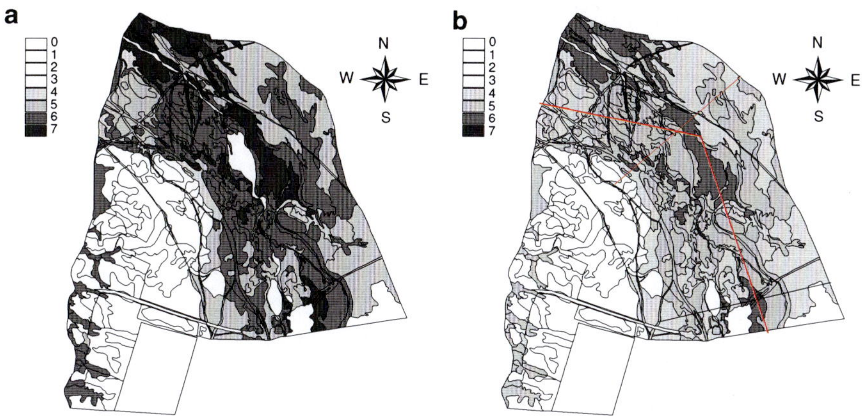

Fig. 12.1 Gypsum content (%) in soils of Dzhizak experimental station in 1987 (**a**) and 2008 (**b**). Legend: 0 (data not available); 1 (<2%); 2 (2–5%); 3 (5–10%); 4 (10–20%); 5 (20–40%); 6 (40–60%); 7 (>60%)

communities of ephemeral plants, including annual (*Carex pachydactylus*, *Poa bulbosa*) and perennial species. The groundwater depth is more than 10 m on the interfluve; at the bottom of the sai valley, the groundwater may rise to the surface. The groundwater salinity is very variable (from slightly saline to strongly saline).

During the fieldwork, the soil map of the territory was developed, and the major types of soils were sampled. Gypsum accumulations in the soil profiles were studied with the use of macro-, meso-, and micromorphological methods.

Several approaches are used to characterize gypsiferous soils. According to the FAO recommendations (FAO 1979), the gypsum content and the forms of gypsum accumulations are taken into account. Kerzum (1974) considers the gypsum content and the depth of the upper boundary of gypsiferous horizons as the main criteria. Minashina and Shishov (2002) suggest that the gypsum content, the depth and thickness of gypsic horizons, and the size and shape of gypsum crystals should be taken into account. The particular criteria for the studied area were developed by Pankova and Mazikov (1985). In this chapter, we follow the scheme suggested by these authors with some modifications. The following criteria of gypsiferous soils are used: (1) gypsum content in the horizon of its maximum accumulation within the 2-m-deep layer, (2) depth of the horizon with maximum gypsum content, (3) depth of the upper boundary of the gypsic horizon (with the gypsum content >10%), and (4) morphology of gypsum pedofeatures in the horizon with the maximum gypsum content.

The chemical analyses of soil water extracts and groundwater samples were performed by standard methods. The gypsum contents in the soils were determined by two methods: (1) standard method of dissolving $CaSO_4$ and further precipitation of SO_4^{2-} and (2) hydration loss method (which can be used in the samples with the high content of gypsum that is more that 10%), based on the loss of crystalline water of gypsum when heating a soil sample. Standard statistical and GIS procedures were used to generalize the results and develop the map of gypsum pedofeatures in the soils of the Dzhizak experimental station.

12.3 Results and Discussion

12.3.1 Gypsum Contents and Depth of Maximum Gypsum Accumulation

With respect to the gypsum content, the soils are divided into (a) the low-gypsiferous soils of the interfluve and (b) the highly gypsiferous soils of the sai valley. In the former group, the gypsum content varies from 0.2 to 1–2% at a depth of 80–150 cm. On the slopes of the interfluve and in the sai valley, the content of soluble salts and gypsum in the soil profiles increases. At the groundwater depth of 2–3 m, the soils contain up to 20–30% of gypsum in the middle horizons. Meadow-sierozemic soils also contain 20–30% of gypsum in the middle and deep horizons.

Soils of the valley bottom contain high amount of gypsum. Soils are represented by slightly saline and highly gypsiferous dark meadow soils with the groundwater level of about 2 m. Only small areas are occupied by nonsaline and non-gypsiferous peat meadow-swampy soils with the groundwater level of less than 1 m.

The high gypsum content (up to 50–60%) is typical of the floodplain soils. The gypsum content in these soils is highly variable and depends on the degree of soil hydromorphism. On the natural levees and on the high floodplain, combinations of dark meadow soils and crusty solonchaks are formed.

In the meadow soils of the high terrace, the gypsum content increases down the soil profile up to 30–40% at the depth 50–75 cm. Meadow solonchaks and crusty solonchaks near the bed of the temporary stream are especially rich in gypsum (up to 60% in the topsoil).

The statistical analysis of data on the gypsum content in the studied soils and the groundwater depth shows a reliable correlation. At the same time, there is no direct correlation between the gypsum content in the soils and the groundwater salinity, as well as between the gypsum content and the concentration of Ca^{2+} and SO_4^{2-} ions in the groundwater, although there is a certain range of concentrations of Ca^{2+} in the groundwater (20–30 mmol l^{-1}) closely correlated to the maximum accumulation of gypsum in the soils. The relationship between these parameters is approximately ($R=0.62$) described by the normal distribution curve.

On the basis of these regularities, the soil map of the territory and the DEM, a map characterizing the gypsum content in the horizon with the maximum accumulation of gypsum, has been developed using the GIS tool.

12.3.2 Morphotypes of Gypsum-Containing Horizons

As the regime of soil moistening in the studied area is largely controlled by seasonal fluctuations of the groundwater level, different morphological types of gypsum pedofeatures may be present in a soil profile. However, the morphology of gypsum pedofeatures in the horizon of the maximum accumulation of gypsum can be considered indicative of the particular type of the soil water regime (Yamnova 1990; Yamnova et al. 2007).

Three major morphotypes of gypsum pedofeatures in the horizons of maximum gypsum accumulation can be distinguished that is incrustational, nodular, and powdery types. The incrustational type of gypsum is most widespread; it is typical of both hydromorphic and automorphic soils and is formed in the zone of the constant presence of capillary water. In the hydromorphic soils, it is found in the upper part of the profile. In the automorphic soils (sierozems) with the groundwater level deeper than 10 m, it is present in the lower part of the soil profile. Fine-, medium-, and coarse-crystalline subtypes are distinguished. The latter subtype is formed in the hydromorphic meadow soils and solonchaks with the groundwater level of 0.9–0.95 m. Large gypsum crystals are formed in interaggregate pores in the lower zone of the capillary water, immediately above the groundwater table, and the gypsum content may reach up to 60%. The edge zone of some crystals consists of freshly precipitated gypsum pointing to their continuous growth due to the hydrogenic precipitation (Fig. 12.2a).

The nodular type is typical of the meadow-sierozemic and meadow soils with the groundwater level of about 2 m. This type is characterized by the development of complex carbonate-gypsum nodules, whose origin is related to the periodical saturation of the soil with floodwater. The maximum accumulation of gypsum is found in the upper horizons, where it is about 20–30%. The nodules are heterogeneous in their composition: gypsum druses (cavities lined with crystals) are unevenly impregnated with fine-grained carbonate (Fig. 12.2e, g).

The powdery type of gypsum is typical of the meadow solonchaks and meadow-swampy soils with the groundwater depth of 0.8–1.0 m. The origin of this form is explained by the periodically alternating wetting and drying cycles; during the drying stage, the rate of the groundwater evaporation sharply increases, and very fine gypsum crystals (gypsum powder) are formed. Two subtypes are distinguished in this morphotype: (a) the crumb-like gypsum typical of less moistened soils with the groundwater level of about 1.05–1.10 m and (b) the marly gypsum typical of better moistened soils with the groundwater level of about 0.8–1.0 m. The horizon of maximum accumulation of powdery gypsum is found in the middle part of the profile of hydromorphic solonchaks containing up to 60–80% of gypsum. Lens-shaped (lenticular) microcrystals of gypsum have also been observed. In the crumb-like and marly gypsum horizons, the pseudomorphs of carbonates after gypsum are often found. In the marly subtype, this process is more intense; in some loci, the former crystals of gypsum are completely replaced by carbonate. This feature is indicative of the strong degree of soil hydromorphism. In the crumb-like subtype, the development of separate purely carbonate aggregates takes place (Fig. 12.2c).

The morphotypes of gypsic horizons display definite relationships with the types of soils, their position in the landscape, and the groundwater level. Thus, the incrustational form of gypsum is typical of the interfluves with sierozemic soils; fine- and medium-crystalline subtypes predominate. The nodular forms of gypsum are only formed in the soils of the terraces. The powdery form of gypsum is typical of the lower most positions in the sai valley. Data on the areas occupied by the soils with different morphotypes of gypsum within the surveyed territory are presented in the Table 12.1.

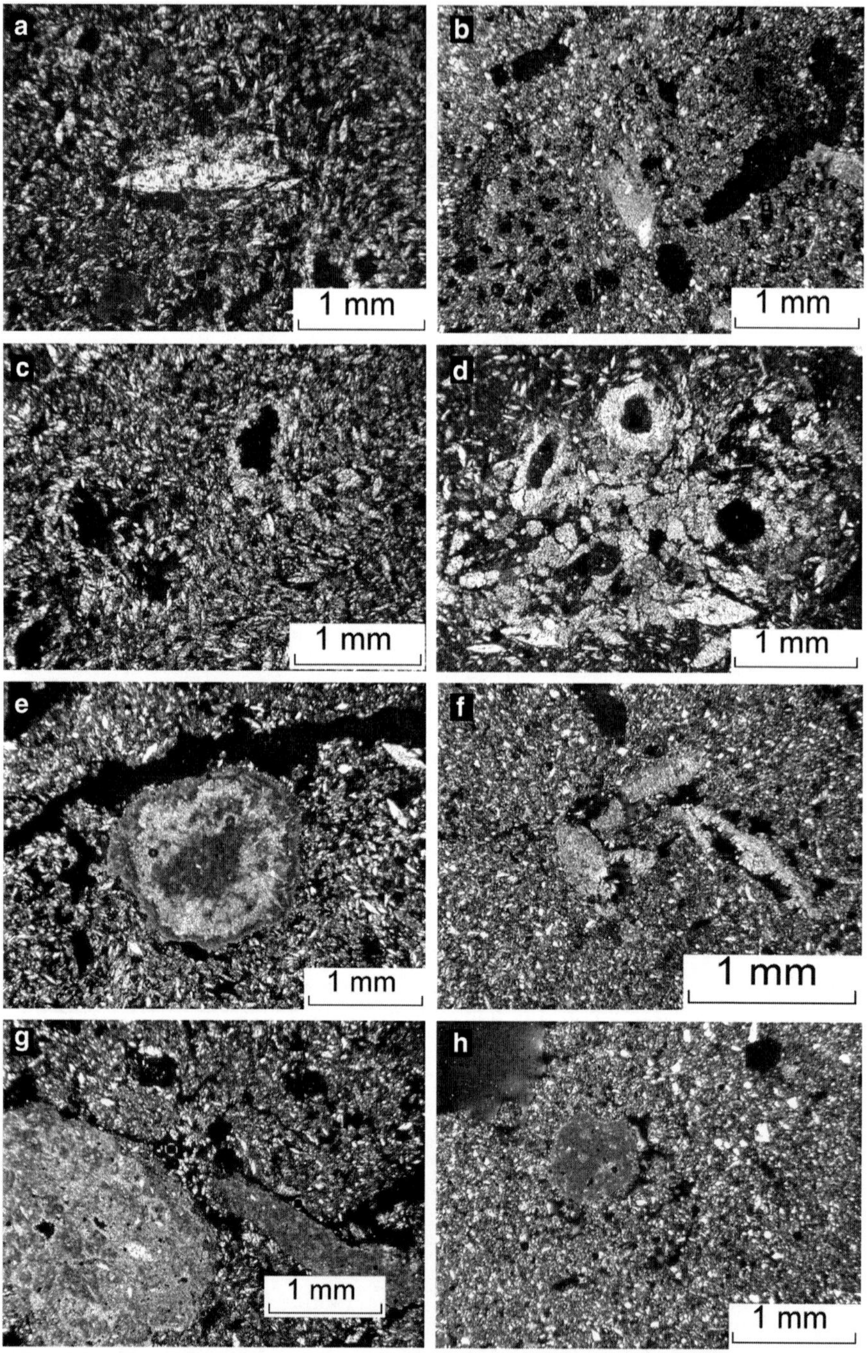

Fig. 12.2 Incrustational gypsum crystals (**a**); micropedofetures changes after melioration (**b**); powdery gypsum (**c**); micropedofetures changes after melioration – replacement of gypsum-bearing neoformations by carbonate pseudomorphoses (**d**); nodular and heterogeneous gypsum (**e**); micropedofetures changes after amelioration – the features of dissolution of the gypsum crystals (**f**); nodular type of gypsum: the big nodular and small crystals of gypsum (**g**); micropedofetures changes after melioration – the absence of gypsum crystals (**h**)

Table 12.1 Changes in the groundwater level and the gypsum content in the horizon of its maximum accumulation at the territory of Dzhizak model polygon for 1987–2008

		Groundwater level (cm)			Gypsum content (%)		Decrease in gypsum content (%)	
Pit No	Soil type (1987)	1987	2008	Difference	1987	2008	Relative	Absolute
I-11	Gleyic fluvisol	100	215	115	–	–	–	–
I-29	Solonchak (gypsum)	105	200	95	61.1	41.0	30.0	21.1
I-41	Solonchak (meadow)	90	120	30	72.5	22.5	70.0	50.0
I-79	Gleyic fluvisol	140	222	82	59.5	24.0	60.0	36.0
I-81	Gleyic fluvisol (meadow steppe-like)	140	222	82	–	–	–	–
I-87	Gleyic fluvisol (meadow steppe-like)	100	250	150	–	–	–	–
I-99	Gleyic fluvisol	188	230	42	36.0	29.0	20.0	7.0
I-109	Gleyic fluvisol (meadow steppe-like)	100	200	100	40.5	24.0	40.0	16.0
II-39	Solonchak (gleyic)	66	180	114	38.0	37.0	2.5	1.0
II-69	Solonchak	72	210	138	55.0	21.0	61.8	34.0
II-79	Solonchak	75	200	125	48.0	28.0	42.0	20.0
II-87	Gleyic fluvisol	100	250	150	32.0	27.0	15.0	5.0
II-105	Gleyic fluvisol	160	250	90	32.7	21.0	35.0	11.7

The mapping of different morphotypes of gypsic horizons can be based on two approaches: (a) the direct mapping of soil features that are not taken into account in the applied soil classification system (Sorokina 2000) and (b) the development of the existing soil classification system with inclusion of such features into it.

As shown in the work of Skvortsova et al. (2006) by the example of Albeluvisols (soddy-podzolic soils), the morphological features of soil horizons that are not taken into account in the existing classification system can be successfully mapped on detailed and large scales; present study has confirmed this assumption. We argue that different morphotypes of gypsic horizons can be mapped on a detailed scale, as they occupy sufficiently large areas with definite position in the landscape. This is important, as such an approach may help us to increase the information capacity of soil maps via showing the features that are not reflected in the soil classification system but can be reliably mapped (Referentiel pedologique principaux sols d'Europe 1992).

According to the second approach, the distinguished morphotypes of gypsic horizons can be included in the refined soil classification system. The horizons of hydrogenic accumulation of gypsum are not separately distinguished in the new substantive-genetic Russian soil classification system. The presence of gypsum is taken into account at the level of diagnostic properties. The results of this study and the review of published materials on gypsic soils of arid territories allow us to suggest that three major morphotypes of hydrogenic gypsic horizons can be included in the list of diagnostic horizons: the incrustational gypsic horizon *CSI*, the nodular (concretionary) gypsic horizon *CSK*, and the marly gypsic horizon *CSM*. Their inclusion in the soil classification system makes it possible to give the following interpretation for the studied soils.

Lately, due to constructing the drainage channel at the territory of model polygon, the groundwater level dropped; the water regime and some valuable soil properties revealed changes including those that have taken place in the composition of water-soluble salts, in the content and forms of gypsum (Table 12.1).

When comparing the results obtained to determine the gypsum content in the horizon of its maximum accumulation in the same soil pits described in 1987 and 2008 (Fig. 12.1a, b), it seemed reasonable to conclude that the gypsum content showed changes being practically decreased in all soil pits under study; its decrease varied in a wide range: absolute decrease accounted for 2.5–70% and relative decrease from 5 to 50%; there is no linear correlation between the initial content of gypsum in 1987 and its content in 2008 ($R^2=0.0002$); at the same time, there exists a clearly expressed dependence between the decrease in gypsum and its initial content ($R^2=0.82$ for absolute and $R^2=0.56$ for relative decrease).

It gives the possibility to assume that the gypsum decrease that resulted from dropping the groundwater depth may be described by using the following exponential dependence:

$$\frac{\mathrm{d}X}{\mathrm{d}t} = -kX, a\ X = X_0 * \exp(-kt) + C,$$

where X_0 is the initial content of gypsum, t is time, and k and C constants determined empirically.

Based upon the data obtained in 1987, this equation allows not only compiling the map of gypsum content in soils for 2008 but also forecasting its content for the next 20 years under the unchangeable conditions (climate, groundwater depth, land-use regime).

Despite the low precipitation and the non-leaching water regime in automorphic sierozems, the studied soils display declining the gypsum content, which is diagnosed not only analytically but also morphologically and micromorphologically in particular. At the same time, the degree and chemical composition of soil salinization do not change in the same way. Probably, this is explained by the fact that due to construction of the drainage channel, the water regime has been changed and became pulsating instead of exudative one. As a result of the groundwater rise during pluvial periods, the water-soluble salts and particularly sodium chloride served as a cause of partial gypsum dissolution because it is always highly dissolved in the presence of chlorides. The decrease in the salt content and especially anions leads to intermediate formation of Na_2CO_3 (at the expense of sodium in the soil solution and CO_2 in soil air), its interaction with gypsum, and subsequent input of sodium sulfate into the groundwaters:

$$Na_2CO_{3\,(solution)} + \underset{(solid\,phase)}{CaSO_{4\,(TB.phase)}} = \underset{(solid\,phase)}{CaCO_{3\,(TB.phase)}} + Na_2SO_{4\,(solution)}$$

Gypsum displays isomorphic replacement by calcite what is clearly seen in thin sections. It is evident that in case of the pulsating water regime and fluctuation of the capillary fringe, the gypsum is dissolved to a considerable extent.

In comparing the pedofeatures caused by the reclamation of gypsiferous soils for 20 years, one should stress that the main pedofeatures of genetic horizons revealed high changes. It allows diagnosing basic trends in elementary soil formation processes due to cardinal changes in the soil-forming factors during 20 years of drainage of hydromorphic soils. In general, the decrease in the gypsum content in horizons of its maximum accumulation, the increase in the organic matter amount, and the enrichment of the topsoil with mica-containing minerals probably at the expense of floodwaters serve as evidence that these gypsiferous soils have become rich in fertility especially potassium contents. A comparative micromorphological analysis of soils with different gypsum morphotypes made it possible to identify the following elementary soil-forming processes conditioned by 20 years of their reclamation: (1) total leaching of gypsum from the upper 0–50-cm layer of all the gypsiferous soils under study (Fig. 12.2b, f); (2) formation of aggregates composing of compacted clayey slightly carbonate material with some gypsum crystals; (3) intensification of the humus-accumulative process, which is manifested through increasing the organic matter content and plant residues in soil voids; (4) oxidogenesis process promoting the occurrence of ferruginous neoformations in the form of small concretions and rings; (5) increase in illuviation of the ferruginous-clayey material, thus forming cutans along the main fissures. A higher content of gypsum-bearing neoformations remains in lower soil horizons; however, the latter reveal traces of dissolution or replacement by carbonate pseudomorphoses (Fig. 12.2d). In soils with nodular (concretions) morphotype of gypsic horizons which have being subject to reclamation for 20 years the amount of fine grains showed their decrease at the microlevel, the content of gypsum crystals became drastically dropped to the depth of 50 cm (Fig. 12.2h), and the interpreted material revealed a higher content of the ferruginous-clayey plasma incorporated by large diamond-shaped gypsum crystals. Downward the profile, the content of gypsum remains rather high, but all the gypsum crystals possess the features of dissolution.

12.4 Conclusions

Typification of gypsum neoformation forms has been performed. As a result, 3 morphotypes of gypsic horizons were identified; (1) incrustative with the following subtypes: fine-crystalline, moderate-crystalline, and coarse-crystalline; (2) concretional; and (3) mealy with spongy and marly subtypes. The micromorphological description of gypsic horizons allowed identifying the specific gypsiferous pedofeatures particularly the different replacement degree of gypsum crystals by carbonates. A clearly expressed correlation has been established between morphotypes of gypsic horizons and the soil hydromorphism pattern and degree. Morphotypes of gypsic horizons enable to be mapped in detail. In a modern soil classification, the hydrogenic diagnostic horizons should be supplemented with hydrogenic-gypsiferous ones. Following the logic of the new soil classification in Russia (Classification and Diagnostics of Soils in Russia 2004), the gypsum content in soil may be taken into complete account at different taxonomic levels depending on the form and formation intensity of gypsum (species, subtype, type, order). In the course of field investigations carried out in 2008, the following results have been obtained: the soils under study reveal decreasing the gypsum content, diagnosed not only analytically but also morphologically and micromorphologically. Being partially dissolved, the gypsum displays isomorphic replacement by calcite, what is clearly seen in thin sections; there is a direct dependence between the gypsum decrease and its initial content. Based upon a great amount of data obtained in 1987, it was possible to compile a map demonstrating the gypsum content in soils in 2008 as well as to forecast it for the next 20 years under unchangeable conditions (climate, groundwater level, land-use regime); and changes in the salt content and composition are different by nature due to a higher spatial-temporary variability of soil salinity within the elementary soil area.

References

Classification and Diagnostics of Soils in Russia (2004) Smolensk, Oikumena, p 342

FAO (1979) A provisional methodology for soil degradation assessment. FAO, Rome

Kerzum PA (1974) Evaluation system of soils in Tajik Soviet Socialist Republic. In: Soil Evaluation of Tajikistan. Dushanbe (Donish), pp 5–108

Kochubey SS (1990) A brief characteristic of natural peculiar features for the territory of Syr-Darya experimental station. In: Proceedings Soil Sci Institute, Moscow, pp 26–30

Minashina NG, Shishov LL (2002) Gypsiferous soils: distribution, genesis, classification. J Pochvovedenie 3:273–281

Pankova YeI, Mazikov VM (1985) Methodical recommendations to use remote spacing data for assessment of soil salinity and salt surveys of the areas under irrigation within the cotton growing zone. In: Proceedings of Soil Sci Institute. Moscow, p 73

Pankova YeI, Muradova ZM (1990) Soil-lithologic-geomorphological regionalization of the Golodnaya Steppe piedmont plain. In: Proceedings Soil Sci. Institute. Moscow, pp 4–12

Pankova YI, Yamnova IA (1987) Forms of gypsum pedofeatures as a factor responsible for reclamative properties of gypsiferous soils. J Pochvovedenie 7:101–109

Referentiel pedologique principaux sols d'Europe (1992) INRA. Paris, p 222

Skvortsova YB, Rukhovich DI, Koroleva PV (2006) Compilation of schematic maps for the threshold space of soils by means of GIS-technologies and micromorphometric methods. J Pochvovedenie 11:1352–1362

Sorokina NP (2000) Elementary soil structures of croplands: mapping experience. J Pochvovedenie 2:158–168

Stoops G, Ilaiwi M (1981) Gypsum in arid soils, morphology and genesis. In: Beinroth FH, Osman A (eds) Proceedings of the 3rd international soil classification workshop, 14–23 Apr 1980. Arab Center for the Studies of Arid Zones and Drylands, Damascus, Syria, pp 175–185

Yamnova IA (1990) Gypsiferous soils in Dzhizak steppe. In: Proceedings of Soil Sci Institute Moscow, pp 37–39

Yamnova IA, Golovanov DL, Lebedeva-Verba MP (2007) Localization of gypsic pedofeatures as a result of genesis and functioning of the soil cover in arid areas. In: Proceedings of the international conference "Spatial-temporary organization of the soil cover: theoretical and applied aspects", St-Petersburg, pp 386–389

Chapter 13
Soils in Arid and Semiarid Regions: The Past as Key for the Future

Bernhard Lucke, Iourii Nikolskii, and Rupert Bäumler

Abstract Growing populations, increasing food demand, and technological advances may soon lead to intensifying land use in semiarid and arid countries through the spread of irrigated agriculture. Improved water harvesting and desalinization technology, coupled with higher efficiency of regenerative energies, might allow to widely extend irrigated areas. While this is a positive development in the light of growing demands for water and food, it presents challenges for land-use planners. Negative examples like Lake Aral make clear that a careful analysis is required before embarking on large-scale irrigation projects.

Soils are central for assessing the impacts of irrigation in the desert. For long-term projects as outlined above, it is insufficient to consider only the present soil distribution. It should also be considered how soils will change under irrigation. In this context, the past is a key for the future, since the modeling of future soil development can be calibrated using reconstructions. Soil surveys which consider the archival role of soils and sediments can partly be used to understand the landscape history and identify risk areas. Paleosols can be evaluated as indicators how changes of moisture availability will affect soil properties and which time frames are involved. This can be coupled with modeling of future soil development. A major methodological challenge for this approach is the use of different parameters and time frames in reconstruction and modeling, which have to be "translated" using experimentally determined relationships.

B. Lucke (✉) • R. Bäumler
Institute of Geography, University of Erlangen-Nürnberg,
Kochstr. 4/4, 91054 Erlangen, Germany
e-mail: bernhard.lucke@web.de

I. Nikolskii
Hidrociencias, Colegio de Postgraduados,
Campus Montecillo, Texcoco, Mexico
e-mail: nikolski@colpos.mx

S.A. Shahid et al. (eds.), *Developments in Soil Classification, Land Use Planning and Policy Implications: Innovative Thinking of Soil Inventory for Land Use Planning and Management of Land Resources*, DOI 10.1007/978-94-007-5332-7_13,

Long-term, large-scale irrigation in arid regions will mean a significant change of the environment and a departure from the conservative idea of sustainability, toward a concept which has been named "progressive development." Its success chances depend largely on our understanding and correct prediction of the consequences of man-made changes of the environment.

Keywords Arid and semi-arid • Dry land • Fertility index • Paleosols • Sustainable development

13.1 Introduction

Most semiarid regions face strong population growth. Modern sanitation needs, industry, and irrigation led to exploding water demands, which can at this time only be met by tapping fossil water reserves like the Disi aquifer between Jordan and Saudi Arabia (Amery and Wolf 2000). The availability of fossil water led to irrigation of the deserts which in the case of the Disi aquifer enabled Saudi Arabia to become the world's sixth largest wheat exporter until 1992 (FAO 1997). In 2008, the Saudi Arabian government had to stop all subsidies to agriculture in order to save the rapidly shrinking groundwater (Reuters 2008), but the fossil irrigation water might in the future be replaced by sewage, more water harvesting, or desalinated water. In any case, it seems very likely that dry lands will face new and significant human land-use pressures, especially from irrigation.

"Sustainable development" is an important keyword regarding these pressures on the environment, which aims to preserve resource availability for future generations by limiting their extraction to the amount which is naturally replenished. However, facing tremendous growth of population and resource needs, it was recently suggested to replace this concept with "progressive development": actively changing the environment in order to improve the human resource basis (Issar 2008).

If this concept is realized, dry lands might not be as marginal as they used to be. As demonstrated by the case of Saudi Arabia, these areas can be very productive as long as enough water is available. If we assume that new methods of water harvesting and desalination can provide this water, it will be important to ensure that large-scale irrigation projects in the dry lands do not lead to salinization and soil degradation. This can be illustrated by the example of Lake Aral, where inefficient irrigation caused a tremendous environmental catastrophe and soil fertility was irreversibly lost due to saline groundwater rising to the surface (Aydarov et al. 1992).

Soil surveys are an important tool to evaluate the land's suitability, for example, for irrigated agriculture. If a sufficiently detailed database is available and consulted before putting areas under new land use, the risk of land degradation can be minimized. However, a soil database can provide even more: It can allow modeling how soil properties will change (Nikolskii et al. 2002). A systematic modeling of soil development could not only provide recommendations to ensure that land use will not lead to soil degradation but could be used to develop strategies to improve soil properties over time. However, models often suffer from uncertainties regarding parameters which can be used as indicators and the involved time frames. These challenges can be addressed by calibrating predictions with reconstructions of past developments. Many dry lands were subject to climate changes during their history which are documented by paleosols. Systematically combining the investigation of these paleosols with the modeling of soil genesis will mutually improve both the precision of reconstructions and the reliability of models aiming to predict future changes of soil properties. However, although theoretically promising, the combination of modeling and reconstruction is a new and challenging approach that requires a careful selection and evaluation of suited parameters.

13.1.1 Reconstructing Past Soil Development

Paleosols deliver long-term records of soil development and are a key for dealing with global warming, since they inform us how soils and landscapes changed under past land use and climate variations. In recent years, important progress was achieved regarding paleosols in semiarid regions of Jordan and the Levant, mostly limestone soils which are often obscured by calcareous material (Cordova et al. 2005; Maher 2005; Lucke 2008). These studies of the past allow better discrimination between human and natural forcing, since the paleosol record covers periods with and without agriculture.

Gvirtzman and Wieder (2001) showed that paleosols are archives that are suited to document the crossing of thresholds in landscape dynamics, which makes them best suited to investigate desertification in the light of climatic variations. Paleosols are usually investigated using genetic parameters, such as color, pH, electrical conductivity, organic matter, $CaCO_3$ content, soil development indices based on iron and manganese oxides, magnetic susceptibility, and texture (Lucke 2008).

13.1.2 Modeling Future Soil Development

It is very important and difficult to predict long-term soil alteration under irrigation because the productivity of agricultural lands strongly depends on soil fertility, which can be deteriorated under irrigation. Additionally, agricultural soils play an important role in the global ecosystem. Soil carbon sequestration can be important to mitigate the extent of global warming and soil degradation causing desertification.

There are well-known methodologies to predict the impact of irrigation on soils regarding the development of its salinity or alkalinity (Nikolskii 1996). These phenomena materialize during relatively short periods (several years or few decades). But in general, the relationship between soil and irrigation is much more complicated and materializes during long terms (at least several decades or centuries). Even with high-quality water, irrigation can have negative impacts on different soil-forming processes, such as microbiological activities, physical and chemical processes, and leaching and loss of useful chemical and physical substances.

A mathematical modeling of such processes during decades or centuries is not very reliable. A possible approach could be based on the geographic law of soil zonality (Dokuchaev 1883; Jenny 1941; Volobuev 1974; Grigor'ev 1954) and the establishment of a quantitative relationship between Budyko's radiative index of dryness (I_{2000}^{v}) and modal values of some regional biological and chemical soil properties (f_{2000}^{v}) (v: for virgin soils) for geomorphologically homogeneous soils (with similar texture and mineralogy of subsoil and with the same slope range of the surface) at the beginning of the twenty-first century. The index (I) is dimensionless and is calculated in the following form:

$$I = Rn / \left[L(Pr - S + Ir)\right] \tag{13.1}$$

where (Rn), (Pr), and (S) are mean annual values of net radiation (Kj cm^{-2} $year^{-1}$), precipitation (mm $year^{-1}$), and surface runoff (mm $year^{-1}$), respectively; (Ir) is mean annual irrigation water requirement (mm $year^{-1}$); (L) is a latent heat of evaporation (2.51 Kj cm^{-2} mm^{-1}). Frequently, in arid zones, the value (S) can be ignored (it means $S=0$). It is obvious that the value (Ir) is considered only under irrigation. $Ir=0$ in formula (13.1) should be considered for rain-fed lands.

This approach is based on assumption that principal soil properties responsible for fertility are in equilibrium with the mean annual climate conditions (considering their rather slow change). These properties of irrigated lands should be also in equilibrium with microclimatic conditions if irrigation is maintained during long periods at the same level (with the same technology, the same crops, and the same mean annual irrigated water consumption).

The f_{2000}^{v} (I_{2000}^{v}) relationship can be applied to predict agricultural soil fertility and its productivity change in future using existing climate change scenarios. According to each climate change scenario, the value ($I_{future\ v}$) is calculated, and the soil fertility index (based on assessment of certain soil properties' change [$f_{future}^{v} - f_{2000}^{v}$] is used to predict the [f_{2000}^{v} (I_{2000}^{v})] relationship. The same relationship can be used to assess the annual irrigation requirements considering crop necessities and long-term soil degradation.

Irrigation has direct and indirect impacts on soils. The direct impact depends on the chemical and biological composition of the water, while the indirect impact is related to an alteration of soil structure and formation processes due to changes of hydrothermal conditions, expressed with the index I in formula (13.1). Using the [f_{2000}^{v} (I_{2000}^{v})] relationship, where $Ir>0$, the indirect impact of irrigation can be predicted and the limited mean annual value Ir be assessed.

The [f_{2000}^{v} (I_{2000}^{v})] relationship can be also used in investigation of paleosols. For example, if some paleosol property (f_{past}^{v} $_{past}^{v}$) is known, its regional modal value can be statistically calculated. Using available [f_{past}^{v} (I_{2000}^{v})] relationships that were established for the investigated region at the present time, it is possible to obtain a climatic index (I_{past}^{v}) for the past. In cooperation with historical studies, information about past climate conditions can be used to analyze and understand the direct and indirect role of agriculture for changes of productivity.

13.2 Methodology

According to Volobuev (1974), the distribution of some properties of main soil types correlates with mean annual values of net radiation (R) and precipitation (Pr) or with Budyko's radiative index of dryness (I) which is expressed with formula (13.1), where Ir = S = 0. The mean annual values of (I) reflect climatic conditions: I<1 – humid, 1 ≤ I<2 – semihumid, 2 ≤ I<3 – semiarid, and I >3 – arid conditions.

It is possible to establish quantitative relationships between (I) and regional modal values of some soil properties (f), for example, organic matter content, ratio between humic and fulvic forms of organic matter, calcium carbonate content, cation exchange capacity, base saturation, pH, exchangeable bases, available P and K, and clay content (focusing on the surface layer 0–10 or 0–20 cm inside the A horizon, since this is the part of the profile which is the most affected and relevant for land use). The relationships f_{2000}^{v} (I_{2000}^{v}) should be established separately for geomorphologically homogeneous groups of virgin soils with similar texture and subsoil mineralogy and within the same range of surface slopes.

The relationships [f_{2000}^{v} (I_{2000}^{v})] can be obtained using sets of modern digital maps showing elevation, land use, hydrology, hydrogeology, and soils (1:50,000 and 1:250,000 scales). The procedure consists of the following steps; it should be noted that the symbols (f) and (I) are used instead of (f_{2000}^{v}) and (I_{2000}^{v}) for the sake of simplicity:

1. Sites without agricultural land use, with similar geomorphologic conditions, and with the same range of surface slopes are selected. The absolute altitude of particular sites plays no role for these relationships because of the dependence of the radiative index of dryness (I) on the altitude and its according variation.
2. The properties (f) of typical soils (with subsoils of similar texture and mineralogy) are extracted. It is desirable to recollect soil sampling points covering different climatic conditions with a wide range of annual precipitation and temperatures and diverse vegetation communities. This permits identifying tendencies in changes of the regional modal values of the property (f) depending on the climatic index (I) and considering natural variations of (f).
3. The radiative index of dryness (I) is calculated for each sampling point using interpolations of data from nearby weather stations.
4. A statistical treatment is applied to all data pairs of (f) and (I) corresponding to each sampling point. For that, the total range of (I) values is subdivided into several intervals. Within these relatively small intervals, the random variation of soil properties due to (I) should be insignificant. Then the mean arithmetic values (f_{av}) of each analyzed soil property are calculated inside each selected interval of (I), and all particular values (f) are divided by (f_{av}) in each selected interval of (I). This approach allows considering normalized data of soil properties (f/f_{av}) within

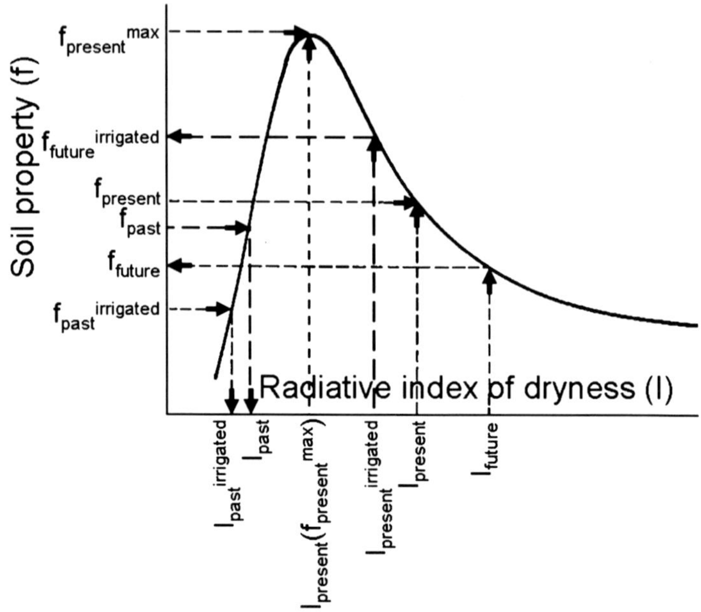

Fig. 13.1 A schematic dependence between a regional modal value of soil property (*f*) and the climatic index (I) at present. I_{past} or I_{future} is the climatic index corresponding to rain-fed conditions in the past or in the future, respectively, at one of the studied sites and calculated by formula (13.1) with Ir = 0 and S = 0. $I_{past}^{irrigated}$ or $I_{present}^{irrigated}$ is the climatic index corresponding to irrigation (in the past or at present, respectively, at one of the studied sites calculated by formula (13.1) with Ir > 0). $I_{present}$ is the climatic index under rain-fed conditions at one of the studied sites today. $I_{present}(f_{present}^{max})$ is the climatic index under rain-fed conditions corresponding to the maximal value of the property ($f_{present}^{max}$) on the graph f(I). $f_{present}$, f_{past}, f_{future}, and $f_{past}^{irrigated}$ are modal regional values of soil property under rain-fed conditions corresponding to the present, past, and future at a studied site. $f_{past}^{irrigated}$ and $f_{future}^{irrigated}$ are modal regional values of soil property under irrigation, corresponding to the past and future at a studied site

every particular interval of (I) as statistically homogeneous sample and then determining the type of statistical distribution of (f/f_{av}) values.

5. Graphical illustrations of absolute modal values of each soil property (f) on the (I) index are developed.

The resulting quantitative relationship between modal regional values of a soil property (f) and the mean annual radiative index of dryness (I) is schematically shown in Fig. 13.1.

Such kinds of relationship can be used for different purposes, for example, to assess past climate conditions on the basis of selected paleosol properties. These could refer, for example, to organic matter content (OM)—not its absolute value (which is rarely preserved in paleosols) but its structure expressed as a ratio (A_h/A_f) between humic (A_h) and fulvic (A_f) forms of organic matter. For this, it is necessary to be sure that:

- The value of the soil property (f_{past}) really corresponds to the studied historical period.

– The relationship between the model regional values of the soil property ($f_{present}$) and climatic index ($I_{present}$) should be firmly established for the present.

The dependence $f_{present}$ ($I_{present}$) should correspond to geomorphologically homogeneous virgin soils formed in conditions similar to the paleosols (e.g., alluvial soils or automorphous soils with deep water table, formed in situ in flat lands). If (f_{past}) is a modal value of paleosol property obtained from analysis of a number of samples taken from a climatically homogeneous region or at least from one studied site, it is possible to determine a climatic index (I_{past}) from the graph f(I) that was established for the present. The index (I_{past}) should reflect mean annual climate conditions in the past. If the mean annual temperature in the past is known, the mean annual net radiation (R_{past}) can be estimated using empiric relationships between (R) and annual temperature (Sellers 1965; Tejeda et al. 1999; Torres 1995).

In case of irrigation, the index ($I^{irrigated}$) in general should be calculated as

$$I^{irrigated} = R^{*} / \left[L \left(Pr + Ir \right) \right] \tag{13.2}$$

where (R^*) is the mean annual net radiation for microclimatic condition under irrigation and Ir is the mean annual amount of irrigation water. It is possible to ignore the difference between (R^*) and (R) and consider (R^*)≅(R) in case of periodic alluvial flooding or relatively rare irrigation applications.

(I) corresponds to irrigated conditions and is expressed by formula (13.2). For rain-fed lands, (I) is calculated by formula (13.1). If the property (f_{past}) corresponds to a paleosol formed under rain-fed conditions, it is possible to obtain the climatic index (I_{past}) on the graph f(I) for the past and to describe it by formula (13.1). If the property ($f_{past}^{irrigated}$) reflects the paleosol formed under irrigation and the graph f(I) corresponds to alluvial soils, the index ($I_{past}^{irrigated}$) is described by formula (13.2). If modal values of property (f_{past}) and ($f_{past}^{irrigated}$) in the same geographical region with the same climate conditions are available, it is possible to assess the amount of irrigated water applied in the past (Ir_{past}) using the following calculation:

$$Ir_{past} = \left(\frac{1}{I_{past}^{irrigated}} - \frac{1}{I_{past}} \right) \frac{R_{past}}{L} \tag{13.3}$$

The relationship f(I) can be used to predict the change of soil property (f) in relation to the changes of the climatic index (I) during the twenty-first century for expected scenarios of climate change. For example, if according to the climatic scenarios for certain periods (2025, 2050, or others) the index (I) in a studied geographical site should be changed from ($I_{present}$) to (I_{future}), the modern regional value of soil property (where the site is located) should change according to Fig. 13.1 from $f_{present}$ to f_{future}. The change of a particular soil property [f(local)] of the studied site during the twenty-first century can be assessed with the following formula:

$$f\left(local\right)_{future} = f\left(local\right)_{present} + \left(f_{future} - f_{present}\right) \tag{13.4}$$

where $f(local)_{present}$ and $f(local)_{future}$ are values of soil property of a studied site at present and in the future, respectively, and $f_{present}$ and f_{future} are regional modal values of the same soil property obtained on the graph f(I) for the present ($I_{present}$) and future (I_{future}) climatic indexes, respectively.

In case of irrigated land characterized by a mean annual amount of irrigation water application ($Ir_{present}$), the graph can be used to verify if the value ($Ir_{present}$) points to future soil degradation due to worse hydrothermal conditions of soil formation under irrigation. For this, it is necessary to remember that the climatic index of irrigated lands ($I_{present}^{irrigated}$) is calculated by formula (13.2). If the climatic index in the site or region ($I_{present}$) under rain-fed conditions or irrigation conditions equals ($I_{present}^{irrigated}$), then the future value of the irrigated soil property ($f_{future}^{irrigated}$), corresponding to ($I_{present}^{irrigated}$), can be greater or smaller than the present value ($f_{present}$). The irrigation requirement can be approved if $(f_{future}^{irrigated}) \geq (f_{present})$; otherwise, it would mean that irrigation will indirectly lead to soil degradation due to a deterioration of hydrothermal conditions.

13.3 Examples

Figure 13.2 shows the present dependence of the organic matter content (OM) in the 0–20-cm soil surface layer on the climatic index (*I*), corresponding to virgin soils of Mexico and the western part of the former Soviet Union (European part of Russia plus Central Asia).

All these soils are automorphous and formed in situ, located on rather flat lands with surface slopes of less than 3%. The sampling points were located at altitudes ranging from 0 to 2,500 m, covered by virgin/native vegetation and characterized by similar geomorphologic conditions and subsoil properties. The variation of mean annual precipitation and temperature is 200–800 mm and 1–17°C in the former Soviet part and 100–3,000 mm and 6–27°C in Mexico. It is assumed that soil properties come to equilibrium when climate change is rather slow. According

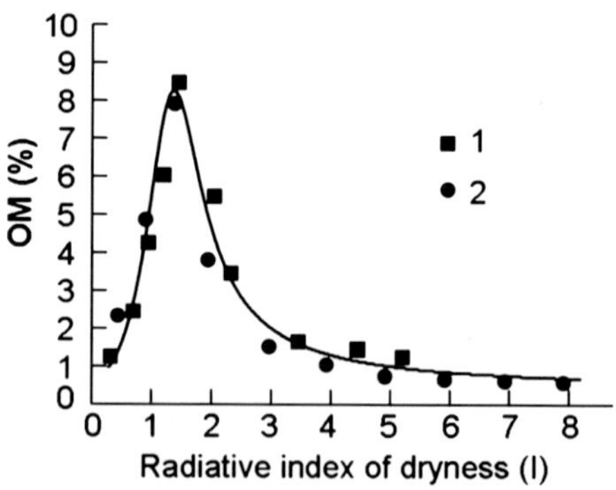

Fig. 13.2 Dependence of regional modal values of organic matter content (OM) in 0–20-cm surface layer on the dimensionless climatic index (I). All soils are automorphous and correspond to lands covered by virgin vegetation, characterized by similar geomorphologic conditions and subsoil properties. 1 Data for the European part of Russia and Central Asia (Aydarov 1985); 2 Data for Mexico (Nikolskii et al. 2002; Castillo-Alvarez et al. 2007)

to Arnold et al. (1990), this applies to organic matter, mineral nutrient contents, and pH. The direct impact of rising CO_2 contents on soil organic matter is rather small and was ignored, taking into account the estimations made by Bazzaz and Sombroek (1996), Reichstein et al. (2005), and Knorr et al. (2005).

Relationships f(I) were obtained also for cation exchange capacity, base saturation, pH, available P and K, the content of clay in Mexico (Nikolskii et al. 2002; Castillo-Alvarez et al. 2007), and content of water stable aggregates (WSA) and clay, pH, and ratio (A_h/A_f) between humic (A_h) and fulvic (A_f) forms of organic matter in the former Soviet part (Aydarov 1985). Confidence intervals (with a 95% probability) for changes of all these soil properties (f) numbered ±25% of the modal values of (f). Unfortunately, f(I) relationships were not yet established for Jordan and the Levant, and the dependences [A_h/A_f(I)] and [WSA(I)] were not investigated in Mexico.

The correlation coefficient (r^2) between all modal values (OM) and the approximating curve in the Fig. 13.2 is equal to 0.97. As we can see, the relationship OM(I) obtained in both countries with quite different climatic conditions is practically the same. It possibly means that this relationship is typical for such kind of soils in the world. The graphs of pH(I) and content of clay (I) relationships published by Aydarov (1985) for European part of Russia and Central Asia and by Nikolskii et al. (2002) and Castillo-Alvarez et al. (2007) for Mexico have also similar forms for each property.

The geometric form of OM(I) graph gives us an idea about the "susceptibility" of this property to climate change. As we can see, the organic matter content varies most noticeably in the interval $0<I<3$. This means that the soils more "susceptible" to climate change are located in zones where the dryness index lies within $0<I<3$, that is, the humid tropical zone ($I\leq 1$), the semihumid zone ($1<I\leq 2$), and the semiarid zone ($2<I\leq 3$). The potential impact of climate change may obviously be estimated mathematically as an absolute value of the partial derivative [$\delta(OM)/\delta I$], meaning the (OM) variation per unit change of (I).

The dependencies of regional modal values of organic matter content and some other virgin soil properties on the climatic index (I) established for present time can be used for prediction of their alteration due to global climate change during the twenty-first century. It is useful to apply the integral index of soil fertility (F) to assess agricultural soil quality alteration in general. The regional integral index of soil fertility (F) can be expressed in a dimensionless form as suggested, for example, by Pegov and Khomyakov (1991):

$$F = 0.46\,\frac{OM}{OM_{max}} + 0.28\sqrt{\frac{P}{P_{max}} * \frac{K}{K_{max}}} + 0.26e^{-\left(\frac{pH-6}{2}\right)^2} \tag{13.5}$$

where OM, P, and K are organic matter, phosphorus, and potassium content, respectively, while OM_{max}, P_{max}, and K_{max} are the maximal values observed in the investigation area. The values of F are varying from 0 (for completely degraded soils) to 1 (for the maximal possible fertility in the considered area).

In order to predict the integral fertility index (F) alteration due to global climate change, the graphs $F_{present}(I_{present})$ should be established separately for virgin and agricultural soils of rain-fed and irrigated lands. If the difference between these graphs is significant, the index of soil fertility (F) can be calculated as

$$F_{future}^{agric} = F_{present}^{agric} + (F_{present}^{max, agric} / F_{present}^{max, nat})(F_{future}^{nat} - F_{present}^{nat}) \quad (13.6)$$

where $F_{present}^{agric}$ and F_{future}^{agric} are local integral fertility indexes of agricultural soil at present and in the future, respectively; $F_{present}^{nat}$ and F_{future}^{nat} are regional integral fertility indexes for virgin soils, corresponding on graph F(I) to climatic indexes at a studied site for the present ($I_{present}$) and future (I_{future}), respectively; and $F_{present}^{max, agric}$ and $F_{present}^{max, nat}$ are maximal values of regional integral fertility indexes observed on the graphs F(I) and established for agricultural and virgin soils, respectively.

If there is no significant difference between graphs F(I) or if now $F_{present}^{nat} \cong F_{present}^{agric}$, the predicted soil fertility index at a certain site F_{future}^{local} will be equal to

$$F_{future}^{local} = F_{present}^{local} + (F_{future} - F_{present}) \quad (13.7)$$

where $F_{present}^{local}$ is the soil fertility index now at the same site.

This approach was applied to predict variations of the virgin soil fertility index (F) in different climatic zones of Mexico at the end of twenty-first century according to selected climate change scenarios assuming a doubling of the CO_2 content in the atmosphere (Gay 2003). The expected soil fertility change varies between moderate decrease (up to −30%) to low increase (up to 20%) depending on the climatic region. Moderate decrease (−20 to −30%) is expected in semiarid zone of Mexico, low decrease (−10 to −20%) and low increase (10–20%) in semihumid zone, and insignificant change (−10 to 10%) in arid zone. It is clear that the reliability of such predictions depends on the reliability of the climatic scenarios.

The integral soil fertility index (I) can be used also to calculate agricultural crop productivity (Y). For example, according to FAO-IIASA (2000) and Pegov and Khomyakov (1991), the crop productivity can be assessed as

$$Y = Y^{max} K F \quad (13.8)$$

where Y^{max} is the maximum possible crop yield (kg ha^{-1}) depending on the predicted monthly values of temperature, global radiation, and biological plant properties; K is a coefficient reflecting soil water availability and depending on predicted monthly values of temperature, global radiation, and precipitation; and F is the predicted integral agricultural soil fertility index. K and F are dimensionless and vary between 0 and 1. Some calculations of the vulnerability of wheat productivity in Mexico to climate change were conducted using formula (13.8). The calculations showed that irrigated crops are more vulnerable to climate change than in rain-fed agriculture, which is due to an expected change of water availability especially in arid zones. Consideration of the soil fertility alteration is very important in assessments of crop vulnerability to climate change. Ignorance of the fertility index alteration in predictions of crop productivity can cause errors up to 50% or more.

In discussing "progressive development" in arid and semiarid zones, it is important to consider the indirect impact of irrigation on soil fertility. Evaluating only

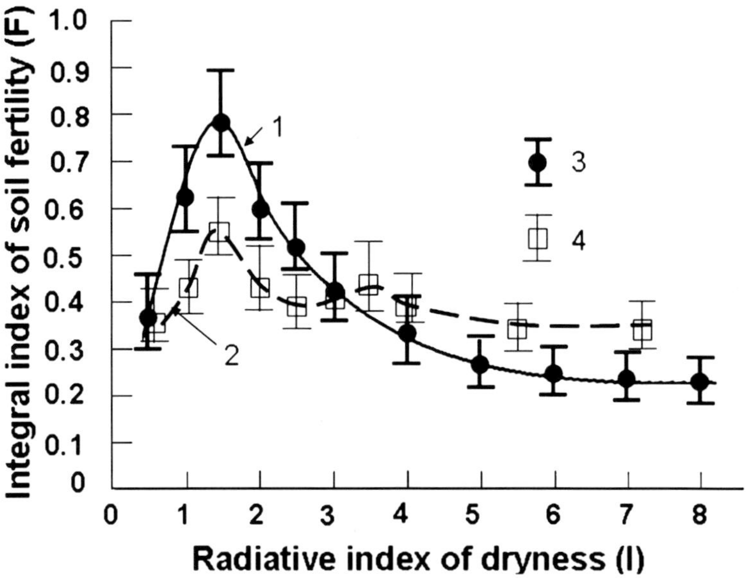

Fig. 13.3 Dependence of dimensionless integral index of soil fertility in the 0–20-cm surface layer on the dimensionless climatic index (I) in Mexico. 1 Virgin (with natural vegetation) and agricultural soils of rain-fed lands under maize and wheat; 2 Agricultural soils of irrigated lands under maize and wheat; 3 and 4 confidence intervals with a 95% probability

agricultural crop requirements may lead to soil degradation: Figure 13.3 shows relationships between regional modal values of the integral soil fertility index (F) and the radiative index of dryness (I), separately for irrigated and rain-fed agriculture as well as for virgin soils. All lands have surface slopes less than 3%. Agricultural land use lasted for more than 50 years.

The comparison of modal values of the integral index of soil fertility (F) for virgin and rain-fed agricultural soils showed insignificant differences. This is possibly due to the extensive use of the majority of rain-fed lands in Mexico in the case of maize and wheat production (without artificial fertilizers and with insignificant manure). Thus, in this figure, only one graph F(I) is shown for virgin and rain-fed agricultural soils.

As it can be noticed, the impact of irrigation on agricultural soils in arid zone ($I>3$) in Mexico is positive in general. The fertility index of irrigated soils is higher than that of nonirrigated and virgin soils. However, the irrigation practice in semiarid ($2<I\leq 3$), and semihumid ($1<I\leq 2$) zones has a negative impact on agricultural soils, at least in the case of maize and wheat production. The fertility index of irrigated soils in these zones is lower than that of nonirrigated and virgin soils. These data are representative because they correspond to large areas of irrigated and nonirrigated lands in Mexico (about 3×10^6 ha for each one). The observed difference between fertility of irrigated and nonirrigated soils can be explained by the improvement of soil formation processes under irrigation in arid zones and

Table 13.1 Examples for a comparison of mean annual irrigation requirements (Ir^{crop}), applied today in practice, and recommendations for soil conservation (Ir^{soil}) in different climatic conditions of Mexico. Crop: wheat

Pr (mm year^{-1})	T (°C)	Ir^{crop} (mm year^{-1})	I (dml)	$I^{irrigated}$ (dml)	Ir^{soil} (mm year^{-1})
528.5	24.6	628	2.7	1.22	409
735.0	21.3	367	1.6	1.11	250
400.0	19.5	540	2.8	1.19	346
422.8	28.1	798	3.5	1.20	552

Note: Pr and T – mean annual precipitation and temperature; I and $I^{irrigated}$ – natural climatic and local microclimatic indexes calculated by formulas (13.1) and (13.2), respectively

deterioration of soil formation process under irrigation in semiarid and especially in semihumid conditions, due to over-irrigation leading to excessive leaching of organic matter and mineral nutrients from the soils.

The difference between fertility indexes of irrigated and virgin soils can be used to evaluate the mean annual irrigation requirements, considering not only agricultural crop productivity (Ir^{crop}) but also long-term soil conservation (Ir^{soil}). Such considerations should be part of sustainable agriculture concepts. Regarding irrigated soil conservation, the irrigation requirement (Ir^{soil}) can be assessed using index ($I^{irrigated}$), described by formula (13.2) as graph in Figs. 13.1, 13.2, and 13.3. Obviously, in arid, semiarid, and partially semihumid zones where $I > 1.5$, the index ($I^{irrigated}$) should be less than 1.5.

In this case

$$1.5 \geq R / \left[L \left(Pr + Ir^{soil} \right) \right] \text{ or } Ir^{soil} \leq \left(R / 1.5\, L \right) - Pr \text{ for zones where } I > 1.5 \qquad (13.9)$$

According to Figs. 13.2 and 13.3, in semihumid and partially humid zones where $I < 1.5$, the index ($I^{irrigated}$) should not number more than index (I) that corresponds to the natural rain-fed conditions (Table 13.1). This means that in order to prevent soil degradation, the mean annual irrigation requirement should not be more than the mean annual water surface runoff in agricultural lands (S) and less than the difference between potential evaporation (E_0) and precipitation without surface runoff ($Pr - S$):

$$Ir^{soil} \leq S \text{ and } Ir^{soil} \leq (E_0 + S - Pr) \text{ for zones where } I < 1.5 \qquad (13.10)$$

A similar approach was proposed earlier by Aydarov (1985).

Examples for the determination of (Ir^{soil}) values for some irrigated districts of Mexico are presented in n (the term dml means dimensionless).

This example shows that in order to prevent possibly negative indirect impact of irrigation on soil fertility during long periods (at least 10 years), it is necessary to reduce mean annual irrigation requirement (Ir^{crop}) but which traditionally only consider crop demand. The final decision about mean annual irrigation requirements which are then applied in practice ($Ir^{practice}$) depends on the climatic conditions and should be based on an economic justification considering the cost of water and soil resources as well as direct benefits related to agricultural production. With such an approach, the irrigation requirement ($Ir^{practice}$) is supposed to stay between (Ir^{crop}) and (Ir^{soil}): $Ir^{soil} < Ir^{practice} < Ir^{crop}$.

13.4 Challenges and Potentials of Combining Modeling and Reconstruction of Soil Development

The above-mentioned modeling of soil properties can be validated regarding individual parameters by using their actual distribution. For example, the valley Wadi Ziqlab in Jordan can illustrate how temperature changes may be related to changing soil properties. While the mean annual precipitation is basically stable over the whole course of Wadi Ziqlab, temperatures vary strongly and follow an altitude gradient of 1,300 m over a valley length of 20 km (Fisher et al. 1966). The soil and vegetation pattern gives a strong impression of increasing aridity with lower altitude. Precipitation cannot explain this trend: The contrasting character of the landscape is due to temperature and evaporation (Fisher et al. 1966). In this context, the soil and vegetation distribution in Wadi Ziqlab may to some degree allow description of thresholds for permanent soil and vegetation change. Considering the water balance and vegetation, the direct impact of temperature and evaporation seems not as relevant as the indirect role of changing soil properties and local microclimatic controls. For example, it is known that higher temperatures can reduce soil aggregate stability and water retention (Imeson and Lavee 1998). According to Bens et al. (2006), increasing dryness may lead to hydrophobic behavior of soils, reducing infiltration and increasing runoff. As well, the soil's resistance to root growth may increase, leading to reduced biomass. It is however unknown to which degree soil change is reversible and in which time frames it takes place.

One major challenge of modeling changing soil properties is the time factor: Since the model is based on the geographic law of soil zonality, it can only display currently existing relationships. In the previous section, we estimated time frames that could be involved in the modeled changes; however, so far, no systematic study exists that established these figures, and unfortunately, long-term monitoring of soil development was only rarely conducted. It is likely that the speed and intensity of soil properties change will vary according to parameters—for example, organic matter contents usually respond much faster to changing environments and land use than pedogenic iron oxides. But this makes it difficult to assess the process of soil change and to predict soil fertility for a given point of time.

Here, the validation of models of soil development by paleosols could make an important contribution. Paleosols can provide a framework describing possible future soil development, since what already happened in the past may happen again. They also give insight into the time frames of soil development, and although the covered time spans are usually very long, recent research in archaeological contexts allowed targeting relatively short periods of centuries or decades (Khokhlova et al. 2004).

A major challenge for combining the modeling of soil development with reconstruction is the use of different parameters as indicators. However, theoretically, any parameter should be applicable in the model and could be determined in the following procedure: A quantitative relationship of soil properties (SP) and a mean annual climatic index (CI – expressed as relationship between mean annual values of net sun radiation, precipitation, and temperature) will be established. As well, regional modal values of the soil properties (SP) of modern soils (preferably virgin soils not used in agriculture) belonging to the same geomorphological group will be defined. This approach can rely on already established relationships SP(CI) for Mexican and the former USSR (European and Central Asian) soils (Nikolskii et al. 2006). A statistical analysis of the soil properties of modern soils and the related climatic indices will be conducted and applied to the paleosols. The establishment of a statistical function expressing the relationship SP(CI) will allow to transfer the climatic index (CI) to paleosol properties (PSP), making it possible to describe a paleoclimatic index PSP(CI). This will allow to validate the projected soil development expressed by the index SP(CI), since the soundness of the PSP(CI) index and related climatic scenarios can be checked with the climate reconstruction on the basis of the paleosol analysis. The forecasting of soil development will then be conducted and based on genetic modeling.

For example, Khresat and Taimeh (1998) investigated how $CaCO_3$ contents are related to changing soil properties. Paleosols in today arid areas in Jordan documented several climatic changes, and dry periods were connected with accumulation of calcium carbonate up to the formation of a hard crust. This led to a significant reduction of vegetation, too, and many soils in the present transition area to the desert are assessed to be in an incipient state of calcium carbonate accumulation, with related negative consequences for water retention, nutrient availability, and organic matter content (Khresat and Taimeh 1998). Bäumler (2001) showed on a transect of present soils and paleosols of different source rocks and climates in the Himalaya, Central Asia, and Kamchatka how combining physical parameters (such as grain size) with chemical properties (such as iron oxides and weathering indexes) can be used for the reconstruction of soil development. Genetic modeling of soil development can thus be combined with the above-described forecasting of soil fertility and desertification: What are still missing are the connection of soil physical and chemical properties with organic matter and a translation of the more descriptive genetic approach into mathematical expressions.

In this context, validation of the above-described model of soil development based on the radiative index of dryness with paleosols faces the problem that most paleosols do not contain significant amounts of organic matter. Organic matter contents are however directly related to magnetic properties, pedogenic iron and manganese oxides, calcium carbonate contents, and particle sizes (Bäumler 2001; Lucke 2008). As well, isotopes from pedogenic carbonates allow for conclusions about the dominant vegetation type during their formation (Bar-Matthews et al. 1998). Considering virgin modern soils, it is possible to evaluate how the above-mentioned properties are related to organic matter contents. Focusing on young sediments and soils developed out of the debris on archaeological sites, relatively short time frames of soil development can be addressed as well. Finally, recent studies of soil magnetics in semiarid regions indicate that the magnetic susceptibility of soils might be useful as indicator of past organic matter contents (Lucke 2008).

13.5 Conclusions

The relationship of agricultural soil properties, climate condition, and irrigation practice can be modeled according to mathematical formulas, which allow predicting the impact of specific land-use technologies for given climatic scenarios. With such tools, it should be possible to plan "progressive development" well ahead as long as the necessary water supply can be ensured. However, uncertainties exist regarding the time frames of change and the development of individual parameters, since the modeling is based on geographic relationships at the present time. This makes it difficult to predict soil fertility for a given point of time. However, the study of paleosols might allow addressing these difficulties by connecting genetic reconstructions of soil development with mathematical modeling. It requires "translating" the parameters used for reconstructions to those used for modeling and investigating how changes of individual parameters are connected in feedbacks. While the combination of modeling and reconstruction of soil properties change still needs further investigation, soil surveys should address and document paleosols already now since the reliability of models depends largely on data availability.

References

Amery H, Wolf A (2000) Water in the Middle East. University of Texas Press, Austin
Arnold RW, Szabolcs I, Targulian VO (eds) (1990) Global soil change. IIASA, Laxenburg, p 109
Aydarov IP (1985) Water, salt and nutrient soil regime management in arid zones. Agropromizdat, Moscow (in Russian)
Aydarov I, Nikolskii Y, Jachaturyan V, Korolkov A (1992) La crisis ecologica en la cuenca del Mar Aral por el uso agrícola de los recursos hidraulicos. In: Agricultura Sostenible: Un Enfoque Ecologico, Socioeconomico y el Desarrollo Tecnologico. Memorias de/I Simposio y I Reunion Nacional, Guadalajara, Mexico, pp 79–83 (in Spanish)
Bar-Matthews M, Ayalon A, Kaufmann A (1998) Middle to late holocene (6500 yr period) paleoclimate in the Eastern Mediterranean region from stable isotopic composition of speleothems from Soreq Cave, Israel. In: Issar A, Brown N (eds) Water, environment and society in times of climatic change. Kluwer Academic, Dordrecht, pp 203–215
Bäumler R (2001) Vergleichende bodenkundliche Untersuchungen in Hochasien und Kamtschatka. Relief, Boden, Paläoklima 16, Gebrüder Bornträger, Munich
Bazzaz F, Sombroek W (1996) Global climate change and agricultural production: direct and indirect effects of changing hydrological, pedological and plant physiological processes. FAO, Rome
Bens O, Buczko U, Sieber S, Hüttl R (2006) Spatial variability of O layer thickness and humus forms under different pine beech-forest transformation stages in NE Germany. J Plant Nutr Soil Sci 169(1):5–15

Castillo-Alvarez M, Nikolskii-Gavrilov I, Ortiz-Solorio CA, Vaquera-Huerta H, Cruz-Bello G, Mejía-Sáenz E, González-Hernández A (2007) Alteración de la fertilidad del suelo por el cambio climático y su efecto en la productividad agrícola. Interciencia 32:368–376

Cordova C, Foley C, Nowell A, Bisson M (2005) Landforms, sediments, soil development and prehistoric site settings in the Madaba-Dhiban Plateau, Jordan. Geoarchaeology 20(1):29–56

Dokuchaev VV (1883) Russian chernozems (Russkii chernozem). Israel Prog Sci Trans, Jerusalem, 1967. Transl. from Russian by N Kraner. Available from US Dept of Commerce, Springfield

FAO (1997) Irrigation in the near east region in figures. FAO, Rome

FAO-IIASA (2000) Global agroecological zones. Methodology and results of the global agro-ecological zones model. CD-ROM. Versión 1.0

Fisher W, Atkinson K, Beaumont P, Coles A, Gilchrist-Shaw D (1966) Soil survey of Wadi Ziqlab. University of Durham, Durham

Gay C (ed) (2003) Una visión hacia el siglo XXI. El cambio climático en México. UNAM, México, p 220

Grigor'ev AA (1954) Some regularities of geographical zonality, vol 5, Ser Geogr. Izv. Akad. Nauk SSSR, Moscow, pp 5–17 (in Russian)

Gvirtzman G, Wieder M (2001) Climate of the last 53,000 years in the Eastern Mediterranean, based on soil-sequence stratigraphy in the coastal plain of Israel. Quat Sci Rev 20:1827–1849

Imeson A, Lavee H (1998) Soil erosion and climate change: the transect approach and the influence of scale. Geomorphology 23:219–227

Issar A (2008) Progressive development in arid environments: adapting the concept of sustainable development to a changing world. Hydrogeol J 16:1229–1231

Jenny H (1941) Factors of soil formation: a system of quantitative pedology. McGraw-Hill, New York

Khokhlova OS, Khokhlov AA, Chichagova CC, Morgunova NL (2004) Radiocarbon dating of calcareous accumulations in soils of the holocene chronosequence in the Ural river valley (Cis-Ural Steppe). Euras Soil Sci 37(10):1024–1038

Khresat S, Taimeh A (1998) Properties and characterization of vertisols developed on limestone in a semi-arid environment. J Arid Environ 40:235–244

Knorr W, Prentice IC, Holland EA (2005) Long-term sensitivity of soil carbon turnover to warming. Nature 433:298–301

Lucke B (2008) Demise of the decapolis: past and present desertification in the context of soil development, land use and climate. VDM, Saarbrücken

Maher L (2005) The epipaleolithic in context: paleolandscapes and prehistoric occupation in Wadi Ziqlab, Northern Jordan. Dissertation, Department of Anthropology, University of Toronto, Toronto

Nikolskii YN (1996) Ecological consequences of irrigation. ICID J 45:59–73

Nikolskii Yu N, Bakhlaeva OS, Contreraz-Benitez A, Ordaz-Chaparro V (2002) Assessment of changes in soil properties as dependent on hydrothermic conditions of lowlands (by the example of Mexico). Euras Soil Sci 35:1031–1036

Nikolskii YN, Castillo-Alvares M, Bakhlaeva OS, Roman-Calleros XA, Maslov BS (2006) The influence of the possible global climate change on the properties of Mexican soils. Euras Soil Sci 39(11):1164–1169

Pegov SA, Khomyakov PM (1991) Simulation of ecological systems alteration. Gidrometeoizdat, Leningrad (in Russian)

Reichstein M, Kätterer T, Andren O, Ciasis P, Schuize ED, Cramer E, Papale D, Valentini R (2005) Does the temperature sensitivity of decomposition vary with soil organic matter quality? Biogeosci Discuss 2:738–747

Reuters (2008) Saudi Arabia scraps wheat growing to save water. http://www.reuters.com/article/latestCrisis/idUSL08699206. Accessed 31 Jan 2009

Sellers WD (1965) Physical climatology. University of Ontario Press, Chicago

Tejeda MA, Vazquez IL, Rivas DA (1999) Algoritmos simples para estimar datos mensuales de irradiacion solar y horarios termohigrometricos en la Republica Mexicana. Universidad Veracruzana, Veracruz

Torres RE (1995) Agrometeorologia. Editorial Trillas Sa De Cv, Mexico

Volobuev VR (1974) Introduction into the energy of pedogenesis. Nauka, Moscow (in Russian)

Chapter 14
Classification, Characterization, and Suitability Evaluation of the Savanna Soils of Oyo North of Nigeria

J.O. Aruleba and A.S. Ajayi

Abstract A detailed soil survey of about 100 ha of Savanna ecoclimatic region of Oyo North of Nigeria was carried out to characterize and classify the soils and to assess the potential of the area for agricultural suitability. During this study, five soil mapping units were identified and delineated. Selected physical and chemical characteristics were determined to support soil taxonomy, FAO–UNESCO legend, and land fertility capability classification. Five soil pedons were classified according to USDA soil taxonomy as Arenic Kandiudults, Arenic Kanhapludults, Typic Kandiudults, Arenic Kandiudalfs, and Gross Arenic Kandiudalfs and according to FAO–UNESCO system as Ferric Lixisols, Stagnic Lixisols, and Haplic Lixisols, respectively. The studied pedons are assessed for fertility capability classification (FCC), and their FCC (Buol classification) is recognized as sandy topsoil (S), low CEC (e), and K deficient (k) and presented as pedon 1 and 2 as Sek (>35% gravels), pedon 3 as Sek (2–4% slope and >35% gravels), pedon 4 as Sek (4–7% slope and >35% gravels), and pedon 5 as Lek (>7% slope and >35% gravels). Sand-sized particles dominate in the entire profiles. The exchangeable bases are generally low, and the exchange sites are dominated by calcium. Total N and organic matter contents are also low. The soils are marginally suitable for commonly grown crops such as cassava, maize, and oil palm. Limitations to agricultural production include soil fertility (CEC), organic matter, poor texture, and climate (rainfall and length of dry season).

Keywords Classification • Savanna soils • Limitations • Agricultural potentials • Nigeria

---

J.O. Aruleba (✉) • A.S. Ajayi
Faculty of Agricultural Sciences, University of Ado-Ekiti, Ado-Ekiti, Ekiti State, Nigeria
e-mail: arulebaunad@yahoo.com

S.A. Shahid et al. (eds.), *Developments in Soil Classification, Land Use Planning and Policy Implications: Innovative Thinking of Soil Inventory for Land Use Planning and Management of Land Resources*, DOI 10.1007/978-94-007-5332-7_14,

14.1 Introduction

There are growing concerns about the need to provide adequate information on the quality of Nigerian soil resources for sustainable agricultural production and to understand the problem of soil degradation resulting from their misuse. In Nigeria, as in most developing countries, agricultural production has been continually declining over the years leading to increasing food insecurity issues and deteriorating quality of life especially among the rural dwellers. The soil information is an integral part that enables more numerous, accurate, and useful prediction to be made for specific purposes that could not have been possible otherwise. To achieve this purpose, it is necessary to determine the pattern of the soil/land cover and to divide this pattern to relatively homogenous units, so enabling the soil properties over an area to be predicted. It also involves characterization of soil mapping units in a way that is useful to predict land use potential and responses to changes in management.

The extensive area of Oyo North accounts for about 40% of the total land mass of Oyo State covering parts of Shaki, Igboho, Okeho, Ileto, etc. The vast land of the Oyo North coupled with the low population density makes the area attractive for extensive agricultural investment. Moreover, with the present pressure on land resources and agricultural activities and also due to the annual increase in production demand, proper land use planning is essential. Given the constraints that little is known about the soils of the area and their agricultural potentials, this study aims at characterizing and classifying soils of the area to evaluate their suitability for agricultural practices and sustainable production.

14.2 Materials and Methods

14.2.1 Study Area (Area of Interest)

The study area (250,432 km^2) in savannah zone (Shaki west local government area of Oyo State) lies between longitude 5°4′ E and latitude 7°32′ N (Fig. 14.1). It is about 100 km^2 northeast of state capital Ibadan. Topography is gently undulating with dominant slopes of 2–12%. Elevation varies between 180 and 250 m above sea level. The critical climatic factor is rainfall and its distribution. The mean monthly rainfall is variable ranging from 213 mm in October to 300 mm in January. A mean monthly rainfall of 104.8 mm and a mean monthly temperature of about 27 °C prevailed. The solar radiation ranged between 10.35 and 17.15 cal cm^{-2} day^{-1} with a mean of 14.63 cal cm^{-2} day^{-1}.

14.2.2 Field Soil Mapping and Description

An area of 100 ha was delineated representing the entire farming community. The soils were described using the soil survey manual (Soil Survey Division Staff

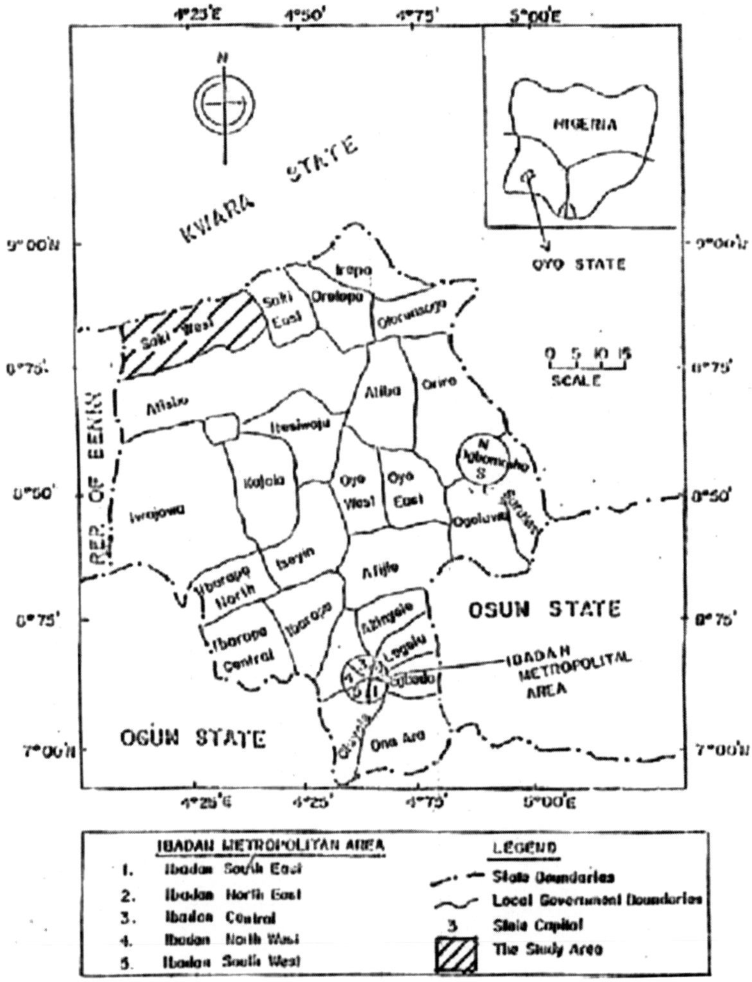

Fig. 14.1 Map of Oyo State showing the study area (Saki west local government area of Oyo State) (Source: Ministry of Lands, Housing and Physical Planning, Oyo State, Department of Geography and Planning Science, Cartographic Unit, University of Ado-Ekiti (UNAD))

1993) and field book (Schoeneberger et al. 2002) and soils classified using the specifications in the keys to soil taxonomy (Soil Survey Staff 2003). Using the grid survey, five soil types were identified and delineated. From each ha, ten samples were collected at 0–25 cm depths and composited; a representative sample was then obtained for laboratory analyses. Based on the morphological features and other characteristics such as texture, structure, color, depth of horizons, and gravel contents, soils with similar properties were grouped together into five soil mapping units (A–E) with different areas (A = 20 ha, B = 25 ha, C = 18 ha, D = 21 ha, and E = 16 ha). Typical soil profiles (1 m × 1m × 1.5 m) representing each map unit (1 from each unit) were exposed. The typical profiles were described, soil samples collected for laboratory characterization. The results were used to name soil taxa using keys to soil taxonomy (Soil Survey Staff 2003), the FAO–UNESCO (1986), and Buol et al. (1982).

Infiltration rate was measured besides each typical soil profile using standard twin ring infiltrometer (Bouwer 1986). The soil moisture content, porosity, and bulk density were also determined using standard procedures. The GPS coordinates of five typical soil profiles were recorded using Garmin Elpex (2000) geographic positioning system (GPS) unit.

14.2.3 Laboratory Analysis

The soil samples were processed for air-drying, sieving (<2 mm) prior to analyses for the selected parameters. Soil texture (particle-size distribution) was determined by standard Bouyoucos hydrometer method (Bouyoucos 1951). Soil pH was measured both in distilled water and 0.1 N KCl in a soil–liquid suspension of 1:2.5 using Beckman's Zeromatic pH meter. Various determinations were made using standard methods, such as organic carbon by Walkley and Black (1934) method, which was converted to organic matter by multiplying with standard conversion factor 1.724; exchangeable cations in 1 N ammonium acetate-NH_4OAC (pH 7) extract using atomic absorption spectrophotometer (ASS); exchangeable acidity in 1 N KCl extract; cation exchange capacity (CEC) as described by Peach et al. (1947); total nitrogen by the micro-kjeldhal apparatus (Bremner 1965); available phosphorous by Bray I methods (Bray and Kurtz 1945); electrical conductivity (EC) as described in USDA Handbook 60 (Richards 1954); effective CEC calculated as the sum of TEA and TEB while base saturation percentage as outlined by Coleman and Thomas (1967); and zinc and manganese by EDTA (complexometric titration) method, and free iron was determined by complexometric titration method (Lakanem and Ervio 1971).

14.2.4 Soil Classification

The soils were classified using USDA Keys to Soil Taxonomy (Soil Survey Staff 2003) and the FAO–UNESCO (1986) system, soil map of the world legend (1989), and Buol et al. (1982) and by describing morphological, physical, and chemical characteristics of the soils investigated.

14.2.5 Suitability Evaluation

The soil suitability evaluation for common crops grown in the area was made using physical and chemical characteristics including soil depth, drainage, texture, and coarse fragments. Soil characteristics were matched with the crop requirements following the Sys (1985) method, and the suitability class was assigned to each soil taxa along with the most limiting factors.

14.3 Results and Discussion

14.3.1 Soil Morphological Characteristics

The morphological characteristics are shown in Table 14.1. Table 14.1 illustrates in-depth differences in Munsell color, soil structure, concretion, and texture. The soils at surface are characterized by dark yellowish brown, black, and very dark brown color. The color difference show that the pedons on the upper and the middle slope of the landscape were reddish to brownish as a result of the presence of iron and manganese concretions which are most prominent in pedons 1, 2, and 3. The soils are deep and well drained. Range of soil structure was observed (single grain, fine crumb, and very fine crumb). Iron and manganese concretions were observed in three pedons (pedons 1, 2, and 3). The pedons are extremely stony to very stony due to the presence of concretions in all horizons; this characteristics no doubt exacerbate the dryness of the soils most of the seasons. The consistence is mainly loose or friable (pedon 1, 3, 4, and 5), occasionally slightly sticky and sticky (pedon 2).

14.3.2 Physical Characteristics

The physical soil characteristics are presented in Table 14.2. The particle-size distribution (PSD) analyses revealed sand being dominant in the primary soil particles, ranging between 63 and 85%. Among the profiles, sand is being distributed uniformly at depth with slight differences. All the pedons are dominant in sand fraction over silt+clay contents; this is an indication that the problems that can be encountered in these soils are high infiltration rate and low water-holding capacity since the groundwater supply may no longer be recharged particularly in the dry period (Stoop 1987). Another problem that can be encountered on these soils if they are left bare is soil compaction due to the battering effect of sun and raindrop. The coarse texture of the soils also confers on its low structural stability, which aid high erodibility. Silt and clay vary within soil profiles. The ratio between silt and clay is generally low. The clay dominates at lower depths where illuviation soil-forming process was observed and the soil horizon was indicated with horizon suffix "t." Soil texture ranges between loamy sand and sandy loam.

The bulk density values within the horizons of a particular soil unit and between soil units did not vary appreciably and were mostly ranges between 1.28 and 1.52 g cm^{-3}. These values were attributed mainly to loamy sand to sandy loam texture. Bulk densities of these ranges are considered suitable for most crops and do not cause hindrance to root penetration and seedling emergence. De Gens (1978) reported a bulk density of 1.63 g cm^{-3} as the critical value causing hindrance to root penetration.

Table 14.1 Morphological characteristics of soils of Oyo North

Horizon	Depth (cm)	Color	Stoniness	Structure	Consistence	Roots	Concretion	Boundary
	Pedon 1							
Ap	0–20	10YR 3/4	Stony	Single grain	Loose	Medium common	Fe–Mn	gs
Bt_1	20–37	5YR 3/4	Stony	Single grain	Loose	Fine common	Fe–Mn	gs
Bt_2	37–52	5YR 4/6	Very stony	Single grain	Loose	Fine common	Fe–Mn	gs
Bt_3	52–80	5YR 5/6	Extremely stony	Single grain	Loose	Few	Fe–Mn	gs
C	80–125	5YR 5/6	Extremely stony	Single grain	Loose	–	Fe–Mn	gs
	Pedon 2							
Ap	0–22	10YR 2/2	Fairly stony	Medium crumb	Friable	Medium many	–	dw
AB	22–34	7.5YR 3/4	Fairly stony	Fine crumb	Slightly sticky	Fine few	–	dw
Bt_1	34–50	5YR 4/6	Fairly stony	Fine crumb	Slightly sticky	Fine few	–	dw
Bt_2	50–75	5YR 5/8	Fairly stony	Fine crumb	Sticky		–	dw
BC	75–109	2.5YR 4/8	Fairly stony	Single grain	Sticky		Fe–Mn	dw
C	109–128	2.5YR 4/8	Fairly stony	Single grain	Sticky		Fe–Mn	dw
	Pedon 3							
Ap	0–20	10YR 2/1	–	Fine crumb	Friable	Medium many	–	cw
AB	20–30	7.5YR 3/4	Stony	Single grain	Loose	Few	Fe–Mn	cw
Bt_1	30–41	7.5YR 3/4	Very stony	Single grain	Loose	Few	Fe–Mn	cw
Bt_2	41–55	7.5YR 3/4	Very stony	Single grain	Loose	Few	Fe–Mn	cw
BC	55–72	5YR 4/6	Very stony	Single grain	Loose	Few	Fe–Mn	cw
C	72–120	2.5YR 5/8	Very stony	Single grain	Loose	Few	Fe–Mn	cw
	Pedon 4							
Ap	0–18	10YR 2/2	–	Fine crumb	Friable	Medium common	–	dw
AB	18–33	7.5YR 3/4	–	Fine crumb	Friable	Fine many	–	dw
Bt_1	33–55	7.5YR 4/6	–	Fine crumb	Loose	Few	–	dw
Bt_2	41–55	5YR 4/6	–	Platy	Firm	-	–	dw
BC	55–72	5YR 6/6	–	Granular	Firm	-	–	dw
C	72–115	5YR 6/8	–	Granular	Firm	–	–	dw

Ap	0–30	10YR 2/2	–	Very fine crumb	Loose	Fine common	–	gs
Bt_1	30–7	10YR 4/2	–	Very fine crumb	Friable	Fine few	–	cs
Bt_2	47–65	10YR 3/6	–	Fine crumb	Slightly sticky	Fine few	–	gs
Bt_3	65–82	10YR 4/4	–	Fine crumb	Friable	Few	–	gs
C	82–120	10YR 5/4	–	Single grain	Friable	–	–	cw

Boundary *c* clear, *d* diffuse, *g* gradual, *s* smooth, *w* wavy

Table 14.2 Physical characteristics of Oyo North soils

Horizon	Depth (cm)	Sand (%)	Silt (%)	Clay (%)	Bulk density (g cm^{-3})	Silt/clay	Textural class
Pedon 1							
Ap	0–20	67.2	13.4	19.4	1.28	0.7	Sandy loam
Bt_1	20–37	85.2	7.4	7.4	1.41	1.0	Loamy sand
Bt_2	37–52	73.2	13.4	13.4	1.51	1.0	Sandy loam
Bt_3	52–80	65.2	15.4	19.4	1.46	0.8	Sandy loam
C	80–125	66.0	16.0	18.0	1.46	0.9	Sandy loam
Pedon 2							
Ap	0–22	63.2	13.4	19.4	1.35	0.7	Sandy loam
AB	22–34	83.2	9.4	7.4	1.44	1.3	Loamy sand
Bt_1	34–50	83.2	11.4	5.4	1.41	2.1	Loamy sand
Bt_2	50–75	73.2	11.4	15.4	1.39	0.7	Sandy loam
BC	75–109	72.1	11.9	16.0	1.38	0.7	Sandy loam
C	109–128	73.2	11.2	15.6	1.39	0.7	Sandy loam
Pedon 3							
AP	0–20	71.2	13.4	15.4	1.31	0.9	Sandy loam
AB	20–30	72.2	11.8	16.0	1.36	0.7	Sandy loam
Bt_1	30–41	72.6	11.4	16.0	1.31	0.7	Sandy loam
Bt_2	41–55	66.6	11.4	22.0	1.27	0.5	Sandy loam
BC	55–72	66.0	16.0	18.0	1.37	0.9	Sandy loam
C	72–120	65.2	15.4	19.4	1.36	0.8	Sandy loam
Pedon 4							
Ap	0–18	66.6	11.4	22.0	1.31	0.5	Sandy loam
AB	18–33	80.6	9.4	10.0	1.34	0.9	Loamy sand
Bt_1	33–41	62.6	11.4	26.0	1.23	0.4	Sandy loam
Bt_2	41–55	64.6	9.4	26.0	1.30	0.4	Sandy loam
BC	55–72	65.0	15.4	19.6	1.36	0.8	Sandy loam
C	72–115	62.6	11.2	26.2	1.34	0.4	Sandy loam
Pedon 5							
Ap	0–35	82.6	7.4	10.0	1.38	0.7	Loamy sand
Bt_1	35–47	82.6	7.4	10.0	1.36	0.7	Loamy sand
Bt_2	47–65	68.6	11.4	20.0	1.52	0.6	Sandy soam
Bt_3	65–82	74.6	7.4	18.0	1.49	0.4	Loamy sand
C	82–119	73.2	11.4	15.4	1.50	0.7	Sandy soam

14.3.3 Chemical Characteristics

The chemical characteristics are given in Table 14.3. The pH of the soils ranged between pH (H_2O) (3.2–5.6) and pH (KCl) (2.7–4.3). The organic carbon is slightly fluctuated with depth indicating continuous deposition of organic material. The total N contents are generally less than 1% ranging between 0.41 and 1.19% following similar trend as for organic carbon. Available P (Bray P) in all the pedons was generally low. The organic matter for most of the pedons is low; this is common to most tropical soils (Faghami and Udo 1982). The low organic matter may be partly due to high temperature and relative humidity which favors rapid mineralization. The OM has positive influence on cation exchange capacity (CEC), buffering capacity, water-holding capacity, structure, pH, and several other physicochemical properties of the soil (Eshett 1985).

The exchangeable bases (K, Ca, Mg, and Na) in all the pedons are generally low. The cation exchange capacity (CEC) was low and fluctuated irregularly, the highest values being recorded at the surface horizon. The low CEC may be attributed to the high leaching intensity, which may have been aggravated by the coarse surface soil texture. The exchangeable acidity (EA) was high in pedons 1, 2, and 4 and moderate in pedons 3 and 5. The base saturation ranged between 4.11 and 90%. The available Fe increased with soil depth in some pedons. Available Mn fluctuated with depth, whereas Zn and Cu distribution down the profile is very erratic. For most tropical soils, the organic matter is the major exchange site for the basic nutrients cations encountered in soils. Thus, the increase of organic matter in these soils will largely increase the CEC and nutrient-holding capacity and influence other soil chemical properties.

14.3.4 Soil Classification

The soil classification is summarized in Table 14.4. Three international soil classification systems were used because Nigeria does not have a national soil classification system. Furthermore, the adoption of the international soil classification systems will among other things facilitate agro-technological transfer. The soil taxa of the investigated pedons using three classification systems are presented in Table 14.4. At the higher category level of USDA soil taxonomy (Soil Survey Staff 2003), only two soil orders have been recognized (Alfisols and Utisols). The pedons where argillic B horizons were identified and have high base saturation (>35%) are classified as Alfisols and Lixisols (FAO-UNESCO 1994). The pedons where an argillic horizon was identified, but have no fragipan, and have base saturation less than 35% are classified as Ultisols and Lixisols (FAO-UNESCO 1994). The Alfisols were also classified as Udalfs (Soil Survey Staff 2003) at the suborder level due to lack of cyric or isofrigid temperature and ustic and xeric moisture regimes. The Ultisols are classified as Udults (Soil Survey Staff 2003) due to udic moisture regime.

14.4 Soil Map and Mapping Units

The soil map of the study area is divided into five soil mapping units based on the suit of soils present in the delineations, soil properties, and landscape positions, as described below (Fig. 14.2). A mapping unit comprises of two or three soil types, but each of the mapping unit is represented by a pedon.

Table 14.3 Chemical characteristics of Oyo North soils

Horizon	pH (H_2O)	pH (KCl)	Org C (g kg^{-1})	Total N (%)	Av. P (mg kg^{-1})	TEB (cmol kg^{-1})	Ca (cmol kg^{-1})	Mg (cmol kg^{-1})	Na (cmol kg^{-1})
Pedon 1									
Ap	3.8	2.9	2.74	0.66	0.35	0.35	0.21	0.09	0.04
Bt_1	3.4	3.5	4.88	1.19	7.51	1.26	0.75	0.27	0.04
Bt_2	3.2	3.0	3.73	0.91	4.24	0.48	0.22	0.12	0.04
Bt_3	3.6	2.9	3.10	0.71	4.68	0.34	0.18	0.11	0.04
C	3.8	3.0	3.00	0.72	4.52	0.47	0.21	0.18	0.05
Pedon 2									
Ap	4.6	2.9	3.10	0.74	2.83	0.26	0.13	0.08	0.04
AB	4.6	3.4	4.40	1.06	4.42	0.48	0.30	0.13	0.04
Bt_1	4.6	3.4	4.10	0.98	9.27	1.00	0.62	0.32	0.04
Bt_2	3.8	3.0	4.01	0.97	8.83	0.38	0.21	0.11	0.04
BC	4.3	3.5	3.00	0.82	7.81	0.44	0.21	0.15	0.05
C	4.5	3.4	2.89	0.80	7.42	0.52	0.31	0.14	0.04
Pedon 3									
Ap	4.6	3.2	3.38	0.81	3.53	0.39	0.23	0.11	0.04
AB	5.6	2.7	3.41	0.83	7.42	0.24	0.12	0.07	0.04
Bt_1	4.4	3.0	3.70	0.95	6.36	0.47	0.28	0.13	0.04
Bt_2	3.8	2.9	3.20	0.84	2.91	0.33	0.20	0.08	0.04
BC	4.5	3.8	3.00	0.79	2.82	0.37	0.20	0.11	0.05
C	4.3	3.8	2.80	2.70	2.70	0.41	0.23	0.12	0.40
Pedon 4									
Ap	3.7	3.0	2.70	0.65	7.90	0.22	0.06	0.09	0.04
AB	4.3	3.4	4.71	1.13	4.42	0.48	0.27	0.15	0.04
Bt_1	5.1	4.3	3.95	1.00	29.40	0.63	0.30	0.27	0.04
Bt_2	5.2	4.3	2.10	0.51	29.32	0.84	0.15	0.63	0.04
BC	5.2	4.2	2.00	0.48	28.41	0.76	0.16	0.54	0.03
C	5.1	4.2	1.89	0.42	26.22	0.67	0.15	0.45	0.04
Pedon 5									
Ap	5.3	4.4	3.11	0.74	56.34	1.51	0.22	1.23	0.04
Bt_1	5.6	4.7	4.10	0.97	55.98	1.77	0.27	1.42	0.04
Bt_2	4.2	3.3	2.81	0.67	9.45	0.42	0.09	8.24	0.08
Bt_3	3.7	2.7	2.35	0.55	8.83	0.28	0.06	0.19	0.02
C	3.8	3.8	2.31	0.41	8.00	0.32	0.09	0.18	0.03

K (cmol kg^{-1})	EA (cmol kg^{-1})	CEC (cmol kg^{-1})	Base saturation (%)	Mn (mg kg^{-1})	Fe (mg kg^{-1})	Cu (mg kg^{-1})	Zn (mg kg^{-1})
0.01	3.8	4.16	8.4	3.3	28.7	0.98	1.19
0.20	0.4	1.66	75.9	19.4	25.3	0.75	0.81
0.10	3.2	3.68	13.0	7.0	20.9	0.76	1.08
0.01	3.6	3.94	8.6	10.0	46.6	0.94	1.35
0.03	3.4	3.84	10.4	12.0	45.5	0.91	1.30
0.01	4.6	4.86	5.4	6.6	39.5	0.86	1.30
0.01	0.6	1.08	44.4	10.6	16.8	0.57	0.85
0.02	0.6	1.06	62.5	33.5	21.7	0.87	3.23
0.02	3.8	4.18	9.1	43.8	62.6	1.28	2.17
0.03	3.9	4.12	10.2	45.7	60.2	1.20	2.11
0.02	3.8	4.05	9.2	41.2	61.1	1.20	2.11
0.01	1.6	1.99	19.6	6.9	25.4	0.93	0.68
0.01	5.6	5.84	4.1	7.5	67.6	1.07	0.69
0.02	1.4	1.87	25.1	9.9	35.9	1.04	1.52
0.01	3.8	4.13	8.0	8.6	45.1	1.11	0.95
0.01	2.8	3.15	8.2	7.4	43.1	1.12	0.85
0.02	2.6	3.00	8.2	7.0	43.1	1.40	0.79
0.03	4.6	4.82	45.6	6.9	66.1	1.20	1.09
0.02	0.2	0.68	70.6	13.0	28.9	0.98	1.25
0.02	0.2	0.83	75.9	36.5	13.4	2.02	9.11
0.02	0.2	1.04	80.8	11.9	19.5	1.51	6.84
0.03	0.2	1.11	80.6	11.7	18.5	1.41	6.91
0.03	0.3	1.09	78.5	10.2	18.5	1.51	5.21
0.02	0.2	1.71	88.3	49.2	23.4	2.22	11.09
0.04	0.2	1.97	89.8	65.6	33.7	3.40	20.66
0.01	2.0	2.42	57.4	14.7	77.0	1.19	1.20
0.01	2.2	2.48	4.3	10.1	39.2	1.00	0.89
0.02	2.1	2.41	42.3	10.0	38.2	1.11	0.74

Table 14.4 Soil classification of Oyo North

Profile	Soil Survey Staff (2003)	FAO/UNESCO	FCC
Pedon 1 (MU1)	Arenic Kandiudults	Ferric Lixisols	Sek >
Pedon 2 (MU2)	Arenic Kanhapludults	Stagnic Lixisols	Sek >
Pedon 3 (MU3)	Typic Kandiudults	Ferric Lixisols	Sek>(2–4%)
Pedon 4 (MU4)	Arenic Kandiudalfs	Haplic Lixisols	Se>(4–7%)
Pedon 5 (MU5)	Gross Arenic Kandiudalfs	Haplic Lixisols	Lek>(>7%)

S sandy topsoil, *L* loam
e = low CEC (applies only to plow layer or surface 20 cm whichever is shallower)
k = K deficient; <10% weatherable minerals in silt and sand fraction
> = >35% gravel or coarse (>2 mm) fraction by volume
% = (slope) where it is desired to show slope within FCC unit

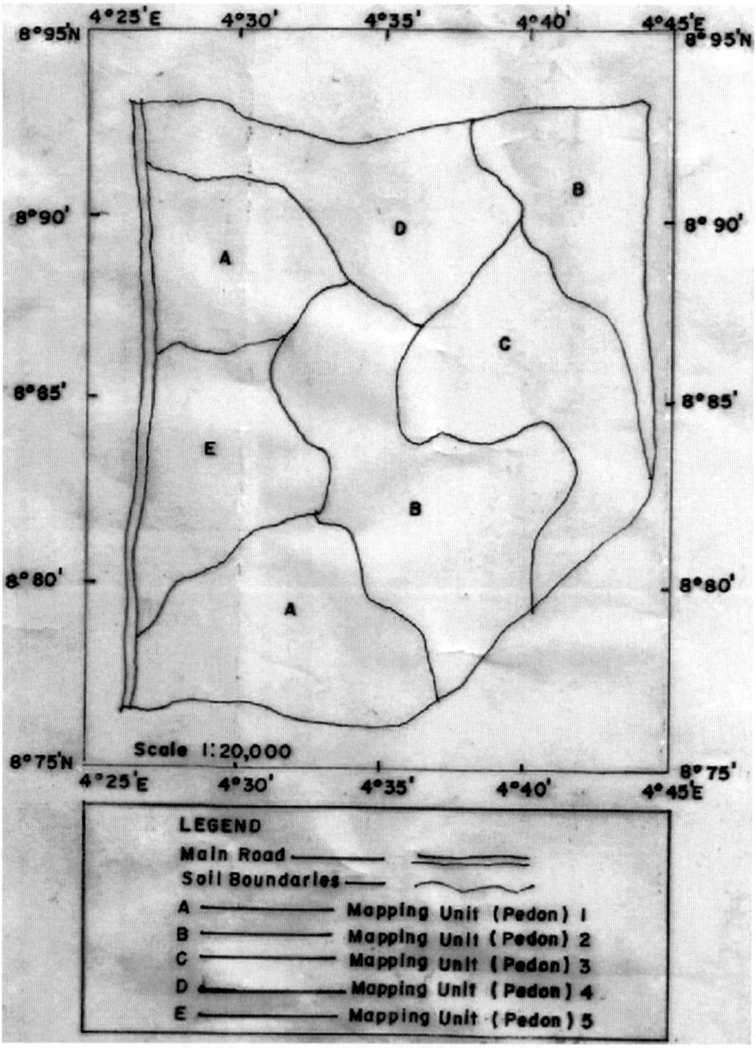

Fig. 14.2 Soil map of part of Saki local government area of Oyo State

14.4.1 Soil Mapping Unit 1 (MU1)

The MU1 is classified as Arenic Kandiudults at the subgroup level (Soil Survey Staff 2003); it presents sandy or sandy skeletal texture from the mineral surface to the top of a Kandic horizon. On the FAO–UNESCO (1994) system, the same pedon is classified as Ferric Lixisols due to ferric properties within 125 cm of the soil surface and an albic E horizon, lacking gleyic properties within 100 cm of the surface. This unit is designated as soil consociation. The soils of this unit are mostly found in the upper terrain, mainly presenting geomorphological units characterized by a subtraction consisting of loamy sand and sandy loam. It is a fallow land since the last 5 years, and quarrying is suspected in the area. Fe–Mn concretion is prominent beneath the B horizon, and nearly all the horizons are very stony.

The mapping unit can be used for arable crop production under intensive irrigation system. The limitations of this unit include excessive drainage, rapid infiltration, sandy texture, and low nutrient status. Based on the limitation of these soils and to counteract the problem of low nutrient and sandy nature of the soil, crop residues and organic manures are to be practiced in the traditional agriculture. It is advisable to supplement soils with inorganic fertilizers to improve nutrient status. Crop residue mulch can help in maintaining soil fertility, prevent raindrop impact, reduce soil compaction, maintain favorable soil temperature and moisture regime, stimulate biological activity of earthworm, and sustain high equilibrium level of organic matter content. Land qualities are given in Table 14.5, and the suitability of the soil for various alternative uses is summarized in Table 14.6. The soil is marginally suitable (S3) for maize, cassava, and oil palm, and the soil can be potentially modified to be moderately suitable for these crops by embarking on the above measures.

14.4.2 Soil Mapping Unit 2 (MU2)

The MU2 is classified as Arenic Kanhapludults at the subgroup level (Soil Survey Staff 2003) due to sandy or sandy skeleton particle-size class throughout a layer extending from the mineral soil surface to the top of a Kandic horizon at a depth of 100 cm and Stagnic Lixisols (FAO-UNESCO 1994) due to stagnic properties within 50 cm of the soil surface, and lacking gleyic and plinthite within 100 cm of the surface. The MU2 is designated as soil consociation. The soils are located in the upper slope/summit of the landscape. The land is fallow and is equivalent to secondary forest. The profiles are very deep (>150 cm) with fine crumb sandy loam topsoil with the Fe–Mn concretions below the B horizons. The soils are generally well drained and slightly acidic. The infiltration rate is high due to high gravel contents. The nutrient status is very low.

Table 14.5 Land qualities/characteristics of Oyo North soils

Pedon no.	Annual rainfall (mm)	Length of dry season (months)	Mean temp. °C	Slope %	Relative humidity (%)	Drainage	Soil depth (cm)	Coarse fragments %	Texture	Exch K (mol kg^{-1})	OC %	ECEC cmol kg^{-1}	BS (%)	pH (KCl)
1	1,800	4½	25	1	76	WD	>125	65–85	SL–LS	0.01–0.20	2.73–4.88	1.66–4.16	8.41–75.9	2.9–3.5
2	1,800	4½	25	3	76	WD	>128	65–83	SL–LS	0.01–0.03	2.89–4.40	1.06–4.84	5.35–62.5	2.9–3.5
3	1,800	4½	25	3	76	WD	>120	65–72	SL	0.01–0.02	2.80–3.70	1.87–5.84	4.1–25.1	2.7–3.8
4	1,800	4½	25	6	76	WD	>115	62–80	SL–LS	0.02–0.03	1.89–4.71	0.83–4.82	45.64–80.7	3.0–4.3
5	1,800	4½	25	7	76	WD	>119	68–82	LS–SL	0.01–0.04	2.31–4.10	1.71–2.48	41.29–89.8	2.7–4.7

WD well drained, ECEC effective CEC, OC organic carbon, BS base saturation

Table 14.6 Suitability class scores and aggregate suitability of pedons for common crops grown in the area

Crops (ratings)		Aggregate suitability (nonparametric)	
Maize		*Actual*	*Potential*
Pedon 1	S2tcf, S3cf	S3sf	S2s
Pedon 2	S2tcf, S3cf	S3f	S2s
Pedon 3	S3tcf, S3cf	S3sf	S2s
Pedon 4	S3tcf, S3cf,	S3cf	S2s
Pedon 5	S2tcf,, S3cf	S3sf	S2s
Oil palm			
Pedon 1	S3tc, S3cf, S2f	S3csf	S2s
Pedon 2	S3tc, S3cf, S2f	S3csf	S2s
Pedon 3	S3tc, S3cf, S2f	S3csf	S2s
Pedon 4	S3tc, S3cf, S2f	S3csf	S2s
Pedon 5	S3tc, S3cf, S2f	S3csf	S2s
Cassava			
Pedon 1	S3cf, S2f	S3sf	S2s
Pedon 2	S3cf, S2f	S3sf	S2s
Pedon 3	S3cf, S2f	S3sf	S2s
Pedon 4	S3cf, S2f	S3sf	S2s
Pedon 5	S3cf, S2f	S3sf	S2s

s soil physical properties, *f* fertility, *t* texture, *cf* chemical fertility
Aggregate suitability class scores:
S1 = highly suitable
S2 = moderately suitable
S3 = marginally suitable

The mapping unit can be used for arable and biennial crop production under intensive irrigation system. Agricultural practices have significant impacts on the status and nutrient reserve of the soil; green manuring is a way of solving the soil fertility problem; legumes like mucuna and centrosema are to be mixed with the soil; this practice would go a long way to improve the soil fertility when the crops are well incorporated into the soil. Appropriate tillage practices in which plowing the soil alone should be done to create microdepressions, which temporarily store water during the rain; these depressions are strategic to control runoffs and soil loss. The use of cover crops and planted fallow can also lead to conservation and restoration of the soil physical and chemical properties and buildup of soil organic matter.

The suitability of the soil is marginal (S3) for the three crops reviewed, and the suitability can be potentially modified to moderately suitable (S2) through the above measures.

14.4.3 Soil Mapping Unit 3 (MU3)

The MU3 is classified as Typic Kandiudults at the subgroup level (Soil Survey Staff 2003) and Ferric Lixisols (FAO-UNESCO 1994) for showing ferric properties within 125 cm of the surface and for lacking gleyic and albic E horizon within

100 cm of the surface. This soil unit is designated as consociation. The soils in this unit are found in the middle slope. The soils have well-developed argillic B horizon and have sandy or sandy skeletal particle size throughout the layer. The Fe–Mn concretions are present nearly in all the horizons. The soils are generally well drained, micronutrients are normal, and the soils are non-sodic but slightly acidic.

The mapping unit can be used for arable and tree crops, but the water requirement has to be met with intensive irrigation. Also as in mapping unit 1 and 2, green manuring, agroforestry, use of cover crops, planting fallow land, efficient fertilizer use, and management would improve the fertility status of the soils.

Due to certain limitations of MU3, the current suitability for most crops is marginal (Table 14.6). The suitability of the soil for crops that are marginally suitable (S3) can also be modified potentially to moderately suitable (S2) with the above management techniques.

14.4.4 Soil Mapping Unit 4 (MU4)

The MU4 is classified as Arenic Kandiudalfs at the subgroup level (Soil Survey Staff 2003) due to sandy particle-size class throughout a layer extending from the mineral surface to the top of Kandic horizon, and Haplic Lixisols (FAO-UNESCO 1994) for lacking an albic E horizon, as well as ferric properties and plinthite within 125 cm of the surface. This soil unit is designated as consociation. The soils are commonly found at the lower slope portion of the site. The soil has medium particle-size class throughout the upper 75 cm of the argillic horizon. The soils are deep and perfectly well drained. They have dark brown sandy loam, weak fine crumb, and friable topsoil underlain by dark red sandy loam. The soil is slightly acidic, low total exchangeable basis, CEC, and fairly low organic carbon and total nitrogen. The soil is fallow without human activity.

The mapping unit can be used for arable and perennial crops. In addition to the methods of soil management in the mapping units 1, 2, and 3, the use of organic fertilizers can improve both the physicochemical and biological properties of the soil; this also has the advantage of supplying additional micronutrients which are not usually available in inorganic fertilizers. The mapping unit is marginally suitable (S3) for maize, cassava, and oil palm, and the marginal suitability can be modified to moderate suitability (S2) by the above system of managements.

14.4.5 Soil Mapping Unit 5 (MU5)

The MU5 is classified as Gross Arenic Kandiudalfs (Soil Survey Staff 2003) due to sandy particle-size class throughout a layer extending from the mineral soil surface to the top of a Kandic horizon within 150 cm of the mineral soil surface and Haplic Lixisols (FAO-UNESCO 1994) for lacking an albic E horizon, and ferric and plinthic properties within 125 cm of the surface.

The soils are located in the lower portion toward the stream at the valley bottom. The soils are deep and well drained and are formed on sedimentary parent material. The soils have dark brown color, weak fine crumb, friable nonsticky, fine common roots and gradual smooth boundary at the plowed layer underlain by red sand, non-sticky and firm irregular boundary. The soils range from loamy sand to sandy loam at the bedrock. The soil is moderately acidic. The fertility status of the soil is moderate for optimum production.

Due to the coarse texture, the silt/clay ratio is high, which is in agreement with the Nwaka and Kwari (2000). In sandy soils, high values of silt–clay indices may be related to skeletal composition of the parent material.

The mapping unit can be used for arable, biennial, and tree crops. Though the soil is marginally suitable (S3) for maize, cassava, and oil palm, the soil can be improved with the integrated methods of soil management that have been discussed in the other mapping units, such as green manuring, agroforestry, use of cover crops and planting fallow land, appropriate tillage practices, efficient use of fertilizers based on soil tests, use of organic fertilizer, and biofertilization.

All the methods of management practices mentioned above when done appropriately would go a long way for soil sustainability management in the study area for crop production.

14.5 Fertility Capability Classification (FCC)

The FCC system developed and modified by Buol et al. (1982) was used in this study. The system consists of three categorical levels "type" (texture of plow layer or top 20 cm, whichever is shallower), substrata "type" (texture of topsoil when there is a marked change in texture from the surface), and the condition modifiers (chemical element consideration). Class designations from the three categorical levels are combined to form the FCC unit. The different FCC units are shown in Table 14.4.

The results of the FCC show that the major differences between the pedons were in texture and slope which are very crucial in evaluating the suitability of soil for most crops. The type shows the texture of plow layer or surface 20 cm which is represented by S in all the pedons except pedon 5 which is L representing loam, while the subtype is used only if there is a textural change from the surface or if a hard root-restricting layer is encountered within 50 cm which is not applicable in all the pedons.

While the condition modifiers indicate the chemical elements that are deficient in the soil, and where more than one criterion are listed for each modifier, only one needs to be met to place the soil in a class. In case of all the pedons which indicate low CEC, k = (k deficient) was common, except pedon 4 which presents gravels (particles >2 mm) >35% by volume which are common to all the pedon. For example, FCC of pedon 1 was Sek>which signifies sandy topsoil, low CEC, and K deficiency with gravel content more than 35%.

The FCC predicts how the soil can be managed, for example, in pedon 1, the sandy topsoil, low CEC, and K deficiency can be improved by the application of organic and inorganic fertilizers to the soil and use of mulching and planting fallow lands to improve physical, chemical, and biological properties of the soil, and where there is slope of more than 4%, erosion measures can be put in place to reduce runoff and soil loss.

14.6 Land Suitability Evaluation

Using the method of Sys (1985), the suitability of soils was evaluated for some crops commonly grown in the area. Table 14.5 shows the detailed land and soil qualities of the area. Conventional approach (FAO 1976) was employed in which pedons were placed in suitability classes by matching their characteristics of the land (Table 14.5) with the requirement of various crops. The suitability class or rating of each pedon is indicated by its most limiting characteristics. The aggregate suitability rating is for both the potential and actual (or current) suitability which also indicates the soil limitation and constraints. These limitations dictate the appropriate technique and technologies to be used to increase the soil quality potentials for increased productivity.

Based on the soil qualities, pedon 1 is currently marginally suitable (S3) due to fertility and moderately suitable (S2) due to texture for maize and marginally suitable (S3) for cassava and oil palm due to fertility status, and the soil can be modified to moderately suitable (S2) for these crops by improving the limitations. The modification of these limitations determines the subsequent suitability of the pedons for crop production.

14.7 Summary and Conclusions

The soils of the study area are unique in their characteristics and classification; these attributes are greatly influenced by the nature of the parent materials derived from the basement complex of southwestern Nigeria. The soils are coarse and sandy in texture and acidic to slightly acidic with low base saturation and low cation exchange capacity. The available Fe increases with soil depth, and available Mn, Zn, and Cu fluctuate with soil depth in most of the pedons. Pedons 1, 2, and 3 are classified as Ultisols and Udults as a result of having udic moisture regime and pedons 4 and 5 as Alfisols and Udalfs as a result of not having cyric or isofrigid temperature and ustic and xeric moisture (Soil Survey Staff 2003). All the soils are classified as Lixisols (FAO-UNESCO 1994) for low CEC and base saturation. Fertility capability classification shows that the soils had sandy topsoils and low CEC at plow layer, K deficiency, and high gravel content. All the soils are currently marginally suitable and potentially moderately suitable for maize, cassava, and oil palm; the most limitations are climate, soil texture, and low soil fertility. The soils can be improved by the addition of organic and inorganic manures, use of extensive irrigation systems, and maintenance of soil cover either by planting fallow land or mulching and liming for sustainability of the soils for crop production.

References

Bouwer H (1986) Intake rate: cylinder infiltrometer method of soil analysis. In: Klute A (ed) Methods of soil analysis. Part I Physical and Mineralogical methods. Soil Science Society of America, Madison, pp 825–844

Bouyoucos GH (1951) A calibration of the hydrometer method for mechanical analysis of soils. Agron J 43:434–438

Bray RH, Kurtz LT (1945) Determination of total organic and available forms of phosphorus in soils. Soil Sci 9:1149–1178

Bremner JM (1965) Total nitrogen. In: Black CA (ed) Methods of soil analysis. Part II, vol 9, American society of agronomy monograph. Soil Science Society of America, Madison, pp 1149–1178

Buol SW, Sanchez PA, Cate RB Jr, Grander M (1982) Soil acidity. Fertility capability classification. A technical soil classification system for fertility management. In: Bornemisza E, Alvarado A (eds) Soil management in tropical America. North Carolina State University, Raleigh, pp 126–145

Coleman NT, Thoman CW (1967) The basic chemistry of soil acidity. In: Pearson RN, Adams F (eds) Soil acidity and liming, vol 12, Agronomy monograph. American Society of Agronomy, Madison, pp 1–41

De Gens JG (1978) Fertilizer guide for the tropics and subtropics. Centre d Etudel Azote, Zurich

Eshett ET (1985) Soil characteristics and farming systems in Northern Cross River State of Nigeria. Unpublished PhD thesis, Department of Agronomy, University of Ibadan, p 225

Faghami A, Udo EJ (1982) The characteristics of two soil toposequences in the basement complex in the Federal Capital Territory of Nigeria. J Agric 4(112):9–24

FAO (1976) A framework for land evaluation, vol 32, Soils Bull. FAO, Rome

FAO-UNESCO (1986) Provisional edition of the revised legend of the FAO-UNESCO soil map of the world. FAO of the United Nation, Rome

FAO-UNESCO (1994) Provisional edition of the revised legend of the FAO-UNESCO soil map of the world. FAO of the United Nations, Rome

Lakanem E, Ervio ER (1971) A comparison of eight extractants in soils Scion. Maatal Scur Juk 123:223–232

Nwaka GK, Kwari JJ (2000) The nature and properties of the soils of Jere Bowl near Maiduguri in Borno State. Samaru J Agric Res 16:25–40

Peach ML, Alexander T, Dean LA (1947) Methods of soil analysis for soil fertility investigation. USDA Agric Circ No 757, USDA, Washington, DC, p 8

Richards LA (ed) (1954) Diagnosis and improvement of saline and alkaline soils. United States Department of Agriculture, Handbook No 60. USDA, Washington, DC

Schoeneberger PJ, Wysocki DA, Benham EC, Broderson WD (eds) (2002) Field book for describing and sampling soils, Version 2.0. Natural Resources Conservation Service, National Soil Survey Center, Lincoln

Soil Survey Division Staff (1993) Soil survey manual. Agric Handb 18. USDA-NRCS. US Govt Print. Office, Washington, DC

Soil Survey Staff (2003) Keys to soil taxonomy, 6th edn. USDA-Natural Resources Conservation Service, Washington, DC

Stoop WA (1987) Variation in soil properties along 3 toposequences in Burkina Faso and implications for the development of improved cropping systems. Agric Ecosyst 19:241–264

Sys C (1985) Land evaluation, vols I, II, III. International Training Centre for Postgraduate Soil Scientist, State University of Ghent, Ghent

Walkley A, Black IA (1934) Determination of organic carbon in soils. Soil Sci 37:29–38

Chapter 15
Use of Soil Survey Database for the Probabilistic Evaluation of Soil Cover Transformation in the Semiarid Zone of Western Siberia

Irina Mikheeva

Abstract The knowledge of soil properties, state of soil cover, and soil-forming processes is necessary for the understanding of landscapes and to plan ameliorative and engineering approaches for reclamation and their uses for agricultural purposes. In the semiarid zone of Western Siberia, it has been shown that the soil survey database can be used for probabilistic estimations and modeling of soil cover changes. Probability distributions were determined for chestnut soil properties, which are suitable for the reliable and holistic evaluation of the effect of soil-forming factors on soil properties and, hence, for comparisons of current and future anthropogenic and natural changes. The divergence of soil properties under deflation, long-term plowing, and irrigation by low-mineralized water with sodium bicarbonate was investigated.

Keywords Anthropogenous • Database • Holistic evaluation • Soil cover • Western Siberia

15.1 Introduction

Ecosystems in semiarid zone of Western Siberia are strongly subjected to anthropogenic pressure owing to agricultural uses. This is promoted by sufficient-leveled relief for agriculture and good thermal resources of such territories. Intensive agricultural use of soils in this zone causes degradation of soil structure, compaction, salinization, and alkalinization.

Obtaining the information for the estimation of soil condition, forecasting of possible changes, and prevention of adverse consequences of the land usage caused

I. Mikheeva (✉)
Institute of Soil Science and Agrochemistry, Siberian Branch of Russian Academy of Sciences, 18, Sovetskaya st., Novosibirsk, 630099 Russia
e-mail: mikheeva@issa.nsc.ru

S.A. Shahid et al. (eds.), *Developments in Soil Classification, Land Use Planning and Policy Implications: Innovative Thinking of Soil Inventory for Land Use Planning and Management of Land Resources*, DOI 10.1007/978-94-007-5332-7_15,

by the changes of natural environment is based on the monitoring of conditions of the environmental components, including soils. Monitoring is connected to acquiring and storing of necessary data for mathematical analyses.

Due to high natural variability of soils and their properties, problem exists to verify the soil estimations statistically. Indeed, soil properties and soil-forming processes are result of functioning of soil as open and complicated system; hence, variability and fluctuations are its inherent property at different organization levels. The concept of deterministic chaos and uncertainty which was developed by Philips (1993) is an adequate conceptual basis for understanding this phenomenon.

Therefore, it is logical to use probabilistic distributions functions (pdf) of properties which reflect two structural levels—an internal structure from elements with different displays of soil properties and determination of system. The theoretical idea of using the pdf to assess the transformation of soils consists of assumption that collective transformation of elements of system can arise from a set of individualistic chaotic changes that are mutually compatible. Integrity is given to this system by the presence of specific and rigid external conditions in which elements of system exist.

Taking these propositions into account, we suggested statistical method of estimating the soil changes according to the change of pdf of the soil properties (Mikheeva 2001). Results have proved law of "wave" of pdf, depending on factors and processes of soil formation. In monitoring practice, pdf of soil properties can serve as diagnostic characteristics or indicators which allow holistic assessment of soil processes and changes of state of soil objects taking into account their variability (Mikheeva 2005). Analysis of pdf dynamics showed that within one soil series, a certain property can often be both weakened and intensified, though, as a rule, a prevailing tendency is observed. However, to give an adequate quantitative estimate of transformation, it is necessary to characterize the soil changes in general. So we proposed holistic estimation of soil transformation by mathematical value of divergence of pdf of soil properties (Mikheeva 2009).

The objectives of this study were reliable and holistic probabilistic estimation of transformation of soil cover in the south of Western Siberia under different agricultural impacts by using this methodic approach on data of soil monitoring survey.

15.2 Materials and Methods

15.2.1 Description of Study Area

The study was conducted over a large part (16,000 km^2 in area, 200 km from north to south, and 80 km from east to west) of the Kulunda steppe located in the south of Western Siberia (Fig. 15.1).

The region is characterized by droughty continental climate, and its relief may be defined as gently undulating plain. The soil cover consists of chestnut soils (70%), meadow-chestnut soils, meadow soils, solonetzes, and solonchaks with different degrees of hydromorphism in Russian classification.

Fig. 15.1 Site location map of the study area

Table 15.1 Soil texture, soil humus, and CEC value as in the upper horizon of chestnut soils

Soil texture variety	Friable sand	Cohesive sand	Coarse-loamy sand	Fine-loamy sand	Sandy loam	Loam
Humus (%)	0.77	0.87	1.24	1.51	1.89	2.65
CEC (cmolc kg^{-1})	Nd.	Nd.	9.13	10.22	13.26	18.69

The soils are formed on ancient Quaternary lacustrine-alluvial deposits, which are responsible for their primarily light granulometric composition and spatial diversity of fractions. Soils significantly vary in texture, from friable sands to loams (Table 15.1). The predominant soil is loamy sand chestnut soil.

The texture of a soil largely determines its physical properties and water regime (moisture-holding capacity), which in turn determine chemical properties of automorphic soils developed on nonsaline parent rocks. The humus content is less than 3% depending on the soil texture (Table 15.1). The cation exchange capacity (CEC) is 9.13–18.69 cmol kg^{-1} soil; calcium is predominant (75–80%), magnesium makes up 20%, and sodium less than 1%. The upper horizons of the chestnut soils have satisfactory physical properties. The porosity is 48–50%, the maximum hygroscopicity is 1–5%, the molecular moisture capacity is 9–12%, and the field capacity is 16–22% (Panfilov 1973).

Data in Table 15.1 are average values of soil properties, but the values in individual soil profiles vary significantly, which is why probabilistic approach to soil assessment is necessary.

The chestnut soils are the dominant soil type of the arid-steppe zone. One can judge the rate and degree of manifestation of desertification and degradation of the soil cover by monitoring the changes of soil properties in these soils. The drier the climate, the more probable is the manifestation of deflation with the increased wind speed, especially in the flat-plain relief, light granulometric composition of soils, and absence of forest, which is typical of that territory. This makes it essential to use the soils properly. The measures on introducing the soil-protecting agriculture (primarily alternation of the strips of grain crops and perennial herbs, plantation of forest shelter belts) taken in the 1960s stopped the deflation caused by the total tillage. Irrigation network was developed between 1970 and 1990 to soften influence of arid climate to agriculture and to meet water requirements of crops. The fresh water is scarce; therefore, low-mineralized water (soluble salts) of deep earth horizon (about 1,000 m) was used for irrigating crops. The mineralization of this water is rather good (about 700–1,000 mg l^{-1} of salts); however, its sodium adsorption ratio is high [$SAR = 20.4$ $(mmol\ l^{-1})^{0.5}$].

15.2.2 Existing Information and Database

The existing data were used in this study as baseline information. In the past, full-scale study (1:25,000) was conducted using standard Russian methods at an area of approximately 1.6 million ha in the southwest of Kulunda steppe. During survey, one full soil profile with six samples was investigated in the grid 350 × 350 m.

During this study, 4,000 soil profiles were investigated; the results are grouped and are used to create a database. Each data sample gives information related to a certain soil variety by one of the soil properties (humus content, content of granulometric fractions, etc.) in a separate soil horizon or layer at certain time or state. Volumes of these data sample range from 20–25 to 650 datasets, depending on the abundance or the soil series. We believe that this is enough to conduct statistical and probabilistic analysis.

The objective of this study was to determine the probability distribution function (pdf) of soil properties that involves selection of the type and parameters of distribution function and its fitting for experimental data. The standard procedures of verification of hypotheses on the closeness of the studied statistical distribution to a theoretical one often fail to give a satisfactory unambiguous answer. So we used the software product of the Department of Applied Mathematics and Informatics of Novosibirsk State Technical University (Lemeshko 1995). It has overcome the problem in the definition of mathematical functions of pdf by using principle, when many hypotheses about coincidence were checked with a set of thirty theoretical functions (Su-Johnson, exponential functions, double power, maximum and minimum value, normal, Ln-normal, lognormal). In each case, only the function with its parameters was chosen, which were evaluated by the method of maximal likelihood.

It is the best approximation coordinated with set of parametric and nonparametric criteria.

We have applied this statistical procedure, which has allowed identification of pdf with very high reached probability value. Usage of big samples ($n=50\ldots600$) gives us confidence that the forms of these pdf are close to their real forms. We have identified pdf of soil properties for each chestnut soil series, and as a result, we received bank of pdf of soil properties of Kulunda steppe at different stages of soil usage (Mikheeva 2001, 2005). An example of part of this pdf bank is given in Table 15.2.

To estimate how much the pdf of soil properties differs in different horizons, at different times, or in different states of soil (arable, deflated, non-deflated, irrigated), we used the value of divergence (*d*) (Gubarev 1992).

$$d = \int \left(W1(x) - W2(x)\right) \ln\left(\frac{W1(x)}{W2(x)}\right) \mathrm{d}x$$

where $W1(x) - W2(x)$—pdf of property in compared objects. This value was calculated in program system Mathcad.

15.3 Results and Discussion

The deflation due to wind erosion destroys ground surface, basically, at droughty territories, especially when the soils are coarse textured. Deflation is connected to climatic rhythms and intensity of land use. It concerns not only to the territory investigated by us but also to other regions of the country and the world. After total plowing of the virgin soils during 1950–1960, and subsequent cycles of wind activity, started the deflationary processes. It has been observed that most parts of the chestnut soils in the investigated territory have been subjected to deflation to some extent.

The deflation represents the physical process made by wind, expressing in a separation, carry, and adjournment of soil particles, so the deflation is spatially widespread process at which sites of removal and accumulations of a deflationary material are formed. Therefore, it is rather difficult to state an exact quantitative estimation of the general losses and infringements of soils in certain territory and influence of deflation on the change of soil properties.

The degree and speed of soil deflation depend on many regional and local factors and conditions, such as climatic, geomorphologic, anthropogenous, and also from the context of physical and chemical characteristics of soils. Stability of soils which is not protected by plants from blowing is determined by their aggregate structure, durability of units, intra- and inter-aggregate connectivity, and surface crust. The listed properties, in turn, are defined by the basic characteristics of soils: their humidity, granulometric composition, the contents of organic substance, and calcium carbonate equivalents. It has been shown by various authors that the stability of soils to deflation can be determined by granulometric composition. In particular, with other things being equal (climatic, position in a relief), the belonging of chestnut soil to certain texture series defines probability of display of a deflation (Kuzmina 1980).

It is already known that soil particles of different sizes, owing to their aerodynamic properties, do not have identical value during a deflation. Deflation is expressed in various forms of movement of particles by the action of wind: fractions of coarse and medium sands are transferred by creeping on soil surface, fractions of fine sand move in steps (saltation), and silt and clay fractions due to light weight stay in suspension in the air. The mechanism of soil particles movement (creeping-saltation-suspension) in the desert environment by the action of wind is described by Shahid and Abdelfattah (2008). Therefore, display of soil deflation is defined by a ratio of all fractions of granulometric composition.

Weak destruction at strong deflationary process is a regional feature of the studied soils. Very big duration of aeolian processes in the given territory (as grounds were formed at constant action of deflation) determines a modern steady condition of studied soils in relation to destroying action of wind. The soil contains minimum of dust particles; nevertheless, a plenty of the fine sand causing spasmodic movement of particles determines high probability of occurrence of deflation. Coarse sand creeps on soil surface, sliding and arranging "ripples," leveling relief, and filling the lakes.

Changes in humus contents of soils under deflational conditions were also observed. After 10 years' use of chestnut soil as arable land under soil-protective agriculture, the tendency of average humus content increase in the upper horizon of sandy and loamy soils is recorded. The changes of soil parameters in these soils represent reorganization of probabilities of the values connected to formation of sandy deposits that reflects specificity of given natural-anthropogenic process more precisely than changes of average values.

The irrigation by low-quality water leads to salinization and alkalinization of soils. Fluctuations of the contents of water-soluble salts show high dynamism of salinization, due to irrigation practices and climatic factors. The stable tendency of the reduction of calcium in soil base-exchange complex is due to the replacement by sodium. Change of granulometric structure in relation to the increase in silt contents and relative reduction of dust and sandy fractions has impulsive and irreversible character.

We modeled and evaluated change of probability distributions of soil properties in loamy sandy chestnut soils, as well as in other texture series of chestnut soils, in different states of grounds: (1) non-deflated, (2) deflated, (3) after long-term plowing without deflation, and (4) irrigation at some irrigated plots at territory of our investigation. These states followed each other in the last 40 years of previous century.

The results of such evaluation of pdf of the contents of coarse sand are visualized in Fig. 15.2. It is shown that central part of pdf of the contents of coarse sand moved to decrease this attribute; moreover, changes of form of pdf was rather essential because of change of probabilities of values of the attribute under influence of deflation. Long-term plowing brings opposite tendency. Irrigation changed pdf of contents of coarse sand dramatically; it was decreased under the influence of sodium bicarbonate water, which destroyed the large soil aggregates.

All these alterations do not change soil taxonomy of the sites investigated, but they influence the status of soil and allow to model long-term trend of soil cover change. It could be evaluated by means of value of divergence (*d*) of pdf. $d = 0.5$ for differences

Table 15.2 Probability distributions of humus content in A_p horizon of chestnut soils

Soil type, status	Type of distribution	Parameters of distribution		α[a]
Cohesive sand, deflated	Double power	$\theta_0=6.71$	$\theta_1=146.60$	0.6
Cohesive sand, non-deflated	Ln-normal	$\theta_0=-0.04$	$\theta_1=0.28$	0.7
Coarse-loamy sand, deflated	Double power	$\theta_0=3.94$	$\theta_1=84.56$	0.6
Coarse-loamy sand, non-deflated	Su-Johnson	$\theta_0=-3.34$	$\theta_1=3.21$	
		$\theta_2=0.48$	$\theta_3=0.59$	0.7
Coarse-loamy sand, protected	Maximum value	$\theta_0=0.23$	$\theta_1=1.18$	0.8
Fine-loamy sand, deflated	Su-Johnson	$\theta_0=-0.42$	$\theta_1=1.64$	
		$\theta_2=0.40$	$\theta_3=1.42$	0.5
Fine-loamy sand, non-deflated	Logistic	$\theta_0=1.42$	$\theta_1=0.30$	0.4
Fine-loamy sand, protected	Nakagami	$\theta_0=0.63$	$\theta_1=0.89$	
		$\theta_2=0.98$		0.5
Sandy loam, deflated	Normal	$\theta_0=1.66$	$\theta_1=0.38$	0.7
Sandy loam, non-deflated	Su-Johnson	$\theta_0=-1.43$	$\theta_1=1.76$	
		$\theta_2=0.48$	$\theta_3=1.40$	0.5
Sandy loam, protected	Maximum value	$\theta_0=0.38$	$\theta_1=1.83$	0.6

[a] Reached level of significance (p-value)

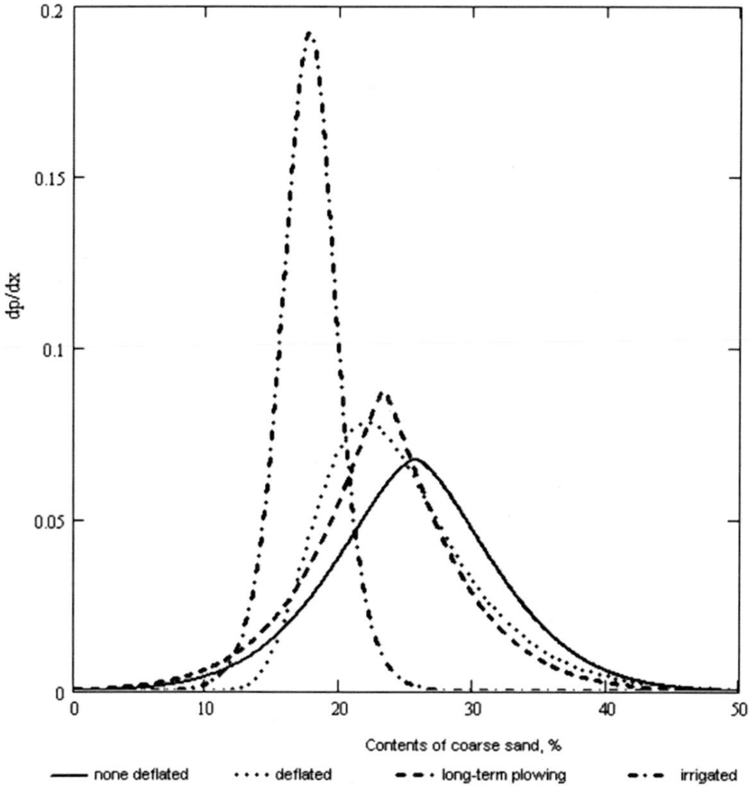

Fig. 15.2 Pdf of the contents of coarse sand in the upper horizon of loamy sandy chestnut soil in different states (non-deflated, deflated, long-term plowing, irrigated)

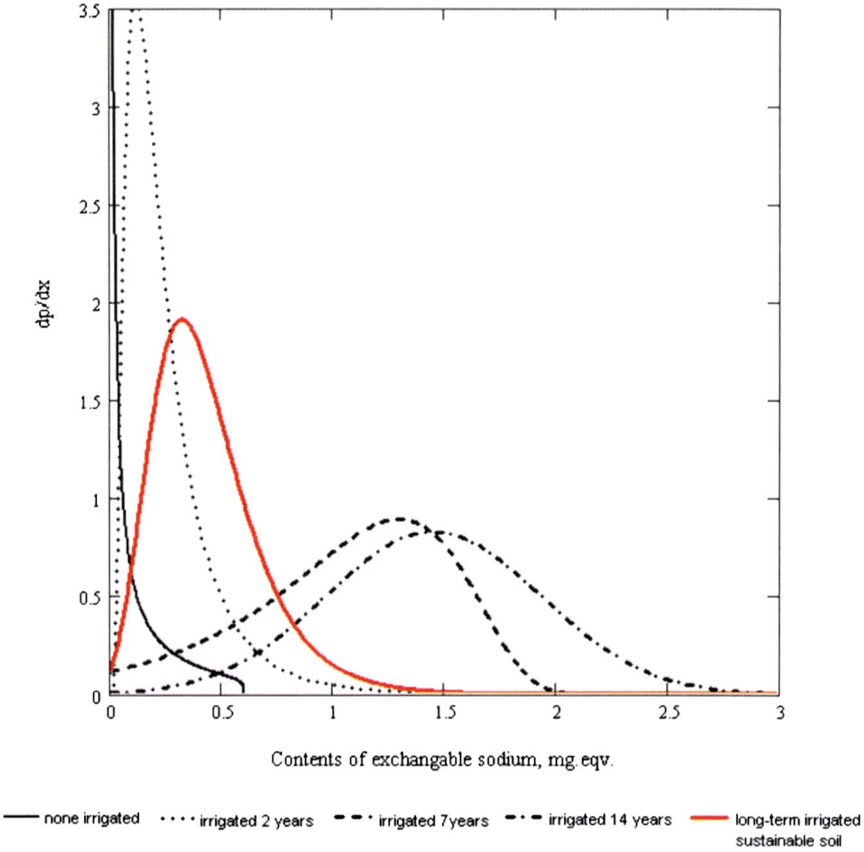

Fig. 15.3 Pdf of exchangeable sodium content in the upper horizon of chestnut soil irrigated by low-mineralized water at different times of irrigation (nonirrigated, irrigated 2, 7, and 14 years, *solid line* is the pdf exchangeable sodium content in 21 years' irrigated soil)

of pdf of the attribute in non-deflated and deflated soils, $d=0.6$ for deflated and long-term plowed soils, and $d=3.4$ for long-term plowed and irrigated soils. Therefore, the influence of irrigation to granulometric composition of soil is stronger than the effect of deflation.

Next example of evaluation and modeling change of soil property pdf is the tendency of irrigation water containing sodium bicarbonate to increase exchangeable sodium on soil exchange complex of chestnut soils. It is shown in *n*.

The irrigation led to significant accumulation of sodium. After 2 years of irrigation, the pdf of sodium contents differ from nonirrigated analog ($d=4.0$). After 7 years of irrigation, the pdf change very strongly ($d=4.6$). After next 7 years of irrigation, the pdf of the attribute continue to alter but in less extent ($d=1.2$). But there are some plots of soils that display sustainability to sodium water, and pdf of sodium content keep its form for a long time. These are the soils with raised content of coarse sand (Fig. 15.3).

15.4 Conclusions

Transformation of soils and a soil cover with the influence of natural processes and anthropogenous influences lead to change of probability structure of values of soil properties. Therefore, it is necessary to use probability distributions for an estimation of transformations. Distinctions of probability distributions can be quantitatively evaluated. It characterizes a degree of influence of distinctions of soil-forming factors and anthropogenous influences on the probability structure of properties of ground. For this purpose, it is necessary to use soil surveys data for monitoring investigations at territory of interest.

References

Gubarev VV (1992) Probability models. Part 2. Novosibirsk State Technical University, Novosibirsk, p 188

Kuzmina YeD (1980) Evaluation of manifestation of deflation and sustainability of dark chestnut soils of western part of Kulunda Steppe. PhD thesis, Tselinograd

Lemeshko BY (1995) Statistical analysis of univariant observations of a random variable: a program system. Novosibirsk State Technical University, Novosibirsk

Mikheeva IV (2001) Statistical probability models of soil properties (with chestnut soils of the Kulunda Steppe). Rossiyskoy Akademii Nauk, Novosibirsk

Mikheeva IV (2005) Statistical probability evaluation of sustainability and variability of nature objects under modern processes (with chestnut soils of the Kulunda Steppe). Rossiyskoy Akademii Nauk, Novosibirsk

Mikheeva IV (2009) Divergence of probability distributions of the soil properties as a quantitative characteristic of the soil cover transformation. Contemp Probl Ecol 2(6):667–670

Panfilov VP (1973) Physical properties and water regime of Kulunda Steppe soils. Nauka, Novosibirsk

Philips JD (1993) Chaotic evolution of some coastal plain soils. Phys Geogr 14:566–580

Shahid SA, Abdelfattah MA (2008) Soils of Abu Dhabi Emirate. In: Perry R (ed) Terrestrial environment of Abu Dhabi. Environment Agency of Abu Dhabi, Abu Dhabi, pp 70–91

Chapter 16
Soil Suitability of Northern State of Sudan to Irrigated Agriculture

Abdelmagid Ali Elmobarak and Fawzi Mohamed Salih

Abstract A semi-detailed soil survey of 390,857 ha on both banks of the river Nile between the third and fourth cataracts, at an extent of approximately 400 km in the Northern State of Sudan, was carried out. The area lies within the desert climatic zone of Sudan. The objectives of the survey are to characterize soils and landscape and assess their suitability for irrigated agriculture and to provide guidance to agronomists and irrigation engineers for future land uses. Thirty-six soil map units were defined and grouped into five main units based on the contents of secondary calcium carbonate and soil texture. In this chapter, different soils, such as recent and sub-recent alluvial and high-terrace soils of Nile, desert plain soils, Nubian formation and soils of recent windblown sand and wadi deposits, are described. The soils are evaluated for irrigated agriculture suitability. The survey revealed recent and middle Nile alluvial soils (R) occupy 12%, Nile high-terrace soils (H) occupy 26%, desert plain soils (D) occupy 29%, Nubian formation soils (N) occupy 17%, and wadi deposits and windblown sand soils (S) occupy 16% of the survey area. Currently about one-third of the survey area is suitable for irrigated agriculture. A further 14% can be made suitable after the remedial work (mostly soil reclamation). Land that might be used in future for LUT 2 and LUT 3 amounts to a further 24% of the survey area, and the remainder 29% is unsuitable for irrigated agriculture.

Keywords Irrigated agriculture • Nile soils • Soil suitability • Soil survey • Sudan

A.A. Elmobarak (✉) • F.M. Salih
Land Evaluation Research Section, Land and Water Centre,
Agricultural Research Corporation (ARC), Wad Medani, Sudan
e-mail: melmobarak2012@gmail.com

S.A. Shahid et al. (eds.), *Developments in Soil Classification, Land Use Planning and Policy Implications: Innovative Thinking of Soil Inventory for Land Use Planning and Management of Land Resources*, DOI 10.1007/978-94-007-5332-7_16,

16.1 Introduction

A semi-detailed soil survey of 390,857 ha was carried out in the Northern State of Sudan. The study area lies between longitudes 30°20 and 31°50 E and latitudes 17°45 and 19°45 N. The area lies within the desert climatic zone of Sudan and is characterized by low humidity, insignificant rainfall, very hot summer and cool winter (Van der Kevie 1973). The climatic data for Karima and Dongola towns is shown in Tables 16.1 and 16.2, respectively. Karima lies in the southern part, while Dongola lies in the northern part of the survey area; they are the only two meteorological stations in that area. The soil temperature and moisture regimes are hyperthermic (Wambeke 1982) and aridic.

The geology of the area is variable according to the published maps (GRAS 1988; GRAS and Robertson 1995; GRAS and BRGM 1981). The solid geology of the area is complex affected by faulting and tectonics, and much of it is obscured by drifting sands. The main geological formations in the order of age are the basement complex, Nubian formations and volcanic intrusive and superficial deposits. The study area is devoid of any vegetation due to high aridity. Previous studies on the ecology of the area classified it either as "desert" or as "riverrain" close to the Nile (Andrews 1948; Smith 1949; Harrison and Jackson 1958; Lebon 1965).

The objectives of the survey are to characterize soils and landscape and land suitability assessment for irrigated agriculture and to provide guidance to agronomists and irrigation engineers for future land uses.

16.2 Materials and Methods

A semi-detailed soil survey was carried out during July to November 2004 on both banks of the river Nile between the third and fourth cataracts, at an extent of approximately 400 km, in the Northern State, Sudan. The study area (390,857 ha) lies between longitudes 30°20 and 31°50 E and latitudes 17°45 and 19°45 N and occurs in the desert climatic zone of Sudan. The present survey was accomplished using various maps and materials, such as topographic map sheets of Northern State, Sudan; geological maps; ortho-photomaps at 1:20,000 scale; aerial photographs; and existing soil survey reports (Lahmeyer International 2005). The field survey was completed on a grid basis. The sites were located using GPS (Global Positioning System). A total of 8,457 augers holes and 137 profile pits were made. The soils were described using FAO standards (FAO 1990).

Field survey strategy was developed to cover various landscapes, e.g. one auger observation was made for every 25 ha for newly surveyed areas, one auger observation for every 100 ha for previously surveyed areas, one auger observation for every 400 ha for sand dunes and one typical soil profile was made for every 3,000 ha survey area.

A total of 8,457 sites were investigated using hand augers. Of which 3,403 sites were sampled at 0–30- and 30–60-cm depths for the establishment of salinity (ECe) and sodicity (SAR) status. Only 137 soil profile pits were exposed and described, and samples were collected for detailed chemical and physical analyses

Table 16.1 Climatic data for Karima meteorological station 1941–2003

Month	Mean relative humidity (%)	Total rainfall (mm)	Daily PE (mm)[a]	Mean wind speed (km day^{-1}) and direction	Air temperature (°C)			Daily sunshine (h)
					Mean daily max.	Mean daily min.	Mean daily	
January	34	0.1	5.6	263-N	27.1	9.0	18.1	10.1
February	30	0.0	6.5	263-N	29.5	10.2	19.8	10.5
March	26	0.0	8.0	277-N	33.8	14.1	23.9	10.3
April	24	0.5	9.6	289-N	38.6	18.7	28.6	10.7
May	19	0.3	10.6	290-N	41.1	23.0	32.5	10.8
June	19	2.9	10.6	275-N	43.2	24.8	34.0	11.2
July	22	9.2	9.5	234-N	42.4	25.4	33.9	10.6
August	24	1.8	9.8	263-N	42.0	26.0	34.0	10.2
September	23	0.2	9.9	292-N	41.6	25.1	33.3	9.7
October	25	0.2	9.1	306-N	38.9	21.5	30.2	10.1
November	31	0.0	6.8	278-N	32.5	15.2	23.8	10.4
December	36	0.0	5.5	250-N	28.6	10.5	19.5	10.2

[a]1941–1970

Table 16.2 Climatic data for Dongola meteorological station 1941–2003

Month	Mean relative humidity (%)	Total rainfall (mm)	Daily PE (mm)[a]	Mean wind speed (km day^{-1}) and direction	Air temperature (°C)			Daily sunshine (hours)
					Mean daily max.	Mean daily min.	Mean daily	
January	29	0.0	5.8	247-N	28.4	12.2	20.3	10.3
February	22	0.0	7.1	275-N	30.5	13.5	22.0	10.7
March	17	0.0	8.5	276-N	34.6	17.1	25.8	10.4
April	15	0.0	9.5	262-N	39.0	21.3	30.1	10.8
May	15	0.6	10.2	260-N	42.2	25.2	33.7	10.9
June	15	0.0	9.7	232-N	43.3	26.6	34.9	10.4
July	22	8.0	8.9	205-N	42.0	27.0	34.5	10.6
August	24	13.9	8.6	206-N	41.4	27.2	34.3	9.9
September	21	2.8	8.7	219-N	41.9	27.0	34.4	9.4
October	22	0.6	8.1	234-N	39.4	24.1	31.7	10.1
November	27	0.0	6.8	249-N	33.6	18.4	26.0	10.5
December	31	0.0	5.8	249-N	29.4	13.8	21.6	10.4

[a] 1941–1970

(719 samples). Infiltration tests were performed on 122 profile pits using double ring infiltrometer and the results used for land suitability assessment. The soil samples were analysed using standard methods (Richards 1954) in the laboratories of Land and Water Research Centre and Agricultural Research Corporation of Sudan, except where otherwise stated. Following is a brief description of the analytical methods:

Particle size analysis was made using the standard pipette method supplemented with wet sieving after giving standard treatments. Using the sand, silt and clay values, soil textural class was determined using the USDA textural triangle. Bulk density was determined on irregular soil clods; coefficient of linear extensibility (COLE) was calculated as the difference between bulk density values at 1/3 bar and oven-dry condition. Available water capacity (AWC) was calculated as the difference between moisture contents at 1/3 (field capacity) and 15 (wilting point) bars as measured by pressure plate extractor. Hydraulic conductivity (cm hr^{-1}) was determined by the constant head method on disturbed soil samples. Organic carbon was determined by modified Walkely and Black method, total nitrogen by micro-Kjeldahl method and available phosphorus by Olsen sodium bicarbonate extraction method. Soil pH was determined on 1:5 soil to water suspension (pH 1:5) and on the saturated soil paste (pHs). The EC was determined in saturated paste soil extract (ECe expressed as dS m^{-1}). Calcium carbonate ($CaCO_3$) equivalents were determined through back titration procedure. Cation exchange capacity (cmoles kg^{-1} soil) was determined by treating soil with 1 N sodium acetate (pH = 8.2) and washed with ethanol (95%). The adsorbed sodium was extracted from the sample using ammonium acetate solution (pH = 7.0). Sodium concentration in the extract was determined by flame photometer and expressed as cmoles kg^{-1} soil.

Exchangeable Na and K were extracted with 1 N ammonium acetate (pH = 7.0) and determined by flame photometer. Soluble cations and anions in soil saturation extract (Ca^{2+}, Mg^{2+}, CO_3^{2-}, HCO_3^-, Cl^-) were measured using standard titration procedures, whereas Na and K measured by flame photometer. Sulphates (precipitated as barium sulphate) were gravimetrically determined. The exchangeable sodium percentage (ESP) was calculated using exchangeable Na and CEC data. Sodium adsorption ratio (SAR) was calculated using soluble Na, Ca and Mg results after substituting in standard equation. Gypsum was determined by standard acetone precipitation method and micronutrients by atomic absorption spectrophotometer in the DTPA + $CaCl_2$ at pH 7.3 extraction solution.

The soils were classified using the USDA system of classification (Soil Survey Staff 1999). The soils were then evaluated for their suitability to irrigated agriculture using the FAO framework for land evaluation and guide for sprinkler and drip irrigation (FAO 1990).

16.3 Results and Discussion

The soil legend was developed using the landforms information, calcium carbonate equivalents, particle size classes and texture of upper soil surface. Based on the landforms information, initially the soils were divided into five main groups

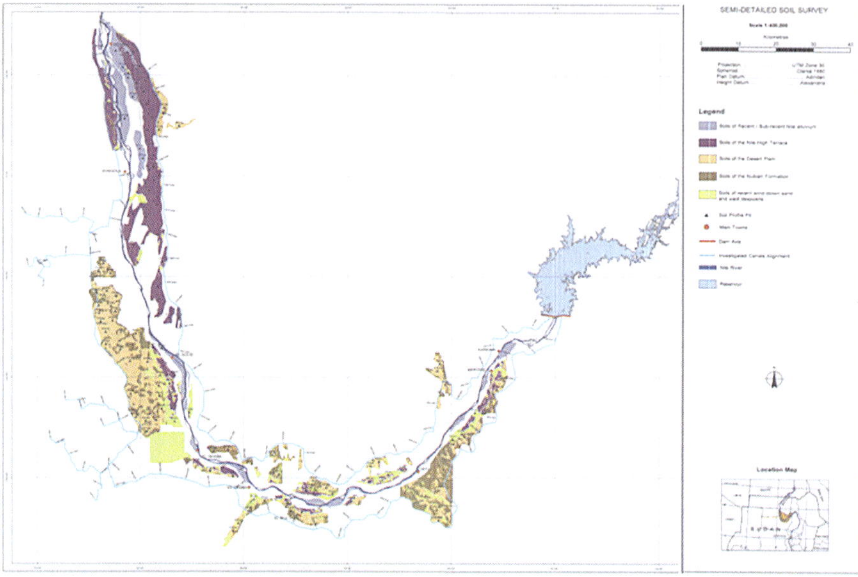

Fig. 16.1 Distribution of soils by landscapes

(Fig. 16.1): recent and middle Nile alluvial soils (R) occupying 12%, Nile high-terrace soils (H) occupying 26%, desert plain soils (D) occupying 29%, Nubian formation soils (N) occupying 17% and wadi deposits and windblown sand soils (S) occupying 16% of the survey area. The soils were classified and 36 soil map units established. The map unit legends reflect defining factors.

At the present scale of survey and mapping, the accuracy or the purity of the mapping units is expected to be about 75% (between 65 and 85%; Landon 1991). When it was not possible to map the soil units separately, they are considered as complexes. The soil complexes comprised several soil units. Two soil complexes were mapped, one in the Nile high-terrace landform having four units and the other one in the desert plain landform with seven units.

The soil classification is defined as arrangement of soils into an orderly distribution of natural grouping. Accordingly, each soil mapping unit has been classified based on the distribution of soils (major and minor soils).

As the survey area is situated in the arid environment, therefore, the landform, time and parent material are considered as the most effective soil formation factors. Using the following properties, soil classification is completed: soil climate; degree of soil profile development; presence, nature and quantity of calcium carbonate or gypsum; presence of subsoil cracking and slickensides; presence of clay skins; levels and distribution of soil sodicity (exchangeable sodium percentage); levels and distribution of electrical conductivity (EC); soil depth to hard rock; thickness of sand or loamy sand and evidence of recent alluvial deposition/distribution and organic carbon. The soil classification is shown in Table 16.3, and the main properties of the different landforms are summarized below.

Table 16.3 Soil classification of the main landforms

Landform	Soil Order	Soil taxonomy (USDA 1999)
Recent Nile alluvial	Entisols	Typic Torrifluvents
	Entisols	Vertic Torrifluvents
Nile high terrace	Entisols	Typic Torripsamments
	Aridisols	Typic Haplocambids
	Aridisols	Sodic Haplocambids
	Aridisols	Petronodic Haplocambids
	Aridisols	Typic Haplocambids
	Aridisols	Sodic Haplocalcids
Desert plain	Aridisols	Typic Haplocambids
	Aridisols	Sodic Haplocambids
	Aridisols	Petronodic Haplocambids
	Aridisols	Typic Petrocambids
	Aridisols	Typic Haplocalcids
	Aridisols	Sodic Haplocalcids
Windblown sand and wadi	Entisols	Typic Torriorthents
	Entisols	Typic Torripsamments
Nubian formation	Aridisols	Lithic Haplocalcids
	Aridisols	Typic Haplocambids
	Aridisols	Sodic Haplocambids
	Entisols	Typic Torriorthents

16.3.1 The Soils of the Recent and Sub-recent Nile Alluvium

They are deep, brown to yellowish brown sand to clay soils of basins, levees, channels and terraces. The soils of the recent alluvium have low $CaCO_3$; they are nonsaline, non-alkali soils. They are subdivided into six soil units based on the surface soil texture. The soils of the sub-recent alluvium have common $CaCO_3$ concretions, slightly to moderately saline-alkali and rarely flooded. They have three soil units based on textural variations in the soil surface.

The soils of the recent alluvium are classified as Entisols (Typic Torrifluvent and Vertic Torrifluvent). They are nonsaline, non-sodic and non-calcareous. These soils are fertile and renewed annually during the flood season. Most of the cropped lands belong to these soils. The soils of sub-recent alluvium are calcareous with some patches of saline and alkali soils; they are flooded only at high-flood seasons. These soils are formed on alluvium deposited by the Nile and do not have enough time to develop; they are classified as Typic Torrifluvent and Vertic Torrifluvent on low-lying areas.

16.3.2 The Soils of the Nile High Terrace

These are deep brown to greyish brown soils formed from old Nile alluvium, usually saline, alkali or saline-alkali. They are divided into three subdivisions. Soils with low $CaCO_3$ concretions or soft segregations, they have three soil units based on their surface soil texture. Soils with high $CaCO_3$ concretions and frequently have (petro) calcic or petronodic horizon below 1.0 m. Based on surface soil textural variations, they have three soil units. Gypsic soils are soils of old channels and basins, they also have high $CaCO_3$, and they are saline-alkali soils.

The soils of the Nile high terrace are deep, brown to greyish brown soils of the old Nile alluvium. They are generally saline and alkali soils and have low to high contents of calcium carbonate equivalents. The texture ranges between clay and sand. In some areas of old basins, these soils are gypsiferous. These soils are developed from the sediment from Nile and have reasonable time for the accumulation of salts, calcium carbonate and clay in subsurface. The soils are classified as Aridisols (Typic Haplocambids, Sodic Haplocambids, Petronodic Haplocambids, Typic Haplocalcids and Sodic Haplocalcids) and have cambic or calcic or petronodic horizons with or without sodium accumulation, while the Nile levee is classified as Entisols (Typic Torripsamment).

16.3.3 The Soils of the Desert Plain

These are flat to gently sloping old fans and wadis. They are colluvial-alluvial soils of variable depths to the Nubian bedrock. They contain three major units based on their contents of $CaCO_3$ and gypsum. These soils contain 15 soil units based on their contents of $CaCO_3$, gypsum and textural variations.

The soils of the desert plain are flat to gently sloping old fans; their depth to the bedrock is variable. They are ranging from sandy to loamy and to clay texture. The calcium carbonate is low to abundant, hence developing a petrocalcic, calcic and petronodic horizons within 100 cm from soil surface. In areas of old lakes and basins, crystallized gypsum is observed. The nature and origin of desert plain soils vary from degradation or aggradations process to fluvial/colluvial deposit and lacustrine sediment. All soils are classified as Aridisols (Typic Haplocambids, Sodic Haplocambids, Petronodic Haplocambids, Typic Petrocambids, Typic Haplocalcids and Sodic Haplocalcids), with cambic, petrocalcic, calcic or petronodic horizons with or without sodium accumulation.

16.3.4 The Soils of the Recent Windblown Sand and Wadi

They are deep sandy and gravelly soils associated with aeolian and wadi depositions. They are divided into two units: sand plains and dunes and recent alluvium of active wadis. They contain about three soil units.

The soils of the recent windblown sand and wadi deposits are deep sandy and gravelly and are associated with aeolian and wadi deposition. These are (1) the sandy soils of the sand plain and sand dunes with a loose cover of 0.5 m and (2) the sandy clay, sandy and gravelly soils of the active wadis recent alluvium. All soils of this group belong to Entisols classified as psamments (Typic Torripsamments) and orthents (Typic Torriorthents).

16.3.5 The Soils of the Nubian Formation

They are rock outcrops and shallow and/or gravelly soils over sandstone pediment. They are composed of two soil units. The soils of the Nubian formation are mainly lithic and extremely gravelly soils of low ridges, mudstone or sandstone. Rock outcrops are common. They are mainly Aridisols (Lithic Haplocalcids, Typic Haplocambids, Sodic Haplocambids); sometimes these are Entisols (Typic Torriorthents).

16.3.6 Soil Suitability Evaluation for Irrigated Agriculture

The soils were evaluated for their suitability to irrigated agriculture. The information on these soils on their permeability, available water-holding capacity and infiltration rate was used. After that the soils were grouped as to their key characteristics in their respective suitability for the anticipated land utilization types which are mainly gravity, sprinkler and drip irrigation systems. The soils were then grouped as to their potential for crop suitability where sandy and coarse-textured soils are suited for sprinkler irrigation, taking into consideration crops grown to be drought tolerant, less water demanding and less fertility demanding, whereas the loamy and clayey soils suit all types of irrigation (gravity, sprinkler and drip irrigation).

The soils of sandy textures on top and clayey or loamy subsoils are suitable for sprinkler, considering growing shallow-rooted, drought-tolerant and less fertility-demanding crops. The soils of the recent alluvium whether clayey, silty or loamy are suitable for all types of irrigation (Table 16.4).

The land suitability is the fitness of an area of land to a defined land use (land utilization type, LUT) and is used to determine or assess the land suitability for the proposed land utilization types. In this study, three general LUTs were defined to assess the potentiality of the soils to irrigated agriculture, they are:

LUT 1. Irrigated agriculture by small farmers with limited resources, irrigated by gravity (G), sprinkler (S) or drip (D) systems for a range of field crops.

LUT 2. Large-scale sprinkler irrigation under commercial/government management on lands unsuitable for LUT 1. If it is suitable for LUT 1, then it will also be suitable, or perhaps more suitable, for larger-scale farming.

LUT 3. Centre pivot irrigated agriculture on land unsuitable for LUT 1 but demonstrated the efficiency of centre pivots. Crop selection may be slightly restricted on somewhat saline land.

Table 16.4 Soil grouping for potential crop suitability

Soil textural group	Irrigation option	Limiting for crops that:	Landform
Sandy	S[a]	Are drought sensitive; water demanding; fertility demanding	Windblown, Nile high terrace sandy soils
Coarse loamy/sandy throughout (+/– loamy topsoil)	S	Are drought sensitive; water demanding; fertility demanding	Desert plain, coarse loamy soils
Loamy throughout	GSD	Prefer sandy or clayey textures	Recent alluvium, high terrace and desert plain loamy texture soils
Loamy/clayey topsoil, clayey subsoil	GSD	Are sensitive to wetness, drainage restriction	Recent alluvium, high terrace and desert plain soils with clayey texture
Sandy, topsoil, clayey subsoil	S	Are shallow rooting and drought-sensitive, water demanding, fertility-demanding; are deep-rooting but sensitive to wetness, drainage restriction	Recent alluvium, high terrace and desert plain sandy on clayey subsoil
Sandy, topsoil, loamy subsoil	S	Are shallow rooting and drought-sensitive, water demanding, fertility demanding	Recent alluvium, high terrace and desert plain sandy topsoil on loamy subsoil
Recent alluvium, clayey/silty clayey	GSD	Are sensitive to wetness, drainage restriction	Recent alluvium, clayey
Recent alluvium, loamy, silty	GSD	Prefer sandy or clayey textures	Recent alluvium having or silty texture

S sprinkler irrigation, *G* gravity irrigation, *D* drip irrigation

[a] Requires large investments in water and fertilizers, therefore more suited to large-scale, centre-pivots irrigation, e.g. fodder, cereals, sugarcane

Table 16.5 Summary of the assessment for land suitability

Suitable for irrigated agriculture, LUT 1	129,817 ha
Suitable for irrigated agriculture (LUT 1) after reclamation, drainage or flood protection	54,103 ha
Potentially suitable for large-scale sprinkler enterprises (LUT 2)	73,969 ha
Possibly suitable for future centre-pivot irrigated agriculture (LUT 3)	18,026 ha
Unsuitable	114,942 ha

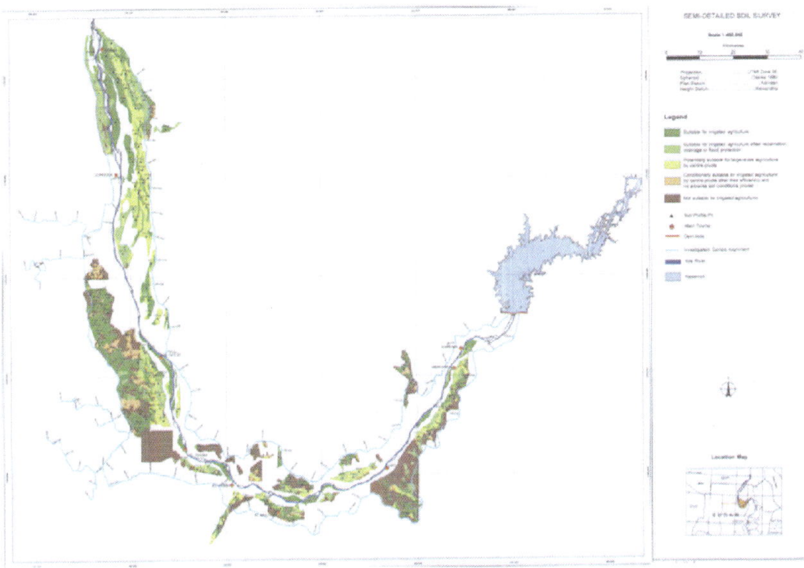

Fig. 16.2 Distribution of land suitability classes

It is assumed that, enough water will always be delivered upon demand at acceptable cost, the sodic and saline soils be reclaimed before agricultural development. The areas did not require flood protection or drainage to be provided and sufficient fertilizers will be available at acceptable cost.

The soils with sandy or other textures in different landforms are suitable for sprinkler irrigation with high investments in fertigation. The loamy texture soils are suitable for all land utilization types, and hence, soils of the recent, high terrace and desert plain of loamy textures fit here. The soils of different landform having clayey subsoil are subject to wetness.

The results for assessment of land suitability are summarized in Table 16.5, whereas Fig. 16.2 shows the distribution of the suitable, conditionally suitable and unsuitable lands. Currently about one-third (33%) of the survey area is suitable for irrigated agriculture. A further 14% is suitable only after remedial work (mostly soil reclamation) has been carried out. Land that might be used in future for LUT 2 and LUT 3 amounts to a further 24% of the survey area and the remainder 29% is unsuitable for irrigated agriculture.

Table 16.6 Percent area in each soil group under different salinity (EC dS m^{-1}) ranges

Soil groups		<4 dS m^{-1}	4.0–11.9 dS m^{-1}	>12.0 dS m^{-1}
R	Recent and sub-recent Nile alluvial soils	7.5	0.9	0.4
DL	Desert plain, low $CaCO_3$ soils	12.3	5.0	2.5
DH	Desert plain, high $CaCO_3$ soils	2.8	1.7	0.7
DG	Desert plain, gypsic clay soils	0.0	0.0	0.0
DX	Desert plain, soil complexes	5.0	3.2	0.8
HL	High terrace, low $CaCO_3$ soils	8.7	3.0	2.5
HH	High terrace, high $CaCO_3$ soils	1.6	0.9	1.6
HG	High terrace, gypsic clay soils	0.1	0.2	0.6
HX	High terrace, soil complexes	2.3	1.4	1.8
N	Nubian formation soils	11.4	5.2	3.8
S	Aeolian sand dunes	10.0	1.3	0.8
	Total	61.7	22.8	15.5

The area of land that is suitable for gravity or sprinkler irrigation, including the land that requires remedial measures, amounts to 157,181 ha. Most of this land is better suited to sprinkler irrigation because soil infiltration rates are only marginally suited to gravity supply. Sprinkler irrigation is also a more effective option for soil reclamation. Only the already-cultivated loamy to clayey, recent and sub-recent alluvial soils close to the Nile are well suited to gravity irrigation, but these amount to only about 30,000 ha.

16.3.7 Management and Reclamation of Soils

About 51,000 ha of land was found be saline (Table 16.6) or sodic (Table 16.7) and need to be reclaimed before cropping. The main measure to be taken is leaching, coupled with good husbandry in an attempt to improve the soil organic matter and hence structure. Depending on the initial level of salinity, leaching over 1–4 years accompanied by cropping (although low yields were anticipated), soils could be reclaimed with an irrigation or leaching regime of 14 mm day^{-1} for sprinkler irrigation.

Alkali soils that cannot be reclaimed by leaching alone but with gypsum and other amendments amounting to 14,000 ha occur on the high-terrace soils containing high calcium carbonate concretions mainly.

A more detailed study to exactly locate these alkali soils is needed, as they occur in patches in an unpredictable way. Most of the soils that need to be reclaimed have moderate or slow permeability; the construction of under drainage system should be considered during the reclamation process if salts are not leached to a sufficient depth. The depth and spacing of drains rely entirely on site determination of hydraulic conductivity.

Table 16.7 Percent sodicity distribution in soil groups (%)

Soil groups		SAR <15	SAR 15–24	SAR >24
R	Recent and sub-recent Nile alluvial soils	8.0[a]	0.4[a]	0.4[a]
DL	Desert plain, low $CaCO_3$ soils	12.3	5.0	2.5
DH	Desert plain, high $CaCO_3$ soils	2.8	1.7	0.7
DG	Desert plain, gypsic clay soils	0.0	0.0	0.0
DX	Desert plain, soil complexes	5.0	3.2	0.8
HL	High terrace, low $CaCO_3$ soils	8.7	3.0	2.5
HH	High terrace, high $CaCO_3$ soils	1.6	0.9	1.6
HG	High terrace, gypsic clay soils	0.1	0.2	0.6
HX	High terrace, soil complexes	2.3	1.4	1.8
N	Nubian formation soils	11.4	5.2	3.8
S	Aeolian sand dunes	10.0	1.3	0.8
	Total	61.7	22.8	15.5

[a]percent area in each soil group

16.4 Conclusions and Recommendations

About one-third (33%) of the survey area is currently suitable for irrigated agriculture. A further 14% area can be made suitable after remedial work (mostly soil reclamation) has been carried out. Land that might be used in future for LUT 2 and LUT 3 amounts to a further 24% of the survey area, and the remainder 29% is unsuitable for irrigated agriculture. The main measure to be taken to reclaim saline soils is leaching, coupled with good land husbandry in an attempt to improve the soil organic matter and soil structure. Alkali soils cannot be reclaimed by leaching alone, these soils can be amended by gypsum and other amendments to rectify soil sodicity. The construction of drainage system should be considered during the reclamation process. A more detailed study is suggested to exactly locate alkali soils, as they occur in patches in an unpredictable way.

References

Andrews FW (1948) The vegetation of the Sudan. In: Tothill JD (ed) Agriculture in the Sudan. Oxford University Press, London, pp 32–61

FAO (1990) Guidelines for soil description, 3rd edn. FAO, Rome

GRAS (1988) Geological atlas of the Republic of the Sudan, scale 1:1,000,000. Geological Research Authority of Sudan, Khartoum

GRAS (Geological Research Authority Sudan), and BRGM (Bureau de Recherches Geologiques et Minieres (1981) Geological map of Sudan, scale 1:2 million. Geological Research Authority Sudan, Khartoum

GRAS-Robertson (1995) Geological notes to 1:1 million scale Geological atlas of Sudan. Bulletin No 40

Harrison MN, Jackson JK (1958) Ecological classification of the vegetation of the Sudan. Forests bulletin No.2 (New series). Forests Department, Khartoum, Sudan

Lahmeyer International (2005) Feasibility study for the Merowe irrigation project. Land resources and sand stabilization. Annex 3.1. Khartoum, Sudan

Landon JR (ed) (1991) Booker tropical soil manual. Longman Scientific and Technical, Thame/Harlow

Lebon JHG (1965) Land use in Sudan. The World Land Use Survey monograph no.4, Geographical Publications, B Cornwall

Richards LA (ed) (1954) Diagnosis and Improvement of saline and alkali soils. Handbook No 60. USDA, Washington, DC

Smith J (1949) Distribution of tree species in the Sudan in relation to rainfall and soil texture. Bulletin No 4. Sudan Ministry of Agriculture, Khartoum

Soil Survey Staff (1999) Soil taxonomy. A basic system of soil classification for making and interpreting soil surveys, 2nd ed. Agriculture handbook no 436. United States Department of Agriculture-Natural Resources Conservation Service, Washington, DC

Van der Kevie W (1973) Climatic zones in Sudan. Soil Survey Department, Wad Medani

Wambeke A (1982) Africa soil moisture and temperature regimes, SMSS Tech Monograph USDA, Washington, DC

Chapter 17
Effects of Plants on Soil-Forming Processes: Case Studies from Arid Environments

William H. Verboom, John S. Pate, Mahmoud A. Abdelfattah, and Shabbir A. Shahid

Abstract Bulk deposits of aeolian sand accumulated over recent timescales provide instructive systems for examining effects of colonising vegetation on soil development. The two contrasting case studies presented here are eucalypt woodland in a dune system in southwest Australia and the rubified sand seas of the United Arab Emirates. In the former, clay pavements forming under the lateral root catchments of the eucalypts are shown to be constructed from iron, aluminium and other mineral elements abstracted from ground waters by deep roots. The pavements concerned have a marked restrictive influence on understorey density and biodiversity while also having an overall role in maximising effectiveness of usage of water and nutrients by the trees in question. Timescales and amounts of iron uplifted in this manner are estimated for the system. In the Arabian example, the occurrence of intense reddening (rubification) of sand towards the mountains of Oman is well known, and abiotic processes have been implicated in the phenomenon. In this chapter, we invoke involvement of a biotic component, having demonstrated a relationship between vegetation density and extent of rubification as seen in a positive

W.H. Verboom (✉)
Department of Agriculture and Food, 10 Doney Street, Narrogin 6312, WA, Australia
e-mail: william.verboom@agric.wa.gov.au

J.S. Pate
School of Plant Biology, The University of Western Australia, Crawley 6009, Australia

M.A. Abdelfattah
Soil and Land Use Management, Environment Agency-Abu Dhabi,
PO Box 45553, Abu Dhabi, UAE

Soils and Water Sciences Department, Faculty of Agriculture,
Fayoum University, Fayoum, Egypt
e-mail: mabdelfattah@ead.ae; mahmoudalia@yahoo.com

S.A. Shahid
International Center for Biosaline Agriculture, P.O. Box 14660, Dubai, UAE
e-mail: s.shahid@biosaline.org.ae

S.A. Shahid et al. (eds.), *Developments in Soil Classification, Land Use Planning and Policy Implications: Innovative Thinking of Soil Inventory for Land Use Planning and Management of Land Resources*, DOI 10.1007/978-94-007-5332-7_17,

correlation between increased reddening and cumulative vegetation encountered as one moves from coast to mountains. We hypothesise that uplift of iron by deep-rooted shrubs/trees might be the agent responsible for progressive reddening. Definitive testing of this hypothesis is required, particularly by analysing for iron in xylem sap flowing up through taproots and looking for evidence of its subsequent release into superficial layers of sand surrounding lateral roots of the trees.

Keywords Desert • Hydraulic lift • Iron • Rubification • Sand • Trees

17.1 Introduction

In arid and semiarid regions of the world, one commonly encounters organisms that actively manipulate abiotic components of their habitat in order to modify local microclimates and resources of water and nutrients. Through time they not only pass on niche-building abilities but also develop adaptive features permitting them to operate highly effectively within the environments which they have created. Within such scenarios, there exist bases for selection of a multitude of extraordinary behaviours and associations (see Odling-Smee et al. 2003 and Boogert et al. 2000).

A very good example of the above is the 'termitarium' building by termites which harvest and ingest local plant and earthen materials and function in symbiotic association with gut flagellates, cellulose-decomposing bacteria and, in certain cases, nitrogen-fixing bacteria. In a similar vein, we have introduced the word 'phytotarium' to denote any of a number of cases in which higher plants modify soils to their advantage and do so with or without mutualistic cooperation by mycorrhizae or rhizosphere bacteria (see review of Verboom and Pate 2006a).

In our studies of pedogenetic phenomena in semiarid ecosystems of Western Australia, we have identified a number of cases in which biota have been suggested to have major influences in modifying soil profiles. Table 17.1 summarises such effects, with each item annotated in terms of classes of biota involved and the likely mechanisms which implement their operation. The 16 items listed have already been documented and discussed by Verboom and Pate (2006a, b), and each item in the Table 17.1 is annotated in respect to the end results to be expected to be gained by continued functioning of the ecosystem in question. It should be emphasised that Table 17.1 is by no means exhaustive. For example, we have also identified possible niche-building activities in the generation of exceptionally acidic rhizospheres under certain acacias and the high accumulations of hydraulically uplifted salt by certain eucalypt species generating sparse understoreys restricted to halophytic Chenopodiaceae. Further study of the edaphic qualities of the above is clearly needed, particularly of the detailed roles which indigenous flora and microorganisms might play in functioning in particular circumstances.

Since most soil profiles encountered across ecosystems of the world today are likely to have supported many generations of evolving biota, it becomes a daunting challenge to specify overall influences through which resident and advected materials have been

Table 17.1 Bioengineering attributes and outcomes suggested as plausible elements of a phytotarium concept

Item No	Soil feature	Formative agents	Mechanisms involved	Function
1	Silicon-lined root channels	Precipitation of Si	Root secretion of biogenic Si by certain eucalypts	Downward channelling of water
2	Ferric-lined root channels	Precipitation of Fe	Processing of Fe by proteoid root exudates and bacteria	Downward channelling of water
3	Water-repelling soil surfaces	Mallet leaf litter, cryptogamic crusts	Chemically based hydrophobic effects	Uneven wetting of upper profile
4	Localised water repellency in subsurface soil	Fungal mycelia and old cluster root mats	Chemically based hydrophobic effects	Uneven wetting of lower profile
5	Texture-contrast seal	Dispersed clay and precipitated Si in certain mallees	Root secretion of biogenic Si and dispersing agents	Hydrologic decoupling of A from B horizon
6	Finely divided carbonate	Precipitation of carbonate	Root secretion of Ca and processing of by bacteria	Enhancement of water storage in B horizon
7	Nodular calcretes	Precipitation of carbonate and Al and Si	Processing of Ca by ectomycorrhizal fungi and bacteria	Sequestration of P and refugia for microbes
8	Hardpan horizons	Precipitation of Si and some Fe	Root secretion of biogenic Si by certain eucalypts	Restricts deep percolation
9	Silcrete horizons	Precipitation of Si	Root secretion of biogenic Si by certain acacias	Restricts deep drainage
10	Pisolithic ferricretes	Precipitation of ferric rinds	Processing of Fe by proteoid root exudates and bacteria	Sequestration of P and refugia for microbes
11	Overprinting of ferricrete by amorphous Si, Al carbonate or clay	Incursion of proteaceous vegetation by certain Myrtaceae	Root secretion of biogenic Al, Si, and Ca	Commandeering of existing niche by other vegetation type
12	Sand-binding roots	Monocotyledon understorey species	Permanent entrapment of sand on lignified root hairs	Biopore for water or occupancy by seedling roots
13	Nutrient enrichment of upper profile	Biota operating in the lower regolith	Phosphorous mining by microbes and absorption by roots	Maintenance of P capital in ecosystem
14	Soil phytoliths	Si accumulation as plant opal	Release of Si from decomposing plant biomass	Contribution of biogenic Si to layers and root pipes
15	Abrupt lateral facies changes from one crete to another.	Transition between vegetation types	Vegetation specific effects on soil profiles	Protection of the adjoining niches
16	Multiple vertical facies changes	Sequential long-term formation and burial of several distinct phytotaria	Phases of aggradation and plant growth under climatic change	Probably none

successively transformed within particular ecosystems. In any event, it has proved to be extremely difficult to examine the effects of any single phytotarium in isolation. We would suggest that the ideal natural laboratory for studying such sequences are situations where plants are colonising recently deposited sands blown in from marine or lacustrine sources. Although such systems may have gone through multiple cycles of weathering, transport and deposition of the parent material there is still likely to be only minimal carry-over of signatures from past pedogenetic activity.

Two contrasting Holocene-based settings of the above ilk are chosen for study in this chapter. The first from southwest Australia comprises a highly uniform dune of quartzitic sand blown out from a playa lake and now colonised by heath and open woodland. The second, involving a much larger study area, embraces the Rub'al Khali (Empty Quarter) sand sea within the borders of the Abu Dhabi Emirate and involves the progressive migration of gulf-derived sand through the region.

17.2 Materials and Methods

17.2.1 Southwestern Australian Study

17.2.1.1 Experimental Site and Principal Study Species

This comprised a lunette of fine quartzitic sand blown out from Lake Chillinup (grid ref: S 34° 32.54 E 118° 04.12) during the late Pleistocene. It is currently colonised by pristine mallee woodland intermixed with myrtaceous-proteaceous heath. Vegetation zones and species listings are as described by Pate and Verboom (2009), who also discuss competitive effects of clay pavement-forming mallees on density and biodiversity of understorey vegetation. The principal study species was the mallee E. incrassata Labill. A cutting bisecting the dune provided an ideal setting for studying long-term effects of different vegetation type on profile development especially in relation to presence or absence of neoformed clay, iron oxides and carbonates.

17.2.1.2 Rooting Morphology and Pavement Formation

This was examined using a combination of pit excavations, air and water spading and evaluations of spatial and mass relationships between roots, mature clay columns and columns currently under construction. Observations were extended to other mallee eucalypts and pavements exposed in the cutting through the dune.

17.2.1.3 Chemical and Physical Composition of Soil Profiles

Elemental composition of profiles in lateral root catchments of mallee eucalypts was assessed by 3-m-deep corings at sites passing directly through pavements.

A matching set of corings was obtained for unmodified (unpavemented) sand under closely adjacent myrtaceous-proteaceous heath. Chemical analyses of samples down profiles involved determinations of total Fe, Al, Ca, K, Th and Mg by inductively coupled plasma (ICP) optical emission spectrometer (OES) or ICP mass spectrometer (MS). Additional chemical analyses of core material from woodland or cutting were pH, exchangeable cations (Ca, Mg, Na, K), oxalate-extractable Al, Fe and Si and % carbonate. Physical analyses included particle size distributions and bulk densities using methodologies as listed by Verboom et al. (2010).

17.2.1.4 Xylem (Tracheal) Sap Collection and Analysis

Sap was obtained by vacuum extraction of clean, bark-free xylem cores cut from freshly excavated 30–40-cm-long sections of healthy lateral root using the techniques described by Pate et al. (1994). Samples related mostly to the principal study species E. incrassata. All samples were membrane filtered and frozen after collection. Analyses were conducted for Al, Fe, Ca, Mg, K, Na and Si using an ICP-OES and organic acids assessed by reversed phase column liquid chromatography.

17.2.2 Arabian Study

17.2.2.1 Area of Investigation

This included the southern and northern areas of sand sea within the borders of the United Arab Emirates (see descriptions of Bresler 1982 and depictions in Fig. 17.1). The bulk of this area was surveyed using 22,000 excavations in conjunction with satellite imagery draped over digital elevation models (EAD 2009). Subsequent investigation for the purposes of this study concentrated on northern sand fields between the Arabian Gulf and the piedmonts of the mountains of Oman. A rainfall gradient currently applies increasing as one approaches the mountains (Fig. 17.2).

17.2.2.2 Measurement of Vegetation Density

This analysis was undertaken along two transects (located in Fig. 17.2) each 1 km wide and stretching for 90 km between the Gulf and mountains of Oman. Large woody species identifiable individually at 2-km eye altitude were assessed by manual counting from Google Earth DigitalGlobe 2010 images. The principal species involved were identified as *Prosopis cineraria, Acacia tortilis and A. nilotica.*

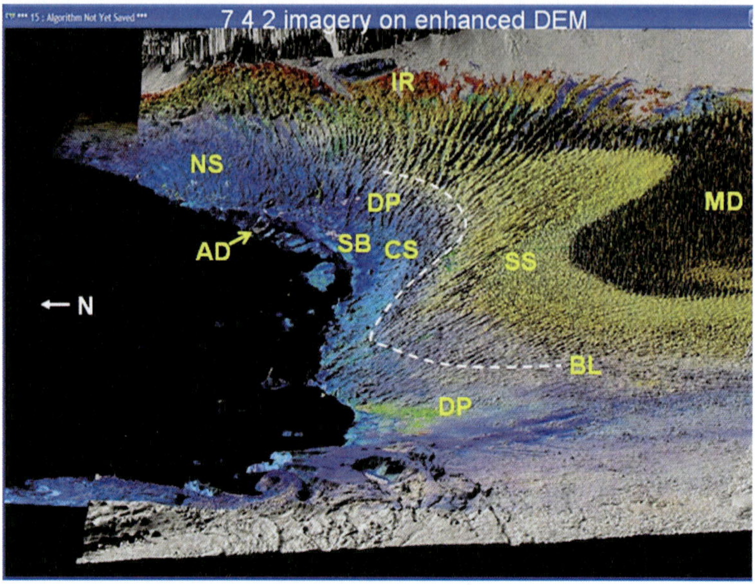

Fig. 17.1 Landsat (bands 7,4,2) imagery draped over enhanced digital elevation model of the Emirate of Abu Dhabi showing deflated sand fanning back from the Gulf and piling up on the higher ground (The mountains of Oman are at *top left* of picture) (**Legend**: AD = Abu Dhabi; BL = Buchanan line marking northern limit of linear dune inversion; CS = pale carbonatic sand sheets; DP = deflation plain; IR = zones of intense rubification; MD = mega barchan dunes; NS = northern sand sea; SB = coastal sabkha; SS = southern sand sea

Graphical representations of vegetation densities were then constructed from measurements obtained from a series of 1 km^2 plots along the 90-km length of the transect.

17.2.2.3 Rubification of Sand

Degrees and general patterning of rubification across the area of investigation were assessed qualitatively from Landsat Thematic Mapper imagery (bands 7, 4, 3, 2 and 1) coupled to ground verification, the latter indicating whether iron was present as simple coats on all grains, only on a proportion of the grains or localised in wind-sorted bands such as encountered on exposed Pliocene-Pleistocene deposits within the region. Investigations of rubification were conducted in parallel with vegetation density using the same series of 1-km^2 plots on the transects mentioned above. Crude quantification of reddening and associated increases in dithionite-extractable iron oxide levels were based on values derived from the Landsat Thematic Mapper imagery (band 3) and data from the study of White et al. (2001), respectively.

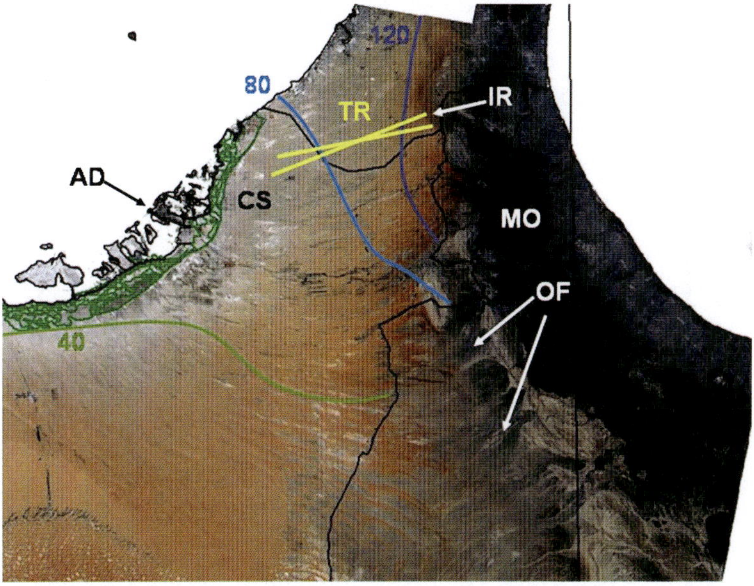

Fig. 17.2 False colour image (Landsat 7 bands 3 = red, 2 = green, 1 = blue) of northern sand sea of the Emirate of Abu Dhabi showing study transects TR along which intensity of vegetation was compared to rubification of sand (Legend: AD = Abu Dhabi; CS = pale carbonatic sand sheets; IR = zones of intense rubification; MO = mountains of Oman; OF = outwash fans from mountains. Annual rainfall isohyets (mm) are designated by colour coded lines and values)

17.3 Results and Discussion

17.3.1 Southwestern Australian Study

Figure 17.3a is a stylistic representation of the study mallee *Eucalyptus incrassata* showing principal morphological features (see legend to plate) in relation to clay columns of the developing pavement. Columns form at discrete foci of clay biogenesis associated with clusters of fine rootlets on lower level (R2) lateral roots (Fig. 17.3b, c). As columns are generated over the life of a tree, they coalesce into a more or less continuous pavement. This was particularly well demonstrated in the man-made cutting driven through the dune (see Fig. 17.4). Parallel studies throughout the region (Verboom and Pate 2006b; Pate and Verboom 2009) suggested that clay column formation was exclusive to certain eucalypts and that depth of location and column morphology varied with species involved. We present the hypothesis that the phenomenon may well have been responsible for generating vast tracts of texture contrast (duplex soils) now utilised extensively for cereal and pasture-based farming across much of Southern Australia. We also have shown that clay column formation in lateral root catchments of mallee has a marked suppressive effect on

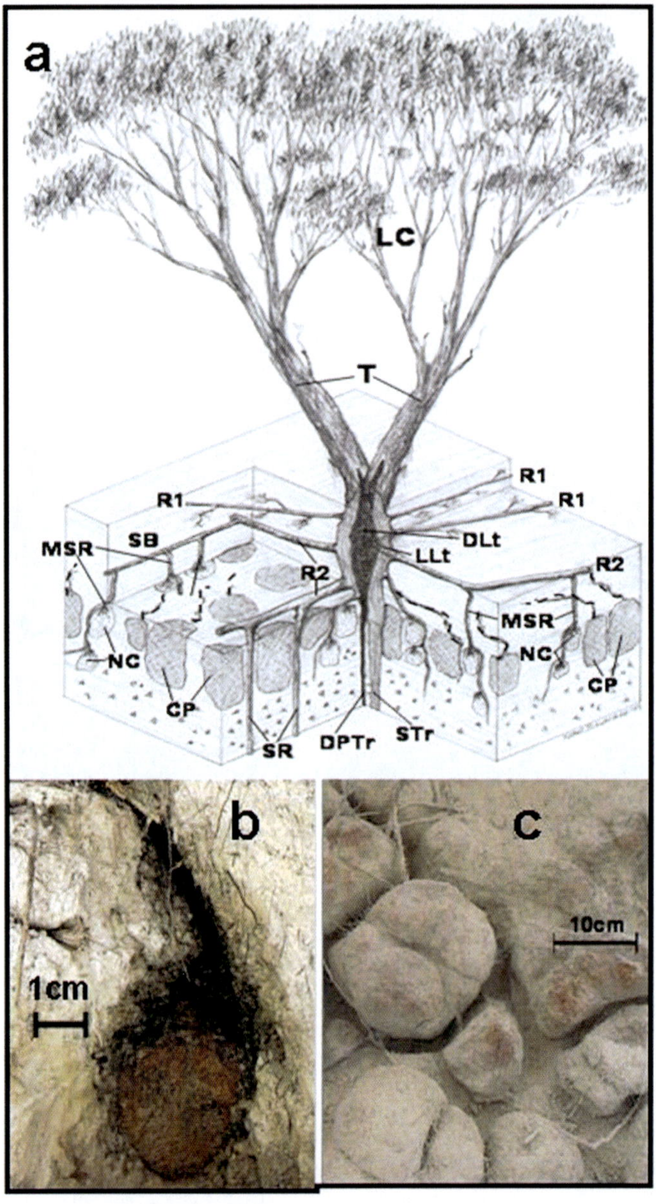

Fig. 17.3 Features relating to clay pavement formation by eucalypts. (**a**) Stylistic representation of typical mature tree of the mallee *Eucalyptus incrassata* showing principal morphological features in relation to clay columns of the pavement. Legend: LC = leafy canopy; T = trunks; LLt = living lignotuber; DLt = decayed first-formed lignotuber; STr = secondary taproots; DPTr = decayed primary taproot; R1 = superficial lateral root lacking sinker roots; R2 = lateral roots which may develop side branches (SB), deeply penetrating sinker roots (SR) and shallow mini 'sinker roots' (MSR), with new columns (NC) on their side branches. Columns of already-formed pavement (CP) are depicted in relation to decayed remains (*broken lines*) of a previous generation of R2 roots (Reproduced from Verboom et al. 2010). (**b**) In situ view of juvenile column forming within a cluster of fine branch roots (Reproduced from Verboom et al. 2010). (**c**) Group of columns in later stages of development which will eventually coalesce into a solid pavement (Reproduced from Pate and Verboom 2009). Note that the red coloration of the clay in Fig. 17.3b, c is due to incorporation of Fe uplifted from perched ground water below the dune

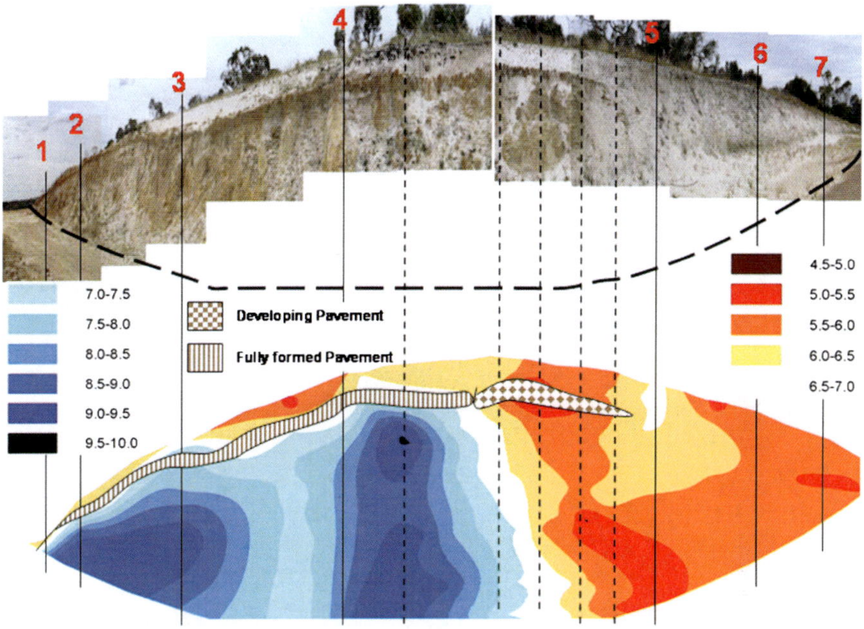

Fig. 17.4 Profile view of the cutting through the dune at Chillinup showing pavement development under mallee woodland (stations 1–4, grading into 5) but not under heath (stations 5–7). A matched Krig-sponsored interpolation of pH (lower half of figure) shows high pH under fully pavemented areas compared to the acidic situation under heath

understorey species in natural uncleared vegetation and does so by limiting their root penetration and limiting access to deeply located stores of water during the dry season.

Chemical analysis of cores through columns of pavemented profiles showed gross elevations of Al and Fe (Fig. 17.5), in the zone where clay columns had been laid down. Bulk densities were elevated in this zone, and substantial deposition of Mg and Ca occurred in and below the columns (see data provided by Verboom et al. 2010). A principal outcome of this uplift is gross elevations of pH in profiles below eucalypts versus that in adjacent heath over unmodified sand (see data for cutting through Chillinup dune shown in Fig. 17.4). A corollary of the above effect is that ground water sources from which these cations have been abstracted by the eucalypt will become progressively acidified (e.g. the highly acidic lakes draining large tracts of mallee-dominated shrubland in southeastern parts of Western Australia).

The xylem sap analyses of E. incrassata (Table 17.2) and other eucalypts (Verboom et al. 2010) have provided evidence of elevated concentrations of the clay-forming mineral elements in column-bearing roots during summer, but not in winter. Thus, as seen in Table 17.1, summer samples averaged at more than 100 times more concentrated in Al ($P=0.008$) and 5–15 times more concentrated in Si, Mg, Fe and S, ($P=0.008$, 0.024, 0.310 and 0.992, respectively) than in winter. However, concentrations of Ca, K and Na were closely similar between the seasons. We also found that concentrations of organic acids particularly citrate were at high

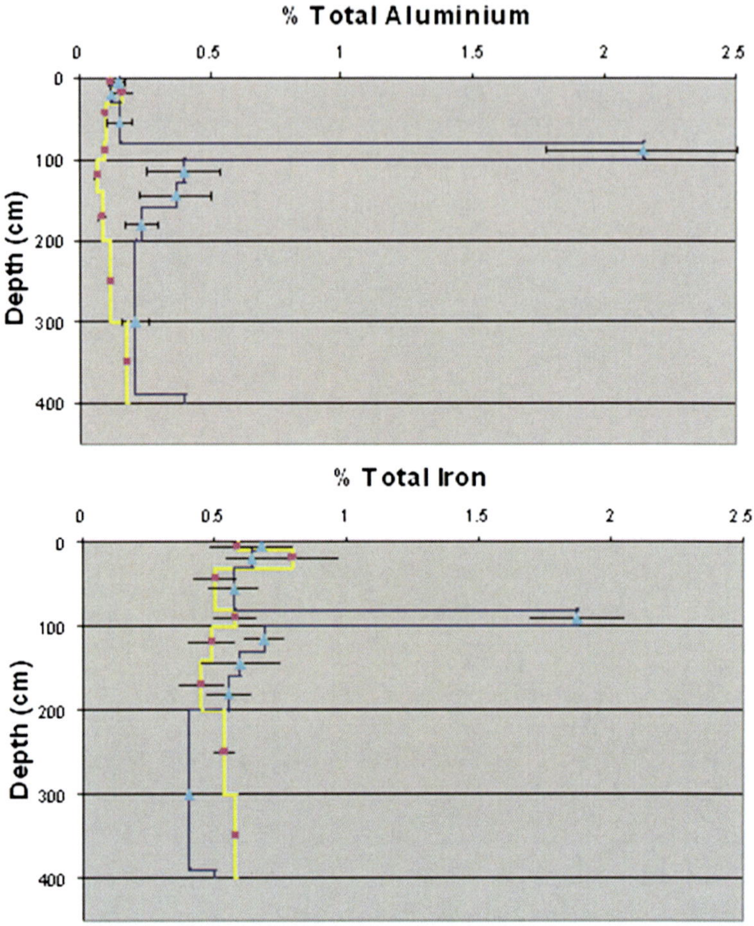

Fig. 17.5 Iron and aluminium concentrations in cores passing through columns of *E. incrassata* (*full lines*) versus sand under adjacent heath (*broken lines*) Columns of *E. incrassata* lie within the depth zone 80–120 cm

Table 17.2 Season comparison of xylem sap analyses of lateral roots of *E. Incrassate*

		Al (mg L^{-1})	Ca (mg L^{-1})	Fe (mg L^{-1})	K (mg L^{-1})	Mg (mg L^{-1})	Na (mg L^{-1})	Si (mg L^{-1})
Wet season ($n=5$)	Mean	0.18	20.7	0.32	59.4	5.0	18.2	0.30
	SE	0.12	3.1	0.13	8.6	0.5	2.0	0.34
Dry season ($n=5$)	Mean	29.7	32.0	1.51	57.4	89.6	20.8	2.40
	SE	16.0	8.6	1.10	8.5	80.0	3.4	0.37

1. All dry season (summer-autumn) samples harvested from authenticated column-bearing roots
2. Xylems particularly high in Fe were also concentrated in organic acids, particularly citrate

level when xylem sap levels of Fe were elevated. These data support our general hypothesis that deep mining of selected mineral elements by taprooted tree species may carry an important function in modifying soil profile composition across a range of ecosystems. In our mallee system, the principal direction of flow and deposition is clearly from deep taproots out into lateral roots utilising the well-known physiological activity referred to in the literature as hydraulic lift or more generally as hydraulic redistribution (see Dawson 1993; Dawson and Pate 1996; Emerman and Dawson 1996; Burgess and Bleby 2006).

An approach used to estimate timescales of clay pavement construction by the resident mallee woodland at the Chillinup site was to consider relationships evident within the rooting catchment of a single-study tree of E. incrassata (see Verboom et al. 2010). The tree in question had a standing biomass of 237-kg dry weight (8.6 g Fe content) while that of the pavement formed over its life was 2,900-kg dry weight (38 kg Fe content). On this basis, one would require the equivalent of almost 4,420 times the current iron pool of the tree to be turned over to meet the observed increment in pavement Fe, assuming that the surrounding sand, highly deficient in Fe, were the only available source. This is clearly an unlikely scenario, whether implemented by internal cycling of Fe via shoots or by shedding and decomposition of biomass. Our alternative hypothesis is that hydraulic uplift of Fe sourced from the lacustrine deposit is our primary input for the iron accumulating in the pavement. Based on a measured maximum Fe concentration of 5.7 mg L^{-1} in xylem sap of column-bearing roots, a targeted uplift of 6.7 million litres of water would have been required during the estimated 300 years of life of our study tree. This equates to uplift of 61 L per day throughout the year, a value within the range of values recorded for hydraulic lift by other woody tree species (e.g. see Emerman and Dawson 1996; Ludwig et al. 2003).

A particularly valuable aspect of the Chillinup site was the timing of construction of the dune during arid Pleistocene time (probably 10–12,000 years ago). Assuming that the mallee woodland and adjacent myrtaceous-proteaceous heathland developed some 5,000 years ago, following a primary phase of colonisation by other taxa, we are able to provide a rough estimate of annual rates of uplift and deposition of Fe across this timescale by measuring current concentrations of the element in fully developed pavements exposed in the cutting through the dune (Fig. 17.4). The analysis shows deposition of Fe at between 0.7 and 5.9 g m^{-2} $year^{-1}$.

17.3.2 Arabian Study

Distributions of aeolian sediment and subsequent diagenetic alterations in the sand sea within the borders of the Emirate of Abu Dhabi have been suggested to have been controlled by a number of interlinked factors including (a) fluctuations in aridity, sea level and sand supply; (b) lateral and leeward accretion as influenced by longshore currents and effects of offshore barriers; and (c) influence of low- or high-wave energy and coastal orientation relative to northwesterly Shamal winds (see

Wagner and Van der Togt 1973; Alsharhan and Kendall 2003). It has also been shown that sea level rise during the Holocene caused dwindling supplies of sand from siliclastic fan deposits fed mostly from rift highlands to the west and south (Garzanti et al. 2003) and perhaps from the Tigris-Euphrates itself (see Al-Sulaimi and Pitty 1995; Glennie 2001). Although sea level rise has now submerged these earlier sources of sand, there remains a substantial supply of mainly white bioclastic and oolitic carbonatic sands emanating from high-energy marine settings of the southern Arabian coast and the immediate offshore. Overall, however, sand supplies have dwindled dramatically, causing deflation to work its way progressively inland (see Fig. 17.1). In the process, the topography of older lithified linear dunes of Pliocene-Pleistocene age has become inverted, upwind of the Buchannan Line, with vast quantities of dunal and interdunal sand piling up on basement high points and old mega dunes towards the south (Fig. 17.1). Interstratifications of rubified quartz and white carbonatic sands in both modern and lithified dunes (see Fig. 17.6a and insets b and c) testify to chemical imprinting and physical reworking of terrestrial and marine sands across the region over multiple cycles.

While it is now well known that rubification is promoted by warm temperatures, oxidising conditions and periodic presence of moisture, the precise mechanisms underpinning the phenomenon have yet to be specified. The northern sands between the Arabian Gulf and the piedmont of the Oman mountains are of particular interest in this respect as they redden dramatically towards the wetter foothills of the Oman mountains, particularly where they sweep over iron-rich metamorphic rocks of the Semail Ophiolite Suite. Herein one finds a possible source of Fe oxides possibly released via subaerial weathering of fan deposits as suggested by White et al. (2001) (also see Anton and Ince 1986).

The density of vegetation across the transect is shown in Fig. 17.7a. Note the virtual absence of large tree/woody species in the drier west and overall increasing density in moving along the transect towards the mountains. However, it must be stressed that distribution of trees at any zone of the transect is generally not uniform and can be restricted to dunes where they abut plains (Fig. 17.6d). It was not possible to identify tree species identities from DigitalGlobe images but general ground observation suggested A. tortilis to be a dominant component.

Rubification indices along the transects and associated dithionite-extractable iron oxide % were indicated in Fig. 17.7b. As to be expected, values were lowest in regions of carbonatic and bleached quartz sand near the coast. Despite limitations as to its reliability, the dithionate treatment comprised the only available option for assessing haematitic fractions associating with pedogenetic iron. Combining the above data on rubification with vegetation density produced a strong positive correlation between reddening and a cumulative plot of vegetation from coast to mountains (see Fig. 17.6c). This fits with the trend of increasing redness running parallel to isohyets for average annual rainfall and water supplied by fan distributaries. Under our hypothesis, iron uplifted by the trees during episodes of hydraulic lift would be the mechanism responsible for reddening, and in a scenario of an easterly drift of sand towards the mountains, one would likely expect the chances of individual grains becoming coated with Fe to increase progressively as sand migrated along the same gradient. We would also suggest that the subaerial weathering referred to

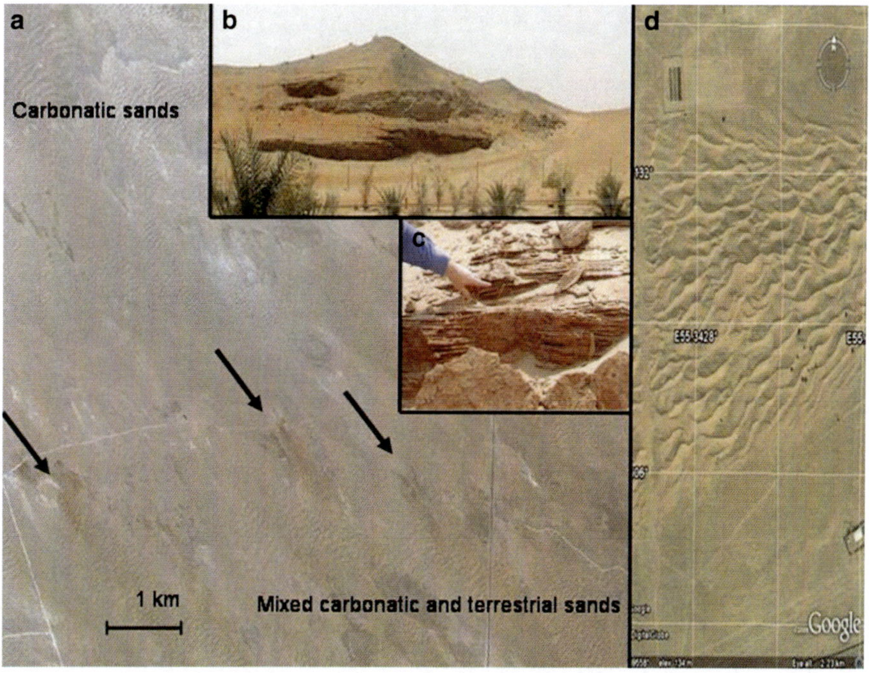

Fig. 17.6 Features relating to clay pavement formation by eucalypts. (**a**) Reddened plumes of sand in the wake of eroding Pliocene-Pleistocene dunes testify to white 'marine' sands mixing with red 'terrestrial' sands towards lower region of photograph. The white carbonatic sand deposits (*top left*) are modern and currently blowing in from the Gulf. *Arrows* indicate both the direction of the Shamal wind and the location of the older red dunes which are carbonate cemented and of Pliocene-Pleistocene age. (**b**) Roadside view of such a red lithified dune showing rubified sand eroding from it and streaking down wind (equivalent to arrowed area shown in satellite view-a). (**c**) Close-up view of an eroded face of red dune showing that it is itself composed of interstratifications of rubified ('terrestrial') grains and non-rubified ('marine') miliolite sand. (**d**) Select strip of part of study transect (TR in Fig. 17.6a) showing very sparse tree scattering across dune field particularly its margins. Trees are typically absent in inter dunal flats with minimal sand covering

by White et al. (2001) may be facilitated by specialised root hairs on deep taproots releasing acids and other scouring chemicals to solubilise important mineral elements such as iron and phosphate (see Yi et al. 2010).

The above interpretation may well prove to be a gross oversimplification of the situation, particularly since it fails to explain currently visible heterogeneities in Fe distribution across the area in the form of reddened embayments and other effects due to winnowing. Under the explanation of in situ 'weathering' presented by Anton and Ince (1986) and White et al. (2001), embayments would become intensely reddened because of lowered sand mobility in such regions. This again concurs with protracted uplift of Fe by the vegetation concentrated in such areas.

Despite the limitations alluded to above, the general correlation between rubification and woody vegetation suggests a causative interrelationship which we would suggest to be implemented by hydraulic uplift of iron by taprooted trees and subsequent release into upper sand. We would thus draw parallels to our Australian

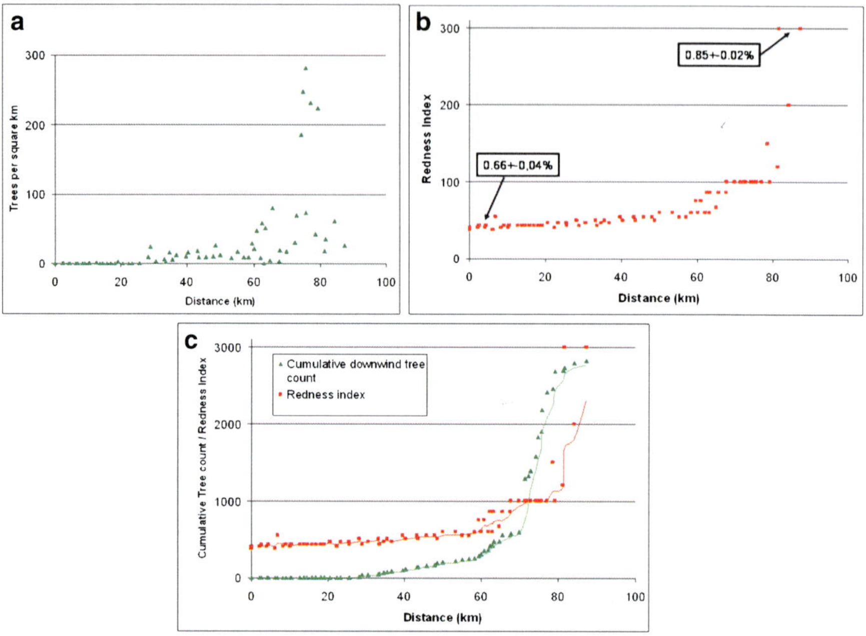

Fig. 17.7 Trends in density of vegetation and intensity of rubification across the transects shown in Fig. 17.2. Data are given as means of measurements for the two transects. Note lack of reddening in coastal zone-lacking trees at start of transects and strong correlation between reddening and cumulative tree count along the remainder of the transects. (**a**) Data for tree counts. (**b**) Assessments of rubification and dithionite-extractable iron %. (**c**) Comparisons of cumulative tree count and rubification with progressive distance along the 90-km length of the transects shown as TR in Fig. 17.2

example at Chillinup despite the fact that the uplifted iron in the latter case becomes deposited in clay layers as opposed to coatings of individual sand grains as applies to the United Arab Emirates. Direct testing of this hypothesis in relation to the Arabian example would require sampling of xylem sap for transported iron from species such as *A. tortilis* and looking for evidence of subsequent release of this Fe into superficial layers of sand surrounding lateral roots of the trees. This species has already been shown to engage in hydraulic lift of water in other African ecosystems but has yet to be studied in the context of mineral transport.

17.4 Conclusions

The mallee ecosystem in the sand dune surrounding a playa lake at Chillinup in Western Australia shows clay pavement formation in the root catchments of its resident species of eucalypts. Evidence is presented that uplift of iron, aluminium and other mineral elements from ground water by deep taproots provides the materials required for clay biogenesis. Possible time courses of pavement construction by successive generations of woodland at the site is presented, including assessment of mean annual rates of iron uplift per unit area in tree root catchments.

Our analysis of the extent and patterning of rubification of sand within the borders of the United Arab Emirates shows that the southern sand sea acquired much of its reddish hue during soil formation in the Pliocene-Pleistocene epoch while similar processes operating during the Holocene probably accounts for the intense reddening of the northern sand sea towards mountains of Oman.

We present the hypothesis that iron uplift by deep-rooted shrubs and trees has been responsible for rubification in both cases and in support of this view show a strong correlation between the extent of reddening along the general flow route for sand in the northern sand sea and vegetation progressively encountered along this path. Definitive studies of iron uptake, translocation and secretion in this ecosystem may well lead to a better understanding of rubification process in current and palaeoenvironments of the region.

Acknowledgements We gratefully acknowledge assistance provided by the Environment Agency-Abu Dhabi (EAD) and Dubai-based the International Center for Biosaline Agriculture (ICBA). Iron oxide data was kindly provided by Kevin White of University of Reading. Andrew Buchanan and Phil Goulding helped prepare Figs. 17.1, 17.2 and 17.4. The Australian component of this work was carried out with the support of the South Coast Natural Resource Management Incorporated. The drawing in Fig. 17.3a was kindly provided by Noel Schoknecht.

References

Alsharhan AS, Kendall CGSC (2003) Holocene coastal carbonates and evaporites of the southern Arabian Gulf and their ancient analogues. Earth Sci Rev 61:191–243

Al-Sulaimi JS, Pitty AF (1995) Original depositional model of wadi Al-Batin and its associated alluvial fan, Saudi Arabia and Kuwait. Sediment Geol 97:203–229

Anton D, Ince F (1986) A study of sand colour and maturity in Saudi Arabia. Z Geomorphol NF 39:339–356

Boogert NJ, Paterson DM, Laland KN (2000) The implications of niche construction and ecosystem engineering for conservation biology. BioScience 56:569–577

Bresler H (1982) The north-eastern Rub al-Khali within the borders of the United Arab Emirates. Z Geomorphol NF 26:495–504

Burgess SSO, Bleby TM (2006) Redistribution of soil water by lateral roots mediated by stem tissues. J Exp Bot 57:3283–3291

Dawson TE (1993) Hydraulic lift and water use by plants: implications for water balance, performance and plant-plant interactions. Oecologia 95:565–574

Dawson TE, Pate JS (1996) Seasonal water uptake and movement in root systems of Australian phraeatophytic plants of dimorphic root morphology: a stable isotope investigation. Oecologia 107:13–20

EAD (2009) Soil survey of Abu Dhabi Emirate. Extensive survey, vol I. Environment Agency Abu Dhabi, Abu Dhabi, p 506

Emerman SH, Dawson TE (1996) Hydraulic lift and its influence on the water content of the rhizosphere an example from sugar maple, Acer saccharum. Oecologia 108:273–278

Garzanti E, Andò S, Vezzoli G, Dell'era D (2003) From rifted margins to foreland basins: investigating provenance and sediment dispersal across Desert Arabia (Oman, UAE). J Sediment Res 73:572–588

Glennie KW (2001) Evolution of the Emirates' land surface: an introduction. In: Hellyer P, Al-Abed I (eds) The United Arab Emirates: a new perspective. Trident Press, London

Ludwig F, Dawson TE, Kroon H, Berendse F, Prins HHT (2003) Hydraulic lift in Acacia tortilis trees on an East African savanna. Oecologia 134:293–300

Odling-Smee FJ, Laland KN, Feldman MW (2003) Niche construction: the neglected process in evolution, vol 37, Monographs in population biology. Princeton University Press, Princeton

Pate JS, Verboom WH (2009) Contemporary biogenic formation of clay pavements by eucalypts: further support for the phytotarium concept. Ann Bot 103:673–685

Pate JS, Woodall G, Jeschke WD, Stewart GR (1994) Root xylem transport of amino acids in the root hemiparasitic shrub Olax phyllanthi R.Br. (Olacaceae) and its multiple hosts. Plant Cell Environ 17:1263–1273

Verboom WH, Pate JS (2006a) Bioengineering of soil profiles in semiarid ecosystems: the 'phytotarium' concept. A review. Plant Soil 289:71–102

Verboom WH, Pate JS (2006b) Evidence of active biotic influences in pedogenetic processes. Case studies from semiarid ecosystems of south-west Western Australia. Plant Soil 289:103–121

Verboom WH, Pate JS, Aspandiar M (2010) Neoformation of clay in lateral root catchments of mallee eucalypts: a chemical perspective. Ann Bot 105:23–36

Wagner CW, van der Togt C (1973) Holocene sediment types and their distribution in the southern Persian Gulf. In: Purser BH (ed) The Persian Gulf. Springer, New York, pp 123–155

White K, Goudie AS, Parker A, Al-Farraj A (2001) Mapping the geochemistry of the northern Rub' Al Khali using multispectral remote sensing techniques. Earth Surf Proc Landf 26:735–748

Yi K, Menand B, Bell E, Dolan L (2010) A basic helix-loop-helix transcription factor controls cell growth and size in root hairs. Nat Genet 42:264–267

Chapter 18
The Sand Land Soil System Placement (Taxonomy) and Society

Ramez A. Mahjoory

Abstract Arid soils of the world require different strategies for optimum land utilization due to the delicate balance between annual climatic cycles and general trends toward desertification. Arid lands are an irreplaceable natural resource covering one-third of the global land surface and harbor within them the potential to elevate the standards of living of more than two billion people. The soils of the arid areas are underutilized due to inadequate public knowledge of the soil system. In a modern age of precise measurements and categorization, it is time to use the available tools to greatly enhance the productivity and use of the soil system in general and arid soils in particular. Quantification of multifunctional and dynamic ecosystems within arid regions can provide opportunities for land users to diversify agroecological systems and for food security based upon well-defined representative soil units in each region. Thus, the soil scientists by considering soil as a system and correlating its taxonomic units would have the ability to extrapolate results of studies and by keeping "balance" between "inputs" and "outputs" in "soil system" would promote national productivity and regional economy. In this study, selected soils in various arid environmental and ecological regions of countries such as Ghana, Jordan, and Iran that have already been investigated (USDA soil taxonomy) to the level of soil series/farmland units on applicable scales will be compared to the soils of the Abu Dhabi Emirate.

Keywords Abu Dhabi • Agroecosystems • Arid soils • Soil system • Taxonomy

---

R.A. Mahjoory (✉)
Institute of International Agriculture, Soil Science,
Michigan State University, East Lansing, MI, USA
e-mail: mahjoor5@msu.edu; rmhjoory@aol.com

S.A. Shahid et al. (eds.), *Developments in Soil Classification, Land Use Planning and Policy Implications: Innovative Thinking of Soil Inventory for Land Use Planning and Management of Land Resources*, DOI 10.1007/978-94-007-5332-7_18,

18.1 Introduction

New concepts in soil utilization in a manner that best meet the demands for diverse, social, economical, and environmental movements are the current challenges faced by the scientists, professionals, and decision makers. Each soil is uniquely different in its strategic role to meet these challenges. This is particularly true in arid soils that are classified within Aridisols, Entisols, Inceptisols, Alfisols, and Vertisols (Soil Survey Staff 2006). These soils cover more than one-third of world resources and have potential to meet the food and fiber demands of the growing population (Lal and Kimble 2000).

Among extensive variations of arid soils, the sand land soils covering more than 900×10^6 ha worldwide are excessively susceptible to degradation/desertification that adversely affect the livelihood of more than a billion people. These soils (>70% sand and <15% clay) are classified as Aridisols and Entisols (Soil Survey Staff 2010) and as Arenosols (IUSS Working Group WRB 2006). Global assessment of land degradation is rather complex, and therefore, a wide range of methods are being implemented, including global estimates of desertification (Dregne and Chou 1994) and of land degradation by the International Soil Reference and Information Centre (Oldeman et al. 1992). It is estimated that 70% of the dry areas, particularly the sand land soils, are being degraded by the combined anthropogenic, physicochemical, biological, and natural factors, causing decline in soil properties and leading to reduction in biomass and aesthetic value of the landscape (Lal et al. 1997).

New technologies, such as satellite imaging, geographic information system (GIS), and greater computing power, can help us hold back the desert, but land degradation is also a people problem, and that must be part of the solution according to land degradation assessment in dry area (LADA) report (FAONews-LADA 2002).

Although the role of soil science in classifying and conserving world land resources is receiving global attention, its role in evaluating potentiality of arid soils with all their importance is neither properly understood nor widely recognized. The perception that enough is already known about soils, so that generalizations can be made for all soils, is incorrect (Eswaran et al. 2001).

Significant amounts of scientific research in classifying and mapping soils have been accomplished in the Abu Dhabi Emirate. Abu Dhabi is the largest emirate of the United Arab Emirates and occupies about 77,000 km^2, of which 3,770 km^2 (5.4%) composes the coastline area. The soil maps of the Abu Dhabi coastline area have been prepared, recognizing the soil taxonomic units at the family/soil association level at a scale of 1:50,000 (Shahid et al. 2004; Abdelfattah and Shahid 2007). In these studies, high amounts of carbonates (>15%), gypsum, and halite and occurrence of more soil diversity and classes have been reported. The saline and sodic soils with extremely high salinity and sodicity located in the coastal salt flats (sabkha) are also reported (Abdelfattah 2009; Shahid and Abdelfattah 2008). The mineralogy of a unique anhydrite soil has recently been reported (Shahid et al. 2007), and gypsum polymorphism in the desert environment (Shahid and Abdelfattah 2009) of the Abu Dhabi Emirate has also been reported in detail.

The main objectives of this study are to compare the classification of some arid area soils from several agroecosystems of the world with characterization and classification of soils in the Abu Dhabi Emirate at the order, great group, and family taxonomic unit levels. This study is also focused on the necessity and importance of developing more soil diversity and classes. It is important to identify those soil classes in detail by attaching increased amounts of information to their mapping units by "updating soil mapping units" and evaluating the composition and/or percent inclusions of updated soil units in survey areas of the emirate, "correlating" soils by naming similar ones under a unique identity in various agroecological regions and "modifying soils legend and survey report." This chapter also emphasizes on "conducting further research" on saline and sodic soils in relation to their morphology, mineralogy, and chemistry and also the soils with high carbonates to evaluate pedogenic carbonates formation and "utilizing soil amendments" for reclamation of degraded lands.

18.2 Materials and Methods

Selected soils such as Laterites, Xeralfs, Xererts, Argids, Orthids, and Ortents and Ochrepts in various arid/semiarid agroecological regions of countries such as Ghana (Mahjoory et al. 2002), Jordan (Awni et al. 2006), and Iran (Gharaee and Mahjoory 1981) were evaluated on the basis of their field (morphological features) and laboratory (physicochemical) characteristics. The soil taxonomic units at the family/association level were called soil mapping units, and the soils were mapped at a scale of 1:50,000, according to the USDA soil taxonomy (Soil Survey Staff 1996).

The data were collected by the author during FAO/UN and World Bank consultancy missions, at the US Fulbright Regional Scholar in the Middle East and Shiraz University academic research and teaching programs. The classification, mapping, and description of these soils and the Aridisols and Entisols in the Eastern Desert of Egypt are compared with characteristics and classification of soils (Soil Survey Staff 2010) in Abu Dhabi Emirate.

The transect methodologies (Mahjoory and Whiteside 1975, 1976) could be utilized for "updating soil mapping units" at the family level and evaluating the percent inclusions in map units in the Abu Dhabi Emirate according to the soil taxonomy (Soil Survey Staff 2010). The "updated" soil units need to be "correlated" and the soil survey report improved through the "modified legend" on the basis of guidelines and methodologies outlined in USDA-National Soils Handbook, No. 430 (Soil Survey Staff 1996).

18.3 Results and Discussion

The soils included in this study are located in different regions of the world, but in arid areas, differ from each other and from Abu Dhabi Emirate. These differences are reflected in taxonomic differences at the order level such as Alfisols, Inceptisols, and Vertisols and family-level taxonomic units such as mineralogy that varies from vermiculitic to smectitic and particle-size distribution that varies from coarse-loamy to very fine. But most soils, including the soils of Abu Dhabi Emirate, are similarly classified within the orders of Aridisols and Entisols and the sandy, mixed/carbonatic/gypsic, and hyperthermic family, exception is with the Entisols which are mixed and hyperthermic. The soils are mapped at the family level of soil taxonomy (Soil Survey Staff 2006) at scales of 1:100,000 and soil series level at 1:25,000 scales (EAD 2009a, b). The soil survey reports and "soil map unit legends" have been developed and map units described. The soil mapping units include several component soil families/series depending on their distribution. Each soil series may also include several "soil types" on the basis of variations in surface features such as texture and slope.

Because of the particular importance of the soils in arid areas of Abu Dhabi Emirate and their excessive susceptibility to erosion, degradation, and also utilization, it is needed to "update" the existing classification of these soils and evaluate percent inclusions of other soils in their "map units" and attach more information to "updated units" for the users at a scale of 1:25,000 or larger.

However, since variations in surface features of soils affect boundary delineation of each "map unit," the surface accumulation of sand dunes and their color, mineralogy, and dynamic process in shaping and height of the dunes as reported in the United Arab Emirates (Howari and Baghdady 2007) may have a higher priority than sandy texture and slope in characterizing the map units. The classification may include variations in sand dunes as special surface features in naming map units. The "updated soil map units" could be correlated by putting together similar units under a unique name in each agroecological/ecosystem region of the Abu Dhabi Emirate (Boer 1998). This would promote the effective transfer of information to and between the land users and also facilitate extrapolation of research data throughout the region and worldwide. The results would definitely help to improve conservation practices, prevent degradation, and meet demands of stakeholders in the emirate.

The soil map "legend" could also be "modified" on the basis of revision, correlation, and descriptions of the "updated soil map units" and Soil Survey Report, prepared with maximum applicability.

Based on the high exchangeable sodium percentage (ESP) (>15) and high clay content of saline and sodic soils of sabkha in the emirate, further studies of morphology, chemistry, and mineralogy are needed to determine the possible presence of a natric horizon and the presence of expandable clays (montmorillonite, smectites, vermiculite). Once these characteristics are confirmed, these soils can then only be reclassified within the Natrargids great group and in the smectitic mineralogy family (Mahjoory 1979). The source of high carbonates in the soils of Abu Dhabi Emirate, in addition to parent materials, might also be related to pedogenic carbonate processes that are natural means of atmospheric carbon sequestration. This suggests the potential of these soils being sources/sinks of CO_2 (Lal and Kimble 2000), impacting carbon credit in world markets. Thus, further research is needed in these aspects related to climate change and global warming.

Land degradation, if not prevented, will remain an important global issue for the twenty-first century due to its adverse impact on the environment, food security, and quality of life. Anthropic and natural factors have major counterproductive impacts on arid soils, particularly the sandy soils.

Mobilizing scientific communities in academic institutions for "adapted teaching" and research according to the needs of the countries (Mahjoory 2005), and also exchange of experiences and research results in internationally held conferences, would become beneficial for stakeholders and decision makers in conserving our lands.

Using naturally occurring soil amendments, such as shampoo-clay (Mahjoory 1996), that might occur throughout the region may improve aggregate stability, stabilize the moving sands, and rehabilitate marginal lands throughout the region. Cultivating "kudzu," a plant that was brought to world markets in 1876 from Japan and has been used in several countries as a form of erosion control and also as legume to increase nitrogen in soil and transfer valuable minerals, might also be beneficial in preventing land degradation (Abbas et al. 2001; Boyette et al. 2006; Tanner et al. 1979).

Additional important task and challenge ahead of us should be to demonstrate the inseparable role of society in sustaining agriculture and the soil in sustaining society.

18.4 Conclusions and Recommendations

The correlation of soils from various regions of arid areas with those from Abu Dhabi Emirate revealed similar soils as Aridisols and Entisols Orders. However, Alfisols, Inceptisols, and Vertisols do not occur in the emirate. It is suggested to "update" and reevaluate soils in the map units by using transect methodologies. The updated soil units could also be "correlated" and the soil map legend "modified" for regional and global extrapolation of research data/database and efficient utilization of the soils.

The unique anhydrite soil mapped in the coastline of Abu Dhabi Emirate is suggested to be included in the USDA soil taxonomy at various levels. The coastline soils of Abu Dhabi Emirate are very highly saline forming salt scalds and, therefore, require costly management for their uses. The sustainable restoration/reclamation of these soils may not be achieved without development of a unique and "integrated classification system." It is, therefore, suggested to add a new soil order "Salsodosols" to classify global saline/sodic soils. This classification could place soils uniquely to the level of mapping/farmland units on a global and larger/workable scale. The criteria for this new order can be developed jointly with soil system scientists. Further investigation is required to confirm the occurrence of Natrargids and vermiculitic/smectitic mineralogy family class for improved understanding, uses, and management of these soils. More research is required on the occurrence of pedogenic carbonates and to evaluate their role in atmospheric carbon sequestering and potential of C-Credit in world markets. The sandy soils are prone to land degradation; these should be protected by using clay and stabilizing the dunes with suitable native plants. A "balanced" soil system science would give the scientists and stakeholders a far larger amount of attention to the soil-related productivities such as food and fiber and also problems such as degradation/desertification than focusing on general patterns of the "soil science."

References

Abbas HK, Tak H, Boyette CD, Shier WT, Jarvis BB (2001) Macrocyclic trichothecenes are undetectable in kudzu (Pueraria montana) plants treated with a high-producing isolate of Myrothecium verrucaria. Phytochemistry 58:269–276

Abdelfattah MA (2009) Land degradation indicators and management options in the desert environment of Abu Dhabi, United Arab Emirates. Soil Surv Horiz 50:3–10

Abdelfattah MA, Shahid SA (2007) A comparative characterization and classification of soils in Abu Dhabi coastal area in relation to arid and semi-arid conditions using USDA and FAO Soil Classification Systems. Arid Land Res Manag 21(3):245–271

Awni YT, Kherast S, Mahjoory RA (2006) Evaluation of representative soils in relation to C-dynamics on a climo-toposequence in arid/semiarid ecoregions of Jordan. Abstracts: American Society of Agronomy Annual Meetings, Indianapolis

Boer B (1998) Ecosystems, anthropogenic impacts, and habitat management techniques in Abu Dhabi. PhD dissertation, University of Paderborn, Germany

Boyette CD, Krishna NR, Hoagland RE (2006) Glyphosate and bioherbicide interaction for controlling Kudzu (Pueraria lobata), Redvine (Brunnichia ovata), and Trumpet creeper (Campsis radicans). Biocontrol Sci Technol 16(10):1067–1072

Dregne HE, Chou NT (1994) Global desertification dimensions and costs. In: Degradation and restoration of arid lands. Texas Technical University, Lubbock

EAD (2009a) Soil survey of Abu Dhabi Emirate-extensive survey, vol I. Environment Agency, Abu Dhabi, pp xx + 506

EAD (2009b) Soil survey of Abu Dhabi Emirate – intensive survey, vol III. Environment Agency, Abu Dhabi, p xviii + 435

Eswaran H, Lal R, Reid PF (2001) Land degradation: an overview. USDA, NRCS, Washington, DC

FAONews-LADA (2002) Dry land degradation: a people problem. www.fao.org/news/2002/020205

Gharaee HA, Mahjoory RA (1981) Characteristics and geomorphic relationships of some representative Aridisols in Southern Iran. Soil Sci Soc Am J 48(5):1115–1119

Howari FA, Baghdady A, Goodell PC (2007) Mineralogical and geomorphological characterization of sand dunes in the eastern part of United Arab Emirates using orbital remote sensing integrated with field investigations. Geomorphology 83:67–81

IUSS Working Group WRB (2006) World reference base for soil resources. World soil resources reports no. 103. FAO, Rome

Lal R, Kimble JM (2000) Pedogenic carbonates and global carbon cycle. In: Lal R, Kimble JM, Eswaran H, Stewart BA (eds) Global climate change and pedogenic carbonates. CRC Press, Boca Raton

Lal R, Kimble JM, Mtimet A, Eswaran H, Scharpenseel H (1997) Global climate change and pedogenic carbonates. An international workshop, Tunis, Tunisia, 13–17 Oct 1997

Mahjoory RA (1979) The nature and genesis of some salt-affected soils in Iran. Soil Sci Soc Am J 43(5):1019–1024

Mahjoory RA (1996) Occurrence and mineralogy of a deposit of Shampoo-clay in Southern Iran. Appl Clay Sci 11(1):69–76

Mahjoory RA (2005) Anthropogenic modifications of soil physico- chemical properties in various agro-eco regions of Iran and the challenging opportunities in future. In: International conference on IUT, Isfahan, Iran. Published in Classifier (Soil Classifiers Association of Michigan), pp 7–10
Mahjoory RA, Whiteside EP (1975) Relationship of modern soil taxonomy to the mineral soils in Michigan. Soil Sci 120(4):295–300
Mahjoory RA, Whiteside EP (1976) Soils of Saginaw County, Michigan, vol I: Soil descriptions and interpretations, vol II: Soil maps. Updating, 1993 soil maps. USDA-NRCS and Michigan State University, Washington, DC and East Lansing
Mahjoory RA, Koohafkan PA, Asiamah RD (2002) The main agro-ecological zones and associated benchmark soils of Ghana. Abstracts: American Society of Agronomy Annual Meetings Indianapolis, USA
Oldeman LR, Hakkeling RTA, Sombrock WG (1992) World map of the status of human-induced soil degradation: an explanatory note. ISRIC-World Soil Information, AJ Wageningen
Shahid SA, Abdelfattah MA (2008) Soils of Abu Dhabi Emirate (chapter 2). In: Perry RJ (ed) Terrestrial environment of Abu Dhabi Emirate. Environment Agency, Abu Dhabi, pp 71–91
Shahid SA, Abdelfattah MA (2009) Gypsum polymorphism in the desert environment of Abu Dhabi Emirate. Eur J Sci Res 29(2):237–248
Shahid SA, Abdelfattah MA, Arshad KR, Al Muhairbi MA, Othman YR, Al Haji A, Al Radha Al, Mansouri M, Ahmed S, Siddiqui A, Azimullah KM, Sekaly AL, Al Qubaisi BS (2004) Soil survey for the coastline of Abu Dhabi Emirate, 2 vols, vol 1: Reconnaissance survey and vol 2: Soil maps. Environment Agency, Abu Dhabi
Shahid SA, Abdelfattah MA, Wilson MA (2007) A unique anhydrite soil in the coastal sabkha of Abu Dhabi Emirate. Soil Surv Horiz 48:75–79
Soil Survey Staff (1996) National soil survey handbook 430, vol 6. Sec 609.06. (Soil Correlation). US Government Printing Office, Washington, DC Available online at: http://soils.usda.gov/technical/handbook/accessed 2010
Soil Survey Staff (2006) Keys to soil taxonomy, 10th edn. US Government Printing Office, Washington, DC
Soil Survey Staff (2010) Keys to soil taxonomy, 11th edn. US Government Printing Office, Washington, DC
Tanner RD, Shahid SH, Lindsey AH, Frederick TW (1979) Kudzu (Pueraria Lobata): potential agricultural and industrial resource. Econ Bot 33(4):400–412

Chapter 19
Digital Soil Mapping Using Spectral and Terrain Parameters and Statistical Modelling Integrated into GIS-Northwestern Coastal Region of Egypt

Fawzy Hassan Abdel-Kader

Abstract This chapter examines digital soil mapping approach for the production of soil maps by using multinomial logistic regression on soil and terrain information from pilot areas in the northwestern coastal region of Egypt. The aim is to reproduce the original map and predict soil distribution in the adherent landscape. Reference soil maps produced by conventional methods at Omayed and Nagamish areas were used. The logit models of the soil classes as expressed by the spectral and terrain parameters were calculated, and predicted soil classes' maps were produced. The IDRISI/SAGA/SATISTCA/SPSS platforms were used in this chapter. The terrain and spectral parameters were found to be significantly influential that the selection of the land surface predictors was satisfactory. The McFadden pseudo *R*-squares ranged from 0.473 to 0.496. The most significant terrain parameters influencing the spatial distribution of the soil classes are found to be elevation, valley depth, multiresolution ridgetop flatness index, multiresolution valley bottom flatness and SAGA wetness index. However, the most influential spectral parameters are the first two principle components of the six Enhanced Thematic Mapper bands. The overall accuracy of the predicted soil maps ranged from 72 to 74% with kappa index ranged from 0.62 to 0.64. The developed probability models were successfully used to predict the spatial distribution of the soil mapping units at pixel resolutions of 28.5 m×28.5 m and 90 m×90 m at adjacent unvisited areas at Matrouh and Alamin. The developed methodology could contribute to the allocation and to the digital mapping and management for new expansion sites in the remote desert areas of Egypt.

Keywords Digital soil mapping • Egypt • Multiple logistic regressions • Northwestern coastal region • Spectral and terrain parameters

---

F.H. Abdel-Kader (✉)
Remote Sensing and GIS Laboratory, Department of Soil and Water Sciences, Faculty of Agriculture, University of Alexandria, El-Shatby, Egypt
e-mail: sfawzy08@yahoo.com

S.A. Shahid et al. (eds.), *Developments in Soil Classification, Land Use Planning and Policy Implications: Innovative Thinking of Soil Inventory for Land Use Planning and Management of Land Resources*, DOI 10.1007/978-94-007-5332-7_19,

19.1 Introduction

Based on the soil mapping activities of the Soil and Water Research Institute, ARC, during 1955–1964 and other works (FAO 1964, 1970; Ahmed 1995) of the soil map of Egypt (Hamdy 1982) and the Land Master Plan (GARPAD 1986), it can be concluded that only 10% of the total area of Egypt has been systematically mapped. Cadastral and topographic maps as well as aerial photographs were used to produce hard copies of the soil and soil potentiality maps at scales of 1:100,000 and 1:200,000. From these studies, only 1% of the total area of Egypt was selected to survey at semi-detailed level at scale of 1:50,000. Based on the existing geological and soil mapping data, hard copies and digital soil maps of Egypt at 1:1,000,000 and 1:100,000 scales were developed using RS/GIS capabilities and using the FAO taxonomy (FAO-UNESCO 1977; Abdel-Kader and Bahnassy 2001; FAO 2003).

The increasing demand for new land in Egypt necessitates an intensive soil mapping activities throughout the unmapped 90% of the total area of Egypt preferably at scales of 1:25,000 and 1:50,000. As the conventional soil surveys at 1:25,000/1:50,000 scales are expensive and time-consuming, there is a growing need for considering digital soil mapping (DSM) concept in Egypt. The DSM is a spatial soil information system created by numerical models that account for the spatial and temporal variations of soil properties based on soil information and related environmental variables (Lagacherie et al. 2006). Remotely sensed data as well as geostatistical models are the basic tools of DSM approach to speed up and refine soil mapping. Maps generated using this DSM procedure may be adequate for extrapolating soil distribution information to areas where no comprehensive soil map is available, but where reference soil maps representing soil diversity and distribution in such regions do exist. The most frequent statistical models used for spatial prediction of soil categorical classes are the multinomial logistic regression model.

The objective of this present chapter is to examine the DSM approach for the production of soil maps by using multinomial logistic regression on soil and terrain information from pilot areas in the northwestern coastal region of Egypt. The main goal of this chapter is to use digital elevation model (DEM) and training soil data from conventionally made soil map to model the relationship between soil mapping units and topographic and spectral attributes so that the model could be used to predict the spatial distribution of the soil groups using spatially continuous likelihood values. The aim is to reproduce the original map and predict soil distribution in the adherent landscape. To achieve the objective the existing reference soil maps produced by conventional methods were used. Relevant parameters derived from an ETM image and an SRTM DEM were considered; these are the following: elevation, multiresolution valley bottom flatness index, multiresolution ridge top flatness index, valley depth, mid-slope position, catchment slope, slope gradient, SAGA wetness index, grain size index, carbonate index and the first and second principle components of six ETM bands. The logit models of the soil classes as expressed by the spectral and terrain attributes were calculated, and predicted soil class's maps were produced. In this chapter, the IDRISI/SAGA/STATISTICA/SPSS platforms were used.

The accuracy of the predicted soil maps was determined by using error matrices where map accuracy indicators were calculated. The probability models were then used to predict the spatial distribution of the soil classes at adherent unvisited sites. Training areas at Omayed and Nagamish Northwest Coast (NWC) were used. Predicted digital soil maps at pixel resolutions of 90 m × 90 m and 28.5 m × 28.5 m for the regional adherent areas at Alamein and Matrouh were processed.

19.2 Materials

19.2.1 Study Area

The northwest coastal zone of Egypt extends over 350 km from west of Alexandria to the Libyan border. The 150-m ASL-coastal area is about 1,427,898 ha (Abdel-Kader 2009), of which only 988,198 ha (69%) were surveyed at 1:100,000 scale and 11,208 ha (7.75%) were surveyed at scale of 1:50,000 (FAO 1970; GARPAD 1986). The soils in these areas are mainly Torripsamments, Torriorthents and Calcic Paleorthids. The human settlements and land uses are entirely dependent on rainfall and on various forms of water harvesting. Annual rainfall is restricted to the winter months and to a narrow 20-km strip along the coast where the 60-year average at Mersa Matrouh is 144 mm with CV 45%.

Potentially productive areas in terms of run-off accumulation and soils are concentrated within 218 wadis and depressions (FAO 1970; Abdel-Kader et al. 2005). For this present chapter, two representative sites were considered to include training and prediction areas at Omayed-Alamin, west of Alexandria, and at Naghamish, east of Matrouh, as shown in Fig. 19.1.

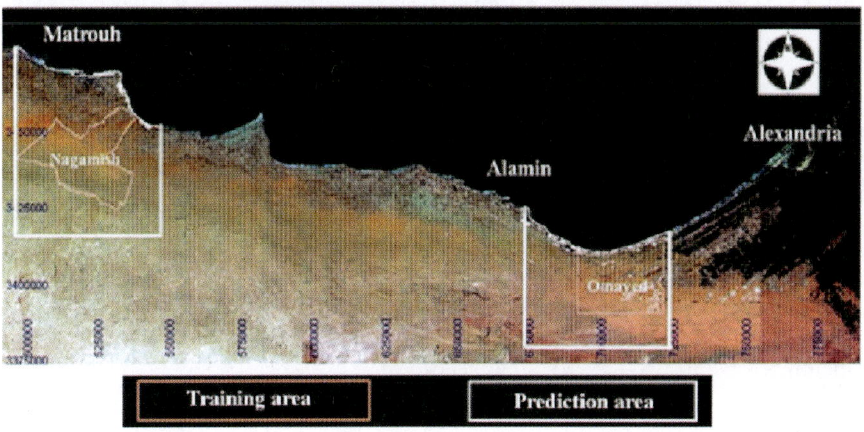

Fig. 19.1 The northwest coast and study sites: training and prediction area at Omayed-Alamin and Nagamish-Matrouh

Table 19.1 Omayed soil mapping units

Mapping units	Description	Hectares	%
MU1	Deep, sandy to sandy loam, nonsaline (Typic Haplocalcids)	12,309.57	20.68
MU2	Deep, sandy to sandy loam, moderately saline (Typic Haplogypsids)	3,606.12	6.05
MU3	Deep, sandy, nonsaline (Typic Torripsamments)	14,669.10	24.65
MU4	Complex of shallow, sandy, nonsaline to very saline (Typic Torriorthents)	18,681.84	31.39
MU5	Stony ridge	10,253.79	17.23
Total		59,520.42	100.00

Table 19.2 Naghamish-Matrouh soil mapping units

Mapping unit	Description	Hectares	%
MU1	Rocky outcrop	3,590.67	5.40
MU2	Very shallow sand to loamy sand, nonsaline to slightly saline (Lithic Torripsamments)	6,102.06	9.18
MU3	Very shallow sandy loam to loam, nonsaline to slightly saline (Lithic Torriorthents)	32,257.85	48.55
MU4	Shallow sandy to loamy sand, nonsaline to moderately saline (Lithic Torripsamments)	13,095.36	19.71
MU5	Shallow sandy loam to loam, nonsaline to moderately saline (Lithic Torriorthents)	7,174.27	10.80
MU6	Moderately deep sand to loamy sand, nonsaline to moderately saline (Typic Torripsamments)	1,132.26	1.70
MU7	Moderately deep sandy loam to loam, nonsaline to moderately saline (Typic Torriothents)	2,606.62	3.92
MU8	Deep sand to loamy sand, nonsaline (Typic Torripsamments)	348.11	0.52
MU9	Deep sandy loam, nonsaline (Typic Torriorthents)	140.25	0.21
Total		6,6447.45	100.00

19.2.2 Data Sources

19.2.2.1 Soil Data

Omayed-Alamin: Soil map of a representative area at Omayed at a grid resolution of 60 m was prepared where five soil mapping units (Fig. 19.1; Table 19.1) were identified and grouped based on the landform units (Abdel-Kader et al. 2005).

Naghamish-Matrouh: Soil map of Wadi Naghamish, east of Matrouh, at a grid resolution of 28.5 m was prepared where nine main soil mapping units (Fig. 19.1; Table 19.2) were identified (Abdel-Kader et al. 2004) based on visual interpretation of SPOT PAN and XS images and DEM as well as using field and laboratory analytical data (soil depth, texture and salinity).

19.2.2.2 Environmental Ancillary Data

Landsat image consists of six 28.5-m bands (GLFC 2006): band 1 (B, 0.45–0.52), band 2 (G, 0.53–0.61), band 3 (R, 0.63–0.69), band 4 (NIR, 0.78–0.90), band 5 (MIR, 1.55–1.75) and band 7 (MIR2, 2.09–2.35).

Digital elevation models (DEM) – DEM25-25 m DEM derived from the topographic maps at 1:25,000 scale and DEM90-90 m DEM derived from the SRTM mission (CGIAR 2005; Jarvis et al. 2006).

All data set layers were projected to 28.5 m × 28.5-m pixel resolution at Nagamish-Matrouh or 90 m × 90-m pixel resolution at Omayed-Alamin using the WGS 84 UTM 35N reference system.

19.3 Methods

19.3.1 Extraction of Land Surface Parameters

Digital image processing: Considering the pedogenesis of the areas following soil-related indices were processed using IDRISI, Clark Labs 2009 (Table 19.3).

1. Carbonate Index (CAI) = (b_3/b_2) (Roecker et al. 2009)
2. Grain Size Index (GSI) = $(b_3 - b_1)/(b_1 + b_2 + b_3)$ (Xiao et al. 2006)
3. Gypsic Index (GYI) = $(b_5 - b_7)/(b_5 + b_7)$ (Nield et al. 2007)
4. Principle components of the six ETM bands: PC1, PC2.

19.3.2 Digital Terrain Analysis

Artificial depressions were removed from the DEM using the method of Planchon and Darboux (2001). The primary and secondary terrain attributes were derived from the smoothed DEM using SAGA (Boehner et al. 2006; Bock et al. 2008). Land surface parameters relevant to the mapping objectives, study area characteristics and scale of application were considered. As the study areas under consideration are of high relief, representative DEM parameters that can explain hydrological, climatic and morphological properties of a terrain were derived and were included, such as the following: elevation, multiresolution valley bottom flatness, multiresolution ridgetop flatness index, slope gradient, catchment slope, mid-slope position, SAGA wetness index and valley depth (Table 19.3).

19.3.3 Statistical Modelling

19.3.3.1 Variable Screening

Considering the soil groups as categorical dependent variables and the terrain attributes as quantitative predictors, the multinomial logistic regression statistical

Table 19.3 Land surface parameters used for spatial prediction by the multinomial logistic regression model

Land surface parameters		Definition	Reference/source
Basic	Elevation	Height above sea level (metres)	Topo-DEM SRTM DEM
	Slope gradient (sg)	Downward gradient (percent)	SAGA
Hydrologic	SAGA Flow accumulation (fa)	Iteratively modified upslope catchment area (meter^2)	SAGA
	Catchment slope (cs)	Average gradient above flow path (percent)	Gallant and Wilson (2000) SAGA
	Mid-slope position	Calculates the extent that each point similar to a ridge or valley position as values 0 through 100	Bohner and Antonic (2009) SAGA
	Multiresolution valley bottom flatness (MRVBF)	Measure of flatness and lowness	Gallant and Dowling (2003) SAGA
	Multiresolution ridgetop flatness index (MRRTF)	Measure of flatness and upness	Gallant and Dowling (2003) SAGA
	SAGA wetness index	Ln (fa/sg)	SAGA
	Valley depth	Metres	SAGA
Multispectral Imagery Enhanced Thematic Mapper (ETM)	Bands 1,2,3,4,5,7	Path178 row 38, path179 row39 L7, July 6, 2000	GLFC (2006)
	Carbonate index	(b_3/b_2)	Roecker et al. (2009)
	Grain size index (GSI)	$(b_3 - b_1)/(b_1 + b_2 + b_3)$	Xiao et al. (2006)
	Gypsic index	$(b_5 - b_7)/(b_5 + b_7)$	Nield et al. (2007)
	Principle components of six ETM bands	PC1, PC2	

model was used. The selection of the continuous quantitative land surface parameters (Table 19.3) was made following the knowledge-driven approach (Kempen et al. 2009) based on the pedogenesis of the northwest coastal zone gained throughout several studies (Abdel-Kader et al. 2004, 2005; Abdel-Kader and Yacoub 2009). To assess the significant importance of the parameters, the raster soil categories and predictors layers were exported to SATISTCA7 (StatSoft Inc. 2007) where feature selection and variable screening module was used. A list of land surface parameters was computed from the respective chi-square values and sorted in ascending order by *p*.

19.3.3.2 Univariate Analysis

The SATISTCA file was saved as SPSS file. Univariate analysis was performed using SPSS Statistics 17.0 (SPSS Inc. 2008).

19.3.3.3 Regression Analysis

The Logistic Model

The logistic model belongs to the family of generalized linear models and is used when the response variable is a categorical variable (Clark Labs 2009). The logit is the logarithmic function of the ratio between the probabilities that a pixel is a member of a class (*P*) divided by the probability that it is not (1 − *P*). Its value can be directly predicted from the predictor values through regression as:

$$\text{Logit i} = \text{Lin}\frac{\text{Pi}}{1-\text{Pi}} = \text{a} + \text{b}_1\text{X}_1 + \text{b}_2\text{X}_2 + \ldots + \text{b}_\text{n}\text{X}_\text{n} + \varepsilon \tag{19.1}$$

The equation shows how to calculate the logit of a category, e.g. soil class *i*, predicted from a number of quantitative factors $X_1 \ldots n$, e.g. terrain attributes. The '*a*' indicates the intercept of the regression curve, the '*b*'s are the coefficients of each predictor, and ε represents random and systematic (if any) error. From this equation, it is possible to derive:

$$\text{Pi} = \frac{\text{e}^{a+b_1X_1+b_2X_2+\ldots+b_nX_n}}{1+\sum_1^{m-1}\text{e}^{a+b_1X_1+b_2X_2+\ldots+b_nX_n}} + \varepsilon \tag{19.2}$$

This equation is used to predict the probability *P* that a soil class *i* is present at a pixel given the levels of the terrain attributes X_1, X_2, …, X_n, by dividing the logit of *i* to that of the total sum of the logits of all other soil classes (except the reference category) plus unity (Menard 2000, 2002). One of the classes, often the first/last in the list, is considered as reference, and its logit is not estimated. However, its probability of existence is given as:

$$\text{Pr} = \frac{1}{1+\sum_1^{m-1}\text{e}^{a+b_1X_1+b_2X_2+\ldots+b_nX_n}} + \varepsilon \tag{19.3}$$

The values of the '*a*' and the '*b*'s have to be determined for each soil class based on empirical data, i.e. land surface parameters data. Once the values have been estimated with statistical significance, models 2 and 3 can be integrated into a GIS tool to map the probability that a given soil class *i* is found at a given pixel based on the values of the terrain attributes $X_1 \ldots n$. Equation 19.3 is used to predict the probability of the reference category.

Multinomial Logistic Regression Analysis

Multinomial logistic regression analysis was carried out using NOMREG module of SPSS Statistics 17.0 (SPSS Inc. 2008). In IDRISI, MULTILOGISTICREG was performed to analyse and visualize predicted images. To reduce the multicollinearity effect within the predictor's variables, principal component analysis was run in IDRISI, and then, instead of using the original predictors, the transformed components were used; those are absolutely independent. The principal component analysis (PCA) is an orthogonal transformation of n-dimensional image data that produces a new set of images (components) that are uncorrelated with one another and ordered with respect to the amount of variation (information) they represent from the original image set. The PCA is typically used to uncover the underlying dimensionality of multivariate data by removing redundancy (evident in intercorrelation of image pixel values as explained by Eastman 2009). Before deriving the predictive components, all input raster layers were linear stretched to the same binary scale (0–255 values) as per Hengl (2007). The multinomial logistic regression (NOMREG) module of the SPSS was chosen since the dependent variable had more than two categories. One of the soil groups, the first or last, was randomly defined as reference category. The chi-square-based maximum likelihood ratio test was used to evaluate the fitness of the overall model (Menard 2002). This statistic tests if the difference between the model prediction and the observed values is minimized enough for the model to fit. The significance of the regression coefficient of each predictor variable for each dependent variable was evaluated using the likelihood statistic.

19.3.4 Accuracy Assessment

The ERRMAT of IDRISI was used to compare the original images of soil units with the predicted ones for the purpose of accuracy assessment. The ERRMAT creates an error matrix that tabulates the different predicted classes to which original soil unit truth cells have been assigned. Output also includes producer's accuracy, user's accuracy, errors of omission and commission, an overall error measure and a kappa index of agreement (KIA).

19.3.5 Predictive Maps for Adhered Unvisited Sites

The land surface predictors for the unvisited sites were processed according to the methods of the sample sites. Based on unvisited land surface predictors, the probability models of the sample site soil groups (Eqs. 19.2 and 19.3) were fed into the raster calculator of IDRISI to produce a map showing the probability of existence at each pixel for each soil class at adhered unvisited sites. The RECLASS/MDCHOICE of IDRISI was used to produce final predictive maps for adhered unvisited sites.

19.4 Results and Discussion

19.4.1 Omayed-Alamin

19.4.1.1 The Land Surface Predictor Relations as Modelled by Logistic Regression

The overall multinomial logistic model was found to be significantly fit at $p<0.01$. Table 19.4 gives the model fitting information for estimation of probabilities of occurrence of soil mapping units at Omayed. The McFadden pseudo R-square is 0.473. The McFadden pseudo R-square measures the reduction in maximized log likelihood. It is conceptually and mathematically close to the ordinary least squares R^2 (Menard 2000). A McFadden pseudo R-square of 1 indicates a perfect fit, whereas a pseudo R-square equal to 0 indicates no relationship. Clark and Hosking (1986) comment that a McFadden pseudo R-square greater than 0.2 can be considered as a relatively good fit. Table 19.5 further shows that the most significant terrain and spectral attributes in influencing the spatial distribution of the soil classes were found to be elevation, multiresolution ridgetop flatness index (MRRTF), multiresolution valley bottom flatness (MRVBF), PC1, gypsic index, grain size index, slope, PC2, SAGA wetness index and carbonate index in descending order. In fact all of the terrain and spectral attributes were found to be significantly influential that the selection of the land surface predictors is satisfactory.

19.4.1.2 Spatial Prediction of the Soil Groups and Its Performance at Omayed

With the reference category MU5, Table 19.6 indicates multiple logistic regressions for estimation of probabilities of occurrence of soil mapping units at Omayed training area. Figure 19.2 gives the predictive soil map versus the training original soil map at Omayed at 1:90,000 scales (90-m pixel resolution).

Table 19.4 Model fitting information, pseudo R-square for estimation of probabilities of occurrence of soil mapping units at Omayed

Model fitting information				
Model	Model fitting criteria	Likelihood ratio tests		
	−2 log likelihood	Chi-square	df	Significance
Intercept only	2.197E5			
Final	1.157E5	1.040E5	40	0.000
Goodness of fit				
Pearson		1.374E7	291,324	0.000
Deviance		115668.559	291,324	1.000
Pseudo R-square				
Cox and Snell	0.760			
Nagelkerke	0.799			
McFadden	0.473			

Table 19.5 The significance of each terrain attribute in the overall model at Omayed (likelihood ratio tests)

Effect	Model fitting criteria	Likelihood ratio tests		
	−2 log likelihood of reduced model	Chi-square	df	Significance
Elevation	1.328E5	1.715E4	4	0.000
MRRTF	1.255E5	9.852E3	4	0.000
MRVBF	1.182E5	2.487E3	4	0.000
PC1	1.172E5	1.538E3	4	0.000
Gypsic index	1.170E5	1.367E3	4	0.000
Grain size Index	1.167E5	1.036E3	4	0.000
Slope	1.166E5	968.705	4	0.000
PC2	1.162E5	524.972	4	0.000
SAGA wetness index	1.162E5	542.840	4	0.000
Intercept	1.161E5	472.496	4	0.000
Carbonate index	1.159E5	205.011	4	0.000

Table 19.6 Parameter estimates of probabilities of occurrence of soil mapping units at Omayed. The reference category: MU5, df = 1

Effect	Model intercepts' *a* and logistic regression (logit) coefficients' *b*			
	MU1	MU2	MU3	MU4
Intercept	18.351**	45.856**	25.749**	21.545**
Carbonate index	−2.829	−16.309**	−7.135**	−11.347**
Grain size index	41.513**	113.982**	89.489**	74.177**
Gypsic index	42.628**	36.564**	19.423**	82.937**
PC1	−0.057**	−0.099**	−0.079**	−0.071**
PC2	−0.061**	0.000	0.029**	−0.073**
Elevation	−0.182**	−0.909**	−0.153**	−0.157**
MRVBF	0.227**	−0.240**	0.344**	0.788**
MRRTF	0.705**	0.928**	0.355**	1.120**
Slope	−0.347**	−1.041**	−0.324**	−0.463**
SAGA wetness index	0.078**	0.107**	0.054**	−0.04**

**indicates $p < 0.01$

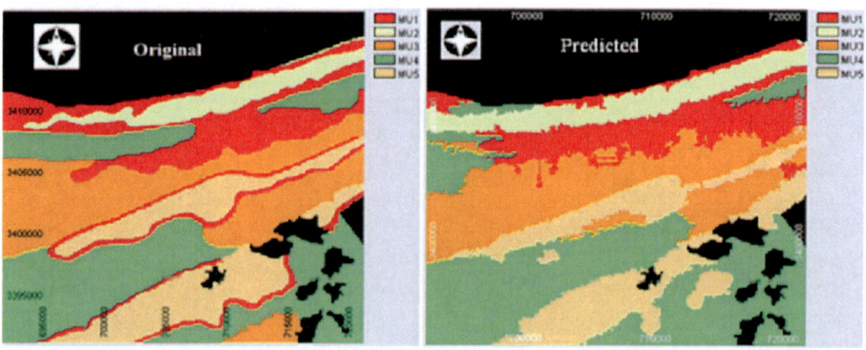

Fig. 19.2 Predicted versus original soil map of Omayed site

Table 19.7 Producer's, user's and overall accuracy and kappa index of agreement of the field-surveyed soil map with predicted soil classes at Omayed

			Kappa index of agreement (KIA)	
Category mapping units	Producer's accuracy %	User's accuracy %	Using predicted soil classes as the reference image	Using the field-surveyed soil map as the reference image
MU1	42.77	59.62	0.4909	0.3280
MU2	97.05	67.10	0.6497	0.9677
MU3	70.64	73.61	0.6498	0.6155
MU4	81.62	74.24	0.6246	0.7194
MU5	83.32	79.43	0.7515	0.7965
Overall accuracy	72.1102%			
Overall kappa	0.6360			

19.4.1.3 Evaluating the Reproducibility of the Original Soil Map at Omayed

Table 19.7 gives error matrices comparing the field-surveyed soil map (columns/truth) with multinomial predicted soil classes (rows/predicted) at Omayed training area. The overall accuracy was 72.11% with kappa index = 0.64.

19.4.1.4 Spatial Prediction of the Soil Mapping Units of Alamin Area

Figure 19.3 and Table 19.8 present the predictive map at 1:90,000 scale (90-m pixel resolution) and mapping units area for the adhered unvisited sites of Alamin area, respectively. The data indicate that Alamin region includes deep, sandy to sandy loam and nonsaline calcareous soils (Typic Haplocalcids) comprising of 12% of the area and deep, sandy to sandy loam, moderately saline gypsiferous soils (Typic Haplogypsids) comprised of 20% of the area. The deep, sandy and nonsaline soils (Typic Torripsamments) comprise of 29% of the area, whereas complex of shallow, sandy and nonsaline to very saline soils (Typic Torriorthents) comprise 25%. The stony ridges comprise 12% of the Alamin area.

19.4.2 *Naghamish-Matrouh*

19.4.2.1 The Land Surface Predictors Relations as Modelled by Logistic Regression

The overall multinomial logistic model at Naghamish was found to be significantly fit at $p < 0.01$. Table 19.9 gives model fitting information and pseudo R-square.

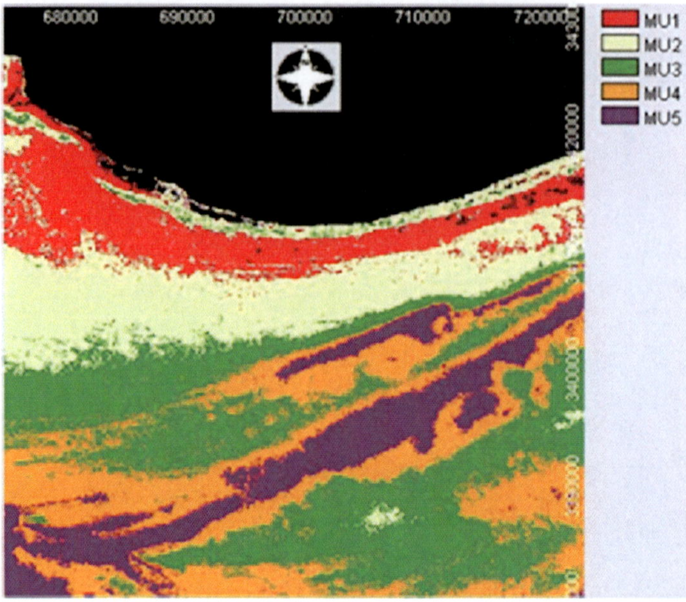

Fig. 19.3 The predicted soil map of Alamin at 90-m pixel resolution

Table 19.8 Areas of the predicted mapping units of Alamin

Mapping units	Description	Hectares	%
MU1	Deep, sandy to sandy loam, nonsaline (Typic Haplocalcids)	21,433.41	12.27
MU2	Deep, sandy to sandy loam, moderately saline (Typic Haplogypsids)	35,853.03	20.54
MU3	Deep, sandy, nonsaline (Typic Torripsamments)	50,882.58	29.15
MU4	Complex of shallow, sandy, nonsaline to very saline (Typic Torriorthents)	44,964.72	25.76
MU5	Stony ridge	21,443.13	12.28
Total		174,576.87	100.00

Table 19.9 Model fitting information, pseudo *R*-square for the estimation of probabilities of occurrence of soil mapping units at Magamish

Model fitting information				
Model	Model fitting criteria –2 log likelihood	Likelihood ratio tests Chi-square	df	Significance
Intercept only	2.494E6			
Final	1.258E6	1.235E6	104	0.000
Goodness of fit				
Pearson		5.401E30	6,541,688	0.000
Deviance		1259083.197	6,541,688	1.000
Pseudo *R*-square				
Cox and Snell	0.779			
Nagelkerke	0.818			
McFadden	0.496			

Table 19.10 The significance of each terrain attribute in the overall model in Nagamish (likelihood ratio tests)

Effect	Model fitting criteria	Likelihood ratio tests		
	−2 log likelihood of reduced model	Chi-square	df	Significance
FDEMM	1.362E6	1.043E5	8	0.000
PC2	1.314E6	5.635E4	8	0.000
WINDEX	1.311E6	5.284E4	8	0.000
VDEPTH	1.300E6	4.221E4	8	0.000
MRRTF	1.297E6	3.940E4	8	0.000
PC1	1.289E6	3.073E4	8	0.000
MSPOSITON	1.286E6	2.797E4	8	0.000
Intercept	1.282E6	2.372E4	8	0.000
GYI	1.273E6	1.491E4	8	0.000
MRVBF	1.273E6	1.520E4	8	0.000
SLOPE	1.264E6	5.801E3	8	0.000
GSI	1.261E6	2.731E3	8	0.000
CAI	1.260E6	2.365E3	8	0.000
CSLOPE	1.260E6	1.741E3	8	0.000

The chi-square statistics is the difference in −2 log likelihoods between the final model and a reduced model. The reduced model is formed by omitting an effect from the final model. The null hypothesis is that all parameters of that effect are 0

McFadden pseudo *R*-square value of 0.496 could be considered a relatively good fit (Clark and Hosking 1986).

Table 19.10 shows that the most significant terrain and spectral attributes influencing the spatial distribution of the soil classes at Nagamish were elevation (FDEM), principal component 2 (PC2), SAGA wetness index (WINDEX), valley depth (VDEPTH), multiresolution ridgetop flatness index (MRRTF), principal component 1(PC1), mid-slope position (MPOSITON), intercept, gypsic index (GYI), multiresolution valley bottom flatness (MRVBF), slope (SLOPE), grain size index (GSI) and catchment slope (CSLOPE) in descending order. These sets of best predictors are related to long-term known relationships between soil-forming factors, landform and soil distribution, which relate soil distribution to erosion processes in steep areas and with water dynamics, and water movement and accumulation in the low-depression areas (Giasson et al. 2008).

The most influential spectral attributes are the first two principle components of the six Enhanced Thematic Mapper bands. The first component accounts for brightness variation among surface samples and generally correlates well with sand content (Palacios-Oureta and Ustin 1998), while second component most often represents variation in vegetative cover (Thiam 1997). In fact all of the terrain and spectral attributes were found to be significantly influential showing that the selection of the land surface predictors is reliable and satisfactory.

19.4.2.2 Spatial Prediction of the Soil Groups and Its Performance at Nagamish

Table 19.11 presents parameter estimates for the estimation of probabilities of occurrence of soil mapping units and the significance of each terrain attribute in the overall model. The reference category is MU1. Figure 19.4 gives the predictive soil map versus the field soil map at Nagamish at 1:30,000 scale (28.5-m pixel resolution).

19.4.2.3 Evaluating the Reproducibility of the Original Soil Map at Nagamish

Table 19.12 illustrates producer's accuracy (PA,%), user's accuracy (UA, %) and kappa index of agreement of the field-surveyed soil map (columns/truth) with multinomial predicted soil classes (rows/predicted) at Nagamish training area. The overall accuracy is 73.56% with kappa index = 0.62. Overall accuracy and kappa index were considered satisfactory, as they are in the same magnitude to the values found by Giasson et al. (2008).

19.4.2.4 Spatial Prediction of the Soil Mapping Units of Matrouh Area

Figure 19.5 shows the predicted map for adhered unvisited sites of East Matrouh area. Table 19.13 presents the areas of the predicted mapping units mapped at 1:30,000 scale (28.5-m pixel resolution) for adhered unvisited sites at Matrouh area. The data indicated that Matrouh region included shallow sandy, nonsaline to moderately saline soils (Lithic Torripsamments) comprising 52% of the area and nonsaline to moderately saline, deep sandy loam to loam soil (Typic Torriothents) comprising 13% of the area. The rocky outcrops comprise 35% of Matrouh region.

19.5 Conclusions

The terrain and spectral attributes were found to be significantly influential showing that the selection of the land surface predictors was satisfactory. The model fitting information for the estimation of probabilities of occurrence of soil mapping units at study sites indicates that the McFadden pseudo R-squares ranged from 0.473 to 0.496. The most significant terrain attributes in influencing the spatial distribution of the soil classes were found to be elevation, valley depth, multiresolution ridgetop flatness index, multiresolution valley bottom flatness and SAGA wetness index. However, the most influential spectral attributes are the first two principle components of the six Enhanced Thematic Mapper bands. With specific reference categories the multiple logistic regressions estimated the probabilities of occurrence of

Table 19.11 Parameter estimates of probabilities of occurrence of soil mapping units in Nagamish. The reference category: MU1, df = 1

	Model intercepts' *a* and logistic regression (logit) coefficients' *b*							
Effect	MU2	MU3	MU4	MU5	MU6	MU7	MU8	MU9
Intercept	18.63**	5.48**	−15.5**	−28.7**	9.75**	−13.6**	−8.24**	12.05**
CAI	0.857	−3.39**	4.31**	2.95**	8.80**	−0.499	7.77**	−0.931
GSI	−1.894	−9.62**	−14.8**	−4.05**	−3.492	43.3**	−10.5**	−29.1**
GYI	−27.5**	20.5**	−53.9**	15.5**	−47.7**	13.98**	−75.3**	−16.9**
PC1	−0.006**	−0.067**	0.033**	0.059**	−0.004**	0.006**	0.028**	−0.019**
PC2	0.072**	−0.133**	0.222**	0.020**	0.247**	−0.196**	0.272	0.105**
FDEM	−0.090**	−0.049**	−0.102**	−0.138**	−0.075**	−0.027**	−0.130	−0.120**
MPOSITON	2.303**	1.72**	5.61**	−1.58**	−0.685**	−0.733**	1.015	3.875**
MRVBF	0.158**	0.404**	0.206**	0.589**	0.747**	0.014**	0.197	0.457**
MRRTF	0.670**	0.561**	1.03**	0.573**	0.692**	0.382**	0.871	0.195**
WINDEX	−0.344**	0.768**	0.959**	1.09**	−0.537**	−0.093**	0.482	0.449**
VDEPTH	0.314**	0.317**	0.268**	0.183**	0.381**	0.275**	0.330	0.260**
SLOPE	8.266**	47.4**	37.1**	32.704	8.78**	18.9**	37.066	33.22**
CSLOPE	−1.656	21.3**	−15.0**	−12.548	−4.99**	13.838	14.705	0.397**

**indicates $p < 0.01$

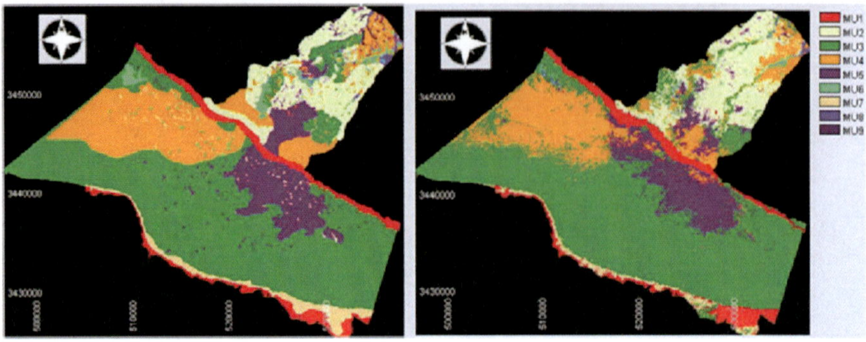

Fig. 19.4 Predictive soil map versus original soil map Nagamish

Table 19.12 Producer's, user's and overall accuracy and kappa index of agreement of the field-surveyed soil map with predicted soil classes at Nagamish

			Kappa index of agreement (KIA)	
Category mapping units	PA producer's accuracy (%)	UA user's accuracy (%)	Using the predicted soil classes as the reference image	Using the field-surveyed soil map as the reference image
MU1	76.00	68.90	0.6713	0.7465
MU2	62.00	52.00	0.4715	0.5766
MU3	88.50	83.84	0.6859	0.7650
MU4	68.00	70.23	0.6292	0.6064
MU5	53.00	58.55	0.5354	0.4842
MU6	12.00	40.02	0.3898	0.1250
MU7	31.50	62.39	0.6085	0.3018
MU8	11.67	34.27	0.3392	0.1151
MU9	3.18	19.10	0.1893	0.0315
Overall accuracy	73.5651%			
Overall kappa	0.6161			

soil mapping units at NWC sample areas. Logit equations were successfully achieved with significant intercepts and *b* values where predictive soil maps were compiled. The overall accuracy of the predicted soil maps ranged from 72 to 74% with kappa index ranged from 0.62 to 0.64. The reproducibility of the original soil maps was undependable on the map resolution of the soil and environmental data used. At Alamin-Omayed area with 90 m × 90-m map resolution, the overall accuracy was 72.11% with kappa index = 0.64. At Nagamish-Matrouh area with 28.8 m × 28.5-m map resolution, the overall accuracy was 73.56% with kappa index = 0.62. Based on the multinomial logistic regression analysis (MOREG) that uses conventional soil maps and available environmental data, this present chapter was able to predict the spatial distribution of the soil mapping units at predicted digital soil maps at pixel resolutions of 90 m × 90 m and 28.5 m × 28.5 m for the regional adherent areas at

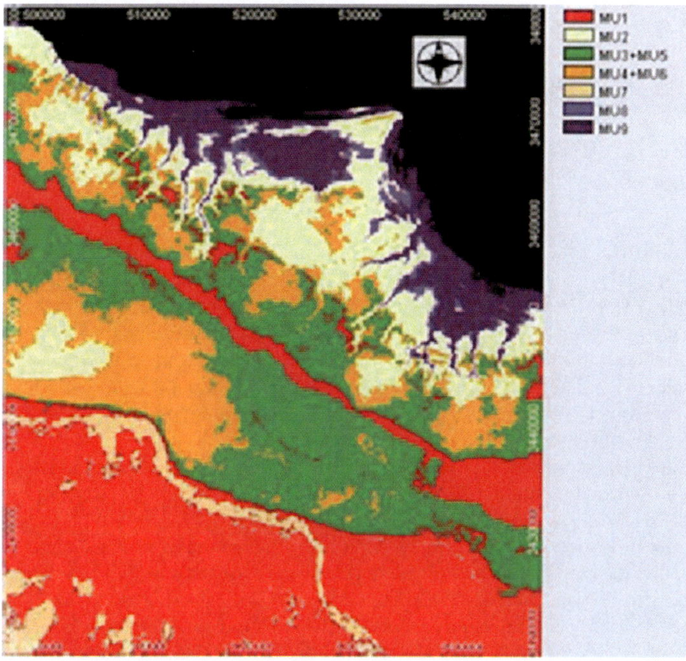

Fig. 19.5 The predicted soil map of Matrouh area at 28.5-m pixel resolution

Table 19.13 Areas of the predicted mapping units map for adhered unvisited sites of Matrouh area

Mapping units	Description	Hectares	%
MU1	Rocky outcrop	91,758.42	34.9879
MU2	Very shallow sand to loamy sand, nonsaline to slightly saline (Lithic Torripsamments)	33713.25	12.8550
MU3 + MU5	Very shallow sandy loam to loam, nonsaline to slightly saline (Lithic Torriorthents)	60,398.16	23.0301
MU4 + MU6	Shallow sandy to loamy sand, nonsaline to moderately saline (Lithic Torripsamments)	42,880.69	16.3506
MU7	Moderately deep sandy loam to loam, nonsaline to moderately saline (Typic Torriothents)	9,970.59	3.8018
MU8	Deep sand to loamy sand, nonsaline (Typic Torripsamments)	5,411.83	2.0636
MU9	Deep sandy loam, nonsaline (Typic Torriorthents)	18,124.67	6.9110
Total		262,257.60	100.0000

Alamien and Matrouh. It is evident that the developed methodology could contribute to the allocation and to the digital mapping and management of new potential expansion sites in the remote desert areas of Egypt.

References

Abdel-Kader FH (2009) Northwest Coast of Egypt. Unpublished data. Department of Soil and Water Sciences, Faculty of Agriculture, University of Alexandria, Egypt

Abdel-Kader FH, Bahnassy M (2001) Digital soil map of Egypt at scale 1:1,000,000. Presented at the first meeting on work progress of the Eastern Mediterranean soil mapping at 1:1 million scale. Lebanese National Council for Scientific Research, Beirut, pp 15–18

Abdel-Kader FH, Yacoub RK (2009) Assessment of land and water resources potentials using RS/GIS analysis and hydrological model for rainfed agriculture, Northwest Coast, Egypt. Egypt J Soil Sci 49(3):319–340

Abdel-Kader FH, Fitzsimons J, Bahnassy M, Moustafa A (2004) Challenges in resource management in rainfed agriculture in Wadi Maghaish, North Western Coastal Region, Egypt. Egypt J Desert Res 54(2):237–258

Abdel-Kader FH, Abdel-Hady AM, Saleh AM (2005) Integration of terrain, thematic mapper, and soil quality data for surface hydrology, water erosion and land suitability modeling: Northwest Coast, Egypt. Egypt J Soil Sci 45(2):225–238

Ahmed AS (1995) Egypt, the country reports. In: Zink JA (ed) Soil survey: perspective and strategies for the 21st century. World soil resources report 80. ITC, FAO, Rome

Bock M, Bohner J, Conrad O, Kothe R, Ringeler A (2008) SAGA: system for the automated geoscientific analysis. Department of Physical Geography, Hamburg, Germany. http://www.saga-gis.org

Boehner J, McCloy KR, Strobl J (eds) (2006) SAGA-analysis and modelling applications. Goettinger Geographische Abhandlungen, Heft 115. Verlag Erich Goltze GmbH, Göttingen, p 117

Bohner J, Antonic O (2009) Land-surface parameters specific to topo-climatology. In: Hengl T, Reuter HI (eds) Geomorphometry: concepts, software, applications. Elsevier, Amsterdam, pp 195–226

CGIAR (2005) Consortium for spatial information. http://srtm.csi.cgiar.org/S

Clark WA, Hosking PL (1986) Statistical methods for geographers (chapter 13). Wiley, New York

Clark Labs (2009) IDRISI Taiga. Clark University, Worcester

Eastman JR (2009) IDRISI Taiga: guide to GIS and image processing. Clark Labs, Clark University, Worcester

FAO (1964) High dam soil survey project United Arab Republic, vols I, II, III. FAO/SF:16/U.A.R

FAO (1970) Pre investment survey of the northwestern coastal region. ESE:SF/UAR 49

FAO (2003) Multipurpose Africover Databases on Environmental resources (MADEFAO 2003). http://www.africover.org/system/user/

FAO-UNESCO (1977) FAO-UNESCO soil map of the World, vol VI, Africa. UNESCO, Paris, p 299

Gallant JC, Dowling TI (2003) A multiresolution index of valley bottom flatness for mapping depositional areas. Water Resour Res 39:1347–1370

Gallant JC, Wilson JP (2000) Primary topographic attributes. In: Wilson JP, Gallant JC (eds) Terrain analysis: principles and applications. Wiley, New York, pp 51–86

GARPAD (1986) Land master plan. Joint project. Kingdom of the Netherlands Ministry of Foreign Affairs, Directorate General for International Co-operation and Arab Republic of Egypt, Ministry of Development, General Authority for Rehabilitation Projects and Agricultural Development (GARPAD)

Giasson E, Figueiredo SR, Tornquist CG, Clarke RT (2008) Digital soil mapping using logistic regression on terrain parameters for several ecological regions in Southern Brazil (chapter 19). In: Hartemink AE et al (eds) Digital soil mapping with limited data. Springer, Dordrecht, pp 225–232

GLFC (2006) Global landcover facility. University of Maryland, College Park. http://glcf.umiacs.umd.edu/index.shtml

Hamdy H (1982) Soil map of Egypt. Final report, Academy of Scientific Research and Technology-ASRT, Cairo, Egypt

Hengl T (2007) A practical guide to geostatistical mapping of environmental variables. EUR 22904 EN ISBN 978-92-79-06904-8. Office for Official Publications of the European Communities, Luxembourg

Jarvis A, Reuter HI, Nelson A, Guevara E (2006) Hole-filled SRTM for the globe. Version 3. Available from the CGIAR-CSI SRTM 90m Database. http://srtm.csi.cgiar.org

Kempen B, Brus DJ, Heuvelink GBM, Stoorvogel JJ (2009) Updating the 1:50,000 Dutch soil map using legacy soil data: a multinomial logistic regression approach. Geoderma 151:311–326

Lagacherie P, McBratney AB, Voltz M (eds) (2006) Digital soil mapping—an introductory perspective, vol 31, Developments in soil science. Elsevier, Amsterdam

Menard SS (2000) Coefficients of determination for multiple logistic regression analysis. Am Statist 54:17–24

Menard SS (2002) Applied logistic regression analysis. Quantitative applications in the social sciences. Sage, Thousand Oaks

Nield SJ, Boettinger JL, Ramsey RD (2007) Digital mapping of gypsic and nitric soil areas using Landsat ETM data. Soil Sci Soc Am J 71:245–252

Palacios-Oureta A, Ustin SL (1998) Remote sensing of soil properties in Santa Monica Mountains: I. Spectral analysis. Remote Sens Environ 65:170–183

Planchon O, Darboux F (2001) A fast, simple and versatile algorithm to fill the depressions of digital elevation models. Catena 46:159–176

Roecker SM, Howell D, Haydu-Houdeshell CA, Blinn C (2009) A qualitative comparison of conventional soil survey and digital soil mapping approaches. https://sharepoint.ngdc.wvu.edu/sites/...soils/.../Roecker_etal2009_Ch29.pdf

SPSS Inc (2008) SPSS statistics 17.0 for Windows. SPSS Inc., Chicago

StatSoft Inc (2007) STATISTICA 8.0. StatSoft, Inc., Tulso, 1984–2007

Thiam A (1997) Geographic information and remote sensing systems methods for assessing and monitoring land degradation in the Sahel region: the case of Southern Mauritania. Dissertation, Clark University, Worcester

Xiao J, Shen Y, Tateishi R, Bayaer W (2006) Development of topsoil grain size index for monitoring desertification in arid land using remote sensing. Int J Remote Sens 27:2411–2422

Chapter 20
Studies on the Micromorphology of Salt-Affected Soils from El-Fayoum Depression, Egypt

Tolba S. Abdel Aal and A.M. Ibrahim

Abstract Salt-affected soils exhibit considerable areas in Egypt. Salt accumulation affects soil's biological, physical, chemical, and micromorphological characteristics. Present study focuses on the identification of micromorphological features related to salinization processes in El-Fayoum area. The salts accumulate as surface crust, subsoil accumulation, and random salt distribution in the profile. The studied soils belong to three soil orders: Aridisols (Haplic Natrargids, Typic Haplosalids, and Typic Haplocalcids), Entisols (Vertic Torrifluents, Typic Torrifluvents, and Typic Torripsamments), and Vertisols (Typic Haplotorrerts). The thin section study under the polarizing microscope revealed spongy microstructure due to surface salt accumulation, occurrence of halite and gypsum crystals in voids and s-matrix, and medium to coarse calcified shell fragments distributed throughout the soil profiles, especially adjacent to saline Qaroun Lake. Other micromorphological features identified are subangular blocky microstructure in the Btz horizon of saline-sodic soil-presenting natric horizon; clay papules with high level of clay orientations in the form of sepic plasmic fabric; false micro-aggregates formed due to high level of salts, precipitated clay particles under sodic environment, which have a natric horizon in the presence of $CaCO_3$ more than 13%; and humus accumulations as highly humified debris and patches randomly distributed in s-matrix, especially in the topsoil.

Keywords Egypt • El-Fayoum depression • Micromorphology • Saline irrigation water • Salt-affected soils

---

T.S. Abdel Aal (✉) • A.M. Ibrahim
Soil and Water Department, Faculty of Agriculture,
El-Fayoum University, Al Fayoum, Egypt
e-mail: tsa00@fayoum.edu.eg

S.A. Shahid et al. (eds.), *Developments in Soil Classification, Land Use Planning and Policy Implications: Innovative Thinking of Soil Inventory for Land Use Planning and Management of Land Resources*, DOI 10.1007/978-94-007-5332-7_20,

20.1 Introduction

Extensive areas in the arid and semiarid regions have become marginal for cultivation due to salinization, the reasons being the fluctuation of shallow saline ground water, poor water management, inadequate drainage systems, and seepage from the systems (Balba 1984). More than 23% of agricultural area in Egypt has become salt affected (Salem 2002); in the El-Fayoum depression, its extent is 9.9% at Sennures and Ibshway districts (Abdel-Motaleb 1997; Abdelfattah 2002). El-Fayoum depression is a closed triangular-shaped basin developed on the Eocene limestone plateau at about 100 km south of Cairo (Fig. 20.1). The soils of El-Fayoum depression slope generally occur from 35 m above sea level at El-Lahoun (southeast) and toward Lake Qaroun at 45m below sea level (northwest) and cover about 380,000 feddan. It has characteristics of a desert having no outlet, while it joined the Nile River through the natural gap of El-Lahoun via the ancient Bahr Youssef canal. So, the composition of different soil deposits belongs to many different sources, that is, Nile fluvial, lacustrine of Qaroun Lake, fluvio lacustrine, old lacustrine shorelines, and eolian desertic deposits (Abdel-Aal 1990).

The vital problem at El-Fayoum depression is the irrigation and drainage system; for example, with the increase of irrigation discharge, the water level of Qaroun saline lake rises and submerges the neighboring cultivated areas. The discharge of Bahr Youssef canal is about 2.3×10^{9} m^{3} $year^{-1}$ which is distributed in controlled system of clear overall weirs in order to keep the water slope in the main canals relevant to land level without water erosion, besides the reuse of drainage water

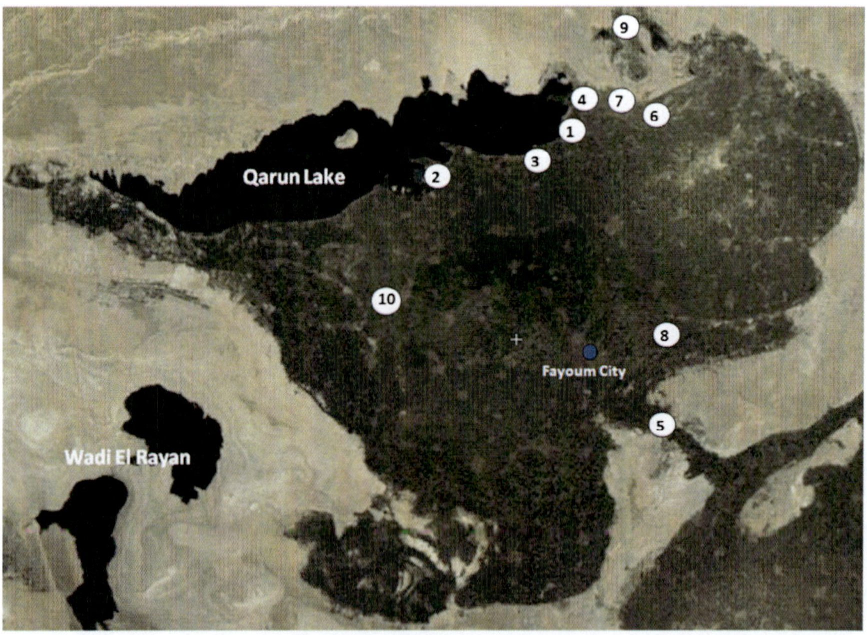

Fig. 20.1 Site location map of soil profiles (El-Fayoum depression)

and recycling of sewage effluent in irrigation. The freshwater salinity values of Bahr Youssef and its secondary canals vary between 262 and 845 mg l^{-1}, (EC_w 0.41–1.32 dS m^{-1}) with SAR of about 1.56 (Farrag 2003). It exceeds the salinity of Nile freshwater which reaches 212–320 mg l^{-1} (EC_w 0.33–0.50 dS m^{-1}) with SAR of about 1.56 (Hegazi 1999). The secondary sources of irrigation water at El-Fayoum are the reuse of drainage water and recycling of sewage effluent water, with or without mixing with fresh canal water. The amount of total drainage water exceeds 714×10^6 m^3 $year^{-1}$ with salinity value of EC_w 0.81–3.13 dS m^{-1} and SAR of about 18.31, while the salinity of sewage effluent varied between EC_w 1.49 and 2.49 dS m^{-1} with SAR of about 8.9 (Hegazi 1999).

In the irrigated agricultural areas, many of the salinity problems are associated with the shallow water table (Ayers and Westcot 1985). A progressive increase in salt accumulation at surface is due to capillary movement and subsequent evaporation of saline ground water when ground water level is above the critical depth (Kisra 1980). According to Balba (1984) and Selassie et al. (1992), the critical depth of water table in heavy soils is closed to 110 cm, whereas other workers (Hensly et al. 1997; Hensly 2000) considered it to be about 60 cm in light (sandy) soils. The quantity of salts accumulated in the topsoil depends on the quality of ground water (ASCE 1996) and rate of evaporation. Low rainfall and high evapotranspiration restricts complete leaching and transportation of salts to lower depths (Ramadan 1986). The misuse of marginal-quality waters for irrigation leads to soil salinization and degradation of soil structure (Ben-Hur et al. 1998). The salt-affected soils show considerable diversity in their biological, physical, chemical, mineralogical, and micromorphological characteristics compared to nonsaline soils. High sodium on soil exchange complex causes sodic properties with consequent degradation of soil structure usually causing crusted surfaces, dispersion of clay and organic matter and their translocation to lower depths develop prismatic, and/or columnar structure in the B horizon (Arshad and Pawluk 1966; Ivanava and Bol'ShaKov 1972). The precipitation of crystalline halite in voids is a common pedofeature of salt-affected soils (Kaewmano 2003; Shahid et al. 1992; Eswaran et al. 1980). Lebedeva-Verba et al. (2009) and Lebedeva-Verba and Gerasimova (2009) showed water table rise to enhance the processes of humus accumulation in topsoil, rearrangement of clay coatings into micromass b-fabric in the natric horizon, and the development of pseudosand fabric, calcite, and gypsum formation in the lower part of the natric horizon.

Owing to the importance of salt-affected soils in irrigated agriculture and the development of special morphological features, the present study was carried out to look into an insight of micromorphological features developed through salinization patters in El-Fayoum area for better understanding and management of these soils.

20.2 Materials and Methods

A field survey was conducted on the salt-affected soils in El-Fayoum depression to distinguish different patterns of soil salinization. Ten soil profiles representing the area were selected (Fig. 20.1) showing different salt distribution patterns. Three to four soil profiles were selected representing each salinization pattern (difference in salinity levels, parent materials, and soil texture classes). The selected soil profiles were described using USDA specifications (Soil Survey Division Staff 2003) and soil classified according to USDA soil taxonomy (Soil Survey Staff 2006). Disturbed soil samples were collected for routine physical and chemical analyses (ECe, pHs, soluble anions, soluble cations, CEC, $CaCO_3$ equivalents, soil texture and exchangeable sodium percentage, organic matter) using the methods as described by Klute (1986) and Page et al. (1982). Undisturbed soil samples were collected in Kubiena boxes for thin section preparation. These samples were collected from different layers of four soil profiles (2, 3, 7, and 9) representing different patterns of salinization processes and texture classes. Bulk density was measured by collecting undisturbed samples in standard cores, following oven drying and weighing, using standard calculation procedure. Ksat was measured by using Guelph permeameter.

The undisturbed soil samples were impregnated with mixture of resin, acetone, and accelerator. After the resin had fully polymerized, the impregnated soil blocks were chopped into standard size using diamond saw. One side of the soil block was polished with aluminum oxide grit (6 μm) and then with diamond pastes (3 and 6 μm) and washed with petroleum spirit. The polished side was stuck to a clean glass slide and put under pressure. Final thin section (25–30 μm) was prepared using lapping and polishing machine observing quartz grains to gray birefringent color under crossed polarized light. The thin section preparation methods are very well described by various workers (Abdel-Hamid 1973; FitzPatrick 1984). Soil thin sections were systematically described (Brower and Sleemen 1969; Bullock et al. 1985; Stoops 2003).

20.3 Results and Discussion

20.3.1 Field Morphological Description

The results from field morphological description of ten soil profiles are presented in Table 20.1. Table 20.1 illustrates that profiles 1, 2, and 3 present surface salt crust of different thickness. Calcium carbonate accumulation is found in the form of concretions, nodules, and lithic contact with calcareous parent material. The light colors accompanying the salt accumulations at the surface of profiles 1, 2, and 3 are dominated with a value more than 5 and a chroma more than 3, through light-textured layers of profiles 4, 6, and 8. While dark colors with value 3 and chroma 1 coupled with the decomposed organic matter are characterized in soil profiles 5, 9, and 10, besides to the natric horizon of profile 3. Dark brown or reddish brown mottles were recorded at the deepest layer of profile 2 (C3gz).

Table 20.1 Morphological description of soil profiles

Profile no.	Soil parent material	Slope gradient	Horizon	Soil depth (cm)	Munsell soil color (dry)	Modified texture class	Soil structure	Soil consistency	Boundary	Field pedological features	Soil taxonomy
1	Fluviolacustrine	Gently sloping	Bzm	0–30	10YR 6/3	C	Mas	Very hard	Clear	Salt crust, shell fragments	Typic Haplosalids
			C1z	30–60	10YR 5/2	C	Cstsbk	Very hard	Diffuse	$CaCO_3$ nodules, decomposed OM	
			C2z	60–80	10YR 4/3	C	Cstsbk	Very hard	–	–	
							Ground water table				
2	Recent lacustrine	Gently sloping	Bzm	0–10	5YR 5/3	LS	Sg	Loose	Clear	$CaCO_3$ nodules, salt crust	Typic Haplosalids
			C1kz	10–20	5YR 6/1	LS	Sg	Loose	Diffuse	Lime concertinos, decomposed	
			C2z	20–40	10YR 6/1	CS	Mas	Slightly hard	Clear	Shell fragments, lime nodules	
			C3gz	40–90	10YR 5/4	CS	Mmsbk	Hard	–	Mottled spots	
							Ground water table				
3	Fluviolacustrine	Almost flat	Bzm	0–5	10YR 4/2	C	Mas	Very hard	Clear	Fine nodules of $CaCO_3$, salt crust	Haplic Natrargids
			Apz	5–20	1 10YR 4/3	C	Mas	Very hard	Clear	Very fine root hair	
			Btz	20–50	10YR 3/1	C	Cstsbk	Very hard	Sharp	Decomposed OM	
			C1z	50–80	10YR 4/3	C	Mmsbk	Very hard	Sharp	Shell fragments, $CaCO_3$ nodules	
			C2z	80–100	10YR 5/3	C	Mas	Very hard	–	–	

(continued)

Table 20.1 (continued)

Profile no.	Soil parent material	Slope gradient	Horizon	Soil depth (cm)	Munsell soil color (dry)	Modified texture class	Soil structure	Soil consistency	Boundary	Field pedological features	Soil taxonomy
							Ground water table				
4	Old lacustrine	Gently sloping	Apz	0–30	10YR 6/3	SL	Mas	Friable	Diffuse	Nodules of $CaCO_3$	Typic Haplosalids
			C1	30–50	2.5YR 6/2	SL	Mas	Friable	Diffuse	Fine root hairs	
			C2	50–80	10YR 6/3	SL	Mas	Friable	–	Nodules of $CaCO_3$	
							Ground water table				
5	Fluvial	Gently sloping	Ap	0–25	10YR 5/2	CL	Mas	Hard	Diffuse	Common worm and root channels	Vertic Torrifluvents
			C1	25–60	10YR 5/3	CL	Msbk	Hard	Clear	Organic matter decomposed	
			C2	60–100	10YR 5/4	CL	Msbk	Hard	Clear	–	
			C3	100–150	10YR 5/3	CL	Mas	Hard	–	–	
6	Fluviolacustrine	Level	Ap	0–20	10YR 7/4	LS	Mas	Friable	Sharp	Root hairs	Typic Torripsamments
			C1	20–80	10YR 5/2	LS	Mas	Loose	Diffuse	–	
			C2	80–120	10YR 5/2	LS	Mas	Loose	–	–	
7	Fluviodesertic	Nearly level	Ap	0–20	10YR 5/2	C	Mas	Hard	Diffuse	$CaCO_3$ nodules	Typic Haplocalcids
			C1	20–80	10YR 6/2	C	Wsbk	Hard	Clear	Root, OM, fine gypsum veins	
			C2	80–120	2.5YR 7/4	C	Wsbk	Hard	–	Segregations of $CaCO_3$	
						Lithic contact limestone basement rock					

8	Eolian desertic	Gently sloping	Ap	0–20	10YR 6/4	S	Sg	Friable	Diffuse	$CaCO_3$ nodules	Typic Haplocalcids
			C1k	20–60	10YR 6/2	SL	Mas	Friable	Clear	Concretions and nodules of $CaCO_3$	
			C2	60–100	10YR 8/3	SL	Mas	Friable	–	–	
9	Fluviolacustrine	Gently sloping	Ap	0–20	2.5YR 6/2	C	Mas	Hard	Smooth	Carbonate concertinos	Typic Haplotorrerts
			C1	20–60	2.5YR 5/2	C	Msbk	Hard	Diffuse	Common slickensides	
			C2	60–100	2.5YR 6/3	C	Msbk	Hard	Diffuse	–	
			C3	100–150	2.5YR 6/2	C	Msbk	Hard	–	–	
10	Fluvial	Nearly flat	Ap	0–25	10YR 4/2	CL	Mas	Hard	Diffuse	$CaCO_3$ nodules	Typic Torrifluvents
			C1	25–60	10YR 4/3	CL	Msbk	Hard	Clear	–	
			C2	60–100	10YR 5/4	CL	Msbk	Hard	Clear	–	
			C3	100–150	10YR 5/4	CL	Msbk	Hard	–	–	

Texture class: *S* sand, *LS* loamy sand, *SL* sandy loam, *SCL* sandy clay loam, *CL* clay loam, *SiC* silt clay, *C* clay
Soil structure: *Sg* single grain, *Mas* massive, *Mmsbk* medium moderate subangular blocky, *Cstsbk* coarse strong subangular blocky, *Cstabk* coarse strong angular blocky, *OM* organic matter

20.3.2 Physical and Chemical Characteristics

The soil profiles are arranged into three groups based on difference in soil texture and to represent three salt distribution patterns (Tables 20.2 and 20.3).

The first group includes four soil profiles where shallow (80–100 cm) water table was observed, two are clayey in texture (profiles 1 and 3). The other two profiles present loamy sand and sandy loam texture (profiles 2 and 4). Increasing and decreasing trends of clay and ECe in term of soil depth are shown in Fig. 20.2a and b, respectively. The salinity is decreasing with depth, whereas clay is increasing with depth, except in profile 3. The increasing trend of salinity (ECe) toward surface reveals net upward flux of ground water and subsequent evaporation (Kisra 1980) and the increase of clay with depth due to the illuviation process (Table 20.1). Organic matter in general is less than 1%, and $CaCO_3$ equivalents are generally between 10 and 20 %. The ionic composition in general follows decreasing trend from Na>Mg>Ca>K. The concentration of soluble magnesium reached many folds as that of soluble calcium. This abnormal pattern of soluble magnesium is rendered to the submerging and seepage effects of Qaroun Lake water upon the more contact areas (recent lacustrine deposits). Soluble anion distribution pattern takes an order of $Cl>SO_4>HCO_3>CO_3$. The ECe (salinity values), clay content, and ESP values are high enough to meet natric horizon requirement (profile 3) and salic horizon (profiles 1, 2, and 4). This group of soils is classified as Typic Haplosalids (profiles 1, 2, and 4) and Haplic Natrargids (profile 3).

The second group included three soil profiles. Their soil texture varied from clay loam, loamy sand to clay (profiles 5, 6, and 7). The increasing and decreasing trend of clay and ECe is shown in Fig. 20.2c and d, respectively. This group of soil profiles presents increasing trend of soil salinity in subsoil revealing the net downward movement of saline solutions. These soils are deep (more than 150 cm) and do not present ground water table; however, a lithic contact was observed at 100 cm depth in profile 7. These soils are presently under cultivation and are located at the end of secondary irrigation canals, therefore receive less water for irrigation. From an ongoing discussion, it can be firmly explained that insufficient water to remove soluble salts from soil profiles is the cause of differential salt accumulation at various depths coincided with high evaporation rate in these arid regions (Ramadan 1986). The soluble ions arranged in the customize trend of Na>Ca>Mg>K and anions $Cl>SO_4>HCO_3$. These soils are classified as Vertic Torrifluvents, Typic Torripsamments, and Typic Haplocalcids (Table 20.1).

The soils of the third group (profiles 8, 9, and 10) are characterized by texture grades of sandy loam, clay, and clay loam textures. The increasing and decreasing trends of clay and ECe (soil salinity) are shown in Fig. 20.2e and f. These soils are irrigated with drainage water (profiles 8 and 9) and sewage effluent (profile 10). The drainage system is nearly sufficient, but the major problem is the salinity of irrigation water. Salts distributed within soil profiles trend similarly, whereas soil texture classes were in different grades. The trend of soluble cations and anions follows the trend in drainage water (Na>Mg>Ca>K and $Cl>SO_4>HCO_3$). The soil profiles 8, 9, and 10 are classified as Typic Haplocalcids, Typic Haplotorrerts, and Typic Torrifluvents, respectively.

Table 20.2 Selected physical characteristics of soil profiles

Profile no.	Depth (cm)	Particle size distribution: Coarse sand (%)	Fine sand (%)	Silt (%)	Clay (%)	Texture class	Bulk density (g cm^{-3})	Particle density (g cm^{-3})	Porosity (%)	Ksat (cm h^{-1})	$CaCO_3$ eq. %	Organ. matter %
Group 1: Salt distribution pattern number 1												
1	0–30	2.3	5.2	22.4	70.1	C	1.35	2.66	49.3	0.23	12.1	0.87
	30–60	1.9	6.6	19.7	71.8	C	1.38	2.66	48.1	0.18	15.3	0.67
	60–80	1.7	5.2	12.2	80.9	C	1.41	2.67	47.2	0.06	15.4	0.25
2	0–10	15.8	70.1	4.4	9.7	LS	1.53	2.64	41.6	3.90	15.7	1.80
	10–20	13.5	75.6	2.5	8.4	S	1.61	2.65	39.2	4.30	20.2	0.80
	20–40	21.6	41.5	9.1	27.8	SCL	1.42	2.65	46.4	2.10	25.8	0.40
	40–90	5.4	44.3	11.3	39.0	SC	1.39	2.66	47.7	1.10	12.2	0.30
3	0–5	2.4	8.3	3 30.6	58.7	C	1.27	2.63	51.7	0.31	13.5	0.71
	5–20	3.2	10.7	26.1	60.0	C	1.29	2.66	51.5	0.04	13.8	1.06
	20–50	1.9	6.3	24.1	67.7	C	1.32	2.67	50.6	0.05	14.4	0.74
	50–80	2.6	18.8	23.4	55.2	C	1.49	2.67	44.2	0.01	14.4	0.55
	80–100	4.0	19.5	21.7	54.8	C	1.46	2.66	45.1	0.01	13.0	0.21
4	0–30	19.3	46.2	19.3	15.2	SL	1.52	2.64	42.4	2.19	10.2	0.90
	30–50	16.4	48.6	18.9	16.1	SL	1.53	2.64	42.0	1.89	7.4	0.65
	50–80	12.7	46.9	16.3	24.1	SCL	1.59	2.65	40.0	0.10	11.2	0.32
Group 2: Salt distribution pattern number 2												
5	0–25	2.5	29.5	29.7	38.3	CL	1.41	2.63	46.4	1.60	6.8	1.03
	25–60	1.8	35.4	27.4	35.4	CL	1.43	2.65	46.0	0.80	5.3	0.76
	60–100	2.1	33.1	26.9	37.9	CL	1.44	2.65	45.7	0.70	4.5	0.48
	100–150	1.3	34.6	25.8	38.3	CL	1.46	2.64	45.1	0.50	3.2	0.59
6	0–20	40.3	37.8	9.2	12.7	LS	1.51	2.64	2.8	4.50	3.5	0.95
	20–80	40.1	35.3	10.9	13.7	LS	1.59	2.66	40.2	4.30	3.2	0.52
	80–120	43.6	32.1	9.1	15.2	LS	1.60	2.66	39.9	1.80	3.3	0.38

(continued)

Table 20.2 (continued)

Profile no.	Depth (cm)	Particle size distribution: Coarse sand (%)	Fine sand (%)	Silt (%)	Clay (%)	Texture class	Bulk density (g cm^{-3})	Particle density (g cm^{-3})	Porosity (%)	Ksat (cm h^{-1})	$CaCO_3$ eq. %	Organ. matter %
7	0–20	7.4	24.9	11.6	56.1	C	1.26	2.65	52.5	0.16	12.4	1.54
	20–80	2.2	27.1	11.8	58.9	C	1.30	2.64	50.9	0.09	8.7	0.79
	80–100	2.3	11.2	41.7	44.8	SiC	1.36	2.65	48.7	0.03	63.1	0.38
Group 3: Salt distribution pattern number 3												
8	0–20	52.1	37.2	3.1	7.6	S	1.55	2.63	41.1	5.10	9.4	1.29
	20–60	45.8	31.9	12.2	10.1	SL	1.58	2.64	40.2	2.50	15.8	1.13
	60–100	48.8	30.3	10.4	10.5	SL	1.58	2.64	40.2	3.00	6.1	0.26
9	0–20	6.5	41.1	6.9	45.5	C	1.53	2.64	42.3	0.93	9.9	1.33
	20–60	1.2	37.4	16.6	44.8	C	1.52	2.64	42.6	0.39	0.6	0.68
	60–100	1.5	37.9	9.8	50.8	C	1.50	2.64	43.4	0.20	0.4	0.51
	100–150	0.5	31.8	8.1	59.6	C	1.41	2.65	46.8	0.07	1.2	0.34
10	0–25	6.4	23.2	36.7	33.7	CL	1.43	2.64	45.8	1.22	11.2	0.98
	25–60	7.7	25.1	31.6	35.6	CL	1.44	2.65	45.7	1.06	9.3	0.75
	60–100	4.2	26.4	32.8	36.6	CL	1.46	2.66	45.1	0.63	8.6	0.31
	100–150	7.3	25.7	30.9	36.1	CL	1.48	2.66	44.4	0.37	7.2	0.25

S sand, *LS* loamy sand, *SL* sandy loam, *SCL* sandy clay loam, *SC* sandy clay, *CL* clay loam, *SC* silt clay, *C* clay

Table 20.3 Selected chemical characteristics of soil profiles

Profile no.	Depth (cm)	pHs	ECe (dS m^{-1})	Soluble cations (meq l^{-1}) Ca^{2+}	Mg^{2+}	Na^{+}	K^{+}	Soluble anions (meq l^{-1}) CO_3^{2-}	HCO_3^{2-}	Cl^{-}	SO_4^{2-}	CEC (cmol (c) kg^{-1})	ESP
Group 1: Salt distribution pattern number 1													
1	0–30	8.38	194.7	46	505	1,413	8	0.3	4.2	1,573	393	41.2	26
	30–60	8.46	61.5	33	118	478	2	0.2	3.2	455	176	40.8	24
	60–80	8.29	30.9	25	71	200	1	0.2	2.1	143	152	42.4	25
2	0–10	7.95	162.5	25	417	1,178	18	–	18.5	1,340	280	6.8	14
	10–20	7.85	87.7	22	213	637	9	–	10.3	668	202	5.8	15
	20–40	8.20	68.1	19	159	492	6	–	6.6	519	150	23.3	15
	40–90	8.10	47.8	15	137	323	6	–	4.3	345	132	31.7	15
3	0–5	8.73	55.3	32	77	348	4	0.2	2.9	362	96	40.9	20
	5–20	8.60	47.6	25	69	370	3	0.2	2.7	258	206	38.7	18
	20–50	8.69	30.1	22	28	175	1	0.2	2.5	148	74	40.2	28
	50–80	8.17	23.7	10	76	145	2	0.2	2.2	127	105	40.8	18
	80–100	8.55	26.9	22	61	141	2	0.2	3.7	63	160	39.6	19
4	0–30	8.51	57.7	39	82	460	2	0.2	3.5	444	139	11.3	23
	30–50	8.29	24.0	18	28	179	2	–	2.5	75	150	12.5	18
	50–80	8.08	16.7	18	15	130	1	–	2.7	47	114	13.7	19
Group 2: Salt distribution pattern number 2													
5	0–25	7.92	5.8	11	9	37	1		3.3	32	23	25.7	18
	20–60	7.96	7.1	12	9	48	1		3.3	40	28	24.8	19
	60–100	8.05	8.2	16	13	53	1		3.4	50	28	28.1	22
	100–150	8.11	8.3	16	13	55	1		3.5	53	27	35.9	18

(continued)

Table 20.3 (continued)

Profile no.	Depth (cm)	pHs	ECe (dS m^{-1})	Soluble cations (meq l^{-1}) Ca^{2+}	Mg^{2+}	Na^{+}	K^{+}	Soluble anions (meq l^{-1}) CO_3^{2-}	HCO_3^{2-}	Cl^{-}	SO_4^{2-}	CEC (cmol (c) kg^{-1})	ESP
6	0–20	7.88	9.4	19	11	64	2		3.1	38	54	11.4	15
	20–80	7.89	10.7	20	17	71	1		3.3	42	64	10.8	16
	80–120	7.94	14.8	19	13	116	3		3.4	80	66	11.6	13
7	0–20	8.04	4.9	9	8	31	1		3.1	20	26	41.0	14
	20–80	8.21	8.6	20	15	53	2		2.9	38	49	47.3	16
	80–100	8.49	11.6	24	19	75	2		1.5	68	50	37.9	24
8	0–20	7.89	11.1	17	18	75	1.1	–	3.8	64	43	5.6	19
	20–60	8.02	12.5	19	20	85	1.1	–	3.2	75	48	6.4	17
	60–100	8.21	12.6	30	19	78	1.0	–	5.7	83	39	7.4	16
9	0–20	8.48	11.8	14	26	76	3.1	–	1.5	60	58	16.1	26
	20–60	8.51	11.6	20	26	70	2.5	–	1.5	50	67	27.7	23
	60–100	8.47	11.6	20	28	68	2.5	–	2.0	52	64	30.6	21
	100–150	8.37	11.7	21	29	67	2.5	–	2.0	56	62	38.3	23
10	0–25	8.01	7.1	12	10	10	1.3	–	3.3	44	24	27.3	20
	25–60	8.09	8.1	16	13	13	1.1	–	3.4	57	22	26.1	22
	60–100	8.15	8.4	17	12	12	0.8	–	3.5	53	28	26.9	25
	100–150	8.24	7.8	10	8	8	0.7	–	3.6	50	25	26.0	28

pHs pH of saturated soil paste, *ECe* electrical conductivity of soil saturation extract, *CEC* cation exchange capacity, *dS* m^{-1} deci Siemens per meter, *ESP* exchangeable sodium percentage

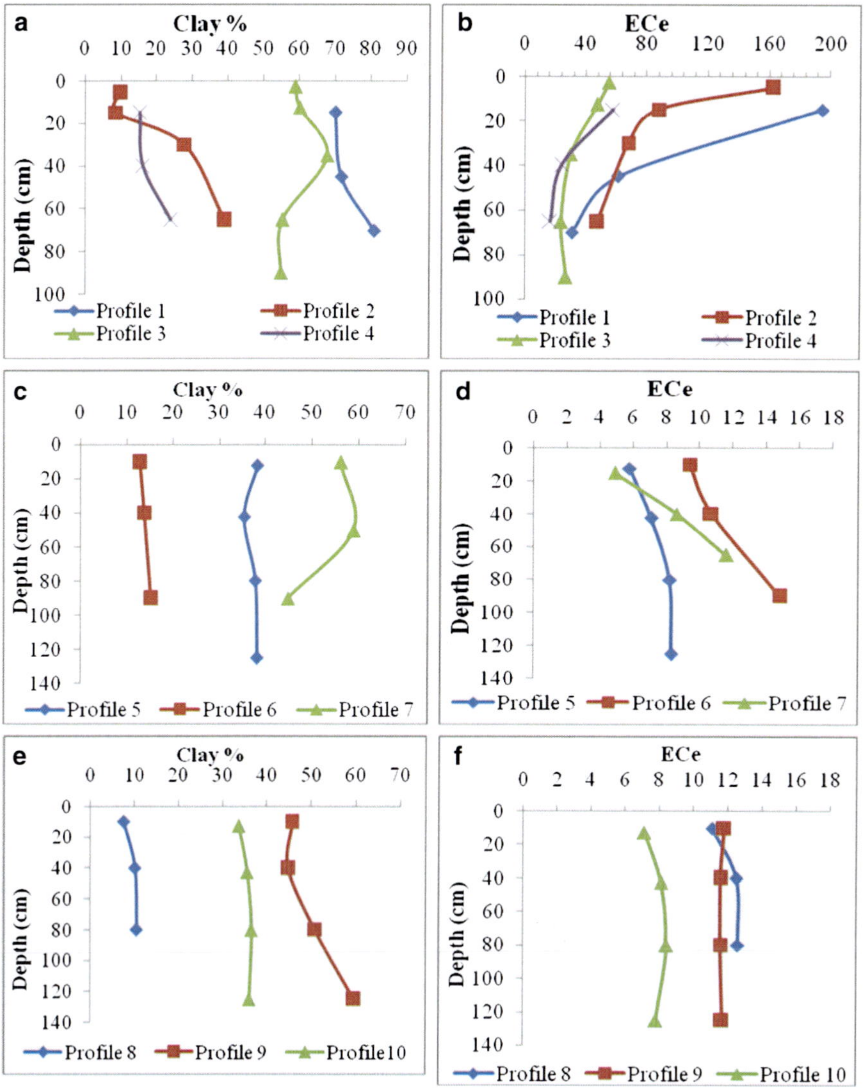

Fig. 20.2 Increasing and decreasing trends of clay and ECe in different profiles

20.3.3 Soil Micromorphology

The prevailing soil forming processes can be inferred directly from the microfabric structure analysis with the condition that sufficient background information is available. The differences in soil microstructure development of the soil materials in soil horizons can reveal extent of these soil formation processes. It is a matter of interpreting the observed changes in the soil material due to differences in soil processes which caused the changes. A number of hypotheses have been put forward concerning the processes involved in the formation of the micromorphological aspects, such as plasma fabric, voids, peds, and pedological features. These hypotheses based on what appear to be logical inferences derived from first principles and related to known properties of the constituents involved and the broad processes for which there is good evidence, such as movement or illuviation of solutions and suspensions through soil profile, diffusion, shrinking, and swelling. The micromorphological description of the studied thin sections (profiles 2, 3, 7, and 9) is revealed in Table 20.4, which enabled the identification of following processes.

20.3.3.1 Salinization and Alkalinization

Thin section description revealed spongy microstructure development at the surface through salt crystallization (Fig. 20.3a). The microstructure (Fig. 20.3b) is predominately apedal s-matrix, with intertextic plasmic fabric (Fig. 20.3c). Vughy-void pattern occupied by salt crystallization is the main micromorphological feature, which is associated with high level of clay orientation in the form of either more developed sepic plasmic fabric (in, mo, ma, vo, and skel sepic) as shown in profile 9 or through coagulation of clay particles in the form of clay papules (profile 2) as shown in Fig. 20.2d and e. Halite (NaCl) crystals are found in a fine joint and skew planes and embedded in s-matrix within profile 3 (Fig. 20.3f). Many medium to coarse completely bio-calcified shell fragments of different types are scattered throughout the soil profiles 2 and 3, especially in the surface layers (Fig. 20.3g) referring to the direct effect of saline water body of Qaroun lake. Also, the void pattern in profile 3 (clayey soils) as shown in Fig. 20.3h and in profile 9 as shown in Fig. 20.4a is dominated by skew, crazy, and joint planes. The latter void pattern is mostly occupied by soluble salts; besides chamber types, most of them occupied with gypsum, as partially weathered cluster of gypsum (Fig. 20.4b) and intercalary gypsum crystals (Fig. 20.4c).

20.3.3.2 Illuviation

The clay papules (profile 2) with surrounded boundaries of either authogenic or orthic pedological features indicating accumulation of clay particles in the coarse media demonstrate the relatively high salinity content, leading to the formation of such feature (Fig. 20.4d). The clay illuviation and their subsequent accumulation in the subsoil layers in sodic environment caused the precipitation of soluble Ca ions as calcium carbonate, thus discounting any role of soluble Ca and the presence of $CaCO_3$ in preventing the movement and accumulation of clay particles (profile 3) and formation of natric horizon in the presence of more than 13 % $CaCO_3$. These findings are in harmony with those reported by Ghassemi et al. (1995), Lebedeva-Verba and Gerasimova (2009), and Lebedeva-Verba et al. (2009). The morphological

Table 20.4 Micromorphological description of the selected soil profiles

Profile no.	Depth (cm)	Distribution pattern	Microstructure	Pedological features	Plasma fabric
2	0–10	Intertextic and granular	Apedal: single grains	Many shell fragments in different weathering stages, spongy secondary salt accumulations, gypsum and halite crystals in different voids, and impregnated within s-matrix	Crystic plasmic fabric
	10–20	Intertextic and granular	Apedal: single grains	Crystals of salts accumulated in different voids, humified OM, orthic medium to fine nodules of $CaCO_3$, gypsum bands	Crystic to silasepic
	20–40	Agglomero	Apedal: soil materials	Diffused clay papules, completely calcified shell fragments of different types, lime concretions, void argillans, and few medium orthic $CaCO_3$ nodules	Skelsepic and argillasepic
	40–80	Agglomero and porphyroskelic	Random micro-false flocculation of clay peds	Skel, and vo-argillans, few medium orthic $CaCO_3$ nodules, humified OM	Skelsepic and argillasepic
3	0–5	Porphyroskelic fabric	Apedal: soil materials	Spongy secondary salt accumulations, spinal gypsum crystals embedded in s-matrix and in elongated vughs, shell fragments	Argillasepic and crystic plasmic
	5–20	Porphyroskelic	Apedal to false primary peds	Calcareous orthic and inherited nodules, some of them stained with ferric oxides, argillans, and salans	Oxides, argillans and salans In, vo, ma-sepic and skelsepic
	20–50	Porphyroskelic	Pedal, primary false subangular blocky	Argillans and vo-calcitans. Patches of decomposed OM stained on some false primary peds	In, vo, ma, and skelsepic

(continued)

Table 20.4 (continued)

Profile no.	Depth (cm)	Distribution pattern	Microstructure	Pedological features	Plasma fabric
	50–80	Porphyroskelic	Apedal to false primary subangular blocky peds	Patches of partially decomposed OM stained with Mn and Fe oxides, vo-argillans. Orthic diffused $CaCO_3$ nodules	In, vo, ma, and skelsepic and argillaspic
	80–100	Porphyroskelic	Apedal: soil materials	Inherited iron oxide nodules, orthic and inherited spinal gypsum crystals embedded in s-matrix	In, mo, ma, and skelsepic and argillasepic
7	0–20	Porphyroskelic	Apedal: soil material	Partially dissolved shell fragments leaving $CaCO_3$ replaced with gypsum bands, $CaCO_3$ nodules, OM decomposed, few fine salt crystals embedded in s-matrix	In, mo, ma, and skelsepic to argillasepic
	20–80	Porphyroskelic	Pedal: fine to medium Primary peds	$CaCO_3$ orthic and inherited in different sizes skel and vo-calcitans	In, mo, ma, and skelsepic to argillasepic
	80–100	Agglomero plasmic and Porphyroskelic	Pedal: in primary peds	Fine lime stone fragments and inherited nodules	Crystic to silasepic and argillasepic
9	0–20	Porphyroskelic	Apedal: soil materials	Few random root tissue structures in reddish color and partially decomposed plant remaining, $CaCO_3$ nodules orthic and inherited and some detritus of Mn or Fe oxides, vo-calcitans, and argillans	Argillasepic and skel-voinsepic
	20–60	Porphyroskelic	Pedal: in subangular blocky primary peds	Vo and skeleton argillans and vo- calcitans, small patches of OM-staining primary peds	In, vo, ma, and skelsepic
	60–100	Porphyroskelic	Pedal in subangular blocky primary peds	$CaCO_3$ nodules, few intercalary gypsum in s-matrix, few dark brown debris of OM in bio-channel	In, vo, ma, and skelsepic and omnisepic
	100–150	Porphyroskelic	Pedal: in subangular blocky primary peds	Highly humified OM scattered in s-matrix, equent inherited Mn and Fe oxides nodules	In, vo, ma, and skelsepic and omnisepic

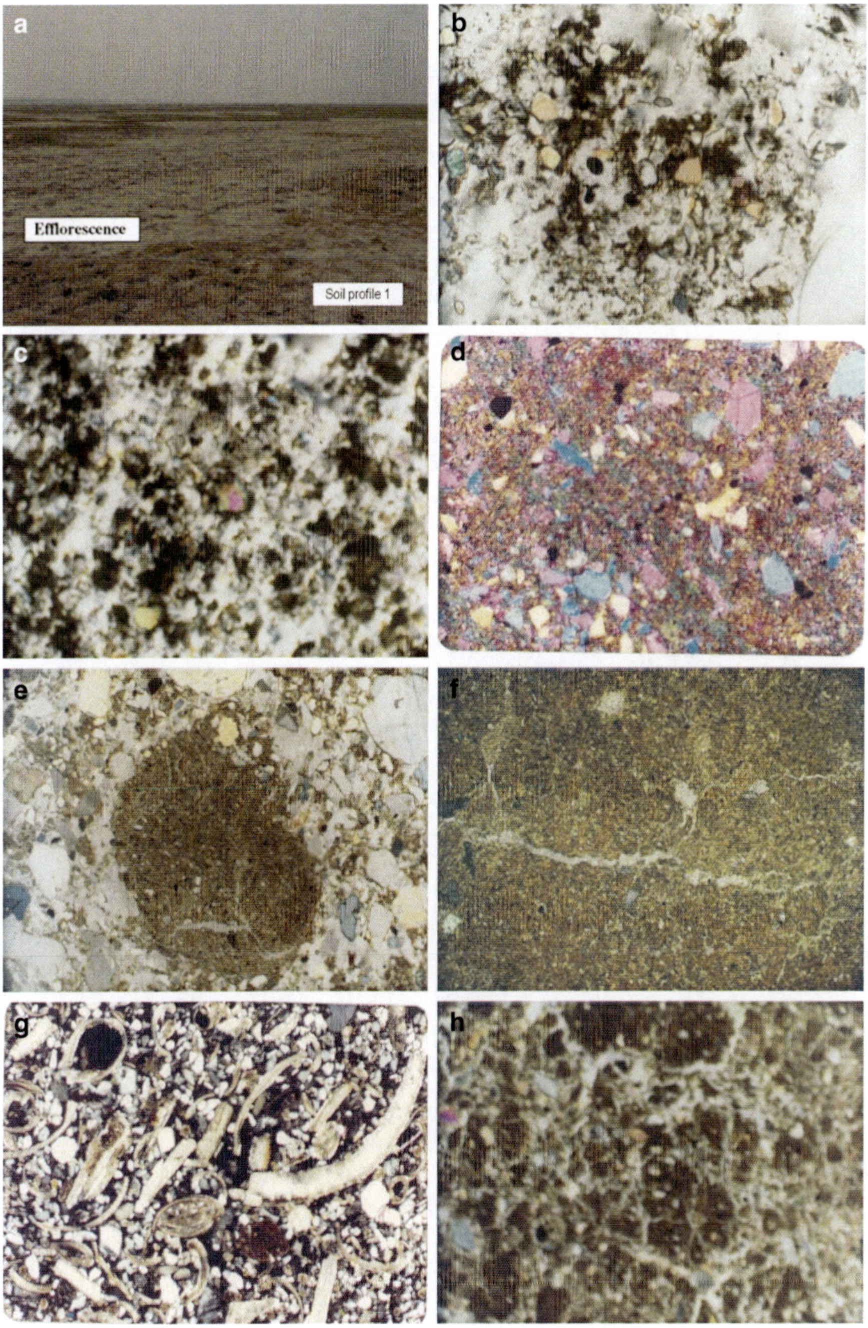

Fig. 20.3 (**a**) Surface salt efflorescence; (**b**) Spongy salty microstructure (×40 profile 2; (**c**) A pedal s-matrix with intertextic plasmic fabric (×40 profile 2); (**d**) N-Mo-Vo-skel and omnisepic plasmic fabric (×100 profile 9); (**e**) Clay papule in salty matrix (×40 profile 2); (**f**) Apedal salty s-matrix in joint and skew planes (×40 profile 3); (**g**) Weathered shell fragments (×40 profile 2); (**h**) S-matrix differentiated into fine primary peds separated by salty crazy plane (×40 profile 3)

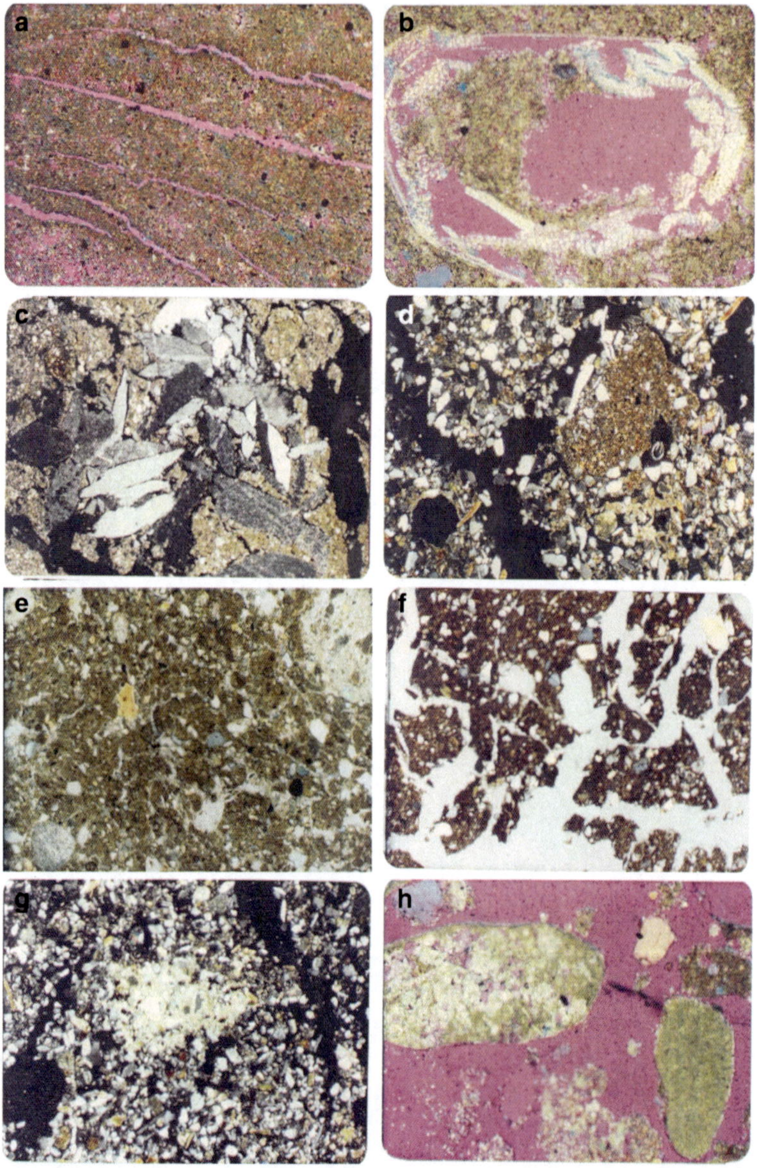

Fig. 20.4 (**a**) Subparallel fine joint planes (×100 profile 9); (**b**) Neo-Vo calcitans and gypsans and clusters of salt crystals (×100 profile 9); (**c**) Intercalary spinal gypsum crystals embedded in chamber and s-matrix (×100 profile 3); (**d**) Clay papules translocated in coarse textured soil (×40 profile 2); (**e**) Pedal subangular blocky structure stained in salty s-matrix (×100 profile 3); (**f**) Well-developed soil peds separated by compound packing voids, (×100 profile 9); (**g**) Spongy calcitic nodule (×40 profile 7); (**h**) Transported orthic and inherited $CaCO_3$ nodules (×100 profile 9)

description and chemical data (Tables 20.1 and 20.3) confirmed the development of subangular blocky microstructure in the Btz horizon within soil profile 3 (Fig. 20.4e) as well as the occurrence of natric horizon.

20.3.3.3 Associated Pedological Features

The comparison between the microstructure developed in the normal clayey soil and in clayey soil that is affected by salinity showed that topsoil of profile 9 is characterized by a microstructure of well-developed peds separated by compound packing voids photo (Fig. 20.4f), and its subsoil is dominated by slickensides phenomenon which is confirmed by the occurrence of subparallel fine joint planes (Fig. 20.4a); on the other hand, the highly affected soils of profile 3 showed a deteriorated microstructure of subangular blocky in the Btz horizon within soil, that is, s-matrix differentiated into fine primary peds separated by salty crazy planes (Fig. 20.4e).

The common pedological features are compacted and spongy calcareous nodules with some spots of iron oxide-stained s-matrix, as well as ferrogeneous nodules photos (*n*). Organic matter in different stages of decompositions and forms (highly humified dibers, eminent, and patches) is also observed.

20.4 Conclusions

Study of soils in thin sections under polarizing microscope clearly revealed soil microstructural development in studied soils from Egypt, for example, spongy microstructure due to surface salt accumulation, coating of voids with halite and gypsum crystal voids and s-matrix, subangular blocky microstructure in the Btz horizon of saline-sodic soil-presenting natric horizon, clay papules, and false microaggregates formed. Micromorphology is the only tool that allows in situ observation of soil features, compared to soil analyses in aqueous solutions, which gives average composition and hence considered most powerful tool in understanding soil development and to formulate better management options of salt-affected soils.

References

Abdel-Aal TS (1990) Cyclic formation of soils on different geomorphic features of Fayoum area, Egypt. PhD thesis, Faculty of Agriculture, Cairo University, Egypt

Abdelfattah MA (2002) GIS evaluation of soil and water management for sustainable agriculture in El-Fayoum Governorate, Egypt. PhD thesis, Faculty of Agriculture, Cairo University, El Fayoum, Egypt

Abdel-Hamid NH (1973) Micromorphology of soil structure and the main properties of some typical soil types of Egypt. MSc thesis, Faculty of Agriculture, Alexandria University, Egypt

Abdel-Motaleb MH (1997) Studies on monitoring of desertification and land degradation processes at El-Fayoum depression, Egypt. MSc thesis, Faculty of Agriculture, Cairo University, Egypt

Arshad MA, Pawluk S (1966) Characteristics of some solonetzic soils in the glacial Lake Edmonton of Alberta. J Soil Sci 17:36–47

ASCE (1996) Agricultural salinity assessment and management. Water Quality Technical Committee of the Irrigation and Drainage Division of ASCE, New York, p 619

Ayers RS, Westcot DW (1985) Water quality of agriculture, irrigation and drainage. Paper 29. FAO, Rome, p 97

Balba AM (1984) Save our soils. Alex Sci Exch 5:1–13

Ben-Hur MM, Agassi R, Keren A, Zhang J (1998) Compaction aging and rain drop impact effect on hydraulic properties of saline and sodic vertisols. Soil Sci Soc Am J 62:1377–1383

Brower R, Sleemen JR (1969) The arrangement of constituents in quaternary soils. Soil Sci 107:435

Bullock P, Fedoroff N, Jongerius A, Stoops G, Tursina T (1985) Handbook for soil thin section description. WIANE Research Publications Wolverhampton UK, p 152

Eswaran H, Stoops G, Abatahi A (1980) SEM morphologies of halite (NaCl) in soils. J Microsc 120:343–352

Farrag FRM (2003) Land degradation as affected by irrigation water quality in El-Fayoum Governorate. PhD thesis, Faculty of Agriculture, Cairo University, El-Fayoum, Egypt

FitzPatrick EA (ed) (1984) Micromorphology of soils. Chapman & Hall, London/New York, p 433

Ghassemi FA, Jakeman J, Nix HA (1995) Salinization of land and water resources. Center for Resources and Environmental Studied. Australian National University, Canberra. ACT 0200, Australia

Hegazi IMA (1999) Factors affecting utilization and development in lands and water resources of the northern area of El Fayoum, Egypt. PhD thesis, Faculty of Agriculture, Cairo University, El-Fayoum, Egypt

Hensly M (2000) Water table and maize production. In: Durand W, due Toit AS (eds) Proceedings of highwild ecoregion workshop. Potchefstroom, South Africa, 21–23 June 1999, pp 67–69

Hensly M, Anderson JJ, Botha JJ, Van Standen PP, Singlels A, Prinsloo M, Dutoit A (1997) Modeling the water balance on benchmark ecotopes. Report to Water Research Commission by the ARC-Institute for Soil, Climate and Water. WRC Project No k5/508

Ivanava EM, Bol'ShaKov AF (1972) Academicia KK Gedroyts "Theory of Solonchaks, Slonetzes and Solod and its subsequent elaboration". Sov Soil Sci 4:156–171

Kaewmano J (2003) Relationship between salt affected soils and landforms in Northwest part of Nakahon Ratchasima province. MSc thesis, Kasesart University, Bangkok

Kisra, AAH (1980) Studies on factors affecting critical depth of ground water in some soils of Egypt. MSc thesis, Faculty of Agriculture, Cairo University, Egypt

Klute A (ed) (1986) Methods of soil analysis. Part-1: Physical and mineralogical methods, 2nd edn. American Society of Agronomy, Madison

Lebedeva-Verba MP, Gerasimova MI (2009) Macro and micromorphological features of genetic horizons in a solonetzic soil complex at the Dzhanybek Research Station. Euras Soil Sci 42(3):237–250

Lebedeva-Verba M, Gerasimova M, Konyushkova M (2009) Micromorphology of solonetzic horizons as related to environmental events in the Caspian Lowland. J Mt Sci 6:132–138

Page AI, Miller RH, Keeney DR (eds) (1982) Methods of soil analysis. Part-2: Chemical and microbiological properties, 2nd edn. American Society of Agronomy, Madison

Ramadan MA (1986) Chemical and mineralogical studies of some salt-affected soils in Egypt. MSc thesis, Faculty of Agriculture, Ain Shams University, Egypt

Salem SYM (2002) Monitoring of environmental pollution for soil and water with some metals and non-metals resulting from fertilizer factories and its remediation. PhD thesis, Institute of Environmental Studies and Research, Ain Shams University, Egypt

Selassie TG, Jurinake JJ, Dudley IM (1992) Saline and saline-sodic soil reclamation: first order kinetic model. Soil Sci 154:1–7

Shahid SA, Jenkins DA, Hussain T (1992) Halite morphologies and genesis in the soil environment of Pakistan. In Proceedings of the international symposium on strategies for utilizing salt-affected soils, Bangkok, Thailand, 17–25 Feb 1992, pp 59–73

Soil Survey Division Staff (2003) Soil survey manual, Agricultural handbook 18. USDA-NRCS. US Government Printing Office, Washington, DC

Soil Survey Staff (2006) Keys to soil taxonomy. US Government Printing Office, Washington DC

Stoops G (2003) Guidelines for analysis and description of soil and regolith thin section. Soil Science Society of America, Madison, p 184

Chapter 21
Correlation of Students' Estimation and Laboratory Determination of Soil Texture

Mohamed S. Alhammadi and Mohamed S. Gheblawi

Abstract Soil texture helps understand physical and chemical properties of soils. Soil texture can be determined accurately in the laboratory, or it can be estimated through finger test in the field, the accuracy of the latter depends on the skills and experience of the person. However, such estimates provide quick results compared to laboratory measurement, which takes ample amount of time for final results. In an attempt to correlate field and laboratory soil texture results, 11 students were trained to develop their skills to determine field soil texture (finger test) and in determining soil texture in the laboratory using standard hydrometer method. When the finger test results were compared with the laboratory results, the students' estimates predicted the actual clay and sand percentages with high degree of accuracy. The results showed a high variation between student's and laboratory determination of soil texture. Estimation of sand fraction is slightly more difficult than clay particles. This study suggests that students require more practice to be able to estimate soil textural class accurately. It is recommended that students should use reference samples (known texture) to improve their skills.

Keywords Feel method • Hydrometer method • Texture correlation • *T*-statistics • UAE University

---

M.S. Alhammadi (✉)
Department of Aridland Agriculture, UAE University,
P.O. Box 17555, Al-Ain, UAE
e-mail: Alhammadi@uacu.ac.ac

M.S. Gheblawi
Department of Agribusiness and Consumer Sciences, UAE University,
P.O. Box 17555, Al-Ain, UAE
e-mail: Gheblawi@uaeu.ac.ae

S.A. Shahid et al. (eds.), *Developments in Soil Classification, Land Use Planning and Policy Implications: Innovative Thinking of Soil Inventory for Land Use Planning and Management of Land Resources*, DOI 10.1007/978-94-007-5332-7_21,

21.1 Introduction

Soil texture is one of the most important soil properties in soil survey (Cooper 1990). Soil Survey Division Staff (2003) defined soil texture based on the percentage of sand, silt, and clay in soil that is less than 2 mm in diameter (fine earth fraction). Soil texture helps to understand many soil physical and chemical properties such as leaching and erosion potential, bulk density and soil consistence, water and nutrient retention, and heat conductance (Burt 2004). Particle-size distribution (PSD) also reflects the relative balance of weathering and pedogenetic processes. Soil texture is a key component of any minimum data set to be used for evaluating soil quality and sustainability of agricultural management practices (Kettler et al. 2001). Soil texture can be determined through detailed particle-size distribution analysis (PSDA), which quantifies clay, silt, and sand fractions in soil sample. The primary soil particles in the USDA classification system (Gee and Bauder 1986) are divided into three major sizes: sand (2.0–0.05 mm), silt (0.05–0.002 mm), and clay (<0.002 mm). Conventional particle-size analysis requires the measurement of clay, silt, and sand fractions, their plotting into standard textural triangle to determine soil textural class.

Soil texture can be determined in the laboratory or it can be estimated through finger test in the field. The accuracy of description is high in the laboratory determination due to the fact that the field estimation depends mainly on the engineer's skill and experience. Post et al. (1999) defined the accuracy of the field estimation based on which soil engineer findings agree with objective criteria that the laboratory determines. Accurate, inexpensive, and rapid field estimation of soil texture is useful to soil engineers to get feelings on the spot. Laboratory analysis for soil texture determination is time consuming and expensive.

During field description, soil engineers use their skills and experience to estimate soil texture. The ability of soil engineers to accurately estimate field soil texture depends on the hands-on experience on a variety of soils (Levine et al. 1989; Cooper 1990). Several studies evaluated students' or soil scientists' ability to determine some soil properties (Post et al. 1986, 1993, 1999, 2001; Levine et al. 1989; Cooper 1990). Post et al. (2001) have studied soil scientists' estimation of soil stickiness and plasticity. Cooper (1990) and Post et al. (1993) have evaluated students' ability to determine soil Munsell color notation. Their results showed that students were able to reach high level of matching soils with Munsell colors. Foss et al. (1975), Levine et al. (1989), and Post et al. (1999, 1986) have evaluated students' and soil scientists' capability to estimate soil texture. They found that sand and clay were most accurately estimated, while 50% of the textural classes were correctly placed in the 12 textural classes.

It is recognized that such field estimations are very important for on-site investigation of certain soil characteristics (soil texture), and therefore, a study was undertaken with the aim to evaluate students' skills in estimating soil texture and to correlate with the results from laboratory determinations. In addition, the study was conducted to observe any variation between student-determined soil texture and the texture determined by using hydrometer by professional scientist in the laboratory. Hydrometer method is sufficiently accurate to be used in research; however, it is not considered the most accurate method to determine particle-size distribution compare to Pipette method. Hydrometer method is a relatively simple and rapid sedimentation procedure, and most laboratories exercise Bouyoucos method (Jacobs et al. 1971).

21.2 Materials and Methods

21.2.1 *Estimation of Soil Texture by Feel Method (Finger Test)*

Eleven students of the UAE University, who were taking regular undergraduate course "Principle of Soil Science" were trained for 3 weeks in developing their feeling skills to estimate soil particle percentages (sand and clay) and to estimate soil textural class for each soil sample. They were provided with several soil samples to practice finger estimation of soil particles' percentage and to follow steps in "Guide to Texture by Feel" diagram (Thien 1979), which was modified later by USDA (Burt 2004). By the end of the training period, a batch of 16 soil samples with diverse unknown textures, collected from widespread areas of the UAE (northern and eastern regions) and Oman (northern region), was given to students to estimate soil particle percentages and textural classes. The same batch of soil samples was also analyzed in the laboratory as reference for comparison and correlation.

21.2.2 *Soil Texture Determination in the Laboratory*

Seven soil samples with diversified properties were given to students to determine particle-size distribution using standard hydrometer method (Bouyoucos 1927), later modified by USDA (Burt 2004). The students' results were compared with the results obtained by professional analyst in the soil laboratory.

To prepare soil samples for laboratory analyses, 50 g of air-dried soil (recorded the weight=S) was transferred to the stirring cup; this was filled (1/2–3/4) with distilled water, and 10 ml of 1N sodium hexametaphosphate $(NaPO_3)_6$ was added as a dispersion agent. In standard soil texture determination, the results are presented on oven-dried soil basis; in this study, we used air-dried samples for correlation with students' texture determination by feel method, who used air-dried soil samples. The soil suspension was stirred for 4–10 min depending on the soil texture (4 min for sandy soil and 10 min for clayey soil). The dispersed soil suspension was transferred to hydrometer jar filled to 1 l with distilled water. The temperature (°C) of the suspension was measured (T1) at each reading. Soil suspension was stirred manually with the plunger (1–4 min of continuous agitation) to unify soil suspension in the jar. First hydrometer reading was recorded after 40 s of shaking (R1). While the second reading (R2) was taken after 4 h and temperature recorded (T2).

The hydrometer readings for temperature (CR) were corrected using the temperature correction factor according to the following equations:

$$\mathrm{CR1} = (\mathrm{R1} - \mathrm{Rb}) + 0.4(\mathrm{T1} - 20)$$

$$\mathrm{CR2} = (\mathrm{R2} - \mathrm{Rb}) + 0.4(\mathrm{T2} - 20)$$

where Rb is the blank reading of solution without soil but with sodium hexametaphosphate solution. Finally, sand, silt, and clay percentages were calculated using equations below:

$$\%\ \mathrm{Sand} = \frac{(\mathrm{S} - \mathrm{CR1}) \times 100}{\mathrm{S}}$$

$$\%\ \mathrm{Clay} = \frac{(\mathrm{CR2}) \times 100}{\mathrm{S}}$$

$$\%\ \mathrm{Silt} = \frac{(\mathrm{CR1} - \mathrm{CR2}) \times 100}{\mathrm{S}}$$

The results from laboratory determination and those determined by the students were evaluated and statistically analyzed.

21.3 Results and Analysis

21.3.1 Estimating the Soil Texture by Feel Method

Table 21.1 illustrates the t-statistics test of average clay and sand estimation data obtained by 11 students, whereas Figs. 21.1 and 21.2 illustrate the regression equations for the clay and sand percentages, respectively. In this study, 16 soil samples were given to students to estimate clay and sand values based on their skills acquired during an intensive 3 weeks training. The percentages of clay and sand determined in the laboratory are considered the population means (μ). In order to analyze the adequacy of students' training and skills, a null hypothesis (H_0) is set up. The H_0 is the difference between the population mean determined in the laboratory and the sample means (x) calculated as the simple mean of the students' results is equal to zero. This implies that there is no difference between the laboratory-determined values and the student-estimated values based on their skills. The alternative hypothesis (H_A) stipulates that there is a difference between these two means, or the students' means are different from the laboratory-determined values.

The *t*-statistics values of the hypotheses testing are shown in Table 21.1. The cutoff *t*-value for a two-tail with 10° of freedom with 5 % significance level is ±2.228. It is evident from the results that a sizable majority of the tests had *t*-statistic<critical *t*-value. This implies that H_0 cannot be rejected, and hence, the

Table 21.1 *T*-statistics test showing the difference between population and sample means

Sample no	Clay	Sand
1	1.428*	−1.445*
2	1.910*	−0.747*
3	2.562	−5.011
4	0.469*	−0.485*
5	0.013*	0.334*
6	1.734*	3.393
7	−0.562*	−0.993*
8	1.875*	−1.203*
9	2.094*	0.562*
10	2.699	2.382
11	−0.337*	−1.954*
12	−3.678	4.709
13	3.479	−2.259
14	4.780	1.856*
15	2.526	−1.446*
16	2.101*	−3.987

*Significantly different at α=0.05

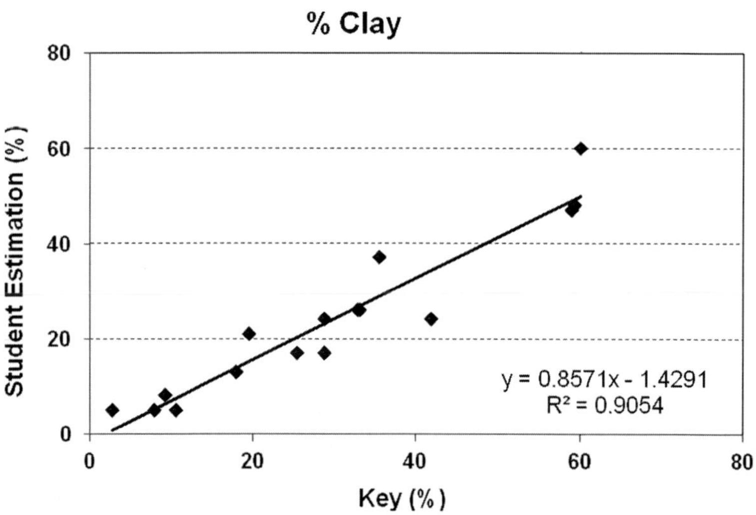

Fig. 21.1 Regression equation for estimated clay %

difference between the sample means and the laboratory values cannot be due to sampling variations.

The success of the students in estimating the values of the samples can be viewed in terms of how many tests were significant in relation to the total number of samples per test category. Table 21.2 lists the success ratios of these tests.

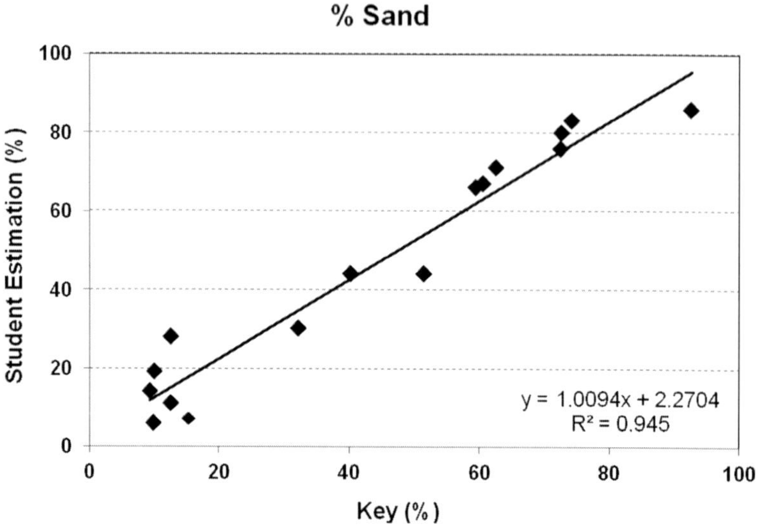

Fig. 21.2 Regression equation for estimated sand %

Table 21.2 Success ratio of students' results

	Clay	Sand
Reject H_0	6	6
Accept H_0	10	10
Success %	62.50	62.50

Table 21.3 Regression results of $y_i = a_i + b_i \overline{x}_i$

	Clay	Sand
R-squared	0.905	0.945
b_0	−1.429	2.270
t-b_0	−0.558	0.681
b_1	0.857	1.009
t-b_1	11.575	15.514

The second part of the statistical analysis utilizes the regression techniques. The regression model in Eq. 21.1 specifies the dependent variable y_i as the laboratory-determined morphological characteristics of the soil sample i ($i = 1, \ldots, 16$). The independent variable $\overline{xi}$ is the mean of students' values for each sample. The intercept is a_i, while the error term is $e_i \sim (0, 1)$.

$$y_i = a_i + b_i \overline{xi} + e_i \tag{21.1}$$

The least-squares method is used to fit the model in Eq. 21.1. The regression results are shown in Table 21.3 and Figs. 21.1 and 21.2 for the two soil properties of samples. The intercept in the two equations is insignificant. However, the goodness of fit is relatively high as indicated by the values of the coefficient of determination

Table 21.4 Clay and sand means (%), standard deviation (SD), and coefficients of variation (CV) for 16 soil samples estimated by 11 students

	% Clay			% Sand			
Sample No	Mean	SD	CV	Mean	SD.	CV	% Correct textural classes[a]
1	18.0	11.1	61.5	62.5	18.5	29.6	13
2	33.2	12.0	35.8	40.2	16.1	40.2	0
3	59.0	14.8	25.1	10.2	5.6	54.6	13
4	9.3	8.6	92.7	72.5	22.5	31.0	16
5	60.1	21.5	35.7	12.7	16.3	128.4	26
6	32.9	12.6	38.3	15.4	7.9	51.0	6
7	19.5	8.2	41.9	59.4	20.8	35.0	13
8	8.0	5.1	63.2	74.2	23.2	31.2	16
9	28.8	7.3	25.2	32.3	12.8	39.6	6
10	25.5	10.0	39.2	51.5	10.0	19.4	3
11	35.6	12.8	35.9	9.4	7.3	77.8	13
12	2.7	1.9	71.6	92.6	4.5	4.8	0
13	10.6	5.1	48.2	72.6	10.3	14.2	3
14	42.0	11.9	28.3	10.0	6.8	68.1	3
15	28.8	14.8	51.3	60.5	14.1	23.3	3
16	59.4	17.1	28.8	12.8	12.0	93.9	23
Overall mean	29.6	10.9	45.2	43.0	13.0	46.4	10

[a] Using 12 textural classes system (Soil Survey Division Staff 2003)

R^2. The slope coefficient for the two equations is significant at $\alpha=0.10$ as indicated by the t-statistics values. The conclusion is that the model may be used with confidence to predict the laboratory value of a given sample with the mean of student's value for a given soil sample.

Table 21.4 shows the mean, standard deviation, coefficients of variation, and correct textural classes for sand and clay estimations of each soil sample. Coefficients of variation were calculated to compare the estimations among students. The mean coefficients of variation agreed closely in both clay and sand soil particles (45.2 and 46.4%, respectively). Standard deviations for clay and sand were calculated to measure the difficulty of clay and sand percentages estimations. Mean standard deviation for sand percentage estimation was slightly higher (SD=13) than clay percentage estimation (10.9) indicating that sand particles were more difficult to estimate compared to clay particles. The overall mean of the textural classes shows that students were able to correctly classify 10% of the textural classes based on 12 textural class classification systems. Students' classification is clearly poor in terms of accuracy compared to the results found by Post et al. (1986) and Levine et al. (1989).

21.3.2 Laboratory Determination of Soil Texture

Figures 21.3, 21.4, and 21.5 show box plots for the mean students' determination of clay, silt, and sand percentages for students using the hydrometer procedure. Students' determination of soil texture in the laboratory was not better than the

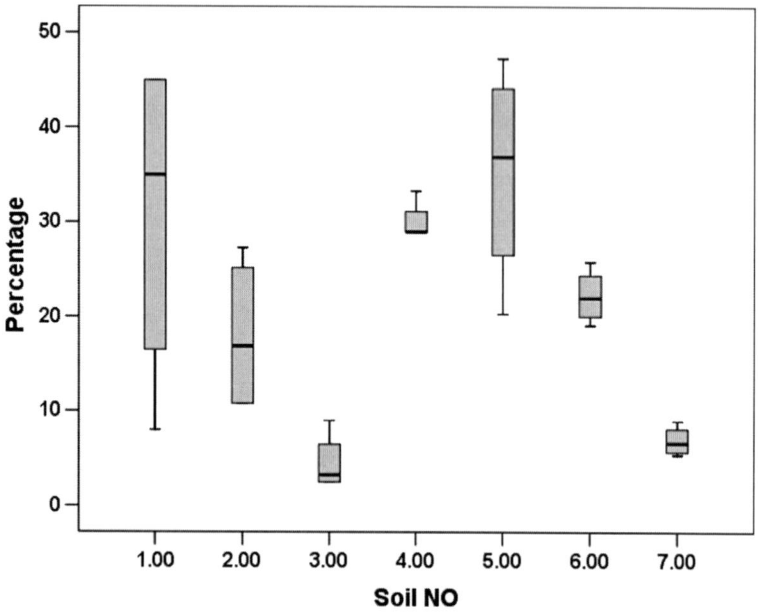

Fig. 21.3 Box plots for the mean students' clay % determination

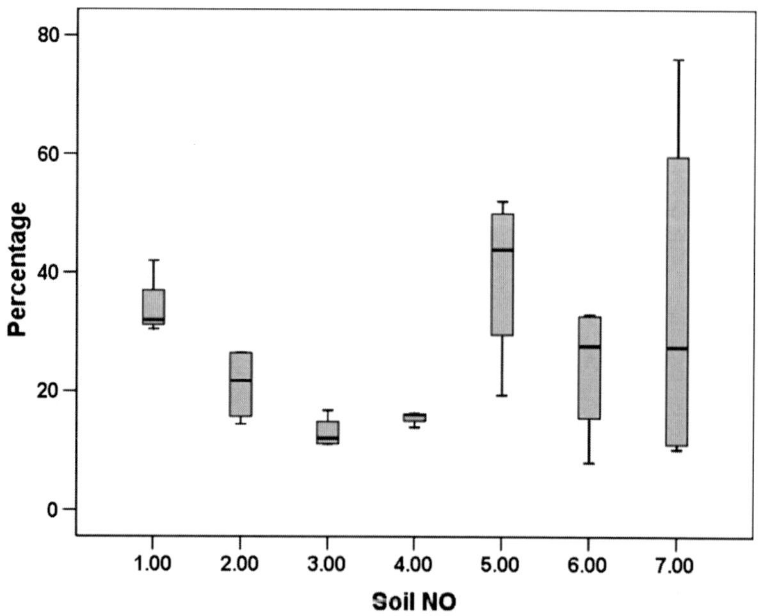

Fig. 21.4 Box plots for the mean students' silt % determination

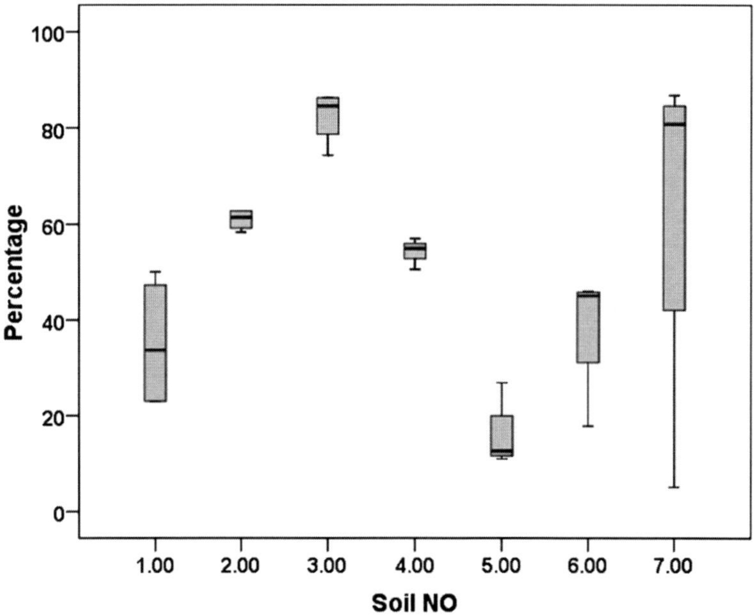

Fig. 21.5 Box plots for the mean students' sand % determination

estimation by feeling skills, even though they were using detailed guiding steps. We observed some variations between the students in determining soil particles in several soil textures. The variation between students in determining clay % differs by soil samples. It was very high for soil samples 1 and 5 indicating that the students were facing some difficulties in determining clay % for these two particular samples. On the other hand, highly clustered values were found in soil samples number 4 and 7 indicating the low differences between students determination of clay %. Normal distribution was clearly found in soil sample number 6 (Fig. 21.3).

Determination of silt (%) fraction is slightly different than the clay fraction. Soil sample number 7 had very high variation between students' determination than soil sample 5 and 6. Whereas soil samples number 4 then 3 had the lowest variations between the students' determination of silt percentage (Fig. 21.4).

Sand percentage determination for soil sample number 7 was highly scattered indicating big variation between the students. Soil samples number 2 and 4 were highly clustered indicating low variation between the students and ease of sand percentage determination (Fig. 21.5).

21.4 Conclusions and Recommendations

The study was focused on testing the ability of students who were given 3 weeks intensive training on how to determine particle-size percentages and textural classes using their feeling skills. They also determined soil texture using standard laboratory procedure. The study revealed that the students were able to estimate the samples' clay and sand contents almost two-thirds of the time (66 % success). However, when the students' results were compared with those obtained from the laboratory measurements, the students' estimates predicted the actual clay and sand percentages with low degree of accuracy. It was observed that the estimation of sand fraction was slightly more difficult than clay fraction. In general, students' classification is clearly poor in terms of accuracy, where they were able to correctly classify only 10 % of textural classes based on 12 textural classes system. The overall results showed a high variation between the students determination of soil texture with those determined in the laboratory using hydrometer method. This study suggests that students require more practice and experience for accurate estimation of soil textural classes. Also it is essential to determine laboratory soil texture by standard Pipette method for better correlation. It is recommended that in such studies, reference soil samples of known particle-size percentages are to be used to improve students' skills. Since field studies or surveys require frequent soil texture determination to have on the site information, therefore, training of students in the feel method of texture determination should be given higher preference in practical classes of soil science, especially in soil physics practical sessions.

References

Bouyoucos GJ (1927) The hydrometer as a new method for the mechanical analysis of soils. J Soil Sci 23:343–352

Burt R (ed) (2004) Soil survey laboratory methods manual. United Stated Department of Agriculture-Natural Resources Conservation Service, Soil survey investigations report 42, version 4.0

Cooper TH (1990) Development of students' abilities to match soil color to Munsell color chips. J Agron Educ 19:141–144

Foss JE, Wright WR, Coles RH (1975) Testing the accuracy of field textures. Scientific article no A2012 of the Maryland Agricultural Experiment Station. Soil Sci Soc Am Proc 39:800–802

Gee GW, Bauder JW (1986) Particle-size analysis. In: Klute A (ed) Methods of soil analysis. Part 1, 2nd edn. Agronomy monograph, vol 9. ASA and SSSA, Madison, pp 383–411

Jacobs HS, Reed RM, Thien SJ, Withe LV (1971) Soil laboratory exercise source book. American Society of Agronomy, Madison, p 126

Kettler TA, Doran JW, Gilbert TL (2001) A simplified method for soil particle size determination to accompany soil quality analyses. Soil Sci Soc Am J 65:849–852

Levine SJ, Post DF, Ellsworth TJ (1989) An evaluation of student proficiency in field estimation of soil texture. J Agron Educ 18(2):100–104

Post DF, Huete AR, Pease DS (1986) A comparison of soil scientist estimations and laboratory determinations of some Arizona soil properties. J Soil Water Conserv 41(6):421–424

Post DF, Levine SJ, Bryant RB, Mays MD, Batchily AK, Escadafal R, Huete AR (1993) Correlations between field and laboratory measurements of soil color. In: Bigham JM, Ciolkosz EJ (eds) Soil color. Soil Sci Soc Am Spec publication 31. SSSA, Madison

Post DF, Ferreira LG, Rapp R (1999) Evaluating the skill of soil scientists to determine soil texture, stickiness, and plasticity. In: Soil Science Society of North Carolina proceedings XLII, Raleigh, pp 94–98

Post DF, Rapp R, Ferreira L (2001) Quantitative field estimations of stickiness and plasticity for predicting WEPP Model parameters. In: Stott DE, Mohtar RH, Steinhardt GC (eds) Sustaining

the global farm: 10th international Soil Conservation Organization meeting, Purdue University, pp 896–902, 24–29 May
Soil Survey Division Staff (2003) Soil survey manual, Agricultural handbook 18. USDA-NRCS. US Government Printing Office, Washington, DC
Thien SJ (1979) A flow diagram for teaching texture by feel analysis. J Agron Educ 8:54–55

Chapter 22
Soil Classification and Genesis in Part of Khorasan Province

Mohammad Hassan Sayyari-Zahan

Abstract Saline and sodic soils occupy an important part of Iran including the vast area in Khorasan province. Also due to secondary salinization, some of the arable lands are becoming salt-affected. It is important to characterize and classify these soils for better management and uses. Preliminary interpretation of aerial photos and using topographic map, the area was divided into different physiographic units, that is, dominantly piedmont alluvial plains and flood plains. In these units, 30 soil profiles were dug and described, and soil samples from each genetic horizon were collected for physical and chemical characteristics using standard USDA methods. The aridic and thermic were recognized as moisture and temperature regimes, respectively. The gypsiferous and saliferous marls are the parent materials, being main factor for soil salinization. Topography, wind erosion and human activities also contributed to soil salinization. Micromorphological studies have revealed the presence of secondary gypsum and calcium carbonate in some pedons, forming gypsic and calcic horizons, respectively. The soils are classified according to the USDA soil taxonomy. Two soil orders (Aridisols, Entisols) and six suborders (calcids, salids, gypsids, cambids, fluvents and orthents) have been identified in the area, and a new subgroup of sodic torrifluvents is proposed.

Keywords Classification • Genesis • Khorasan province • Salinization • USDA

M.H. Sayyari-Zahan (✉)
Department of Soil Science, Agriculture Faculty,
University of Birjand, 97175-331 Birjand, Iran
e-mail: mh.sayyar@gmail.com; msayari@birjand.ac.ir

S.A. Shahid et al. (eds.), *Developments in Soil Classification, Land Use Planning and Policy Implications: Innovative Thinking of Soil Inventory for Land Use Planning and Management of Land Resources*, DOI 10.1007/978-94-007-5332-7_22,

22.1 Introduction

Salinization of land resources is a major impediment to their optimal utilization in many arid and semiarid regions of the world including Iran. Many countries have taken the initiatives to reduce soil salinization, but it is increasing at an alarming rate in the world. In this regard, the characterization, management and reclamation of salt-affected soils are gaining importance in many countries to bring these marginal lands into production to meet ever increasing food demand of the growing population. In Iran, about 25×10^6 ha of lands, representing 15% of the state area, are affected to varying levels of salinity and sodicity. Such lands constitute approximately 30% of Iran's plains and mild slopes and 50% of the lands with potential for irrigated agriculture. In Russia, China, India, Pakistan and Iran, saline soils are occurring on large areas. In these countries, most of the agricultural farms are irrigated with saline/brackish waters, which eventually have affected millions of hectares to become saline, although they have yielded abundantly but non-sustainably. Generally, in the arid and semiarid areas, low precipitation, high evaporation and storm activities bring the salts to the soil surface, a common phenomenon under hot climate.

The average annual rainfall ranges from less than 50 mm in the Central Plateau to more than 1,600 mm on the Caspian coastal plain, with a 250-mm national average. Owing to extremely hot summer, the average annual potential evaporation is very high, ranging from less than 700 mm along the Caspian Sea shore to over 4,000 mm in the deserts (Kavir-e-Loot). The problems of salt-affected soils have serious implications in the irrigated areas where about half of the land is prone to different levels of salinity (Cheraghi 2004) leading to national economic loss of more than 1 billion US$. Consequently, the dependency on irrigated agriculture is at stake in areas where land and water resources degradation has increased over time. Although salt-affected soils exist throughout Iran, slightly and moderately salt-affected soils are mostly found in the northern part, while soils with high salinity levels are prevalent in the central part.

In India it has been reported (Abrol 1972) that about 7×10^6 ha of India's lands were affected by salinity and alkalinity due to insufficient rainfall to leach salts. While studying salt-affected soils in Iraq, AlLayla (1978) has reported the origin of salts in certain parts of Iraqi soils due to salts carried via subterranean or phreatic water as well as salts transported via marine currents. In another study on Iraqi soils, Ayers (1960) described soil salinity development due to residual effects of parent material, slates, shales, limestone and gypsum as well as through drainage and evaporation.

Mahjoori (1975) studied saline and sodic soils in the lowlands of Iran. He collected soil samples from three soil series from the central and southern part of Iran. The study revealed the origin of salts from marly gypseous and saline deposits and their transportation to soils by water. In another study, pedogenesis of saline soils in an area of Roodasht in Esfahan revealed the role of parent material such as alluvium, in conjunction with slow drainage causing soils to become saline and sodic.

The present study was conducted with the objectives to determine physical and chemical characteristics of the zonal soils of Nishaboor plain and to classify soils using pedological taxonomy.

22.2 Materials and Methods

The study area is located in the Khorasan province (northeast of Iran). It is located between 59°, 30″ east longitude and 35°, 45″ north latitude, 85-km southwest of Mashhad (northeast of Iran) and is known as Neishabour Lower Dizbaud. The area is lying across the seasonal wind channel; that is why it is named Dizbaud (baud in Persian language means wind). At the time of study, some of the area was under cultivation.

By means of air photos and topographic maps (1:50,000 scale), the area was delineated into different physiographic units, and 30 profile sites were selected. Soil profiles were dug and photographed to show in-depth horizons. Following to the description of soil profiles (Soil Survey Division Staff 1993), soil samples were collected from different horizons for laboratory analysis. Soil samples were air-dried and passed through 2-mm sieve (fine earth fraction), calcium carbonate ($CaCO_3$) equivalents, gypsum, soluble salts, pH, cation exchange capacity (CEC), exchangeable sodium percentage (ESP), soluble anions and cations, organic carbon and particle size distribution were determined using standard USDA methods (Burt 2004).

22.3 Results and Discussion

The field description of soil profiles and results from laboratory analyses of soil samples (profile numbers 3, 7, 11, 13 and 21 located in the sloppy plain and 1, 4 and 27 located in the flood plain) revealed that the major source of the salts are marly, gypseous and salty formations present in the zone as parent materials which are carried into lower landscapes through the floodways in the region. Tables 22.1 and 22.2 present the results of physical and chemical characteristics of soil profiles.

In Iran, Persia's tectonic formation has caused such soils to form. In a survey of saline and alkali soils of lowlands based on three soil series from the central and southern parts of Iran, Mahjoory (1979) attributed the origin of salts to parent materials of marly and gypseous deposits and their transfer through water and occurrence of salinization.

The increasing and decreasing trends of electrical conductivity of soil saturation extract (ECe) and sodium adsorption ratio (SAR) with respect to soil depth in the soil profiles have been shown in Figs. 22.1, 22.2, 22.3 and 22.4 in piedmont plain and Figs. 22.5 and 22.6 in flood plain. Where both ECe and SAR are increased with depth (Figs. 22.1, 22.2 and 22.3), these are related to the origin of salts in the parent

Table 22.1 Physical and chemical characteristics of soil profiles from piedmont plain

Profiles no	Horizon	Depth (cm)	OC (%)	ESP	SAR (mmoles $L^{-1})^{0.5}$	$CaCO_3$ equivalents (%)	ECe ($dS\ m^{-1}$)	pH	SP	Particles percent Sand (%)	Silt (%)	Clay (%)	Soil texture
3	A1	0–20	0.20	5.6	4.9	21.4	1.5	8.4	38.1	37.6	39.4	23.0	L
	E1	20–40	0.14	12.6	10.7	21.5	7.1	8.2	37.2	39.6	43.4	17.0	L
	Ey1	40–55	0.16	32.8	34.1	21.6	44.9	7.9	36.1	47.6	37.4	15.0	L
	Ey2	55–98	0.10	39.7	45.7	23.4	45.0	7.8	41.5	29.6	49.4	21.0	SiL
	C1	98–107	0.08	41.3	48.8	22.2	46.8	7.8	37.7	71.6	17.4	11.0	SL
	C2	107–130	0.08	48.3	64.3	22.7	61.6	7.9	48.5	15.6	57.4	27.0	SiL-SiCL
	C3	130–150	0.06	42.6	51.3	22.0	42.5	7.9	40.7	37.6	47.4	15.0	L
7	A1	0–20	0.19	0.9	1.4	19.8	1.4	8.0	31.2	39.6	33.4	27.0	L
	B1	20–48	0.15	2.3	2.4	18.2	1.2	8.0	30.8	47.6	29.4	23.0	L
	Bk	48–70	0.12	6.8	5.8	24.4	6.5	7.9	29.0	66.6	15.4	18.0	SL
	C	70–102	0.10	3.6	3.4	19.8	4.1	7.7	30.6	71.6	12.4	16.0	SL
	2Bkb	102–130	0.06	11.5	9.7	25.4	4.5	7.8	55.5	41.6	21.4	37.0	CL
11	A	0–23	0.21	4.0	3.7	18.0	0.8	8.3	30.0	34.2	47.4	18.4	L
	Bk1	23–65	0.17	17.9	15.7	24.8	4.4	8.2	35.0	41.2	36.4	22.4	L
	Bk2	65–90	0.10	7.7	6.5	20.2	7.3	8.1	32.8	68.2	15.4	16.4	SL
	C	90–150	0.09	8.4	6.9	19.2	6.5	8.0	30.0	62.2	18.4	19.4	SL
13	Ap	0–35	0.12	4.3	3.9	24.2	3.5	8.0	32.1	63.2	27.4	9.4	SL
	By1	35–60	0.08	15.4	13.2	23.5	9.0	8.0	30.6	70.2	19.4	10.4	SL
	By2	60–83	0.10	13.5	11.8	22.5	9.8	7.8	30.8	59.2	28.4	12.4	SL
	C1	83–125	0.10	14.3	13.0	22.7	14.7	7.8	26.0	72.2	16.4	11.4	SL
	C2	125–150	0.04	22.1	20.1	23.2	19.4	8.0	35.6	46.2	31.4	22.4	L
21	Ap	0–25	0.22	35.3	38.1	23.2	37.2	8.0	35.5	44.2	39.4	16.4	L
	C1	25–50	0.15	30.2	30.3	22.6	30.1	8.0	33.5	50.2	31.4	18.4	L
	C2	50–70	0.12	14.4	12.2	20.8	9.8	7.9	27.0	74.2	13.4	12.4	SL
	C3	70–140	0.08	10.5	8.8	17.7	9.1	8.0	26.1	80.2	4.4	15.4	SL

OC organic carbon, *ESP* exchangeable sodium percentage, *SAR* sodium adsorption ratio, *ECe* electrical conductivity of soil saturation extract, *SP* saturation percentage, *L* loam, *SiL* silty loam, *SiCL*, silty clay loam, *LC* loamy clay, *SL* sandy loam, *CL* clay loam

Table 22.2 Physical and chemical characteristics of soil profiles from flood plain

Profiles no	Horizon	Depth (cm)	OC (%)	ESP	SAR (mmoles $L^{-1})^{0.5}$	$CaCO_3$ equivalents (%)	ECe (dS m^{-1})	pH	SP	Particles percent Sand (%)	Silt (%)	Clay (%)	Soil texture
1	Ap	0–37	0.38	26.3	23.0	23.1	13.0	7.8	31.6	55.7	32.0	12.3	SL
	C1	37–60	0.23	20.1	21.0	20.4	3.9	8.1	23.9	79.7	10.0	10.3	LS-SL
	C2	60–80	0.15	22.5	20.0	22.1	12.3	7.8	42.5	15.7	56.0	28.3	SiCL
	2By	80–120	0.36	19.5	13.0	23.5	11.1	7.7	42.6	20.7	49.0	30.3	CL
	2C	120–145	0.15	13.3	6.0	24.8	5.0	7.9	38.4	35.7	38.0	26.3	L
4	A1	0–15	0.32	19.8	31.7	22.6	33.5	7.6	37.5	35.7	42.0	22.3	L
	By	15–45	0.23	22.9	59.7	25.2	50.6	7.9	39.3	46.2	27.4	26.4	L-SCL
	Cz	45–75	0.31	27.4	91.7	24.0	79.5	8.0	34.0	55.7	24.0	20.3	SCL-SL
	2Byzb1	75–108	0.28	29.1	93.4	24.5	88.8	8.0	47.9	16.2	52.4	31.4	SiCL
	2Byb2	108–140	0.19	27.5	32.8	25.0	29.1	7.7	43.2	20.2	58.4	21.4	SiL
27	Ap	0–20	–	2.3	2.5	20.0	5.0	7.9	48.4	20.2	48.0	31.8	CL-SiCL
	By1	20–50	–	19.7	17.5	21.7	13.0	8.0	50.0	24.2	50.0	25.8	SiL
	By2	50–75	–	28.4	27.8	23.6	27.9	8.0	50.0	34.2	37.4	28.4	CL
	C1	75–95	–	30.1	31.1	24.8	34.8	7.9	39.2	40.2	37.4	22.4	L
	C2	95–115	–	31.0	31.4	23.1	36.9	7.9	40.8	36.2	42.4	21.4	L
	C3	115–140	–	31.4	32.1	24.8	38.0	7.8	38.1	52.2	29.4	18.4	SL-L

OC organic carbon, *ESP* exchangeable sodium percentage, *SAR* sodium adsorption ratio, *ECe* electrical conductivity of soil saturation extract, *SP* saturation percentage, – = not reported, *SL* sandy loam, *LS* loamy sand, *SiCL* silty clay loam, *CL* clay loam, *SCL* sandy clay loam, *L* loam, *SiL* silty loam

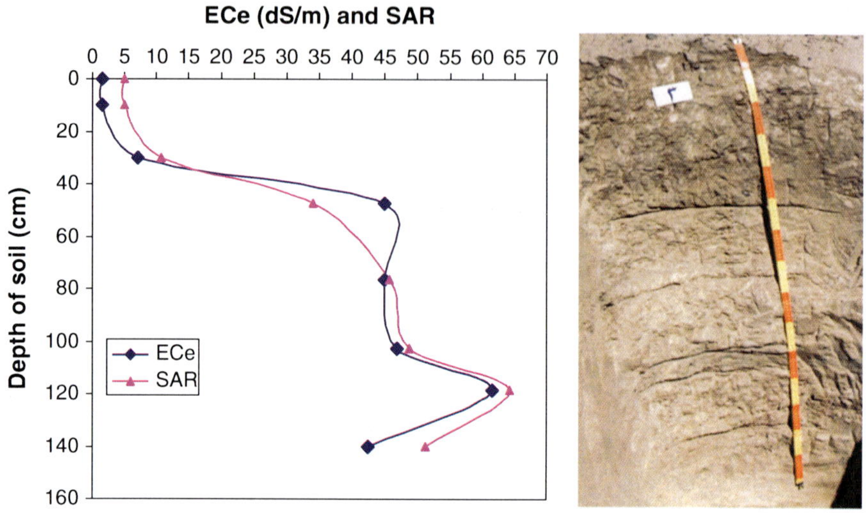

Fig. 22.1 In-depth trends of ECe and SAR in profile no. 3 (piedmont plain)

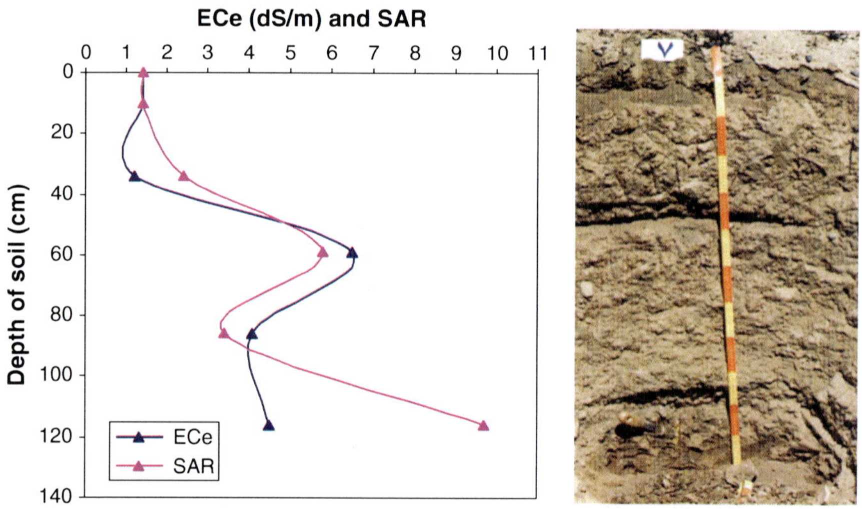

Fig. 22.2 In-depth trends of ECe and SAR in profile no. 7 (piedmont plain)

material on which profiles are developed. In another study of Hamadan plain of Asadabad, Asadiyan (1989) has cited geological formations (e.g. schist and calcareous stones) being responsible for salinity and alkalinity development.

It has been observed that the regional topography has played a crucial role in salt transfer and accumulation. The results revealed that the soils with lower salinity and coarse textures are lying at higher altitudes of the region (profiles nos. 3, 7 and 13),

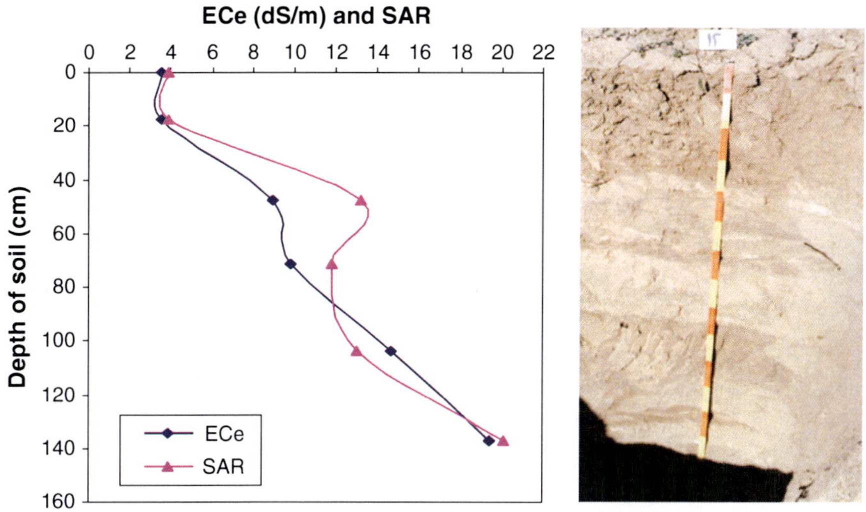

Fig. 22.3 In-depth trends of ECe and SAR in profile no. 13 (piedmont plain)

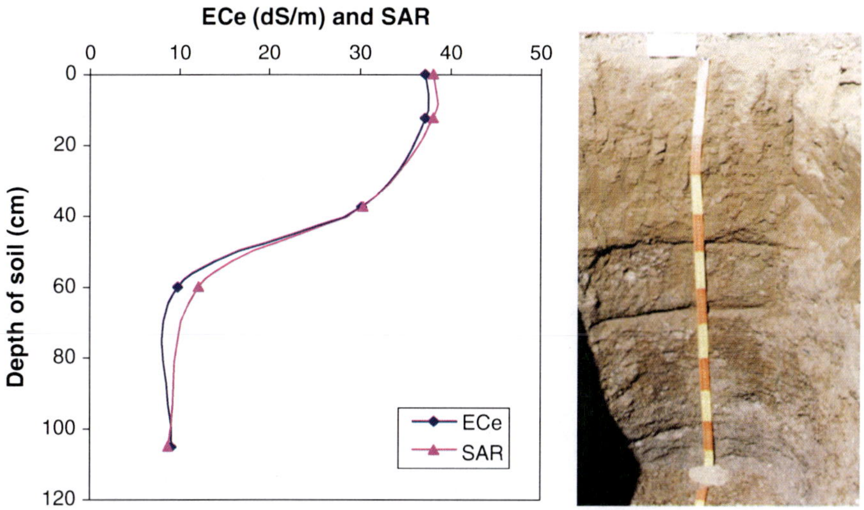

Fig. 22.4 In-depth trends of ECe and SAR in profile no. 21 (piedmont plain)

whereas the soils higher in salinity and fine texture lie at lower altitudes of the region (profile nos. 4 and 27). Having investigated the causes of soil salinity and alkalinity, Szabolcs (1985) came to the conclusion that these two phenomena were affected by various landscapes. High salinity at 75-cm depth (profile no. 4, Fig. 22.6) has formed as salic horizon. Taimeh (1992) has demonstrated the constitution of salic horizon in the absence of subterranean and phreatic water effects and

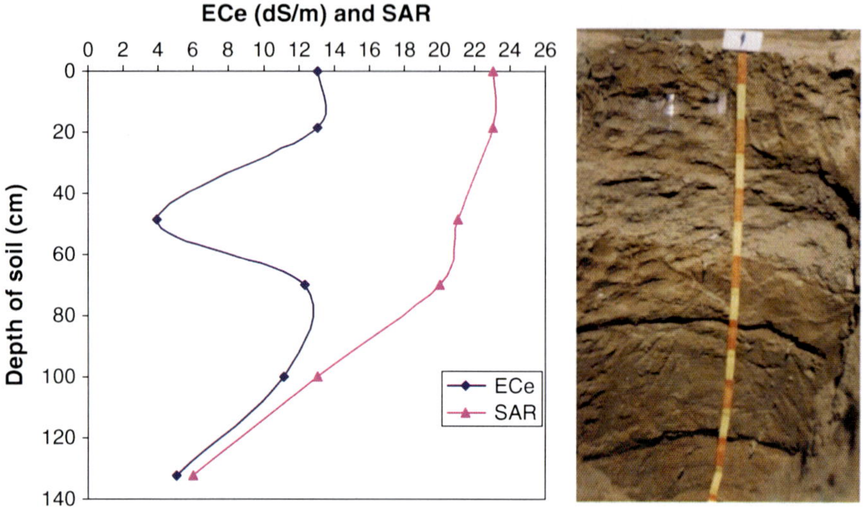

Fig. 22.5 In-depth trends of ECe and SAR in profile no. 1 (flood plain)

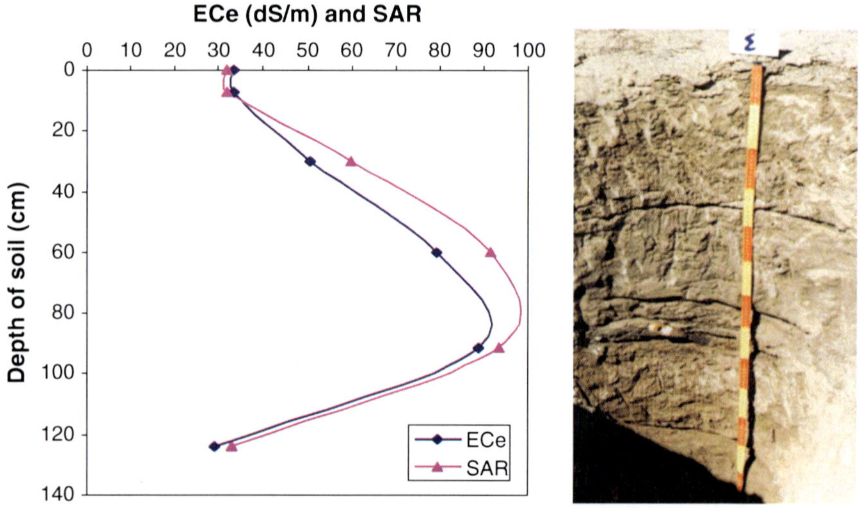

Fig. 22.6 In-depth trends of ECe and SAR in profile no. 4 (flood plain)

concluded that this horizon has been shaped due to wind activity and alluvial deposits, and the maximum surface salinity is due to wind-borne (aeolian) deposits.

Besides the running waters, wind is the next important agent to transport salt crystals or salty materials from the upper parts to the low-lying lands. Barren surface and scarce vegetation cover enhance the wind action. This is evident from accumulated coarse fractions around the bushes in the area of pedon number 3, which lies on the

Fig. 22.7 Soil salinity map of Iran indicating area under salt-affected soils and other types of landscape (Modified from Banie 2001)

central part of the plain. It has been estimated that nearly 50 cm of this profile is filled with wind-deposited materials within an interval of 10 years.

The performance of plant species used as a management strategy for salt-affected soils can be improved by seed improvements: (1) pre-soaking seed treatment, that is, soaking of seeds in freshwater before sowing; (2) pretreatment of seeds for improved germination; (3) applying farm manure to the soil, where available; and (4) changing cropping patterns to avoid extreme stresses from hot and dry weather and high concentrations of salts (Qadir and Oster 2004; Qadir et al. 2008). Soil salinity map of Iran showing different salinity zones as well as different landscapes is presented in Fig. 22.7 (Banie 2001). The soil salinity map of Iran reveals 27% area affected to various level of salinity. Of 27%, 13% area is severely (<16 dS m^{-1}) affected, 11% area is moderately (8–16 dS m^{-1}) affected, and 3% area is slightly (4–8 dS m^{-1}) affected (Fig. 22.7). These affected areas need careful management and reclamation for the optimum use for various purposes such as irrigated agriculture.

Table 22.3 Classification of soil profiles in piedmont plain (USDA 1994)

Profile no.	Family	Subgroup	Order	Proposed
3	Coarse loamy mixed thermic	Sodic Haplocambids	Aridisols	Sodic Torrifluvents
7	Fine loamy mixed thermic	Typic Haplocalcids	Aridisols	
11	Fine loamy mixed thermic	Typic Haplocalcids	Aridisols	
13	Coarse loamy mixed (calcareous)thermic	Sodic Haplocalcids	Aridisols	
21	Coarse loamy mixed (calcareous)thermic	Typic Torriorthents	Entisols	Sodic Torriorthents

Table 22.4 Classification of soil profiles in flood plain (USDA 1994)

Profile no.	Family	Subgroup	Order	Proposed
1	Fine loamy mixed thermic	Typic Torriorthents	Entisols	Sodic Torriorthents
4	Fine loamy mixed thermic	Gypsic Haplosalids	Aridisols	
27	Fine loamy mixed thermic	Sodic Haplogypsids	Aridisols	

22.3.1 *Classification of Sloppy and Flood Plains Soils*

Profiles nos. 7 and 13 are classified as calcids due the presence of calcic horizon. Profile no. 7 is classified as Typic Haplocalcids and profile no. 13 as Sodic Haplocalcids due to high ESP at a depth of 100 cm from soil surface. The soil classification results are presented in Tables 22.3 and 22.4 for piedmont and flood plains, respectively.

The sodic torrifluvents is a newly proposed subgroup, with the following definition: Sodic torrifluvents are torrifluvents that have an ESP > 15 or an SAR > 13 $(\text{mmoles L}^{-1})^{0.5}$ in all layers within the upper 50 cm which decrease with depth below 50 cm.

22.4 Conclusions

The development of irrigation systems in the area with deep water table had an ameliorative effect on the salinity and sodicity of the soils as the salts from the upper soil layers have been moved to the deeper layers. With variable levels of success, different approaches, such as salt leaching and drainage interventions, crop-based management, chemical amendments and fertilizers, and integrated application of these approaches, have been used to enhance the productivity of salt-affected soils in the country. The soils revealed to fit with Aridisols and Entisols orders and calcids, salids, gypsids, cambids, fluvents and orthents suborders, accordingly. The new subgroup of sodic torrifluvents has been proposed.

References

Abrol IP (1972) Saline and alkali soils in India, their occurrence and management. Central Soil Salinity Research Institute (CSSRI), Karnal

AlLayla AM (1978) Effect of Salinity on Agriculture in Iraq. J Irrig Drain Div 104(2):195–207

Anis Al-Layla M (1978) Effect of salinity on agriculture in Iraq. J Irrig Drain Div 104(2): 195–207

Asadiyan MS (1989) A survey of the soil salinity and alkalinity of Hamadan Plain of Asadabad. MSc thesis, Pedology, College of Agriculture, Tehran University

Ayers AD (1960) Saline and sodic soils of Spain. Soil Sci 90(2):133–137

Banie MH (2001) Soil map of Iran: land resources and potentialities. Soil and Water Research Institute (SWRI), Tehran

Burt R (ed) (2004) Soil survey laboratory methods manual. SSIR No 4., Ver 4.0. US Government Printing Office, Washington, DC

Cheraghi SAM (2004) Institutional and scientific profiles of organizations working on saline agriculture in Iran. In: Taha FK, Ismail S, Jaradat A (eds) Prospects of saline agriculture in the Arabian Peninsula: proceedings of the international seminar on prospects of saline agriculture in the GCC countries 18–20 March 2001, Dubai, United Arab Emirates. Amherst Scientific Publishers, Amherst, pp 399–412

Mahjoory RA (1975) Clay mineralogy, physical and chemical properties of some soils in arid region of Iran. Soil Sci Soc Am Proc 39:1157–1164

Mahjoory RA (1979) The nature and genesis of some salt-affected soils in Iran. Soil Sci Soc Am J 43:1019–1024

Qadir M, Oster JD (2004) Crop and irrigation management strategies for saline-sodic soils and waters aimed at environmentally sustainable agriculture. Sci Total Environ 323:1–19

Qadir M, Qureshi AS, Cheraghi SAM (2008) Extent and characterization of salt-affected soils in Iran and strategies for their amelioration and management. Land Degrad Dev 19:214–227

Soil Survey Division Staff (1993) Soil survey manual. USDA handbook no 18. USDA, Washington, DC

Szabolcs I (1985) Assessment of soil salinity and alkalinity. Taxonomy and management of vertisols and aridisols. In: Proceedings of the fifth international soil classification workshop, Sudan, 2–11 November 1982. Part 1(198):253–261

Taimeh AY (1992) Formation of salic horizon without water table in an arid region. Soil Sci 154(5):399–409

USDA (1994) Keys to soil taxonomy, 6th edn. USDA-NRCS. US Government Printing Office, Washington, DC

Chapter 23
Classification, Characterization, and Management of Some Agricultural Soils in the North of Egypt

Sabry M. Shaheen, Mohamed E. Abo-Waly, and Rafaat A. Ali

Abstract Present study is completed on soils developed from different sources and types of materials (fluvial, lacustrine, marine, sandy and fluvio-sandy, and calcareous deposits). These soils belong to orders entisols and aridisols. The diverse geological nature of the deposits on which these soils are developed is reflected in the wide variation of soil characteristics (morphologies, clay, cation-exchange capacity, carbonate equivalents, and oxides of Fe, Al, Si, and Mn). All soils are alkaline in reaction, ECe (salinity) ranges between 0.66 and 8.0 dS m^{-1}, clay (6.2–57.5%), cation-exchange capacity (3.0–79.1 cmolc kg^{-1}), organic matter (0.29–2.68%), and calcium carbonate equivalents (0.07–55.62%). Total Fe, Al, and Mn concentrations differed greatly between soils, and the majority of the Fe and Mn occurred in crystalline form. Aluminum oxides are amorphous especially in clayey soils. Total free Si was similar to or exceeded to those of amorphous Si in all soils except calcareous ones. The soils developed from different materials are classified, such as fluvial (Vertic Torrifluvents, Typic Torrifluvents, Typic Fluvaquents, and Typic Ustifluvents), lacustrine (Typic Xerofluvents and Typic Fluvaquents), marine (Typic Xeropsamments and Typic Psammiaquents), sandy (Typic Quartzipsamments and Typic Torripsamments), and calcareous (Typic Haplocalcids).

Keywords Agricultural soils • Classification • Egypt • Macromorphology • Sesquioxides

S.M. Shaheen (✉) • M.E. Abo-Waly • R.A. Ali
Department of Soil & Water Sciences, Faculty of Agriculture,
Kafrelsheikh University, 33516-Kafr El-Sheikh, Egypt
e-mail: smshaheen1973@yahoo.com; mabowaly@yahoo.com;
rafaataali1950@yahoo.com

S.A. Shahid et al. (eds.), *Developments in Soil Classification, Land Use Planning and Policy Implications: Innovative Thinking of Soil Inventory for Land Use Planning and Management of Land Resources*, DOI 10.1007/978-94-007-5332-7_23,

23.1 Introduction

The Nile Delta and the Nile River Valley of Egypt is one of the oldest agricultural areas in the world. These areas are under continuous cultivation for at least 5,000 years. Egypt is situated at the northeast extremity of Africa; it lies between the latitudes 31° N and 22° N and between the longitudes 25° E and 34° E. Its gross area is 1,001,450 km^2, of which about 3.6% is suitable for agricultural production. Physiographically, Egypt is divided into three main regions, the delta, the Nile Valley, and the deserts (Amer and Abo-Zeid 1989).

Soil properties and classifications are influenced by the type of materials from which the soils are developed and the conditions of depositional environments. Pedologists intend to establish relationships between the deposition environments and soils morphological, physical, chemical, and mineralogical characteristics as well as soil classifications (Perttjhon 1984).

The soil parent materials could be from sources such as rocks and mineral mixtures, sediments from floods, glacial deposits, and aeolian deposits (Raymond and Roy 1995). Fluvial deposits include all sediments deposited by the running waters. These deposits are generally poorly sorted and reveal evidence of rapid but interrupted deposition. The alluvial soils are developed through the deposition of alluvium brought by streams to the land (Worcoster 1969). The lacustrine deposits are settled out of stagnant lakes, whereas the marine deposits originate from stream action and deposited in the sea (FAO-SF 1964; Selley 1982).

It is well established that the type and source of the soil materials and the depositional environments control the soil properties and classifications due to particle-size distribution and the mineralogical composition. Present study was completed with the objective to investigate the influence of different sources of soil materials on soil properties and classifications and based on the basis of these results to discuss management issues of these soils.

23.2 Materials and Methods

23.2.1 *Field Description and Sample Collection*

Eighteen soil profiles from cultivated entisols and aridisols representing different geological deposits of Egypt (fluvial, lacustrine, marine, sandy and fluvio-sandy, and calcareous) were selected for the present study. Fluvial soils were selected from the Nile Delta, lacustrine soils (Burullus lake, north of the Nile Delta), sandy marine soils (north of Nile Delta by the Mediterranean sea), and calcareous soils (north of the western desert of Egypt). Soil profiles were dug, various genetic horizons and layers described using standard FAO procedures (FAO 1970), as well as using American system of soil classification (Soil Survey Staff 2006). The horizons were described for soil color (moist, dry), soil texture, structure, and consistence (wet, moist, dry), and effervescence test was made to check calcium carbonates. Soil samples were collected from representative depths of the soil profiles.

23.2.2 *Laboratory Procedures*

The soil samples were processed prior to their analyses in the laboratory. The samples were air-dried and ground to pass through a 2-mm sieve to obtain fine earth fraction. The soil samples were analyzed for routine physical and chemical characteristics using the procedures referred by Sparks et al. (1996). EC of the soil saturation extract (ECe) was measured by standard EC meter. Soluble cations (Na, K, Ca, Mg) and anions (CO_3, HCO_3, Cl, SO_4) were measured using standard titration method and atomic absorption spectrophotometer (AAS) as appropriate. Soil reaction (pH) was measured in a 1:2.5 soil-deionized water suspension (Thomas 1996). Calcium carbonate equivalents were determined by calcimeter. Organic matter content was determined by the Walkley-Black method (Walkley 1947). Cation-exchange capacity (CEC) was measured by saturating the soil exchange complex with 1N ammonium acetate solution at pH 7 (Sumner and Miller 1996); the ammonium was replaced and measured to represent CEC. Extractable cations (Na, K, Ca, Mg) were determined in 1N ammonium acetate extract; the difference between the extractable and soluble cations represented as exchangeable cations. Particle-size distribution analysis was accomplished by standard pipette method (Gee and Bauder 1986) following by wet sieving to quantify five sand subfractions (very coarse, coarse, medium, fine, and very fine). Soil texture was determined using USDA textural triangle (Soil Survey Division Staff 1993). Total Fe, Al, and Mn were extracted by using mixture of concentrated nitric acid, concentrated hydrochloric acid, and 30% hydrogen peroxide according to USEPA method (USEPA 1995). Total free iron oxides (FIO) were extracted with 3 M sodium citrate, 1 M sodium bicarbonate, and 1 g sodium dithionite in a water bath at 85 °C (Mehra and Jackson 1960); Al, Mn, Si, and Ca were also measured in the same extract. Amorphous Fe oxides as Fe (Feox) were extracted with 0.175 M ammonium oxalate + 0.1 M oxalic acid adjusted to pH 3.0. Calcareous soils were pretreated with 1N sodium acetate (pH 5.5) to remove carbonates according to Loeppert and Inskeep (1996). The P, Al, Mn, Si, and Ca were measured in this extract. The Fe, Al, Mn, Si, and Ca were measured by atomic absorption spectrometry (Varian, SpectrAA-400 Plus, Australia).

23.3 Results and Discussion

In this section, soils developed from different sources of materials are described.

23.3.1 *Fluvial Soils*

23.3.1.1 Field Description, Soil Classification, Physical, and Chemical Characteristics

Table 23.1 illustrates general aspects and soil classification of the fluvial soil profiles. Generally, the soil color is very dark grayish brown (moist) and dark grayish brown (dry); massive structure and soil consistency is sticky and plastic (wet), firm (moist), and hard (dry). These properties are due to the type of sediments with varying clay contents, cemented materials, and type of clay minerals. Based on the genetic horizons development and the sequence in the profiles, the fluvial profiles are classified as Typic Torrifluvents, Typic Fluvaquents, Typic Ustifluvents, and Vertic Torrifluvents (Table 23.1).

Silty clay (heavy) is the common soil texture in various soil horizons of all soil profiles (Table 23.2), except at few depths where other textures were found (clay loam, silty clay loam). The ranges of primary soil particles are clay (17.0–57.5%), silt (14.1–53.8%), and sand (<1–68.9%). The sand was irregularly distributed with depth, the dominant being in the fine and very fine fractions (250–50 μm).

Data in Table 23.2 showed that fluvial soils (P1, P2, P4, and P5) are either nonsaline (ECe < 2 dS m^{-1}), or they are (P3 and P6) slightly saline (ECe 4–8 dS m^{-1}). There is slight but an insignificant in-depth difference of salt distribution within the profile. The highest ECe was recorded at the subsurface (P6); this could be attributed to the mobility of salts downward with descending irrigation water or from below through rising water table. Relatively water soluble Na^+ is dominant over the Ca^{2+}, Mg^{2+}, and K^+. General trend for cations (Na > Ca + Mg > K) and anions ($Cl > SO_4 > HCO_3 > CO_3$) was observed. In P1, different trend exists for cations (Ca > Mg > Na > K) and anions ($Cl > HCO_3 > SO_4 > CO_3$). Increasing and decreasing trend of other properties is shown in Table 23.2. The soil reaction (pHs) in general is in the moderately alkaline range (7.9–8.4).

The organic matter is low (0.88–1.26%). The relatively high value at the surface is due to the residual effect of cultivation on these soils and the addition of organic manure and plant residues to improve soils physical and fertility aspects. All profiles indicate different trends in distribution of total $CaCO_3$ with depth. However, most of them present less than 3%, other between 3 and 5%. From such low quantities, it is difficult to confirm the calcification process. Higher quantities at surface (Sparks 1995) can be due to secondary carbonates.

The CEC values are due to higher clay contents in the profiles. The CEC shows irregular trend of distribution in the profiles; this could be due to the type of sediments, clay minerals, and organic matter and perhaps due to fine clay illuviation; however, this has not been confirmed in the present study. The values of the exchangeable Na and CEC were used to determine soil sodicity, which shows soils to be in general highly sodic with ESPs ranging between 27.6 and 60.6.

Table 23.1 The location, general description, and soil classification of fluvial soil profiles

Profile no.	Site	Depth (cm)	Soil description: Color – Moist	Color – Dry	Texture-class[a]	Structure	Soil consistencies – Wet	Moist	Dry	General features	Soil CLASSIFICATION
P1	Soils of an island in Disuq district, Kafr El-Sheikh Governorate, 100 m to the west of Disuq Damanhur road	0–20	Very dark grayish brown 10YR 3/2	Grayish brown 10YR 5/2	SiC	Massive	Sticky and plastic	Firm	Hard	Few fine roots of Zea mays, few shell fragments, and weak effervescence with HCl. Ten-cm deep and 1-cm wide surface cracks	Typic Torrifluvents
		20–80	Very dark grayish brown 10YR 3/2	Dark grayish brown 10YR 4/3	SiC	Massive	Very sticky and very plastic	Firm	Hard	Few fine and very fine roots, few small shell fragments especially in 50–80 cm layer, and weak effervescence with HCl	
		80–120	Olive brown, 2.5Y 4/4	Light olive brown, 2.5Y 5/6	SiC	Granular	Slightly sticky and slightly plastic	Friable	Slightly hard	Very few small shell and very weak effervescence with HCl	

(continued)

Table 23.1 (continued)

Profile no.	Site	Depth (cm)	Soil description: Color – Moist	Color – Dry	Texture-class[a]	Structure	Soil consistencies – Wet	Moist	Dry	General features	Soil CLASSIFI-CATION
		120–150	Light olive gray 5Y 6/2	Light gray 5Y 7/1	SiC	Massive	Slightly sticky and slightly plastic	Firm	Slightly hard	Very few dead roots, few small shell fragments, and weak effervescence with HCl	
P2	El-Banawan village, El-Mahalla El-Kubra district, El-Gharbia Governorate	0–120	Very dark grayish brown 10YR 3/2	Dark grayish brown 10YR 4/2	SiC	Massive	Sticky and plastic	Firm	Hard	Level land, cultivated for cotton crop, fine and medium roots diffused in the surface cracks (>10 cm deep and 1 cm wide), slight to moderate effervescence with HCl. The soil profile is homogeneous. Upper layers show clear morphological features	Typic Torrifluvents

P3	Kafr Dukhmeis village, El-Mahall El-Kubra district, El-Gharbia Governorate	0–120	0–120 Very dark grayish brown 10YR 3/2	Dark grayish brown 10YR 4/2	SiC SiCL at 6–90 cm	Massive	Sticky and plastic	Firm	Hard	Level land used for vegetable cultivation. Fine and medium roots diffused in the surface cracks (>10 cm deep and 1 cm wide). Some scattered broken shells occur throughout showing moderate effervescence with HCl. The soil profile is homogeneous. Upper layers show clear morphological features	Typic Torrifluvents

(continued)

Table 23.1 (continued)

Profile no	Site	Depth (cm)	Soil description: Color Moist	Color Dry	Texture-class[a]	Structure	Soil consistencies Wet	Soil consistencies Moist	Soil consistencies Dry	General features	Soil CLASSIFI-CATION
	El-Khadmia village, Kafr El-Sheikh district, Kafr El-Sheikh Governorate	0–20	Very dark grayish brown 10YR 3/2	Dark grayish brown 10YR 4/2	SiC	Massive	Sticky and plastic	Firm	Hard	Level land, cultivated by cotton and vegetable crops. Fine and medium roots diffused in this layer, moderate effervescence with HCl	Typic Fluvaquents
P4		20–80	Very dark grayish brown 10YR 3/2	Dark grayish brown 10YR 4/2	CL	Massive	Slightly sticky and slightly plastic	Firm	Slightly hard	Few fine roots diffused. Few shell fragments and weak to moderately effervescence with HCl	
		80–105	Dark grayish brown 10YR 4/2	Pale brown 10YR 6/3	SiC	Massive	Very sticky and very plastic	Very firm	Hard		

P5	El-Banawan village,	0–120	Very dark grayish brown 10YR 3/2	Dark grayish brown 10YR 4/2	SiC	Massive	Sticky and plastic in all layers and very sticky and very plastic in the deepest layer	Firm in all layers and very firm in the deepest layer	Hard	Level land, cultivated by cotton with some vegetable crops. Fine and medium roots diffused in the surface, slight to moderate effervescence with HCl. The soil profile is homogeneous. Upper layers show clear morpho logical features	Typic Ustifluvents

(continued)

Table 23.1 (continued)

Profile no	Site	Depth (cm)	Soil description								
			Color		Texture-class[a]	Structure	Soil consistencies			General features	Soil CLASSIFI-CATION
			Moist	Dry			Wet	Moist	Dry		
P6	Faculty of Agriculture farm. 500 m South Kafr El-Sheikh City	0–160	Very dark grayish brown 10YR 3/2	Dark grayish brown 10YR 4/2	SiC	Strong, coarse to medium, sub-angular blocky	Sticky and plastic	Firm	Hard	Fine and medium roots diffused in the surface, cracks (>20 cm deep and 1 cm wide). Slight to moderate effervescence with HCI The soil profile is homogeneous throughout its layers	Vertic Torrifluvents

[a]Apparent soil texture
Texture – *SiC* silty clay, *SL* sandy loam, *SCL* sandy clay loam, *SiCL* silty clay loam, *CL* clay loam

Table 23.2 Selected physical and chemical characteristics of fluvial soil profiles

Profile no.	Depth (cm)	pHs	ECe (dS m^{-1})	Soluble cations (meq l^{-1})[a] Na	K	Ca	Mg	Soluble anions (meq l^{-1})[a] CO_3	HCO_3	Cl	SO_4	OM (%)	$CaCO_3$ (%)	CEC (cmolc kg^{-1})	ESP	Particle-size distribution (%) VCS	CS	MS	FS+VFS	Sand	Silt	Clay	Texture class
P1	0–20	7.87	0.79	2.6	0.2	2.9	2.5	0.0	3.2	4.0	1.0	1.92	1.6	79	30	0.5	0.4	2.7	8.8	12.4	40.4	47.2	SiC
	20–50	7.87	0.62	2.1	0.1	2.2	2.3	0.0	2.5	3.1	1.1	1.64	1.5	65	28	0.1	0.2	0.2	1.2	1.7	46.5	51.8	SiC
	50–80	7.93	0.56	1.7	0.1	2.1	2.0	0.0	2.5	2.3	1.1	1.23	3.5	69	31	0.1	0.1	0.2	0.8	1.3	46.4	52.3	SiC
	80–120	7.98	0.63	1.7	0.1	2.0	2.7	0.0	1.7	3.4	1.4	0.48	1.1	38	39	1.7	2.8	6.9	57.5	68.9	14.1	17.0	SiC
	120–150	7.96	0.68	1.9	0.1	2.3	2.4	0.0	0.9	3.3	2.5	1.02	2.3	59	29	1.2	2.0	8.4	3.26	44.1	27.3	28.6	SiCL
P2	0–30	8.11	1.56	11.4	0.2	2.3	2.1	1.8	2.7	9.0	2.5	1.67	2.4	44	32	0.2	0.2	0.3	1.7	2.4	50.2	47.4	SiC
	30–60	8.31	1.62	12.3	0.2	2.0	2.4	1.3	2.7	7.2	5.7	1.11	1.8	44	33	0.3	0.1	0.1	2.2	2.7	47.5	49.8	SiC
	60–90	8.39	1.65	12.1	0.2	2.4	2.0	0.8	3.0	6.4	6.5	0.70	2.7	42	34	1.2	0.6	0.2	1.6	3.6	52.2	44.2	SiC
	90–120	8.42	1.61	11.8	0.1	2.0	1.8	0.4	2.0	5.6	7.7	0.90	1.6	43	34	0.4	0.8	0.2	1.5	2.8	48.7	48.5	SiC
P3	0–30	7.92	7.06	45.0	0.6	15.0	11.0	0.8	2.0	39.0	29.8	1.58	4.2	43	33	0.5	0.5	3.2	11.4	15.6	41.9	42.5	SiC
	30–60	7.88	6.77	45.0	0.5	14.0	9.7	0.4	2.6	29.2	37.0	0.97	3.9	44	37	0.4	0.4	2.8	7.6	11.2	44.8	44.0	SiC
	60–90	8.17	5.75	47.5	0.3	6.0	9.9	0.0	2.2	31.8	29.7	0.62	3.1	39	42	0.5	0.5	2.9	9.8	13.7	47.6	38.8	SiCL
	90-120	8.29	4.79	37.5	0.2	4.3	4.7	0.0	2.6	25.8	18.3	0.35	4.7	41	41	0.3	0.4	1.9	6.4	9.0	45.8	45.2	SiC
P4	0–20	8.01	2.00	11.7	0.2	3.7	4.2	1.4	2.2	12.8	3.4	2.00	3.9	41	33	0.4	0.4	2.0	6.5	9.3	40.2	50.5	SiC
	20–50	7.93	1.68	9.1	0.1	4.3	3.3	0.0	3.4	11.2	2.2	1.04	4.8	36	30	0.7	0.8	4.0	15.3	20.8	48.6	30.6	CL
	50–80	8.25	1.93	10.1	0.1	4.7	3.7	0.0	3.0	14.9	0.7	0.62	2.3	34	30	0.7	0.6	2.8	11.7	15.8	47.8	36.4	CL
	80–105	7.39	1.62	8.6	0.2	4.4	2.8	0.0	3.0	11.3	1.7	1.31	0.6	52	38	0.3	0.2	0.3	1.4	2.2	44.9	52.9	SiC
P5	0–30	8.24	1.65	10.9	0.1	3.7	2.7	1.2	3.2	9.6	3.4	2.10	3.7	52	28	0.4	0.6	0.9	1.3	3.2	49.5	47.3	SiC
	30–60	8.04	2.23	14.4	0.2	4.0	3.2	0.6	2.5	12.4	6.3	1.24	3.4	47	28	0.2	0.1	0.3	1.2	1.7	53.8	44.5	SiC
	60–90	8.19	1.25	8.5	0.1	2.3	1.7	0.0	3.4	6.8	2.4	0.82	1.6	57	31	0.2	0.2	0.4	1.1	1.8	50.0	48.2	SiC
	90–120	8.35	1.21	8.8	0.2	1.8	1.4	0.5	2.9	6.8	2.0	0.82	1.2	59	34	0.1	0.1	0.1	0.4	0.7	48.5	50.8	SiC

(continued)

Table 23.2 (continued)

Profile no.	Depth (cm)	pHs	ECe (dS m^{-1})	Soluble cations (meq l^{-1})[a] Na	K	Ca	Mg	Soluble anions (meq l^{-1})[a] CO_3	HCO_3	Cl	SO_4	OM (%)	$CaCO_3$ (%)	CEC (cmolc kg^{-1})	ESP	Particle-size distribution (%) VCS	CS	MS	FS + VFS	Sand	Silt	Clay	Texture class
P6	0–30	8.93	4.00	25.7	0.4	6.1	7.9	0.0	4.2	30.2	5.7	2.22	3.1	57	58	0.7	0.8	0.9	3.2	5.5	49.2	45.2	SiC
	30–60	8.81	5.10	37.2	0.4	4.2	10.1	0.0	2.1	35.1	14.6	1.07	2.3	58	57	0.4	0.7	0.7	2.2	4.1	50.3	45.6	SiC
	60–90	9.14	5.10	37.9	0.3	4.9	10.9	0.0	1.8	40.0	12.2	1.02	2.2	52	51	0.3	0.4	0.4	1.6	2.8	47.7	57.5	SiC
	90–120	9.07	6.00	45.0	0.3	6.0	11.8	0.0	1.5	45.2	16.4	0.95	2.3	52	60	0.2	0.3	0.3	1.9	2.8	49.2	48.0	SiC
	120–140	8.86	6.90	52.2	0.4	8.4	12.7	0.0	1.6	50.0	22.4	0.91	1.4	59	61	0.1	0.2	0.2	0.7	1.1	48.8	50.1	SiC
	140–160	8.70	8.00	67.6	0.4	9.3	16.7	0.0	2.8	54.9	36.3	0.91	1.8	63	60	0.1	2.0	0.2	0.7	1.1	49.8	49.1	SiC

[a]In soil saturation extract

ECe EC of soil saturation extract, *VCS* very coarse sand, *CS* coarse sand, *MS* medium sand, *FS* fine sand, *VFS* very fine sand. Texture class – *SiC* silty clay, *SiCL* silty clay loam, *CL* clay loam, *CEC* cation-exchange capacity, *OM* organic matter, *pHs* pH of saturated soil paste, *ESP* exchangeable sodium percentage

23.3.2 *Lacustrine Soils*

23.3.2.1 Field Description, Soil Classification, and Physical and Chemical Characteristics

General aspects and soil classification of four lacustrine profiles are presented in Table 23.3. Field description showed very dark grayish brown color in the moist state of most layers in all profiles, while in dry state, the dominant color was dark grayish brown. The structure in all layers of four tested profiles was massive and massive blocky structure except for the surface layer in P1 and P2, where it was coarse to medium blocky structure. The soil consistence was sticky and plastic in surface layers and very sticky, very plastic in the deepest ones in the wet state, firm in moist state, and hard in dry state.

These features are mainly due to the nature of sediment, the increase of clay content, and cemented materials as well as the type of clay minerals. The classifications of all lacustrine profiles were Typic Fluvaquents (P2 and P4) and Typic Xerfluvents (P1 and P3).

Silty clay is the common soil texture, followed by silty clay loam and in one depth (P4), it is clay loam (Table 23.4). Table 23.4 illustrates heavy texture at surface and subsurface layers and also in the deepest layers, while the texture was relatively light in the middle layers in profile nos. 2, 3, and 4. Such variations in soil texture are due to multi-depositional cycles. The clay (21.4–56.3%) and sands (0.98–55.05%) were distributed variably. The fine and very fine sand (250–50 μm) was the dominant fractions. The differences in soil texture, $CaCO_3$ content, and CEC values are closely related to the matter of sedimentation mode and the nature of parent materials.

The ECe is relatively higher at deeper layers due to leaching. The difference is due to type of irrigation water used, clay content, and drainage conditions. Salt distributions throughout the profiles were irregular and differ in profiles. The soil reaction (pHs) ranges between 7.93 and 9.13 (moderately to strongly alkaline reaction), and ECe is generally less than 4 dS m^{-1} (Table 23.4). Water soluble sodium was dominant among cations, other cations trend $Mg^{2+} > Ca^{2+} > K^+$, and among anions, Cl^- was dominant over others trending $SO_4 > HCO_3$. Lacustrine soils exhibited more soluble Mg^{2+} and K^+ relative to other soils. Similar trends are reported earlier (Ali 1985).

The organic matter content is low. The relatively higher but insignificant values at surface are due to cultivation and addition of organic manure and plant residues. Surface and subsurface layers (P2, P3, and P4) have high $CaCO_3$ (17.3%), while the deepest layers of P4 have the lowest (0.07%). The high values of $CaCO_3$ are due to the presence of broken and complete shells.

The CEC values vary greatly and have irregular trend due to sedimentation regime, clay, and organic matter contents. The soils (P1 and P2) are sodic where ESP > 15; other profiles are non-sodic (ESP < 15). The sodic soils are to be amended with gypsum based on gypsum requirement to improve properties of these profiles.

Table 23.3 The location, general description, and soil classification of lacustrine soil profiles

Profile no.	Site	Depth (cm)	Soil description: Color – Moist	Color – Dry	Texture class[a]	Structure	Soil consistencies – Wet	Moist	Dry	General aspects	Soil classification
P1	Khalej kepli, Mutubis district, Kafr El-Sheikh Governorate	0–120	Very dark grayish brown 10YR 3/2	Dark grayish brown 10YR 4/2	SiC	Coarse to medium blocky in surface layer and massive in deeper layers	Sticky and plastic in all layers, very plastic in the deepest layer	Firm to very firm	Hard	Level land, plowed after *Zea mays* crops. Fine and medium dead roots in the surface. Scattered shell fragments at surface and strong effervescence with HCl. Few shells and moderate effervescence with HCl in the subsurface layers. Some red mottling in the fluctuating water zone	Typic Xerfluvents
P2	El-Thamaneen village, El-Hamul district Kafr El-Sheikh Governorate	0–120	Very dark grayish brown 10YR 3/2	Dark grayish brown 10YR 4/3	SiC all layers SiCL at 60–90 cm	Coarse to medium blocky in surface layer and massive in deeper layers	Sticky and plastic in all layers and very plastic in the deepest layer	Firm to very firm	Hard	Level land, plowed after *Citrullus* sp. crops. Fine and medium dead roots in the surface. Scattered shell fragments in all layers except the deepest one showing strong effervescence with HCl in all layers, except the deepest one. Red mottling in the fluctuating water table zone due to oxidation of iron	Typic Fluvaqents

P3	El-Khasha, 500 m to the east of Burulus lake, 7 km to the west of Baltim-El-Hamul road	0–27	Very dark grayish brown 10YR 3/2	Dark grayish brown 10YR 4/2	SiC	Massive	Sticky and slightly plastic	Firm	Slightly hard	Scattered shell fragments band of broken and non-broken shells, few fine roots, strong effervescence with HCl	Typic Torrifluvents
		27–37	Very dark grayish brown 10YR 3/2	Dark brown 10YR 4/3	SiC	Massive	Sticky and plastic	Firm	Hard	Scattered shell fragments, very fine roots, strong effervescence with HCl	
		37–60	Very dark grayish brown 10YR 3/2	Dark brown 10YR 4/3	SiC	Massive	Sticky and plastic	Firm	Hard	Few scattered shell fragments, very fine roots, weak effervescence with HCl	
		60–70	Dark grayish brown 10YR 4/2	Pale brown 10YR 6/3	SiCL	Massive	Slightly sticky and slightly plastic	Slightly firm	Slightly hard	Many scattered shell fragments, strong effervescence with HCl	

(continued)

Table 23.3 (continued)

Profile no.	Site	Depth (cm)	Soil description: Color Moist	Color Dry	Texture class[a]	Structure	Soil consistencies Wet	Moist	Dry	General aspects	Soil classification
		70–90	Very dark grayish brown 10YR 3/2	Dark grayish brown 10YR 4/2	SiCL	Massive	Sticky and plastic	Firm	Hard	Very few scattered shell fragments, weak effervescence with HCl. Some red mottling observed	
		90–110	Very dark grayish brown 10YR 3/2	Dark grayish brown 10YR 4/2	SiC	Massive	Vey plastic and very sticky	Very firm	Very hard	This layer reveals the fluctuating water table zone. Very few shells and very weak effervescence with HCl	
P4	2 km to the south of Baltim city, 500 m to the east of Burulus lake, 7 km to the west of Baltim-El-Hamul road	0–20	Very dark grayish brown 10YR 3/2	Dark grayish brown 10YR 4/2	SiCL	Massive and blocky	Sticky and slightly plastic	Firm	Slightly hard	Scattered shell fragments, very fine roots, cracks (10 cm deep and 0.5 cm wide), strong effervescence with HCl	Typic Torrifluvents

20–30	Very dark grayish brown 10YR 3/2	Dark grayish brown 10YR 4/2	SiCL	Massive and blocky	Sticky and plastic	Friable	Hard	Scattered shell fragments, band of broken and non broken shells, very fine roots, strong effervescence with HCl
30–72	Dark grayish brown 10YR 4/2	Pale brown 10YR 6/3	SCL	Massive and blocky	Slightly sticky and slightly plastic	Firm to very firm	Hard	Very few scattered shell fragments, very fine roots weak effervescence HCl
72–90	Very dark grayish brown 10YR 3/2	Dark grayish brown 10YR 4/2	SiCL	Massive and blocky	Sticky and slightly plastic	firm	Slightly hard	Scattered shell fragments, very fine roots, cracks (10 cm deep and 0.5 cm wide) strong effervescence with HCl

[a]Apparent soil texture – *SiC* silty clay, *SiCL* silty clay loam, *SCL* sandy clay loam

Table 23.4 Selected physical and chemical characteristics of lacustrine soil profiles

Profile no.	Depth (cm)	pHs	ECe ($dS\ m^{-1}$)	Soluble cations ($meq\ l^{-1}$)[a] Na	K	Ca	Mg	Soluble anions ($meq\ l^{-1}$)[a] CO_3	HCO_3	Cl	SO_4	OM (%)	$CaCO_3$ (%)	CEC ($cmolc\ kg^{-1}$)	ESP	Particle-size distribution (%) VCS	CS	MS	FS + VFS	Sand	Silt	Clay	Texture class
P1	0–30	8.06	1.96	9.4	0.3	2.9	6.6	0.8	2.8	11.6	4.0	2.68	8.60	64	30	0.3	0.2	0.4	2.0	2.9	44.3	47.2	SiC
	30–60	8.66	2.06	15.1	0.4	1.9	3.7	0.7	1.8	14.0	4.5	1.04	3.90	60	14	0.2	0.2	0.3	0.4	1.0	44.4	51.8	SiC
	60–90	8.68	1.99	14.1	0.4	1.7	3.4	0.7	1.7	14.0	3.3	1.04	2.90	69	24	0.1	0.2	0.2	0.4	1.0	42.7	52.3	SiC
	90–120	8.91	2.41	18.9	0.3	1.9	3.9	0.8	1.9	16.3	6.0	0.83	6.50	60	38	0.4	0.4	0.7	v3.3	4.89	44.3	17.0	SiC
P2	0–30	8.16	2.81	19.9	0.3	3.3	3.6	1.6	3.6	20.5	1.7	1.79	7.80	51	30	0.3	0.2	0.3	0.4	1.2	55.1	47.4	SiC
	30–60	8.18	4.77	35.5	0.7	3.9	8.1	0.4	2.9	31.6	13.3	1.17	5.70	56	34	0.2	0.2	0.4	0.2	1.0	53.2	49.8	SiC
	60–90	8.11	7.88	60.0	1.2	5.3	9.7	0.4	2.1	57.4	16.3	0.97	7.20	51	25	0.9	1.9	2.8	9.9	15.4	45.3	44.2	SiCL
	90–120	7.99	11.20	90.5	1.7	6.6	17.4	0.0	2.0	83.8	30.4	0.97	1.20	48	35	0.1	0.2	0.7	0.8	1.9	55.3	48.5	SiC
P3	0–27	8.12	2.23	13.5	0.3	4.3	5.3	0.8	3.2	14.8	4.6	1.45	17.30	45	7	0.4	0.4	1.8	3.1	5.7	48.0	46.3	SiC
	27–37	8.02	2.59	14.7	0.5	5.3	4.3	0.0	2.4	17.9	4.4	0.62	6.60	50	8	0.3	0.4	0.9	0.7	2.1	48.9	49.0	SiC
	37–60	8.00	2.59	14.7	0.7	3.9	4.9	0.0	2.2	17.9	4.1	1.24	3.40	54	8	0.1	0.2	0.3	1.0	1.6	46.6	51.8	SiC
	60–70	8.21	3.14	16.4	0.8	4.3	8.1	0.0	1.5	23.0	5.1	0.90	6.80	34	8	0.9	1.9	2.9	12.3	17.9	46.0	36.1	SiCL
	70–90	7.93	2.98	17.3	0.9	3.6	9.4	0.0	0.9	21.3	9.0	0.76	0.60	36	7	0.7	1.5	2.7	6.3	11.5	48.7	39.8	SiCL
	90–110	8.11	3.10	15.6	1.0	4.0	9.4	0.0	1.3	22.7	6.0	1.11	0.70	37	8	0.8	1.5	3.2	5.2	10.7	48.3	41.0	SiC
P4	0–20	8.24	3.30	20.8	0.7	3.5	8.5	0.0	1.5	15.2	16.8	1.75	9.31	53	5	1.5	1.8	2.0	6.9	12.2	50.1	37.7	SiCL
	20–30	8.04	2.80	20.9	0.8	2.4	6.6	0.6	3.4	17.5	9.8	1.38	9.85	45	5	1.4	2.1	2.8	8.5	14.8	46.7	38.5	SiCL
	30–72	8.19	2.70	16.7	1.2	2.3	7.2	0.0	2.1	14.9	10.4	0.41	0.94	24	6	2.7	4.2	5.1	43.2	55.0	23.6	21.4	SiCL
	72–wt	8.35	2.40	20.0	0.8	1.8	4.2	0.0	3.0	16.2	7.6	1.08	0.09	46	10	0.0	0.1	0.1	1.4	34.6	25.3	40.1	CL
	wt	8.93	2.50	22.7	0.8	1.9	4.1	0.0	4.2	14.8	10.5	0.84	0.07	44	9	0.0	0.1	0.2	2.6	2.8	58.4	38.8	SiCL

[a] In soil saturation extract

ECe EC of soil saturation extract, *VCS* very coarse sand, *CS* coarse sand, *MS* medium sand, *FS* fine sand, *VFS* very fine sand; Texture class – *SiC* silty clay, *SiCL* silty clay loam, *CL* clay loam, *CEC* cation-exchange capacity, *OM* organic matter, *pHs* pH of saturated soil paste, *ESP* exchangeable sodium percentage, *wt* water table

23.3.3 *Marine Soils*

23.3.3.1 Field Description, Soil Classification, and Physical and Chemical Characteristics

Table 23.5 presents soil properties and classification. The color is variable dominantly light yellowish brown (dry) and light olive brown (moist). The dominance of light color, structureless, nonsticky, nonplastic and friable properties, and in depth difference in sand, silt, and clay are due to the nature of sandy sediments. The soil classification is Typic Xerpsamments (P1) and Typic Psammiaquents (P1 and P3). Sand is the texture in all profiles at all depths, very fine and fine sands together dominate in sand subfractions, other fraction ranges in their mean values as 2.0–4.6 (silt) and 4.23–5.5 (clay).

Table 23.6 presents physical and chemical characteristics, revealing generally organic matter and $CaCO_3$ (<1%); CEC ranges between 3.1 (P3) and 10 cmol kg^{-1} (P1), and ESP is lowest 6.3 in P3 and the highest 62.9 in P1. The ECe is very low (<0.73 dS m^{-1}) in P1, <3.2 dS m^{-1} in P2, and <7 dS m^{-1} in P3. This shows that P1 and P2 are sodic and P3 is saline at subsurface (below 30 cm). Such a development is due to the use of irrigation water and high drainage conditions in sandy texture that leached the salts.

Soluble Na and chlorides dominate among other cations and anions, respectively. The pH is in the range of moderately to strongly alkaline (P1 and P2) and very strongly alkaline (P3). The pH is above the optimum range (pH 6.6–7.3) where most of the plant nutrients become available to plants. It is not clear why the pH in P1 and P2 is lower than P3, the former are sodic profiles, which usually give pHs>9.00 (Richards 1954).

23.3.4 *Sandy and Fluvio-Sandy Soils*

23.3.4.1 Field Description, Soil Classification, and Physical and Chemical Characteristics

Variable color, structure, and consistence observed during profile description (Table 23.7). Table 23.8 shows sandy (P1) and variable texture (clay loam, loam, and sandy loam) in P2. Sand being dominant over silt and clay in both profiles. Profile 2 represents the zone between fluvial and sandy deposits, and there is large difference in texture due to multi-depositional regime. Similar findings were reported by El-Shahawi (1994). The difference in color and other characteristics is due to type of sediment and the management of these soils. The $CaCO_3$ is low (less than 3%), organic matter (<1%), soil reaction (pHs) is very strongly alkaline in all layers of both profiles, except upper layer of P2 (strongly alkaline), and hence will restrict nutrient availability to plants. The CEC is relatively lower in P1 than P2, this is due to high clay content and perhaps type of clay minerals in P2. Soluble Na and

Table 23.5 The location, general description, and soil classification of marine soil profiles

Profile no.	Site	Depth (cm)	Soil description: Color, Moist	Color, Dry	Texture class	Structure	Soil consistencies: Wet	Soil consistencies: Moist	Soil consistencies: Dry	General aspects	Soil classification
P1	Khaleg Bahary, Mutubis district, Kafr El-Sheikh Governorate	0–120	Brownish yellow 10YR 6/8	Yellow 10YR 7/6	Sandy	Structure less	Nonsticky and non-plastic	Very friable	Loose	Level soil, cultivated by peach trees. Prolonged application of organic fertilizers turned the surface darker than below. The profile is homogeneous	Typic Xerpsamments
P2	Ezbet Hammad, Baltim district, Kafr El-Sheikh Governorate	0–90	Light olive brown 2.5Y 5/3	Light yellowish brown 2.5Y 6/4	Sandy	Structure less	Nonsticky and non-plastic	Very friable	Loose	Level cultivated land (cotton, vegetable, and trees). The profile is homogeneous. Thin roots observed at 0–60 cm. Decayed palm roots found in the water table fluctuating zone. Few shell fragments and weak effervescent with HCl	Typic Psammiaquents

P3	0.5 km to the south of Mediterranean Sea. EI-Sheikh Mubarak village	0–90	Light olive brown 2.5Y 5/3	Light yellowish brown 2.5Y 6/4	Sandy	Structureless	Nonsticky and non-plastic	Very friable	Slightly hard	The profile is homogeneous. Thin roots observed at 0–60 cm. Decayed palm roots found in the water table fluctuating zone. Few shell fragments and weak effervescent with HCl	Typic Psammiaquents

Table 23.6 Selected physical and chemical characteristics of marine soil profiles

Profile no.	Depth (cm)	pHs	ECe (dS m^{-1})	Soluble cations (meq l^{-1})[a] Na	K	Ca	Mg	Soluble anions (meq l^{-1})[a] CO_3	HCO_3	Cl	SO_4	OM (%)	$CaCO_3$ (%)	CEC (cmolc kg^{-1})	ESP	Particle-size distribution (%) VCS	CS	MS	FS + VFS	Sand	Silt	Clay	Texture class
P1	0–30	8.36	0.73	4.3	0.3	1.4	1.9	0.0	2.1	3.9	1.9	0.15	0.6	10.0	59	0.5	0.9	22.8	62.1	86.2	6.2	7.6	S
	30–60	8.63	0.66	3.4	0.4	1.0	1.6	0.0	1.5	3.4	2.1	0.07	0.3	8.5	57	0.4	0.8	21.8	66.6	89.6	5.2	5.2	S
	60–90	8.74	0.50	2.8	0.4	0.9	1.4	0.0	1.0	2.9	1.6	0.04	0.2	7.5	62	0.3	0.9	20.5	70.0	91.6	4.1	4.3	S
	90–120	8.78	0.99	4.9	0.3	2.0	2.9	0.0	0.9	5.6	3.6	0.03	0.2	70	63	0.3	1.0	23.6	68.1	82.9	3.1	4.1	S
P2	0–30	8.25	2.67	13.9	0.5	5.3	5.7	0.0	0.6	16.3	8.5	0.42	0.6	9.4	39	0.4	0.9	18.4	69.0	88.6	4.6	6.8	S
	30–60	8.58	3.16	20.0	0.6	4.4	6.9	0.0	2.9	27.4	1.6	0.21	0.4	7.5	38	0.3	0.8	19.9	69.5	90.4	4.7	4.9	S
	60–90	8.49	2.95	18.1	0.5	4.3	5.8	0.0	3.1	22.1	3.5	0.16	0.3	7.9	45	0.3	1.0	21.8	68.7	91.7	3.5	4.8	S
P3	0–30	9.15	2.80	20.4	0.7	2.2	5.9	0.0	5.1	21.9	2.3	0.57	1.10	5.4	7	0.2	0.6	21.9	1.6	91.8	3.5	4.7	S
	30–60	9.29	6.40	59.0	0.9	5.7	7.8	0.0	3.0	50.2	20.2	0.24	0.90	3.1	7	0.0	0.7	31.5	0.6	95.0	1.2	3.8	S
	60–90	9.26	6.90	60.7	0.8	5.8	9.7	0.0	3.5	55.0	18.6	0.25	0.50	4.9	6	0.0	0.4	33.6	0.6	94.5	1.3	4.2	S

[a] In soil saturation extract

ECe EC of soil saturation extract, *VCS* very coarse sand, *CS* coarse sand, *MS* medium sand, *FS* fine sand, *VFS* very fine sand; Texture class – *S* sand, *CEC* cation-exchange capacity, *OM* organic matter, *pH*s pH of saturated soil paste, *ESP* exchangeable sodium percentage

Table 23.7 The location, general description, and soil classification of sandy and fluvio-sandy soil profiles

			Soil description								
			Color				Soil consistencies				
Profile no.	Site	Depth (cm)	Moist	Dry	Texture class	Structurewet	Wet	Moist	Dry	General aspects	Soil classification
P1 Sandy	South El-Tahreer, El-Beheira Governorate	0–30	Brown	Yellow 10YR 7/5	S	Structure less	Nonsticky and non plastic	Very friable	Loose	Prolonged use of organic fertilizers converted the surface as back. The profile is homogeneous	Typic Xerpsamments
P2 Fluvio-sandy	El-Zapharany village Kom Hamada, city El-Beheira Governorate	0–36	Very dark grayish brown 10YR 3/2	Grayish brown 10YR 5/2	SCL	Massive and blocky	Slightly sticky and plastic	Firm	Hard	Few fine roots of *Zea mays*, some very fine shell fragments and weak effervescence with HCl	Typic Psammiaquents
		36–61	Olive brown 2.5Y 4/4	Light olive brown 2.5Y 5/6	LS	Massive and blocky	Nonsticky and nonplastic	Friable	Slightlyhard	Tongue of clay through the layer	
		61–88	Olive yellow 2.5Y 6/6	Yellow 2.5Y 7/6	S	Structure less	Nonsticky and nonplastic	Very friable	Loose	Tongue of decayed OM observed	
		88-water table	Very dark grayish brown 10YR 3/2	Dark grayish brown 10YR 4/2	CL	Massive	Very sticky and very plastic	Very firm	Veryhard	This layer reveals fluvial nature so it has dark color	

Texture class – *S* sandy, *SCL* sandy clay loam, *LS* loamy sand, *CL* clay loam

Table 23.8 Selected physical and chemical characteristics of sandy and fluvio-sandy soil profiles

Profile no.	Depth (cm)	pHs	ECe (dS m^{-1})	Soluble cations (meq l^{-1})[a]				Soluble anions (meq l^{-1})[a]				OM (%)	$CaCO_3$ (%)	CEC (cmolc kg^{-1})	ESP	Particle-size distribution (%)							Texture class
				Na	K	Ca	Mg	CO_3	HCO_3	Cl	SO_4					VCS	CS	MS	FS + VFS	Sand	Silt	Clay	
P1 Sandy	0–30	9.30	1.2	5.7	0.1	1.7	1.8	0.0	2.3	9.8	0.2	0.15	2.1	3.6	5	5.9	14.9	19.7	54.9	95.4	2.5	2.1	S
	30–60	9.25	1.3	9.6	0.1	2.0	1.8	0.0	2.2	11.0	0.3	0.14	2.4	3.0	5	7.8	13.6	20.7	54.8	97.0	1.6	1.4	S
	60–90	9.14	2.4	15.5	0.3	3.6	5.4	0.0	2.0	19.2	3.5	0.12	2.0	3.6	4	6.2	16.8	23.6	49.1	95.7	3.1	1.2	S
P2 Fluvio-sandy	0–30	8.86	4.1	8.9	0.4	4.5	6.5	0.0	4.5	13.5	2.3	1.62	3.3	22.7	3	3.6	5.8	6.2	51.4	37.3	23.7	39.0	CL
	30–60	9.33	3.6	11.8	0.4	4.2	6.3	0.0	2.2	15.0	5.5	0.42	1.6	17.0	4	5.7	9.7	17.6	51.7	49.2	27.8	23.0	SCL
	60–90	9.16	3.6	17.6	0.3	4.0	3.0	0.0	3.1	14.5	7.3	0.41	1.5	14.7	7	7.1	11.4	15.9	61.0	56.8	29.0	14.1	SL
	8–wt	9.17	2.1	20.8	0.3	4.1	4.5	0.0	4.5	16.6	8.6	0.29	2.6	16.3	7	1.0	4.5	3.8	27.2	46.6	32.9	20.5	L
	Wt	9.06	2.1	10.7	0.2	2.0	5.2	0.0	4.2	12.4	1.5	0.30	2.5	18.2	9	0.8	1.9	2.9	21.1	42.5	28.7	28.8	CL

[a] In soil saturation extract

ECe EC of soil saturation extract, *VCS* very coarse sand, *CS* coarse sand, *MS* medium sand, *FS* fine sand, *VFS* very fine sand; Texture class – *S* sand, *CL* clay loam, *SCL* sandy clay loam, *L* loam, *CEC* cation-exchange capacity, *OM* organic matter, *pHs* pH of saturated soil paste, *ESP* exchangeable sodium percentage, *wt* water table chlorides dominate among other cations and anions, respectively. All layers (P1 and P2) are nonsaline and non-sodic (ECe < 4 dS m^{-1} and ESP < 15), except upper surface in P2 which is saline. Profiles are classified as Typic Quartzipsamments (P1) and Typic Torripsamments (P2). In a previous study, Shaheen (1999) has reported similar results.

23.3.5 Calcareous Soils

23.3.5.1 Field Description, Soil Classification, and Physical and Chemical Characteristics

Variable color, texture, and consistence observed during profile description (Table 23.9). Table 23.10 shows silty loam being the dominant texture (P1 and P2) and variety of textures in P3. Compared to all other profiles investigated, these soils are heavier in soil texture, and silt being dominant. The trend of primary particles is silt > clay > sand. This can be attributed to the type of parent material developed under marine and lagoonal conditions high in carbonate contents (Moustafa 1987).

The difference in color and other characteristics is due to type of sediment and the management of these soils. All profiles are highly calcareous, and the $CaCO_3$ is relatively higher in P1 compared to other two profiles. The high-$CaCO_3$ contents correspond to high silt content (El-Gamal 1992). The organic matter in general is <1%, and at surface of all profiles, it is present up to 1.61%. Soil reaction (pH) is strongly alkaline in all layers (P1 and P2) and very strongly alkaline (P3), and hence will restrict nutrient availability to plants. The mean CEC is higher (11.5–17.8 cmolc kg^{-1}) due to high clay content. The entire profiles (P1, P2, P3) are nonsaline and non-sodic (ECe < 4 dS m^{-1} and ESP < 15). All profiles are classified as Typic Haplocalcids. The trend of soluble cations (Na > Ca > K) and soluble anions is same (Cl > SO_4 > HCO_3) as for other profiles.

23.4 Sesquioxides in the Tested Profiles

Table 23.11 reveals sequioxides of profiles, showing the dominance of total Fe and Al when expressed on g kg^{-1} basis. Total Fe being the lowest in marine soils (6.1 g kg^{-1}), and the highest in lacustrine soils (58.4 g kg^{-1}). Total aluminum follows same trend as for total Fe (lowest 5.9 g kg^{-1} in marine and highest 64.9 g kg^{-1} in lacustrine soils). The general trend of citrate-bicarbonate-dithionite extractable Fe, Mn, Al, Si, and Ca is found as Ca dominates over the others, Mn being the lowest, whereas Fe and Al codominate over other ammonium oxalate extractable contents (Mn, Si, Ca). The variations of sequioxides with respect to profiles developed from different sediments can be seen in Table 23.11.

Table 23.9 The location, general description, and soil classification of calcareous soil profiles

Profile no.	Site	Depth (cm)	Description: Color Moist	Color Dry	Texture class	Structure wet	Soil consistencies Wet	Soil consistencies Moist	Soil consistencies Dry	General aspects	Soil classification
P1	Mariut at about 1.5 km to the west of the km 30 of Alexandria-Cairo desert road	Yellowish brown 10YR 6/4	Very pale brown 10YR 7/4	SiL, L and CL	Massive and blocky		Firm	Hard to very hard	The soil profile is homogenous. The profile is deep where water table was not observed	Typic Haplocalcids	
P2	Mariut at about 1 km to the west of the km 30 of Alexandria Cairo desert road		Yellowish brown 10YR 6/4	Very pale brown 10YR 7/4	Silt loam, and clay loam	Massive and blocky		Firm	Hard to very hard	The soil profile is homogenous	Typic Haplocalcids

P3	Ezbet El-Adlia 1 km to the north of Nubaria canal. Abo El-Matameir, El-Beheira Governorate	0–28	Dark brown 10YR 5/3	Brown 10YR 5/3	Clay loam	Massive and blocky	Sticky and plastic	Firm	Hard	This layer has relatively dark color as a result of mixing with clay fraction, few fine fibrous roots and strong effervescence with HCl	Typic Haplocalcids
		28–50	Strong brown 7.5YR 5/8	Reddish yellow 7.5YR 6/6	Sandy Clay loam	Massive and blocky	Slightly sticky and slightly plastic	Very firm	Very hard	Few fine roots, few white soft lime nodules, and very strong effervescence with HCl	

(continued)

Table 23.9 (continued)

Profile no.	Site	Depth (cm)	Description									Soil classification
			Color		Texture class	Structure wet	Soil consistencies			General aspects		
			Moist	Dry			Wet	Moist	Dry			
		50–80	Yellowish brown 10YR 6/4	Very pale brown 10YR 7/4	Sandy loam	Massive and blocky	Slightly sticky and slightly plastic	Firm	Very hard	Very few and very fine roots, white soft lime nodules, common white lime nodules and concretions, and very strong effervescence with HCl		
		80–wt	Dark brown 10YR 4/4	Brown 10YR 5/3	Loam	Massive and blocky	Slightly sticky and slightly plastic	Firm	Very hard	Common black hard gravels and concretions, very strong effervescence with HCl		

Table 23.10 Selected physical and chemical characteristics of calcareous soil profiles

				Soluble cations (meq l^{-1})[a]				Soluble anions (meq l^{-1})[a]								Particle-size distribution (%)							
Profile no.	Depth (cm)	pHs	ECe (dS m^{-1})	Na	K	Ca	Mg	CO_3	HCO_3	Cl	SO_4	OM (%)	$CaCO_3$ (%)	CEC (cmolc kg^{-1})	ESP	VCS	CS	MS	FS + VFS	Sand	Silt	Clay	Texture class
P1	0–30	8.53	5.2	33.5	0.8	8.2	10.0	0.0	4.5	26.5	21.8	1.61	33.3	13.0	4.3	1.6	2.7	5.2	23.7	33.2	39.5	27.3	CL
	30–60	8.75	3.0	20.8	0.5	4.6	4.4	0.0	2.0	17.1	11.2	0.51	48.6	10.8	4.3	1.1	1.9	4.6	13.1	20.8	58.1	21.1	SiL
	60–90	8.80	2.9	20.4	0.5	4.4	4.2	0.0	2.1	15.5	11.9	0.43	55.6	9.6	2.9	1.6	2.2	4.8	16.1	24.7	54.3	21.0	SiL
	90–120	8.89	2.5	19.0	0.5	3.8	2.8	0.0	2.3	15.0	8.9	0.50	40.3	12.4	4.1	1.5	2.4	6.2	16.6	26.7	50.6	22.7	SiL
	120–150	8.94	2.3	18.5	0.4	2.2	3.0	0.0	2.4	13.8	7.9	0.49	39.2	11.6	5.2	1.5	2.9	5.4	20.0	29.9	45.3	24.8	L
P2	0–30	8.75	2.9	13.4	0.7	6.2	9.0	0.0	3.5	12.5	3.3	1.03	32.7	9.8	2.4	1.4	2.4	5.5	24.9	34.2	38.5	27.3	CL
	30–60	8.78	2.9	11.9	0.5	6.2	10.3	0.0	3.3	20.0	5.6	0.37	34.5	18.4	3.3	1.6	1.8	4.4	14.3	22.2	57.0	20.8	SiL
	60–90	8.49	2.7	12.1	0.5	5.1	9.3	0.0	3.4	20.0	3.6	0.52	37.0	18.0	2.8	1.8	2.2	4.2	18.8	26.9	50.6	22.5	SiL
	90–120	8.58	2.3	12.8	0.4	4.2	6.4	1.3	3.8	15.0	3.7	0.37	35.1	16.6	2.3	1.7	2.6	6.5	17.8	28.6	50.3	21.1	SiL
P3	0–28	8.86	4.1	28.9	0.6	7.2	8.1	0.0	3.5	23.0	18.3	1.60	20.7	22.7	3.1	1.4	3.4	1.2	31.2	37.3	23.7	39.0	CL
	28–50	9.33	3.6	27.9	0.2	5.6	4.4	0.0	3.3	18.5	16.3	0.42	32.2	17.0	5.1	2.9	3.8	5.0	38.4	49.2	27.8	23.0	SCL
	50–80	9.16	3.6	25.8	0.1	5.9	8.1	0.0	2.3	18.4	19.2	0.41	49.5	14.7	5.4	5.8	3.4	4.7	4.29	56.8	29.0	14.1	SL
	80–wt	9.17	2.1	16.4	0.2	4.0	3.2	0.0	2.5	15.0	6.3	0.29	36.7	16.3	7.3	2.9	2.8	5.2	35.8	46.6	32.9	20.4	L
	uwt	9.06	2.1	15.4	0.3	4.0	2.3	0.0	2.4	14.3	5.3	0.30	27.8	18.2	4.5	1.8	4.8	2.9	33.0	42.5	28.7	28.8	CL

[a]In soil saturation extract

ECe EC of soil saturation extract, *VCS* very coarse sand, *CS* coarse sand, *MS* medium sand, *FS* fine sand, *VFS* very fine sand, Texture class – *CL* clay loam, *SCL* sandy clay loam, *L* loam, *SiL* silt loam, *CEC* cation-exchange capacity, *OM* organic matter, *pHs* pH of saturated soil paste, *ESP* exchangeable sodium percentage, *wt* water table, *uwt* under water table

Table 23.11 Amounts of sesquioxides in the surface layers of tested soil profiles

Profile no.	Depth (cm)	Soil classification	Fe_t	Fe_d	Fe_o	Mn_t	Mn_d	Mn_o	Al_t	Al_d	Al_o	Si_d	Si_o	Ca_d	Ca_o
Fluvial soils		g kg-1 soil													
P1	0–20	Typic Torrifluvents	57.8	10.3	2.49	1.05	0.75	0.52	51.7	0.79	2.02	1.12	1.01	13.5	0.016
P2	0–30	Typic Torrifluvents	63.2	7.9	2.02	0.85	0.69	0.48	69.9	1.03	2.09	1.79	0.83	16.4	0.025
P3	0–30	Typic Torrifluvents	47.9	6.2	1.72	0.74	0.51	0.33	47.1	0.70	1.72	1.28	0.81	17.1	0.014
P4	0–30	Typic Fluvaquents	59.7	9.0	3.35	0.81	0.59	0.41	61.4	0.96	2.48	0.89	1.05	13.0	0.016
P5	0–30	Typic Ustifluvents	62.0	7.6	1.85	1.04	0.67	0.44	69.2	1.07	2.54	1.28	0.90	20.0	0.015
P6	0–30	Vertic Torrifluvents	58.1	5.9	2.03	0.94	0.64	0.46	55.8	0.64	2.15	1.33	0.94	20.5	0.018
Lacustrine soils															
P1	0–30	Typic Xerofluvents	61.9	7.4	3.94	1.00	0.63	0.47	72.0	1.19	3.06	1.37	1.42	26.1	0.012
P2	0–30	Typic Fluvaquents	61.7	6.6	3.76	1.19	0.89	0.62	67.4	0.86	2.15	1.59	0.98	27.8	0.015
P3	0–30	Typic Xerofluvents	62.4	8.2	4.86	0.94	0.63	0.49	71.0	0.78	2.16	1.51	1.14	26.8	0.019
P4	0–30	Typic Fluvaquents	47.8	6.0	3.50	0.65	0.39	0.29	49.1	0.54	1.84	1.23	1.00	26.6	0.023
Marine soils															
P1	0–30	Typic Xeropsamments	7.9	0.9	0.31	0.13	0.07	0.05	6.9	0.29	0.27	0.24	0.20	2.1	0.026
P2	0–30	Typic Psammiaquents	4.6	0.6	0.22	0.13	0.11	0.07	4.9	0.23	0.24	0.30	0.17	3.8	0.025
P3	0–30	Typic Psammiaquents	6.6	0.7	0.26	0.15	0.08	0.06	6.0	0.19	0.27	0.27	0.23	9.0	0.022
Sandy and fluvio-sandy soils															
P1	0–30	Typic Quartzipsamments	3.6	0.7	0.11	0.08	0.10	0.06	4.1	0.32	0.30	0.10	0.14	5.2	0.016
P2	0–30	Typic Torripsamments	25.2	3.2	1.23	0.41	0.31	0.21	27.8	0.60	0.87	0.57	0.54	18.1	0.031
Calcareous soils															
P1	0–30	Typic Haplocalcids	18.3	4.1	1.06	0.34	0.26	0.14	23.3	0.50	0.90	0.22	0.36	37.7	0.025
P2	0–30	Typic Haplocalcids	22.0	5.0	1.48	0.39	0.26	0.14	32.1	0.53	1.02	0.23	0.40	48.5	0.007
P3	0–30	Typic Haplocalcids	22.7	3.9	1.14	0.41	0.33	0.21	28.6	0.54	0.98	0.45	0.52	43.0	0.009

Fe_t, Al_t, and Mn_t = Total Fe, Mn, and Al; Fe_d, Al_d, Mn_d, Si_d, and Ca_d = Citrate-bicarbonate-dithionate extractable (Fe, Al, Mn, Si, Ca); Fe_o, Al_o, Mn_o, Si_o, and CaO = Ammonium oxalate extractable (Fe, Mn, Al, Si, and Ca)

23.5 Conclusions and Recommendations

The classification of profiles and their basic properties are investigated. The profiles vary considerably in physical and chemical properties depending on the geological nature of the deposits (fluvial, lacustrine, marine, calcareous, sandy, and fluvio-sandy). The fluvial, lacustrine, marine, and sandy and fluvio-sandy soils are classified as entisols, whereas calcareous soils as aridisols. The soils exhibited quite different physical and chemical properties. The pH is above optimum range of nutrient availability; therefore, the soil needs careful management to improve nutrient efficiency in soil. The sodic soils need to be amended by gypsum application to reduce sodicity. The diverse geological nature of these deposits is reflected in the wide variation of clay content, carbonates content, and in the form of Fe, Al, Si, and Mn oxides.

References

Ali RA (1985) Mineralogical and chemical composition of some Egyptian soils in relation to potassium status. PhD thesis, Tanta University

Amer MH, Abo-Zeid M (1989) History of land drainage of Egypt. In: Amer MH, Ridder NA (eds) Land drainage in Egypt. DRI, Cairo

El-Gamal AA (1992) Different approaches for evaluation of soil development and uniformity at the area between Nubaria and El hager canal. MSc thesis, Tanta University

El-Shahawi MI (1994) Pedochemical studies on some Egyptian soils. PhD thesis, Tanta University

FAO (1970) Guidelines to soil description. FAO Bull, Rome

FAO-SF (1964) High dam soil survey. UAR, United Nations Development Program, FAO, Rome

Gee GW, Bauder JW (1986) Particle size analysis. In: Klute A (ed) Methods of soil analysis: physical and mineralogy methods, part 1, 2nd edn. ASA/SSSA, Madison, pp 383–412

Loeppert RH, Inskeep WP (1996) Iron. In: Sparks DL (ed) Methods of soil analysis: chemical methods, part 3. ASA/SSSA, Madison, pp 639–664

Mehra OP, Jackson ML (1960) Iron oxides removal from soils and clays by dithionate-citrate system buffered with sodium bicarbonate. Clays Clay Miner 7:317–327

Moustafa AM (1987) Pedological studies on new formations in some calcareous soils of the western desert, Egypt. MSc thesis, Alexandria University

Perttjhon EJ (1984) Sedimentary rocks, 3rd edn. CBC Publishers and Distributors, Delhi

Raymond WA, Roy LD (1995) Soils in our environment. Prentice Hall, Englewood Cliffs

Richards L (ed) (1954) Diagnosis and improvement of saline and alkali soils, vol 60, USDA handbook. US Salinity Lab Staff, Washington, DC

Selley RC (1982) An introduction to sedimentology, 2nd edn. Academic, London

Shaheen SM (1999) Some pedological and geochemical studies of some Egyptian soils. MSc thesis, Tanta University

Soil Survey Division Staff (1993) Soil survey manual. Handbook No 18. USDA, Washington, DC

Soil Survey Staff (2006) Key of soil taxonomy, 10th edn. USDA-NRCS, Govt Printing Office, Washington, DC

Sparks DL (1995) Environmental soil chemistry. Academic, New York

Sparks DL, Page AL, Helmke PA, Loppert RH, Soltanpour PN, Tabatabai MA, Johnston CT, Summner ME (1996) Methods of soil analysis: chemical methods, part 3. ASA/SSSA, Madison

Sumner ME, Miller WP (1996) Cation exchange capacity and exchange coefficients. In: Sparks DL, Page AL, Helmke PA, Loppert RH, Soltanpour PN, Tabatabai MA, Johnston CT, Summner ME (eds) Methods of soil analysis: chemical methods, part 3. ASA/SSSA, Madison, pp 1201–1230

Thomas GW (1996) Soil pH and soil acidity. In: Sparks DL, Page AL, Helmke PA, Loppert RH, Soltanpour PN, Tabatabai MA, Johnston CT, Summner ME (eds) Methods of soil analysis: chemical methods, part 3. ASA/SSSA, Madison, pp 475–490

United States Environmental Protection Agency (USEPA) (1995) Test methods for evaluating solid wastes. USEPA SW 846. US Govt Printing Office, Washington, DC

Walkley A (1947) A critical examination of a rapid method for determining soil organic matter. Soil Sci 63:654–661

Worcoster GP (1969) A text book of geomorphology, 2nd edn. Affiliated East–West Pvt Ltd, New Delhi

Chapter 24
Semiarid Soils of Eastern Indonesia: Soil Classification and Land Uses

Anny Mulyani, Adi Priyono, and Fahmuddin Agus

Abstract Despite the high annual rainfall (2,000–5,000 mm) of Indonesia, 3.3×10^6 ha area of East Nusa Tenggara (NTT) province receives annual rainfall between 1,000 and 2,000 mm with 5–8 dry months (<100-mm rainfall), and about 1×10^6 ha area receives <1,000 mm annual rainfall with 8–10 dry months. About 1.7×10^6 ha land is mountainous (>30% slope), and 1.5×10^6 ha is hilly (15–30% slope). The steep slopes and high-intensity rainfall in rainy season cause high erosion and sedimentation. This results in >1-m soil depth at the valley bottom dominated by Haplustepts and Haplustolls, and <50-cm soil depth in the upper slope dominated by Lithic Ustorthents or Lithic Haplustepts. The soil reaction is acidic to slightly alkaline (pH 4.1–7.8), low to high organic carbon (1.2–5.4%), high to very high (25%-HCl extractable) P and K, moderate to high exchangeable cations (18–41 cmol(+) kg^{-1}), and high base saturation percentage. About 2.4×10^6 ha land is suitable for agriculture, of which 0.5×10^6 ha occurs in the <1,000 mm; 1.7×10^6 ha in the 1,000–2,000 mm; and 0.2×10^6 ha in the >2,000 mm annual rainfall areas. Due to low rainfall, the recommended commodities were tailored to the availability of water. Based on soil characteristics and rainfall, recommendations are made for crop selection. Annual crops are recommended for the flat areas and valley bottoms, while perennial crops are recommended on the sloping areas. It is recommended, for example, for annual crops, adjustment of the planting time is to be made with the rainfall distribution, whereas for perennial crops, selection of those crops is to be made which require several months of dry period, such as candlenut, cashew nuts, kapok, and Jatropha.

Keywords Annual crops • Annual rainfall • Eastern Indonesia • Land suitability • Perennial crops

---

A. Mulyani (✉) • A. Priyono • F. Agus
Centre for Agricultural Land Resources Research and Development, Bogor, Indonesia
e-mail: anny_mulyani@ymail.com

S.A. Shahid et al. (eds.), *Developments in Soil Classification, Land Use Planning and Policy Implications: Innovative Thinking of Soil Inventory for Land Use Planning and Management of Land Resources*, DOI 10.1007/978-94-007-5332-7_24,

24.1 Introduction

The climate of East Nusa Tenggara province (NTT) is untypically dry relative to the rest of Indonesian archipelago. Of 4.6×10^6 ha land area of the province, 1.0×10^6 ha receives annual rainfall (<1,000 mm), and 3.3×10^6 ha receives annual rainfall (1,000–2,000 mm) with more than 6 dry months (<100 mm monthly rainfall) and <4 wet months (>200 mm monthly rainfall) (Balitklimat 2004). With the short duration of wet months, rainfall tends to be very high (200–500 mm monthly rainfall) during the wet months and so is rainfall intensity.

About 71% area are hilly (15–30% slope) to mountainous (>30% slope). The high-intensity rainfall during the rainy season and steep slope topography cause high soil erosion. The long-term impact of erosion process can be observed from the shallow (<50 cm) soil depth at the interfluves, the upper and middle slopes positions, and deep (sometimes reaching more than 1 m) soil depth at the valley bottom. Therefore, the selection of crops suitable with the ecosystem and their arrangement in the landscape to control erosion and determine agricultural sustainability in the area is essential.

In the present study, attempt has been made to investigate the spatial distribution of soils and land suitability evaluation for various agricultural commodities for NTT province. Besides soil properties, positions of points in the landscape, the slope, and rainfall amount and distribution are among the important factors considered in determining land suitability.

24.1.1 The Setting of Semiarid Areas of Indonesia

Indonesia is an archipelago (a sea abounding in islands) with a land area of about 188.2×10^6 ha, with a variety of soil orders (10 orders), parent materials, climate types, elevation, and reliefs. The diversity of these land resources provides benefits and opportunities for Indonesia in terms of development of various agricultural commodities.

Western Sumatra, Java, Bali, Kalimantan, Sulawesi, and Papua are mostly humid with rainfall of more than 2,000 mm per annum. This moisture partly originates from the mountains that trap damp air and partly brought by oceanic wind. The city of Bogor, near Jakarta, is a place of the world's highest number of rainstorms of around 322 events annually. About 45.3×10^6 ha of Indonesian areas receive rainfall (<2,000 mm); 43.2×10^6 ha (1,000–2,000 mm) and 2.4×10^6 ha (<1,000 mm) per annum (Balitklimat 2004). The areas which receive annual rainfall more than 2,000 mm are dominated by soils belonging to soil orders, such as Inceptisols (31.6%), Ultisols (29.0%), and Oxisols (12.4%) (Subagyo et al. 2000; Puslitbangtanak 2000).

It is to be noted here that the eastern part of the Indonesia, which is under the influence of Australian Continent, is drier. Of 2.4×10^6 ha Indonesian land (annual rainfall <1,000 mm), approximately 1.0×10^6 ha is located in the province of East Nusa Tenggara (NTT).

24.2 Materials and Methods

Present study on the land identification and evaluation was conducted in three major islands of East Nusa Tenggara province, namely, Flores, Sumba, and Timor, covering 4.6×10^6 ha area. The intensive field observations were made on 7 representative sites including 3 in the Flores Island, 1 in the Timor Island, and 3 in Sumba Island.

The distribution of soil, parent material, physiography and relief, and climate in the study area (NTT province) was evaluated from the existing soil and climate database at 1:1,000,000 scale (Puslitbangtanak 2000), including agricultural climatic resource map (Balitklimat 2004) and land use recommendation map (Puslitbangtanak 2001). The soil maps provided soil, physiography, landform, and elevation data; the climatic map provided information on the distribution of climate type, rainfall, and numbers of wet and dry months.

The soil, lithology, parent materials, and climatic maps were overlaid using the geographical information system (GIS) to generate interim land suitability. These base maps and the overlay map were used as the references for ground truthing.

Identification and characterization of the land were carried out by observing the land morphological characteristics in the field based on soil pit (1 m × 1 m × 2 m depth), mini pit (0.6 m × 0.6 m × 0.5 m depth), and through soil auger observations and morphological description (FAO 1990; Soil Survey Division Staff 1993). Soil morphological information include soil depth, soil color, texture, structure, consistency, drainage, pH, cementation, the concentration of coarse material (stoniness and rockiness), and plant root distribution.

After the description of soil pits and auger holes, soil samples from representative horizon depths were collected. The soil samples were processed (drying, sieving) in the laboratory and analyzed for texture (sand, silt, clay), organic C, total N and C/N ratio, soil reaction (pH), 25%-HCl extractable P and K contents, exchangeable Ca, Mg, K, and Na (1 N NH_4-acetate extraction), cation exchange capacity (CEC), and base saturation (BS) using standard procedures (Sulaeman et al. 2005; Burt 2004). Soil analytical results were used to evaluate the rainfall and landscape positions, influence on soil fertility, as well as inputs for land suitability evaluation and land management system development. Land suitability evaluation was made by matching the land characteristics with crop growth requirements as explained in Djaenudin et al. (2003).

24.3 Results and Discussion

24.3.1 Climate Characteristics

Based on climatic classification (Balitklimat 2004), NTT province has a region which includes "semiarid" areas with rainfall patterns IA and IC covering 1.05×10^6 ha (Table 24.1), with annual rainfall <1,000 mm and 7–10 dry months (monthly rainfall

Table 24.1 Rainfall pattern and climate type in the East Nusa Tenggara province

Rainfall pattern	Climate type	Annual rainfall (mm)	Months with rainfall <100 mm	Months with rainfall 100–150 mm	Months with rainfall 150–200 mm	Months with rainfall >200 mm	Total area (ha)
IA	Dry	<1,000	7–10	≤4	≤3	≤2	644,040
IB	Dry	<1,000	8–12	<3	0	0	–
IC	Dry	<1,000	8–9	<2	≤2	≤2	409,412
IIA	Dry	1,000–2,000	5–8	<3	≤2	≤4	2,316,822
IIB	Dry	1,000–2,000	≤4	<5	≤5	≤4	–
IIC	Dry	1,000–2,000	≤5	<5	≤6	≤5	954,363
IIIA	Wet	2,000–3,000	≤6	<4	≤5	≤6	140,840
IIIB	Wet	2,000–3,000	≤4	<4	≤5	5–6	–
IIIC	Wet	2,000–3,000	≤4	<4	≤5	6–8	48,788
IVA	Wet	3,000–4,000	≤2	<3	≤4	7–9	47,944
Total							4,562,210

Source: Balitklimat (2004)

<100 mm) and ≤2 wet months (monthly rainfall >200 mm). For example, the climate station located in the Alor Island recorded an annual rainfall of about 925 mm, with 1 wet month and 7 dry months (Table 24.2).

The largest area in NTT covering about 2.3×10^6 ha is the area with rainfall pattern IIA, where the annual rainfall is in the 1,000–2,000 mm range, 5–8 dry months and ≤4 wet months. Regions with IIA rainfall pattern are distributed in Flores and Timor Islands. Data from Ende station in Flores Island and in Kupang and Belu in Timor Island shows annual rainfall of 1,000–2,000 mm, with 3–4 wet months and 5–7 dry months. Rainfall distribution in NTT province is shown in Fig. 24.1.

Table 24.3 shows that most (1.7×10^6 of the 2.3×10^6 ha) land with rainfall 1,000–2,000 mm are located in the hilly to mountainous areas with slopes >15%; thus, these areas are vulnerable to erosion, especially those with four wet months. In a recent study, Widiyono et al. (2006) found soil loss as high as 11 t ha^{-1} year^{-1} in Kupang District of Timor Island.

At some stations in Kupang and Belu, the wettest period from December to February are marked by rainfall reaching 300–500 mm. The remaining areas of around 0.75×10^6 ha have rainfall patterns IIC, IIIA, IIIC, and IVA with 5–9 wet months, and they are located in hilly to mountainous (15–40% slope) relief.

24.3.2 Soil Characteristics and Classification

Of the total area (4.6×10^6 ha), the Inceptisols (Haplustepts) cover 2.4×10^6 ha. Most of the soils are derived from sedimentary rocks and volcanic materials, limestone, alluvium, and coral rock (Table 24.4 and Fig. 24.2).

Table 24.2 Rainfall distribution from five climatological stations in East Nusa Tenggara province

Month	Ende (Flores Island)		Alor (Alor Island)		Kupang (Timor Island)		Belu (Timor Island)		West Sumba (Sumba Island)	
	Rainfall (mm)	Rainy days (Day)	Rainfall (mm)	Rainy days (Day)	Rainfall (mm)	Rainy days (Day)	Rainfall (mm)	Rainy days (Day)	Rainfall (mm)	Rainy days (Day)
January	269	17	328	22	322	22	324	12	401	16
February	220	19	189	22	509	22	333	11	386	17
March	156	9	83	18	277	18	130	9	336	18
April	75	2	26	8	63	8	170	9	377	16
May	22	1	2	3	19	3	186	7	81	5
June	10	0	1	3	8	3	86	7	33	2
July	0	0	1	2	5	2	33	4	51	2
August	8	0	1	1	0	1	19	2	17	1
September	0	0	1	0	4	0	7	0	37	2
October	31	1	7	3	37	3	18	2	119	7
November	107	11	117	13	142	13	108	7	273	12
December	241	16	169	20	358	20	208	12	245	14
Total	1,138	77	925	114	1,745	114	1,622	80	2,357	111

Source: BMG NTT Province (1996–2007)

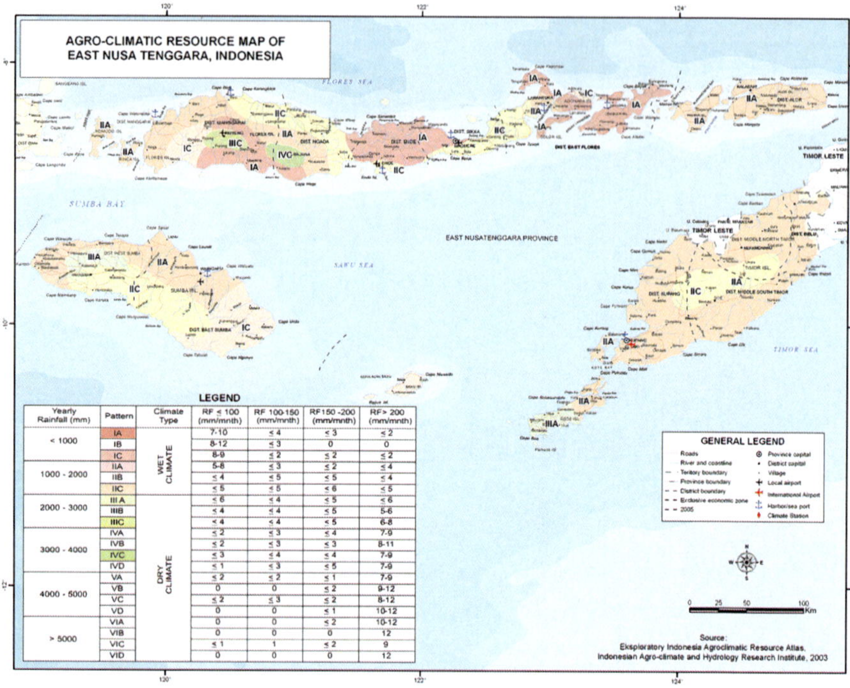

Fig. 24.1 Distribution of rainfall pattern in NTT province

Table 24.3 Rainfall pattern and reliefs in East Nusa Tenggara province

Rainfall pattern	Relief (area in hectares)[a]					Total (ha)
	Flat	Undulating	Rolling	Hilly	Mountainous	
IA	37,595	36,202	19,237	166,439	384,567	644,040
IC	27,044	127,065	29,162	136,595	89,546	409,412
IIA	124,851	267,059	206,226	758,102	960,584	2,316,822
IIC	66,293	96,780	131,791	390,242	269,257	954,363
IIIA	20,405	33,383	60,274	26,193	585	140,840
IIIC	–	–	6,778	19,432	22,578	48,788
IVC	627	15,880	3,329	18,747	9,362	47,945
Total	276,815	576,369	456,797	1,515,750	1,736,479	4,562,210

[a] Flat = 0–3% slope; Undulating = 3–8% slope; Rolling = 8–15% slope; Hilly = 15–30% slope; Mountainous = >30% slope

Haplustepts are derived from the sedimentary material and are generally associated with Ustorthents (Entisols) covering about 1.3×10^6 ha area and are distributed mostly in hilly and mountainous areas with slopes >15%. The volcanic soils, Haplustepts, are associated mostly with Haplustalfs (Alfisols) and Haplustolls (Mollisols), and mainly located in mountainous areas with slope >30% (Table 24.5).

Besides Haplustepts, wide distribution of soils belonging to Mollisols (Haplustolls) cover about 1.3×10^6 ha and are predominantly derived from volcanic, limestone,

Table 24.4 Distribution of parent material, landform, and soil types based on exploration map in NTT province

Parent material	Landform	Soil type		Area (ha)
Alluvium	Alluvial	Endoaquepts	Udifluvents	39,139
		Haplusterts	Endoaquepts	68,568
			Haplustepts	149,146
	Marine	Endoaquepts	Halaquepts	19,962
Limestone	Karst	Haplustepts	Haplustalfs	32,728
			Ustorthents	287,580
		Haplustolls	Haplustepts	113,708
			Ustorthents	106,294
Coral stone	Marine	Haplustolls	Ustorthents	268,836
Sediment	Tectonic	Haplustepts	Haplustalfs	191,041
			Ustorthents	1,315,644
		Haplustolls	Argiustolls	195,269
		Paleustults	Haplustepts	45,936
Lime sediment	Tectonic	Hapludolls	Eutrudepts	34,447
		Haplustolls	Haplustepts	64,702
Volcanic	Volcano	Eutrudepts	Endoaquepts	11,411
		Hapludands	Dystrudepts	55,460
			Endoaquepts	30,415
			Eutrudepts	110,475
		Hapludolls	Argiudolls	52,281
		Hapludults	Dystrudepts	42,530
		Haplustepts	Haplustalfs	387,461
			Haplustolls	103,555
			Ustorthents	96,607
		Haplusterts	Haplustepts	111,752
		Haplustolls	Argiustolls	182,542
			Haplustalfs	62,634
			Haplustepts	320,585
		Ustipsamments	Haplustands	61,502
Total				4,562,210

and coral reefs parent materials. The Haplustolls are mostly associated with Haplustepts, Argiustolls, and Ustorthents in hilly to mountainous areas with slopes >15%, while the association between Haplustolls and Ustorthents is also found in undulating areas with slopes <8% and primary parent materials of sedimentary rock, limestone, and coral rocks (Table 24.5).

In Flores Island, Haplustepts are associated with Ustorhents while Haplustolls are associated with Argiustolls. Haplustepts and Haplustolls are generally located on the foot and lower slopes, while Ustorthents and Argiustolls occur at the summit and upper slopes. These Ustorthents and Argiustolls have shallow soil and rock fragments/lithic property (Lithic Ustorthents, Lithic Haplustepts, and Lithic Argiustolls). Ustorthents in the volcanic cone and tectonic hills are found in Flores Island. They have good drainage and coarse to medium texture generally mixed with porous gravel and are slightly acidic (Hikmatullah and Chendy 2003).

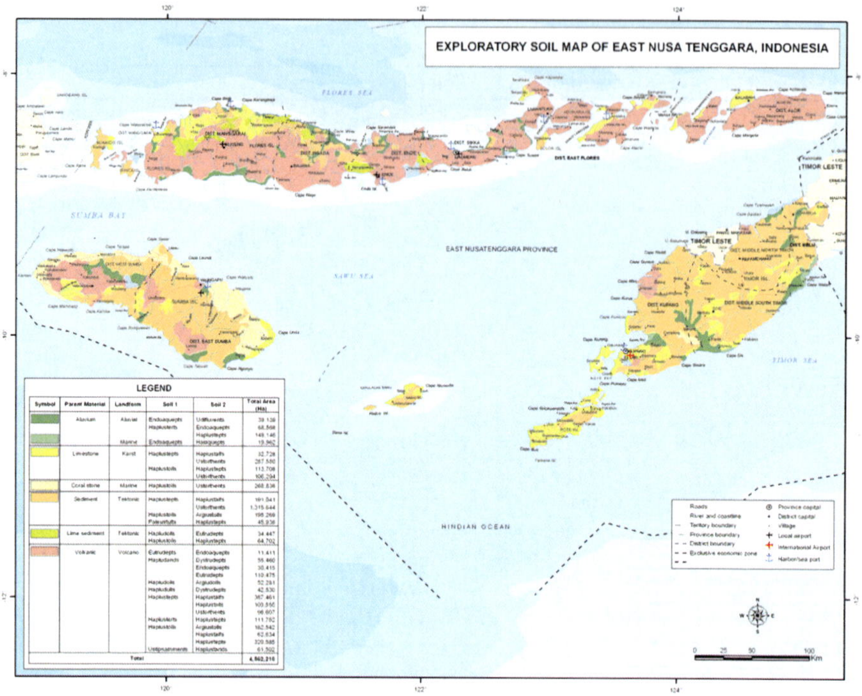

Fig. 24.2 Exploration soil map in NTT province

In the Ende District, the Haplustolls and Haplustepts occurring in the granite intrusion areas varied in solum depth and soil texture. On the steep slopes, the soil is shallow and gravelly/rocky, except those developed from volcanic tuff or ash, which has the deep soil (>100 cm). The soils developed from lava mostly contain rocks or rock fragments with shallow soil depth (lithic properties) (Chendy et al. 2003).

24.3.3 Physical and Chemical Soil Characteristics

The site location map presenting seven points of observation is shown in Fig. 24.3. It has been found that soil nutrient content is highly correlated with the parent material, rainfall, and vegetation cover. Table 24.6 illustrates the Haplustepts (derived from volcanic material) in Flores Island (Pedon 1) are neutral in reaction (pH 6.5–6.7), while the Haplustepts of Sumba Island (Pedon 6) derived from sedimentary parent material are very highly acidic (pH 4.1–4.3). Besides the sedimentary materials, the soil acidity was also influenced by high rainfall of >2,000 mm (Table 24.2), resulted in intensive leaching of bases (Subagyo et al. 2000). The Pedon 6 presents low P, K, exchangeable bases, CEC, and base saturation, leading to low overall soil fertility (Mulyani et al. 2008).

Table 24.5 Soil great groups (area in ha) and relief of Nusa Tenggara province, based on exploration map

Soils and soil association		Flat	Undulating	Rolling	Hilly	Mountainous	Area (ha)
Endoaquepts	Halaquepts	16,917					16,917
	Udifluvents	39,139					39,139
Eutrudepts	Endoaquepts		11,411				11,411
	Udipsamments		9,833				9,833
Hapludands	Dystrudepts			27,907		17,720	45,627
	Endoaquepts			30,415			30,415
	Eutrudepts					110,475	110,475
Hapludolls	Argiudolls				52,281		52,281
	Eutrudepts					34,447	34,447
Hapludults	Dystrudepts					42,530	42,530
Haplustepts	Haplustalfs		32,728	9,884	226,825	341,793	611,230
	Haplustolls					103,555	103,555
	Ustipsamments		827				827
	Ustorthents			368,371	589,710	740,923	1,699,004
Haplusterts	Endoaquepts	68,568					68,568
	Haplustepts	149,146	111,752				260,898
Haplustolls	Argiustolls		95,046		282,765		377,811
	Haplustalfs				62,634		62,634
	Haplustepts			20,220	214,217	264,558	498,995
	Ustorthents		268,836		87,318	18,976	375,130
Paleustults	Haplustepts		45,936				45,936
Udipsamments	Endoaquents	3,045					3,045
Ustipsamments	Haplustands					61,502	61,502
Total		276,815	576,369	456,797	1,515,750	1,736,479	4,562,210

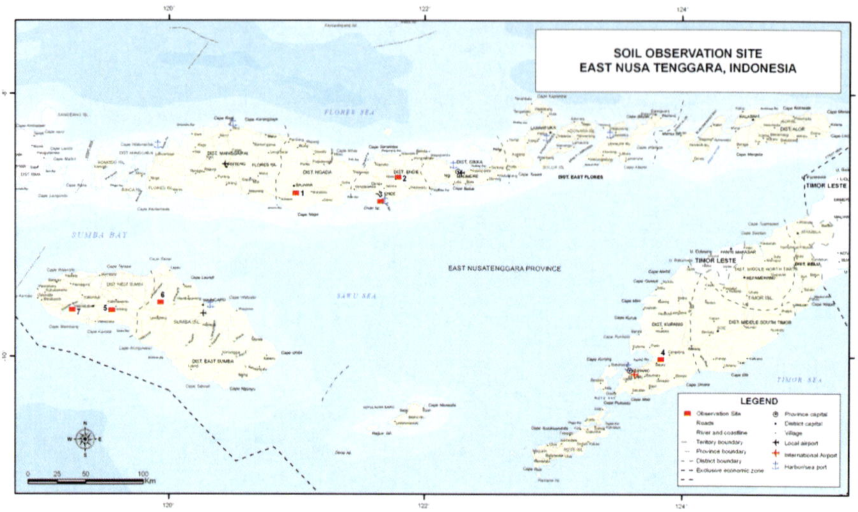

Fig. 24.3 Site location map in East Nusa Tenggara (NTT) province. The *numbers* are soil observation points

The Haplustepts in Kupang, Timor Island (Pedon 4), located in a flood plain (alluvium) are slightly alkaline (pH 7.1–7.8). This is due to the reason that the surrounding area is hilly karst region feeding basic cations to the area, resulting in high exchangeable Ca (45–53 cmol (+) kg^{-1}), high CEC and base saturation (Table 24.6). Total P and K contents (25%-HCl extractable) of soils derived from limestone parent materials are generally higher than those developed from volcanic and sedimentary materials.

The Mollisols derived from volcanic materials are coarse textured (>50% sand) compared to those derived from limestone parent materials with <20% sand. Nutrient contents and general soil fertility levels are better for the soils derived from volcanic and limestone parent materials compared to those derived from sedimentary parent material.

Soil depths in the upper and middle slopes decrease with the increase in the land slope. Table 24.6 shows that the soils in the hilly or mountainous areas generally have a shallow solum (lithic properties) and a coarser texture in the topsoil layer. Soil organic C is generally high due to addition of litter from bush and other natural vegetation cover.

24.3.4 Existing Land Use

Table 24.7 presents area under different land uses including lowland rice covers (2.5%), annual upland agricultural crops (16%) and the plantation-woodland (tree crops, 8%), in addition to nonagricultural areas (73%) of the total land area. With the low proportion of area allocated for agricultural, especially for lowland rice, the province is partially dependent on other provinces for food supplies. Parts of the nonagricultural areas have the potential for agricultural development as explained in the following section.

24.3.5 Land Use Planning for Agricultural Development

Of the total 4.6×10^6 ha area of the province, 2.5×10^6 ha is suitable for agricultural development, 0.3×10^6 ha is suitable for conservation forest, and 1.7×10^6 ha is unsuitable for agriculture; because due to some constraints like steep slopes >30%, coarse texture, rock outcrops, and lithic properties, such unsuitable areas are recommended for regreening with locally adaptable vegetation.

The suitable crops for the area receiving <1,000 mm annual rainfall, with <50 cm soil depth, in the lowland include upland rice (*Oryza sativa*), maize (*Zea mays*), sorghum (*Sorghum bicolor*), soybean (*Glycine max*), peanut (*Arachis hypogaea*), mungbean (*Phaseolus radiatus*) and cotton (*Gossypium hirsutum*), the perennial crops include cashew nut (*Anacardium occidentale*), kapok (*Ceiba pentandra*), candlenut (*Aleurites moluccana*), and Jatropha (*Jatropha curcas*).

Table 24.8 shows that the lands suitable for lowland rice and annual crops are distributed in all districts. Land suitable for fish ponds is available in the districts of Belu and Kupang. Pasture land can be developed in Kupang, Timor Tengah Selatan, Sumba Barat, and Sumba Timur districts.

Comparing the findings from the present study with the existing land uses (Table 24.7), it is recommended that there are ample opportunities for the development of paddy field, annual upland crops, ponds, and perennial crop plantation in the study area. This is supported with the figures that suitable land for lowland rice is 194,091 ha (Table 24.8), while the existing area under lowland rice is 115,600 ha (Table 24.7). For annual upland crops, land available for development is about 784,385 ha.

The province receives limited but irregular rainfall, therefore, it is recommended to rotate the annual crops in a way that high water-requiring crops are planted in the wet season, and those requiring less water planted in the transition period between the rainy and dry seasons. For irrigated lowland rice areas, the suggested cropping pattern is flooded rice in the beginning of rainy season, followed by maize in the late rainy season and fallow in the dry season or rice cultivation in the early rainy season followed by two crops of mungbean in the late rainy and dry seasons. For the rain-fed areas, rice is recommended as the first crop in the beginning of rainy season, while the second crop is mungbean, or avoid second crop to minimize the drought risk. In some locations, water pumping systems from surface waters (river or water retention ponds) are used for irrigation, and these can prolong the cropping period.

Annual and low-standing crops are recommended for lands with <15% slopes, such as maize, rice, soybeans, mungbeans, and cotton (Table 24.9). If water and irrigation system is available, cotton is a good choice as a dry season crop in rotation

Table 24.6 Selected soil characteristics of seven pedons from East Nusa Tenggara province

Soil horizons[a] and depth (cm)	Sand (%)	Silt (%)	Clay (%)	Texture classes	EC (dS m^{-1})	pH H_2O	Org C (%)
Pedon 1 Typic Haplustepts, Ngada District, Flores Island, 8% slope, volcanic plateau							
0–27 (Ap)	32	30	38	CL	–	6.7	1.77
27–54 (Bw1)	27	31	42	C	–	6.5	1.29
54–70 (Bw2)	29	29	42	C	–	6.6	0.81
70–90 (Bw3)	35	24	41	C	–	6.6	0.64
90–120 (BC)	29	26	45	C	–	6.6	0.54
Pedon 2 Typic Haplustolls, Ende District, Flores Island, 8–15% slope, volcanic ridge							
0–26 (Ap)	60	26	14	SL	–	6.7	0.91
26–47 (Bw1)	61	13	26	SCL	–	6.9	0.41
47–78 (Bw2)	60	26	14	SL	–	6.8	0.19
78–120 (BC)	56	32	12	SL	–	6.7	0.09
Pedon 3 Lithic Ustorthents, Ende District, Flores Island, >30% slope, volcanic mountain							
0–14 (Ap)	41	39	20	L	–	6.2	3.47
14–42 (AR)	25	28	47	C	–	6.4	2.01
> 42 (R)							
Pedon 4 Oxiaquic Haplustepts, Kupang District, Timor Island, 3% slope, flood plain							
0–21 (Ap)	3	37	60	C	0.198	7.1	1.67
21–44 (Bw1)	16	32	52	C	0.161	7.6	0.85
44–62 (Bw2)	5	37	58	C	0.145	7.7	0.49
62–100 (Bw3)	21	27	52	C	0.156	7.8	0.34
Pedon 5 Oxic Haplustoll, Sumba Barat Daya District, Sumba Island, 2% slope, karst plateau							
0–24 (Ap)	20	71	9	SiL	0.059	6	5.06
24–35 (Bw1)	12	76	12	SiL	0.044	6	3.05
35–60 (Bw2)	10	71	19	SiL	0.032	6	1.45
60–90 (Bw3)	4	62	34	SiCL	0.026	6	1.2
Pedon 6 Oxic Haplustepts, Sumba Tengah District, Sumba Island, 7% slope, tectonic plain							
0–19 (Ap)	59	8	33	SCL	0.022	4.3	2.2
19–41 (Bw1)	48	3	49	SC	0.012	4.3	0.99
41–72 (Bw2)	58	4	38	SCL	0.013	4.2	0.54
72–110 (Bw3)	35	9	56	C	0.02	4.1	0.35
Pedon 7 Lithic Argiustolls, Sumba Barat Daya District, Sumba Island, >15% slope, hilly karst							
0–17 (Ap)	0	60	40	SiC	0.123	6	5.41
17–32 (Bt)	0	18	82	C	0.074	6.1	2.27
>32 (R)							

[a]A = A master horizon; B = B master horizon; C = C master horizon; R = strongly cemented to other disturbance; t = accumulation of silicate clay; w = development of color or structure; – = not *SCL* = sandy clay loam

P_2O_5	K_2O	Exchangeable cations						BS
		Ca	Mg	K	Na	Sum	CEC	%
HCl 25% (mg kg^{-1})		cmol(+) kg^{-1}						
100	130	11.58	4.29	0.16	0.19	16.22	16.31	99
110	120	14.27	4.31	0.1	0.27	18.95	18.71	100
–	–	14.96	5.7	0.1	0.33	21.09	21.13	100
–	–	14.81	5.82	0.1	0.39	21.12	21.63	98
–	–	17.72	6.75	0.08	0.59	25.14	24.03	100
6	19	9.75	2.82	0.25	0.3	13.12	15.94	100
6	13	12.45	3.54	0.16	0.73	16.88	19.85	76
8	14	12.79	3.82	0.16	0.61	17.38	20.31	100
8	21	11.75	3.53	0.33	0.49	16.1	18.16	100
28	118	15.4	5.2	1.59	0.11	22.3	21.29	100
14	159	15.95	5.33	1.58	0.25	23.11	25.49	55
117	145	45.16	2.07	0.86	0.17	48.26	24.73	100
98	113	51.9	1.8	0.61	0.29	54.6	22.39	100
92	65	53.31	2.57	0.31	0.4	56.59	21.2	100
96	52	49.8	5.23	0.22	0.57	55.82	19.36	100
73	87	17.47	2.89	1.06	0.27	21.69	21.7	100
52	37	15.41	2.2	0.5	0.35	18.46	19.6	94
50	23	12.73	1.29	0.31	0.22	14.55	16.61	88
45	19	10.99	1.05	0.28	0.3	12.62	15.37	82
8	1	1.93	0.92	0.03	0.05	2.93	10.63	28
4	0	1.52	0.56	0	0.05	2.13	8.59	25
9	1	1.11	0.47	0	0.05	1.63	7.74	19
12	2	1.69	0.55	0.03	0.02	2.29	10.19	22
64	77	54.4	3.55	0.65	0.21	58.81	40.86	100
34	30	51.13	1.92	0.17	0.16	53.38	35.31	100

indurated bedrock; *BS* = base saturation; *CEC* = cation exchange capacity; p = plow layer-tillage or available; C = clay; *CL* = clay loam; *L* = loam; *LS* = loamy sand, *SiL* = silt loam; *SL* = sandy loam;

Table 24.7 Existing land use types and area in East Nusa Tenggara province

Land use	Area	
	ha	%
Lowland rice	115,596	2.53
Upland annual crops	738,254	16.18
Temporary fallow land	731,890	16.04
Estate crop	355,885	7.80
Meadows	903,422	19.80
Pond and dyke	2,638	0.06
Woodland	371,961	8.15
Settlement area	201,648	4.42
Forest and others	1,140,915	25.01
Total	4,562,209	100.00

Source: Biro Pusat Statistik (2008)

with maize. However, in areas where there is no irrigation system and the rainfall is <1,000 mm, cotton is not recommended because of high chance of water deficiency.

The selected perennial plantation crops should be those that are resistant to 6–7 dry months; these are cashew nut, candlenut, and Jatropha. Flores Island has a good prospect for the development of dry land plantation crops such as cashew nut, candlenut, kapok (*Ceiba pentandra*), and mango (Hikmatullah et al. 1999). The cashew and candlenut have been planted widely, while Jatropha is drawing attention as a biofuel crop. The Jatropha can be used for the rehabilitation of degraded lands with shallow and stony solum that are largely found in Ende, Sikka, Sumba Timor, and Sumba Barat districts. Jatropha can be adapted in marginal lands and has the potential to grow on stony, sandy, clayey, as well as eroded lands (Bhag Mal and Joshi 1991). Jatropha has been reported on the rocky and hilly areas or along the canals and farm borders (Heller 1996; Rivaie et al. 2006).

Of the total 4.6×10^6 ha of NTT province, about 3.5×10^6 ha are located in the lowlands (<700 m above sea level—masl) and the remaining 1.2×10^6 ha in the high land areas (>700 masl). Of the total land in the highlands, only 273,527 ha are suitable for the development of food and perennial crops. In these highland areas, recommended annual crops are wheat (*Triticum aestivum*), tobacco (*Nicotiana tobacum*) and vegetables such as potatoes (*Solanum tuberosum*), cabbage (*Brassica oleracea*), carrots (*Daucus carota*); and from the perennial crops, these are coffee (*Coffea arabica*) and orange (*Citrus* sp.) (Table 24.9). The recommendation land use map in the NTT province is presented in Fig. 24.4.

24.4 Conclusions

From the present study, following conclusions are drawn. The eastern archipelago of Indonesia exposed to the southerly wind of Australia is relatively dryer than most parts of Indonesia. The promising crops for the area are recommended considering

Table 24.8 Land use recommendations (ha) for in East Nusa Tenggara province

District	Rice-field	Annual crop	Perennial crops	Pasture	Ponds	Forest conservation	Not suitable[a]	Total (ha)
Alor	–	20,162	49,270	–	–	16,596	206,932	292,960
Belu	18,488	35,381	70,932	–	14,038		54,097	192,936
Flores Timur	2,246	34,141	47,559	–	–	19,111	79,808	182,865
Kota Kupang	1,289	11,438	4,512	–	–		7,375	24,614
Kupang	18,290	98,282	145,966	12,060	2,879	39,897	199,984	517,358
Lembata	854	1,512	–	–	–	19,475	106,081	127,922
Manggarai	56,172	62,303	178,593	–	–	118,040	272,170	687,278
Ngada	47,966	24,763	119,797	–	–	16,019	87,933	296,478
Rote Ndao	–	95,338	23,812	–	–			119,150
Sikka	9,366	26,344	98,458	–	–	33,286	194,795	362,249
Sumba Barat	7,237	146,512	90,307	95,476	–	26,477	28,769	394,778
Sumba Timur	7,332	84,932	249,406	191,448	–	18,312	140,554	691,984
Timor Tengah Selatan	23,873	74,814	63,599	1,072	–	40,688	155,520	359,566
Timor Tengah Utara	978	68,463	36,291	–	–	3,742	202,598	312,072
Total	194,091	784,385	1,178,502	300,056	16,917	351,643	1,736,616	4,562,210

[a] Recommended for rehabilitation or regreening with pioneer crops

Table 24.9 Crop recommendation for agricultural land use in East Nusa Tenggara province

Land use recommendation	Alternative crops/commodities	Areas	
		ha	%
Lowland rice (paddy system)	Rice, maize, mungbean	194,091	4.3
Annual upland crops	Maize, upland rice, soybean, mungbean, cotton, [a]wheat, tobacco, potatoes	784,385	17.2
Perennial or plantation crops	Cashew nut, candlenut, cacao, jatropha, [a]coffee, orange	1,178,502	25.8
Meadows/grassland	Cattle	300,056	6.6
Ponds	Shrimp and milkfish	16,917	0.4
Conservation forest	Secondary forest	356,619	7.8
Unsuitable land	Rehabilitation/natural vegetation	1,731,640	38.0
Total		4,562,210	100.0

[a]Commodities for high land agriculture

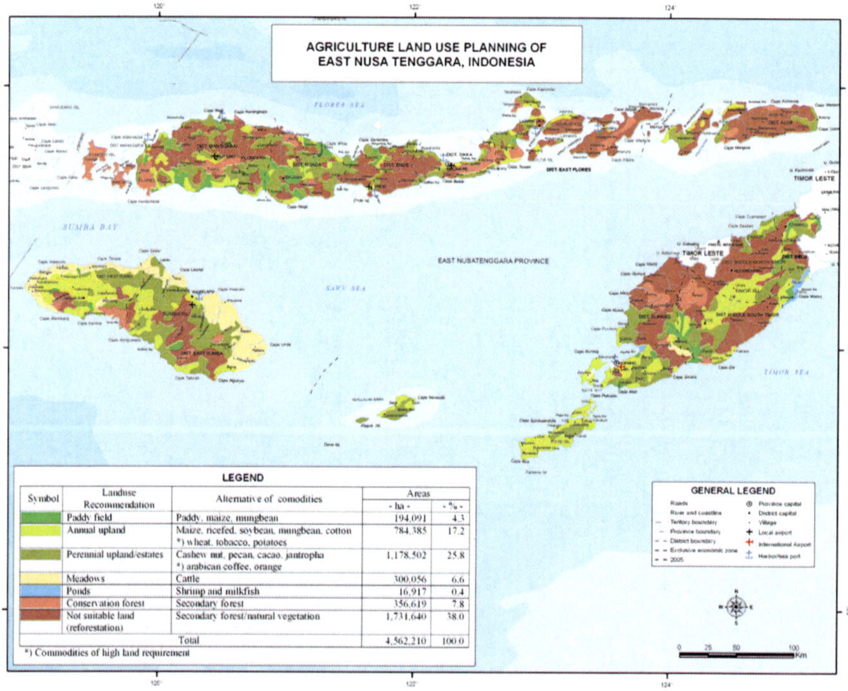

Fig. 24.4 Land use recommendation map for NTT province

the soil properties, slope, and rainfall patterns. In the hilly and mountainous topographic areas and where the torrential rainfall occurs in the rainy season, it is recommended to grow annual crops in relatively flat areas (mainly in the valley bottom) and relatively drought-resistant perennial crops in the hilly areas. The stony and the lithic properties areas are recommended for regreening with pioneer trees, while some existing forests are to be conserved. In order to achieve agricultural sustainability, it is important to adopt conservation measures and improve water management.

References

Balitklimat (2004) Atlas Sumberdaya Iklim/Agroklimat untuk Pertanian (The Atlas of Climate Resources/Agro-climate for Agriculture). Balai Penelitian Agroklimat dan Hidrologi, Bogor

Bhag Mal, Joshi V (1991) Underutilized plant resources. In: Paronda RS, Arora RK (eds) Plant genetic resources – conservation and nanagement. Malhotra Publishing House, New Delhi, pp 211–229

Biro Pusat Statistik (2008) Land utilization by provinces in Indonesia. Central Bureau of Statistics, Jakarta. www.bps.go.id. Accessed 1 Feb 2008

BMG NTT Province (1996–2007) Data of rainfalls and rainy days averaged from 1996 to 2007 in NTT Province. Badan Meteorologi dan Geofisika, NTT, Bogor

Burt R (ed) (2004) Soil survey laboratory methods manual. Soil Survey Investigations Report No. 42, Version 4.0, November 2004. USDA-NRCS, Washington, DC

Chendy TF, Kosasih SE, Umat DK, Tanjung S, Hikmatullah (2003) Penyusunan Peta Pewilayahan Komoditas Pertanian berdasarkan Zone Agro-Ekologi skala 1:50.000 di Kabupaten Ende, Provinsi Nusa Tenggara Timur. Balai Penelitian Tanah, Bogor

Djaenudin D, Hendrisman M, Subagyo H, Hidayat A (2003) Petunjuk Teknis Evaluasi Lahan Untuk Komoditas Pertanian (Guidelines of land suitability evaluation for agriculture commodities). Balai Penelitian Tanah, Bogor

FAO (1990) Guidelines for soil description. 3rd edn. (Revised). FAO/UNESCO, Rome

Heller J (1996) Physic Nut (*Jatropha curcas L.*). Promoting the conservation and use of underutilised and neglected crops. 1. Institute of Plant Genetics and Crop Plant Research. Gatersleben/ International Plant Genetic Resources Institute, Rome, p 66

Hikmatullah, Chendy TF (2003) Klasifikasi dan Sifat-sifat Tanah (Classification and soil properties). In: Sumberdaya Tanah Pulau Flores Nusa Tenggara Timur: Karakteristik dan potensinya untuk Pertanian (Land resources of Flores East Nusa Tenggara: characteristics and potential for agriculture). Balai Besar Litbang Sumberdaya Lahan Pertanian, Bogor, (in Indonesian), pp 37–91

Hikmatullah, Chendy TF, Sukarman, Sunyoto Rb, Subagyo (1999) Kondisi Sumberdaya Lahan Pulau Flores Provinsi Nusa Tenggara Timur (Land resource condition of Flores, East Nusa Tenggara Province). Pusat Penelitian Tanah dan Agroklimat, Bogor, (in Indonesian)

Mulyani A, Pramudia A, Sukarman, Hartomi H, Allorerung D (2008) Laporan Akhir Identifikasi dan Evaluasi Kesesuaian Lahan Jarak Pagar (*Jatropha curcas L.*) (Final report on identification and evaluation of land suitability for Jatropha (*Jatropha curcas L.*). Balai Besar Penelitian dan Pengembangan Sumberdaya Lahan Pertanian, Bogor, (in Indonesia)

Puslitbangtanak (2000) Atlas Sumberdaya Tanah Eksplorasi Indonesia. Skala 1:1,000,000 (Exploration land resource atlas of Indonesia 1:1,000,000 scale). Pusat Penelitian dan Pengembangan Tanah dan Agroklimat, Bogor

Puslitbangtanak (2001) Atlas Arahan Tata Ruang Pertanian Indonesia. Skala 1:1,000,000 (Referral atlas of agriculture spatial planning of Indonesia. 1:1,000,000 scale). Pusat Penelitian dan Pengembangan Tanah dan Agroklimat, Bogor

Rivaie AAR, Allorerung D, Mahmud Z, Effendi DS, Sumanto F, Isa (2006) Karakteristik fisik lingkungan daerah pertanaman jarak pagar (*Jatropha curcas L.*) di Cikeusik-Banten (Physical characteristics of Jatropha (*Jatropha curcas L.*) planting area in Cikeusik-Banten). Makalah Lokakarya II Status Teknologi Tanaman Jarak Pagar. Hotel Pangrango-Bogor, 29 November 2006 (in Indonesian)

Soil Survey Division Staff (1993) Soil survey manual. USDA Handbook No 18. USDA, Washington, DC

Subagyo H, Suharta-dan N, Siswanto AB (2000) Tanah-tanah Pertanian di Indonesia (Agricultural soils of Indonesia). In: Sumberdaya Lahan Indonesia dan Pengelolaannya (Land resources of Indonesia and their management system). Pusat Penelitian Tanah dan Agroklimat, Bogor (in Indonesian), pp 21–66

Sulaeman, Suparto, Eviati (2005) Petunjuk Teknis Analisis Kimia Tanah, Tanaman, Air dan Pupuk. Balai Penelitian Tanah, Bogor

Widiyono W, Abdulhadi R, Lidon B (2006) Kajian erosi dan pendangkalan 'Embung' di Pulau Timor-NTT: Studi kasus 'Embung' Oemasi dan 'Embung' Leosama (The study on erosion and water sedimentation pond in Timor Island, East Nusa Tenggara: case study Oemasi and Leosama). Limnotek, vol XIII, No 2, Lembaga Ilmu Pengetahuan Indonesia, Jakarta (in Indonesian)

Part II
Land Use Planning and Policy Implications

Chapter 25
Land Use Planning and Policy Implication: Bridging Between Science, Politics and Decision Making

Winfried E.H. Blum

Abstract We describe land use planning based on the six main functions and uses of soil and land. Policy input is mainly needed to control the competition between different land uses, avoiding or minimising irreversible impacts. For achieving this, we describe an indicator framework, which allows for planning and controlling complex land use systems.

Keywords Decision making • Land use planning • Policy implication • Politics • Science

25.1 Introduction

Land use planning is a complex issue in view of the six different functions and uses of soil and land, delivering goods and services to humans and the environment. Policy implication is mainly concerned with the control of competition between the main uses in space and time, aiming at specific targets. In this context, indicators can help decision making by explaining the complexity of the many different land use systems.

W.E.H. Blum (✉)
Institute of Soil Research, University of Natural Resources and Life Sciences (BOKU), Vienna, Austria
e-mail: winfried.blum@boku.ac.at

S.A. Shahid et al. (eds.), *Developments in Soil Classification, Land Use Planning and Policy Implications: Innovative Thinking of Soil Inventory for Land Use Planning and Management of Land Resources*, DOI 10.1007/978-94-007-5332-7_25,

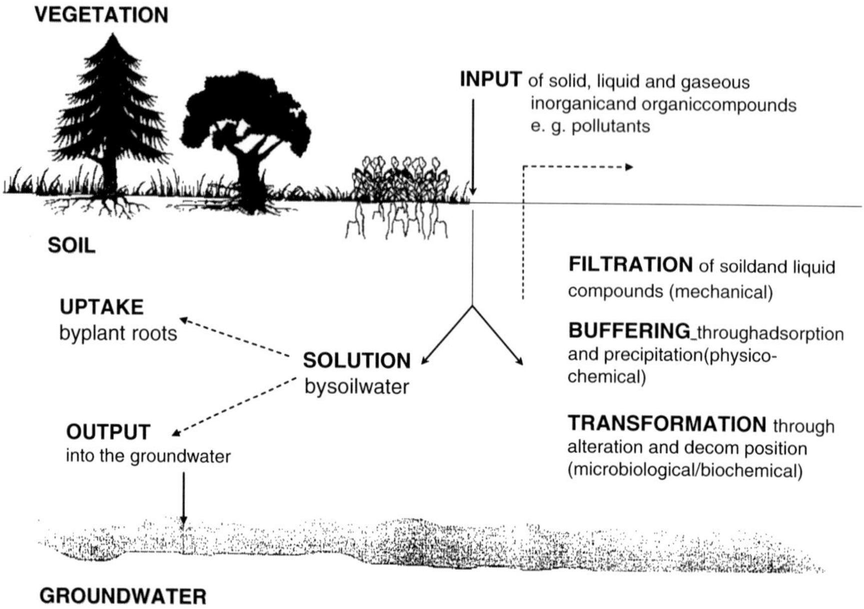

Fig. 25.1 Filtering, buffering and transformation through soil

25.2 Land Use Planning and the Six Main Functions of Soil and Land

Soil has six main functions for the cultural, social and economic development of human societies, which can be distinguished into three more ecological functions and three others, directly linked to human activities in the sense of technical, industrial and socio-economic uses (Blum 2005).

The three ecological functions are:

1. Production of biomass, ensuring food, fodder, renewable energy and raw materials. These well-known functions are the basis of human and animal life.
2. Filtering, buffering and transformation capacity between the atmosphere, the ground water and the plant cover, strongly influencing the water cycle at the land surface as well as the gas exchange between terrestrial and atmospheric systems and protecting the environment, including human beings against the contamination of ground water and the food chain (Fig. 25.1). This function becomes increasingly important because of the many solid, liquid or gaseous, inorganic or organic depositions, due to urban development, industrialisation, transport processes and others. As long as these filtering, buffering and transformation capacities can be maintained, there is no danger to the ground water and to the food chain. However, these capacities of soil and land are limited and vary according to the specific conditions.

3. Biological habitat and gene reserve, with a large variety of organisms. Soils contain three to four times more species in number and quantity than all other above-ground biota together. Therefore, soils are the main basis of biodiversity, on which human life is extremely dependent. Moreover, genes from the soil become increasingly important especially for biotechnological and bioengineering processes.

In addition to these three ecological functions, soil and land have three other functions, linked to technical, industrial and socio-economic uses:

1. They are the physical basis for technical, industrial and socio-economic structures and their development, e.g. industrialisation and urbanisation. One of the main problems in this context is the exponential increase of urban and peri-urban areas, including transport facilities between them. This is also true for the Arabian peninsula.
2. Soils are a source of raw materials, delivering clay, sand, gravel and minerals in general, as well as energy and water. Raw materials from soils are the basis for technical, industrial and socio-economic development.
3. Last but not least, soils are important as a geogonic and cultural heritage, forming an essential part of the landscape in which we live and concealing and protecting palaeontological and archaeological treasures of high importance for the understanding of our own history and that of the earth.

In view of the fact that soil and land are an absolutely limited resource, which cannot be extended or enlarged, the use of these six main functions of soil and land, which are often concomitant in the same area, becomes a key issue of sustainable development, based on land use planning. Land use planning under this aspect can be defined as the spatial harmonisation of a simultaneous use of all these functions in a given area.

25.3 Land Use Planning and Policy Implication

For understanding soil and land use planning under the aspects of a sustainable development, it is necessary to define the interactions and competitions which exist between the different uses of soil and land (Fig. 25.2). In this context, three different categories of interaction and competition can be distinguished as below:

- Exclusive competition between the use of land for infrastructural development, as a source of raw materials and as a geogonic and cultural heritage on the one hand and the use of soil for biomass production, filtering, buffering and transformation activities and as a gene reserve on the other hand.

This becomes evident by the sealing of soil through urban and industrial development, e.g. the construction of roads, industrial premises, houses, sporting facilities and others, well known as the process of urbanisation and industrialisation, thus excluding all other uses of soil and land. The process of sealing is at the moment the most important challenge for land use planning and a great threat to sustainable soil use (Figs. 25.3, 25.4, and 25.4).

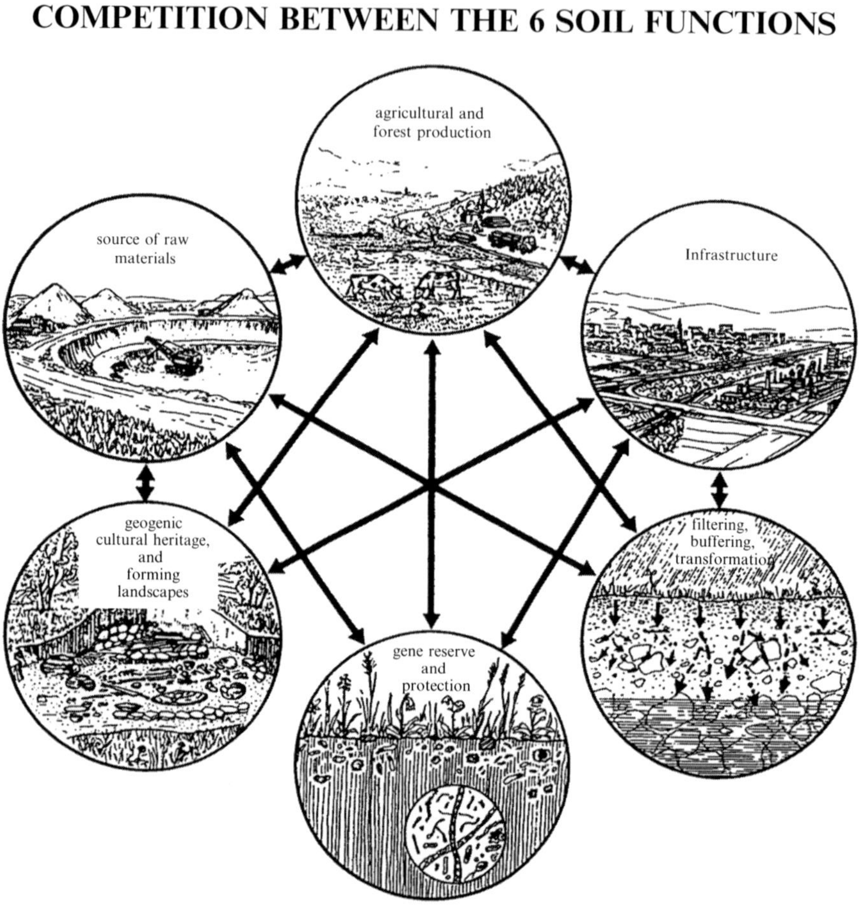

Fig. 25.2 Competition between the six uses of soil and land

- A second category of competition exists through intensive interactions between infrastructural soil and land uses and the use of land for urbanisation and industrialisation, due to the problem of soil contamination and pollution through the atmospheric pathway, the waterway and terrestrial transport (Fig. 25.6).
- A third form of competition also exists among the three ecological soil uses themselves, e.g. when we use waste and sewage sludge on agricultural land, in addition to the deposition of air pollutants, which can have a negative influence on the ground water and the food chain, surpassing the natural capacity of soils for dealing with these adverse contaminants. This is specifically true for high-input agricultural systems (Fig. 25.7), because each drop of rain falling on agricultural land has to pass soil before it becomes ground water underneath. Therefore, the type of soil use is very important for the production of ground water or water filtering for ground water production.

Fig. 25.3 Europe's natural environment at daylight

Fig. 25.4 Europe's built environment at night

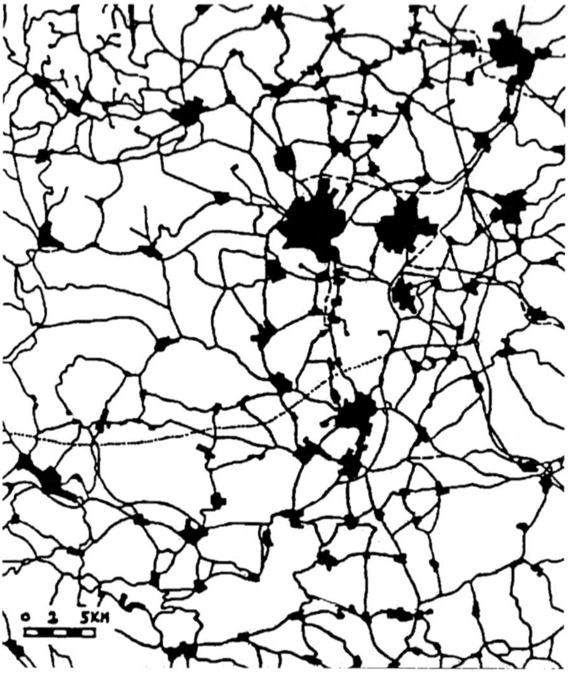

Fig. 25.5 Sealing of soils in southern Germany (observe the scale of 5 km)

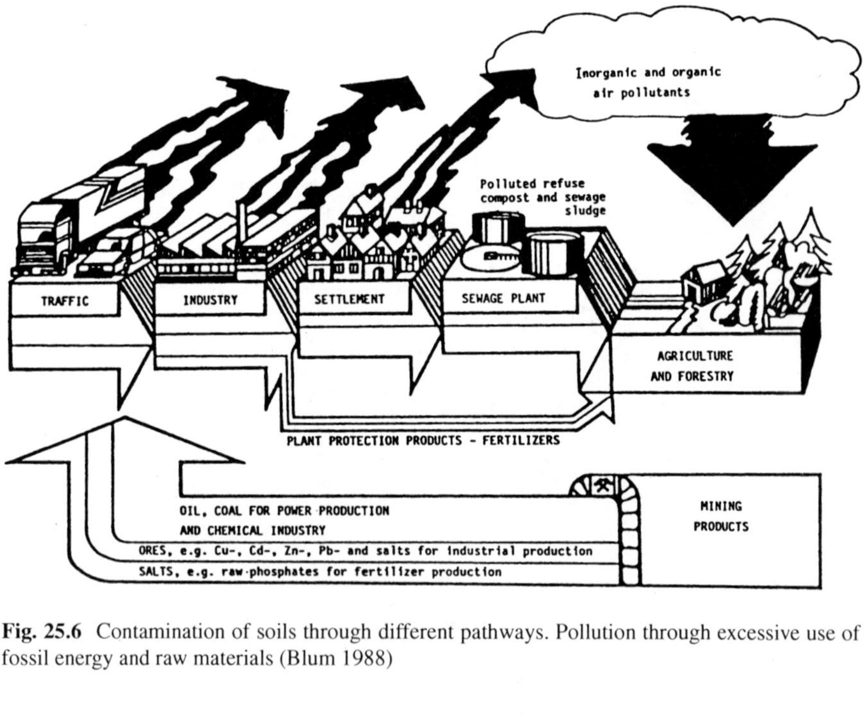

Fig. 25.6 Contamination of soils through different pathways. Pollution through excessive use of fossil energy and raw materials (Blum 1988)

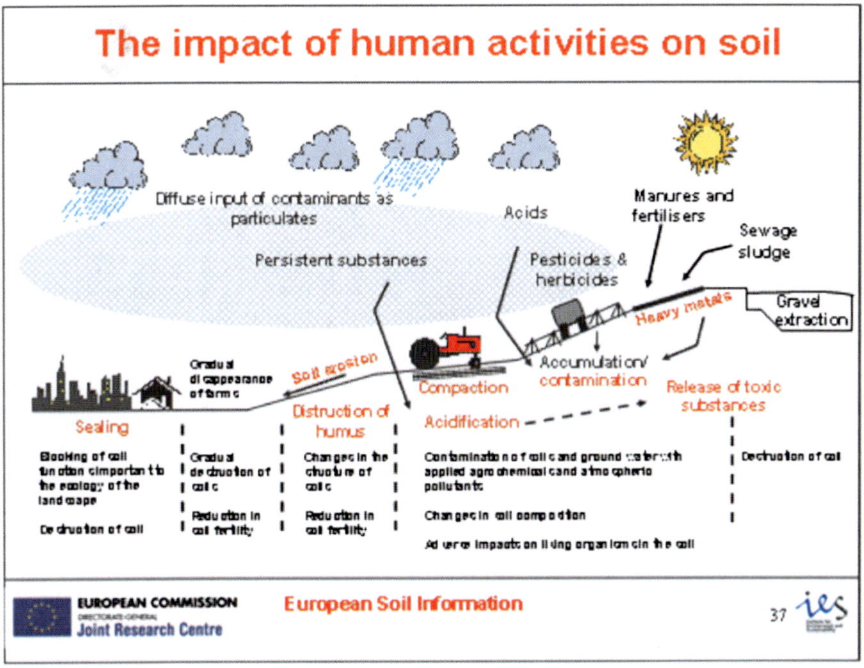

Fig. 25.7 Impact of human activities by urban and rural land use

Summarising, it can be stated that there exist eight main threats to land and soil: sealing through urbanisation and industrialisation, local and diffuse contamination, erosion by water and wind, compaction and further forms of physical degradation, decline in soil organic matter; loss of biodiversity, salinisation and alkalisation and floods and landslides. These main threats were politically defined by the European Strategy of Soil Protection (European Commission 2006).

However, it is not only important to know the threats and damages to land and soil but also to classify them in order of urgency. Under this aspect, we can distinguish between irreversible and reversible damages and threats. The definition of irreversibility is based on a time span of 100 years (about four human generations). Under this definition, irreversible damages and threats are soil loss through sealing; extraction of materials; mining and erosion by water and wind; intensive pollution by heavy metals, xenobiotics and radioactive compounds; advanced acidification and salinisation and deep-reaching compaction. Reversible damages or threats are soil pollution by biodegradable organic compounds, which can be mineralised and metabolised; compaction; glazing and other physical deterioration of the top soil.

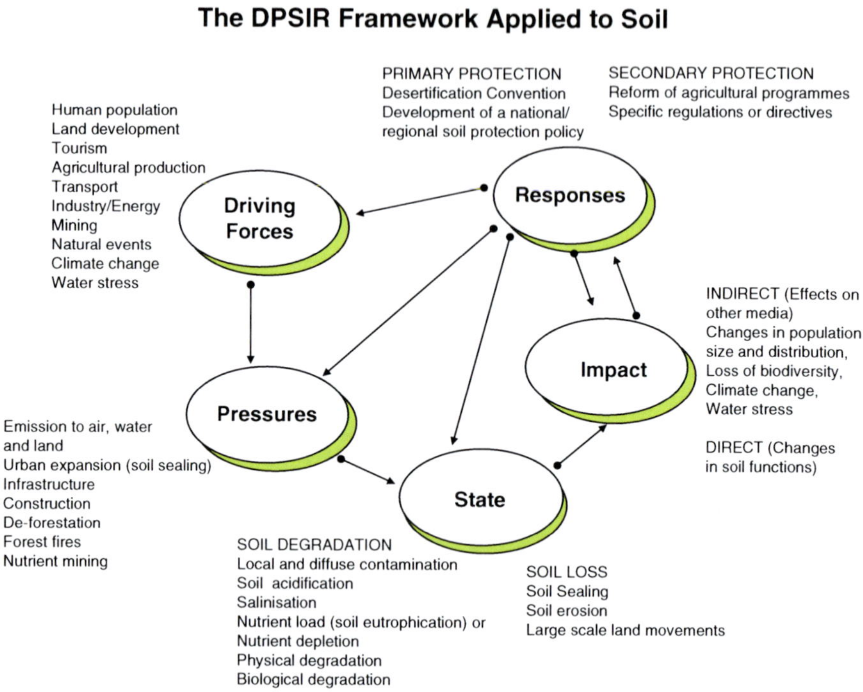

Fig. 25.8 The DPSIR framework approach applied to soil

25.4 Indicators for Land Use Planning: Bridging Between Science, Politics and Decision Making

Land use planning aiming at a sustainable use of land and soil must harmonise the different uses in such a way that irreversible uses are minimised and the six main soil functions are maintained (Blum 2009). However, this is not a scientific question but a political one, which means that people living in a given area or a certain space need to decide which soil functions they may use at a given time or in a given space, using a top-down-bottom-up approach. In this context, scientists only have the possibility to develop scenarios and explain which causes and impacts may occur, when different options are exercised.

For action, politicians and decision makers as well as stakeholders need information. This information can be given by indicators, which are following the cycle of political decision making, e.g. the DPSIR approach (Blum 2004) as shown in Fig. 25.8. Indicators in this sense are information for understanding and managing complex systems. They can be cultural, social, economic, ecological or technical information. Scenarios based on such indicators can help politicians and decision makers as well as land use planners to understand the complexity of the system and to choose the right options.

For this, the criteria for indicators must be policy relevant; focus on real demand and less on the supply of data; are analytically sound, based on science and revealing a clear cause-response relationship; are easy to interpret and understandable for stakeholders at the grass-root level as well as for decision makers and politicians and, finally, are easily measurable and therefore feasible and cost effective in data collection, processing and dissemination.

The most used indicator framework for bridging between science on one side and politics and decision making on the other side is the DPSIR framework approach (Fig. 25.8). From this figure, it becomes clear that a certain state (S) of a soil, e.g. a degraded soil or an eroded soil, is caused by driving forces (D), e.g. agricultural land use, developing pressures (P), such as compaction or erosion. On the other side, the state (S) has an impact (I) which can be direct, e.g. changes in important soil functions, like soil fertility, or indirect, when the changes are affecting the local population which, e.g. cannot produce enough food and therefore cannot live any longer in that area. For mitigating the driving forces or pressures, the state or the impact, responses (R) are needed, which can reach from information, subsidies, until legal regulations at different levels.

In order to make this approach operational, concepts for integrated research, especially on soils, were developed, distinguishing between main research goals, research clusters and sciences involved (Table 25.1) (Blum et al. 2004).

From Table 25.1, it becomes clear that it is necessary (1) to understand the main processes in the soil, induced by threats; (2) to know where these processes occur and how they develop with time; (3) to know the driving forces and pressures behind these processes, as related to cultural, social, economic, ecological or technical, local, regional or global developments; (4) to know the impacts on the ecoservices provided by soils to other environmental compartments and finally (5) to develop operational tools for the mitigation of the threats and impacts.

For this, 5 research clusters were developed, showing the analysis of processes; the analysis of the state and its changes with time (soil monitoring); the relating of the threats to driving forces and pressures, cross-linking with cultural, social and economic drivers, to analyse the impacts on other environment compartments, like air, water, biomass production, human health and culture and finally the development of strategies and operational procedures for the mitigation of the threats (responses) as shown in Fig. 25.9. The relationship between the explained eight threats to driving forces and pressures through cross-linking with policies is shown in Fig. 25.10, where the agricultural policy is made responsible for erosion, losses of organic matter, decline in biodiversity, compaction, salinisation and contamination.

An example of the analysis of impacts by relating those to the soil deliverables into other ecological compartments is shown in Fig. 25.11, where the impact of erosion on the air, the water, the biomass production, human health and biodiversity is shown.

Finally, looking into the necessity to involve different sciences in different operational steps aiming at developing scenarios for politicians and decision makers, Table 25.1 shows that for most of the priority research areas, different sciences are necessary for developing sound results.

Table 25.1 Concept for integrated research in ecology – example soil

	Main research goals	Research clusters	Sciences involved
1	To understand the main processes in the eco-subsystem soil; induced by threats	Analysis of processes related to the 8 threats to soil and their interdependency: erosion, loss of organic matter, contamination, sealing, compaction, decline in biodiversity, salinisation, floods and landslides	Interdisciplinary research through cooperation of soil physics, soil chemistry, soil mineralogy and soil biology
2	To know where these processes occur and how they develop with time	Development and harmonisation of methods for the analysis of the state (S) of the 8 threats to soil and their changes with time = *soil monitoring* in Europe	Multidisciplinary research through cooperation of soil sciences with geographical sciences, geo-statistics, geo-information sciences (e.g. GIS)
3	To know the driving forces and pressures behind these processes, as related to cultural, social, economic, ecological or technical, local, regional or global developments	Relating the 8 threats to *driving forces (D)* and *pressures (P)* = cross-linking with EU and other policies (agriculture, transport, energy, environment, etc.)	Multidisciplinary research through cooperation of soil sciences with political sciences, social sciences, economic sciences, historical sciences, philosophical sciences and others
4	To know the impacts on the ecoservices provided by the subsystem soil to other environmental compartments (eco-subsystems)	Analysis of the *impacts (I)* of the 8 threats, relating them to soil ecoservices for other environmental compartments: air, water (open and ground water), biomass production, human health and biodiversity	Multidisciplinary research through cooperation of soil sciences with geological sciences, biological sciences, toxicological sciences, hydrological sciences, physiogeographical sciences, sedimentological sciences and others
5	To have operational tools (technologies) at one's disposal for the mitigation of threats and impacts	Development of operational procedures for the mitigation of the threats = *responses (R)*	Multidisciplinary research through cooperation of natural sciences with engineering sciences, technical sciences, physical sciences, mathematical sciences and others

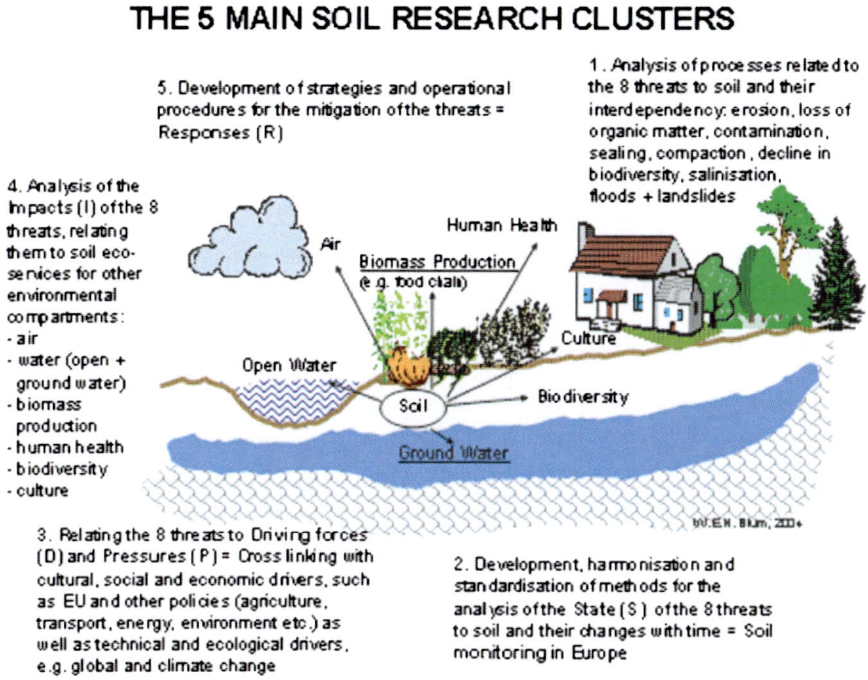

Fig. 25.9 The five main research clusters (Blum 2004)

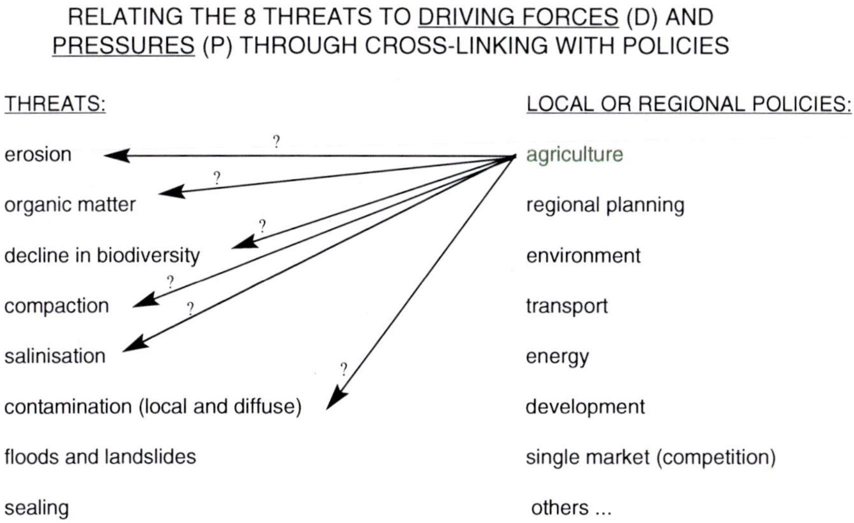

Fig. 25.10 Relating the eight threats to driving forces (D) and pressures (P) through cross-linking with policies

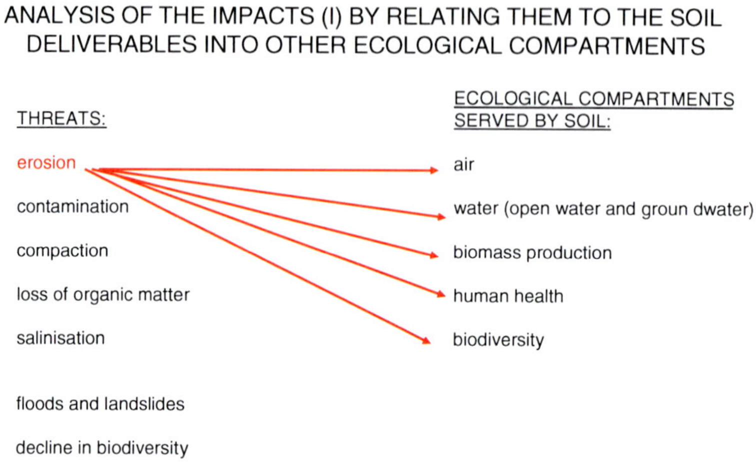

Fig. 25.11 Analysis of the impacts (I) by relating them to the soil deliverables into other ecological compartments

This indicator framework was developed by the European Environment Agency (EEA), on the basis of frameworks developed by the Organisation of Economic Co-operation and Development (OECD).

25.5 Summary and Conclusions

Land use planning should be based on scenarios which are scientifically sound and contain clear criteria, e.g. indicators. The DPSIR approach allows for understanding and managing complex ecological (technical) systems through the development of such scenarios. For the definition of indicators, new concepts are needed, including interdisciplinary and multidisciplinary approaches, bringing together technical, ecological, cultural, social and economic sciences. Indicators based on this approach can bridge between science and technology on one side and stakeholders, decision making and politics on the other side, thus sharing knowledge between those who have it and those who need it.

References

Blum WEH (1988) Problems of soil conservation, vol 39, Nature and environment series. Council of Europe, Strasbourg

Blum WEH (2004) Soil indicators for sustainable land management. Soil Sci Annu. 55(1):19–24 (Warschau, ISSN 0080–3642)

Blum WEH (2005) Functions of soil for society and the environment. Rev Environ Sci Biotechnol 4:75–79

Blum WEH (2009) Reviewing land use and security linkages in the Mediterranean region. In: Rubio JL, Safriel U, Daussa R, Blum WEH, Pedrazzini F (eds) Water scarcity, land degradation and desertification in the Mediterranean Region. Springer, Dordrecht, pp 101–117

Blum WEH, Büsing J, Montanarella L (2004) Research needs in support of the European thematic strategy for soil protection. Trends Anal Chem 23:680–685

European Commission (2006) Thematic strategy for soil protection COM (2006) 231 final

Chapter 26
Application of Soil Survey in Land Use Planning and Policy Development

Noel Schoknecht

Abstract Soils have a crucial role in addressing some of the key issues of our time, such as food security and climate change, and provide a key natural resource asset underpinning sustainable development. Until now, this role has largely been ignored in the policy debate. This situation is changing internationally, as the importance of the land in supporting our future survival and prosperity is increasingly realised. A soil management strategy providing a clear purpose and direction for policy development and a framework to coordinate activities is essential. A policy vision and a set of guiding principles to meet this challenge are proposed, and the value of good underpinning soil resource information is demonstrated. The soil survey of the Abu Dhabi Emirate, which delivers a soil and land resource dataset at a scale, accuracy and consistency required to support land use planning and policy development, is a significant achievement that should be applauded. The opportunities provided by this new comprehensive information to achieve sustainable development in the Abu Dhabi Emirate are immense.

Keywords Land use planning • Soil management • Soil policy • Soil strategy • Soil survey

N. Schoknecht (✉)
Department of Agriculture and Food, 3 Baron Hay Court, Western Australia, South Perth 6151, Australia

National (Australian) Committee on Soil and Terrain, 3 Baron-Hay Court, South Perth 6151, Australia
e-mail: noel.schoknecht@agric.wa.gov.au

S.A. Shahid et al. (eds.), *Developments in Soil Classification, Land Use Planning and Policy Implications: Innovative Thinking of Soil Inventory for Land Use Planning and Management of Land Resources*, DOI 10.1007/978-94-007-5332-7_26,

26.1 Introduction

Many countries have a long history of soil survey and soil and land classification. This information is very useful in identifying opportunities and risks of land use and can help direct government policy towards sustainable and profitable use of the soil and land resource. The scale, quality and consistency of the information play a major role in its value in planning and policy applications. This information, however, is often underutilised. A challenge for government is to utilise the wealth of useful information that a soil survey provides for developing appropriate land management and soil policy at a national level. To be successful, this process needs to engage government, academia, private sector and the community. This is an undertaking that Australia and many other countries such as United Arab Emirates are currently pursuing. This chapter investigates the current trends in the use of soil and land information, drawing on experience from within Australia and neighbouring countries and developments globally, and discusses the opportunities offered by the new comprehensive soil survey information now available for the Abu Dhabi Emirate.

26.1.1 Need for Soil Policy

In all the debates around climate change and water scarcity, one of the most important elements has been missing from the discussion – the soil. Soil management is fundamental for food security, for water security and for storing carbon and reducing global greenhouse gas emissions, yet it has been largely ignored in the policy debate to date.

This situation is changing internationally, as the importance of the land in supporting our future survival and prosperity is increasingly realised. A soil management strategy providing a clear purpose and direction for policy development and a framework to coordinate activities is essential.

The need for a coordinated policy agenda to underpin the sustainable use of soils was clearly outlined in agenda 6 (providing guidance to develop and implement national soil policies) of an International Union of Soil Sciences (IUSS) publication A World Soils Agenda. Discussing International Actions for the Sustainable Use of Soils (Hurni and Meyer 2002). International interest in policy frameworks continues. The Commission of the European Communities published a thematic strategy for soil protection in 2006 (Commission of the European Communities 2006a), with an associated impact assessment (Commission of the European Communities 2006b) and a proposal for establishing a framework for the protection of soil (Commission of the European Communities 2006c). Unfortunately, this framework has yet to be implemented due to the complexities of gaining approval from the numerous member states. The importance of soil and improved soil management in mitigating climate change is increasingly realised, such as the Review of existing information on the interrelations between soil and climate change (Commission of the European Communities 2008). In Australia, a forum on development of a soil policy and a national soil strategy was held at the 19th World Congress of Soil Sciences in Brisbane in August 2010.

The fourth UNEP Global Environment Outlook – environment for development (GEO-4) assessment report was published in October 2007. This report addresses the key environmental challenges threatening our existence on this planet, highlights emerging environmental issues that require policy attention and makes recommendations for decision makers to act.

The findings were a wake-up call, and key findings included:

- Since 1987, an average of 73,000 km^2 of forest has been lost annually (mostly to farmland, with farmland converted to urban).
- The intensity of land use for agricultural production has increased dramatically since 1987.
- Regionally, land degradation is a fundamental and persistent problem, which is driven by unsustainable land use or inappropriate land management.
- Chemical contamination and pollution pose hazards to human health and the environment.
- Depletion of soil nutrients is the most significant biophysical factor, limiting crop production over large areas in the tropics.
- Desertification – land degradation in dry lands – is most sharply expressed in poor countries, affecting the livelihoods of rural people.
- Demands on land resources and the risks to sustainability are likely to intensify.

The Soil Science Society of America, acknowledging the critical need for soils information and knowledge in the policy and political arena, in 2006 formed a Congressional Soils Caucus with the following goals:

1. Increase appreciation and support in Congress for the importance of soils and soil science through continuing education and timely communication of contemporary issues in soil science.
2. Identify and prioritise soil research and education needs critical to advancing soil science to solve problems related to world food security, natural resource and environmental protection and land use.
3. Provide timely information and input to assist in the development of farm bill programmes and other legislation involving soil research and education priorities.
4. Enhance connection and interaction among public and private organisations, federal and state agencies and constituencies to accomplish the mission.

A recent report – Food Futures: Rethinking UK Strategy (Royal Institute of International Affairs 2009) – underlines the need for good soil management and policy in the light of the real potential that growth and demand will outstrip our increases in global food production.

There are many other examples. The message is clear – the need for national and international action on soils policy is unequivocal.

26.1.2 A Soil Policy Vision

Managing the land so that we meet our current needs as well as proudly passing it on to future generations to satisfy their needs requires a significant and long-term commitment. A policy vision for Abu Dhabi, based on principles proposed for Australia by Campbell (2008), could see "landscapes in which soil is conserved for its ecological values and the ecosystem services it provides, and soil health is enhanced for sustainable production and the betterment of its population". In this vision, soil is just one component of an integrated and complementary natural resource system.

26.1.3 The Policy Challenge

Thus, for the current generation, our challenge is to develop more sustainable ways of:

- Managing our soils in the face of environmental change and increasing demands upon soil resources
- Restoring the productive capacity of degraded soils
- Putting in place robust and resilient systems of land use and management that prevent the further degradation of our soils and landscapes

26.1.4 Principles Underpinning Soil Policy

To achieve this vision and meet this challenge, and before a detailed action plan can be devised and implemented, a set of guiding principles is required. A following set of principles is suggested, but ultimately, these should be the product of government, industry and community discussion.

Guiding principles for a national soil policy:

1. Soil is a crucial natural asset, and sustainable management and protection of the soil resource is fundamental to our future prosperity.
2. Degradation of our soil resource is an ongoing issue resulting in partial or total loss of productivity and biodiversity (reducing capacity to provide ecosystem services from the land) and creating significant off-site impacts.
3. Prevention of soil degradation is nearly always substantially cheaper than the cost of restoration and in most cases is a much better investment.
4. It is the responsibility of individuals, communities, industries and governments to not knowingly degrade soil and/or water resources.
5. Soil management and policy decisions at all levels should be based on the best available knowledge, including knowledge from traditional owners, and should be evidence and science based.

6. Sustainable soil management is most likely to be achieved through integrated approaches to sustainable agriculture and natural resource management (NRM) where long-term condition of the resource is built in as a core consideration.
7. Governments have a responsibility to provide an institutional framework that encourages and supports sustainable management and discourages unsustainable management of soil resources.
8. Industries that depend on the land have a responsibility to inform themselves about their impact on soil condition and to promote and support sustainable soil management practices within their industry.
9. With the right to own, manage and use land and soil, landholders accept a duty of care to prevent soil degradation that affects others and to implement management practices that maintain or improve soil condition and productive capacity.
10. Sustainable management of soil resources across the country requires coordination, cooperation and collaboration among all levels of government in partnership with industry, land managers and the community, regardless of land tenure.

26.1.5 Land Use Planning and Soil Policy

Land use planning based on good soil information is an important process to help achieve sustainable use of soil and land resources – an ultimate aim of the policy process. The concepts of land capability and land suitability have been around for a considerable time (e.g. FAO 1976). Fundamental to these concepts is the importance of managing land within its capability to avoid degrading it. This requires an understanding of the inherent properties of the land and their spatial distribution at appropriate scales and accuracies. This is where high-quality soil and land information is a key underpinning dataset and where the importance of the Abu Dhabi soil survey (EAD 2009a, b) is most pronounced.

26.2 Role of Soil Information in Informing Policy and Planning Decisions

High-quality land is a finite resource, and our demands on this resource are increasing. The average area of agricultural land per capita across the world has halved from 1961 to 2005 (Fig. 26.1). This trend is expected to continue.

The bulk of the world's most productive lands has already been utilised, and there is likely to be an increasing move to more marginal land of lower quality. This marginal land is also usually more at risk from land degradation processes.

As the world moves towards using more land for agricultural production, it is essential that we know where the best land is, the opportunities it offers and its risks. This is where fundamental land resource information, such as that collected in the Abu Dhabi soil survey (EAD 2009a, b), is critical for planning future developments in a sustainable way.

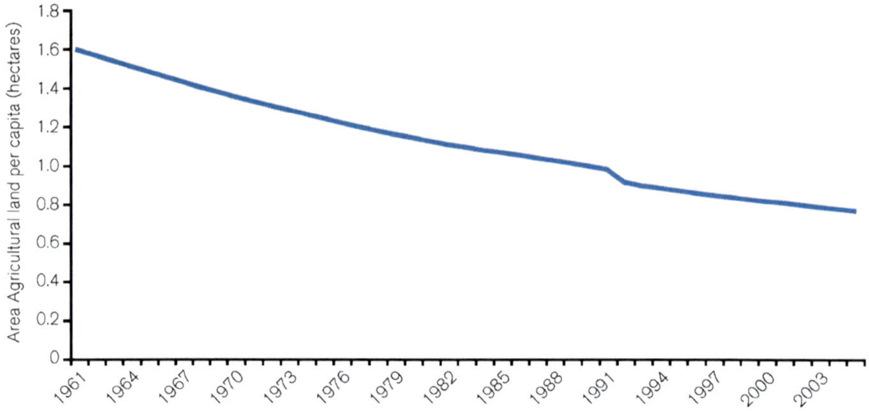

Fig. 26.1 World hectares of agricultural land per capita, 1960–2005 (From: Food Futures: Rethinking UK Strategy (Royal Institute of International Affairs 2009))

Land capability assessment, based on soil surveys and mapped spatially, has formed the basis of most approaches used to identify high-quality agricultural land and land suitability for other uses around Australia. In Australia, this information is used as layers in the planning process to determine optimum land uses. In an era of rapid growth and increasing pressure on land resources, agricultural land is increasingly threatened. Once the land with the greatest significance for agriculture is identified, it can be protected through planning controls. Land capability mapping, based on good soil information, also underpins identification of areas most suitable for urban nodes, industrial development, rural wedges and areas at degradation risk and requiring special protection.

Soil information that is consistent, standardised and captured at an appropriate scale is paramount. The Abu Dhabi soil survey ticks all these boxes, and this information will provide many benefits into the future. If the experience in Australia is any guide, it will also provide unforeseen benefits far beyond those originally planned.

As an example, a recent report on finding new opportunities for economic development in northern Australia has identified opportunities for development for a range of industries across northern Australia. Land resource information was a critical input to this study and will underpin planning to future developments. Full report is available at www.regional.gov.au/regional/ona/nalwt.aspx. An example of these findings is shown in Fig. 26.2.

An important learning in the southwest agricultural areas of Western Australia has been the need to provide planners with user friendly information interpreted from land capability mapping to suit their needs. We have learnt that planners do not have the capacity or time to do this. A good example is the identification of high-capability land for irrigated agriculture from land capability mapping, water resource information and industry knowledge. This land identified as high capability for irrigated horticulture will be identified for protection from subdivision and changed land use in state planning policies.

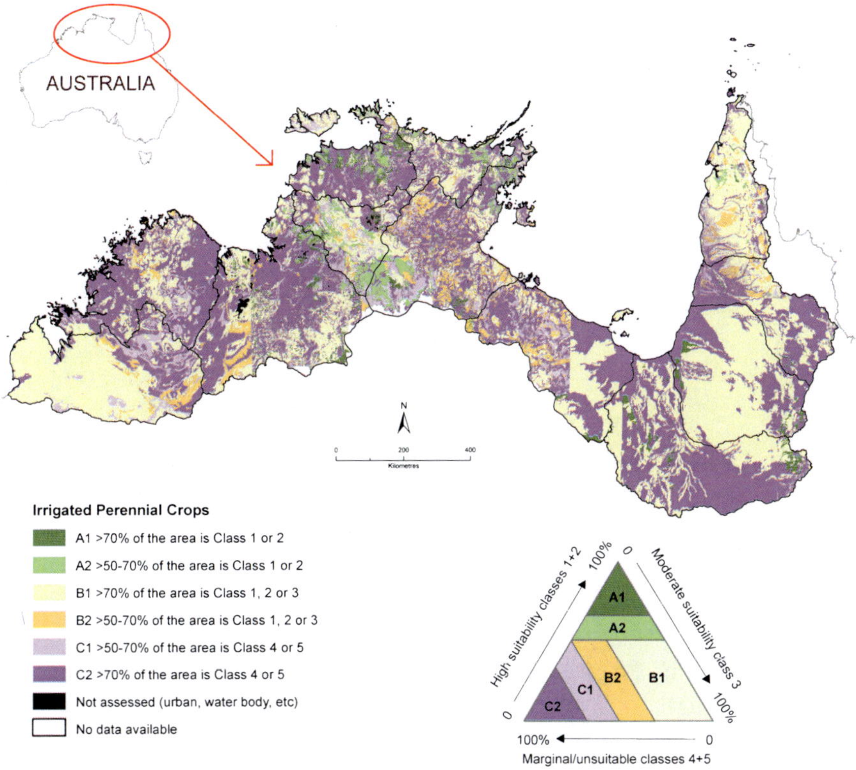

Fig. 26.2 Case study – the potential for irrigated perennial crops in northern Australia (Wilson et al. 2009). Definitions: class 1 – land with very high suitability for the purpose (in this case irrigated perennial crops), class 2 – high suitability, class 3 – moderate suitability, class 4 – low suitability and class 5 – very low or no suitability

26.3 The Need to Integrate Across Disciplines

The benefits of well-managed soils and landscapes go far beyond healthy soils and a profitable agricultural sector. Well-considered and well-integrated policy at agriculture, environmental and health levels leads to healthier foods and lifestyle and ultimately a healthier population (Fig. 26.3). Hence, high-quality soil information, by informing and influencing land management and policy, is a key to this outcome.

The logic is simple-healthy farms, healthy landscapes, healthy food, healthy people and healthy communities are all interconnected (Campbell 2009).

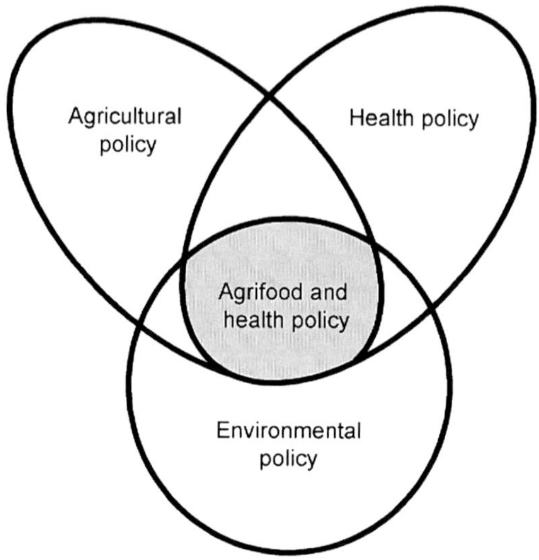

Fig. 26.3 Diagrammatic representation of the interconnection between agriculture, environment and health (Source: Tyrchniewicz and McDonald 2007)

26.4 Conclusion and Next Steps

The soil survey of Abu Dhabi is a fundamental dataset for land use planning and achieving the sustainable use of the soil and land resource in Abu Dhabi. Assembling this dataset at a scale, accuracy and consistency required to support land use planning and policy development is a significant achievement that should be applauded. Further steps are required to maximise its value. The survey tells us where the resource occurs and its properties. There is a need to interpret this information to meet the needs of land use planners and other clients in readily accessible products. A long-term programme to determine the status and trend in soil condition is also required with targets and indicators that define the resource condition we need to achieve to maintain productivity with all the flow-on ecosystem services the soil provides. We need to report on our success at achieving these targets and, if necessary, be prepared to change the way we manage the land to keep it within these targets.

This information needs to be integrated into policy and decision making in other disciplines. The next challenge for Abu Dhabi, and indeed all countries, is to ensure that we pass on our land to the next generation in as good or better condition than we received it and to build healthy communities in the process.

References

Campbell A (2008) Managing Australia's soils: a policy discussion paper. Prepared for the National Committee on Soil and Terrain (NCST) through the Natural Resource Management Ministerial Council (NRMSC), Australia. www.clw.csiro.au/aclep/publications/reports.htm

Campbell A (2009) Converging insecurities: the water, energy, carbon and food nexus. Public lecture at the Australian Academy of Science, Canberra, 4 Nov 2009

Commission of the European Communities (2006a) Thematic strategy for soil protection. http://ec.europa.eu/environment/soil/pdf/com_2006_0231_en.pdf

Commission of the European Communities (2006b) Impact assessment of the thematic strategy on soil protection. http://ec.europa.eu/environment/soil/pdf/sec_2006_620_en.pdf

Commission of the European Communities (2006c) Establishing a framework for the protection of soil and amending Directive 2004/35/EC. http://ec.europa.eu/environment/soil/pdf/com_2006_0232_en.pdf

Commission of the European Communities (2008) Review of existing information on the interrelations between soil and climate change. http://ec.europa.eu/environment/soil/pdf/climsoil_key_messages.pdf

EAD (2009a) Soil survey of Abu Dhabi Emirate-extensive survey, vol 1. Environment Agency-Abu Dhabi, Abu Dhabi, p xx+506

EAD (2009b) Soil survey of Abu Dhabi Emirate-extensive survey, vol 2. Environment Agency-Abu Dhabi, Abu Dhabi, p vi+226 map sheets

FAO (1976) A framework for land evaluation, Soils bulletin 32. Food and Agriculture Organization of the United Nations, Rome

Hurni H, Meyer K (eds) (2002) A world soils agenda. Discussing international actions for the sustainable use of soils. Prepared with the support of an international group of specialists of the IASUS working group of the International Union of Soil Sciences (IUSS). Centre for Development and Environment, Berne

Royal Institute of International Affairs (Chatham House) (2009) Food futures: rethinking UK strategy. www.chathamhouse.org.uk/files/13248_r0109foodfutures.pdf

Tyrchniewicz E, McDonald B (2007) Finding common ground: food for a healthy population and a healthy agri-food sector. CAPI Food and Health Project

United Nations Environment Programme (2007) Global Environment Outlook - environment for development (GEO-4), UNEP, Nairobi, Kenya. www.unep.org/geo/geo4.asp

Wilson P, Ringrose Voase A, Jacquier D, Gregory L, Webb M, Wong M, Powell B, Brough D, Hill J, Lynch B, Schoknecht N, Griffin T (2009) Land and soil resources in Northern Australia, Northern Australia land and water science review. CSIRO, Canberra

Chapter 27
General Framework for Land Use Planning in United Arab Emirates

Khalil A. Ammar

Abstract An integrated framework is presented for analyzing and evaluating land use planning to address growing land use pressures, taking into account the socio-economic and environmental aspects and the effect of land use change on water resources in United Arab Emirates. In land use planning, it is essential to identify the current problems and then to find proper solutions and their implementation with the aim of planning toward long-term conservation and sustainable use of land resources. This chapter identifies areas of land with high value for conserving water resources to help public sector for their planning activities. A general spatial modeling framework using geographic information system (GIS) capabilities and based on land use suitability units is used for evaluating how planning alternatives could affect water resources and best satisfies defined policies. The framework is applied to United Arab Emirates as a case study. This framework is important for making decisions about land and water resource use, managing growth, growing land use pressures and cumulative effects, reconciling competing demands for land, and integrating land use policies. It is anticipated that the policy makers, land use planners, decision-makers, and engineers can benefit from the framework.

Keywords Framework • Land use planning • Decision-makers • Resource degradation • UAE

27.1 Introduction

The United Arab Emirates (UAE) is located in the southeastern part of the Arabian Peninsula (Fig. 27.1). Due to its geographic location, the UAE has limited arable land, harsh climate, and poor renewable water resources. These natural factors

K.A. Ammar (✉)
International Center for Biosaline Agriculture, P.O. Box 14660, Dubai, UAE
e-mail: kaa@biosaline.org.ae

S.A. Shahid et al. (eds.), *Developments in Soil Classification, Land Use Planning and Policy Implications: Innovative Thinking of Soil Inventory for Land Use Planning and Management of Land Resources*, DOI 10.1007/978-94-007-5332-7_27,

Fig. 27.1 Location of United Arab Emirates

have limited the use of available land in UAE. In addition, as a consequence of population growth and economic development in the UAE, since 1970s, the pressure on land use has gradually been increased.

This has affected the land, water, and environmental resources. In this context, it is necessary to set a land use planning framework for optimum use of resources to ultimately achieve sustainable development in the country. This framework will be crucial to assist in decision-making on the location, nature, and control of investment activities to achieve sustainable development.

Land use planning is an interdisciplinary concept intersecting different themes like biodiversity and environment, health, natural resources, water resources and minerals, energy, and engineering infrastructure. It is an iterative process by learning from experience and new findings and adapted for planning according to these new outcomes. Land use planning framework can manage growth, decrease land use pressure and cumulative effects, reintegrate competing demands for land, ensure sustainability of land use for current and future generations, and integrate land use policies.

Best international management practices in land use planning recognize the importance of the setting of a framework for land use planning. For example, Canada-Alberta sets a general framework (Kennett and Schneider 2008) that helped the following: balancing social, economic, and environmental objectives in participatory, inclusive, and transparent approach; management of strategy that combines cumulative effects modeling and policy analysis; institutional capacity; good governance-effective stakeholder involvement, transparency, and accountability in decision-making and monitoring; and strong legislation implementation.

Another important and specific example is from Ireland (Department of the Environment and Local Government 1999). Ireland developed land use framework to delineate land areas in terms of groundwater vulnerability to pollution (combined aquifer map and vulnerability map). The Ireland framework assisted in decision-making; it is used for land surface zoning, created groundwater protection responses, and addressed source water protection issues and groundwater monitoring.

The objective of the present study is to develop a general land use planning framework for UAE to help different stakeholders involved in land use and planning to ultimately achieve sustainable development.

27.2 Land Use Change

The examination of the previous information in the UAE revealed major changes have taken place in land uses. Most of these changes have been in the coastal areas and are due to the main reasons discussed in the following section:

Expansion of urban areas, due to high population growth (6% per annum) in the UAE (MOEW 2010), which is the highest in the world, and due to high economic development.

Expansion in artificial residential and recreation areas, such as building new man-made islands, like the Jumeirah Nakhla Island (Fig. 27.2).

Expansion in agricultural areas due to the government policy to support farmers through subsidies to grow forages. Compared to 1991 (40,000 ha), the year 2008 data show total cultivated areas in the UAE have been increased to over 200,000 ha (Fig. 27.3). In the beginning, the expansion was close to the existing agricultural farming area in Al Ain; later, expansion occurs to the other surrounding arable areas. The trend of gradual increase is shown in Fig. 27.3.

In the UAE, forestry plantation was made following the government policy to combat desertification and moderate climate affect and to clean the environment with the slogan "greening the desert." Natural or wildlife forests do not exist in the UAE. In addition, these man-made forests increased the aesthetic and recreational value of the desert landscapes. Most of the forest areas are located in Abu Dhabi (305,000 ha) and Dubai (47,000 ha) emirates. The remaining five emirates (Ajman, Fujairah, Ras al-Khaimah, Sharjah, and Umm al-Qaiwain) have limited forest areas. Most of these areas are irrigated with brackish groundwater. The total groundwater use for irrigating forestry was about 694 million cubic meters (MCM) in the year 2008 (MOEW 2010).

The urban amenity and landscaping areas represent the parks and irrigated planting areas along the roads. Most of these areas are irrigated with 483 MCM reclaimed water (MOEW 2010). Industrial zones are the areas where factories and oil production activities are common.

Fig. 27.2 Jumierah Nakhla Island (Source: http://www.trekearth.com)

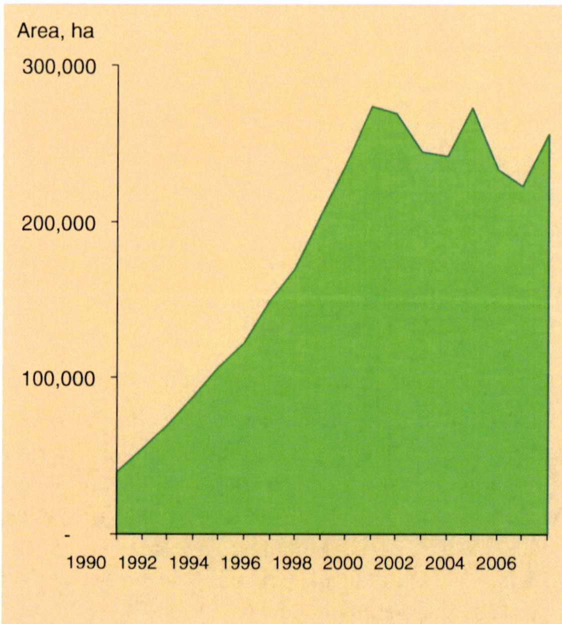

Fig. 27.3 Growth in agricultural areas in UAE (Source: Water Strategy 2010 (MOEW 2010))

27.3 Impacts of Land Use Change on Water Resources and the Environment

The land use change caused several impacts on both water resources and the environment. The main impacts included the following.

27.3.1 Water Resources Degradation

The main degradation was in groundwater. Groundwater is the major source for irrigation. The irrigation is accomplished using both the fresh and brackish groundwater sources. Overpumping, unplanned use, and poor groundwater recharge caused the decline of groundwater levels in most of the agricultural and forestry areas and subsequently depleted this valuable resource in several areas of UAE. The overpumping also caused saltwater intrusion in the coastal areas and migration of poor-quality water from adjacent aquifers in inland areas (Fig. 27.4).

In addition, the intensive agricultural activities degraded the water quality due to overuse of chemical fertilizers and pesticides; this practice has increased the groundwater contamination, for example, nitrate (NO_3) level has reached in some places to more than 650 mg L^{-1} (Robins et al. 2006). This has exceeded the World Health Organization (WHO) standards of NO_3 (50 mg L^{-1}) for drinking water.

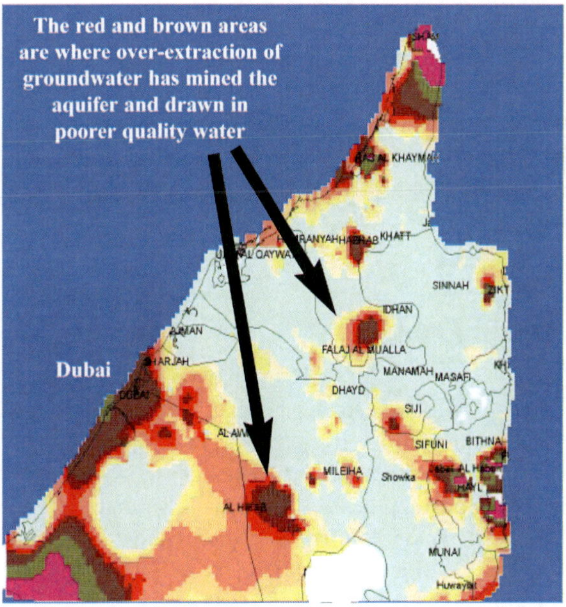

Fig. 27.4 Examples of saltwater intrusion in Northern Emirates (Source: Water Strategy 2010 (MOEW 2010))

27.3.2 Overgrazing

Due to harsh climatic conditions, the UAE rangelands are very poor in vegetation to meet the demands of grazing animals. The rangelands have been overgrazed above the carrying capacity, and these rangelands have very poor regeneration capacity. The poor management and uses of the rangelands and gradual increase of livestock caused overgrazing which ultimately resulted into degradation of several areas in the UAE.

27.3.3 Loss of Biodiversity and Marine Ecology

Many species are classified as endangered species. These endangered species need protection and conservation for the future generations. Priority species, habitats, and ecosystems include intertidal mudflats, mangroves, vegetated sandy beaches, marine (sea grass, coral), wadis, sand sheets and low dunes, interdunal plains and high dunes, alluvial plains, and jebels.

27.3.4 Air Pollution

The total CO_2 emission from power and desalination plants is about 20×10^6 t a year for the United Arab Emirates. The total emissions in Abu Dhabi Emirate from power and desalination plants is 13.5×10^6 t of gases and particulates per year, and carbon dioxide contributes to 99.65% of total emissions. In addition, nitrous oxide and nitrogen dioxide emissions are about 34×10^3 t per year. While the volume of nitrous oxide is relatively small, it is 200 times more effective as a greenhouse gas than CO_2 and is equivalent to 6.8×10^6 t of CO_2.

27.4 Abu Dhabi Estidama

Mitigating the impacts of land use changes is a big challenge. A remarkable effort was made in Abu Dhabi Emirate to overcome or mitigate the negative impacts of land use changes through developing Abu Dhabi sustainable plan. This plan aimed to achieve sustainable development in Abu Dhabi Emirate named Estidama which is the Arabic name of sustainability to address most of the major challenges and land use impacts mentioned above. The example from Abu Dhabi is discussed in this chapter to highlight the positive aspects taken into account that can guide other emirates and the federal government to achieve sustainable development in the UAE.

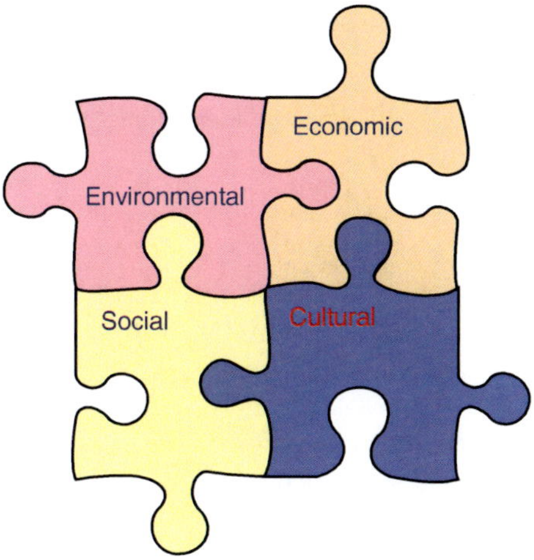

Fig. 27.5 Estidama bases and concepts (Source: Abu Dhabi Urban Planning Council, ADUPC (2010))

Estidama was based on four pillars: environmental, economic, cultural, and social (Fig. 27.5). Aspiration of Estidama was incorporated into Plan Abu Dhabi 2030 and other Urban Planning Council policies such as Development Code. Plan Abu Dhabi 2030 incorporated environmental framework, open-space framework, and land use framework in addition to transportation framework (ADUPC 2007).

The environmental framework incorporated:

1. Preserving mangroves and tidal flats and sea grass beds which are considered the most important ecological resources in UAE.
2. Establishing national park system in order to preserve vital ecologies in both terrestrial and marine environment.
3. Initiate green gradient between the natural core of the park and the urbanized core of the city.
4. Sand belt to ring the city through desert and define the outer limit of growth.
5. Desert fingers to provide undeveloped buffer between the city and coastal towns giving wildlife corridors to the protected coastal areas.

The open-space framework aimed to create:

1. Interconnected network of parks and open space
2. Active urban environments to encourage active lifestyles by providing building occupants and users with recreational public open spaces
3. Regionally responsive planning to reflect the unique climatic, social, and historical influences of the site and region in the community planning

The land use framework included:

1. Central business district (Al Sowwah Island).
2. Capital district.
3. Define the limits of growth for the city.
4. Industrial lands are strategically allocated close to the port.

27.4.1 Water Resources Categories

In addition, Estidama defined two water resource categories out of the seven categories that are fundamental to more sustainable development; these include (ADUPC 2010) the following.

27.4.1.1 Preserving the Region's Critical Natural Environments and Habitats: Natural Systems

1. Natural system assessment, protection, design, and management strategy to ensure assets are protected and the impacts are either mitigated or compensated
2. Natural system design and management strategy to minimize demand for resources and ensure the long-term survival and management of landscaped habitat areas
3. Reuse of land
4. Remediation of contaminated land
5. Ecological enhancement to improve the ecological value of the site by planting native or adaptive species
6. Habitat creation and restoration to maintain habitats that are connected to other similar habitats and to increase the ecological value of the site
7. Food systems to create a more localized approach to food with sustainable food production

27.4.1.2 Reducing Water Demand and Encouraging Efficient Distribution and Alternative Water Sources: Precious Water

1. Community water strategy to minimize the overall water consumption and establish water balance
2. Building water guidelines to promote water conservation in buildings within the community
3. Water monitoring and leakage detection to reduce loss of water
4. Community water use reduction to encourage water-efficient landscape design through plant selection, irrigation technology and management, reduced potable water use for heat rejection, minimized evaporative loss, and the use of recycled water
5. Stormwater management to minimize peak stormwater discharge and protect the stormwater drainage system from pollution
6. Water-efficient buildings to promote reductions in the water consumption of buildings

27.5 Proposed Conceptual Framework for Land Use Planning

A framework is developed to help in analyzing and evaluating land use planning and the effect of land use change on water resources (Fig. 27.6). This framework integrates the natural system vulnerability and human intervention into unified

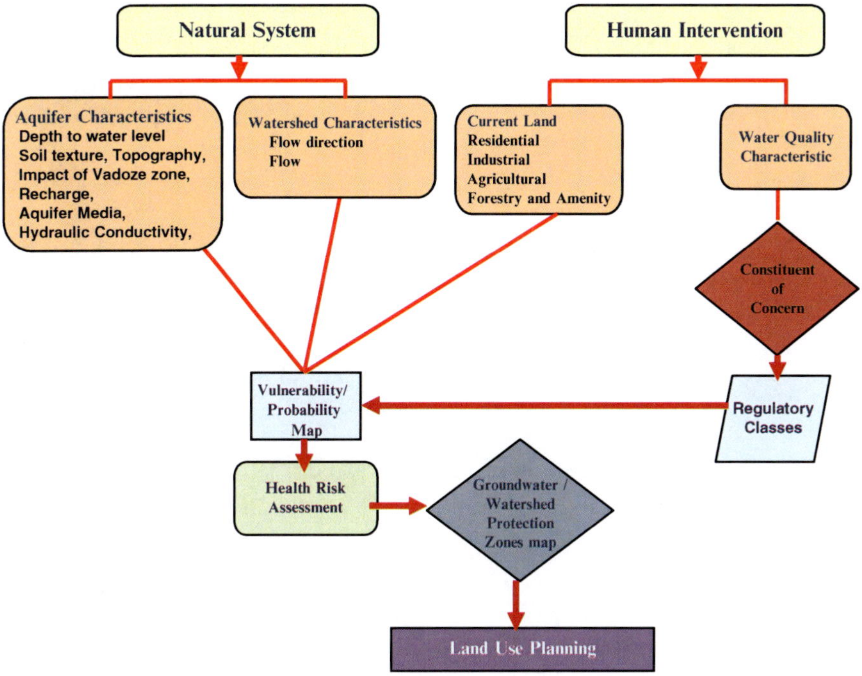

Fig. 27.6 Proposed conceptual framework for land use planning

framework. The natural system characteristics could include depth to water level, topography, soil, recharge, aquifer media, flow direction, and accumulation. Each theme can be detailed according to whatever data is available. On the other hand, human intervention includes all possible parameters that might affect the water resources and the environment. Most importantly are the water quality parameters and accordingly the constituent of concern, for example, heavy metals. Usually, there are certain regulatory limits that define or categorize different concentrations into regulatory limits either according to national standards or international standards (WHO standards). Then, probability of occurrence of certain concentration into one of these regulatory classes can be obtained using either simple probability distribution function (pdf) or state-of-the-art classification methods like artificial neural network, support vector machines, and relevance vector machines. Risk can be assessed based on the probability of occurrence and the consequences of certain pollutant. Finally, the created vulnerability/probability map and water availability and quality zoning map can be used to delineate groundwater and/or watershed protection zones. These delineated groundwater protection zones can be used thereafter as the bases for land use mapping.

Integration of the above-mentioned themes could be accomplished using GIS analysis tools. For example, groundwater quality and availability map can be integrated with current land use map, irrigated/agricultural areas, forest areas, and built-up areas as shown in Fig. 27.7. More themes or layers could be added like

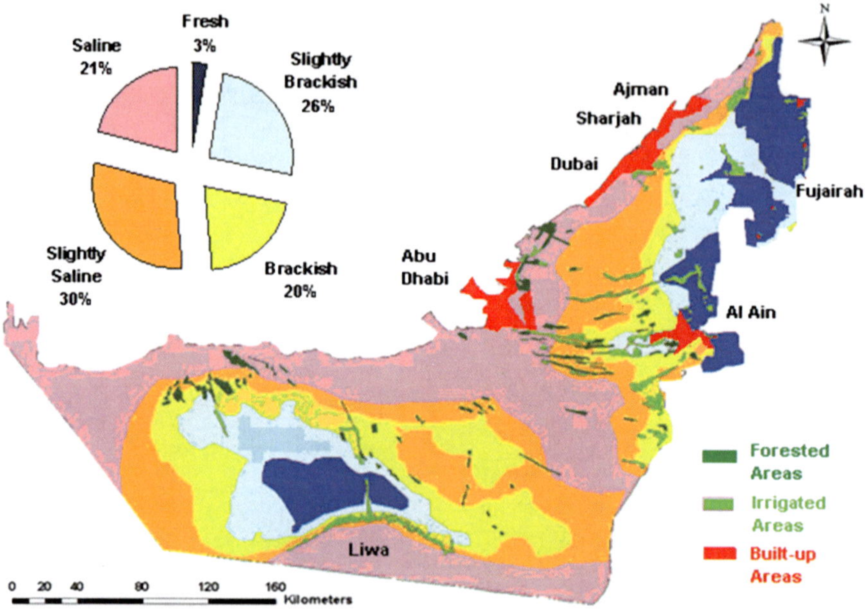

Fig. 27.7 Current land use map and groundwater quality zones (Source: Water Strategy 2010 (MOEW 2010))

land suitability for agriculture map and land salinity. All these detailed maps can help in delineating the land use map.

27.6 Conclusions and Recommendations

The following conclusions and recommendations are drawn from the present study.

It is recognized that the land use planning framework is essential for conservation and stewardship on private and public lands. It is essential to promote efficient use of land to reduce negative impacts of human activities. In order to achieve the goal successfully, the involvement of relevant stakeholders in land use planning is necessary. The impact of land development on water resources and the environment should be managed properly.

In order to integrate all land use planning efforts on national scale, it is recommended to establish federal planning council; establish information system and implement regular monitoring program to support land use decision-making; identify appropriate limits at regional and local levels; define pollution limits for groundwater, wadis, and the environment; define certain types of development in ecologically sensitive areas; and adopt best practices to mitigate development impacts. Land use planning should be linked to other types of planning and integrated into development planning at the emirate and federal level; it should be linked to national resources planning programs, to strategies in the field of land resources planning, and to land law.

References

Abu Dhabi Urban Planning Council (ADUPC) (2007) Plan Abu Dhabi 2030, urban structure framework plan, Abu Dhabi

Abu Dhabi Urban Planning Council (ADUPC) (2010) The pearl rating system for Estidama, community rating system design and construction-version 1, Abu Dhabi

Department of the Environment and Local Government, Environmental Protection Agency, and Geological Survey of Ireland (1999) Groundwater protection scheme, Dublin

Kennett S, Schneider R (2008) Making it real, implementing Alberta's land-use framework. Canadian Parks and Wilderness Society and PEMPINA Institute, Drayton Valley

Ministry of Environment and Water (MOEW) (2010) UAE strategic plan for water conservation, Dubai

Robins N, Benham A, Mitchell C (2006) The geology and geophysics of the United Arab Emirates, vol 3: Economic geology and hydrogeology. Ministry of Energy, Abu Dhabi

Chapter 28
Spatial Mapping and Analysis of Integrated Agricultural Land Use and Infrastructure in Mhlontlo Local Municipality, Eastern Cape, South Africa

Charles Nhemachena and James Chakwizira

Abstract The study spatially mapped and analysed agriculture land use and infrastructure requirements in Mhlontlo Local Municipality linked to ASGISA Eastern Cape's agrarian transformation and rural development initiatives. Emphasis was placed on informing ASGISA Eastern Cape's broader rural infrastructure programme in support of agrarian transformation and rural development for the Eastern Cape province. The study approach was participatory, extensively involving discussions with stakeholders, visits to project areas, internal and external workshops and document analysis. The results show that the municipality has a huge potential for agriculture and improved utilisation of available arable agricultural land that still needs to be realised. The poor state of basic infrastructure for economic and social service delivery remains a key constraint to sustainable and productive agricultural land use and rural development in Mhlontlo Local Municipality. This finding tends to hamper the contributions of the rural labour force to productive agricultural enterprises as well as limiting the knowledge base of rural people. This study recommends an integrated approach to rural agricultural transformation in Mhlontlo which requires infrastructure investments with a broader scope that transcends agricultural land use developments. For example, direct agricultural infrastructural investments and activities need to be complemented

C. Nhemachena (✉)
Human Sciences Research Council,
134 Pretorius Street, Pretoria 0002, South Africa
e-mail: CNhemachena@hsrc.ac.za; cnhemachena@gmail.com

J. Chakwizira
School of Environmental Sciences, University of Venda,
P/Bag X5050, Thohoyandou 0950, South Africa
e-mail: james.chakwizira@univen.ac.za; jameschakwizira@yahoo.com

S.A. Shahid et al. (eds.), *Developments in Soil Classification, Land Use Planning and Policy Implications: Innovative Thinking of Soil Inventory for Land Use Planning and Management of Land Resources*, DOI 10.1007/978-94-007-5332-7_28,

by investments in social services aimed at reducing poverty and stimulating socio-economic growth and development of the local municipality.

Keywords Agriculture land use • Infrastructure • Mhlontlo Local Municipality • Rural development • South Africa

28.1 Introduction

Few countries and regions have experienced sustained growth at a high level without growth in agricultural productivity (the search for biofuels will even make it significantly profitable to invest in agriculture) (World Bank 2008, 2009; FAO 2004). Although agriculture alone does not constitute the only ingredient in Mhlontlo's rural development, it certainly is a significant contributor to the municipality's cash flows. The poor state of basic infrastructure for economic and social service delivery remains a key constraint to sustainable agricultural and rural development in the Eastern Cape province, locking out the contributions of the rural labour force to productive enterprises and limiting the knowledge base of the rural people (Nhemachena et al. 2009; Eastern Cape PGDP 2004).

To realise the agricultural potential of Mhlontlo Local Municipality, it is important to seek to put in place an integrated approach along the production, processing and marketing continuum. In this regard, integrated agricultural infrastructure and services can be crucial for developing links with agricultural markets and services and in the process enhancing rural growth and development. For example, to strengthen access of small farmers to markets, there is a need to improve and maintain as well as expand road networks to open up these areas of high agricultural production potential and facilitate the development of village market centres linked to the mainstream district, provincial and national markets (Chakwizira et al. 2010; Nhemachena et al. 2009). The need to improve linkages between collection points, terminal markets and agro-industry cannot be overemphasised.

Rural infrastructure is not only a key component of rural development but it is also an important ingredient in ensuring sustainable poverty reduction. Some of the advantages of investing in infrastructure as the foundation for unlocking the potential for local economies include the following: infrastructure reduces transaction costs and lower input costs used in production; it enables economic actors (individual, government, business, etc.) to come and invest in the economy due to the wide economic base provided by infrastructure; it facilitates trade and entrepreneurship (e.g. investments in infrastructure open opportunities for wider thinking, entrepreneurship and opportunities for trading with other surrounding economies and integrating in the larger national economy); it creates employment, e.g. public works models in providing infrastructure, and it enhances human capital and infrastructure could attract better service providers (Chakwizira et al. 2010; Nhemachena et al. 2009; Mashiri et al. 2008).

The main objective of the study was to assist ASGISA Eastern Cape with the analysis and spatial mapping of integrated infrastructure requirements linked to its rural development initiatives in Mhlontlo Local Municipality. The assignment was to inform a broader rural infrastructure programme in support of agrarian transformation for the Eastern Cape province. The specific objectives of the study were to (a) identify infrastructure requirements that would enhance economic growth in the agricultural sector and the rural economy of Mhlontlo Local Municipality; (b) provide, based upon analysis of existing infrastructure, a spatial mapping of infrastructure requirements that ASGISA Eastern Cape can systematically invest in to address the major infrastructure requirements and constraints to agricultural and rural development in Mhlontlo Local Municipality; and (c) contribute to an improved operational framework to support more effective investments for agricultural and rural development infrastructure in Mhlontlo Local Municipality.

28.2 Study Area and Methods

28.2.1 Study Area

Mhlontlo Local Municipality is located on the north-east side of the Eastern Cape provincial border alongside the N2 national road between Mthatha and Mt Frere. The municipality falls within the jurisdiction of the OR Tambo District Municipality (Fig. 28.1).

Mhlontlo Local Municipality is predominantly rural incorporating two main urban centres, Tsolo and Qumbu. The rural settlements of the local municipality are characterised by large uneven and low levels of services. However, some settlements, especially the peri-urban, near major intersections and on major routes, have developed into rural service nodes accommodating various community facilities (Mhlontlo Local Municipality IDP 2009; Mhlontlo Local Municipality LED Strategy 2007).

28.2.2 Study Methods

The study methods included extensive discussions with stakeholders, field visits, internal and external workshops as well as literature review. The approach involved two phases; the initial was a comprehensive analysis of the existing agricultural and other rural infrastructure. This provided baseline information on the current infrastructure situation and needs for the mapping of integrated agricultural infrastructure. The second phase was spatial mapping of integrated agricultural infrastructure requirements for strengthening agricultural transformation, food security and integrated rural development alongside the desired spatial form and land uses of Mhlontlo Local Municipality.

Fig. 28.1 Location of Mhlontlo Local Municipality (Source: Adapted from Mhlontlo Local Municipality Local Economic Development Strategy 2007)

28.3 Results and Discussion

28.3.1 Analysis of the Agricultural Sector and Existing Agricultural Infrastructure

The good climatic conditions, soils and land abundance point to the richness of the municipality and potential in high value crops and fruit production (Mhlontlo Local Municipality LED Strategy 2007). The municipality has a huge potential for agriculture and improved utilisation of available arable agricultural land that still needs to be realised. Lack of existing infrastructure and other supporting structures has constrained productive utilisation of a number of areas and implementation of agricultural and rural development projects (Mhlontlo Local Municipality IDP 2009; Mhlontlo Local Municipality LED Strategy 2007). Clearly, availability of infrastructure provides a foundation and basis for productively utilising agricultural land and implementing developmental projects and programmes. Figure 28.2 presents the key agricultural zones in the municipal area. The key agricultural zones include cultivated temporary semicommercial/subsistence dry land, cultivated commercial sugarcane and forest and woodland.

Table 28.1 summarises the key agricultural opportunities and infrastructure requirements in Mhlontlo Local Municipality. Although we focus on agriculture, other sectors such as tourism, forestry and community services are important in the municipality.

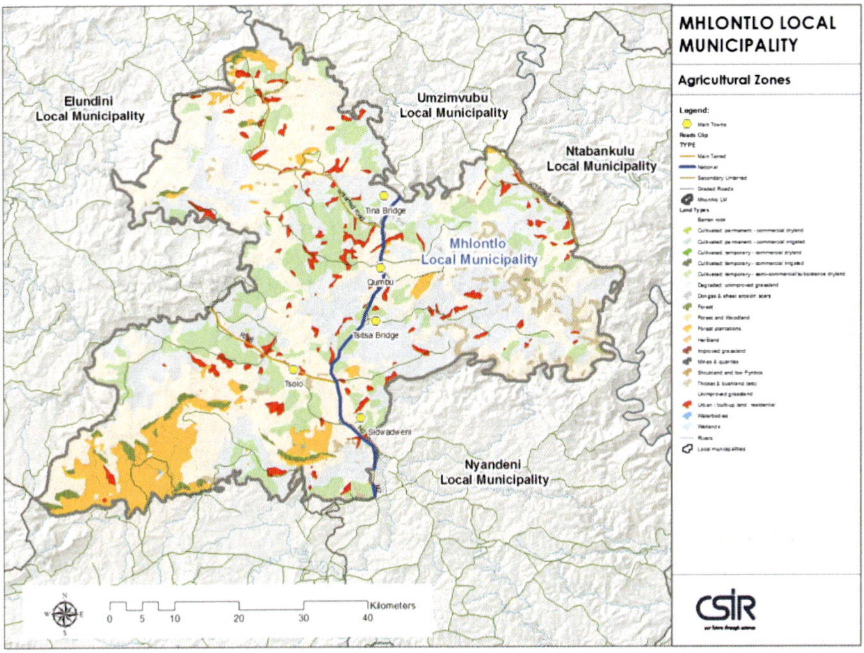

Fig. 28.2 Key agricultural zones in Mhlontlo Local Municipality

28.3.2 Spatial Assessment and Mapping of Integrated Agricultural Infrastructure

Integrated rural and agricultural infrastructure includes all of the basic services, facilities, equipment and institutions needed for the economic growth and efficient functioning of the rural, food and fibre economies. In many rural areas of South Africa, the availability of rural and agricultural infrastructure is not uniform across the country. Whereas the agricultural infrastructure is readily available in rural areas that are inhabited by the so-called organised agriculture/commercial farmers, the rural areas occupied by emerging and developing farmers as well as communal lands, like Mhlontlo Local Municipality, lack such infrastructure (Nhemachena et al. 2009). This has hindered all facets of development in such rural areas, where the most likely key development driver is agriculture.

Having realised the importance of agricultural infrastructure to achieve faster rate of economic growth and poverty reduction, ASGISA Eastern Cape intends to close the gap through intense investment in rural and agricultural infrastructure in its endeavours to reduce poverty, improve food security and stimulate sustainable development. Infrastructure for irrigation, crops handling and storage, value-adding processing plants for agricultural produce, minor roads and bridges, farm access road network, information communication technology, fencing, animal housing, among others, are required in the Eastern Cape to offer the basic services

Table 28.1 Key agricultural opportunities and infrastructure requirements

Items	Opportunities/infrastructure requirements
Land and water availability	Install irrigation schemes, erect fencing for both arable and grazing land, build dams and water reservoirs to enable all-year-round crop production, maximise crop and animal production
Good climatic conditions	Maximise land utilisation, implement production of high value crops
Agricultural institute in place	Upgrade TARD and link it with other agricultural institutions nationally and internationally, maximise technology transfer to local farmers, ensure that TARD develops tailor-made courses to suit local and rural "clientele" within the area and also to suit development needs of MLM
Lack of basic infrastructure	Improve infrastructure such as roads, markets, agro-processing, telecommunications, sanitation and electricity
Unfenced grazing and arable land	Fence all arable and grazing land
Poor access roads to arable land	Upgrade feeder roads to enable transportation of production inputs and produce to markets
Underutilisation of water	Install irrigation schemes and build reservoirs to enable all-year-round crop production
Agro-processing infrastructure	Invest in agro-processing infrastructure
Establishment of agricultural information centres	Invest in agricultural information centres
Availability of marketing infrastructure	Invest in agro-based marketing infrastructure

necessary for sustainable rural and agricultural development and growth (Eastern Cape PGDP 2004).

Figure 28.3 summarises key integrated agricultural and rural infrastructure mapping for Mhlontlo Local Municipality. The following sections discusses infrastructure mapping for crop and livestock production as well as irrigation development. The last section discusses proposed bankable agricultural infrastructure projects for the municipality.

28.3.2.1 Infrastructure for Crop Production

Crop production infrastructure requirement depends on the type of crops produced in an area. In Mhlontlo Local Municipality, crop production plans under the proposed projects primarily focus on the production of maize, soybean, etc., both under dry land and irrigation farming. This assessment of infrastructural requirement is therefore inclined towards the production of these crops. Table 28.2 presents visited crop production projects and infrastructure requirements.

The field visit revealed that apart from a few damaged storage structures, there are no other structures in the municipality for crop production. The area does not

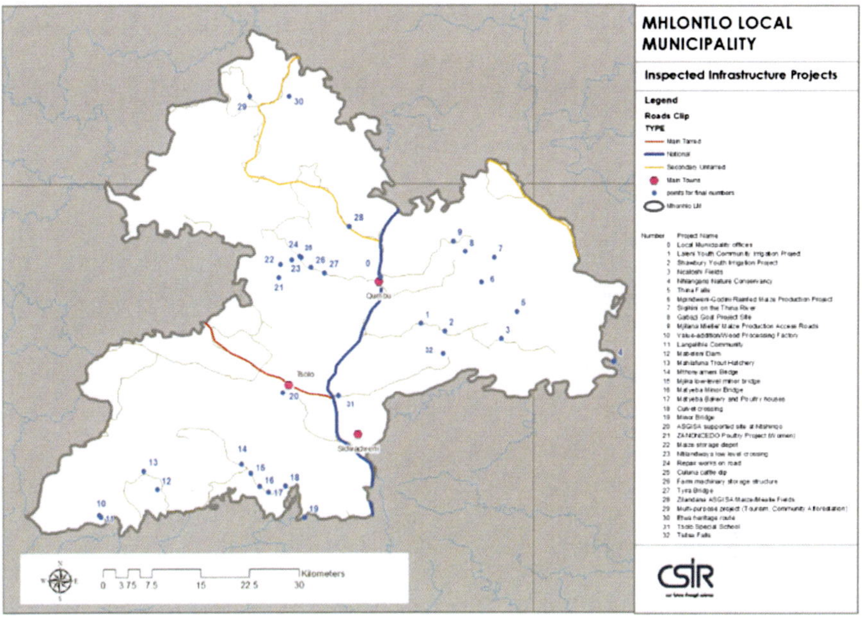

Fig. 28.3 Spatial mapping of integrated agricultural and rural infrastructure for Mhlontlo Local Municipality

have any supplier of farm inputs such as fertilisers and seeds. In addition, agricultural mechanisation services, which are crucial for successful crop production, and value-addition processes, such as animal feed preparation, are not yet developed within the municipality. Rural electrification is another important infrastructural requirement for the value-addition aspect of crop production that is not well developed in the municipality. The telephone network, especially cell phone coverage, was noted to be particularly well with strong signal strength.

Infrastructure that needs immediate attention for sustained crop production in the municipality includes development of mechanisation services, construction and maintenance of an access road network and depots for the storage of farm inputs and crop produce and agro-processing facilities at suitable locations within the municipality.

28.3.2.2 Infrastructure for Animal Production

The infrastructure required for animal production is many and varied. The variety depends on the diversity of animals and the type of animal farming system being practised in an area. In Mhlontlo Local Municipality, animal infrastructure is required for cattle, goats, sheep and poultry production. The infrastructure required for profitable animal farming includes fencing, cattle dips and spraying races, appropriate housing and access roads for delivery of inputs and accessing the markets. Table 28.2 presents inspected animal production projects and infrastructure.

Table 28.2 Summary of the inspected crop production infrastructure

Point no	Project location	Description of the project and infrastructure status	Proposed requirements
28	*Ward*:	*Physical infrastructure*	
	GPS location: S31 05.332 E28 49.422	Access road: the fields are adjacent to a tarred road providing easy access both for transporting inputs and outputs to and from the fields.	The key notable infrastructures on site were the fence and the access tarred road. They were both in excellent condition and exemplify what needs to be done in terms of fencing and developing access roads (all-weather access roads) for the other areas that have been earmarked for agricultural production within the municipality
	Type: maize production	Fencing: the fence around the farm is not only new but in good condition and adequate.	It is important to provide support for farmers along the whole agricultural value chain from inputs, production, output storage, processing and marketing. For example, assessments showed that the only nearest input source is Mthatha which is about 50 km away and there are no ready output marketing structures within the local municipality
	Name of project: Zilandana ASGISA Maize/mealie fields	*Social infrastructure*	
		Community involvement structures: it is key to identify the community involvement structures in the project as this is crucial for sustainability after funding is terminated	It was striking to find that a bag of maize costs a mere R70 and a bag of 12.5-kg maize meal costs R60, which is expensive, compared to R45 for the same bag in Pretoria. Clearly, marketing and agro-processing infrastructure will help increase earnings for local farmers and are crucial for successful agricultural production in the municipality

9	*GPS location*: various points along the route (see map on Fig. 28.3) *Type*: Mjilana Valley Access Road and Maize Production Project *Name of project:* rural access road leading to an area earmarked for ASGISA Maize Production Project	The total area that can be brought under agricultural crop production is 400 ha but can be expanded as there is suitable land adjacent to the earmarked site *Physical infrastructure* Access road: the road is in a very poor state, and this has hindered development of the maize and soybean production project Bridges, culverts, low-level crossing, etc., on this road are in an equally poor condition. For example, Flags no 166 and no 176 are low-level bridges that had been washed away by rain Fencing: the area is not at all fenced. Other Infrastructure: the markets are very far away and there are no storage facilities	The access road needs urgent upgrading including the crossing levels and bridges along the road Other infrastructural requirements: Fencing Handling and storage structures for the produce and farm inputs Mechanisation services Watering points for animals along the river that pass through this area
6	*GPS location*: S31 10.057 E29 00.442 *Type*: N/A *Name of project*: Mpindweni-Godini Rainfed Maize Production Project	*Physical infrastructure* Access: the access road leading to these fields is very bad There is no low-level crossing to connect wards 11 and 12 and Godini village. There are many maize fields on the other sides of the Thina river, and people from Mpindweni and Mzoboshe cannot cross to their fields and have to go via Mt Frere which is a very long road Fencing: there is no fence around the fields Other infrastructure There are no: storages facilities, mechanisation services and agricultural input supplies	All the bridges, low-level crossing and culverts on this road were in a poor condition and need upgrading, replacements and reconstruction A low-level crossing to connect wards 11 and 12 and Godini village is needed Other required infrastructure: Fencing Storages facilities Mechanisation services Agricultural input supplies

(continued)

Table 28.2 (contiuned)

Point no	Project location	Description of the project and infrastructure status	Proposed requirements
3	*GPS location*: S31 14.625 E29 01.983	*Physical infrastructure*	Further improvement of the road is necessary to open up this area and to improve accessibility
	Type: maize fields *Name of project*: Ncaloshi Fields (ASGISA maize field project)	Access road: while the road up to this point was generally accessible, it could not be classified as good	
		Fencing: the fields are already fenced, and the condition of the fence is good	
1	*GPS location*: S31 13.385 E28 55.479	*Physical infrastructure*	An access road needs to be constructed
		Access road: there is no access road to the project site	Irrigation infrastructure needs to be reviewed
	Type: irrigation scheme	Fencing: the fields are already fenced, and the condition of the fence is good	The revival of this project would be like starting anew, and it would therefore have to undergo all the irrigation design stages
	Name of project: Laleni Youth Project	Irrigation: while there is irrigation infrastructure (water pump), its condition needs to be reviewed	The successful implementation of this project would greatly depend on addressing the problems that led to its abandonment by the community
		Social infrastructure	There is a need to train the beneficiaries, in advance, on the management of the irrigation systems to be installed, if they are to manage it
		The irrigation project was initiated for the community, but due to lack of accessibility, it was abandoned at very early stages	Engage the services of an experienced irrigation systems manager (e.g. Tsolo Agricultural and Rural Development Institute) at the beginning to manage the scheme and train the beneficiaries over an agreed period of time. Thereafter, the beneficiaries can take charge of the scheme
		The project was to get water from the nearby Tsitsa river	
		There was no documentation about the project, and design details are therefore unknown	

26	*Ward:* 16 *GPS location*: S31 08.853 E28 46.450 *Type*: farm machinery storage structure *Name of project*: N/A	*Physical infrastructure* Buildings: the structure is in a dilapidated condition It is also not being used for its intended purpose that has led to the accelerated rate of dilapidation Though meant for machinery storage, particularly storage of tractors, it is currently used for shearing of sheep The site where it is located is however suitable, as it is centrally located, relative to the land earmarked for crop production projects by ASGISA	Upgrade the building Social infrastructure: need to identify who would be responsible for looking after the building
22	*Ward*: 16 *GPS location:* S32 08.626 E28 43.993 [Note: Flag 203 coincided with Flag 204] *Type*: storage depot for maize *Name of project*: N/A	*Physical infrastructure* Buildings: this storage structure measuring about 40-m length by 12-m width is in a fairly good condition However, it was not being used because of a lack of crop produce to store in it From the discussion with the councillor, it was meant to be a depot for maize, but maize production has since dwindled to very low levels in the area	The structure requires minor upgrading Social infrastructure: need to identify who would be responsible for looking after the building Who looks after the building (ownership and usage) needs to be cleared through tripartite agreement between the municipality, community and funders of the project Build capacity of local farmers to produce more agricultural crops so that they can employ the structures
		An adjacent structure used for keeping horses. This second structure did not appear to have been designed and constructed for animals, and its use for horses has accelerated its rate of deterioration	Consider using the structure as a satellite service node (see details on bankable projects)

(contiuned)

Table 28.2 (continued)

Point no	Project location	Description of the project and infrastructure status	Proposed requirements
30	*Ward*: 19 *GPS location*: S30 55.112 E28 41.456 *Type*: low-level bridge *Name of project*: Multipurpose project (tourism, community afforestation, crop production and value-addition processing plants) – Etwa Heritage Route	The area is suitable for maize production, and 2,500 ha of land has already been earmarked for production in order to reduce food insecurity *Physical infrastructure* Access: the access road is not maintained, and the bridge on the Etwa heritage route is in bad condition Fencing: there is no fence around the fields *Other infrastructure* There are no storages facilities, mechanisation services, agricultural input supplies, value-adding processing plants for forestry, electricity, water, health facilities, school	Upgrading access road and low-level crossing and bridges Other required infrastructure: Fencing Storages facilities Mechanisation services- School Agricultural input supplies Value-adding processing plants for forestry Electricity Water Health facilities School

Table 28.3 Summary of the inspected animal production infrastructure

Point no	Project location	Description of the project and infrastructure status	Proposed requirements
8	*Ward*: 13 *GPS location*: S31 07.634 E28 59.098 *Type*: community goat project	This GPS location was taken close to the area that has been earmarked for a commercial goat project The project, if implemented, would benefit more than 400 households from Godini, Ngxakolo, Qunu and Newstead. Goat production in the areas is not new The goats are used for rituals among the Xhosa community. The demand for goats from this area was reported to be good because this is the only area in the entire municipality and province that produces big goats	Provide access road to the goat site Provide fence around the proposed area for the goat project Other required infrastructure:
	Name of project: ASGIAGabazi Goat Project	*Physical infrastructure* Access: there is no access road to the goat site Fencing: there is no fence around the proposed area for the goat project *Other infrastructure* There are no watering points along the river	Watering points along the river
25	*Ward*: 16 *GPS location*: S31 08.099 E28 45.682 *Type*: cattle dip *Name of project*: Community Cattle Dip	*Physical Infrastructure* Cattle dip: the structure was dilapidated, and the floor needs to be reworked Drainage rage: the rage was short, meaning that fewer cattle can be dipped per hour Footbath: there was no footbath Water trough: there was no water trough Fencing: it was unmanned and not ring fenced *Other issues:* Chemical: the dipping chemical in the dipping tank section appeared old, and its efficacy in tick eradication was therefore doubtful	Upgrade the dip focusing on the identified infrastructure issues: Rework the floor Drainage rage Footbath Water trough Fencing

(continued)

Table 28.3 (contiuned)

Point no	Project location	Description of the project and infrastructure status	Proposed requirements
21	*Ward:* 16	*Physical infrastructure*	The physical infrastructure is in good condition
	GPS location: S31 09.701 E28 43.873	Building: a poultry house with capacity for 1,000 birds for the production of broiler and layers	There is need to strengthen social infrastructure
	Type: poultry project	The structure is in very good condition and is appropriately designed	Project can be revived if the issues that lead to its failure can be addressed, such as:
	Name of project: ZANONCEDO Poultry Project (Women)	Access: the road is generally well maintained, and the project is easily accessible	Skills training (production and management) of the beneficiaries on poultry production
		Fencing: it is fenced, and the condition of the fence is good	Community involvement structures from early planning stages
		Social infrastructure	
		The project faced operational problems and currently is not in use	
		The local chief reported that it was not well managed owing to lack of management skills	
		The collapse of the project is blamed on training prior to handing over of the project by the funders to the earmarked beneficiaries	
		Fencing: there is no fence around the fields.	
		Other infrastructure	
		There are no storages facilities, mechanisation services and agricultural input supplies	

There are a number of cattle dips and poultry production structures in the municipality. The cattle dips, though operational, are poorly managed. This lack of proper management has accelerated the deterioration of some of these structures, and all three cattle dips visited and inspected can be categorised as being in a poor condition. For these dips to function properly, a proper management system needs to be set up.

Although a number of poultry production structures were found in different wards in the municipality, none was operational. All the units visited and inspected were however found to have been properly designed and well constructed. These projects are still in a good state and can easily be revived. The locations of some of these units are however difficult to access, and therefore, any revival plans must include the improvement of access roads. Improved access roads will improve market accessibility and delivery of feed and other services required in poultry production.

During the field visit, it was found that the land in the municipality does not have adequate fencing. The impact of lack of adequate fencing in agricultural areas has many drawbacks particularly in those areas where mixed farming is practised. The lack of fencing implies additional labour requirements, in terms of herders, to keep the animals away from crops and cropping lands. Fences are also used to control diseases; manage livestock; control overgrazing, which can lead to soil erosion and environmental degradation; afford degraded lands adequate time to recover by keeping the animals away from such land parcels; etc.

Near the roads, fencing is a cheap and effective means of making the roads safe by keeping the animals away from the roads. Fencing also controls the movement of people in and around farms, thereby controlling theft and pilferage. Since there is inadequate fencing in the municipality, the proposed projects must ensure the construction of appropriate and adequate fencing for the different areas earmarked for development. Fencing of the roads in the municipality will have a double benefit – safer roads for drivers and the elimination of animal deaths from road accidents.

28.3.2.3 Infrastructure for Irrigation Development

Smallholder irrigation for communities is a promising vehicle for rural development. It offers farmers in the community increased crop production thereby helping reduce poverty and improving food security and nutrition of households in rural setups. Irrigation development can make a meaningful contribution towards productive utilisation of land. Thus, the problem of underutilisation of land by the communities in the municipality (and the Eastern Cape at large) can partly be addressed by investing in irrigation infrastructure. This is particularly true given the fact that two permanent rivers, viz., Thina and Tsitsa rivers transverse the municipality, and there are large parcels of community land that are suitable for irrigation (the municipality has a business plan for the development of irrigation in the Tsitsa river basin).

Two youth irrigation schemes (Laleni and Shawberry) were visited. The Laleni youth irrigation scheme had been abandoned while the Shawberry youth irrigation project, though operational, had difficulties with maintaining a steady number of members. From the discussions with the municipal officials, it was apparent that irrigation technology is not well understood by the farmers in this municipality. Traditionally, the furrow and flood irrigation system has been used to introduce irrigation technology to farmers who are unfamiliar with irrigation. However, the land terrain, apart from the river valleys, is undulating thus making the use of these two types of irrigation technically unfeasible. The sprinkler system is the only viable option that can be used to introduce irrigation technology to the farmers in the municipality.

A sprinkler irrigation system, particularly the hand-move type, is not new in the area. Already, the Shawberry youth group with the assistance of the church and other development partners in the municipality is using this type of irrigation system. However, discussions with members of the youth group who met during the field visit revealed that the youth lacked the technical know-how for managing this type of irrigation. This was not totally unexpected, given the low levels of formal education and farmers found in such rural areas.

Successful introduction of irrigation technology in this municipality has to be accompanied or preceded by a well-thought-through training programme. This is because the available training materials in the market may be unsuitable for use in training the farmers in the municipality and the Eastern Cape as a whole. Some of the reasons that make the available training material on irrigation management for farmers in Mhlontlo Local Municipality unsuitable may include language, target group, levels of education and applicability. Given the low level of education of farmers in this municipality area, there is a need to make simpler training material, preferably in the farmers' own language (mother tongue). Such training material must be made available in the local language and should be in text, pictorial and audio visual (video or DVD) formats.

28.4 Conclusions and Recommendations

The study spatially mapped and analysed agriculture land use and infrastructure requirements in Mhlontlo Local Municipality linked to ASGISA Eastern Cape's agrarian transformation and rural development initiatives. The study approach was participatory, extensively involving discussions with stakeholders, visits to project areas, internal and external workshops and document analysis. The results show that the municipality has a huge potential for agriculture and improved utilisation of available arable agricultural land that still needs to be realised. The poor state of basic infrastructure for economic and social service delivery remains a key constraint to sustainable and productive agricultural land use and rural development in Mhlontlo Local Municipality. This finding tends to hamper the contributions of the rural labour force to productive agricultural enterprises as well as limiting the knowledge base of rural people. Because local municipalities are largely unable to fund their infrastructure requirements, external intervention is often required. The study recommends an integrated approach to rural development in Mhlontlo Local Municipality.

This approach can be adopted in other rural areas of South Africa and elsewhere in developing countries. This recommendation requires infrastructure investments with a more broader scope that transcend agricultural developments, where direct agricultural infrastructural investments and activities are complemented by investments in social services aimed at reducing poverty and stimulating economic, social growth and development of the local municipality. Related infrastructure that should also be considered includes rural housing, water supply for drinking and commercial purposes, soil conservation and watershed development, forestry development, educational infrastructure including village knowledge centres, public health institutions such as mobile clinics, mini hydropower generation, harvesting indigenous knowledge systems and infrastructure for information technology.

References

Chakwizira J, Nhemachena C, Mashiri M (2010) Connecting transport, agriculture and rural development: experiences from Mhlontlo Local municipality integrated infrastructure atlas. In: 29th Southern African transport conference, walking together. CSIR International Convention Centre, Pretoria, South Africa

Eastern Cape PGDP (2004) Eastern Cape provincial growth and development plan (2004–2014). Office of the Premier, Eastern Cape, East London

FAO (2004) Socio-economic analysis and policy implications of the roles of agriculture in developing countries. Summary report, Roles of agriculture project. FAO, Rome, Italy

Mashiri M, Chakwizira J, Nhemachena C (2008) "Rejecting the inevitability of poverty": empowering women for sustainable rural livelihoods through community-based employment intensive rural infrastructure maintenance projects. In: Science real and relevant: 2nd CSIR Biennial conference. CSIR International Convention Centre, Pretoria, South Africa

Mhlontlo Local Municipality IDP (2009) Integrated development plan 2009/10. Mhlontlo Local Municipality, Qumbu

Mhlontlo Local Municipality LED Strategy (2007) Local economic development strategy. Mhlontlo Local Municipality, Qumbu

Nhemachena C, Mashiri M, Chakwizira J, Maritz J, Norswrothy J, Marenya M (2009) ASGISA-Eastern Cape: Mhlontlo Local Municipality integrated infrastructure atlas. CSIR Built Environment, Pretoria, South Africa. CSIR/BE/RIS/ER/2009/0035/C

World Bank (2008) Agriculture for development. World development report 2008. The World Bank, Washington, DC

World Bank (2009) Implementing agriculture for development. World Bank group agriculture action plan: FY2010-2012. The World Bank, Washington, DC

Chapter 29
Agricultural Land Conversion: Application of Land Capability Classification in Land-Use Planning of Embaderho Village in Eritrea

Tewoldemedhin D. Rustu

Abstract Land capability classification has been used for sensible land reallocation and distribution for settlement purposes. A case study has been conducted to assess how much fertile agricultural land is converted to urban use between 1995 and 2003. This conversion has been assessed in terms of total as well as under each land capability class to examine whether competition between land use for agriculture and housing is leading to landlessness. Assessment is made by superimposing the maps on grid transparent papers using 1995 capability map. Land area possession per family is calculated by dividing total cropland area with total number of households in the village. Interviews and discussions with inhabitants witnessed their feelings about land conversion and subsequent impact on farmers' possession of arable land and livelihood. The assessment demonstrated that 234.9 ha (23.7% of the highly and moderately potential cropland) has been converted to urban use, including 146.5 ha (38.5% of the capability class I land) being reallocated for settlement. Cultivable land area possession per household is reduced (1–0.35 ha) causing land fragmentation and landlessness in the village affecting the livelihood in general and specifically women-farmers-led families. Based on the present study, it has been concluded that the available land capability classification system has not been properly practiced in the reallocation and distribution of land for settlement purpose.

Keywords Agricultural land • Arable land • Land capability classification • Land reallocation • Land-use planning and urban use

---

T.D. Rustu (✉)
Department of Land Resources and Environment, Hamelmalo Agricultural College, P.O. Box 7807, Asmara, Eritrea
e-mail: tewolded@yahoo.com

S.A. Shahid et al. (eds.), *Developments in Soil Classification, Land Use Planning and Policy Implications: Innovative Thinking of Soil Inventory for Land Use Planning and Management of Land Resources*, DOI 10.1007/978-94-007-5332-7_29,

29.1 Introduction

The conversion of prime agricultural land into built-up area (agriculture to urbanization) is a historical phenomenon associated with the economic modernization in the past and a sociopolitical issue in the recent years (Leland 2001). Though humans have been modifying land for thousands of years, current rates, extents, and intensities of land change are far greater than ever recorded in the history, driving unprecedented changes in ecosystems and environmental processes at local, regional, and global scales. These changes encompass the greatest environmental concerns of human populations today (Ellis 2007). The land conversion usually results in the loss of agricultural output and the positive externalities associated with farming and wide-open space (Leland 2001).

Recently, land conversion assessments have been made by many researchers. In China, the dynamic changes and optimal allocation of arable land conversion has been studied by Hengzhou et al. (2007). Later, Liu et al. (2009) reported most of the arable land is lost to urbanization and rural settlements development. In a socio-economy study, Xie et al. (2005) identified socioeconomic driving forces responsible for arable land conversion. Whereas, Oli (2001) expressed that city areas will continue to grow on good agricultural land unless intervened by planning process by land-use planners and decision makers.

The competition for land between urbanization and agriculture is becoming acute and complex (FAO 1999). In many densely populated regions, land demand for residential purpose is the major driving force to transform prime agricultural land out of agriculture production (Ellis 2007); therefore, the protection of prime agricultural land has become the world focus. Government of Eritrea (2004) has realized the urgent need for wiser use of the scarce rural land resources through land-use planning. In the developing countries, land capability classification is used to protect prime agricultural lands from urbanization, in the village land reorganization and for new settlements (Dent and Young 1981). Government of Eritrea (1995) has already acknowledged that there is a high land demand for urban expansion, as well as urgent need to protect agricultural land loss in the highlands of Eritrea.

Total population of Eritrea is about 4.0 million. It is estimated that population will grow on an average of about 2.7% per year (World Bank 2002). The land (12,432,000 ha) is the most important natural resource of Eritrea, of which 439,000 ha (3.5% of the total land area) is arable land (FAO 1994).

Eritrea is predominantly a rural society, and about 80% of the population lives in rural areas depending on subsistence agriculture for their livelihood (FAO 1994). In general, the demands for both arable land and urban development are increasing every year. The current cropland is under heavy pressure owing to the increasing population densities. According to Ministry of Agriculture (2002) estimates, average rural population density is 7.2 persons per ha of cropland, or 0.14 ha cropland per capita. With population increase, per capita cultivable land will decline. The figure may be even less as land becomes alienated from agriculture to urbanization. Moreover, natural and human impact on the land may degrade it, and hence this resource may become marginal for cultivation (Bojo 1995).

All land in Eritrea is owned by the state. The state allocates land to community or individuals for purposes such as agriculture, housing, and other developments (Castellani 2000; Government of Eritrea 2004). Recently, the government of Eritrea proclaimed that land be reallocated and distributed to the needy population.

Customarily, villages and towns in Eritrea occur in hilly and mountainous area or on barren and affected land. This is for security reasons and principally to secure potential agricultural land for crop and livestock production. However, recently, it has been observed that fertile agricultural lands have been encroached by the villages and towns. The reason being that reallocation and distribution of land for construction purposes to the needy has long been restrained to meet the land demand of growing population. Besides population growth in small towns, villages and Asmara (capital city) has been critical and intense. The dramatic population growth in Asmara and the subsequent paucity of houses are acutely affecting the periphery villages for the purpose of housing demand. The rate at which these are taking place is not allowing alternatives for the needed urban expansion other than engulfing arable land. Furthermore, nearly all villages and small towns are located in the vicinity of agricultural land, even though the village boundary is confined to barren land. Naturally, when a village or town grows, it begins to encroach arable land.

In the past, the villagers were extremely careful to consider land capability when deciding where to grow crops and where to construct houses. The local knowledge and the good practices seem to have disappeared in the modern society. Landlessness has never been a problem in Eritrea, but now may become one as hundreds of hectares of fertile agricultural land are being converted to construction sites. The loss of land in subsistence agriculture must be equated with the loss of livelihood. Thirteen villages in the periphery of the Asmara city were recently incorporated into the municipal boundary. Therefore, concentrating on areas around Asmara would be the most logical approach when studying conversion of agricultural land to human settlements. One of these villages around Asmara recently incorporated under Asmara city town planning is Embaderho. Its cultivable area is under the shadow of loss due to urbanization.

This chapter provides a case study of the conversion of agricultural land to urban uses between 1995 and 2003 in Embaderho village of Eritrea. The aim of the study is to determine how much fertile land is converted to residential area and whether a system of land capability classification is employed in the reallocation of land, and whether competition for land use between agriculture and housing is causing landlessness in the village.

29.2 Objective

The objective of the present study is to examine if an appropriate system of land capability classification is practiced in the process of land reallocation and distribution for housing purpose. Thus, the assessment will be made to see how much total agricultural land has been converted to urban use in the village and principally the conversion in terms of land capability classes. The study also looked into the effect of conversion on individual household's land possession and whether the conversion, competition for land use between agriculture and housing is causing landlessness in the village.

29.3 Materials and Methods

29.3.1 Description of the Study Area

The Embaderho village (biggest in the highlands) is located in the Maekel zone and Serejeka subzone, approximately 12 km North East of Asmara (capital city) along the main road to Keren town at latitude of 1,704,753.20 m N and longitude of 488,623.18 m E. Its altitude is about 2,200 m above sea level (asl). Total village area is about 2,687 ha, of which 1,124.63 ha is arable. Primarily the area consists of undulating, gently sloping terrain and plains. The gentle slopes and the plains in between are the arable areas. Steeper hills also occur in the area.

The village has mild to fairly cold climate with a rainy season from June to August, mean annual rainfall of about 500 mm, and mean annual air temperature of 17.5°C.

The dominant soils belong to three soil orders (Lithosols, Cambisols, and Luvisols). Generally, the soils are very shallow; however, in the lower plains where deposition of eroded soil from higher areas has taken place, the soil depth varies between 0.8 and 1.5 m. The soils are poorly managed and therefore present low organic matter content and soil fertility.

The major crops grown are barely wheat and beans in rainfed farmland and potatoes and cabbages in irrigated land. Total area under irrigation is 32.17 ha; rest of the area is rainfed.

The natural vegetation includes *Rumex nervosa* spp., *Euphorbia* spp., *Echinops giganteus*, *Aloe*, *Cynodon dactylon*, and *Hyparrhenia species*. Vegetation cover is very low due to trees clearance for firewood. Recently, efforts have been made in planting eucalyptus and acacias in limited area.

Total population of Embaderho village is 6,446 based on the census conducted in early 2004. This includes native people and outlandish people permanently residing in the village. There are about 1,310 households, which give an average size of five persons per household.

The land tenure system in Embaderho village is called "diessa." In a diessa system, the land is owned communally by the whole village. Every eligible person (native permanent resident) gets an equal share of land for cultivation. Land is redistributed every 5–7 years by elected representatives, normally community elders. In the past, only married men above the age of 18 years were eligible to get land, but very recently, government introduced a new system allowing women to obtain the land. Agricultural land shares of migrated and died persons are reserved for reallocation to newly married couples.

29.3.2 Land Classification

Land is not the same everywhere and there has to be a key to equal and fair division of agricultural land to families. This is addressed in the village through a local land classification system. In this traditional classification system, land is classified according to the potential or fertility of the land and distance from the village. Land is then again subdivided into individual farms capable of producing similar amount. Every family has to receive a share of fertile land as well as poor land in different locations. This process of redistribution of land is carried out through the drawing of lots, called "echa."

Another land classification system practiced in the Embaderho village is the modified USDA capability classification (Haile et al. 1998). Based on this system, Ministry of Land, Water, and Environment (MoLWE-Department of Land, Asmara, Eritrea) classified agricultural land into highly and moderately potential cropland comprising capability class I, II, and III land; less potential cropland as capability class IV land; and fallow and grazing land to capability class IV, VI, and VII land. This modern capability classification is used by the government. Land-use map of Embaderho is shown in Fig. 29.1.

29.3.3 Data Collection and Analysis

Land-use cover change is assessed and mapped between 1995 and 2003, for which maps are available with the MoLWE. The maps are prepared by superimposing the existing maps. Table 29.1 illustrates land-use cover changes between the years 1995 and 2003.

Land capability classification system is the most widely used method for evaluating and identifying best agricultural land (Van Lanen 1991). Land capability map prepared by MoLWE is used to analyze area of land converted from agriculture to urban use in each capability class. When analyzing, the land-use cover and the land capability maps are redrawn into transparent paper using colored pens. Each land-use cover map is then separately overlaid with the capability map for area calculation. Change in area (hectares) of the land that falls into the different capability classes in the study area as a result of urban encroachment between 1995 and 2003 is given in Table 29.2. Table 29.3 shows distribution of land-use cover change (hectares) against each land capability classes of the study area between 1995 and 2003.

Area of land owned by single family household is calculated by dividing total cropland area by the total number of households in the village. Discussion and interviews are held with the village leading farmers, local government, and agriculture experts to look into the farmers' perception on agricultural land conversion and its impact on farmers' arable land possession and on their livelihood.

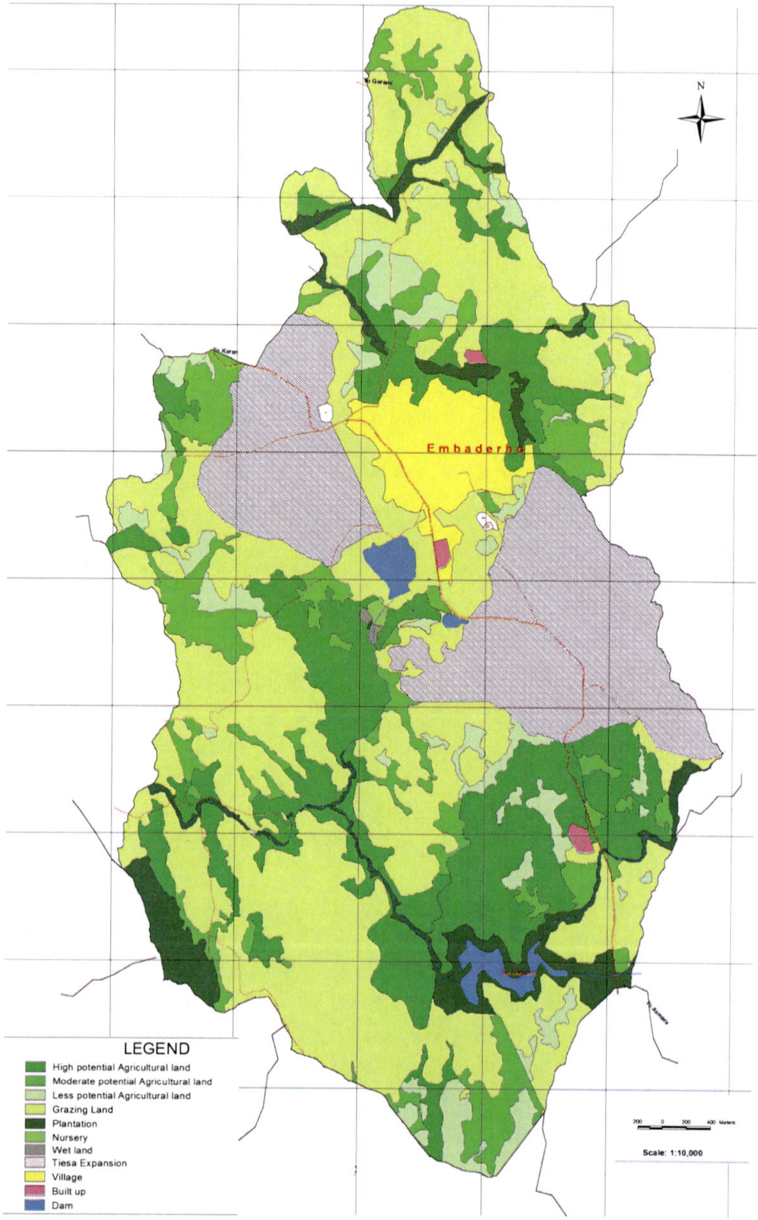

Fig. 29.1 Land-use map of Embaderho (Source: MoLWE)

Table 29.1 Land-use cover changes between 1995 and 2003

Land-use cover	Area (ha) 1995	Area (ha) 2003	Change[a] (ha)
Highly and moderately potential cropland	980.63	754.80	−225.83
Less potential cropland	144.00	144.00	0
Fallow and grazing land	1,417.08	1,049.10	−367.98
Tree plantation	46.94	157.0	+110.06
Wet land	1.90	1.90	0
Nursery	2.50	2.50	0
Dam	23.40	14.30	−9.1
Settlement	67.95	560.80	+492.85
Cemetery	2.60	2.60	0
Total	2,687.00	2,687.00	0

[a] – shows decrease in land-use cover; + shows increase in land-use cover

Table 29.2 Land capability classes' changes depicted between 1995 and 2003

Land capability class	Area (ha) 1995	Area (ha) 2003	Change[a](ha)
I	380.16	233.68	−146.48
II	88.88	65.21	−23.67
III	511.59	455.91	−55.68
IV	1,263.41	968.16	−295.25
VI	276.93	204.20	−72.73
VII	20.74	20.74	0.00
VIII (wet land)	1.90	1.90	0.00
Tree plantation	46.94	157.00	110.06
Nursery	2.50	2.50	0.00
Dams	23.40	14.30	−9.10
Cemetery	2.60	2.60	0.00
Settlement	67.95	560.80	+492.85
Total	2,687.00	2,687.00	0

[a] – shows decrease in land-use cover; + shows increase in land-use cover

29.4 Results and Discussion

29.4.1 The Process of Land Reallocation and Distribution

The village community fully supports and is the key player in the land reallocation and distribution for residential development. It selects where, within its demesne territory, it wants its housing development to be and sets the criteria a permanent resident of the village has to fulfill to be eligible to apply for land for housing. With the approval of the government, it then distributes the land to the eligible person.

Four main bodies influence the process of rural agricultural area conversion to residential land use. These include the village "tiessa" committee, the local government, the village community members, and the Ministry of Land, Water, and Environment.

Table 29.3 Land-use cover change against land capability classes between 1995 and 2003

Land-use cover	Capability class	Area (ha) 1995	Area (ha) 2003	Change[a] (ha)
Highly and moderately potential cropland	I	380.16	233.68	−146.48
	II	88.88	65.21	−23.67
	III	511.59	455.91	−55.68
	Total	980.63	754.80	−225.83
Less potential cropland	IV	144.00	144.00	0
Fallow and grazing land	IV	1,119.41	824.16	−295.25
	VI	276.93	204.20	−72.73
	VII	20.74	20.74	0.00
	Total	1,417.08	1,049.10	−367.98
Tree plantation	VI	46.94	157.00	+110.06
Wet land	VIII	1.90	1.90	0.00
Nursery	–	2.50	2.50	0.00
Dam	–	23.40	14.30	−9.10
Cemetery	–	2.60	2.60	0.00
Settlement	–	67.95	560.80	+492.85
Grand total		2,687	2,687	0

[a] – shows decrease in land-use cover; + shows increase in land-use cover

All these activists have different roles. The main promoter is the village "tiessa" committee, a special committee which is selected by the village community with the responsibility of administering the reallocation and distribution of land for housing purpose. In the process of reallocation and distribution, the village "tiessa" committee locates and selects area for residential development. Upon approval by the local government and the village community members, the selection of the area is communicated to the ministry. The ministry then makes all the necessary preparation: land leveling, roads and parcels cutting, land reserving for infrastructural and potential mining development, and other government interests. Concurrently, the village "tiessa" committee registers applications and selects eligible individuals. The result of the selection is then conversed again with the ministry for approval.

29.4.2 Potential Agricultural Land Conversion

The study shows that 492.85 ha (20.19% of the total agricultural area) inclusive of 234.93 ha (23.74% of the highly and moderately potential cropland) is converted to settlement (tiessa expansion) between 1995 and 2003 (Table 29.1). Of this figure, 146.48 ha (38.53% of the land capability class I) represents prime farmland, land that is ideal for agriculture, converted to urban settlement (Table 29.3). These figures demonstrate that the amount of total agricultural land lost to urbanization in the village is small relative to the total supply of this land. However, the percentage of prime farmland, capability class I land, lost to urbanization is significant. The results support the estimation and the concern of land conversion made by the ministry in 2002. Despite the 2002 estimates of the ministry that only a tiny fraction of Eritrea's arable land is likely to be converted to settlement over the next few decades, there is a concern of potential problems and with special minority groups like women which may be adversely affected by conversion of agricultural land to urbanization.

Table 29.3 indicates 225.83 ha (23.03% of the highly and moderately potential agricultural land) which comprises capability class I, II, and III land is allocated to urban use. These figures suggest that available land capability classification might not have been used properly if used at all in the land-use planning process, land reallocation, and distribution for settlement. According to the 2002 estimates of the ministry, land-use planning is a new practice in Eritrea, and its application through land capability classification in managing land is stalled by the current land tenure system. Much arable land could be preserved in agricultural production if development is more wisely managed using capability classification (Oli 2001). A concern for meeting present needs might have led to short-term objectives and inappropriate land-use practices in the highlands of Eritrea (Zemenfes 2002).

Arable land area possession per family was on average 1 ha in the village. The study shows a reduction from 1 to 0.35 ha between 1995 and 2003. This represents a 65% reduction in single household's land area possession because of land conversion to settlement. There is a long-standing tradition of family farming in the village as is true with all rural areas of Eritrea. Conversion of agricultural land to urban uses together with population growth has resulted in fragmentation of the available arable land. Arable land holdings have now become too small to sustain the livelihood of farming families. These families have lost their cultivable land and now have to find an alternative means of livelihood. The effect of loss in arable land to urban use on the livelihood of the farmers is clearly reflected in the reduction of total and per capita crop production. As obtained from interviews and discussion with farmers, total crop production from the village has reduced by more than 50%.

The village arable land is under further threat of loss due to urbanization. According to discussion with village community members, it is expected that more arable land will be converted to settlement soon. Any permanent resident of the village above the age of 18 years is entitled to get land for residential development. Reallocation and distribution of land to eligible individuals is not a onetime endeavor and so does the threat from urbanization as there will be many people who will turn on 18 every year and have to get land too.

Currently, conversion of agricultural land to residential settlement is supported by the whole community despite the fact that it has caused a reduction in their arable land and loses of their livelihood. This has been found mainly for the following reasons. First, in the current land tenure system, people have only the right to use the arable land as long as they or their family stayed in the village and not migrated or died and is returned to the village thereafter. The system allows only married male and divorced female above the age of 18 years to get land for cultivation. Thus, all unmarried ones are disadvantaged by the agricultural area deriving their preference for land for housing.

Second, by law, the government owns all lands in Eritrea; therefore, it can engage any size of land wherever it is located and for any purpose. For example, during the study period, about 37 ha of fertile and potential irrigable land is utilized by the Ministry of Defense for a project of growing flowers and vegetables. This creates a competition for land between the government and the community and pushes the community to prefer land for housing than for agriculture.

Third, even though families fully depend on subsistence agriculture for their livelihood, they do not produce enough. The agricultural land cannot be sold. Thus, the allocation of agricultural land for residential development is perceived as an attractive opportunity for the community to sell the land at a very attractive price. The amount gained is incomparable with the annual income per capita to be derived from agricultural production on the same land. Currently, value or price of land allocated for housing is exaggerated in Eritrea.

Fourth, the conversion of agricultural land to residential development has given the farmers an economic opportunity. They can generate more income from house rent than from subsistence agriculture. Of the urban population of Eritrea, about 55% (487,000) live in Asmara with estimated urban growth rate of about 5% per year. The urban primacy and thus the demands for housing are very much marked. For the village included in the municipal boundary of the city, its transportation infrastructure is upgraded to enjoy the Asmara city bus services making it attractive to reside there. This results in influx of urban people to look for a rental house in the village. This has increased the value of houses in the village giving selection advantage to land for housing over land for agriculture.

Societies have some concern about potential negative social influences that urbanization entails. Communities perceive inflow of new residents as a cause for loss of their belongings to the village or erosion of their social cohesion. This has now changed in the village according to discussion and interviews with village community members. The discussion and interviews reveal that village community members would like to see inflow of residents. The study suggests that urbanization and a shift in the livelihood of the farmers from agriculture have changed the attitude of the community about inflow of residents to their territory.

29.5 Conclusions

This chapter, through a case study, has examined if available system of land capability classification is being used in land reallocation and distribution for settlement purpose in villages around Asmara city, Eritrea. It has analyzed if the decisions on land use are affecting families' arable land possession and farmers' livelihood. The study suggests that reallocation and distribution of land for residential use is causing the transformation of significant portion of the most fertile agricultural land out of production. Because of agricultural land conversion, arable land is being fragmented; families are losing their cultivable land, and this is causing a shift in the long-held traditional farming system of livelihood of many farmers.

References

Bojo J (1995) Land degradation in Eritrea. Mimeograph paper. World Bank, Washington, DC

Castellani LG (2000) Recent developments in land tenure law in Eritrea, Horn of Africa. Working paper no 37. Legal Tenure Center, Institute for Research and Education on Social Structure, Rural Institutions, Resources Use and Development. University of Wisconsin, Madison

Dent D, Young A (1981) Soil survey and land evaluation. George Allen & Unwin, London, p 278

Ellis E (2007) Land-use and land-cover change. In: Cleveland CJ (ed) Encyclopedia of earth. Environmental Information Coalition, National Council for Science and the Environment, Washington, DC

FAO (1994) Agricultural sectors review and project identification for the Government of Eritrea. FAO, Rome

FAO (1999) The future of our land: facing the challenge, Guidelines for integrated planning for sustainable management of land resources. FAO and UNEP, Rome

Government of Eritrea (1995) National environmental management plan for Eritrea. Eritrean Agency for the Environment, Asmara

Government of Eritrea (2004) The land reform proclamation (no 58/1994). Asmara, Eritrea

Haile A, Gebretatios I, Omer MK, Ogbazghi W, Arai W, Gebremariam T, Gebreselassie G (1998) Rehabilitation of degraded land in Eritrea, 2nd edn. Ministry of Agriculture and University of Asmara, Asmara, p 162

Hengzhou X, Futian Q, Zhongxing G (2007) Dynamic changes and optimal allocation of arable land conversion in transition of Jiangsu Province. China Popul Resour Environ 17(5):54–58

Leland JR (2001) Estimating the determinants of land conversion. Graduate thesis, Department of Economics, Ateneo de Manila University

Liu YS, Wang JY, Long HL (2009) Analysis of arable land loss and its impact on rural sustainability in Southern Jiangsu Province of China. J Environ Manag 91(3):646–653

Ministry of Agriculture (2002) The national action program for Eritrea to combat desertification and mitigate the effects of drought. Asmara, Eritrea

Oli PP (2001) Expansion of urban areas in agricultural land. In: International conference on spatial information for sustainable development. Nairobi, Kenya, 2–5 Oct 2001

Van Lanen HAJ (1991) Qualitative and quantitative physical land evaluation: an operational approach. Doctoral thesis, Wageningen Agricultural University, the Netherlands

World Bank (2002) African development indicators. The World Bank, Washington, DC

Xie Y, Mei Y, Guangjin T, Xuerong X (2005) Socio-economic driving forces of arable land conversion in China. J Glob Environ Change A 15(3):238–252

Zemenfes T (2002) Highland farming in Eritrea: its potential and constraints for sustainable production. In: Abstracts of international research symposium, University of Asmara, Eritrea, 15–18 June 2002

Chapter 30
Assessment of Land Use Planning and Development in Nigeria: Challenges and Policy Implications on Agriculture

Abiodun E. Obayelu

Abstract The world is facing problems of poor land, water, and waste products management. This has attracted the international, national, and local attentions. This chapter presents a comprehensive land use plan based on the past and present land use planning and management practices in Nigeria and highlights the challenges and implications on agricultural production. It is revealed that land transformation is a common phenomenon in Nigeria. The rising land costs and accessibility to urban land has become a serious issue affecting agricultural production with over 70% of the citizens living on less than US$1 a day. The land use management has been wholly concerned with the granting of statutory right of occupancy and approval of plans to use land for different purposes without adequate monitoring of its outcomes. The lack of monitoring is attributable to a number of factors including lack of interests to adopt alternative land use systems. To create conducive environment for the present and future generations, it is essential to fundamentally rethink on land use control mechanism, policy, and action. To achieve such a favorable environment, it is prerequisite that land use laws should be enforced by the policy makers. It is also important to involve relevant stakeholders in the process of environmental planning to share their interests and opinions.

Keywords Development • Land management • Land planning • Land use • Policy implications

A.E. Obayelu (✉)
Department of Agricultural Economics and Farm Management, Federal University of Agriculture Abeokuta (FUNAAB), Abeokuta, Ogun State, Nigeria
e-mail: obayelu@yahoo.com

S.A. Shahid et al. (eds.), *Developments in Soil Classification, Land Use Planning and Policy Implications: Innovative Thinking of Soil Inventory for Land Use Planning and Management of Land Resources*, DOI 10.1007/978-94-007-5332-7_30,

30.1 Introduction

The word "land" has different meanings under different disciplines. In its simplest form, it means the surface part of the planet earth. This definition, however, appears to be restrictive since it recognizes only the immediately visible aspect of the land. In a cultural context, land means a measurable entity divisible into thing like "parcel" by means of mathematical and technical processes of surveying and cartography (Bohanna 1966). In a geographical context, land is seen as a specific area of the earth's surface. This includes its characteristics, all reasonably stable or predictably cyclic attributes of the biosphere that are vertically above and below this area including those of the atmosphere, the soil and underlying rocks, the topography, the water, the plant, and animal populations (Vink 1975). In economic terms, land is seen as the sum total of the natural and man-made resources over which possession of the earth's surface gives control (Barlowe 1963). Within this broad conception, land includes the ground, water, ice, forests, and mineral deposits. It also includes natural phenomena such as sunlight, wind, rain, and temperatures as well as man-made improvements like farm fixtures attached to the surface of the earth.

Demand for this resource is derived demand and is not an end in itself but a means to an end. Land is a major factor of production and a vital element in the socioeconomic development of any country or society. It is the most important resource input in Nigerian agriculture. The policies affecting the land ultimately affect the farming population more than other members of the community. Land use determines whether a resource could be conserved or not and the level of conservation attainable for natural resources. Thus, as the nation grew in size and rural areas become urban centers and urban centers become large metropolitan areas, there is always increased competition as well as demand for land for different purposes. As a result of the population growth, the Nigerians are competing with land, and there is increasing evidence of deterioration of man-land relationship. In rural areas, declining soil fertility and the impact of soil erosion threaten the ability of land to produce sufficient food. In urban areas, competition for productive agricultural land has led to its conversion to urban uses.

The land use conflict and the design of effective and efficient management systems to combat crisis, degradation resulting from uncontrolled and unguided use of land, have attracted international, national, and local attentions. For instance, land reforms and home ownership constitute a prominent item in President Yar'Adua's seven-point agenda, and it is pivotal to achieving Vision 2020. To understand the land use pattern in Nigeria and its shortcomings, it is important to examine the characteristics of past and present land use structure.

The questions this chapter attempts to address are the following: What is land, land use, and land tenure? Why do land use problems still persist despite the various government interventions and actions? Under what conditions or situation or circumstances are the policy packages introduced? Do the land use management practices, conditions, and circumstances prevailing in the country permit the kind of policy introduced? What are the situational factors that hinder effective or efficient management of land use in Nigeria? What are the effects of the land use policies?

30.2 Objectives

This chapter has the following objectives: Firstly, to review the various concepts relating to land and land tenure in Nigeria. Secondly, to examine the concept of land use and control measures in Nigeria. Within the framework provided by the second section, the third part of this chapter presents the challenges of the existing land use policies in Nigeria and the implications on public acquisition and agricultural production.

30.3 Land Use Planning and Control Measures in Nigeria

This section describes the situation of land in Nigeria, the past and present land use policies, and land tenure system.

30.3.1 Nigeria Land and Land Use Dynamics

Land is taken as a commodity for daily use for many purposes by over 140 million people (representing 20% of the entire population of Africa) of Nigeria. For several decades, land has continued to influence the lives of Nigerians socially, economically, and politically. In the process of using the land, complex set of relationships has emerged among groups showing that land is a sensitive asset whose administration must be based on meaningful policy decisions to benefit most people.

In this study, we have observed that the mere presence of land to cultivate has never been a serious problem of agricultural development in Nigeria with a total land area of about 98.3×10^6 ha (77.4×10^6 ha in the savanna zones, 7.57×10^6 ha in the derived savanna zones, and 13.4×10^6 ha in the forest zone) (Omofonmwan and Osa-Edoh 2008). Farming is a very resourceful activity in Nigeria, where about 71.2×10^6 ha of land is cultivable; however, in 2004, only one-third of the land was under cultivation (Daramola 2004), and 30.7×10^6 ha was under cultivation in 2009 (Saletti 2009) contrary to 34×10^6 ha reported in 2006 (ICARRD 2006). The analytical review revealed that 66% of Nigerian land is under various forms of food production (crop and animal) and forest plantation. About 42 and 21% are under crop production and extensive grazing, and livestock projects, respectively, while less than 2 and 1% are under plantation crop and forest, and agriculture in denuded areas, respectively (Asadu et al. 2004). It is now clear to state that 35% of the Nigerian land is devoted to permanent crop cultivation and 56% to forestry and permanent pastures. Apart from the actual land currently under cultivation, it is estimated that another 31% of 56% classified under forestry and permanent pastures is considered potential agricultural land, because this land is neither under forest reserves and nor under permanent pastures.

Nigeria has a high proportion of actual (35%) and potential (31%) agricultural land (Areola and Ofomata 1978). The major problems are the availability of these lands when needed and accessibility in urban areas. Land under cultivation is

Table 30.1 Land use categories and annual rate of change per ha/year in Nigeria (FAO 2001)

Land use category	Annual rate of change (ha/year^{-1})
Agriculture (intensive/extensive)	554,657.10
Floodplain agriculture	67,616.10
Grassland	131,224.60
Dominantly trees/woodlands/ shrubs	−858,720.40
Dominantly shrubs and grasses	−104,974.30
Dominantly grasses	6,166.16
Forest	−104,231.00
Freshwater marsh/swamp	−69,453.05
Forested freshwater swamp	1,707.86
Mangrove forest	−14,982.77
Water	45,474.02
Bare surface	129,113.70

– means decrease in ha

Table 30.2 Comparative land use pattern in Nigeria 1995–2020 (ha) (FAO 2001)

Land use category	Base year 1995	Year 2010	Year 2020	Steady state
Agriculture cropland	61,900,000	68,063,519	70,652,157	87,408,773
Floodplain agriculture	2,400,000	3,141,000	3,390,062	479,783
Grassland	3,150,000	4,398,238	4,765,522	1,071,156
Dominantly trees/ woodland and shrub	9,000,000	3,866,596	2,276,170	7,309
Dominant shrubs and grass	7,100,000	4,290,518	3,017,151	102,583
Dominantly grasses	1,100,000	1,065,057	1,040,003	227,530
Forest	2,650,000	1,436,848	938,066	929
Freshwater marsh/swamp	620,000	181,072	100,943	3,534
Forested freshwater swamp	1,800,000	1,820,089	1,834,930	3,134,486
Mangrove forest	1,190,000	978,706	845,074	701
Water	680,000	1,065,986	1,144,630	172,308
Bare surface	1,892,000	3,174,371	3,477,292	872,908
Total	93,482,000	93,482,000	93,482,000	93,482,000

increasing at an average rate of 554,657 ha per annum while the land under high forests is decreasing by 105,865 ha per annum (Table 30.1). It is estimated that the agricultural cropland in Nigeria will increase from its level of 61.9×10^6 ha in 1995 to 70.6×10^6 ha by 2020, whereas forest land will decrease from 2.6×10^6 ha in 1995 to 0.938×10^6 ha by 2020 (Table 30.2).

30.3.2 Land Use Classification in Nigeria

Land use is the application of human controls in a relatively systematic manner to the key elements within an ecosystem to derive benefits (Vink 1975). The Vink's definition recognizes the dynamic nature of land and land use and emphasizes the importance of man within the ecosystem. The efforts of man according to Vink are crucial determinants of what happens to land. Man's influence may be favorable or adversely affect the land. The use to which land is put differs from one location to another since the immediate needs of the society or community concerned are not the same. The different uses of land revolve around agriculture (crop and animal production), industry (both oil and non-oil), and social needs such as the provision of infrastructure. Land use classifications vary from one society to another. Using the criterion of decree of intensity, Boserup (1966) adopts a fivefold classification: firstly, the forest fallow cultivation in which 1–2-year cultivation is followed by periods of fallow for a duration of at least 20–25 years; secondly, the bush fallow system in which periods of cultivation may vary from 1 to 2 or 6 to 8 years and then the land is left to revert to bush for another period of 6–8 years; thirdly, the short fallow system in which fallow lasts only for a few years to be invaded by grasses; fourthly, the system of annual cropping in which the land is left uncultivated for some months between the harvesting of one crop and the cultivation of the next one; lastly, it is the system of multiple cropping which is the most intensive form of land use. Under this system, the same plot of land carries two or more successive crops every year (Boserup 1966).

In all cases, there are evidences that most of these land use systems are practiced in many parts of Nigeria. However, the most prevalent is agriculture land use, and it is based upon the bush fallow system. The forms of cropping in Nigeria are sole cropping, double cropping, multiple cropping (in irrigated areas), mixed cropping, and intercropping.

The land use pattern is exacerbated by drought, forest fires, overgrazing, and flooding which led to severe environmental degradation, loss of biodiversity, and diminished forest productivity. The institution of land use planning in the cities of Nigeria significantly affects the development. The delays in securing land and development rights affect plot sizes and impose unrealistic standards on both urban agriculture housing developments.

30.3.3 Land Tenure System in Nigeria

Land tenure is dynamic and a social institution in Nigeria. The concept of land tenure has been described as a systematization of the rules which functions by specifying what different classes of persons must or must not do with reference to the occupancy, use, abuse, or disposition. The tenure streams evolved through war and settlement but based on local administration of natural resources. Over the years, land tenure had been the decisive factor in resource management at local level. Unfortunately, the impact of tenure on natural resources allocation and exploitation is often ignored in public land policy. Yet land tenure issues contribute to deforestation, degradation of the environment, lowering of carrying capacities of soils, and poaching and extinction of wild biotic resources. Land tenure systems influence the use to which land is put for economic and social development. Yet land use determines whether a resource could be conserved or not and the level of conservation attainable for natural resources. Land tenure is a tool for conservation, and it involves sets of rules and regulations used to control and manage natural resources: soils, water, wild living resources, and the environment. The systems in Nigeria respond to socioeconomic and political changes put in place for resource utilization. Tenure systems are however not monospecific: They vary from one rural community to another but pivoted by three broad systems of communal, individual, and family ownership.

Traditional land tenure throughout Nigeria was based on customary laws under which land was considered community property. Access to land is based on membership of a landholding community by birth. Right to natural resources such as land, plants, animals, and water is often communal (Osemeobo 1991). An individual had no statutory rights to the land he only farm in his lineage or community area. He could possess the land as long as he used it to his family's or society's benefit and could pass the land on to heirs and pledge its use to satisfy a debt, but could not sell or mortgage it. The right of disposal belongs only to the community which is acting through traditional authorities to exercise this right in accordance with customary law. The communal tenure enjoys strong proprietary and security rights to biotic resources in rural areas between and among two or more indigenous settlements, respectively. The breakdown in traditional tenure and disputes on boundary marks have made Nigerian government to be involved in boundary adjustments in administrative units throughout the country for communities to have exclusive rights to discrete areas of land.

There are evidences that when the tenure system is stabilized at the local level, it responds to changes in land use from communal to individual. But at the village level, the tenure system involves some elements of societal control of land use (Bruce 1988).

At the local level, natural resources are held as a common property. This implies common exploitation and management of resources with respect to hunting; collection of firewood; harvesting of fruits, nuts, and leaves; and even farming or grazing of livestock. It also relates to group interest and control of resources, with each group composed of extended family, a lineage, or a village defined by common descent or common residence. The control of land is vested in council of chiefs and elders who hold them in trust for all members of the community. Land is allocated to individual indigenous farmer or household for farming where land has no conflicting rights. With increasing population pressure, access to land is carried out on equity basis until land reallocation loses its points. Land is passed from generation to generation with customary rules of succession. Nevertheless, where agricultural cultivation is stabilized, dominant user rights are acquired on the land by the individual farmer and the land reverts to individual ownership within the family land or the village land.

30.3.4 Formal Planning and Control of Land Use in Nigeria: The Journey So Far

From the planning perspective, land represents a mosaic that ought to be regulated to ensure conformity and balance of the built environment (Ratcliff 1976). Land policies are the rules, customs, and regulations guiding the use, allocation, distribution, planning, and management of land resources, while land administration is the process of determining, recording, and disseminating information about ownership, value, and use of land when implementing land management policy. Land planning/administration is a process and an instrument for government to offer security of tenure on land, regulate land market, implement land reforms, protect environment, levy taxes, etc., as well as serve the peculiar development needs of citizens security.

All the three tiers of government (federal, state, and local) are involved in land management in Nigeria in most cases through their agencies. Formal land use planning and management in Nigeria began in 1863 with the enactment of the Town Improvement Ordinance by the colonial government (Aribigbola 2008). The ordinance was meant to control development and urban sanitation in Lagos then the federal capital of Nigeria. However, modern land use planning began in the country in 1946 when the Nigerian Town and Country Ordinance was enacted. This law among other provisions in most parts of the country led to the establishment of Town Planning Authorities (TPAS), which were to control and guide the orderly development of land within their jurisdiction by approving proposals for physical development and the preparation of development schemes and land use plans (Olujimi 1993). At present, the planning, control, and management of land in the cities by most state governments in Nigeria is undertaken by the Departments of Urban and Regional Planning, Land Services and Survey, Ministry of Works, Lands and Housing.

Before the promulgation of the Land Use Decree of 1978, there were differences in the administration of land between different parts of the country as already explained under the land tenure system. This implies that the country (Nigeria) has no coherent and comprehensive land policies before 1978 Decree. At this period (before Land Use Decree), government and individuals approached communities for land for farming, housing or planning services, etc., with all the sociocultural implications. But in response to a potential crisis in land distribution, the Federal Military Government promulgated the Land Use Decree of March 1978, establishing a uniform tenure system for entire Nigeria and subsequently incorporated this in the country constitution of 1979. The decree effectively nationalized all land by requiring certificates of occupancy from the government for land held under customary and statutory rights and the payment of rent to the government. The decree stipulated that anyone in a rural or urban area who normally occupied land and developed it would continue to enjoy the right of occupancy and could sell or transfer his interest in the development of the land.

The Land Use Decree of 1978 was instituted to achieve a number of objectives; prominent objective among them was to solve the problems of immediate lack of availability of land for use when required by all the concerned and curb the activities of land speculators. In summary, the decree was introduced to remove bitter controversies resulting at times in the loss of lives which the land is known to be generating through land dispute, streamline and simplify the management and ownership of land in the country, assist the citizens irrespective of his/her social status to realize their ambition and aspiration of owning the place where his/her family will live a secured and peaceful life, and enable the government to bring under control the use to which land can be put in all parts of the country, thus facilitating planning and zoning programs for particular use. By the promulgation of the decree in Nigeria in 1978, ownership and management of land was transferred to the government from individuals and communities. By the decree, land in the urban areas of each state is vested to the hands of the Governor of that state, while rural lands are vested in the local governments exercising jurisdiction in the particular areas. In the urban areas, the decree provides the establishment of a Land Use Allocation Committee to advise the Governor generally on matters relating to control and management of urban lands. The committee is also to advise on issues such as revocation of rights of occupancy, compensation, and resettlements.

In the rural areas, a Land Allocation Advisory Committee advises the local government generally on issues relating to effective management of land. At the state level, the Governor is empowered to grant statutory rights of occupancy to any person for all purposes and to issue certificates of occupancy, levy rents, impose penalties, and extend, curtail, or waive conditions pertaining to statutory rights of occupancy in the urban areas. Owners of developed land were to receive certificates of occupancy, while owners of land undeveloped before March 29, 1978, were limited to half a hectare of land (1.25 acres) as a maximum. In the rural areas, local governments are empowered to grant customary rights of occupancy to individuals or organizations for agricultural, residential, and other purposes. Customary grants are however limited to 5,000 ha for grazing land.

Today, the law guiding land administration in Nigeria is the Land Use Act, Cap L.5 of 2004 (originally promulgated as Land Use Decree No. 6 of 1978). It is a national land policy that seeks to streamline the ownership, acquisition, and disposition of land just like the Land Use Decree. This law is mainly concerned with use and allocation of land in the country. The law adopted the nationalization of all lands in Nigeria as introduced by the Land Tenure Law of 1962 (of the Northern Region). The Act harmonized the different land tenure laws which are operated in the Southern and Northern parts of the country in order to reduce the bottlenecks in land acquisition to encourage rapid national development. Under the law, all lands within a state (except land belonging to the federal government) vest in the hands of the Governor of that state who holds the land in trust for all Nigerians. It abolished freehold interests in land nationwide. By the Land Use Act, government is the trustee of land and an investor as well as the administrator. The citizens have limited interests in land.

The Department of Land Services is responsible for allocating public lands, issuance of Certificate of Occupancy, and management of government lands and estates in the city. In most states, this department is named the Nigerian Urban and Regional Planning Department formed through Nigerian Urban and Regional Planning Decree of December 1992. The Department of Urban and Regional Planning is responsible for the preparation of government layout plans and development schemes, approval of development plans (i.e., building plans, layout plans, etc.), and general physical development control through its Area Urban and Regional Planning Office in the city. Whereas, the Survey Department is in charge of mapping government lands, vetting, and approval of survey plans prepared by private consultants in the city.

The Town Planning Ordinance gave the planning authorities the power to expropriate lands for the preparation of planning schemes (and to compensate for the expropriation) and wide powers too to regulate the manner and type of developments on the schemes in order to ensure that lands are used and developed in the best planning principles.

One of the major government efforts on land development to encourage agricultural production in Nigeria was the establishment of the National Agricultural Land Development Authority (NALDA) in 1992 charged with the responsibility of intervening in the agricultural development of Nigeria. During the period of operation, remarkable progress was noticed in the performance of agricultural sector, but the program was eventually scrapped in 2000 without replacement owing to misappropriation of fund.

30.3.5 Major Challenges of Land Use in Nigeria and Policy Implications for Public Acquisition and Agricultural Production

Land is by far the most important production input in Nigeria. Ownership affects farming systems, institutional structures, ecological conditions, adoption and use of technology, food production and self-sufficiency, and overall well-being of the rural and urban population. Poverty and resource misuse is linked because of the pattern of land distribution which often favors the rich class. Observation has shown that the rich have access to land which is less prone to degradation or erosion. In addition, the rich class has the economic resources to invest in and improve the land. But poor farmers continue to till a marginal resource base despite increase in their number (Medugu 2006). The increment of land charges to over US$120 per square meter in most urban areas in Nigeria is seriously affecting urban agriculture. Nigeria is still a country where over 70% of its citizens live on less than US$1 a day. This is capable of derailing the achievement of the seven-point agenda or Vision 2020 on land use. The increasing population and suburban expansions in Nigeria also have many effects on both land acquisition and land use. Large cities like Lagos (Lagos State Nigeria), Ibadan (in Oyo State, Nigeria), and Kano (Kano State, Nigeria) demand huge quantities of land, water, energy, and other resources. It is evident that formal land use planning and the accompanying policies have not been used to promote and enhance improved people's access to land for use in Nigeria. A number of factors are responsible for this state of inadequate land use planning and management. These constraints include:

30.3.5.1 Inadequate Spatial Information/Data on Land Use

Inadequate information/data is one of the major problems facing land use planning and management in Nigeria. Field investigation reveals lack of data on land use and no comprehensive and up-to-date plan or map showing land use pattern and structure of ownership in most areas. Lack of base map was the major problem faced during land use planning. Various attempts made by the state and federal governments of Nigeria to map the entire country using modern techniques including satellite imagery are yet to be materialized. Thus, land use planning in most part is still based on the available base map. Individual plans generated by families and agencies involved in land use planning and management are not related and coordinated. No land census or land register exists showing the ownership of each parcel of land and the total acreage by ownership and type of use by which each parcel is committed. Thus, planning of land use is undertaken in piecemeal through family and community layout without adequate and up-to-date information about the available land and their location. Lack of vital land information made it difficult to monitor the allocation and use of each parcel of land in most areas resulting into consistent and inefficient allocation and development on land for various development purposes.

30.3.5.2 Non-adoption and Utilization of Modern Planning Approaches and Techniques

Based on the international community, a number of concepts and approaches such as sustainable urban development that emphasized sound environmental management including land management have been devised and adopted. Field investigations show that these new approaches and methodologies have not been incorporated into land use planning and management in most parts of Nigeria. As shown by Aribigbola (2006), land use planning in the city is still undertaken mainly by government officials in the ministry without involving the public who are the stakeholders to use such land. Therefore, non-adoption and incorporation of the new approaches is a major constraint that needs to be surmounted to ensure better management of land in Nigeria.

30.3.5.3 Outdated and Outmoded Land Use Planning Policies, Laws, and Regulations

Land use and management in Nigeria generally is still based on the Land Use Decree of 1978 changed to Land Use Act, 2004. This law was mainly concerned with the use and allocation of land for different purposes. The Land Use Act mainly deals with allocation and acquisition and confirmation of title on owners; it does not indicate the vital aspect of management which is the control of development on the land. The Urban and Regional Planning Act of 1992 that was meant to improve land use planning activity and land management in Nigeria is yet to be implemented in most states and local government areas of the federation, 19 years after enactment. Thus, land use planning and management in most urban areas in Nigeria is still based on the 1946 Act which itself was based on the 1932 Town and Country Planning Act of the United Kingdom that has been changed several times in the countries where the Act were borrowed from. Consequently, one of the major constraints to effective and efficient planning and management of land in Nigeria is the absence of up-to-date and dynamic laws and regulation to guide and control land use activity and management.

30.3.5.4 Inadequate Manpower to Control and Monitor Land Use

Closely associated with the above constraints is inadequacy of qualified land use planners. In Nigeria, most urban planning office is saddled with the responsibility of carrying out the planning and control of development on land for over 140×10^6 people. Majority of people in most of the planning offices are nonprofessionals with only very few people registered as town planners by the Town Planners Registration Council of Nigeria (TOPREC), the body responsible for regulating and registering land use planners in the country.

30.3.5.5 Rising Cost

Rising costs of land and accessibility to urban land in Nigeria like most developing countries have become a serious issue in urban areas of the country. Conventionally, policy makers are blamed for this problem. The rising cost has been attributable to a number of factors including corruption and lack of interests to adopt alternative land use systems.

30.3.6 Effects of Compulsory Acquisition of Land on Smallholder Farmers in Nigeria

The process of compulsory land acquisition emanating from the Land Use Act has been defined as the coercive taking of private lands (individual or communal) for public purposes (Umeh 1973). The outcome of the Land Use Act in Nigeria empowered the federal and state government to hold land in trust and transfer the usage only to those people who want to use land after due approval. This compulsory acquisition of land is however found to have some effects on smallholders' agricultural production. It dislocates their production programs on the land; compulsory land acquisition is disruptive on the social life of farmers' families; most farmers have to be moved to less-fertile lands thus affecting their level of productivity per hectare. The delay in payment of compensation money to community or individual where the lands are forcefully taken by government is detrimental to the welfare of farmers. In most cases when compensations are paid, its inadequacy constitutes a major income problem for the farmer.

Lands acquired from farmers in the oil-producing areas (Niger-Delta region) and given to oil companies have had the following effects on smallholder farmers: unemployment of aged farmers; depletion in soil fertility; accelerated perishability of crops; resultant high cost of living of farmers; inadequate maintenance of lands carrying pipelines, oil spillage affecting fish production, and other dangers emanating from oil wells.

30.4 Conclusions and Recommendations

Agriculture remains the backbone of Nigerian economy, but in spite the country's endowment in agricultural resources, the potential of the sector is yet to be fully realized. Land use is a complex phenomenon, which cannot be discussed separately from land and land tenure. The close relationship between land tenure and land use shows itself in various aspects such as in geographical, sociological, economic, political, and ecological aspects. This chapter has highlighted the importance of land as well as the complexity of land use in a developing country like Nigeria. In the absence of planning, the use of land and the exploitation of natural resources have been virtually unguided and uncontrolled. This is evidenced in Nigeria where there is an increasing rate of deterioration in man-land relationship. Land use planning has an impact on the efficiency of economic and agricultural production, social activities, and physical development. An increasing incidence of urban flooding, for instance, indicates a fundamental disharmony of land use in watershed areas and in the city.

Land use management in the city has been wholly concerned with the granting of statutory right of occupancy and approval of plans to use land for different purposes, without adequate monitoring the outcomes. This implies that land management and control tools in Nigeria are still weakly implemented, disjointed, and uncoordinated. Several organizations and agencies are involved without a coordinating agency or an overall land use plan within which effective land use management can be undertaken.

The inability of land users to acknowledge traditional tenure rights in project design and implementation results in conflicts which lead to high rate of de-reservation, deforestation, excessive poaching, and overexploitation of resources within the reserves.

On private lands, tenure regimes have led to abuse and misuse of resources due to dominant user rights existing within individual land ownership. But no meaningful agricultural revolution will be possible without focusing on availability of land. Positive legislation in this area is necessary to achieve the desired goal of proper land use management.

Land use planning and policies in Nigeria reveals that mechanisms on land use planning are in place in the city; they are not yet fully implemented and do not have any significant effect on land accessibility. Acquisition of land in most urban areas is still out of reach by the poor. To assure successful rural and urban agriculture in the country, effective land management that controls cost of land acquisition is required, including efficient soil and water conservation measures to negate environmental deterioration.

Sustainable conservation of land must be based on public education, mutual agreement on their utilization, and the recognition of the rights of the rural people who own, live with, and earn a living on them within ecological limits.

Land should be listed and registered to ensure the security of tenure of the holder. It is only when the quantity and quality of lands are known that land could be classified for various uses, and other policies applied to them. A comprehensive coding of land should facilitate easy referencing as well as define property boundaries, thus eliminating unnecessary disputes and litigations associated with urban land use in the city.

References

Areola O, Ofomata GEK (1978) Land use control and natural resource management in Nigeria. Geo J 2(5):429–436

Aribigbola A (2006) Public participation in planning in developing countries: the example of Akure, Ondo State, Nigeria. In: Monjur M (ed) Global built environment: towards an integrated approach for sustainability. Proceedings of GBEN 2006 conference held 11–12 Sept 2006, University of Central Lancashire, Preston, UK

Aribigbola A (2008) Improving urban land use planning and management in Nigeria: the case of Akure. Theoretical and empirical researches in urban management, year 3, number 9, 2008. Available online at http://um.ase.ro/no9/1.pdf

Aside CAL, Ezek PI, Nandi GU (2004) Land use and soil management situations in Nigeria: an analytical review of changes. USA J Outlook Agric 33(1):27–37

Barlowe R (1963) Land resource economics. Prentice-Hall Inc., Englewood Cliffs

Bohanna P (1966) Land and land tenure. In: Blebuyck D (ed) African agrarian systems. Published for the International African Institute by the Oxford University Press, London

Boserup E (1966) The conditions of agricultural growth. George Allen & Unwin Ltd, London, Chapters 1 & 2

Bruce JW (1988) Land tenure issues in project design and strategies for agricultural development in sub-Saharan Africa. LTC Research paper 128. Madison, Land Tenure Centre, University of Wisconsin, p 190

Daramola GA (2004) Competitiveness of Nigerian agriculture in a global economy: any dividends of democracy. Inaugural lecture series 36, delivered at the Federal University of Technology, Akure, on 2 Mar 2004

FAO (2001) Country report – Nigeria. FAO Corporate Document Repository

International Conference on Agrarian Reform and Rural Development (ICARRD) (2006) National report-Nigeria, Porto Alegre, 7–10 Mar 2006

Medugu IN (2006) Achieving sustainable agriculture in Nigeria: a land-use policy perspective. Tokyo Academic, Industry & Cultural Integration Tour 2006, 10–19 Dec 2006, Shibaura Institute of Technology, Japan

Olujimi BAJ (1993) The administration of physical planning in Nigeria: the Ondo state experience. Landsc Urban Plan 25:115–126

Omofonmwan SI, Osa-Edoh GI (2008) The challenges of environmental problems in Nigeria. J Hum Ecol 23(1):53–57

Osemeobo GJ (1991) Effects of common property resources utilization on wildlife conservation in Nigeria. Geo J 23(3):241–248

Ratcliff J (1976) Land policy: an exploration of the nature of land in society, The built environment. Hutchinson, London, pp 1–27

Saletti A (2009) The secrets of Nigerian agriculture, 25 Dec 2009 EzineArticles.co. Available online at http://ezinearticles.com/?The-Secrets-of-Nigerian-Agriculture!-Naija-Face&id=3478784

Umeh JA (1973) Compulsory acquisition of land and compensation in Nigeria. Sweet & Maxwell, London

Vink APA (1975) Land use in advancing agriculture. Springer, New York

Chapter 31
Soil Use Planning and Decelerating Crop Productivities: Policy Implications for Indian Punjab

Davinder K. Grover

Abstract Soil is the prime natural and economic resource of any nation. Soils are heterogeneous in composition due to difference in parent material, soil texture, physical and chemical properties, climatic conditions and occurrences on different topographic positions. Soils of the Indian Punjab have been developed on alluvium in flood plain (alluvial soils). Others are loamy, sandy, desert and kandi soils. These soils are different in their inherent soil fertilities, presenting options to cultivate various crops. The combination of major breakthrough in Mexican wheat and Filipino rice cultivars, availability of well-developed irrigation network, adequate marketing infrastructure and price support policy led to predominantly monoculture oriented rice-wheat rotation in the Indian Punjab. This has resulted in manifestation of several adverse effects on soil use efficiency and fast deceleration of crop productivities. To improve soil fertility, annually 1.3×10^6 t of nitrogen, 0.354×10^6 t of phosphorus and 0.039×10^6 t of potash fertilizers are added to soils. The use of chemical fertilizers in the state has risen from 0.213×10^6 t in 1970–1971 to 1.698×10^6 t in 2007–2008. Even though the soil fertility (macro- and micronutrients) is depleting continuously, calling for a pragmatic soil/land use planning for crop diversification is based on soil suitabilities in different areas of the state for particular crop(s).

Keywords Crop diversification • Crop productivity • Deceleration • Fertilizers • Indian Punjab

---

D.K. Grover (✉)
Agro-Economic Research Centre, Department of Economics and Sociology, Punjab Agricultural University, Ludhiana 141004, India
e-mail: dkg_59@rediffmail.com

S.A. Shahid et al. (eds.), *Developments in Soil Classification, Land Use Planning and Policy Implications: Innovative Thinking of Soil Inventory for Land Use Planning and Management of Land Resources*, DOI 10.1007/978-94-007-5332-7_31,

31.1 Introduction

Soils are the prime natural and economic resource of any nation. Soil fertility is the nutrient status or ability of soil to supply nutrients for plant growth. Whereas soil productivity is the capability of soil to produce a specified quantity of plant yield per unit area as well as the ability to produce sequence of crops under a specified management system.

Soils are formed on different parent materials under different climatic conditions over a period of time; therefore, they are heterogeneous in composition (texture, pH, salinity, fertility). Soils of the Indian Punjab developed on alluvium in the flood plain are alluvial soils – suitable for the cultivation of paddy, wheat, sugarcane and vegetables. Loamy soils are fertile and productive and intensively cultivated for wheat and paddy crops. Sandy soils are capable of producing cotton, citrus, oilseeds, wheat and fodder crops. Desert soils are suitable for the cultivation of cotton, moth, citrus, wheat, bajra and other kharif fodder. The kandi soils – highly eroded and less productive – are used for dry farming (Anonymous 2003). The soils due to different properties are heterogeneous in inherent fertilities and productivities presenting options to cultivate various and specific crops. In addition to soil properties, climatic variations are also important in crop diversification in the Punjab. Soils of the southwestern Punjab are calcareous, which include desert and sierozem soils. Soils of the central Punjab are different in texture ranging from sandy loam to clayey and hence are highly vulnerable to the effects of alkalinity and salinity. The alluvial soils are described as arid, brown or tropical arid brown soils. The soils in the Eastern Punjab are loamy to clayey. Based on physiographic variations, substratum composition, ground water quality, annual rainfall and pattern and soil moisture, the Indian Punjab is divisible into six distinct regions.

Five of these regions are further divisible due to variations in soil characteristics. The regions thus delineated have homogeneous agricultural problems and potentials and hence provide stronger base in the soil planning for agricultural development.

The soils of submountain undulating region in the north and south are medium to heavy in texture, and in the central area they are light to medium in texture. Important kharif season crops of this region are maize, groundnut and rabi season crops are sugarcane and wheat. In the undulating plain region, the soils are medium to heavy in texture with paddy as the main kharif crop following by maize, sugarcane and groundnut, while wheat is important crop in the rabi season following by oilseeds crop. In the central plain region, the soils are medium to heavy in texture with mild to serious alkali problems. The paddy, maize, cotton, groundnut and sugarcane are the suitable crops on such soils in the kharif season. During rabi season, wheat, gram and barley are the suitable crops in this region.

In the western plain region of the state, the soils are medium to heavy in texture with serious alkali and waterlogging problems. Cotton is the most suitable crop of the region with paddy in certain pockets during kharif, and wheat is the principal rabi crop following by oilseeds, gram and barley. The western region has medium to heavy soil texture in north and light to medium in south. The soil in north suffers from alkali problems and waterlogging whereas in the south, soils are suffering from shifting sand dune and wind erosion. Cotton is the dominant crop following by paddy in few selected pockets during kharif season, while wheat, gram, oilseeds and barley in rabi season in this region. The soils of flood plain region are fertile with medium to heavy texture in most part. The dominant crop enterprise in the region throughout the state is paddy in kharif and wheat in the rabi season. The soils in the state are heterogeneous between various agroclimatic regions with different soil fertility level suitable for cultivating various and specific crops, yet rice and wheat are dominant crops of the state, and over the years traditional crops have been vanished from the crop map of the state.

31.2 Objectives

The objectives of this chapter are as follows: (1) to examine the radical changes taken place in the cropping pattern over the years in the favour of rice and wheat and its impact on the state agriculture in decelerating crop productivities and (2) to review various agroclimatic regions delineated on the basis of soil characteristics, physiographic variations, substratum type and water quality and amount of rainfall and moisture, which are more suitable for specific crop rotations, and to trace various policy implications and accordingly formulate recommendations for sustainable agriculture in the state.

31.3 Materials and Methods

To accomplish the above-stated objectives, the study has been completed using secondary data from various sources. To examine the temporal changes in the cropping pattern, area under major crops has been collected from various publications of the state government. To establish decelerating crop productivity trends, time-series data on area and productivity of major crops grown in different districts in various agroclimatic regions of the state from 1970–1971 through 2007–2008 were collected from various issues of the Statistical Abstract of Punjab. The data so collected were divided in to four periods, namely, period I, post-Green Revolution phase (1970/1971–1985/1986); period II, stabilized Green Revolution phase (1986/1987–1999/2000); period III, decelerating productivities phase (2000/2001–2007/2008); and period IV, whole Green Revolution phase (1970/1971–2007/2008). Based on the variations in physiographic, substratum composition and ground water quality, amount of rainfall and moisture, Punjab is divisible into six agroclimatic regions (Fig. 31.1). Five of these regions are further divisible on the basis of variations in soil characteristics. The soil information for region delineation has been based on the report entitled agroclimatic regions of Punjab, prepared jointly by the multidisciplinary team including agronomists, soil scientists, soil and water engineers, agricultural economists and extension educationists of the Punjab Agricultural University, Ludhiana. Land utilization/crop pattern from 1960 to 2008, fertilizer consumption in terms of NPK and emergence of macro/micronutrients deficient soils over the years in the state have been extracted from various secondary

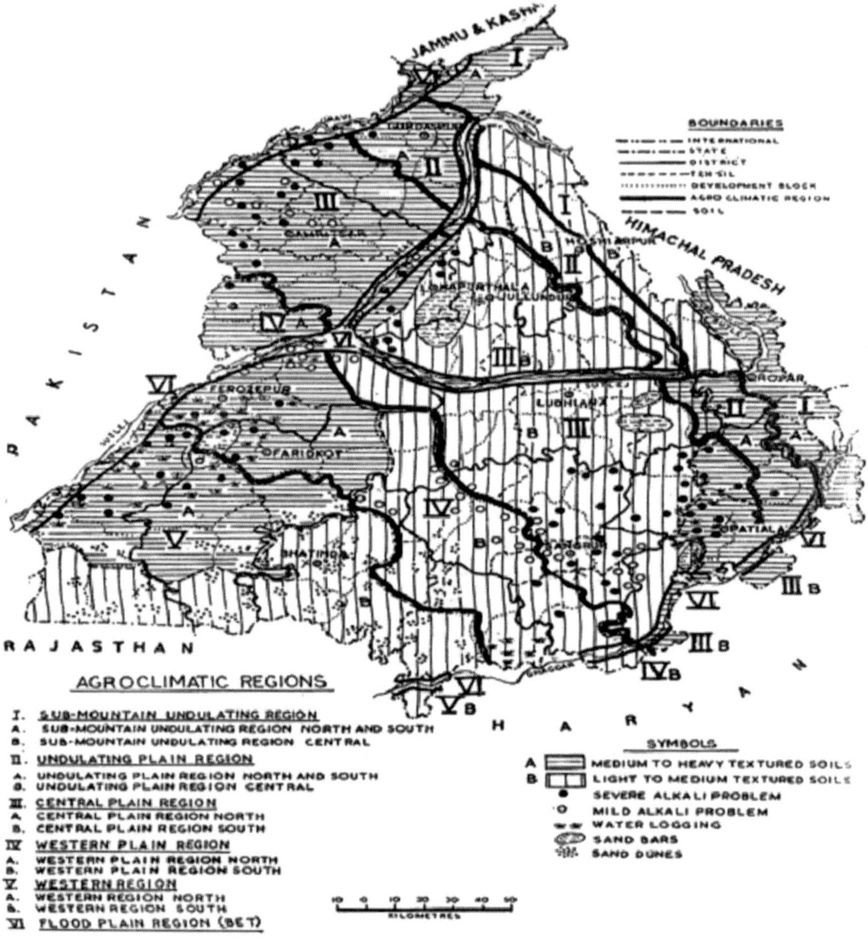

Fig. 31.1 Delineation of agroclimatic regions of Punjab

sources published by the state government. The data was analysed using simple statistical tools for mean, percentage, compound growth rates and student's *t*-test to make the results more explicit. The area and productivity trends in various districts/agroclimatic regions of the state have been depicted from graphical presentations. The compound growth rates were calculated by employing the power function as given below:

$$Y = abt$$

Where Y = dependent variable

a = Constant term

$b = (1 + r)$, regression coefficient

$r = (b - 1) * 100$, compound growth rate in percentage

t = time variable

31.4 Results and Discussion

31.4.1 Emergence of Monoculture

Different soils are suitable for cultivating various specific crops. Owing to the major technological breakthrough in terms of Mexican wheat and Filipino rice cultivars along with the availability of well-developed irrigation network, adequate marketing infrastructure and price support policy led to predominantly monoculture rice-wheat rotation in the state. A rapid shift has been observed in the cropping pattern that favoured rice and wheat crops only. Consequently, the area under rice which was only 6.06% of the cultivated area in kharif season during 1960–1961 increased to about 63% in 2007–2008 at the cost of the area under crops like maize, millets, groundnut and recently cotton. Similarly, in rabi season, cultivated area under wheat increased from 37.33% in 1960–1961 to about 84% in 2007–2008 at the expense of the area under oilseeds, gram and other pulses (Table 31.1). This shift in crop pattern in favour

Table 31.1 Temporal shifts in the cropping pattern in Indian Punjab, 1960–1961 to 2007–2008 (% cultivated area)

Crop	1960–1961	1970–1971	1980–1981	1990–1991	1997–1998	2007–2008
Paddy	6.06	9.62	28.22	47.77	54.20	62.53
Maize	8.72	13.69	7.25	4.34	3.99	3.68
Bajra and jowar	3.43	5.23	1.66	0.28	0.11	0.14
Groundnut	1.78	4.29	1.98	0.27	0.21	0.11
Cotton	11.92	9.79	15.49	16.61	17.30	14.51
Sesamum	0.21	0.36	0.40	0.43	0.34	0.23
Sugarcane	3.55	3.17	1.69	2.38	3.07	2.37
Kharif pulses	0.67	0.81	1.39	1.74	1.55	0.60
Wheat	37.33	56.72	67.10	77.60	78.52	83.56
Barley	1.77	1.41	1.55	0.87	0.94	0.45
Gram	22.35	8.84	6.16	1.42	0.32	0.09
Rapeseed and mustard	2.85	2.54	3.24	1.64	1.67	0.98
Linseed	0.08	0.07	0.05	0.02	–	0.005
Lentil	0.81	0.32	0.48	0.23	0.11	0.10
Sunflower	–	–	–	0.12	1.33	0.37
Tobacco	–	–	1.66	–	–	–
Potato	0.24	3.03	0.95	0.55	1.24	1.82
Chillies	–	0.42	0.31	0.05	0.09	0.03
Other vegetables	0.01	0.57	0.56	0.73	2.02	0.81
Fruits	1.12	1.23	0.69	1.64	2.15	1.36
Fodders & other crops	23.29	17.97	20.51	19.16	18.07	17.89
Net cultivated area ('000 ha)	3,750	4,053	4,191	4,218	4,204	4,174

Source: Based on the data collected from various issues of Statistical Abstract of Punjab

Table 31.2 Consumption of chemical fertilizers in Indian Punjab: 1960–2008 ('000 Metric tonnes)

Year	N	P	K	Total NPK	kg ha^{-1} cropped area	
					Punjab	India
1960–1961	5	0	0	5	1.1	–
1970–1971	175	31	7	213	38.2	–
1980–1981	526	207	29	762	112.7	–
1990–1991	877	328	15	1,220	162.7	–
2000–2001	1,008	282	24	1,314	165.3	88.0
2001–2002	1,070	308	29	1,407	177.2	91.5
2002–2003	1,111	299	31	1,441	184.1	84.8
2003–2004	1,170	335	38	1,543	195.2	89.8
2006–2007	1,299	354	39	1,692	215.2	90.4
2007–2008	1,316	344	38	1,698	215.8	92.2

Source: Statistical Abstracts of Punjab: Various issues; – = data not available

of wheat and rice is mainly due to these crops being more stable and remunerative. However, the two crop systems repeated year after year on intensive scale made the agro-ecosystem of the state extremely fragile in the context of water depletion, pest and crop diseases, soil health, human health and overall living environment. Agricultural growth rate in Punjab has also slowed down from 5.94% in 1970s, 5.16% in 1980s, 2.4% in 1990s and 0.83% in 2000s due to stagnation in the productivity levels of major crops, namely, wheat, rice, maize, oilseed and cotton in the state.

31.4.2 Deteriorating Soil Fertility

Due to high nutritional requirements of rice and wheat, the major crop rotation in the state has exhausted soil nutrients. Resultantly, Punjab state which has just around 3% of cultivated area accounts for about 10% of total chemical fertilizer consumption in the country. The state is adding annually about $1,316 \times 10^3$ t of nitrogen, 344×10^3 t of phosphorus and 38×10^3 t of potassic fertilizers to soil. The use of chemical fertilizers in the state has gone up many times from 213×10^3 t in 1970–1971 to $1,698 \times 10^3$ t in 2007–2008 (Anonymous 2008). The nutrient consumption has been recorded as 215.8 kg ha^{-1} cropped area in Punjab as against the national average of only 92.2 kg ha^{-1} cropped area (Table 31.2). Most of the basic elements have been extracted from soil than added. The fertility of Punjab soils has diminished over the years with deficiency in nitrogen and phosphorus. This was soon followed by deficiency of zinc during 1970s and other nutrients like potash, manganese and sulphur during 1980s. Above all, the deficiency of copper was also visualized since 1990s (Table 31.3).

Thus, it is obvious that present farming system is not sustainable to maintain nutrient status of soils leading to deficiency of all micro- and macronutrients. Maintenance of soil fertility is essential to sustain agricultural production. The soils

Table 31.3 Soil nutrients deficiency in Punjab during various decades

Decade	Deficiency
During 1960s	Nitrogen
During 1970s	Nitrogen + Potash + Zinc + Phosphorus
During 1980s	Nitrogen + Potash + Zinc + Phosphorous + Manganese + Sulphur
During 1990s	All above + Copper
During 2000	All above

Source: Department of Soils, Punjab Agricultural University, Ludhiana, India

of the Punjab are generally low in N content, low to medium in P and medium to high in K except in kandi belt which has low to medium K content (Brar and Chhibba 1994). In the past two to three decades, intensive agricultural practices have exerted huge pressure on soils and resulted in steady decline in soil fertility (micro- and macronutrients). Both rice and wheat have high nutritional requirements, and the double cropping of this system has heavily depleted the nutrient contents of soil. The rice-wheat sequence yields 7 t ha^{-1} of rice and 5 t ha^{-1} of wheat; this removes more than 300 kg N, 30 kg P and 300 kg K ha^{-1} from the soil. Even with the recommended rate of fertilization in this cropping pattern, a negative balance of primary nutrients still exists (Benbi et al. 2006). Moreover, partial factor productivity of NPK in Punjab has also dropped from 80.9 in 1966–1967 to 16.0 in 2003–2004. Hence, farmers in the state have been applying higher and higher doses of major nutrients, especially nitrogen for sustaining adequate production levels (Department of Agriculture, Govt. of Punjab: Adapted from Hira et al. 2004).

31.4.3 Red Alert on Soil Fertility and Soil Productivity

The productive gains in the rice-wheat cropping pattern system have slowed down and reached to a plateau. Advisory Committee on "Agriculture Policy and Restructuring" in 2002 suggested suitable production pattern adjustments replacing 1×10^6 ha area under rice-wheat rotation with other crops like oil seeds and pulses with a view to conserve the natural resources of the state in terms of soil fertility degradation and protecting ground water (Johal and Sidhu 2002). A study by International Food Policy Research Institute (IFPRI) has alerted the farm sector in the Punjab and warned that Punjab should take immediate action to diversify crops by reducing wheat and rice to avoid India of turning to a begging bowl by 2030 (Gulati et al. 2006). Agriculture of Punjab has lost its place among the fastest growing economy in the country and surpassed by other states including Karnataka, Madhya Pradesh, Maharashtra, Rajasthan and West Bengal. In order to maintain state's agriculture sector, it is essential to make significant changes to regain its leadership.

Land is the fundamental basis for most of the human or natural activities and is one of the major natural resources on the earth. Agricultural productivity is entirely dependent on the availability of suitable land and management. The availability of vast tracts of fertile land was a major factor responsible for the agricultural success of Punjab. Further, 83.4% (4,200 × 10^3 ha) of the total land is under agricultural activities with cropping intensity of 189%. This is double than the average percentage (43%) of land under agriculture in the country as a whole. The State has the highest percentage of net sown area in India. The continuous emphases on increasing the cropping intensity and yield and extensive agriculture practices coupled with exhaustive cropping patterns over the years have severely affected soil physical and chemical health and fertility status in the state. Nutrient deficiency, loss of fertility and decline in organic matter content are the serious challenges in the agriculture sector of Punjab. As a consequence of declining soil, quality area under total pulses and oilseeds has been reduced sharply. The area under total pulses recorded a sharp decline from 19% in the early 1960s to 0.45% of total gross cropped area of the state. The total area under oilseeds also declined by more than 70% in the last over four decades. This includes groundnut, lentils and guar. The pulses are the leguminous crops which have special characteristics of roots having nodules responsible for the conversion of atmospheric nitrogen in the plant roots (biological nitrogen fixation) which is used by plants.

It is important to state that after the Green Revolution, the farmers of Punjab abandoned their traditional cropping practices in the favour of the government-supported wheat-rice cropping system. However, this has caused soil degradation including nutrient imbalance, depletion of ground water table (up to 75 cm $year^{-1}$), soil and water contamination due to the abuse of pesticides and fertilizers that has created several environmental and health hazards, besides creating economic and social imbalances.

31.4.4 Decelerating Crop Productivities

The production of crops like wheat, rice, cotton and oilseeds and maize in the state has increased significantly over the last few decades, especially in the post-Green Revolution period. Wheat and rice have been playing a major role in pushing up agricultural production in the state over the years. The production of wheat which was 5,145 × 10^3 metric tonnes (MT) in 1970–1971 rose to 15,720 × 10^3 MT during 2007–2008 registering an increase of about 185%. Similarly, the production of rice, which was 688 × 10^3 MT in 1970–1971, increased to 10,489 × 10^3 MT in 2007–2008 showing an increase of around 140%. The production of oil seeds and pulses from 233 × 10^3 MT and 305 × 10^3 MT in 1970–1971 to about 90 × 10^3 MT and 25 × 10^3 MT in 2007–2008 registering decrease in production by 60 and 90%, respectively. The production of other cereal crops in the state like maize, bajra, barley and jowar has been decreasing sharply mainly due to decrease in the area under these crops.

31.4.5 *Decelerating Rice Productivities*

Temporal analysis revealed that rice productivity in the state grew at 1.60% per annum in the whole Green Revolution phase (1970/1971–2007/2008). The compound annual growth rate of rice productivity in period I – post-Green Revolution phase (1970/1971–1985/1986) was recorded as high as 4.09%. The rice yield growth decelerated in the period II – stabilized Green Revolution phase (1986/1987–1999/2000) with annual compound growth rate of just 0.39%. Similarly, in all agroclimatic regions-districts, the productivity growth has significantly fallen in period II as compared to period I. In the districts, where rice has been growing for quite a long period, the negative productivity growth has been observed, due to rice being macro/micronutrient exhaustive crop. Despite the continuous increase in the use of NPK and micronutrients, the soils have not been able to maintain their fertility level essentially required to improve yield level. The districts with minus trend in rice productivity in period II were Rupnagar, Ludhiana, Patiala, Faridkot and Sangrur. The rice yield growth rate that ranged between about 54% in Faridkot and about 1.80% in Amritsar during the period I, decelerated to just −1.07% in Ludhiana with maximum 1.69% in Amritsar. During period III (2000/2001–2007/2008), though owing to varietal improvement in paddy cultivation, the growth rates improved, yet remained far lower than the average yield growth rate achieved during the period I (Table 31.4). Rice acreage and productivity trends in the state have been depicted in Fig. 31.2.

31.4.6 *Decelerating Wheat Productivities*

An alarming deceleration and stagnation has been observed in wheat productivity in the state. The annual compound growth rate for wheat productivity was 2.59% in period I – post-Green Revolution phase (1970/1971–1985/1986), reduced to just 0.39% in the period II – stabilized Green Revolution phase (1986/1987–1999/2000) and 0.38% during period III (2000/2001–2007/2008) with annual compound growth rate of just 1.13% during whole Green Revolution phase. Consistent wheat productivity deceleration has been experienced in the districts Rupnagar (3.88% in period I, 2.50% in period II and only 0.85% in period III), Hoshiarpur (2.38% in period I, 1.71% in period II and only 0.99% in period III), Amritsar (3.02% in period I, 2.85% in period II and minus 1.09% in period III), Kapurthala (2.47% in period I, 1.94% in period II and minus 0.44% in period III), Jalandhar (2.60% in period I, 2.08% in period II and minus 0.88% in period III) and Ludhiana (1.66% in period I, 1.58% in period II and minus 1.08% in period III). Similarly, the other districts (Ferozepur, Faridkot and Sangrur) where the wheat productivities grew in the range of 4–55% during period I have eventually ended up with negative growth rates in the wheat productivities during period III (Table 31.5). Wheat acreage and productivity trends in the state have been depicted in Fig. 31.3. The consistent decline in the wheat productivity in the state has raised a serious question on the sustainability of the farm sector in the state.

Table 31.4 Trends (annual compound growth rates) in area and productivity of rice in various agroclimatic regions/districts of Punjab

State/Region/District	Period I 1970/1971–1985/1986	Period II 1986/1987–1999/2000	Period III 2000/2001–2007/2008	Period IV 1970/1971–2007/2008
Percent trends in area				
Punjab	11.40**	2.89**	0.478*	5.16**
Undulating plain region				
Rupnagar	12.65**	4.35**	−5.500*	5.79**
Gurdaspur	4.99**	1.56**	0.28NS	2.45**
Hoshiarpur	2.64**	0.59NS	−1.08NS	1.70**
Central plain region				
Amritsar	7.21**	1.98**	−9.76*	2.83**
Kapurthala	9.19**	0.99**	1.04**	3.84**
Jalandhar	15.66**	−1.37NS	1.93**	5.16**
Ludhiana	29.91**	2.01**	0.97**	9.41**
Patiala	10.60**	−0.52NS	−1.05*	2.96**
Southwestern region				
Ferozepur	9.60**	1.64**	−0.04NS	3.44**
Faridkot	46.21**	−2.87NS	0.93NS	7.64*
Bathinda	30.54**	5.30*	−1.27NS	13.55**
Sangrur	24.63**	3.74**	−4.04*	9.30**
Percent trends in productivity				
Punjab	4.09**	0.39NS	1.90**	1.60**
Undulating plain region				
Rupnagar	6.69**	−0.81*	2.01**	2.11**
Gurdaspur	3.70**	1.15*	1.41**	1.56**
Hoshiarpur	3.39**	1.61*	0.68NS	1.62**
Central plain region				
Amritsar	1.80*	1.69*	0.36NS	1.19**
Kapurthala	2.96**	0.80NS	1.96*	1.33**
Jalandhar	4.00**	0.14NS	0.39NS	1.17**
Ludhiana	4.45**	−1.07NS	2.09*	1.27**
Patiala	4.72**	−0.71NS	3.41**	1.73**
Southwestern region				
Ferozepur	2.61**	1.51*	2.17**	1.54**
Faridkot	53.67*	−0.31NS	2.31*	9.22*
Bathinda	6.09**	0.62NS	2.79**	2.04**
Sangrur	6.30**	−0.20NS	3.02**	2.08**

**Significant at 1%; *Significant at 5%; NSNot Significant

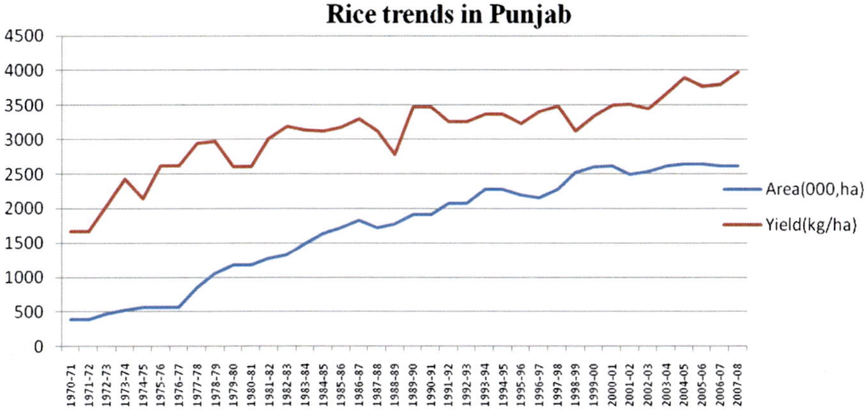

Fig. 31.2 Rice area and productivity trends in Indian Punjab, 1970/1971–2007/2008

Table 31.5 Trends (annual compound growth rates) in area and productivity of wheat in various agroclimatic regions/districts of Punjab

State/Region/District	Period I 1970/1971–1985/1986	Period II 1986/1987–1999/2000	Period III 2000/2001–2007/2008	Period IV 1970/1971–2007/2008
Percent trends in area				
Punjab	2.59**	0.39**	0.38*	1.13**
Undulating plain region				
Rupnagar	1.72**	0.88**	−6.41*	0.67*
Gurdaspur	2.50**	0.44^{NS}	1.23*	1.10**
Hoshiarpur	1.51**	−1.84**	0.49^{NS}	−0.14^{NS}
Central plain region				
Amritsar	2.40**	0.34*	−11.238*	0.38^{NS}
Kapurthala	3.82**	−0.08^{NS}	−0.26^{NS}	1.22**
Jalandhar	−1.01^{NS}	−2.87**	0.01^{NS}	−0.57^{NS}
Ludhiana	0.93**	−0.28*	0.04^{NS}	0.11*
Patiala	1.98**	−2.16**	−1.44*	−0.13^{NS}
Southwestern region				
Ferozepur	1.38^{NS}	0.27^{NS}	0.63*	0.78**
Faridkot	42.88*	−11.76**	0.88**	3.01^{NS}
Bathinda	2.25*	−4.28**	0.23^{NS}	0.12^{NS}
Sangrur	2.33**	0.32*	−4.58*	0.77**
Percent trends in productivity				
Punjab	2.98**	2.73**	−0.26^{NS}	2.22**
Undulating plain region				
Rupnagar	3.88**	2.50**	0.85^{NS}	2.19**
Gurdaspur	2.90**	3.26**	−1.07^{NS}	2.36**
Hoshiarpur	2.38**	1.71*	0.99^{NS}	2.33**

(continued)

Table 31.5 (continued)

State/Region/District	Period I 1970/1971–1985/1986	Period II 1986/1987–1999/2000	Period III 2000/2001–2007/2008	Period IV 1970/1971–2007/2008
Central plain region				
Amritsar	3.02**	2.85**	−1.09NS	2.20**
Kapurthala	2.47*	1.94*	−0.44NS	2.28**
Jalandhar	2.60**	2.08**	−0.88NS	2.00**
Ludhiana	1.66**	1.58*	−1.08NS	1.51**
Patiala	4.35**	1.73**	−0.21NS	2.32**
Southwestern region				
Ferozepur	4.02**	2.00**	−0.26NS	2.11**
Faridkot	55.10*	2.89**	−1.13NS	10.40*
Bathinda	2.89**	3.27**	0.73NS	2.33**
Sangrur	3.22**	1.74**	−1.08NS	2.19**

**Significant at 1%; *Significant at 5%; NSNot Significant

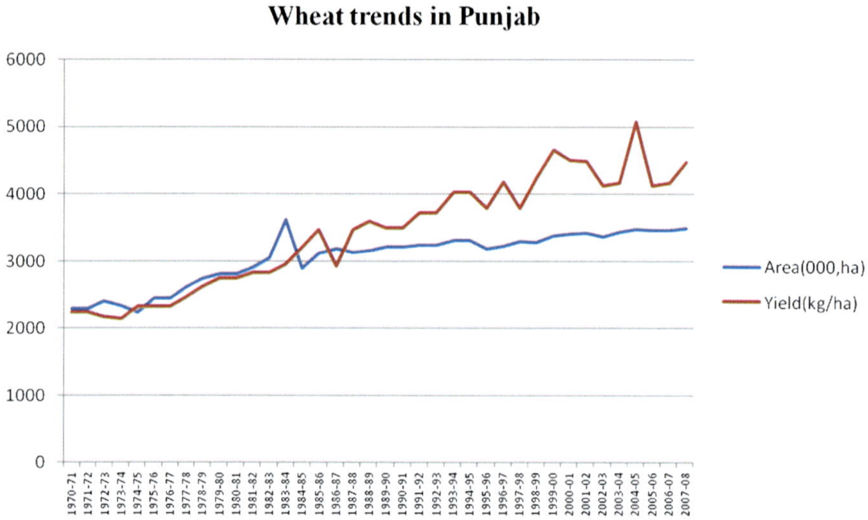

Fig. 31.3 Wheat area and productivity trends in Indian Punjab, 1970/1971–2007/2008

31.5 Conclusions

With the emergence of monoculture of rice-wheat rotation during the Green Revolution for producing more and more food grains, the natural soil and water resources have been degraded and depleted. If this trend continues, it will be impossible to sustain the production level achieved so far. The soils of Punjab have been degraded due to over-exploitation to achieve maximum food production. Excessive use of chemical fertilizers, insecticides and pesticides has destroyed the soils physical and chemical properties that have led to decreased soil water-holding capacity, friable and loose structure resulted into poor fertility and production capacity of the soils. The extensive rice cultivation has been identified as the major problem for productivity stagnation or even productivity deceleration of rice as well as wheat in the state. The rice, a nontraditional crop in the state has been extensively grown that has depleted underground water resources and exhausted soil nutrients status. The land suitable for cotton cultivation in the southwestern region of the state has also been shifting for rice cultivation mainly because of its better profitability and easy marketing, encouraged by the government through consistent hike in minimum support price and effective procurement. Therefore, there is an urgent need to reorient soil use planning in the state keeping in view the declining soil fertility/soil health trend, so that the general deceleration trends emerged in state farm sector in general and crops productivities in particular may be reversed without any further harm. District-block wise crop adjustment programs need to be prepared taking into account crop diversification and cropping intensity better suited to the soils fertility level in a particular pocket.

References

Anonymous (2003) Department of land resources, Wastelands Atlas of India. Prepared by Department of Land Resources, Government of India

Anonymous (2008) Economic and statistical advisor, statistical abstract of Punjab. Published by Directorate of Economics and Statistics, Government of Punjab

Benbi DK, Nayyar VK, Brar JS (2006) The green revolution in Punjab: impact on soil health. Indian J Fertil 2(4):57–66

Brar JS, Chhibba IM (1994) N, P and K status of Punjab soils. Indian J Ecol 21(1):34–38

Gulati A, Cummings R Jr, Joshi PK (2006) Withering Punjab agriculture. International Food Policy Research Institute (IFPRI), India, p 162

Hira GS, Jalota SK, Arora VK (2004) Efficient management of water resources for sustainable cropping in Punjab. Department of Soils, Punjab Agricultural University (PAU), Ludhiana

Johal SS, Sidhu RS (2002) Future of agriculture in Punjab. Centre for Research in Rural and Industrial Development, Chandigarh

Chapter 32
Land-Use Planning for Sustainable Development Using Geoinformatics: Policy Implications for Drylands

R.B. Singh and Dilip Kumar

Abstract The land is the basic natural resource of the planet earth. Continuing population growth has definitely increased food, fuel and fodder demands, and this has put pressure on the judicious use of land resources. Therefore, proper and rational planning is essential to achieve long-term benefits from this resource. The proper planning for conservation, optimum utilisation and management of these resources is not only vital for sustenance of life but also to meet the growing needs of agriculture, expanding urbanisation, increasing industrialisation and for overall socioeconomic development of the country. The Agenda 21 ratified by over 170 nations at the Earth Summit in Rio de Janeiro in 1992 highlights that the land-use planning plays a key role in natural resource management. The science of sustainable development requires setting up of strategies based on the accurate assessment of the earth's carrying capacity. This requires an integrated approach towards harnessing land resources after taking into account the vulnerable environmental condition. In this study, a dryland part of middle Ganga plain, known as Son-Karamnasa interfluve, in India is selected. Based on the results, alternative land-use systems and integration of livestock enterprises with the agriculture system have been suggested for land resources management. The objective of this chapter is to present a land-use plan to increase the productivity of land for sustainable development. The present study is likely to contribute required inputs to help policymakers to improve the socioeconomic and environmental conditions of the drylands.

R.B. Singh (✉)
Department of Geography, Delhi School of Economics, University of Delhi,
New Delhi 110 007, India
e-mail: rbsgeo@hotmail.com

D. Kumar
Department of Geography, Shivaji College, University of Delhi,
New Delhi 110 027, India
e-mail: dilipgis@gmail.com

S.A. Shahid et al. (eds.), *Developments in Soil Classification, Land Use Planning and Policy Implications: Innovative Thinking of Soil Inventory for Land Use Planning and Management of Land Resources*, DOI 10.1007/978-94-007-5332-7_32,

Keywords Drylands • Geoinformatics • Land-use planning • Policy implication • Sustainable development

32.1 Introduction

Land resources are the important base for biomass production and constitute the core of ecological systems. The continuing pressure of human and livestock population growth and ever intensifying development demands caused great stress on the existing land resources as these cannot be easily extended to new areas. Therefore, there is an increasingly urgent need to match land types and land uses in the most possible rational way to optimise production and satisfy the diverse needs of the society while at the same time conserving and maintaining fragile ecosystems for future generations and sustainable development. This requires an integration of production and natural resources conservation strategies to enhance productivity and to maintain for long-term basis. Land-use planning is the systematic assessment of land and water potential, alternatives for land use and economic and social conditions in order to select and adopt the best land-use options. Its purpose is to select and practise land uses that will best meet the needs of the people while safeguarding the resources for the future. To achieve this it is essential to have authentic information about land resources, a prerequisite for proper land-use planning (Food and Agricultural Organisation 1993). Amongst others, the geoinformatics is one of the best technologies providing the reliable land resources information by combination of remote-sensing, global positioning system (GPS) and geographical information system (GIS) tools. The integration of these three tools provides accurate information leading to promote sustainable development of a region.

32.2 Materials and Methods

32.2.1 Description of the Study Area

In this study a dryland part of middle Ganga plain, known as Son-Karamnasa interfluve, in India is selected. The study area has a well-defined natural boundary, stretching between the river Son in the east and south, river Karamnasa in west and the river Ganga in north, covering an area of about 1,098,393 hectares (ha). The area is situated in four districts administration, i.e. Bhojpur, Buxar, Kaimur (Bhabua) and Rohtas of Bihar state in India. The physiography of Son-Karamnasa interfluve is broadly divided into alluvial plain and plateau. The alluvial plain is productive land for agriculture. Almost all the plain area is conducive to holding water and producing good crops. It receives silt deposits from the Ganga almost every year and is extremely fertile. This region is considered the best wheat growing area in the state. The plateau and hilly areas are rich in mixture of coarse sandy soils and stone pieces and are not considered suitable for food crops abundantly. Soil is the basic determinant of land use in any region. The soil pH, texture and depth to hardpan or water table are key soil features that determine suitability of soil types for land use. The Indrapuri barrage on the river Son enhances has played a key role in increasing agricultural prosperity of the southern part of the Bhojpur plain. There are three underground water sources, i.e. surface well, deep-tube well and springs. The Rohtas plateau is a relatively more dissected and least-developed region. The foothill plains are mostly used for summer pastures. The Kaimur limestone ($CaCO_3$) and Amjhor pyrites (FeS_2) are precious mineral resources providing opportunities for economic development (Singh 1989).

The months of April and May are extremely hot, monsoon sets in by the end of June to end of September. The cold weather begins from the month of November and remains till March. The average rainfall in the normal condition, recorded during monsoon period is in the proximity of 1,000 mm. Other rainy months are October, January and February. The irrigated and nonirrigated areas except the hills and forest are being exploited for cultivation. Rice, wheat, barley, grams and pulses are the main crops of the districts. The winter rice forms the greater part of this crop, and the central part of the study area is called the 'rice bowl'. The study area is well connected with road and railways to all the major towns in the state as well as country. According to the Census of India (2001), the total population of the study area is 7.37×10^6 that is 8.9% of the total population of Bihar state. The population density is 696 persons km^{-2}.

32.2.2 Procedural Matters

The physical parameters like physiography, geomorphology, climate, soil characteristics and slope are used to evaluate the land capabilities of the study area. The Indian Remote Sensing Imagery IRS-P6 (RESOURCESAT-1), LISS III satellite data, Path 103 and 104 and Row 54, dated 31 October 2004 and 5 November 2004 for kharif season (summer crop); 28 February 2005 and 5 March 2005 for rabi season (winter crop) season are used for land-use/cover mapping along with survey of India topographical sheet. The digital image classification procedures were used to classify multispectral pixels into different land-use/cover classes in ERDAS Imagine software. The critical environmental issues are collected from the NBSSLUP. This plan has taken into consideration the contemporary technology and the resources, climate and terrain parameters. Generation and Integration of various thematic maps like soil, slope, land use/cover, hydro-geomorphology and groundwater potential coupled with agroclimatic and socioeconomic data was made, and suitable combination of practices has been suggested for various land parcels in GIS environment by using ArcGIS software (Fig. 32.1). In the development plan, existing land use/cover has been taken as a base for potential land-use

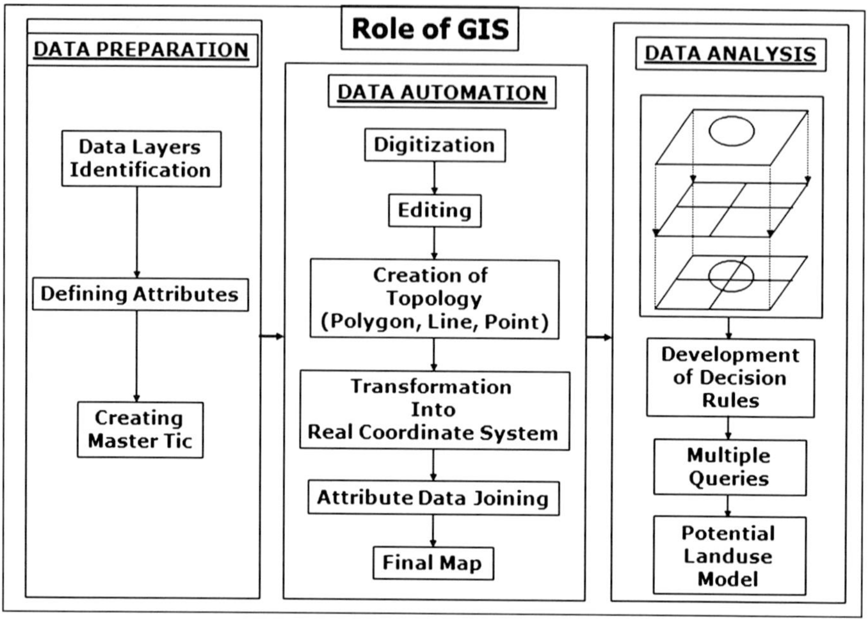

Fig. 32.1 Flow chart of GIS application

practices. In order to understand the cause and effect in respect of problems and potential of the resources, the individual and combination of thematic maps were studied to make the spatial assessment of the relative variations of the resource potentials and a wide range of terrain conditions and the logical association of various parameters in the thematic maps. The combination of different thematic maps resulted into a composite map having all the parameter of land resources. This composite map is used in generating the decision rule for optimum land-use planning in the study area (Fig. 32.2).

32.3 Results and Discussion

The total area of the Son-Karamnasa interfluve is 1,098,393 ha. An area of 52,059 ha is built up and water bodies like river, pond and canal constituting approximately 4.7% of the study area, and these areas cannot be used for the alternate land use. A land-use model has been prepared for an area of 1,046,333 ha constituting 95.3% of the study area. In land-use planning, three land aspects are used, i.e. agricultural land, forest land and wasteland. Amongst these, agricultural land-use planning is the most important aspect for increasing crop production, soil conservation as well as sustainable development. Proper crop management controls soil erosion, increases water infiltration rate, improves soil fertility and moistures retention capacity, which would improve land productivity. Insecticides, pesticides and other related disease

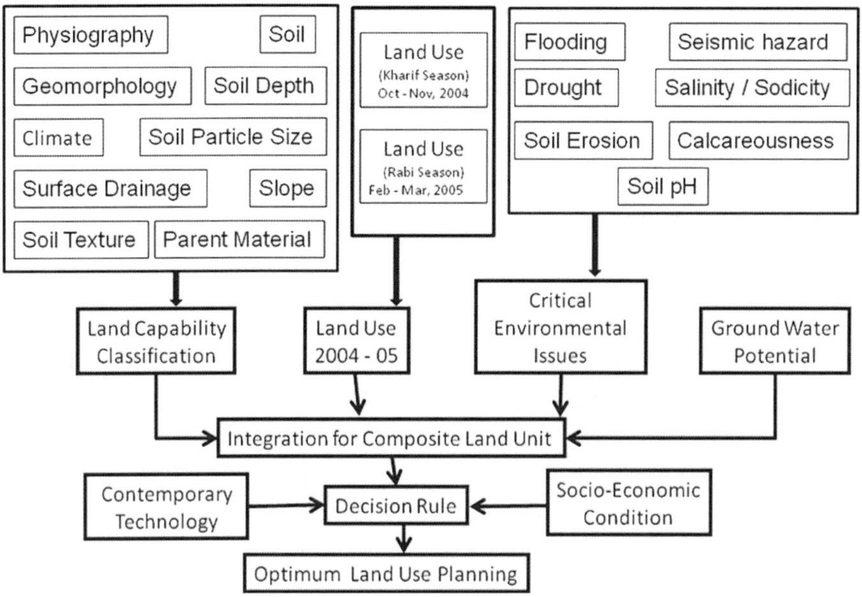

Fig. 32.2 Flow chart of land-use planning

control operations can be accomplished as needed. Forest land is important as it provides forest resources, controls land degradation and conserves wildlife and environment. Due to the lack of proper management practices, many lands are not properly used and hence are not providing expected benefits to the human beings. There are various methods and tools to make decision about land uses; it is also well known that land use is the function of geo-environmental condition, available technology and human resources. Amongst these, geo-environmental condition is most important factor. According to these factors, the optimum land use has been suggested (Figs. 32.3 and 32.4).

32.3.1 Double Crop/Double Crop with Special Effort

The continuing population growth pressure and subsequent increasing multiple demands have necessitated the optimum use of land. To meet these demands, double crop is recommended in areas where the single crops are grown either as rabi or kharif, irrigation facility is available or the underground water quality is safe for irrigation purpose. The multiple criteria analysis reveals that 728,999 ha of land has the capability for double crops constituting 66.4% of the study area. Amongst these, 624,025 ha of land constitutes 56.8% of the study area, which can be used as double crop without any limitation. The watershed-wise analysis reveals that more than

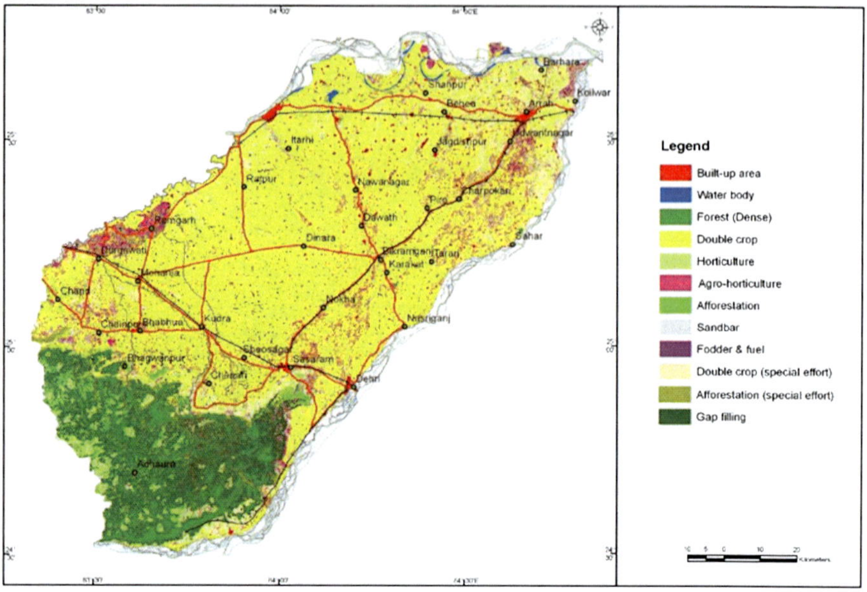

Fig. 32.3 Map showing distribution of different optimum land uses

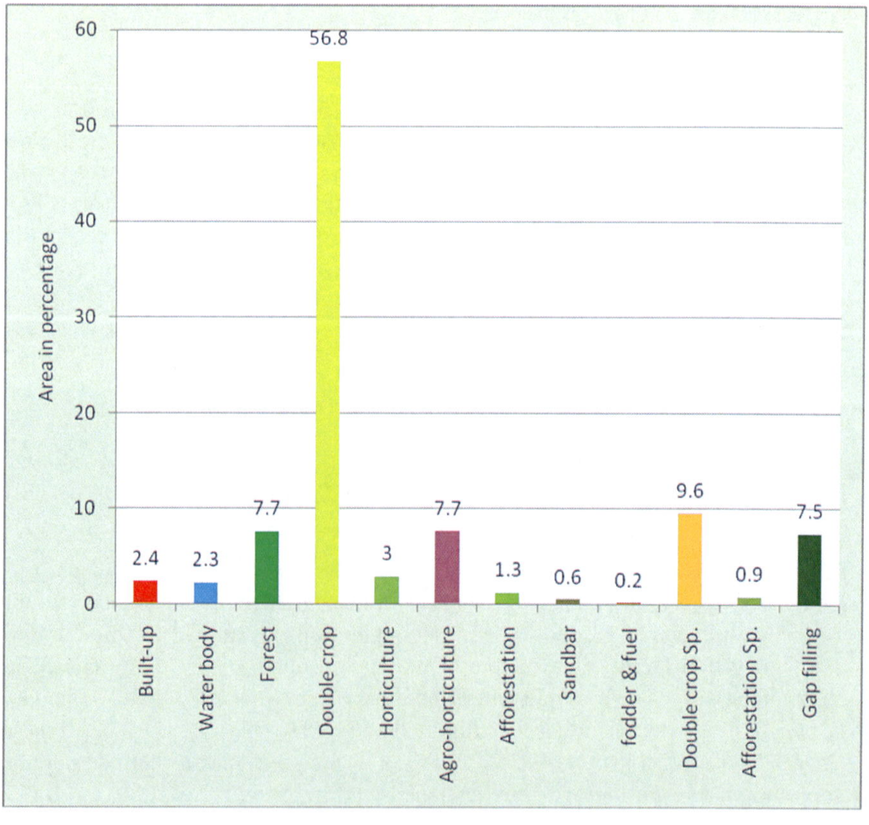

Fig. 32.4 Percent area of optimum land uses in difference categories

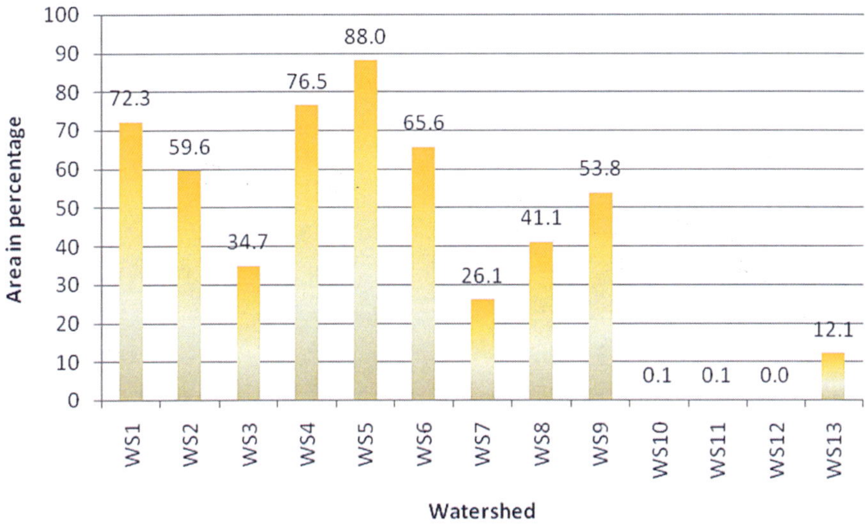

Fig. 32.5 Watershed-wise potential area for double crop

50% of the watershed area in WS1, WS2, WS4, WS5, WS6 and WS9 are capable of double crop without any limitation, and the maximum area (88%) is in WS5 (Fig. 32.5).

The analysis also reveals that 104,973 ha, constituting 9.6% of the study area has the capability for double crops; however, there are some limitations for uses such as waterlogging due to the frequent floods. The WS1, WS2, WS3, WS6 and WS13 are prone to these problems which is maximum (20.8%) in WS2 (Fig. 32.6). If these areas are to be exploited for double cropping, they require special management (installation of drainage system) for successful double cropping.

32.3.2 *Agro-horticulture*

In the arid and semiarid regions of India, much of the agricultural lands are mismanaged, and this increases the risk for crop production. Adoption of appropriate new, innovative and efficient alternative methods can make these lands more productive on sustainable basis. Hence, the concept of agro-horticulture, i.e. growing fruit trees with agricultural crops, has been suggested. Such a combination has added advantage, as it controls soil erosion, improves soil fertility and soil moisture retention capacity that ultimately leads to sustainable productivity of the land. It also opens new opportunities for small farmers to raise income without jeopardising agriculture and improving existing land use rather than land transformation. The areas where

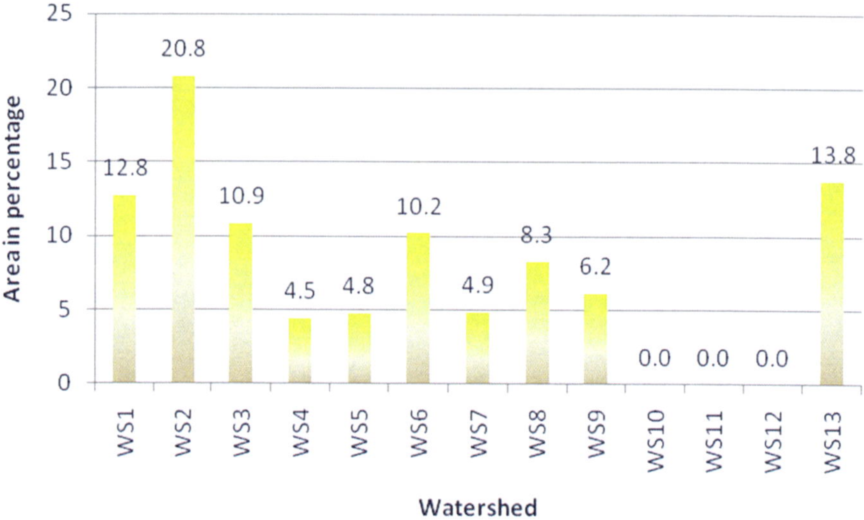

Fig. 32.6 Watershed-wise potential area of double crops with special effort

rabi crops are cultivated and land is not capable for kharif crop, fruit trees can be planted along agriculture crops. The analysis reveals that 84,243 ha of land has potentiality of agro-horticulture, which constitutes 7.7% of the study area. The watershed-wise analysis shows that WS2, WS6 and WS9 have more than 10% area of the watershed under this category and maximum (19.9%) being in WS9. The WS10 to WS12 has less than 2% watershed area under agro-horticulture because these areas are Kaimur and Rohtas plateau, and maximum areas are covered by the forest (Fig. 32.7).

32.3.3 Horticulture

Horticulture is the growing of fruit trees. It has advantages in improving the natural resource base and the economic condition of the farmers. The fruit trees have deep rooting system and thus have capacity to draw water from deeper soil layers. The fruit trees possess enormous resilience to survive under harsh climatic conditions. They are source of firewood and can be used for other wood works. The land which is lying fallow for long period and is considered poor area for rabi crops, this area can be exploited for horticultural crops, including fruit trees, vegetables, spices and floriculture. The analysis reveals that 33,072 ha of land has the potential for horticulture, constituting 3% of the study area. The analysis also reveals that WS10 to WS12 has more than 10% area under horticulture and maximum area being (23.6%) in WS10 (Fig. 32.8).

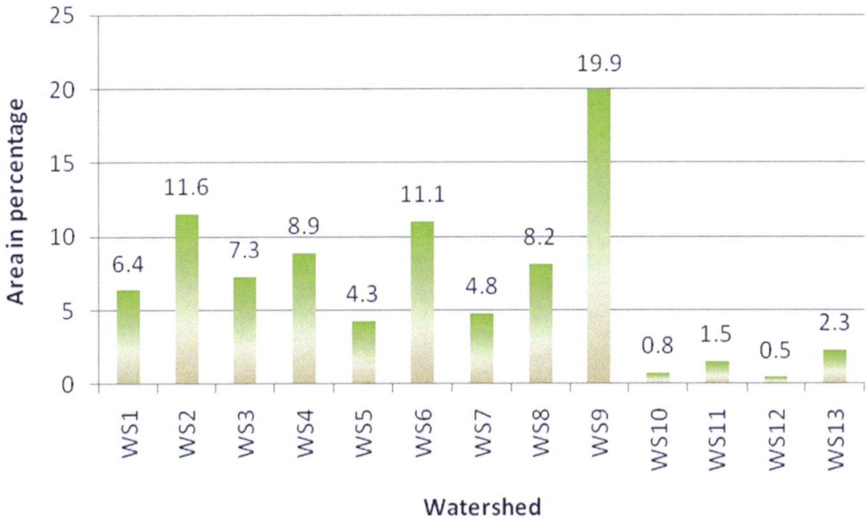

Fig. 32.7 Watershed-wise potential area of agro-horticulture

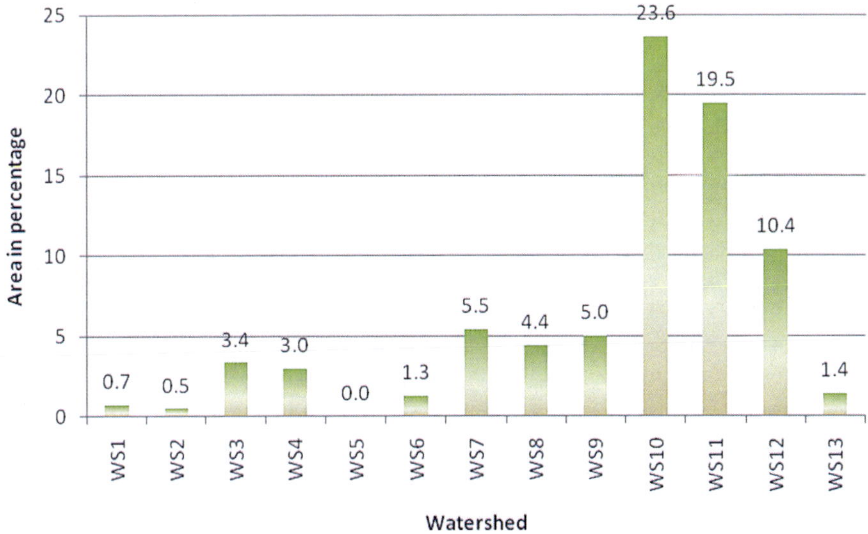

Fig. 32.8 Watershed-wise potential area for horticultural crops

32.3.4 Afforestation and Gap Filling

Perennial vegetation, such as trees and grasses, is the alternatives for the economic utilisation of degraded lands unsuitable for agriculture. Such plantations reduce soil erosion and run-off considerably from such marginal lands. They provide useful basic products (wood, fodder, etc.), enhance soil organic matter to maintain soil

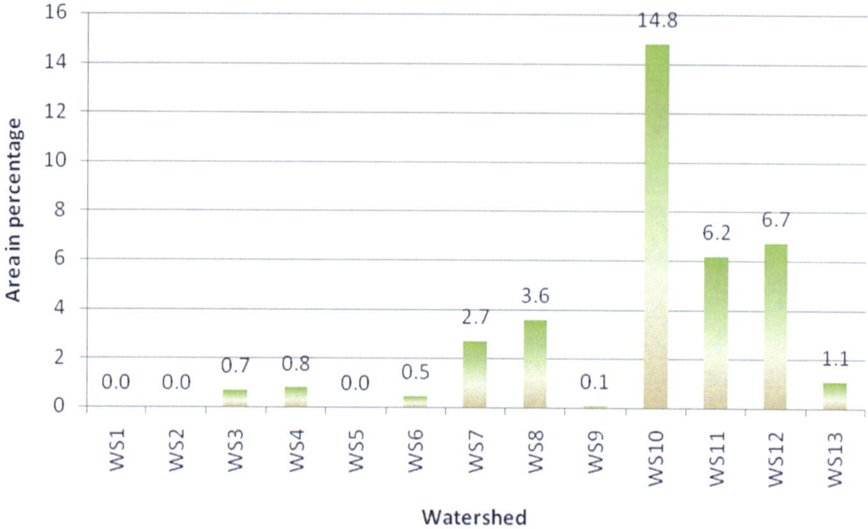

Fig. 32.9 Watershed-wise potential area of afforestation

fertility, bring up nutrients from deeper soil layers and create favourable microclimate and improve the environment quality. The land unsuitable for agriculture can be used for afforestation.

The analyses of the remote-sensing images obtained through 2004–2005 revealed more than 50% forest land degraded as open or scrub forests. According to the geo-environmental condition of the study area, 24,338 ha of land constituting 2.2% of the study area needs afforestation. The analysis reveals that more than 5% of the watershed area needs afforestation in WS10 to WS12 and the maximum (14.8%) in WS10 (Fig. 32.9). Amongst the total afforestation area, 10,233 ha of land, constitutes 0.9% of the study area, needs special or more efforts for afforestation due to the moderate to very severe soil erosion problem. The analysis also reveals that 6.6% of the watershed area in WS10 also requires special efforts for afforestation. The data show that less area is under the forest in the alluvial plains (Fig. 32.10). Apart from scrub forest, the study area does present open forest land, and this has the potential for gap filling. The analysis reveals the entire gap filling area existing in the plateau region and the maximum (57.3%) area being in WS12 (Fig. 32.11).

32.3.5 Silvi-Pasture/Fodder and Fuel

The silvi-pasture is one alternative land-use system available for improving the fodder resources of the study area. The total area under fodder and fuel is 1,539.7 ha covering 0.2% of the study area.

This system offers an extra yield of grasses during the rainy season and browse material in long dry spell. The demand for fuel wood and fodder is ever increasing in

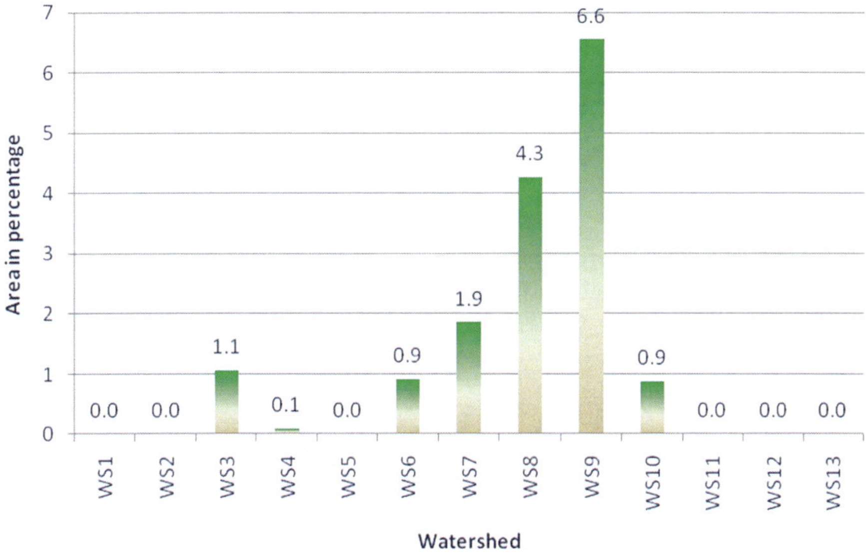

Fig. 32.10 Watershed-wise potential area of afforestation with special effort

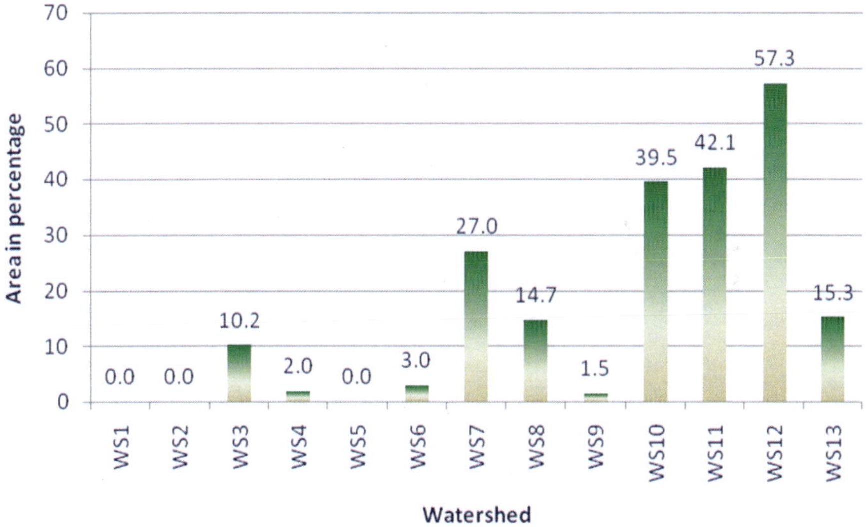

Fig. 32.11 Watershed-wise potential area for gap filling

the rural sector due to the increase in livestock population resulting in merciless cutting of trees all around, especially in the forests. In order to make the region ecologically sustainable and to reduce the pressure on forest land, there is a need to develop silvi-pasture especially near the villages. It reduces the soil erosion, improves moisture conservation, improves soil fertility through selection of forage legumes and nitrogen fixing trees, provides extra fuel wood or charcoal and sparing the nutrient rich animal manure for agriculture purpose and meeting the fodder requirement for the farmers throughout the year in general and during the lean periods in particular.

32.3.6 Levelling and Cropping

This method is useful where the land is lying waste after shifting of brick-kilns. These lands should be brought under cultivation after removing bricks pieces and levelling. To improve soil quality, deep ploughing is required as well as mixing of farmyard manure. The farmyard manure should be spread thoroughly with a thin layer on the surface area followed by deep ploughing with cultivator.

32.3.7 Gully Plugging

The erosion gullies are formed through localised surface run-off affecting the fragile soil material to form channels resulting in undulating terrain. The gullies are the first stage of excessive land dissection followed by their networking, which lead to the development of ravines land. The ravine is usually associated not with an isolated gully but a network of deep gullies formed generally in thick alluvium and entering a nearly river, flowing much lower than the surrounding high grounds. The ravines are extensive system of gullies developed along river course. Some gullies are developed in the study area along all the rivers. Therefore, to control the further erosion, check dams/obstacles, growing of grass species and other measures should be taken.

32.3.8 Integration of Livestock Enterprise with Alternate Land-Use Systems

In dry or semi-drylands, livestock enterprise is the most important from the point of view of income generation, nutrition and drought power; adequate production of fodder crops and pastures is fundamental to increased and sustained production of livestock. The farmyard manure produced from the livestock enterprise can be ploughed back into the soil to improve the soil fertility and crop yields.

32.3.9 Fish Culture

Currently fish culture is gaining popularity amongst panchayats (groups) and individuals and becoming an attracting source of income in the study area. Fisheries Department of the Government of India is opening fish seed farms and supplying fish seeds of major variety to farmers at concessional rates. In the study area, approximately 2,558 ponds or lake occur covering 4,292.1 ha constituting 0.4% of the total area. Amongst these, 2,867.8 ha is perennial and 1,424.4 ha is non-perennial ponds. The seasonal ponds, which retain water for 3–6 months, can be identified, renovated for better impounding of rainwater and can be used for the fish culture. Similarly, a large number of small check dams, farm ponds, impoundments, etc, proposed in the watersheds, can also be exploited for fish culture by adopting seasonal cropping techniques (National Remote Sensing Agency 1995). Fisheries can help the landless or the unemployed persons to find a new job opportunity in view of the fact those villages with small ponds are quite common and in village where no one exists now, new ponds can be developed for this purpose.

32.4 Conclusions

The Son-Karamnasa interfluve has rich potential for land resource development. There exist 75% of the lands that can be used either for agriculture, or horticulture, or both uses to increase the productivity of the land on sustainable manner and would reduce the wasteland from 24 to 0.6%. The analysis reveals that the double crop area can be increased up to 66%, and the maximum potential area lies in the alluvial plains in WS1, WS2, WS4, WS5 and WS6 and lowest area in WS10, WS11 and WS13 due to plateau region. Horticulture land could be increased from 0.4 to 3%, and maximum potential land is in WS10, WS11 and WS12. About 7.7% land is capable for agro-horticulture, and maximum potential area exists in WS2, WS6 and WS9. Forest cover would increase from 7 to 16% by afforestation and gap filling of new plantation. The afforestation is required in WS8 to WS12 due to scrub forest, and gap filling is required in WS7, WS10, WS11 and WS12 due to open forest. We recommend integration of livestock with their alternate land-use system to increase the economic status. The Government of India has initiated various long-term programmes, and currently, project "Hariyali" and "From Hariyali to Neeranchal" are initiated for the land resource development in the country. The findings of the present study can be used by the policy makers and other agencies as guidelines for the best planning of the available resources to improve the socioeconomic and environmental conditions of the region as well as developing new policies and strategies for sustainable development.

References

Census of India (2001) Primary census abstract for Uttranchal, Bihar and Jharkhand, vol 4. Office of the Registrar General, Government of India, New Delhi

Food and Agricultural Organisation (1993) Guidelines for land-use planning. Food and Agriculture Organisation of the United Nations, Rome

National Remote Sensing Agency (1995) Integrated mission for sustainable development. National Remote Sensing Agency, Department of Space, Government of India, Hyderabad

Singh RL (ed) (1989) India: a regional geography. National Geographical Society of India, Varanasi

Chapter 33
Remote Sensing and Geographical Information as an Aid for Land Use Planning and Implications to Natural Resources Assessment: Case Study, South India

N. Jayaraju and J. Abdullah Khan

Abstract Anthropogenic changes in land use are being increasingly recognized as critical factors influencing global change. Land use is often shaped by human and socio-economic and political influences on the land. Remote sensing (RS) integrated with geographical information system (GIS) provides an effective tool for analysis of land use and land cover changes at a regional level. The geospatial technology of RS and GIS holds the potential for timely and cost-effective assessment of natural resources. These techniques have been used extensively in the tropics for generating valuable information on forest cover, vegetation type and land use changes. In the present study, RS and GIS have been used to assess land cover patterns in Pulivendula-Sanivaripalli area of south India. With this in view, an assessment has been made on some of the natural resources and environmental potential of Pulivendula-Sanivaripalli area of south India. To achieve these, three thematic maps (land use and land cover, drainage and slope) were prepared through image interpretation and limited checks. The land use-land cover pattern falls under the broad categories of agricultural land, forest land and wasteland. The agricultural land is further subdivided into dry and wet agricultural land. Forest land has been classified based on the United States Geological Survey (USGS) land use and land cover classification system, using remote sensing data as dry and wet forest land. Further, the forest land has been classified into reserve forest and wasteland vegetation. Social forestry

N. Jayaraju (✉)
Department of Geology and Geoinformatics, Yogi Vemana University,
Kadapa 516 503, India
e-mail: nadimikeri@gmail.com; naddimi_raju@rediffmail.com

J. Abdullah Khan
State Remote Sensing Applications Centre (APSRAC), Nagarjuna Hills,
Punjagutta, Hyderabad 500 482, India
e-mail: abd_jolly@yahoo.co.in

S.A. Shahid et al. (eds.), *Developments in Soil Classification, Land Use Planning and Policy Implications: Innovative Thinking of Soil Inventory for Land Use Planning and Management of Land Resources*, DOI 10.1007/978-94-007-5332-7_33,

programme includes plantations and development of pasture land; control of soil erosion by afforestation has been suggested. In the study area, most of the land (60%) is unused, and this can be used for industries and urban planning.

Keywords GIS • Land use planning and implications • Natural resources • RS • South India

33.1 Introduction

Land cover is the physical material at the earth surface. The agriculture fields, lakes and concrete highways, etc., are all examples of land cover types. The terrain land use relates to the human activity associated with specific piece of land use and can also be described as urban use or single-family residential use. The terms land use and land covers are strongly correlated and in many cases have been used interchangeably. The knowledge of land use and land cover is very important for land planning and land management activities (Angi Reddy 2002). Land is the most important natural resource on which most developmental activities are based. The use of remotely sensed data for land use mapping is considered as one of the most suitable techniques since the last almost seven decades. Over the past few decades, there has been raising concerns from the researchers and planners about the increasing anthropogenic influence on land resources.

More recently, high-resolution satellite data and aerial photography have enhanced and opened new fields for detailed planning of land use and land cover mapping for various developmental and scientific schemes (Burrough 1986; Sabins 1997; Chauhan 2003). Land use refers to the human activities directly related to land, whereas land cover denotes the natural features and artificial constructions covering the land use surface (Kunwar et al. 2010). The land use practices of a region are influenced by a number of parameters, namely, physical and chemical environments, socio-economic factors and needs for the masses. Ever increasing demand due to continued population growth has put heavy pressure on the natural resources of the country in general, and the study is in specific. This necessitates the planning of the area for optimal use of the natural resources on sustainable basis. To achieve this objective, it is imperative to gather information on existing natural resources scenario, their physical/terrain features, climate parameters, ecological conditions, current practices of planning and management and the contemporary technologies to be used for the sustainable use and management of natural resources. The RS and GIS have played an important role in the present study to assess the natural resources. Anthropogenic changes in land use and land cover are being increasingly recognized as critical factors influencing global change.

Information on land and water resources for their proper management is the most important components for the planning of area-specific development activities. To achieve this, it is essential to integrate the data on various natural resources. Such data can be obtained more accurately and reliably by using RS and GIS techniques (Navalgud et al. 2007). It is due to this reason the present study was conducted to characterize the study area with respect to land use-land cover (LULC), slope and drainage, etc., to inventory detailed soil resources and to assess the problems and potentials of the land use. In addition to the above, alternative practices have also been recommended for sustainable development.

33.2 Materials and Methods

33.2.1 Study Area

The study area (360 km^2) lies between Pulivendula and Sanivaripalli and situated between 78°00′0″–78°15′0″ E longitude and 14°15′0″–14°30′0″ N latitude with intended boundary falling in the Survey of India (SOI) topographic sheet number 57J 03 on 1:50,000 scale (Fig. 33.1a). The study area includes five revenue mandals (subdivision of districts), namely, Pulivendula, Lingala, Udumakurti, Krishnamagaripalli and Sanivaripalli. The central and southern part of the study area is occupied by high hills, ridges and valleys. The Maddaleru is the only river in the study area which is a seasonal river fed by monsoon rain (June–October), which is flowing in the NW direction in the southern part of the imagery. Some major tanks also exist in the same area. The climate is dry with mean annual rainfall of 100–150 cm and mean annual temperature of 32°C. May is the hottest (45°C) and December the coldest (25°C) month. The duration of any crop in the area is more than 120 days. The value of contours in the study area ranges between 200 and 600 m and the slope is within 1–15 degrees.

33.2.2 Methodology

The broad methodologies are used in the present study, which involve an innovative three-tier approach, namely, image interpretation, field surveys and laboratory investigations and cartography, and Geographical Positioning System (GPS). The geocoded imagery number 57J 03 (IRS-P6) obtained from the National Remote Sensing Agency (NRSA), Hyderabad, was visually interpreted based on textural and tonal variations (Table 33.1). The topographical maps (1:50,000 scale) were used for the delineation of landform units and land use together with physical verification in the field. The information was then transferred to the topobase topographical maps prepared by the Survey of India (SOI) and used as base maps for field surveys. The base map is prerequisite for any mapping and field ground truthing. The base map is used for the collection and representing the thematic (resource) information on a common uniform scale. The major features like important town or

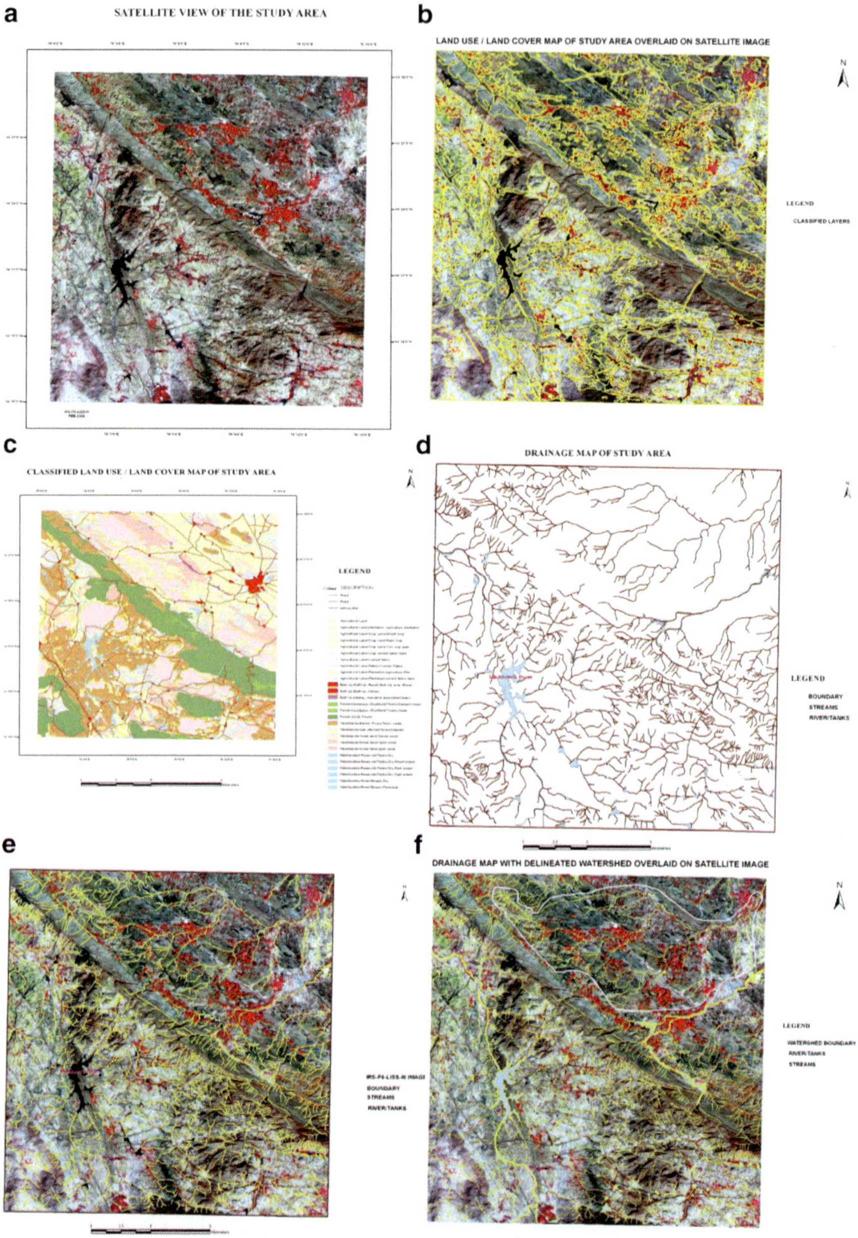

Fig. 33.1 Satellite view (**a**), land use and land cover (**b**, **c**) and drainage maps (**d**, **e**, **f**) of the study area

villages, rivers, reserved forest area, and transport network were traced on tracing sheet to prepare base map at 1:50,000 scale.

Table 33.1 Land use-land cover image interpretation keys developed from LANDSAT remote sensing data

S. No.	Land use class	Image characteristics: Band 5	Tone band 7	Band 4, 5 and 7	Texture	Pattern	Size	Location
1	Built-up land	Lighter	Medium to light grey	Blue mixed with red and white	Fine	Semicircular to circular	Small	In plains and plateaus along the rivers, roads and railways
2	Agricultural land							
	Crop land	Dark grey	Light to medium	Red	Fine	Distinct	Varying sizes	Plains and valley land
	Fallow land	Light grey	Light grey	Yellowish	Rough	Irregular		
3	Forest land Fairly dense	Very dark grey	Light to medium	Light red to brown	Fine	Irregular with distinct Irregular	Varying irregular sizes	Highland hills
	Sparse	Medium dark grey	Light grey	Light brown	Medium to rough			
4	Water bodies River/stream	Medium to dark grey	Dark grey	Medium blue to dark	Fine	Linear	Long	Natural stream
	Lake/tank	Dark grey	Very dark grey	Light to dark blue	Fine	Triangular	Medium	Surrounded by hills
5	Uncultivable land							
	Barren land	White grey	Light grey	White to yellowish	Fine	Distinct	Large	Hilly terrain along rivers/streams
	Wasteland	Light to medium	Medium grey	Brown to grey	Medium to rough	Distinct	Small	

Table 33.2 Land use-land cover classification system

Level 1	Level 2
1. Urban or built-up land	Residential, industrial Transportation, communication and utilities
2. Agriculture land	Crop land—wet cultivated land dry cultivated land Fallow land
3. Forest land	Fairly dense forest land Sparse forest land
4. Water bodies	Streams and canals Lakes Tanks or reservoirs
5. Uncultivable land	Barren land Wasteland

33.2.3 Preparation of Thematic Map

Thematic maps are the true representation of the earth's phenomena such as the spatial distribution of natural resources at the time of survey (Gupta 2003). The satellite image (IRS P6) which is a true record of the various environmental resources on the base map is used. The thematic maps presenting spatial distribution of forest, agriculture, soil, water resources, etc., are prepared by visual interpretation of the satellite imagery. Visual interpretation is carried out based on the image characteristics like tone, size, shape, pattern and texture, in conjunction with the existing map/literature. These pre-field thematic maps are modified and confirmed with limited field checks. The image interpretation keys used for land use-land cover classification are shown in Table 33.1 (Lillisand and Keifer 1979). The land use and land cover system adopted is derived from the USGS land use and land cover classification system (Table 33.2). The details of land use-land cover of the study area are given (Table 33.3; Fig. 33.1b, c). Any differences observed from pre-field interpretation to post-field checks are suitably incorporated at this stage and map finalized after necessary modification.

33.3 Results and Discussion

The urban or built-up land is covered intensively by structures, including cities, towns, villages, industrial and commercial complexes and institutions. In the study area, major towns or villages are Pulivendula, Lingala, Ippatla, Udumukurti, Nallapureddipalli, Dorigallu, Diguvaplli and Sanivaripalli.

Mining The industrial mining of asbestos and serpentine minerals is carried out in the villages of Brahmanapalli, Ippatla, Chinna Kuddala, Ramanutlapalli and Lapatnutala, which are trending in NW-SE direction, accounting for 219 ha (0.29%).

Table 33.3 Land use and land cover of the study area

S. No.	Description	Area (ha)	Percentage of total study area
1	Agricultural land	34.69	0.05
2	Agricultural land plantation, agriculture plantation	770.95	1.04
3	Agricultural land, crop current fallow land	118.29	0.16
4	Agricultural land, cropland, kharif crop	4,215.08	5.68
5	Agricultural land, cropland, rabi crop	6,099.06	8.21
6	Agricultural land, cropland, two-crop area	4,471.89	6.02
7	Agricultural land, current fallow	314.80	0.42
8	Agricultural land, fallow, current fallow	5,638.62	7.59
9	Agricultural land, plantation, agriculture plantation	1,908.15	2.57
10	Agricultural land, plantation, current fallow land	108.09	0.15
11	Built-up, built-up (rural), built-up area (rural)	484.08	0.65
12	Built-up, built-up (urban)	316.13	0.43
13	Built-up, mining/industrial area, mine/quarry	219.02	0.29
14	Forest, deciduous (dry/moist/thorn), dense/closed	110.59	0.15
15	Forest, deciduous (dry/moist/thorn), open	153.56	0.21
16	Forest, scrub forest	12,232.58	16.47
17	Wastelands, barren, rocky/stony waste	15,545.54	20.94
18	Wastelands, salt-affected land, moderate	163.01	0.22
19	Wastelands, scrubland, dense scrub	12,334.46	16.61
20	Wastelands, scrubland, open scrub	6,457.64	8.70
21	Water bodies, reservoirs/tanks, dry	75.10	0.10
22	Water bodies, reservoirs/tanks, dry, kharif extent	7.27	0.01
23	Water bodies, reservoir/tanks, dry, rabi extent	727.90	0.98
24	Water bodies, reservoir/tanks, dry, zaid extent	243.93	0.33
25	Water bodies, river/stream, dry	1,489.60	2.02
26	Water bodies, river/stream, perennial	1.43	0.00
	Grand total	74,250.46	100.00

Agricultural land All the cultivated land with or without crops, orchards and plantations are considered agricultural land. This land use class is further subdivided into two subclasses, namely, wet land (cropland) and dry land (fallow land) (Table 33.2).

Cropland Croplands are the agricultural lands under crops. Due to the large size of land holdings, the farmers are using modern agricultural tools and implements. In the study area, the croplands have wet cultivation and dry cultivation. Wet cultivation includes food crops mainly paddy and wheat. The wet cultivated area (20.15%) is situated in Ippatla, Murarichintala, Kottapalli, Ambakapalli, Udumakuthi and Siddugarippali. The irrigated fields located in the valleys are cultivated throughout the year. Dry cultivation is practised in Nallapureddipalli, Diguvapalli, Dorigallu and Indukuru areas and includes orchard trees and groundnut.

Fallow land The land intended for cultivation but temporarily out of cultivation for a particular period. The fallow land accounts for 10.58% of study area and distributed in villages (Krishanamgaripalli, Peddarangapuram-Lingala, Indukuru) four km right east of Siddugaripalli and west of Velumvaripalli. The area is fallow mainly due to non-availability of water for irrigation purpose (Kunwar et al. 2010).*Uncultivated Land*

*Barren land*Barren land covers all lands which are uncultivable like mountains, deserts, bare exposed rock, strip mines, gravel pits and quarries. The areas which comprise barren lands are surrounded by villages Dondi, Diguvapalli and Lapatnutala in the northeastern part and in southwestern part where the major land is bare rock (rock outcrop) and has not been used. These lands (20.94%) are suggested to be used for industrial and urbanization purposes (Table 33.3).

*Wasteland*Wasteland is degraded, currently uncultivated, or unutilized and exists outside the notified forest land due to several constraints. Wastelands are nonagricultural areas which cannot be brought under cultivation due to high cost. In the study area, the wastelands occur nearby villages such as Dondi, Murarichintalakanampalli and 2 km south of Velumvaripalli. The wasteland accounts for 15.45% of the study area.

*Forest land*Forest land represents areas that have a tree crown aerial density of 10% or more, are stocked with trees capable of producing timber or other wood products and exert an influence on the climate or water regime. In the study area, forest land is classified into two levels: (1) fairly dense forest land covered with the hills trending NW-SE adjacent to villages Matnutalpalli, Ambakapalli and Murarichlintalapallietc and (2) sparse forest land located on undulating terrain, uplands and slopes of the hills. The sparse forests are situated near to Velumvaripalli, Nakkalapalli and Bandamidapalli villages. Forest blanks are the notified forest land completely devoid of any vegetation cover. This category occupies 16.83% land of the study area.

33.3.1 Recommended Alternative Use of Land Use and Land Cover (LULC)

One of the objectives is to identify the problem areas and to generate site-specific LULC developmental plans highlighting alternate land use practices leading to optimal utilization of available resources for sustainable development (Kunwar et al. 2010). To achieve this objective, various thematic maps (LULC, hydrometric, drainage and slope) have been integrated along with filed information and collateral data. This investigation is based on decision rule or key which has been developed in discussions with a multidisciplinary team and district officials and local people. The LULC map has been taken as base layer, and other resource maps were integrated in GIS environment. All the suboptimally utilized and wasteland classes were then identified, and suggestive measures were recommended to transform/alternative LULC. The transformed land use plan thus prepared was verified in the field (Murthy et al. 1997). Integrated approach using GIS and RS techniques has been successfully applied earlier for transformation of LULC. Alternative land use practices suggested for the study area are as follows.

33.3.1.1 Double Cropping/Intensive Agriculture

Double cropping has been recommended where existing agriculture land is cultivated for a single crop. The capability of such land could be improved for double crop with suitable sources of irrigation water (Kunwar and Kachwala 2001).

33.3.1.2 Agro-horticulture

Agro-horticultural practices are suggested for existing waste and barren land. Due to the uncertainty of crop survival, growing suitable fruit trees with existing crop in such areas would not only supplement the income to the farmers but also control soil erosion. The tolerance of fruit trees towards monsoon is much better than short-duration crops; thus, these can utilize off season rains and soil moisture from deeper layers. Type selection of horticulture plants to be planted in these areas should be based on site conditions and the choice of the local people as well as market demand.

33.3.1.3 Agroforestry

Agroforestry system has been suggested to meet the needs of local people for fuel and fodder in addition to agriculture crops. Agroforestry has been suggested on hilly terraces and barren land. Some wasteland areas are also recommended for agroforestry after providing suitable source of irrigation water. The food and fodder trees could be planted along hilly slopes. The soil on these terraces is relatively poor and has relatively low soil moisture. Once the trees are established, they have greater drought tolerance and therefore will cater towards fuel and fodder requirements besides controlling soil erosion and enhancing water recharge.

33.3.1.4 Horticulture

Wasteland and long fallow land with suitable soils and slopes in the vicinity of road network have been recommended for horticulture trees. Depending on the site conditions and choice of the local people, suitable fruit tree could be planted in these areas.

33.3.1.5 Afforestation

The left over area, after using for horticulture, agro-horticulture and agroforestry, has been recommended for afforestation. However, selection of trees should be made at the time of the implementation of action plan selecting from local species and based on the actual site conditions.

33.3.1.6 Gap Filling

Gap filling is tough task and could be done with selective tree species. Tree species preferred may be selected according to the suitability of the site for plantation in the gaps. Moreover, where the forest has high crown density and it is difficult to do gap filling, in such areas it is proposed to totally protect them from biotic interference by suitable means to allow natural regeneration.

33.3.2 Water Resources

The water category includes streams, lakes, canals and reservoirs. The major tanks/ reservoirs occur around Lakkasamudram, Reddivaripalli, Sanivaripalli and Gollapalle in the southern part of imagery; however, streams and canals occur in the northern part of imagery. Maddaleru River is flowing towards NW direction in the southwestern part of imagery. The villages adjacent to the river are Sanivaripalli, Adavibrahmanpalli, Nakkalapalli, Dorigallu, Indukuru and Maartadu.

33.3.2.1 Watershed Modelling (Drainage Map)

The watershed area occurs between Eguvapalli in the west and Krishnamgaripalli in the east. The watershed covers an area of 332.5 km^2 (Fig. 33.1d, e, f). The demarcated watershed is named after rishnamgaripalli village. It enjoys a tropical climate with rainfall between 100 and 150 mm annually. A drainage basin is the topographic region from which a stream receives runoff, through flow and groundwater flow. Drainage basins are divided from each other by topographic barriers called a watershed. A watershed represents all of the stream tributaries that flow to some location along the stream channel. Watershed is a basic unit for water resource survey. For hydrologic analysis, the physical structure and geometry of watershed is of prime importance for the prediction of water and sediment yield. Land cover parameters are valuable direct inputs and also as indirect inputs by inferring basin characteristics such as drainage area, drainage density, drainage texture and bifurcation ratio. They are described below.

Linear Aspects

1. Stream order (U): The first step in drainage basin analysis is the designation of stream orders. The stream order is a measure of the degree of stream branch within a watershed. Each length of the stream is indicated by its order, for example, first order, second order and third order. In the given watershed, five stream orders have been calculated.
2. Stream number (Nu): The count of stream channel in its order is known as stream number. The number of streams decreases as the stream order increases. The demarcated watershed has first-order stream orders (Nu = 100), second-order stream orders (Nu = 20) and third-order stream orders (Nu = 6).

3. Bifurcation ratio (Rb): The Rb is defined as the ratio of the number of streams one order to the number of the next higher order. It is given below:

Bifurcation ratio (Rb)	= Nu / (Nu + 1)
Bifurcation ratio for first order	= No. of 1st order streams/No. of 2nd order streams +1
	= N1/(N2 + 1)
	= 100/(20 + 1) = 4.76
Bifurcation ratio for second order	= N2/(N3 + 1)
	= 20/(6 + 1) = 2.85
Bifurcation ratio of third order	= N3/(N4 + 1)
	= 6/(1 + 1) = 3

Bifurcation ratio characteristically ranges between 3 and 5. This indicates that in watershed, the geologic structures do not distort the drainage pattern.

4. Stream length (Lu): Mean length of a stream channel segments of order U is a dimensional property revealing the characteristic size of components of a drainage network and contributing basin area. To obtain the mean length channel Lu of order U, the total length is divided by the number of segments Nu of that order:

Lu = Total length of particular order stream/Nu	
Length of 1st order stream L1	= 98.75/100 = 0.987
Length of 2nd order stream L2	= 33.25/20 = 1.663
Length of 3rd order stream L3	= 29.75/6 = 4.958
Length of 4th order stream L4	= 10.25/1 = 10.25

5. Stream length ratio (RL): The RL is the ratio of main length (Lu) of segments of order U to main length of segments of the next lower order (Lu − 1). Here, the stream length ratio tends to be constant throughout the successive orders of a watershed. It indicates that stream lengths are decreasing with increase in the order of stream:

RL = Lu/(Lu − 1)	
Stream length ratio for fourth order	= 10.25/4.958 = 2.06
Stream length ratio for third order	= 4.958/1.663 = 2.98
Stream length ratio for second order	= 1.663/0.897 = 1.85

33.3.3 Drainage Analysis

33.3.3.1 Drainage Area

This area contributes water to a particular channel or set of channels. The basin area is one of the important parameters like the length of the draining basin. In the demarcated watershed, the total drainage area or basin area is about 332.5 km^2 (Fig. 33.1f).

33.3.3.2 Drainage Net

It is related to the hydraulic geometry of stream channel. If the streams are relatively far apart, then the drainage net is less dissected. If the streams are relatively low spaced, then the drainage net is intensively dissected. In the Krishnamgaripalli watershed, the drainage net is less dissected because the streams are relatively far apart.

33.3.3.3 Drainage Density (Dd)

This is the total length of all the streams in the basin to the area of whole basin. It is calculated as:

$$
\begin{aligned}
Dd &= \varepsilon Lu / Au \\
&= 162.95 / 332.5 \\
&= 0.505\ \text{km}^2
\end{aligned}
$$

The low values of the drainage density indicate that the region is highly resistant of highly permeable subsoil materials, dense vegetation cover and low relief.

33.3.3.4 Drainage Texture

The drainage texture is the measure of the closeness of the channel spacing and related directly to the drainage density. The drainage density values of the watershed are 0.448 km km^{-2}, that is, less than two. Therefore, the drainage area has very coarse drainage texture.

33.3.3.5 Slope Map

Knowledge of the slope angles is necessary to study the present-day processes and to understand the development of relief. The two main fields of application for slope studies are engineering and agriculture. In both cases, the relevant material includes aspects of surface processes, the form of land surface and the properties of the regolith. Expressions of ground slope, that is, slope angles, can be made in degrees and gradients either as a ratio 1:10 or percentage 10% or in parts of thousands (45). In the study area, the total northeastern part is having slope angle decreasing from 5 to 1 degrees, that is, strongly inclined to plane towards NE, and the rest of the imagery in the southeastern part has strong inclined slopes in areas of Sanivariplalli situated in northwest corner, and Lakkasamudram (central part) and Udukulakurti and Siddugaripalli are in northeast corner of the study area.

33.4 Conclusions

Remote sensing and GIS are useful techniques to provide practical, robust and cost-efficient output for environmental protection and management. The remote sensing technique is useful for mapping and assessment of natural resources to take immediate remedial measures for sustainable development. The worst damage could have been avoided by recognizing suitable methods and adhering to some simple environmental guides and standards (Ramachandran et al. 2005). The data-gathering capability of space-borne remote sensors has now been well established for continuous monitoring and management of natural resources. Since the region is very poor in water resources, top priority should be given for its development. The groundwater distribution in the study area is not uniform, and it depends on the drainage pattern. This watershed, near the village Krishnamgaripalli, is demarcated so as to fulfil the needs. The land which is not cultivated or not used for settlements is suggested to be used for industrial and urban planning. The forests are drastically decreasing in the last few years; therefore, the social forestry programme is suggested.

Different components of watershed, such as drainage, physiography and land use, assessed through remote sensing data in combination with SOI topographic sheets helped to assess their potential land uses. The database thus generated can be digitized, stored and processed for the generation of thematic and interpretative maps for watershed development plan using GIS (Walia et al. 2010). The management plans, which have been prepared by suggesting alternate use of the land, would lead to optimal utilization of land resources and ultimately sustainable development of the area.

Acknowledgements We would like to thank Prof. A.R. Reddy, Vice-Chancellor, Yogi Vemana University, Kadapa, India, for the encouragement and technical assistance for the completion of the present study. The financial assistance provided by the APCOST, Hyderabad, is highly acknowledged and appreciated.

References

Angi Reddy M (2002) Remote sensing and geographical information system. BS Publication, Hyderabad

Burrough PA (1986) Principles for geographical information system for land resources assessment. Oxford University Press, New York

Chauhan TS (2003) Geographical information system and remote sensing for sustainable development, vol 2. Universal Scientific publishers, Jaipur

Gupta PR (2003) Remote sensing geology. Springer International Edition, New Delhi

Kunwar P, Kachwala TS (2001) Transformation of land cover land use for optimal utilization of natural resources on sustainable basis in Mandawara Block, Lalitpur District, Utter Pradesh, using remote sensing and GIS techniques. In: Proceedings of the National Geographical Society of India BHU, Varanasi, Environment and Development, pp 151–156

Kunwar P, Kachwala TS, Kumar A, Agarwal AK, Singh AN, Nisha M (2010) Use of high resolution IKNOS data and GIS technique for the transformation of Land use and Land cover for sustainable development. Curr Sci 98(2):204–212

Lillisand TM, Keifer RW (1979) Remote sensing and image interpretation. Willey, New York
Murthy YVS, Bhaskar N, Uday R, Kameswara SV, Perumal A (1997) GIS approach to suggest land use alternative for sustainable rural development. In: Proceedings of the ICORG, Hyderabad, India, pp 44–47
Navalgud RR, Jayaraman V, Roy PS (2007) Remote sensing applications: an overview. Curr Sci 93:1747–1766
Ramachandran S, Anitha S, Valamuragan V, Dharaniranjan K, Ezhil Vendhan MK, Divien MIP, Senthil VA, Sujjahad HI, Vdayaraj A (2005) Ecological impact of tsunami on Nicobar Islands. Curr Sci 89(1):195–200
Sabins FF (1997) Remote sensing principles and interpretation. W.H Freeman and Company, New York
Walia CS, Singh SP, Dhankar RP, Ram JH, Kamble KH, Katiyar DK (2010) Watershed characterization and soil resources mapping for land use planning of Moolbari watershed, Shimla District, Himachala Pradesh in lesser Himalayas. Curr Sci 98(2):176–182

Chapter 34
Common Land Resources: The Present Status and Need for Their Conservation in North India

Mohd Sadiq Salman and Abdul Munir

Abstract The common land resources (CLRs) as the name implies have common access to all for various economic gains. The CLRs include forests, pastures, barren land, uncultivated land other than current fallow land and cultivable wastelands. The forests provide timber and pastures support livestock. The uncultivated and barren lands support industrial and urban development. Agroforestry and social forestry are also practised. The "common access" to these resources has led to unchecked and rampant use, leading to their degradation. In general, they account for a substantial share in income, socio-economic development and sustainable livelihood of the landless people, marginal and small farmers. Over the last 50 years, the population growth, urbanization and industrialization have led to overexploitation of the resources having "common access". The share of CLRs in the Uttar Pradesh during 1950–1951 was 34.28%; since then CLRs continue declining. Considering the declining trend of CLRs and their role in socio-economic development of the unprivileged, there is a need to manage them in a judicious way through the formulation of suitable and effective policies by the government to prevent degradation and extinction of CLRs.

Keywords Common land resources • GIS • Land use planning • Natural resources • North India

34.1 Introduction

India is a developing country, and its major population lives in rural areas, depending mainly on agriculture. Thus, the livelihood highly depends on their land resource. With continuing population growth, there is an immense pressure on the land

M.S. Salman (✉) • A. Munir
Department of Geography, Faculty of Science, Aligarh Muslim University, Aligarh, Uttar Pradesh 202002, India
e-mail: msadiqsalman@gmail.com; Abdulmunir.amu@gmail.com

S.A. Shahid et al. (eds.), *Developments in Soil Classification, Land Use Planning and Policy Implications: Innovative Thinking of Soil Inventory for Land Use Planning and Management of Land Resources*, DOI 10.1007/978-94-007-5332-7_34,

resources specially the agricultural land. The share of marginal and small farmers is maximum among the total landholders. The present era of competition and globalization has led to decline in the economic benefits through agriculture. Thus, small pieces of land give very low remunerations or have become uneconomical (Mohammad 1981). Thus, the marginal and small landholders having small piece of land rely on the "common land resources" for their livelihood. The landless people also get economic benefit through utilizing these CLRs. The CLRs are common to all, and no one has exclusive rights on them. They are generally utilized by the landless poor, marginal and small farmers for economic gains. The forests provide timber, the pastures support the livestock, and the uncultivated and barren lands are utilized for infrastructures such as construction of houses, poultry farms and animal husbandry (Munir et al. 2008). In India and especially in the Uttar Pradesh (the most populous state of India), the CLRs are declining due to increasing population pressure, "open access" and rampant use.

34.1.1 Conceptual Base

The definition of resources has been different to different people in different ages. To state briefly we can define resource as "features of the environment which are considered to be capable of serving man's needs". They are given utility by the capabilities and wants of man. The "resource" to one community may not be a resource to another, and to someone else, it may be a "neutral stuff" only. Earlier the natural resources were abundant so there was no competition among the users. But gradually conflict started with the increase in the number of users. The natural resources have been thus classified into four categories by Sharma (1984) on the basis of whether the use does or does not lead to conflict between the users. The four categories are (1) resources used for individual's benefit which do not involve conflict, (2) resources used only for social ends without conflict, (3) resources used by individuals and society with conflict if the resource used is in limited supply and (4) resources used both by individuals and society without conflict if the resource used are in abundant supply.

With the increase in the number of users, people turn towards the common property resources (CPRs). The CPRs are resources owned by an identifiable group of people who have a "common access" to it, and the resource is regulated by social conventions and legally enforceable rules (Singh 1994). These include land, water, grasslands, wildlife and forests which are regulated by social conventions and legally enforceable rules (Burges and Gochfield 1998). Traditionally, the CPRs include land, water, grass, wildlife and forest (Berkes 1989). The CPRs have been studied by various scholars since the publication of Hardin's paper "The Tragedy of Commons" in 1968 where he has stated "ruin is the destination towards which all men reach, each pursuing his own interest in a society that believes in the freedom of the Commons" (Hardin 1968).

The "CLRs" are a part of CPRs and is used to refer to land owned and defended by a community of resource users, to property owned by no one and to property owned by a government to which the people have "common access". In India, there are variety of CLRs, such as forests, pastures and grazing lands, threshing grounds, manure pits, cemeteries, cremation grounds, fallow lands, barren land and uncultivated lands. The CLRs are common to all, and no one has exclusive rights upon them.

34.1.2 Objectives

The CLRs have significant effect upon the livelihood of the landless people, small and marginal farmers. Therefore, there is a need to manage these resources in a judicious way. The present study was thus conducted with the following objectives:

1. To study the spatial distribution of the CLRs in various districts of Uttar Pradesh
2. To study the decadal change in CLRs in the various districts of Uttar Pradesh
3. To identify the problems in managing CLRs
4. To suggest possible ways for proper management of the CLRs

34.2 Database and Methodology

The present study is based on the primary and secondary data. The secondary data regarding land use of Uttar Pradesh (1995 and 2005) was obtained from State Land Use Board, Uttar Pradesh and Directorate of Economics and Statistics, Uttar Pradesh, Lucknow, India. This data was used to study the spatial distribution and decadal change in the share of CLRs in different districts of Uttar Pradesh. The primary data regarding different CLR utilization modes and problems associated with their management was collected through field survey of a sampled village of a sampled district during 2009.

The secondary data of land use was used to study the spatial distribution of the CLRs. The districts were divided into five categories on the basis of the percentage of area under the CLRs. A district from the medium category was selected for the field survey and a sample village for the collection of data regarding the utilization modes of CLRs and the associated management problems. The selection of the district was made on the basis of the percentage share of area under CLRs in the district, diversity in physiography and land use pattern and the accessibility for field survey.

The district level data regarding geographical background of the area, climatic conditions including soil characteristics, rainfall, vegetation and land use pattern including general and agricultural land use was obtained from the Statistical Booklet of the selected district. A sampled village was selected from the district selected on the criteria given above. Household level field survey was conducted in the sampled village. The data regarding the population of the sampled village and the social structure was obtained from the village pradhan (political head). The selection of the village was based on population size, distance from the nearest city/town and accessibility.

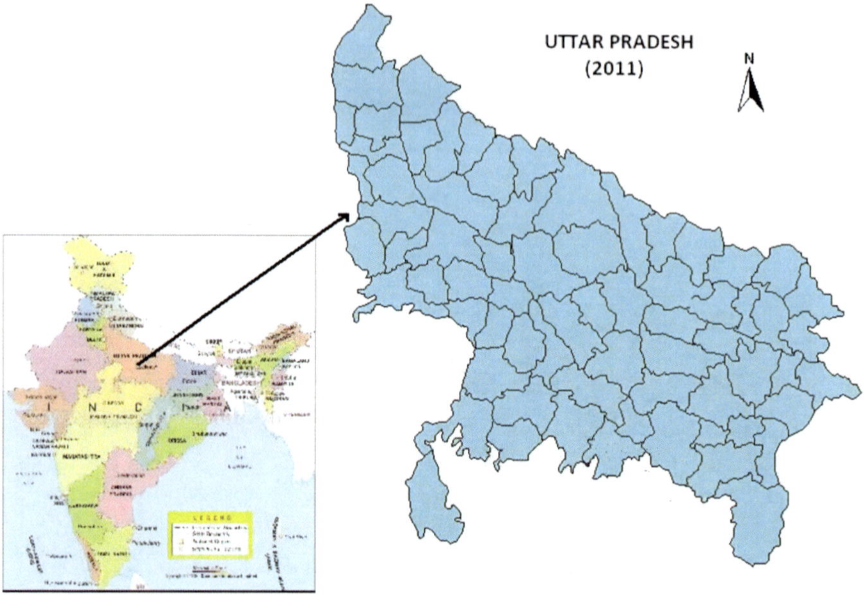

Fig. 34.1 Study area—Uttar Pradesh

Out of the total households, a 50% sample was taken for detailed field survey. The questionnaire includes the mode of CLR utilization and their associated management problems. The field survey data was thoroughly checked and processed using simple statistical and GIS techniques. The data is then presented as maps, tables and diagrams.

34.2.1 Study Area

The state of Uttar Pradesh (UP) is situated between 23°52′ N and 31°28′ N latitudes and 77°3′ and 84°39′ E longitudes in South Asian region. The location of the study area is shown in Fig. 34.1. The UP is the most populous and fifth largest state in India having 199,581,477 population (Census of India 2011) and an area of 24,201,586 km^2. The UP can be divided into three distinct hypsographical regions:

1. The Himalayan region in the North
2. The Gangetic plain in the centre
3. The Vindhya Hills and plateau in the south

The Himalayan region comprises of high mountains formed from sedimentary rocks. The perpetual snows in the higher reaches are the source of perennial water which forms a big river system watering the entire plain. The hilly areas are sparsely populated. Only few trees can grow in this terrain, where soil is subject to heavy erosion. Irrigation facilities are scarce and only a small area is under artificial irrigation. The soils in valley areas are fertile, and there is intensive cultivation on terraced hill slopes. The Siwalik Range forming the southern foothills of the Himalayas has rich forests, cutting across it are innumerable streams which swell into raging torrents during the monsoon. The main crops are wheat, rice and sugarcane and jute is also grown. Tea is grown in the submountain area of the Dehradun.

The Gangetic plain is the most important agricultural area of the country stretching across the entire length of the state from east to west. The Gangetic plain is watered by the Yamuna, the Ganga and its major tributaries, the Ramganga, the Gomti, the Ghaghra and Gandak. This is alluvial and very fertile plain. The major crops are rice, wheat, pearl millet, gram, and barley. Sugarcane is the main cash crop of the region. The southern fringe of the Gangetic plain is demarcated by the Vindhya Hills and plateau. It comprises the four districts of Jhansi, Jalaun, Banda, and Hamirpur. These districts are part of the Bundelkhand division of Uttar Pradesh. The Betwa and Ken rivers join the Yamuna from the south-west in this region. Of four distinct soils found in the Bundelkhand region black, cotton soil is difficult to manage. Due to scarce rain, dry farming is generally practised in the Bundelkhand region of the state.

34.2.2 Climate and Seasons

The climate of UP is tropical monsoon; however, variations exist with altitudes. The Himalayan region is cold. The average temperature varies in the plains from 3–4°C in January to 43–45°C in May and June. Climate is marked by three distinct seasons:

> *Summer* (March–June): hot and dry (temperature rise to 45°C, sometimes 47–48°C), low relative humidity (20%), and dust-laden winds, *Monsoon* (June–September): 85% of average annual rainfall (990 mm). Fall in temperature (40–45°C) on rainy days, and *Winter* (October–February): cold (temperature drop to 3–4°C, sometimes below −1°C), clear skies, and foggy conditions in some tracts

34.2.3 Forest and Wildlife

Most of the forests occur in the Himalayan region and the Terai and Bhabhar area in the Gangetic plain. The Vindhyan forests consist mostly of scrub. The hilly forests also have a large variety of medicinal herbs. Corresponding to its variegated topography and climate, the state has a wealth of animal life. Animals that can be found in the jungles of Uttar Pradesh include the tiger, leopard, wild bear, sloth bear, chital, sambhar, jackal, porcupine, jungle cat, hare, squirrel, monitor lizards and fox. The most common birds include the crow, pigeon, dove, jungle fowl, black partridge, house sparrow, peafowl, blue jay, parakeet, kite, mynah, quail, bulbul, kingfisher and woodpecker.

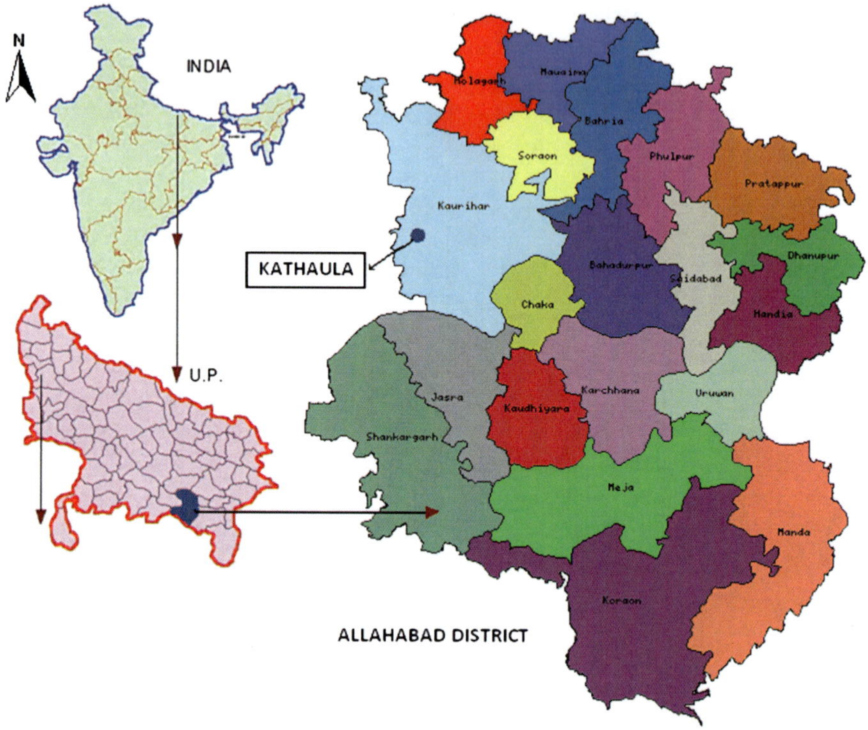

Fig. 34.2 Sampled village of Allahabad district

34.2.4 Sampled District and Village

Allahabad district is selected as study area. It is located in the eastern part of Uttar Pradesh between 24°47′ N and 25°43′ N latitudes and between 81°31′ E and 82°21′ E longitudes. It covers an area of 5,482 km^2. The total rural area is 5,339.28 km^2, and the urban area is 142.72 km^2. The River Ganga and Yamuna flow through the district. From administrative point of view, the district has been divided into 7 Tehsils and 20 blocks which include 3,065 villages (District Statistical Booklet 2005). On the basis of physiography, the district can be divided into three parts, that is, Gangapaar, Yamunapaar and Doab regions. The Gangapaar region has "Khadar" soil formed by floods, Doab has fertile alluvial soil, and "Yamunapaar" possesses gravelly light sandy soil. All the blocks of the district are well connected to other parts of the district and the state through the network of rail and road. The Grand Trunk Road connecting the capital city of New Delhi to other parts of the country cross the district from west to east.

The sampled village Kathaula selected for the field survey lies in the Kaurihar Block falling in the Doab region of Allahabad district (Fig. 34.2). The village is located at a distance of 11 kms from the Allahabad city which is also the headquarters

Table 34.1 Decadal population change in Uttar Pradesh (1951–2011)

Year	Population (In 000's)	Population growth (In 000's)	Percentage change
1951	60,274	–	–
1961	70,144	9,870	14.07
1971	83,849	13,705	16.34
1981	1,05,137	21,288	20.24
1991	1,32,062	26,925	20.38
2001	1,66,198	34,136	20.53
2011	1,99,581	33,384	20.09

Source: Directorate of Statistics and Economics, Uttar Pradesh, Lucknow

of the Allahabad district. The village is well connected to nearby town, Allahabad and the nearby villages with metalled and unmetalled roads. Total population of the village is 1,700 with approximate 300 households. Hindus and Muslims are two major communities in the Kathaula village. Bhartiya, Brahman, Chamar, Khurmi, Lohar, Madari, Nai and Pasi are the main Hindus castes living. The people are generally illiterate, and a large number of the residents are landless A large number of people have small or marginal farms, and very few have big landholdings.

34.2.5 Temporal Change in Population of Uttar Pradesh

The population of Uttar Pradcsh is continuously increasing over the last few decades. This continuing population growth has led to immense pressure on CLRs of the state. The area being the same, increasing population has led to rampant use of CLRs which have an "open access". Since independence the decadal population change is inconsistent (Table 34.1).

34.3 Common Land Resources (CLRs) of Uttar Pradesh

The CLRs in UP are comprised of 3,460,826.8 ha area, which is 14.30% of the total state area, divided into different CLR categories (Table 34.2) varying in size from 0.28% (pastures and grazing land) to 6.98% (forest).

34.3.1 Spatial Distribution of Common Land Resources

The CLRs in 70 districts of UP vary from 2.07% in Moradabad district to 59.09% in Sonbhadra district. The CLR difference is being due to difference in physiography, land use pattern, population and urbanization. The spatial distribution of CLRs in

Table 34.2 Share of different categories of common land resources in Uttar Pradesh (2001)

Category	Area (ha)	Percentage of total area of UP
Forest	1,689,270.7	6.98
Barren and unculturable	575,997.74	2.38
Culturable wasteland	498,552.67	2.06
Pasture and grazing land	67,764.44	0.28
Other fallow land	629,241.24	2.60
Total CLR	3,460,826.79	14.30
Total area of UP	24,201,586.00	100.00

Source: State Land Use Board, Uttar Pradesh, Lucknow (2005)

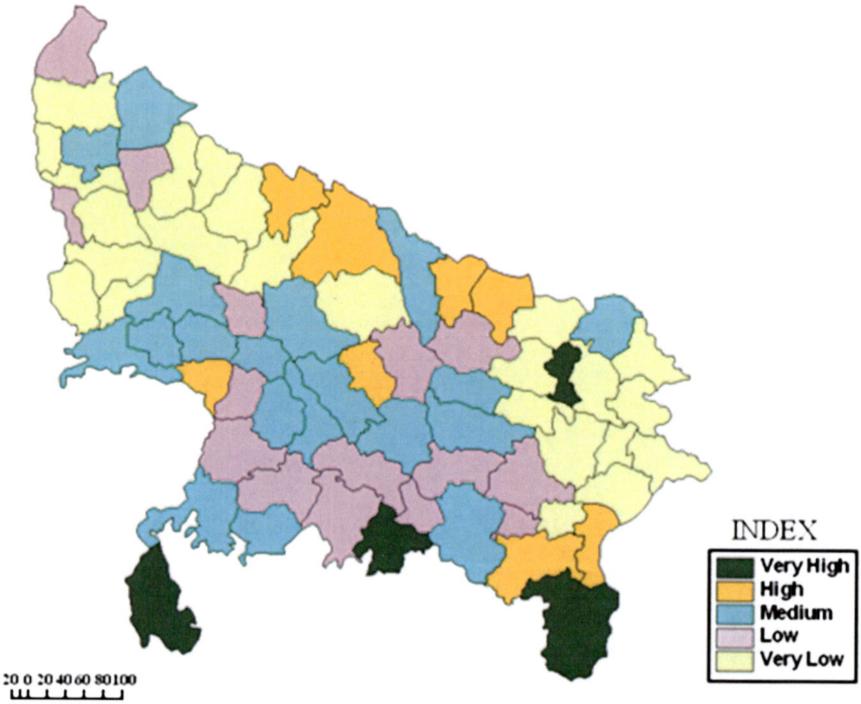

Fig. 34.3 Common land resources of Uttar Pradesh (2005)

different districts and the number of districts in different categories are shown in Fig. 34.3 and Table 34.3, showing largest number of districts in the north-western and eastern part falling in very low category (24 districts). The western and eastern parts are formed from fertile alluvium and have dense population. Thus, most of the forests and grazing lands have been cleared for agriculture or urbanization. There are 15 districts under the low category, generally lying in the south-central part of UP. The next largest share is the medium category with 19 districts mostly located in the central part of the state. This region has a large share of barren and uncultivable land, culturable wasteland and other fallow land. The numbers of districts in high and very

Table 34.3 Number of districts under different category and share of CLRs

Category	Range (share of CLRs to total area)	No. of districts
Very high	(37.05–64.05)	04
High	(18.74–37.05)	08
Medium	(11.41–18.74)	19
Low	(7.94–11.41)	15
Very low	(0–7.94)	24
Total		70

Source: Directorate of Statistics and Economics, Uttar Pradesh, Lucknow

high categories are 8 and 4, respectively. The districts under high category lie mostly in the northern part of the state, whereas those in very high categories are in the southern part of the state which forms the part of the Vindhya Hills and plateau.

34.3.2 Temporal Change in Common Land Resources

The CLRs in the UP are changing continuously. The forest cover and the pastures are continuously decreasing at a fast rate; however, the change in uncultivable land, fallow land and the cultivable wasteland has been observed at different rates. Thus, the CLRs are in a dynamic state.

To investigate the temporal change in CLRs, their spatial distribution was analysed for the years 1995 and 2005. The share of CLRs in UP during 1995 was 14.88% which decreased to 14.30% in 2005. Thus, a decline of 20,765 ha has taken place. The share of CLRs during 2005 among various districts varied between 4.12% (Deoria district) to 64.06% (Sonbhadr district) in 1995 (Fig. 34.4). The data revealed a decrease in the CLRs. A variation recorded in the number of districts under various categories in 1995 and 2005. There is a shift in the number of districts from very high and high categories to medium and from low to very low categories (Fig. 34.5).

Further analysis of the district wise decadal change in CLRs was made, and an increase of CLRs in only seven districts was observed, in others CLRs declined. The increase (Muzaffarnagar and Maharajganj districts) was due to the increase of barren and unculturable land area and creation of new districts (Ambedkar Nagar and Baghpat). In other districts, the increase in barren and unculturable land, culturable wasteland and fallow land is the main reason of CLR increase.

The significant decrease in CLRs is recorded in Moradabad (−68.52%), Gorakhpur (−53.69%), Varanasi (−75.07%), Saharanpur (−44.56%) and Gonda (−45.98%) districts; this is due to the increase in population and urbanization, the high price of land in the National Capital Region being the reason of agricultural land transformation to urbanization, industry and even to institutional area and other factors being decrease in pastures, uneconomic animal rearing, and low economic return from agricultural practices, leading to investment in other sectors and selling of land for nonagricultural purposes.

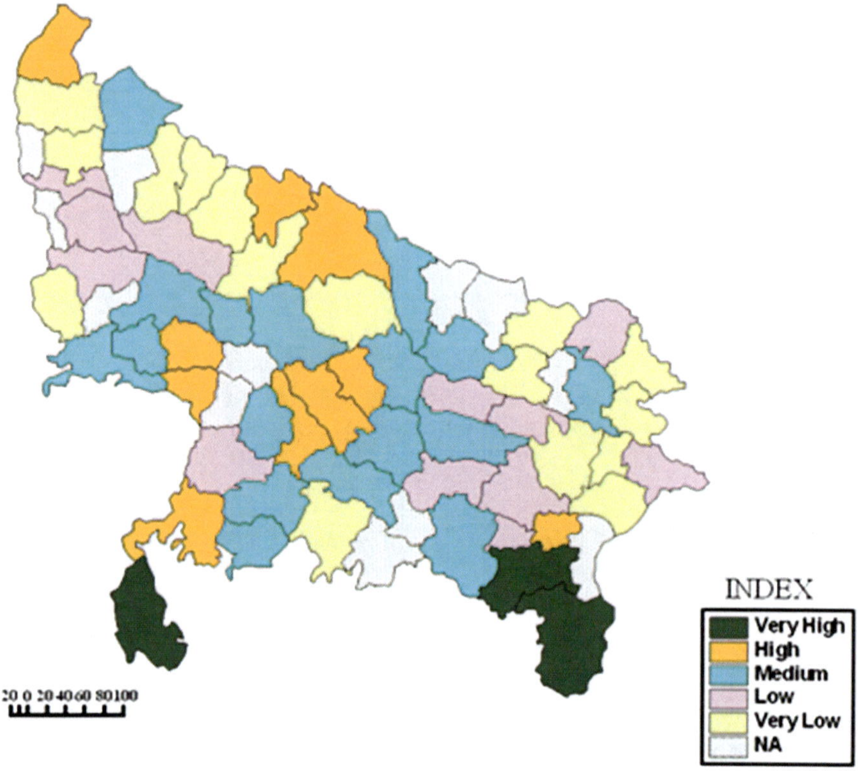

Fig. 34.4 Common land resources of Uttar Pradesh (1995)

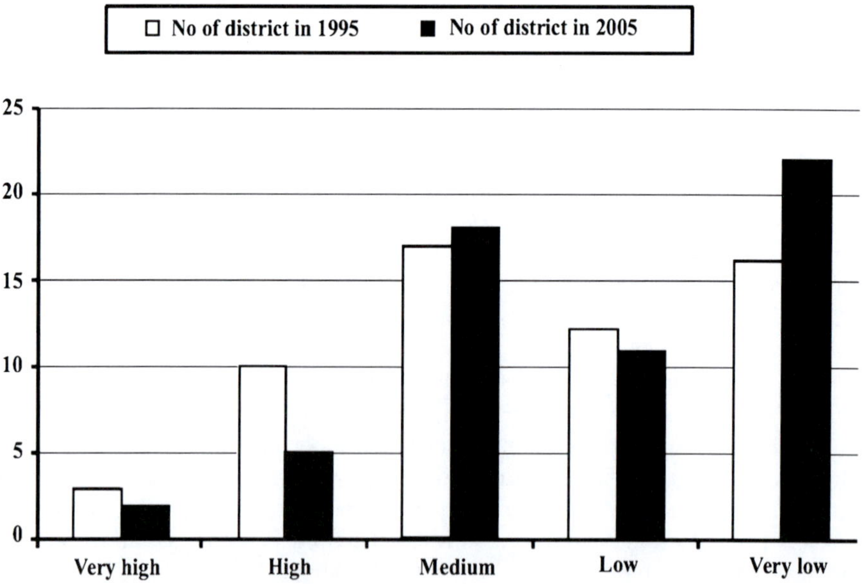

Fig. 34.5 Categorical change in the districts of Uttar Pradesh (1995–2005)

Table 34.4 Landholding wise share of total respondents using CLR

Category	Size of landholding (ha)	Respondents		Respondents using common land resources	
		Number	Percent	Number	Percent
Landless	–	55	36.67	44	38.26
Marginal	<1	73	48.67	62	53.91
Small	1–2	14	9.33	6	5.22
Medium	2–10	8	5.33	3	2.61
Total	–	150	100.00	115	100.00

34.3.3 Socio-economic Profile of the Respondents

The respondents include Bhartiya, Brahman, Chamar, Khurmi, Lohar, Madari, Nai, Pasi and Muslims. The largest share being of Khurmi (15.00%) followed by Muslims (13.97%) and Brahman (11.02%). The respondents have joint family system and variable family size (3–14 persons). People are generally illiterate and not inclined to education for earning money. Thus, the educated are not interested to live in village, but prefer to work as office clerks or teachers to supplement income from agriculture. The landless are engaged in rickshaw pulling (taxi driver), labour, agricultural labour, private business and shopkeeping, etc.

The families with more than 10 members have 3–4 earners, and the families with 10–15 persons have 5–6 earners. The families with very large landholdings enjoy the benefits of agricultural mechanization, and sufficient income is generated from very large landholdings; thus, there is no social or economic pressure on other family members.

34.3.3.1 Size of Landholding

Table 34.4 illustrates the status of landholdings by 150 respondents. Only 95 (63.33%) respondents have landholdings, others are landless, and 115 (76.67%) were actually using the CLRs of different sizes. Of 115 households using CLRs, the largest share was of marginal farmers (53.91%) followed by the landless people (38.26%) and small category (5.22%). A continuous decrease in the total number of households with the increase in the size of landholdings was observed (Table 34.4).

34.4 Utilization Mode of Common Land Resources

The respondents were using CLRs in various ways, the most common being grazing and pastureland (92.17%) followed by crop cultivation (78.26%) and other uses (77.39%). The CLRs were also used for agroforestry and as manure pits, cemeteries, storage grounds, playgrounds, temporary construction of sheds for animals, storage of fodder and agricultural produce, etc. (Table 34.5).

Table 34.5 Landholding wise share of the Respondents under various modes of Common Land Utilization

Category of landholders	Respondents using CLR	Number of respondents utilizing common land resources under different modes			
		Grazing/pasture	Agroforestry	Crop cultivation	Other
Landless	44 (38.26)	44 (41.51)	12 (29.27)	36 (40.00)	36 (40.45)
Marginal	62 (53.91)	55 (51.89)	23 (56.10)	51 (56.67)	48 (53.93)
Small	6 (5.22)	6 (5.66)	5 (12.20)	3 (3.33)	3 (3.37)
Medium	3 (2.61)	1 (0.94)	1 (2.44)	0 (0.00)	2 (2.25)
Total	115 (100.00)	106 (92.17)	41 (35.65)	90 (78.26)	89 (77.39)

Source: The figures in bracket shows percentage

34.4.1 *Grazing and Pastures*

This is most common use by 92.17% respondents. This mode is widely practised among all categories of CLR users. The landless people and all landholders especially marginal and small farmers generally keep cows and buffaloes for agricultural, domestic and business purposes. The landless people depending upon drought animals generally utilize the CLRs as grazing and pasture land.

34.4.2 *Crop Cultivation*

Crop cultivation is next important use of CLRs (78.26%) among the total modes of utilization of common land resources. These lands are either Gram Smaaj land or surplus land acquired during land ceiling act. This land is sometimes given on lease by village administrative bodies to the landless villagers for a fixed period of time; often, the land is grabbed by rich and politically influential farmers, should it occur adjacent to their agricultural fields. If such land is taken by rich farmer, these CLRs are very well utilized with high inputs. The marginal (56.67%) farmers are the largest beneficiaries from this mode of CLR utilization.

34.4.3 *Other Uses*

The miscellaneous uses of CLRs are housing, poultry farms, playgrounds, quarrying and resting grounds for animals. The proper use of CLRs can provide many socio-economic benefits.

34.4.4 Agroforestry

About 35.65% respondents use CLRs for agroforestry. There are two types of agroforestry: (1) planting eucalyptus or Babool trees around the cultivated areas as shelter belts and as source of fuelwood and fodder, and (2) plantation of fruit trees on the borders of agricultural fields, providing timbre, small quantity of fuelwood and fruits.

34.5 Problems in CLR Management

The role of CLRs in providing economic benefits to landless people, small and marginal farmers, is evident. There is a need for optimum use of CLRs for improving livelihood; however, there are many major constraints as described below:

Open access: The CLRs due to open access to all lead to the problem of their preservation and management.

Lack of suitable laws: Lack of laws for those who degrade or misuse the CLRs. The government should take action against those doing undue activities and rampant use.

Awareness among masses: The people are unaware of the environmental problems caused due to soil resources degradation and depletion, and the processes and factors causing loss of land resource.

Lack of participation by local people: The lack of people interest has adversely affected many government schemes of land reclamation, afforestation and soil conservation.

Social injustice: The CLRs are not given to the needy people and often encroached by the wealthy and large farmers who hardly care for these resources.

Political problems: There are many political issues related to the allotment, management and control of the CLRs. Thus, the proper management is lacking.

34.6 Conclusion and Recommendations

Considering the present situation, the urgent need to preserve and manage CLRs and to address associated problems following recommendations are made. A thorough survey of the common land resources should be undertaken by the government through State Land Use Boards and other agencies to know the actual situation. There is an urgent need to formulate laws to protect the CLRs and livelihood of the unprivileged people. Environmental awareness should be encouraged through electronic media and by other suitable means. The encroachment of CLRs should be controlled by local administration, and the landless people should be allotted these lands for a small period of time. The local people should be involved for protection of forests and pastures by giving some incentives. Agroforestry should be encouraged as modes of CLR utilization to protect them from depletion.

References

Berkes F (1989) The benefits of the commons. Nature 340:91–93

Burges J, Gochfield M (1998) The tragedy of the commons thirty years later. Environment 40:4–27

Census of India (2011) Provisional population totals. Office of the Registrar General and Census Commissioner, India, Ministry of Home Affairs, 31 Mar 2011

District Statistical Booklet, Allahabad 2005 and 1995. Directorate of Statistics and Economics, Lucknow, Uttar Pradesh

Hardin G (1968) The tragedy of the commons. Science 162:1243–1248

Mohammad N (1981) Perspective in agricultural geography-framework for agriculture, vol 3. Concept Publications, New Delhi

Munir A, Salman MS, Toufique M (2008) Impact of common land resources on the livelihood of landless, small and marginal farmers: a study of Allahabad district. The Geographer 55(2):1–15

Sharma PR (1984) Resource: its concept, classification and appraisal. In: Rameah A (ed) Resource geography. Heritage Publishers, New Delhi

Singh K (1994) Managing common pool resources. Oxford University Press, New Delhi

State Land Use Board (2005) Department of planning. Government of Uttar Pradesh, Lucknow

Chapter 35
Participatory Soil and Land Evaluation Mapping: An Alternative Approach to Improve Soil and Land Evaluation Information for Decision Makers

Ivan Luiz Z. Bacic, Juniele R. Pivetta, and Roberta P. Martins

Abstract Results of soil and land evaluation survey and mapping, which were expected to be essential to rural land use and management planning, most of the time are not adequately used by their potential users and rarely reach the decision makers. This may reflect the poor communication between the main actors of the rural land use planning process and inappropriate language of the presented results. When following traditional methods, there is a risk that questions may be answered that have no relevance and/or those questions may not be answered properly according to the community expectation. Therefore, it is important to change the strategies of communication and language of the results to make information more useful to the decision makers. In this context, the objective of this work was to test a participatory soil survey and land evaluation methodology to make the information more useful and consequently more used by the decision makers. This work shows the results obtained up to now in Barra Bonita municipality, Santa Catarina State, southern Brazil. The study started with meetings, interviews and questionnaires with local community, and the main demands for information raised were the need for area expansion and management improvement of pastures for milk production. But, instead of giving them information showing soil types with technical language and a general information about land evaluation, the presented information shows what they want to know,

I.L.Z. Bacic (✉)
Centro de Informações de Recursos Ambientais e de Hidrometeorologia de Santa Catarina (CIRAM), Empresa de Pesquisa Agropecuária e Extensão Rural de Santa Catarina (EPAGRI), Florianópolis, SC, Brazil
e-mail: bacic@epagri.sc.gov.br; ivan.zilli@gmail.com

J.R. Pivetta • R.P. Martins
Conselho Nacional de Desenvolvimento Científico e Tecnológico (CNPq), Centro de Informações de Recursos Ambientais e de Hidrometeorologia de Santa Catarina (CIRAM), Empresa de Pesquisa Agropecuária e Extensão Rural de Santa Catarina (EPAGRI), Florianópolis, SC, Brazil
e-mail: batateiraju@hotmail.com; robervalda@hotmail.com

S.A. Shahid et al. (eds.), *Developments in Soil Classification, Land Use Planning and Policy Implications: Innovative Thinking of Soil Inventory for Land Use Planning and Management of Land Resources*, DOI 10.1007/978-94-007-5332-7_35,

namely, spatial location of potential new areas for pasture, what pasture type is better for different conditions and what management is recommended for each area. In the final stage of the study, participants positively evaluated the presented information.

Keywords Alternative approach • Decision makers • Soil mapping • Rural land use • Land evaluation

35.1 Introduction

Land evaluation is the process of predicting land performance over time according to specific land use types (Van Diepen et al. 1991; Rossiter 1996). These predictions should then be used to guide strategic land use and management decisions. However, results from land inventories and soil and land evaluation surveys and mapping, which were expected to be essential to rural land use and management planning, most of the time are not adequately used by their potential users-stakeholders and rarely reach the decision makers (Rossiter 1996; Bouma 1999; Bacic 2003; Bacic et al. 2003). This may reflect the low relevance of many land evaluation and soil survey reports, the poor communication between the main actors of the land use planning process (e.g. soil surveyors, farmers and rural extensionists) and perhaps inappropriate language of the presented results. When following traditional methods, there is a risk that questions may be answered that have no relevance and/or those questions may not be answered properly according to the community expectation.

Bacic (2003) suggests that the process should begin with a careful analysis of the environment where the stakeholders live and, consequently, where decisions are made, following a participatory and demand-driven methodology. Besides, it is important to change the strategies of communication and language of the results. Therefore, it is expected that the decision makers will be given the information they consider more relevant and with more appropriate language.

In this context, the objective of this work was to test a participatory soil survey and land evaluation methodology to make the information more useful and consequently more used by the decision makers.

35.2 Materials and Methods

This work is part of the project "Demand-driven land evaluation" that is being developed in Santa Catarina State, southern Brazil, by the "Empresa de Pesquisa Agropecuária e Extensão Rural de Santa Catarina-EPAGRI", with financial support from the "Conselho Nacional de Desenvolvimento Científico e Tecnológico-CNPq", the "Ministério do Desenvolvimento Agrário/Secretaria de Agricultura Familiar-MDA/SAF" and the "Ministério do Desenvolvimento Social e Combate a Fome/Secretaria Nacional de Segurança Alimentar e Nutricional-MDS/SESAN". The project started on December 2007 and lasts for 30 months. Three municipalities were selected according to the following main criteria: (1) location on three different regions in Santa Catarina State and (2) acceptance and support of local teams (extensionists), local government and primarily the communities. Following these criteria and after several meetings, the following municipalities were selected: Lontras (Upper Itajaí River Valley), Luzerna (Mid-West) and Barra Bonita (West), where traditional soil and land evaluation maps and reports are available and were barely used.

The work presented in this chapter shows the results obtained up to May 2010 in Barra Bonita municipality. The study area, "Pomba Branca Microcatchment", was selected in a meeting with local leaders.

The first step was to raise the demands for interpreted information (e.g. land evaluation, climatic risks, production systems, economic information), according to the local decision makers (farmers, extensionists and local leaders) needs. The demands were raised in meetings, after a presentation, questionnaire application and interviews. At this time, in order to understand better the local conditions, the farmers also answered general questions about farm size, number of people working in the farm, main rural activities, if they wanted to improve or even abandon these activities, as well as if they had the intention to start a new activity and which information would be necessary to make this decision. The demanded information was then organized and ordered following the priorities, according to the urgency and frequency. This chapter deals with the main demands raised, namely, the need for area expansion, recommended pasture species and management improvement of pastures for milk production.

The criteria to evaluate potential expansion areas for pasture and to indicate the better species and management for different areas were defined based on literature and from meetings with experts. The main criterion was defined as the slope degree. In areas with slope degrees lower than 20%, it is possible to recommend both annual and perennial pastures, and no special conservation practices are needed. In this case, the land cover produced by the pasture is considered to be enough to control the erosion. In areas with slope degrees ranging from 20 to 30%, as the erosion susceptibility is higher, recommended species are with the following characteristics: erosion control potential, cattle trampling resistance, resistant to environmental adversities, rapid growing and good soil covering. In areas where the slope degrees vary from 30 to 45%, it is recommended to use perennial pastures due to the even higher erosion susceptibility and difficulties to use agricultural mechanization. In areas where the slope degrees are higher than 45%, it is not recommended to use pasture as land use, due to the high erosion susceptibility, difficult workability and bad conditions to the animals. Another criterion to define potential expansion areas was the land use. It was considered as potential areas only those with annual crops, meadow and reforestation. Regarding the pasture management improvement potential, the areas with continuous grazing (identified as pastures in the land use map) associated to slope degree <30% are considered.

As primary information, the soil and land use survey and mapping were used. The final results about potential expansion areas and recommended species were obtained overlaying the information described above.

Based on the criteria explained above, the following four classes were defined for the study area:

- Class 1: <20% slope degree associated to annual crops, meadow or reforestation
- Class 2: 20–30% slope degree associated to annual crops, meadow or reforestation

- Class 3: 30–45% slope degree associated to annual crops, meadow or reforestation
- Class 4: <30% slope degree associated to pasture

The classes 1, 2 and 3 are considered the classes with potential for pasture area expansion; class 4 is related to the potential for pasture management improvement.

Next, other meeting was organized to present the demanded information as well as to assess the relevance and quality of the demanded information, to evaluate the tools and methods used and to re-evaluate the priorities according to the participants' opinions.

35.3 Results and Discussion

35.3.1 The Study Area

The study area is the Pomba Branca Microcatchment (460 ha), which is representative of most of the Barra Bonita municipality (107 km^2) in western Santa Catarina State, Brazil (Fig. 35.1).

Barra Bonita city centre is located at 730 m above sea level (masl). Applying the Köppen system, Barra Bonita is included in Cfa climate, mesothermal humid, without defined dry season and warm summer (temperatures above 22°C in the warmest month) (Santa Catarina 1986). The underlying geologic formation is the "Serra Geral", composed of dark greyish to black basalt (Silva and Bortoluzzi 1987). There are, according to IBGE (2000), 1,862 people living in rural areas and 256 in urban areas. The principal land uses are annual crops (mainly maize, cassava, tobacco and beans), pasture for milk production and swine (Santa Catarina 2005; IBGE 2000, 2006).

Pomba Branca Microcatchment is characterized by small family farms, as the farms are mostly smaller than 20 ha and most of the labour is from family members. The main commercial agricultural activities are milk production, tobacco, maize and beans. The predominant soils in the study area are Cambissolos, Neossolos Litólicos and Neossolos Regolíticos according to the Brazilian system (EMBRAPA 2006), corresponding to Inceptisols and Entisols in Soil Taxonomy (Soil Survey Staff 1999). Table 35.1 shows the Physiographic units distribution in the microcatchment, as well as the main land evaluation classes according to the Santa Catarina System (Uberti et al. 1991) and predominant slope degrees. The land use distribution in Pomba Branca Microcatchment is shown in Table 35.2. The physiographic and land use maps are shown in Figs. 35.2 and 35.3, respectively.

35.3.2 Land Evaluation Using Traditional Santa Catarina Method

Land evaluations maps and reports in Santa Catarina State are prepared according to the "Metodologia para Classificação da Aptidão de Uso das Terras do Estado de Santa Catarina" (Methodology for Land Suitability Classification in Santa Catarina

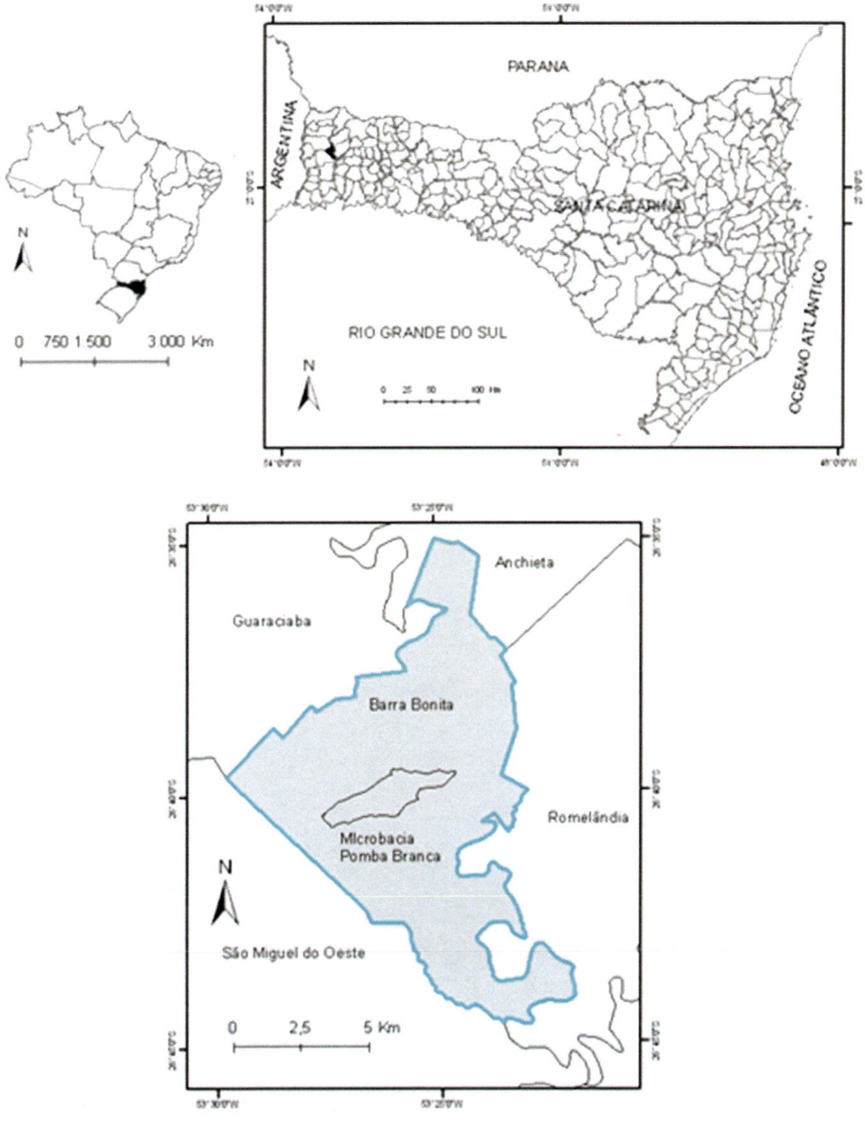

Fig. 35.1 Location of Pomba Branca Microcatchment in Barra Bonita municipality, Santa Catarina State, southern Brazil

State) (Uberti et al. 1991). It classifies land into one of five classes of physical suitability, which are first defined in descriptive terms as follows:

- Class 1—Good suitability for climatically adapted annual crops
- Class 2—Moderate suitability for climatically adapted annual crops
- Class 3—Restricted suitability for climatically adapted annual crops, moderate suitability for fruit production and good suitability for pasture and reforestation

Table 35.1 Physiographic units distribution, slope degrees and land evaluation classes in Pomba Branca Microcatchment

Physiographic units	Area (ha)	Area (%)	Slope degrees (%)	Land evaluation classes according to Uberti et al. (1991)
Ee—erosional hillslope	216	46.9	>30	4e (3dp)
Eec—erosional-colluvial hillslope	109	23.7	20–30	3pe (3prp) (3dp)
Ce—erosional summit	10	2.2	0–8	3prp
Ece1—colluvial-erosional hillslope 1	96	20.9	0–8	2pr (2p)
Ece2—colluvial-erosional hillslope 2	19	4.1	8–20	2de
Ep—steep hillslope	10	2.2	8–20	2p (3prp)
Total	460	100.0	–	–

Table 35.2 Land use in Pomba Branca Microcatchment

Land use	Area (ha)	Area (%)
Buildings	0.7	0.2
Fruit trees	0.8	0.2
Water reservoir	6.0	1.3
Pasture	250.0	54.3
Meadow	15.0	3.3
Annual crops	63.5	13.8
Forest	116.0	25.2
Reforestation	8.0	1.7
Total	460.0	100.0

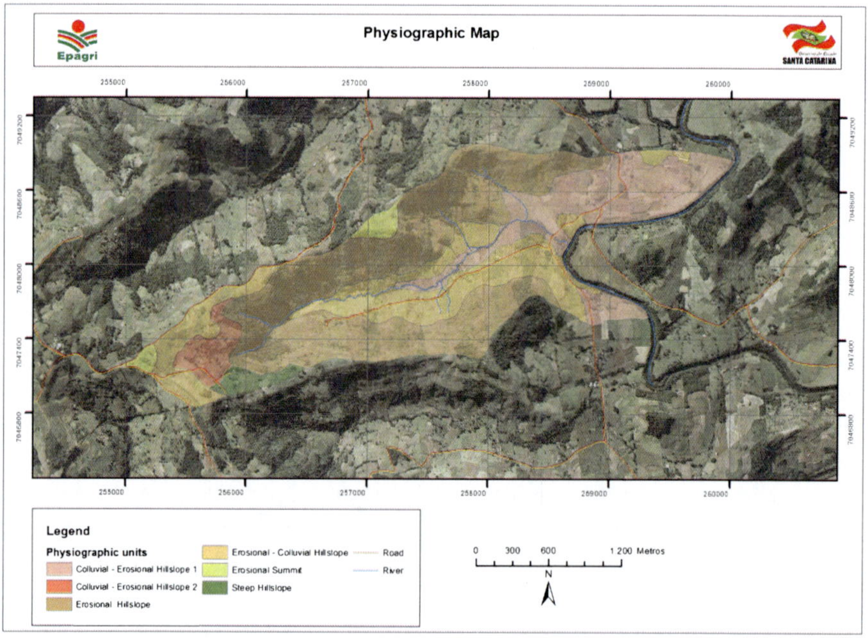

Fig. 35.2 Physiographic Map (Cartographic Base Anchieta, IBGE 1:50,000; Projection Universal Transverse Mercator; Datum SAD-69; Zone 22 South)

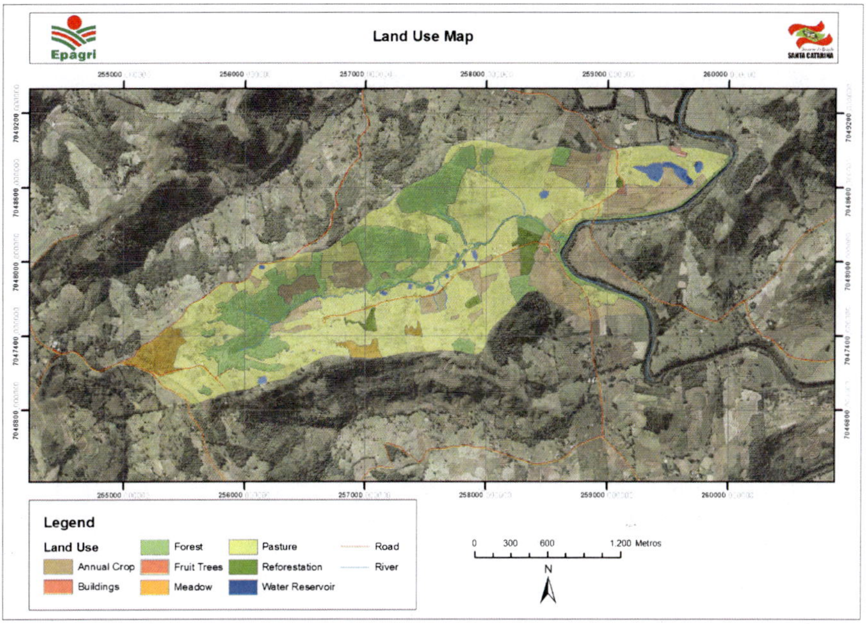

Fig. 35.3 Land Use Map (Cartographic Base Anchieta, IBGE 1:50.000; Projection Universal Transverse Mercator; Datum SAD-69; Zone 22 South)

- Class 4—Restricted suitability for fruit production and moderate suitability for pasture and reforestation
- Class 5—Permanent conservation areas

To make the description operational, the following specific land characteristics are measured for each map unit: slope gradient (d), effective soil depth (pr), stoniness (p), susceptibility to erosion (e), fertility (f) and drainage conditions (h). These are compared with defined limits for each suitability class (Table 35.3), using the maximum limitation method (Sys et al. 1991). There is no separate evaluation for different annual crops or for fruit, horticulture, pasture or forestry, and in addition, there is no differentiation between management levels or techniques.

Following this methodology and considering the land evaluation classes found in the study area (Table 35.1), the recommendations for pasture would be that some areas have good suitability (classes 2 and 3) and others moderate suitability (class 4). Therefore, it is clear that those recommendations are not enough to answer to the farmers' demands.

35.3.3 Potential Areas for Pasture Expansion and Recommended Species and Management

Observing and analysing information in Fig. 35.4 and Table 35.4, it is possible to identify the low potential of the study area to pasture expansion. Only 24.6 ha (5.3% of the study area) is suitable to pasture expansion. Thus, this can just be an alternative for a

Table 35.3 Santa Catarina land use suitability classification rating table

Land suitability classes	Slope gradient (%) (d)	Effective soil depth (cm) (pr)	Stoniness (p)	Susceptibility to erosion (e)	Fertility (ton lime ha^{-1}) (f)	Drainage conditions (h)
Class 1	0–8	>100	No stones	No/slight	0–6	Well drained
Class 2	8–20	50–100	Fairly stony	Moderate	6–12	Imperfectly drained
Class 3	20–45	<50	Stony/very Stony	Strong	>12	Any
Class 4	45–75	Any	Very stony	Very strong	Any	Any
Class 5	>75	Any	Exceeding stony	Any	Any	Any

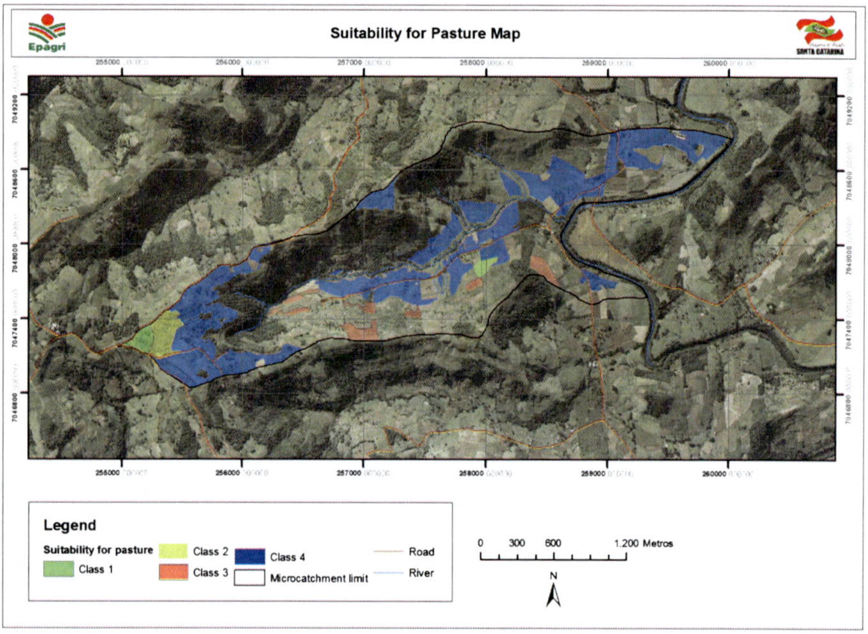

Fig. 35.4 Potential areas for pasture expansion and recommended species and management (Cartographic Base Anchieta, IBGE 1:50.000; Projection Universal Transverse Mercator; Datum SAD-69; Zone 22 South)

small number of farmers. But, considering that the main reason the farmers wanted to increase their pasture area was to increase their profit, improving their financial conditions, another possibility is to improve the pasture management. The potential in the area to improve management is higher, as 30.4% (140 ha) of the study area currently with pasture under continuous grazing could be changed to Voisin grazing management. Following this recommended management, it will be possible to increase number of animals, milk productivity, production and consequently the income. Table 35.5 shows the recommended species and management for different conditions.

Table 35.4 Potential areas for pasture expansion and management improvement

Class	Area (ha)	Area (% of the study area)
Class 1	5.2	1.1
Class 2	7.8	1.7
Class 3	11.6	2.5
Class 4	140.0	30.4

Table 35.5 Recommended pasture species and management

Slope degree (%)	Grass pasture	Legume pasture	Recommended management
0–20%	*Avena sativa* L. (Aveia branca); *Avena strigosa* Sereb (Aveia preta); *Secale cereale* L. (Centeio); *Lolium perenne* L. (Azevém); *Braquiaria decumbens* Stapf. Prain (Braquiária); *Paspalum notatum* (Grama comum); *Cynodon nlemfuensis* Vanderryst (Tifton); *Zea mays* L. (Milho); *Sorghum vulgare* Pers (Sorgo); *Pennisetum purpureum* Schum (capim elefante); *Festuca arundinaceae* Schreb (Festuca); *Braquiaria mutica* (Angola); *Braquiaria humidicola* (Humidicola); *Panicum maximum* (Colonião)	*Vicia sativa* L. (Ervilhaca); *Lotus corniculatus* (Cornichão); *Trifolium repens* L. (Trevo branco); *Trifolium pratense* L. (Trevo vermelho); *Arahis pintoi* (Amendoim forrageiro)	Voisin grazing management
20–30%	*Braquiaria decumbens* Stapf. Prain (Braquiária); *Cynodon nlemfuensis* Vanderryst (Tifton); *Pennisetum purpureum* Schum (Capim Elefante); *Festuca arundinaceae* Schreb (Festuca); *Braquiaria mutica* (Angola); *Braquiaria humidicola* (Humidicola); *Panicum maximum* (Colonião)	*Lotus corniculatus* (Cornichão); *Arahis pintoi* (Amendoim forrageiro)	Voisin grazing management
30–45%	*Panicum maximum* (Colonião); *Braquiaria humidicola* (Humidicola); *Braquiaria mutica* (Angola); *Pennisetum purpureum* Schum (Capim Elefante); *Cynodon nlemfuensis* Vanderryst (Tifton)	*Arahis pintoi* (Amendoim forrageiro)	Continuous grazing

But it is also important to emphasize that even the management improvement will not be the best option to all the farms. As the labour in the study area is based on the family labour, some farms will not have enough people to carry on the additional work demands. In these cases, other land use alternatives should be considered in following stages of the project.

35.4 Conclusions

The presented information showing clearly what the farmers wanted to know, instead of giving them the traditional information showing soil types with technical language and general information about land evaluation, was positively evaluated by them. They easily understood and considered the information useful to help them to make their decisions.

As the specific conditions of each farm are different, there are no recommendations that fit every individual situation. This chapter showed alternatives like the potential for area expansion and the management improvement of pastures, as possible options for some farmers, but they should also receive a set of other realistic land use alternatives from which they could choose from.

The methodology is based on several negotiation rounds, and up to the end of the project, it is expected to find in a collective way, different land use and management techniques, new options to increase incomes without environmental damages, improving the rural families' social inclusion and their life quality.

Acknowledgements This work was supported by CNPq, MDA/SAF, MDS/SESAN and Epagri.

References

Bacic ILZ (2003) Demand-driven land evaluation: with case studies in Santa Catarina, Brazil. Wageningen University/ITC, Wageningen/Enschede

Bacic ILZ, Rossiter DG, Bregt AK (2003) The use of land evaluation information by land use planners and decision-makers: a case study in Santa Catarina, Brazil. Soil Use Manag 19:12–18

Bouma J (1999) Land evaluation for landscape units. In: Sumner ME (ed) Handbook of soil science. CRC Press, Boca Raton, pp E393–E412

EMBRAPA-Empresa Brasileira de Pesquisa Agropecuária (2006) Centro Nacional de Pesquisa de Solos. Sistema brasileiro de classificação de solos, 2nd edn. Rio de Janeiro, p 306

IBGE (2000) Censo Demográfico do Brasil 2000 – Características da População e dos Domicílios: Resultados do universo. Rio de Janeiro, IBGE. Available at: http://www.ibge.gov.br/home/estatistica/populacao/censo2000/universo.php?tipo=31o/tabela13_1.shtm&paginaatual=1&uf=42&letra=B. Verified on Mar 2010

IBGE (2006) Censo Agropecuário 2006. IBGE, Rio de Janeiro. Available at: http://www.sidra.ibge.gov.br/bda/pesquisas/ca/default.asp?o=2&i=P. Verified on Mar 2010

Rossiter DG (1996) A theoretical framework for land evaluation (with discussion). Geoderma 72:165–202

Santa Catarina (1986) Atlas de Santa Catarina. Aerofoto Cruzeiro, Rio de Janeiro, p 173

Santa Catarina (2005) Secretaria de Estado da Agricultura e Desenvolvimento Rural. LAC – Levantamento agropecuário de Santa Catarina: 2002–2003. Florianópolis: Instituto Cepa-SC, Available at: http://cepa.epagri.sc.gov.br/Publicacoes/lac_fev_2005.pdf. Verified on Mar 2010

Silva LC, Bortoluzzi CA (1987) Mapa geológico do Estado de Santa Catarina escala 1:500000: texto explicativo. DNPM – 11o Distrito/Secr. Ciência Tecnologia, Minas e Energia - Coordenadoria de Recursos Minerais, Florianópolis

Soil Survey Staff (1999) Soil taxonomy: a basic system of soil classification for making and interpreting soil surveys, 2nd edn, Agriculture handbook 436. United States Department of Agriculture, Natural Resources Conservation Service, Washington, DC

Sys C, Van Ranst E, Debaveye J (1991) Land evaluation, part 1: principles in land evaluation and crop production calculations. General Administration for Development Cooperation, Brussels

Uberti AAA, Bacic ILZ, Panichi JAV, Laus Neto JA, Moser JM, Pundek M, Carriao SL (1991) Metodologia para classificação da aptidão de uso das terras do Estado de Santa Catarina. EMPASC/ACARESC, Florianópolis

Van Diepen CA, Van Keulen H, Wolf J, Berkhout JAA (1991) Land evaluation: from intuition to quantification. In: Stewart BA (ed) Advances in soil science. Springer, New York, pp 139–204

Chapter 36
Land Suitability Assessment of the Proposed Uranium Mining Area in North-East Botswana

Oagile Dikinya

Abstract Land suitability assessment of the proposed uranium mining site was made to mitigate the possible environmental hazards arising from its potential use. The FAO framework of land evaluation and guidelines for land use planning were employed to assess the land suitability for arable, pasture/grazing and forestry uses. Typical soil textures are sandy loam to sandy clay loam, generally characterized by high base saturation and good fertility. The soils are deep enough for plant growth. Water-holding capacities are favourable for most of the agronomic crops and pastures. The mining area is assessed as marginally to moderately suitable for arable farming and grazing. Poor drainage, soil salinity and rainfall are the limiting factors for arable crops, whilst poor quality of vegetation species and rainfall are the major constraints to pasture and grazing. The area is marginally suitable to forestry due to rainfall being the main constraint. Overall, the area is suitable for grazing/pastures and arable farming, with few pockets suitable for forestry use.

Keywords Arable farming • Forestry use • Land suitability • Pasture and grazing • Uranium mining area

36.1 Introduction

Knowledge of soil properties is essential to understand environmental conditions and factors that can affect land uses. It is well established that soil is a long-term sink for environmental pollutants including heavy metals. These polluted soils can

O. Dikinya (✉)
Department of Environmental Science, University of Botswana,
Private Bag 0022, Gaborone, Botswana
e-mail: dikinyao@mopipi.ub.bw

S.A. Shahid et al. (eds.), *Developments in Soil Classification, Land Use Planning and Policy Implications: Innovative Thinking of Soil Inventory for Land Use Planning and Management of Land Resources*, DOI 10.1007/978-94-007-5332-7_36,

seriously affect agronomic crops, pastures and forestry. Therefore, for their proper use, the mining sites are to be cleaned and rehabilitated to mitigate the mining impact after mine closures. In order to achieve this, it is necessary to characterize the mining site first to determine the feasibility of remedial alternatives at the hazardous waste sites (Moon et al. 2006). This is especially important for sites where uranium is mined, which is radiotoxic and chemotoxic heavy metal, with a potential to pollute soils and accumulate in plants that can affect human health (Stojanoviæ et al. 2009). In other studies, Ibrahim and Whicker (2005) and Straczek et al. (2006) investigated the content of uranium in soils and water draining from land near mine sites and in native flora grown around mines.

Different environmental factors and soil types present specific soil properties that determine ecological functionality. Fundamental to assessing the soils biophysical productivity is the understanding of basic physical and chemical characteristics of soils (e.g. soil texture, soil pH, water infiltration, compaction). Moreover, soils play a crucial role in any ecological system by providing a medium for plants' growth, habitat for soil microorganisms, quarry and recycling of raw materials as well as foundation for engineering work and construction of infrastructures including buildings (Brady and Weil 1996). The changes in soil properties can impact the biodiversity, hydrogeological behaviour and biophysical functionality of the ecosystem. Therefore, the role of soils in the sustenance of the humans' livelihoods cannot be overemphasized especially in the rural environments of the developing world where the majority still practise both pastoral and arable farming.

The significance and magnitude of the biophysical productivity of an area depends on soil status and other environmental factors including climate, vegetation and socio-economic settings. These factors are needed to adequately assess capability of the land for various uses. Forest lands, lands with natural or planted forest or that can be afforested through planting, assist natural regeneration or protection of vegetation for forestry purposes; (b) arable/crop land is used for growing crops; and (c) pasture and rangelands are native lands used for haying or grazing as a source of forage for livestock. Rangeland refers to expansive, mostly unimproved lands on which a significant proportion of the natural vegetation is native grasses, grass-like plants and shrubs.

Land evaluation assessment is a process of estimating the potential of land for a specific or alternative use. Land classification is a method of grouping land or elements of land into classes based on their characteristics. Land suitability is the fitness of a given piece of land for a defined land use. Land suitability classification is the process of appraisal and grouping of land for a specific use. For example, the FAO framework for land evaluation (FAO 1976) recognizes four levels of generalizations in classification of land suitability (Driessen and Kronijn 1992): (a) land suitability orders, reflecting kinds of suitability, i.e. suitable (S) or not suitable (N); (b) land suitability classes, indicating the degree of suitability within an order; (c) land suitability subclasses, specifying kinds of limitation or kind(s) of required improvement measures within classes; and (d) land suitability units, indicating differences in required management with subclasses. The general concept of land

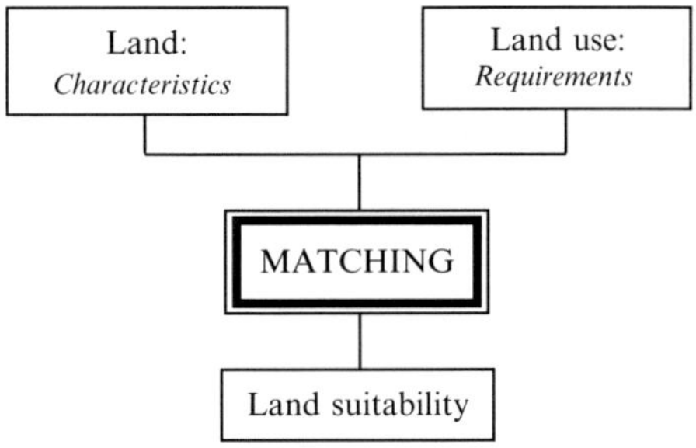

Fig. 36.1 Land evaluation rationale/scheme

evaluation is presented in Fig. 36.1, where land characteristics is an attribute of land that can be measured or estimated, e.g. slope, texture and rainfall; land quality is a complex attribute of land which acts differently in its influence on the suitability of land for a specific kind of use, e.g. moisture availability and erosion hazard; land use requirements are a set of land qualities that determine the production and management conditions of land use; and limitations are land qualities which adversely affect a kind of land use.

The objective of this chapter is to establish the pre-mining environmental conditions as well as to determine the effects of physico-chemical properties of soils on the biophysical productivity and land suitability assessment of the proposed uranium mining area.

36.2 Materials and Methods

36.2.1 Description of the Study Area

The study area, earmarked for uranium mining, is situated 7 km north-west of Serule and 5 km south of Gojwane villages in the North-East Botswana. The estimated total population of the survey area (or the two villages) in 2001 was estimated at 2,756 (Central Statistics Office 2001), predominately rural engaged in livestock and arable farming. The climate is semiarid with rainfall in summer months of October to April (Sims 1981; Bhalotra 1987; Dambe 1987), averaging 440 mm per annum. The rainfall is characterized by isolated and often localized showers, with some dry spells in between, and this is particularly important for rainfed arable farming. Summers are hot, and winters are cool during the night with occasional frost.

Five sites have been investigated, and soil samples representing these sites were collected for further characterization:

Site 1: latitude 21°52′37.5″ S and longitude 027°17′40.0″ E
Site 2: latitude 21°52′37.5″ S and longitude 027°17′40.1″ E
Site 3: latitude 21°50′51.0″ S and longitude 027°18′06.1″ E
Site 4: latitude 21°50′50.0″ S and longitude 027°17′49.6″ E
Site 5: latitude 21°50′50.0″ S and longitude 027°17′49.6″ E

36.2.2 Geology, Physiography and Vegetation

The study area is underlain by the basement complex gneisses of Precambrian age containing local intrusions of metamorphosed mafic and felsic igneous rocks. Topographically it consists of a gently undulating plain with incised, sometimes eroded, valleys with structurally controlled ridges with hills rising up to 200 m above the plain level. The incised valleys have well-defined drainage lines (e.g. Masokobale river) through which water flows intermittently during the rainy season. The area is predominately tree and bush savanna vegetation typified by species of *Acacia, Combretum, Boscia albitrunca, Eragrostis and Aristida.*

36.2.3 Survey Methods and Data Collection

Prior to the field survey, initially the identification of the survey area (using the existing information such as previous survey maps, ortho maps) was carried out to acquaint with the potential problems if they exist in the survey area. The recent (year 2000) areal or ortho-photographs at 1:50,000 scale were used for the interpretation of the soil-earth surface features during survey. The information was complimented with the geological and topographic maps (1:50,000) used for the purpose of land-form identification and navigation or to access roads/routes in the project survey area. In addition, standard survey equipment (GPS, picks, soil augers, spades, sampling bags, soil proforma, etc.) were used for field monitoring and measurements.

36.2.4 Soil Analysis and Measurements

Soil samples were analysed for selected physical and chemical characteristics and properties using the standard methods (Van Reeuwijk 1993). The pH and EC were measured using the pH and EC meter probes in soil to water suspension ratio of 1:2, whilst the organic carbon was measured using the modified Walkley and Black oxidation method. The cation exchange capacity (CEC) was determined using ammonium acetate method (pH 7.0). The exchangeable bases Ca, Mg, Na and K were determined using the atomic absorption spectrophotometer (AAS). The percentage base saturation (PBS) was estimated from the exchangeable bases and the CEC using standard calculation procedure. The particle size distribution analyses (PSDA) were made using the hydrometer method (Breitbart 1988). The soil bulk density, infiltration rate, compaction and water-holding capacity were interpolated and estimated from data of soil properties conducted in similar environments (Dikinya 1999).

36.2.5 Soil Classification System

Together with ortho-photo interpretation, soil analytical data were used to classify soils and to determine their distribution (soil map) in the proposed mining uranium area. An

Table 36.1 Requirements for arable farming, pasture and rangeland and forest uses in the proposed site

Land use	Requirements
Forests	Rainfall, temperature, topography, grass or tree cover, type of grasses
Pasture and rangelands	Soil type, slope, soil depth, tree density or grass cover, altitude, rainfall, suitable temperature and frost, moisture, rooting depth, aeration (soil drainage), nutrients, salinity/sodicity
Arable farming	Depth effective, soil drainage, soil texture, moisture availability, salinity/sodicity, soil erosion

Table 36.2 Schematic structure of land suitability classification

Order	Class	Subclass	Unit
S (Suitable)	S1	S2me	S2e-1
	S2	S2e	S2e-2
	S3	S2me	
N (not suitable)	N1	N1m	
	N2	N1c	

adopted and revised FAO/UNESCO-Botswana soil legend (Verbeek 1990) and FAO/UNESCO system (FAO 1988) was used for classification. Soil types or classes were generalized at second level of classification and subsequently mapped in the area.

36.2.6 Land Suitability and Capability Classification

Land that is allocated to any particular capability/suitability class has the potential for the use specified for that suitability class and for all classes below it. For example, suitability class 1 land, whilst excellent for arable use, can be equally put to any other uses. The capability or suitability class indicates the range of uses to which each soil could be put (Dent and Young 1981). For instance, each proposed major kind of land use has its own requirements (Table 36.1).

The FAO framework for land evaluation (FAO 1976) was employed as a land suitability assessment method or procedure. According to the framework, there are four classes of land suitability: highly suitable (S1), moderately suitable (S2), marginally suitable (S3) and not suitable or very limited (N1). Suitability subclasses (Table 36.2) indicate the degree of suitability as well as the nature of limitation that makes the land less than completely suitable. For example, S3e indicates limitation in S3 due to erosion hazards, (e) whilst "m" and "c" refers to limitations due to moisture and calcium carbonate, respectively.

36.3 Results and Discussions

36.3.1 Environmental Soil Physical-Chemical Characteristics and Properties

Soil texture of the soil samples (0–30 cm) ranged between sandy clay loam to loamy sand (Table 36.3) suggesting their better potential for arable farming as well as the soil natural attenuation to potential uranium contamination on soil.

All soils have low exchangeable Na and thus low exchangeable sodium percentage (Table 36.4); this indicate that the soils have high potential for aggregation and high aggregate stability; therefore, the soils will be less vulnerable to structural collapse due to slight erosion. The organic carbon is relatively high to sustain aggregate stability and soil fertility. The high base saturation and cation exchange capacity clearly demonstrates the high fertility status, a condition satisfactory to both arable and pasture and grazing. The electrical conductivity (EC) is somewhat high suggesting a potential problem of salinity toxicity with subsequent detrimental effects to sensitive crops, plants and vegetation species. The soil pH is in the slightly acid to neutral range (Soil Survey Division Staff 1993) where the nutrients are optimally available to many crops and vegetation species.

Table 36.5 presents physical and hydraulic soil properties. The porosity, compaction and water-holding capacity are favourable for root anchorage, tillage or physical soil manipulation, drainage and aeration. With the exception of site 2 (Gleyic Luvisol), other sites are unlikely to be waterlogged. The infiltration rates are

Table 36.3 Particle size distribution analyses of the proposed mine sites

	Particle size fractions in (mm)				
	Sand %	Coarse silt %	Fine silt %	Clay %	Textural class (Soil Survey Division Staff 1993)
Site number	0.05–2.0	0.02–0.05	0.002–0.02	<0.002	
1	78.3	9.3	7.4	5.0	Loamy sand
2	54.8	7.4	4.9	32.9	Sandy clay loam
3	81.3	7.9	1.9	8.9	Loamy sand
4	75.3	6.4	2.1	16.2	Sandy loam
5	68.9	2.7	5.4	23.0	Sandy clay loam

Table 36.4 Chemical characteristics of soils from the proposed mine site

	pH	EC	OC	BS	CEC	Na	K	Ca	Mg
Site number	(1:2)	dS m^{-1}	%	%	cmole(+) kg^{-1}				
1	6.7	13.3	0.98	74.0	16.62	1.63	2.86	6.11	1.70
2	7.4	41.0	0.74	84.7	30.40	0.81	4.64	8.58	11.72
3	6.6	7.3	1.09	71.0	12.63	0.26	2.75	4.97	0.99
4	6.2	6.7	2.28	66.6	9.05	0.35	2.67	1.76	1.25
5	7.1	21.5	0.33	81.1	36.66	0.46	5.59	19.8	3.89

EC electrical conductivity, *BS* base saturation, *OC* organic carbon, *CEC* cation exchange capacity

Table 36.5 Some selected physical properties of the proposed mine sites

Site number	Soil texture	Bulk density (g cm^{-3})	Porosity (%)	[a]Surface compaction	AWHC (mm m^{-1})	Infiltration rate (cm h^{-1})
1	Loamy sand	1.52	42.6	Compacted	125	3.1
2	Sandy clay loam	1.12	57.7	No compaction	176	2.2
3	Loamy sand	1.64	38.1	Moderately compacted	91	20.1
4	Sandy loam	1.49	43.8	Moderately compacted	112	9.4
5	Sandy clay loam	1.57	40.8	Moderately compacted	168	11.0

AWHC available water-holding capacity (mm m^{-1})
[a]Based on the surface clay content

Table 36.6 Typical mapping units and associated soil types in the proposed mine sites

Site number	Soil mapping units	Soil classification
1	A14-13-9	Orthic Luvisol
2	A7-9-4b	Gleyic Luvisol
3	D1a-1b	Eutric Regosol, partly lithic
4	A13	Chromic Luvisol
5	D10	Eutric Nitosol

relatively favourable to water entry and circulation within the soil in general and in particular in enhancing water-holding capacities of the soils.

36.3.2 Soil Description, Classification and Mapping

Table 36.6 shows soil mapping units and associated soil classes (FAO 1988; Verbeek 1990). The soils are classified (Table 36.6) using the FAO classification system (FAO 1988) and the revised legend of Botswana (Verbeek 1990). The analyses of the soil samples from the investigated sites are presented in Tables 36.3 and 36.4. Relative to other soils, the Gleyic Luvisol and Eutric Nitosol possess high clay, CEC and BS which is an indication of high soil fertility status of these soils, and therefore, these two soil types have relatively better potential to reduce nutrients leaching as well as natural attenuation to withstand further potential contamination.

Prior to soil morphological description of the selected sites, the landscape features including elevation, landforms, land element, position of the site, topography, surface sealing and cracking, land use, vegetation (type and trees/shrubs), grass cover, parent material, drainage, moisture condition, surface stones or rock outcrop, erosion and human influence to site were recorded on the standard data recording sheets. The spatial distribution of soils in the study area is shown in Fig. 36.2. Generally, the effective soil depth (Verbeek 1990) was more than 100 cm (deep soil)

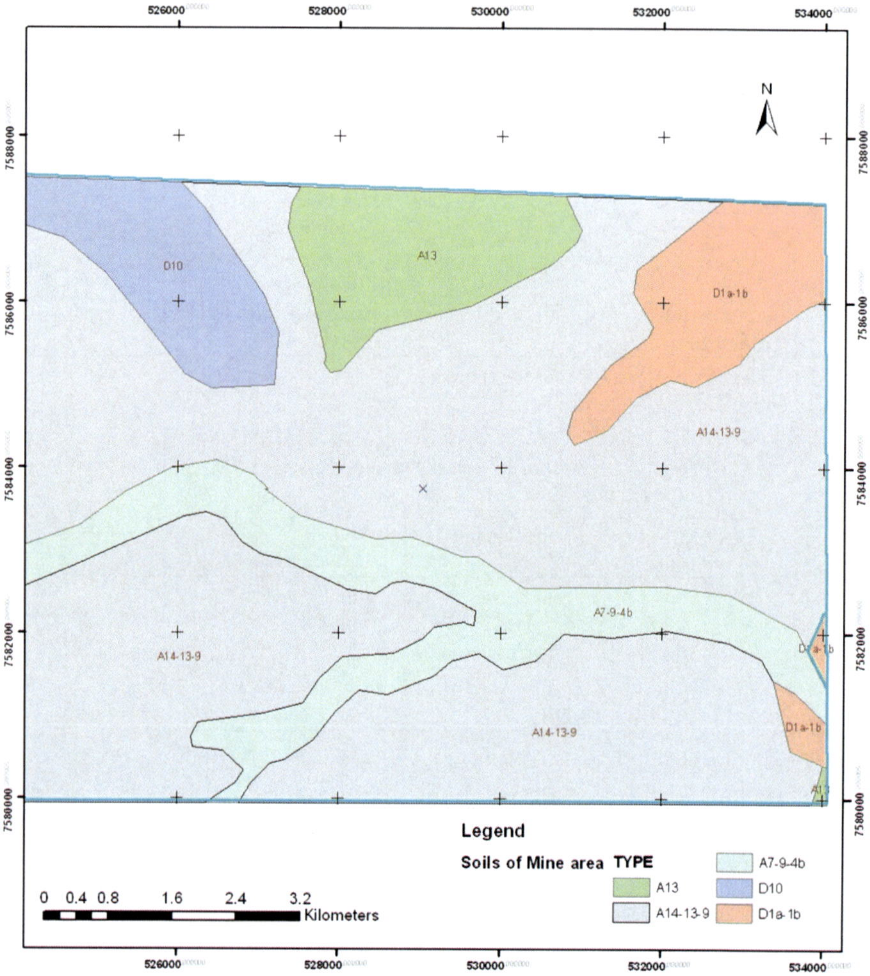

Fig. 36.2 Soil map of the proposed uranium mining area

in all soils, except soil unit D1a–1b (Eutric Regosol, partly lithic) where a lithic contact was found at 70-cm depth (moderately deep). Therefore, all soils are suitable for arable crops, pastures and forestry. The following soil types or mapping units are predominant in the proposed area.

36.3.2.1 Description of Soil Mapping Units

The assessment for the land suitability was carried out in an approximate area of 55.7 km^2 (5,570 ha). The following soil types or mapping units (see Fig. 36.2) are predominant in the proposed area.

Orthic Luvisol (A14-13-9)

Orthic Luvisol covers an area of 41.2 km^2. These soil types are developed in alluvial deposits or alluvial reworked materials. They are moderately deep to very deep, moderately well to well drained, dark brown to yellowish brown, sandy loam to sandy clay, flat to gently undulating soils.

Gleyic Luvisol (A7-9-4b)

Gleyic Luvisol covers an area of 5.6 km^2. These soil types are developed in alluvial deposits or alluvial reworked materials and also formed in colluvium of mixed origin or colluvially reworked material. They are deep to very deep, poorly to imperfectly drained, dark grey to greyish brown, sandy clay loam to clay and slightly flat soils.

Eutric Regosol (D1a-1b)

Eutric Regosol covers an area of 3.2 km^2. These soils are developed on fine-grained sedimentary rocks predominately consisting of shales, siltstone, grits and feldspathic sandstone. They are very shallow to moderately deep, well drained, yellowish brown to reddish brown and sandy loams to clay loam soils.

Chromic Luvisol (A13)

Chromic Luvisol covers an area of 2.9 km^2. These soil types are developed in alluvial deposits or alluvial reworked materials. They are moderately deep to deep, moderately well to well drained, strong brown to dark red, sandy loam to sandy clay loam and almost flat to gently undulating soils.

Eutric Nitosol (D10)

Eutric Nitosol covers an area of 2.8 km^2. These soils are developed on fine-grained sedimentary rocks predominately consisting of shales, siltstone, grits, and feldspathic sandstone. They are very deep, moderately well to well drained, red, sandy loams to sandy clay loam, and flat to gently undulating soils.

36.3.3 Land Suitability Assessment

36.3.3.1 Land Suitability Assessment: Concepts

The land suitability assessment was carried out to identify the potential areas suitable for forestry, agriculture (arable) and grazing and pastures. Systematically the following were considered in land suitability assessment:

Table 36.7 Land quality rating of some selected soil-land use requirements

Suitability rating	Textural class	AWHC (mm m^{-1})	Effective rooting depth (cm)	Effective depth (cm)
1	Sandy clay	>110	>75	>100
2	Fine sandy clay	70–110	50–75	50–100
3	Sandy clay loam	40–70	25–50	25–50
4	Silty clay loam	<40	10–25	10–25
5	Silty clay			<10

1. Delimitation of relevant zones based on the actual land use.
2. Definition of land utilization types (LUT) based on land use including forestry, agriculture, grazing and pasture.
3. Identification of land use requirement for each specific LUT—the conditions of land necessary or desirable for the successful and sustained practice of a given LUT. This can be expressed by land qualities determined and evaluated according to how they affect land characteristics (Sys 1985).
4. Description of land units based on the soil units.
5. Matching land units with land use requirements and with the land quality of land unit under consideration. Land qualities are rated on a scale from 1 to 5 for the corresponding requirements. The suitability ratings (Table 36.7) are defined to reflect the range of conditions and related land use requirements (see Table 36.1).

The ideal soil for suitability purpose is characterized by pH (6–8), EC (<0.4 dS m^{-1}), texture (loamy sand or sandy loam), drainage (well), BS (>50%), ESP (<15) and bulk density (1–1.6 g cm^{-3}). The ideal soil requirements for sorghum are mean temperature (18–35°C), AWHC (45–65 mm m^{-1}), pH (5.2–8.2), salinity or EC (<10 dS m^{-1}), nutrients (medium) and limited ponding (Rhebergen 1988).

36.3.3.2 Land Suitability for Arable Farming, Pasture and Rangeland and Forestry

To develop a suitability classification, soil data (Tables 36.3, 36.4 and 36.5) from arable farming, forestry, pasture and grazing are used and classified on the basis of limitations. That is, if any one limitation is of sufficient severity to lower the land to a given class, it is allocated to that class, no matter how favourable all other characteristics might be. Limitations adversely affect a kind of land use. The overall suitability is defined by limitations or the lowest suitability class(es) of land qualities. The process involved the following steps: (a) determining ratings of the land qualities, (b) establishing the relationship between land quality ratings and suitability class by comparing specific land use requirements and ratings, and (c) the lowest suitability class(es) of land qualities determines the final suitability of the land or soil unit.

Table 36.8 shows the suitability of various soil mapping units for arable farming in the proposed area. The mapping unit A14-13-9 is moderately suitable, with poor

Table 36.8 Land suitability assessment for arable farming

		Mapping unit				
	Site characteristics	A14-13-9	A7-9-4b	Dla-1b	A13	D10
Land qualities and requirements	Depth effective	S1	S1	S2	S1	S1
	Drainage	S2[a]	S3[a]	S2	S1	S2[a]
	Soil texture	S1	S2	S1	S1	S1
	Moisture availability	S1	S2	S1	S1	S1
	Nutrient availability	S1	S1	S2	S1	S1
	Absence of toxic substances (salinity, sodicity)	S1	S1	S3[a]	S2[a]	S1
	Soil erosion	S2	S2	S1	S2[a]	S1
	Soil temperature and frost	S1	S1	S1	S1	S1
	Overall suitability	S2	S3	S3	S2	S2

S1 suitable, *S2* moderately suitable, *S3* marginally suitable, *N* unsuitable
[a]Suitability as determined by the limiting factor

drainage as the limiting factor. The unit A7-9-4b is marginally suitable as limited by drainage as well. Units D1a–1b and A13 are marginally suitable with higher salinity (or EC) as a limiting factor. Land unit D10 is moderately suitable but limited by poor drainage. The suitability map for rainfed arable farming is presented in Fig. 36.3. Overall the study area is rated either as moderately suitable or dominantly as marginally suitable for arable farming.

Similarly Table 36.9 and Fig. 36.4 show suitability of soil mapping units for pasture and grasses in the proposed area. The mapping unit A14-13-9 is moderately suitable, with rainfall, topography, poor quality vegetation and nutrient availability as the main constraints. The unit A7-9-4b is marginally suitable and is limited by poor quality of vegetation (grasses) species. Mapping units D1a–1b and A13 are marginally suitable due to constraints by poor nutrient availability. Mapping unit D10 is moderately suitable but limited by poor nutrient and moisture availability. Overall the study area is dominantly moderately suitable, and relatively less area is marginally suitable.

Table 36.10 shows suitability of various soil or land units for forestry in the proposed area. All the mapping units are marginally suitable for forestry with rainfall as a limiting factor in particular during the winter season. However, the current land use (traditional-communal grazing) suggests that the proposed area is sustainably utilized for its maximal and potential use. It is, therefore, imperative to engage on proper land management in order to sustain its use. The utilization of the proposed area for forestry will undoubtedly have consequential detrimental effects. For instance, in unit A7-9-4b, the soils are suitable for arable farming.

Further, land capability assessment was carried out based on the existing soil or land units following land suitability assessment of the area to form an aggregated land capability classes. Soil and land units or delineated areas of land with specified environmental conditions were employed as the basis for land evaluation

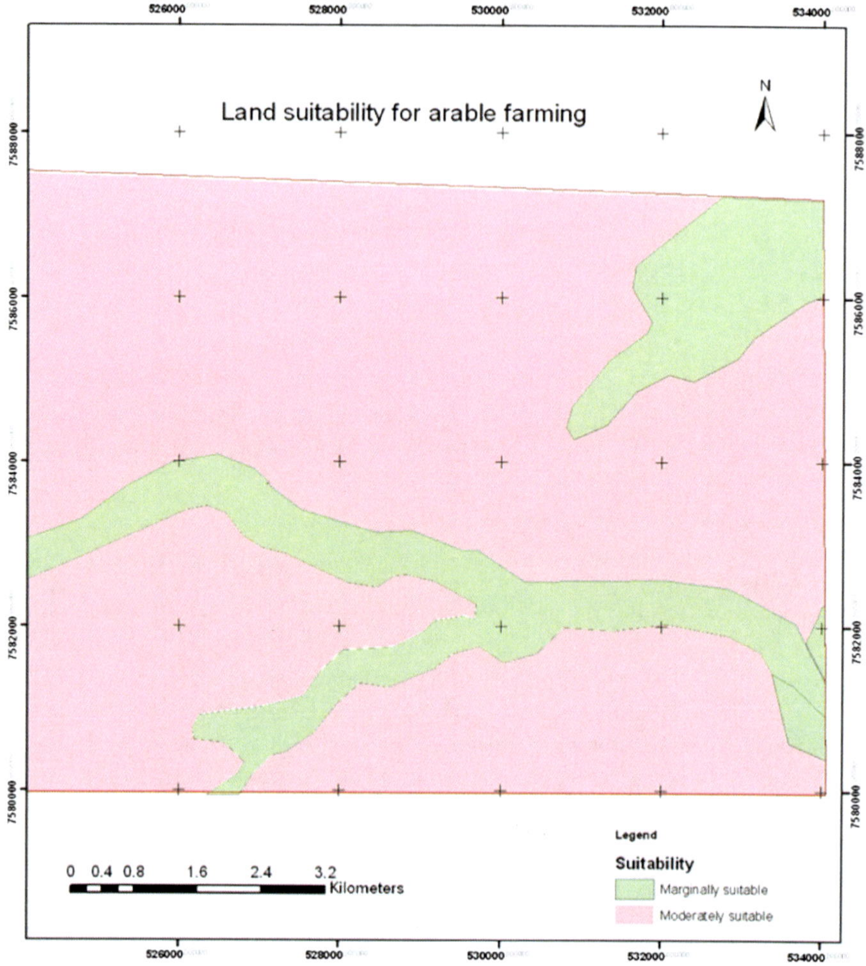

Fig. 36.3 Land suitability for rainfed arable farming

(Table 36.11). Using the above classification criteria, it was clear that preference will be given to arable farming followed by grazing pasture and forestry in accordance with the proposed area's land suitability and capability.

36.4 Conclusions

The study was undertaken to assess the land suitability or capability assessment of the proposed uranium mine site in Serule. It is concluded that various soil types occur in the proposed area with variable soil textures and relatively higher base

Table 36.9 Land suitability assessment for pasture and rangeland use

		Mapping units				
	Site characteristics	A14-13-9	A7-9-4b	Dla-1b	A13	D10
Land Qualities and requirements	Topography or slope	S2	S1	S1	S1	S1
	Soil depth	S1	S2	S1	S1	S1
	Vegetation, tree density and quality of grass species	S2[a]	S3[a]	S2	S2	S2[a]
	Elevation or altitude	S1	S1	S1	S1	S1
	Rainfall	S2	S2	S2	S2	S2
	Temperature and frost	S1	S1	S1	S1	S1
	Moisture availability	S1	S2	S2	S2	S2
	Nutrients availability	S2[a]	S2	S3[a]	S3[a]	S2[a]
	Aeration (soil drainage)	S1	S2	S1	S1	S1
	Salinity/sodicity	S1	S1	S2	S2	S1
	Overall suitability	S2	S3	S3	S3	S2

S1 suitable, *S2* moderately suitable, *S3* marginally suitable, *N* unsuitable
[a]Suitability as determined by the limiting factor

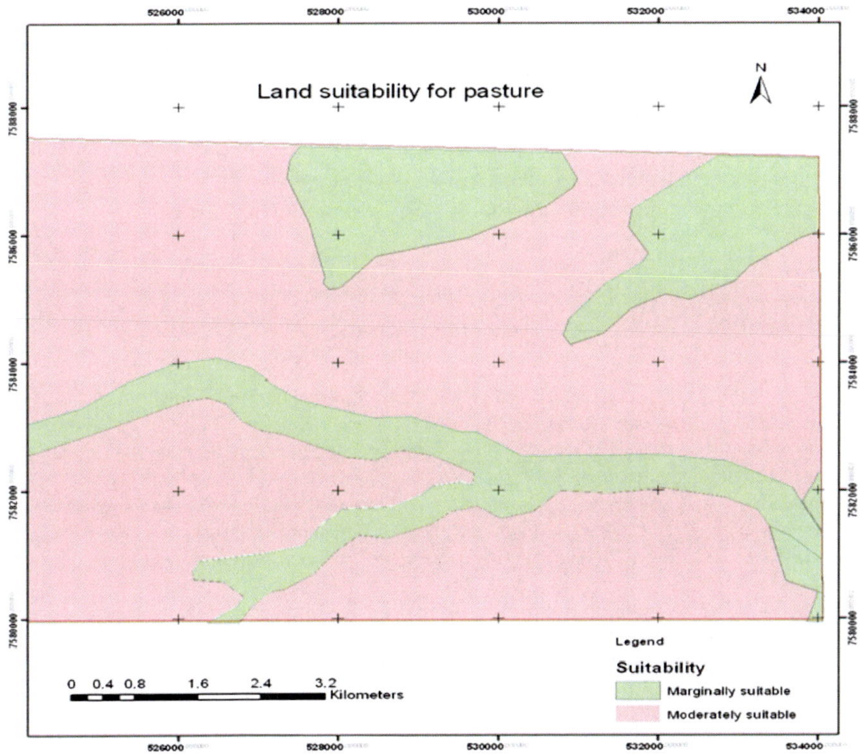

Fig. 36.4 Land suitability of the proposed uranium mining area for pastures

Table 36.10 Land suitability assessment for forest use

		Mapping units				
	Site characteristics	A14-13-9	A7-9-4b	Dla-lb	A13	D10
Land qualities and requirements	Rainfall	S3[a]	S3[a]	S3[a]	S3[a]	S3[a]
	Temperature	S1	S1	S1	S1	S1
	Topography	S1	S1	S1	S1	S1
	Grass or tree cover	S2	S3[a]	S2	S2	S2
	Nature of vegetation species	S1	S2	S2	S2	S2
	Overall suitability	S3	S3	S3	S3	S3

S1 suitable, *S2* moderately suitable, *S3* marginally suitable, *N* unsuitable
[a]Suitability as determined by the limiting factor

Table 36.11 Land capability assessment and classification for various land uses

	Mapping units				
Major land use	A14-13-9	A7-9-4b	Dla-lb	A13	D10
Arable farming	Moderate	Low	Low	Moderate	Moderate
Pasture and rangeland	Moderate	Low	Low	Low	Moderate
Forestry	Low	Low	Low	Low	Low

Moderate = moderate use of land; low = low potential use of land

saturation, thus indicating good fertility of the soils. The overall land capability classification shows that most area is occupied by grazing pasture and rangelands and few pockets for forestry use. Present study supports that the area in general is suitable for arable farming and grazing pasture and rangeland use.

References

Bhalotra YPR (1987) Climate of Botswana. Part II. Elements of climate. Department of Meteorological Services, Gaborone

Brady N, Weil RR (1996) The nature and properties of soils, 11th edn. Prentice Hall International Inc., Upper Saddle River, New Jersey, USA

Breitbart R (1988) Soil testing procedures for soil surveys, vol 3. FAO/UNDP/Government of Botswana project BOT/85/011. Field document no 3, Gaborone, Botswana

Central Statistics Office (CSO) (2001) Government of Botswana, Government Printers, Gaborone, Botswana

Dambe D (1987) Agro-climatic zones in Botswana. Department of Meteorological Services, Botswana

Dent D, Young A (1981) Soil survey and land evaluation. Allen & Unwin, London, p 278

Dikinya O (1999) Reference guide: physical properties of Botswana soils, Gaborone, Botswana, p 13

Driessen M, Kronijn NT (1992) Land use systems. INERS Publishers, Wageningen

FAO (1976) A framework for land evaluation, Soils bulletin no 32. FAO, Rome, p 72

FAO (1988) Revised legend of the FAO-UNESCO soil map of the world. World soils resources report no 60. FAO/UNESCO/ISRIC, Rome

Ibrahim SA, Whicker FW (2005) Comparative plant uptake and environmental behaviour of U-series radionuclides at a uranium mine-mill. J Radioanal Nucl Chem 156:253–267

Moon JW, Roh Y, Phelps TJ, Phillips DH, Watson DB, Kim Y, Brooks SC (2006) Physicochemical and mineralogical characterization of soil-saprolite cores from a field research site, Tennessee. J Environ Qual 35:1731–1741

Rhebergen GJ (1988) A system of land evaluation for arable farming in Botswana. FAO/UNDP/ Government of Botswana project BOT/85/011. Field document no 4

Sims D (1981) Agroclimatological information, crop requirements and agricultural zones for Botswana. DLU, Ministry of Agriculture, Government of Botswana, Government Printers, Gaborone, Botswana

Soil Survey Division Staff (1993) Soil survey manual, Agriculture handbook no 18. United States Department of Agriculture, US Government Printing Office, Washington, DC

Stojanoviæ M, Stevanoviæ D, Iles D, Grubišiæ M, Milojkoviæ J (2009) The effect of the uranium content in the tailings on some cultivated plants. Water Air Soil Pollut 200:101–108

Straczek A, Dams J, Vandenhove H (2006) Localisation of uranium in roots by chemical extractions and by a short term uptake study. Influence of phosphate. Uranium in Environment, 525–530. DOI:10.1007/3-540-28367-6_52

Sys C (1985) Land evaluation. ITC Gent, Belgium

Van Reeuwijk LP (1993) Procedures for soil analysis. International Soil Reference and Information Centre (ISRIC) technical paper, no 9 ISRIC, p 19

Verbeek K (1990) Revised general soil legend of Botswana. FAO/UNDP/Govt of Botswana project BOT/85/011. Field document no 32, Gaborone, Botswana

Chapter 37
Sustainable Development and Management Policies for Soil and Water Conservation in Egypt

A.Z. El Bably and S.A. Abd El-Hafez

Abstract Sustainable development and management policies for soil and water conservation in Egypt pose a dilemma to develop irrigation strategies for irrigated agriculture to reduce negative environmental impacts, an inevitable consequence of irrigation. For proper irrigation management, it is necessary to (1) improve the accuracy of soil water balance components to calculate a reliable estimate of the leaching fraction, (2) estimate the leaching requirements and add to irrigation requirement, (3) consider the water distribution uniformity to decide which part of the field should receive at least the leaching fraction for salinity control, (4) take into account that leaching salts periodically is more practical than every irrigation, (5) consider that there is no need to increase irrigation frequency to control salt concentration except for drip irrigation, and (6) monitor the root zone salinity, especially prior to periodic leaching. Adoption of appropriate irrigation system for specific soil types is necessary, e.g. sprinkler system is well adapted to sandy and loamy soils but less so to heavy or clayey soils. The drip or trickle irrigation system is better adapted to loamy or clayey soils and release water through many emitters at rate of 2–4 l h^{-1}. Moreover, drip irrigation provides a greater opportunity for using saline water. The policy of the Egyptian Government is to use drainage water up to EC of 4.5 dS m^{-1} after blending with fresh Nile water and to assure that the resultant EC of the blended water do not exceed 1.0 dS m^{-1}.

Keywords Blending • Conservation • Drainage • Egypt • Leaching fraction

---

A.Z. El Bably (✉) • S.A. Abd El-Hafez
Soil, Water and Environment Research Institute, ARC, Giza, Egypt
e-mail: azelbably@yahoo.com

S.A. Shahid et al. (eds.), *Developments in Soil Classification, Land Use Planning and Policy Implications: Innovative Thinking of Soil Inventory for Land Use Planning and Management of Land Resources*, DOI 10.1007/978-94-007-5332-7_37,

37.1 Introduction

Soil and water salinization inevitably occur with irrigation. Salinity control measures, therefore, must be implemented if irrigated agriculture is to be sustained over the long term. Ideally, such measures should be compatible with the processes in the natural geohydrological system. Salinity management practices offer different levels of control. Some are aimed at controlling salinity within the root zone in a particular field; others control salinity over a much larger scale, such as an irrigation project (El-Mowlhi and Abd El-Hafez 1999). The practice should protect the environment, as well as surface and groundwater. Usually, no single method is sufficient to control salinity of an irrigated land, and thus, many practices are integrated. The development of integrated salinity control strategy depends on the purpose, economic conditions, and climatic, social, edaphic, and hydrologic factors.

Management practices for the control of salinity, sodicity, and toxicity in the root zone include (1) selecting varieties (rootstocks) that produce satisfactory yields under the existing conditions of salinity, sodicity, and toxicity; (2) using planting procedures that minimize salt accumulation around the seed; (3) using land preparation methods that facilitate uniform infiltration of irrigation water; (4) adopting irrigation methods that maintain sufficient available soil moisture and cause periodic leaching of the soil profile; (5) installing and maintaining drainage systems that facilitate leaching, rooting, and trafficability; and (6) using treatments, e.g. chemical amendments and organic matter, to maintain soil permeability and tilth.

In the arid and semiarid region, the scarcity of fresh water resource is the main factor limiting the expansion of irrigated agriculture area to meet the necessary food demands of the progressively growing population. Therefore, to overcome the shortage of food demands and to reach satisfactory level of food production, the utilization of alternate water resources besides the fresh one is must. In the recent few years, more attention has been given to the utilization of alternate water resources for irrigation, which are already practised in many places, especially where available fresh water resources are scarce. This certainly results in greater amounts of water for irrigation but to the detriment of its quality. If these resources are not properly used, it is likely that prolonged use could seriously affect crop production and degrade physico-chemical soil characteristics. To avoid build-up of soil salinity, proper methods and technologies must be used to manage crops irrigation with saline water for sustainable production. This requires an intelligent assessment and understanding of climatic, edaphic, hydrological, cultural, as well as institutional and social conditions where such saline water is intended to be used for irrigation. In irrigated agriculture, the hazard of salt water is a constant threat. Poor-quality irrigation water is generally of more concern as the climate changes from humid to arid conditions. Salinity is not normally a threat in high rainfall area where salts are leached with high rainfall, making the root zone less saline and conducive for crop production. The irrigation water not used by plants moves through root zone and builds up water table if subsoil consists of a dense or hard layer. This percolating process flushes (leaches) soluble salts.

Most of crop species are glycophytes and are not adapted to high salt concentrations, and there are large differences among these species in their growth and yield

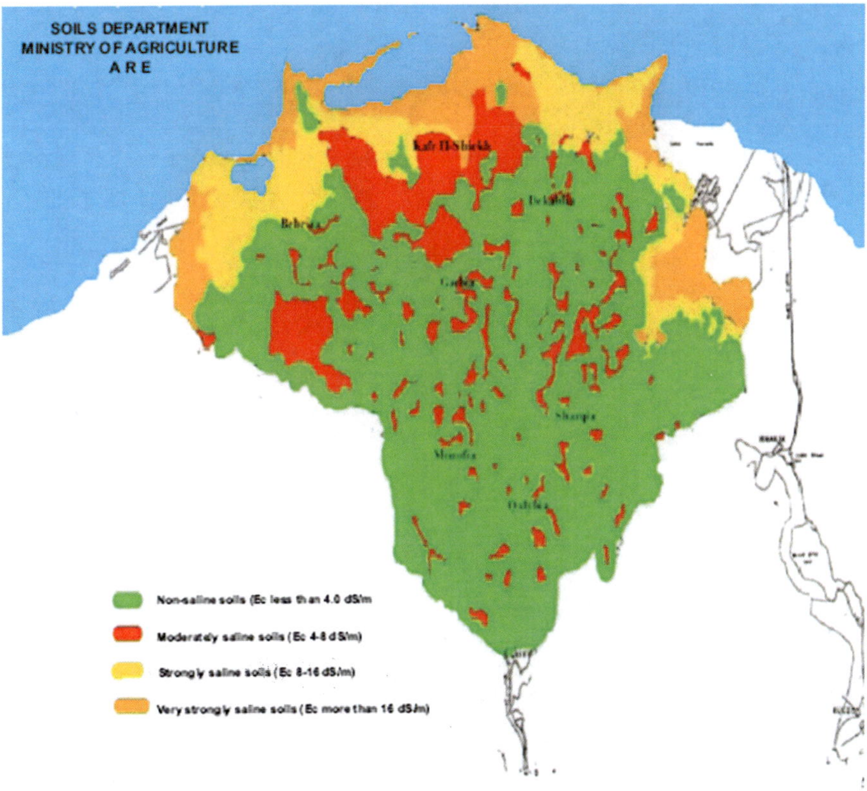

Fig. 37.1 Salinity status in the Nile Delta soil (surface soils)

response to salinity (Mass and Hoffman 1977). A few salt-tolerant crops (e.g. barley, date palm, and sugar beet) are standard crops for saline areas.

37.1.1 Soil Survey and Salinity Status

General status of soil salinity in the Nile Delta is shown in Fig. 37.1. The soils of the southern part of the delta are generally nonsaline, as the electric conductivity of their saturation extracts (ECe) is blow 4 dS m^{-1}. The salinity problem becomes more severe when approaching the Mediterranean coast and around the north lakes where soils are extremely saline. The total soluble salts in general increase with depth due to the effect of the groundwater (Abdel-Aal 1971). Soluble magnesium exceeds soluble calcium at this site due to seawater intrusion. Sulphates are relatively higher near Manzala Lake due to secondary gypsum formation usually found in SIRW area (Abdel-Aal 1971). In the southern part, salt accumulation has resulted from irrigation system in the last century without the construction of an adequate drainage system. But in the northern part, salt accumulation has resulted in from seawater intrusion and the northern lakes and tidal marshes. This has been accelerated by the flat topography and low land level

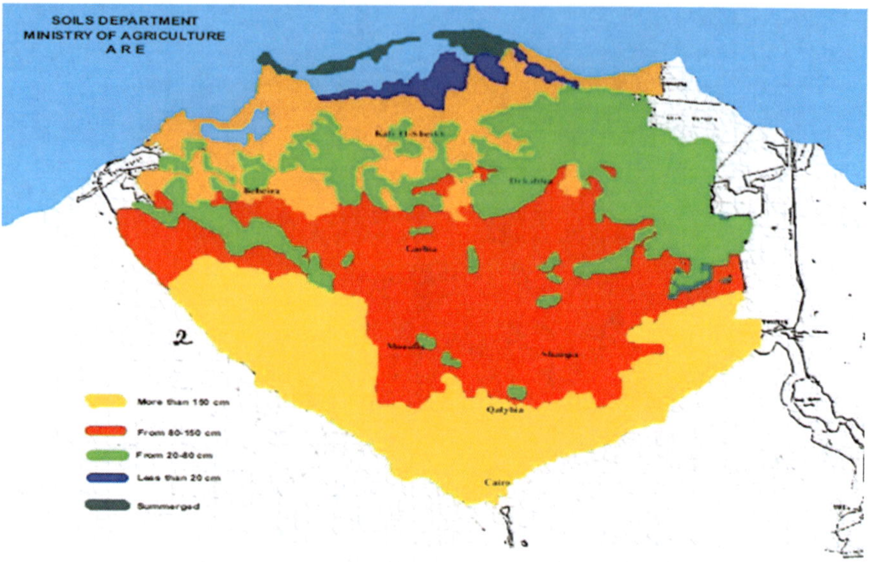

Fig. 37.2 Water table depth in the Nile Delta soils

in the area bringing the water table close to the soil surface (El-Nahal et al. 1977). pH is generally above 7 and mostly below 8.5 being generally higher in the subsoil layers.

37.1.2 Groundwater Table

Figure 37.2 shows the water table deeper than 150 cm in the majority of the southern part; however, it is shallower in the northern part near the coastal lakes approaching to 20 cm from soil surface; in some areas, the soil is submerged in water. In all cases, however, the subsoil in the northern areas was quite wet and showing evidence of poor drainage and aeration. The presence of reduced iron and manganese concentration and mottling in the subsoil indicates poor drainage and anaerobic conditions. However, the depth of the groundwater table in Fig. 37.2 reflects only the situation at the sampling time and does not represent seasonal variations. In some areas the measurement of groundwater depth as well as other soil surveys was carried out during high groundwater seasons. These records are considered as descriptive and could be used to give indication of the range in which the groundwater in a certain area lies (El-Nahal et al. 1977).

Shallow water table in the southern part of the delta was found in areas subjected to water seepage from irrigation canals and where drainage was lacking or inadequate. Groundwater table was deep in areas where no seepage occurred or where drainage system was constructed. Some researchers (Khater et al. 1997; Zein et al. 2000) have found variable EC and sodium adsorption ratio (SAR) of groundwater in various locations. Table 37.1 illustrates higher values of EC and SAR of water table in locations where drainage water was used for irrigation purposes. Such high values

Table 37.1 The EC and SAR values of water table in El-Hamoul and Balteim sites

Location	Water sources	EC dS m^{-1}	SAR (mmoles $L^{-1})^{0.5}$
El-Hamoul	Drainage	11.70	31.27
	Mixed	5.69	14.39
Baltiem	Drainage	12.50	23.34
	Mixed	6.50	11.55

Adapted from Zein et al. (2000)

are due to seawater intrusion and high levels of salinity in the drainage water used in irrigation compared to blended irrigation water.

37.1.3 Soil Productivity

El-Nahal et al. (1977) showed the productivity classes of the Nile Delta soils compared to soil survey work as shown in Fig. 37.3.

The class 1 (excellent) soils cover an area of only 4% of the Nile Delta. These 4% soils occur mainly in the southern part of Menofia and Qalybia Governorates. These soils are suitable for all crops; the crops grown in these soils gave the highest yields. The soils have deep profiles and are of medium texture, have efficient irrigation system, and are well drained. The soils are nonsaline and free of alkalinity. The class 2 (good) soils make up nearly 5% of the delta area. The soils of this class differ from those of the class 1; these soils have heavier texture and are slightly saline. They are located in large areas in Menofia, Qalybia, and Garbia Governorates and in small scattered areas in the rest of the delta. The class 3 (medium) soils make up 33.5% of the delta area. They are concentrated in large areas in Sharkia, Dekahlia, Kafr El-Sheikh, and Beheira Governorates which occupy most of the northern part of the delta. The class 4 (poor) soils make up 7.8% of the delta area. They are concentrated mainly in the newly reclaimed soils along the desert fringe, south of the northern lakes and along the sea coast. The class 5 soils (flooded land) make up 22.5%, and the public utilities make up 7.4% of the delta area.

Generally, the high salt content, high water table, and the heavy or coarse texture are the main factors that restrict the productivity of class 3 and 4 type of soils.

37.1.4 Water Salinity

In Egypt, rainfall is the most critical environmental factor. Rainfall before and during the irrigation season makes it possible to use more saline irrigation water because salts are diluted in the root zone and leaching is increased. To calculate the average water salinity, consider both rainfall and irrigation water. The average salinity of the applied water (C_a) can be calculated from

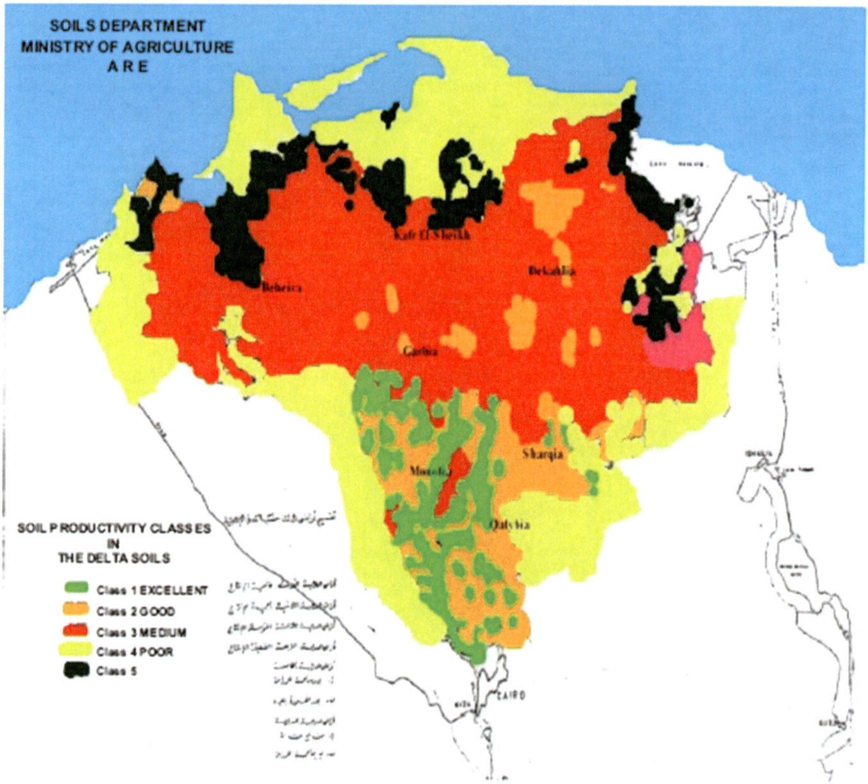

Fig. 37.3 Soil productivity classes

$$\text{Equation: } [Ca = Cr\,Dr + Ci\;\;Di] \div [Dr + Di] \tag{37.1}$$

The variable C can be expressed as concentration (mg l^{-1} or ppm) or electrical conductivity (dS m^{-1} or mmhos cm^{-1}), D is depth (cm), and the subscripts a, r, and i indicate average applied, rain, and irrigation water, respectively. For example, in the Northern Delta, if 6.7 cm of rainfall was totalled and 60 cm of irrigation water with a salt content of 2,500 mg l^{-1} was applied during the growing season, the resulting average salt concentration of the applied water would be

$$C_a = [(0 \times 6.7\text{cm.}) + (2,500\text{mgl}^{-1} \times 60\text{cm.})] \div [6.7\text{cm} + 60\text{cm.}]$$
$$C_a = 2,249\text{mgl}^{-1}$$

37.1.5 Could Soil Salinity Cause a Loss of Yield?

To convert these salt concentrations to electrical conductivity, divide Ca by 640; this results in an ECa of 3.5 dS m^{-1}. These values can be converted to soil salinity; assuming a leaching fraction of 0.15, by multiplying ECa by 1.5, soil salinity,

Table 37.2 Salt tolerance of herbaceous crops

Crop		Threshold ECe ($dS\ m^{-1}$)	Slope %	
Common name	Botanical name	*t*	*s*	Rating
Fibre, grain, and special crops				
Barley	*Hordeum vulgare*	8.0	5.0	T
Bean	*Phaseolus vulgaris*	1.0	19.0	S
Broad bean	*Vicia faba*	1.6	9.6	MS
Cotton	*Gossypium hirsutum*	7.7	5.2	T
Cowpea	*Vigna unguiculata*	4.9	12.0	MT
Flax	*Linum usitatissimum*	1.7	12.0	MS
Groundnut	*Arachis hypogaea*	3.2	29.0	MS
Guar	*Cyamopsis tetragonoloba*	8.8	17.0	T
Kenaf	*Hibiscus cannabinus*			MT
Maize	*Zea mays*	1.7	12.0	MS
Millet, foxtail	*Setaria italica*			MS
Oats	*Avena sativa*			MT
Rice, paddy	*Oryza sativa*	3.0	12.0	S
Rye	*Secale cereale*	11.4	10.8	T
Safflower	*Carthamus tinctorius*			MT
Sesame	*Sesamum indicum*			S
Sorghum	*Sorghum bicolor*	6.8	16.0	MT
Soybean	*Glycine max*	5.0	20.0	MT
Sugar beet	*Beta vulgaris*	7.0	5.9	T
Sugarcane	*Saccharum officinarum*	1.7	5.9	MS
Sunflower	*Helianthus annuus*			MS
Triticale	*X Triticosecale*	6.1	2.5	T
Wheat	*Triticum aestivum*	6.0	7.1	MT
Wheat (semidwarf)	*T. aestivum*	8.6	3.0	T
Wheat, Durum	*T. turgidum*	5.9	3.8	T
Grasses and forage crops				
Alfalfa	*Medicago sativa*	2.0	7.3	MS
Alkaligrass, Nuttall	*Puccinellia airoides*			T
Alkali sacaton	*Sporobolus airoides*			T
Barley (forage)	*Hordeum vulgare*	6.0	7.1	MT
Bent grass	*A. stolonifera palustris*			MS
Bermuda grass	*Cynodon dactylon*	6.9	6.4	T
Bluestem, Angleton	*Dichanthium aristatum*			MS

After Maas and Hoffman (1977)

expressed as ECe, is 5.3 $dS\ m^{-1}$ (Table 37.2), and the threshold value (upper limit with no yield loss) for sugar beet yield is 7.0 $dS\ m^{-1}$. From this information, if leaching was 0.15, the sugar beet yield would be 90%:

$$\text{Equation: Yr} = 100 - s(\text{ECe} - \text{t}) \tag{37.2}$$

where Yr is crop yield relative to the same conditions without salinity, *t* is the threshold salinity, *s* is the linear rate of yield loss with increasing 1 unit of EC above the threshold (slop of the line), and ECe represents the average root zone salinity measured as the electrical conductivity of a soil saturation extract, as shown in Table 37.2.

$$\text{Yr} = 100 - 5.9(5.3 - 7.0)$$

37.1.6 How Much Salt Is Added to the Soil Without Leaching Requirements?

To convert this salt concentration to kg during growing season, divide Ca by 1,000,000 and then multiply by 1,000 to get kg of salt per m^3; since ET of sugar beet was 2,134 m^3 $acre^{-1}$ using modified Penman (Eid 1999), the salt added to soil with irrigation water will be amounted 853.6 kg $acre^{-1}$ $season^{-1}$. It means that 853.6 kg of salt will be added when concentration of irrigation water is 2,500 mg l^{-1}.

37.1.7 Leaching Requirements

To prevent salts from building up to levels that are detrimental to crop production, water must drain through the crop root zone to keep the root zone salinity below threshold level. In most instances, natural drainage is sufficient to leach salts from the crop root zone. If natural drainage is not adequate, however, a drainage system must be installed. Where salinity is a hazard, the length of time before productivity is reduced depending on water management, drainage, and the hydrogeology of the area.

The necessary leaching requirement (LR) can be estimated from the following Eq. (37.3) as described by Rhoades (1972):

$$\text{LR} = \text{ECw} / (5(\text{ECe}) - \text{ECw}) \tag{37.3}$$

where

LR = the minimum leaching requirement needed to control salts within the tolerance (ECe) of the crop with ordinary surface methods of irrigation

ECw = salinity of the applied irrigation water in dS m^{-1}

ECe = average soil salinity tolerated by the crop as measured on a soil saturation extract. Obtain the ECe value for the given crop and the appropriate acceptable yield from Table 37.2. For water in the moderate to high salinity range (1.5 dS m^{-1}), it might be better to use the ECe value for maximum yield potential (100%) since salinity control is critical to obtaining good yields.

Given:

ECw = 2.5 dS m^{-1}

ECe = 7.0 dS m^{-1} (from Table 37.2 for sugar beet at a 100% yield potential)

LR = 2.5/[5 (7.0)–2.5] = 0.077 (for a 100% yield potential)

Table 37.3 Relationship between applied water (AW m^3 $acre^{-1}$) and leaching requirements (LR) using polynomial and cubic equations for different crops in Egypt

Crops	Polynomial and cubic equations	R^2
Filed crops		
Cotton	$AW=4{,}732.7[LR]^2+3{,}005.33[LR]+3{,}189.5$	1.0
	$AW=3{,}390.2\{LR\}^3+3{,}492\{LR\}^2+3{,}139.9\ [LR]+3{,}185.3$	1.0
Sugar beet	$AW=3{,}214.4[LR]^2+2{,}005.4[LR]+2{,}138$	1.0
	$AW=3{,}078.9[LR]^3+2{,}087.6[LR]^2+2{,}127.7[LR]+2{,}134.3$	1.0
Wheat	$AW=2{,}447.1\{LR\}^2+1{,}533.6\{LR\}+1{,}637.5$	1.0
	$AW=3{,}845.3\{LR\}^3+1{,}039.9\{LR\}^2+1{,}686.2\{LR\}+1{,}632.7$	1.0
Corn	$AW=8{,}985.4[LR]^2+1{,}086.3[LR]+2{,}378.2$	1.0
	$AW=100{,}920[LR]^3\pm 27{,}948[LR]^2+5{,}093.4[LR]+2{,}252.6$	1.0
Vegetables crops		
Tomato	$AW=4{,}334.4[LR]^2+2{,}705.1[LR]+2{,}886.8$	1.0
	$AW=3{,}563.4[LR]^2+3{,}030.3[LR]^2+2{,}846.6[LR]+2{,}882.4$	1.0
Potato	$AW=1{,}984.98[LR]^2+1{,}295.4[LR]+1{,}363$	1.0
	$AW=2{,}755.5[LR]^3+976.6[LR]^2+1{,}404.8[LR]+1{,}359.5$	1.0
Pepper	$AW=2{,}369.8[LR]^2+1{,}471.8[LR]+1{,}574.4$	1.0
	$AW=2{,}756.7[LR]^3+1{,}361[LR]^2+1{,}581.2[LR]+1{,}570$	1.0
Onion	$AW=4{,}765.8[LR]^2+3{,}089.6[LR]+3{,}259.9$	1.0
	$AW=5{,}409.0[LR]^3+2{,}786.3[LR]^2+3{,}304.3[LR]+3{,}253.2$	1.0
Forage crops		
Alfalfa	$AW=8{,}866.2[LR]^2+5{,}358.1[LR]+5{,}735.6$	1.0
	$AW=-19{,}756.0[LR]^3+16{,}096.0[LR]^2+4{,}573.7[LR]+5{,}760.2$	1.0
Clover	$AW=3{,}605.0[LR]^2+2{,}245.7[LR]+2{,}399.0$	1.0
	$AW=4{,}537.0[LR]^3+1{,}944.6[LR]^2+2{,}425.8[LR]+2{,}393.4$	1.0
Fruit crops		
Date palm	$AW=7{,}442.6[LR]^2+4{,}727.0[LR]+5{,}020.1$	1.0
	$AW=7{,}474.7[LR]^3+4{,}707.1[LR]^2+5{,}023.8[LR]+5{,}010.7$	1.0
Orange	$AW=7{,}442.6[LR]^2+4{,}727.0[LR]+5{,}020.1$	1.0
	$AW=7{,}474.7[LR]^3+4{,}707.1[LR]^2+5{,}023.8[LR]+5{,}010.7$	1.0
Grape	$AW=4{,}716.7[LR]^2+2{,}993.9[LR]+3{,}176.3$	1.0
	$AW=5{,}509.6[LR]^3+2{,}700.3[LR]^2+3{,}212.7[LR]+3{,}169.4$	1.0

37.1.7.1 Applied Water

Calculation of applied water for above example:

The author found that polynomial and cubic equations achieved a convenient relation between applied water (m^3 $acre^{-1}$) and leaching requirements. Applied water of each crop could be calculated using leaching requirements as shown in Table 37.3. Applied water could be equalled and expressed as evapotranspiration when leaching requirements (LR) = 0. In this case, ET (evapotranspiration) of sugar beet becomes 2,138 m^3 $acre^{-1}$ $season^{-1}$ in Eq. (37.4) and 2,134.3 m^3 $acre^{-1}$ $season^{-1}$ in Eq. (37.5), while ET of sugar beet was 2,134 m^3 $acre^{-1}$ $season^{-1}$ using modified Penman:

$$\text{Applied water} = 3{,}214.4\left[LR\right]^2 + 2{,}005.4\left[LR\right] + 2{,}138 \tag{37.4}$$

$$\text{Applied water} = 3{,}078.9[\text{LR}]^3 + 2{,}087.6[\text{LR}]^2 + 2{,}127.7[\text{LR}] + 2{,}134.3 \quad (37.5)$$

In case leaching requirements=0.077, the applied water becomes 2,311.5 m^3 $acre^{-1}$ $season^{-1}$ (Eq. 37.4) and 2,311.9 m^3 $acre^{-1}$ $season^{-1}$ (Eq. 37.5). Therefore, both equations could be validated to interpret the relationship between applied water and leaching requirements in one hand and applied water and evapotranspiration (ET) on the other hand. So it has been validated to calculate applied water using leaching requirements for each crop. This relation could be showed in guidelines form to be easy in the application.

37.1.7.2 Sodicity

When the concentration of exchangeable sodium becomes excessive in proportion to exchangeable calcium plus magnesium and the exchangeable sodium percentage (ESP) reached more than or equal 15, or SAR more than or equal 13 (mmoles $l^{-1})^{0.5}$, the soil is said to be sodic. Ramadan et al. (1989) and Zein et al. (1997) indicated that excessive sodium causes soil mineral particles to disperse, translocate in soil profile, and deposit in water conducting pores, and thus water mobility in soil is decreased. High ESP becomes a problem when infiltration rate is reduced to the extent that crop is not adequately supplied with water or when the hydraulic conductivity of the soil profile is too low to provide adequate drainage. The ESP excess of 15 may add difficulties for crops through crusting seed beds, temporary saturation of the soil surface, high pH, and the increased potential for diseases, weeds, soil erosion, lack of oxygen, and inadequate nutrient availability.

The permissible value of the SAR is a function of salinity. High salinity levels reduce swelling and aggregate breakdown (dispersion), and promote water penetration. High proportions of ESP, however, produce the opposite effect. Regardless of the sodium content, water with an electrical conductivity less than about 0.2 dS m^{-1} causes degradation of the soil structure, promotes soil crusting, and reduces water penetration. Both EC and SAR of the applied water must be considered when assessing the potential effect of water quality on soil water penetration. Application of gypsum to the soil surface after tillage, or incorporation of gypsum into the surface soil to 10 cm depth, is an effective method to improve infiltration rates. The application of gypsum to irrigation water to solve water-related infiltration problem usually requires less gypsum per acre than dose for soil application. Gypsum is particularly effective when added to water if the water salinity is low (EC less than 0.5 dS m^{-1}). It is much less effective for higher salinity water because of the difficulty in applying and getting sufficient calcium into solution to counter the sodium present effectively. Finely ground gypsum (less than 0.25 mm in diameter) dissolves much more rapidly. Therefore, the finely ground, usually pure grades of gypsum are generally more satisfactory for water application, (FAO 1985). However, the grinding cost must be taken into consideration while making decision on gypsum particle size.

37.1.8 Salinity Tolerance of Crops

Excess salinity within the plants root zone has a general deleterious effect on plant growth which is manifested as nearly equivalent reduction in the transpiration and growth rates (including cell enlargement and synthesis of metabolites and structural compounds). This effect is primarily related to total electrolyte concentration and is largely independent of specific solute composition. The hypothesis that best seems to fit observation is that excessive salinity reduces plant growth primarily because it increases the energy that must be expended to acquire water from the soil of the root zone and to make the biochemistry adjustments necessary to survive under stress. This energy is diverted from the processes which lead to growth and yield. Crops differ greatly in their response to salinity. The most distinct signs of injury from salinity are reduced crop growth and loss of yield. Crops can tolerate salinity up to certain levels without a measurable loss in yield (this is called the salinity threshold). The more salt tolerant the crop, the higher the threshold level. At salinity levels greater than the threshold, crop yield reduces linearly as salinity increases (Eq. 37.2). Crops differ greatly in their values of both threshold (t) and slope (s). Values of threshold and slope for many crops are presented in Table 37.2.

37.1.9 Irrigation Management Under Saline Conditions

For the proper management of irrigation scheduling for crops, it is essential to (1) improve the accuracy of the soil water balance components to calculate a reliable estimate of the LF, (2) estimate the LRs and add that to the irrigation requirement, (3) consider the water distribution uniformity to decide which part of the field should receive at least the LF for salinity control, (4) take into account that leaching salts periodically is more practical than every irrigation, (5) consider that there is no need to increase irrigation frequency to control salt concentration (as explained before) except for drip irrigation, and (6) monitor the root zone salinity periodically, especially prior to the times of periodic leaching. This would result in optimum salt control with minimum losses to deep percolation.

Irrigation more frequently is a simple and effective approach especially in soils having an initially high infiltration rate but for which the rate drops rather quickly due to low adequate water at all times without secondary problems developing such as waterlogging and poor aeration. Irrigating more frequently maintains a higher average soil water content and reduces the possibility of water stress that might result if irrigations were spaced further apart. If the crop is not stressed for water between irrigations, increasing the irrigation frequency does little good.

Pre-planting irrigation can be relied upon to fill the rooting depth to field capacity at a time when there is little chance of causing crop damage. In some difficult soils, a pre-planting irrigation is the only opportunity to wet the deeper part of the crop root zone.

Table 37.4 Quantity of drainage water, salinity levels, and estimated reuse in years 1988 and 1992

	Quantity of drainage water in MCM							
	Water EC levels (dS m^{-1})						Estimate reuse	
Regions	<1	1–2	2–3	3–4	>4	Total	Year 1988	Year 1882
Eastern Delta	949	1,565	1,055	310	433	4,312	1,130	2,000
Middle Delta	330	1,421	1,832	273	1,191	5,047	686	1,400
Western Delta	473	412	1,291	901	1,914	4,991	554	1,050
Total	1,752	3,398	4,178	1,484	3,538	14,350	2,370	4,450

Adapted by Mashali (1989) based on data reported by Amer and Ridder (1988) and Rady (1990)

Selection of an appropriate irrigation system is necessary for specific soils. Sprinklers are well adapted to sandy and loamy soils but less so to heavy or clayey soils. Drip or trickle irrigation system are better adapted to loamy or clayey soils and apply water through many small outlets (emitters) at a rate of 2–4 l h^{-1}. At these low rates, they do not disperse the soil particles, as do sprinklers (FAO 1985). Drip irrigation (Ragab et al. 1984) provides a greater opportunity for using saline water. Sprinkler irrigation may cause surface sealing (Ragab 1983) and leaf burn of sensitive crops. Leaf burning can be reduced by night irrigation and by irrigating continually rather than intermittently.

Basin irrigation has greater potential for uniform application than other methods of flooding such as border irrigation or wild-flooding irrigation provided that the basins are levelled and sized properly.

Furrow irrigation tends to accumulate salts in the seedbeds because leaching occurs primarily below the furrows. The length of the furrow, the slope, size of the stream, and time of application are factors that govern the depth and uniformity of application. Leaching and salinity control require a proper balance among these factors.

37.1.10 Egyptian Strategy to Use Drainage Water

Egypt is a predominantly arid country, and the scattered rain showers in the north can hardly support any agricultural crops. Agriculture thus depends mainly on irrigation from the River Nile (55.5 billion cubic metres (BCM) per annum). The needed increase in food production to support the acceleration of population growth (2.7%) compels the country to use all sources of water (i.e. drainage water, groundwater, and treated sewage water) for the expansion of irrigated agriculture. The policy of the Egyptian Government is to use drainage water (up to EC of 4.5 dS m^{-1}) after it is blended with fresh Nile water to form blended water of EC equivalent to 1.0 dS m^{-1}. The drainage water presently used for irrigation amounts to 4.7 BCM per annum, and it is likely to increase to 7 BCM per annum by the year 2020 (see Table 37.4).

In fact, direct use of drainage water for irrigation with EC varying from 2 to 3 dS m^{-1} is common in the districts of Northern Delta where there are no other alternatives or in areas with limited good water quality supply. Farmers in Beheira, Kafr El-Sheikh,

Table 37.5 Yields of dominant crops in Kafr El-Sheikh and Beheira Governorates using drainage water for irrigation

Irrigation water	Average yields Rice tonnes ha^{-1}	Clover (berseem) tonnes ha^{-1}	Barley tonnes ha^{-1}	Cotton tonnes ha^{-1}	Squash kg ha^{-1}
Drainage water (Kafr El-Sheikh, EC = 2–2.5 dS m^{-1})	8.0	150	–	–	–
Drainage water Beheira	8.2	155	3.7	1.9	330
Fresh Nile water (EC = 0.4 dS m^{-1})	8.5	160	3.7	2.0	350

After Mashali (1985)

Damietta, and Dakhlia Governorates have successfully used drainage water directly for periods of 25 years to irrigate over 10,000 hectares (ha) of land, using traditional farming practices. The soil texture ranges from sand and silt loam to clay with calcium carbonate equivalent contents of 2–20% and very low organic matter. The major crops include clover "berseem", rice, wheat, barley, sugar beet, and cotton. Yield reductions of 25–30% are apparently acceptable to local farmers. Yield reductions observed are attributed to waterlogging and salinization resulting from over-irrigation and other forms of poor agricultural, soil, and water management practices.

Pilot studies carried out in Kafr El-Sheik and Beheira Governorates showed that by applying appropriate management practices (i.e. crop selection, use of soil amendments, deep ploughing, tillage for seedbed preparation, land levelling, fertilization, minimum leaching requirements, mulching, and organic manuring), drainage water of EC 2–2.5 dS m^{-1} can be safely used for irrigation without long-term hazardous consequences to crops or soils (see Table 37.5).

In Fayoum Governorate, the annual average volume of drainage water available amounts to 696 million cubic metres (MCM), of which 350 MCM $year^{-1}$ is used at present after blending with canal water. Results of pilot demonstrations in Ibshwai District during the period 1985–1987 on direct and cyclic use of drainage water (EC = 2.8 dS m^{-1}) with fresh Nile water are presented in (Table 37.6).

The following strategy emerges from above demonstrations, i.e. to irrigate sensitive crops (maize, pepper, onion, alfalfa, etc.) in the rotation with fresh Nile water and salt-tolerant crops (wheat, cotton, sugar beet, etc.) directly with drainage water, and moderately sensitive crops (tomato, lettuce, potato, sunflower, etc.) can be irrigated with drainage water but after seedling establishment with fresh Nile water. Based on these results, the governorate is planning to reclaim 4,000 ha of area using the drainage water.

The estimated present annual abstraction from groundwater resources in the Nile Valley and Delta is about 2.6 BCM (for agricultural, municipal, and industrial use) with an average EC of 1.5 dS m^{-1} but ranging far higher, at least to 4.0 dS m^{-1} (the estimated use of this groundwater resource by the year 2010 is 4.9 BCM). Saline groundwaters ranging EC 2.0–4.0 dS m^{-1} have been successfully used for decades to irrigate a variety of crops in large areas of scattered farms in the Nile Valley and Delta.

Table 37.6 Effect of irrigation with different salinity levels on principal crops grown in the area

Source of irrigation water (EC in dS m^{-1})	Wheat grain dry tonnes ha^{-1}	Onion tonnes ha^{-1}	Maize tonnes ha^{-1}	Summer tomato tonnes ha^{-1}	Winter tomato tonnes ha^{-1}	Pepper tonnes ha^{-1}
Drainage water (EC 2.8 dS m^{-1} with SAR 22)	5.0	6.5	1.8	2.5	8.0	12.5
Fresh Nile water for seedling establishment and then drainage water	3.0	6.5	2.0	4.0	8.7	20.0
Fresh Nile water (EC 0.5 dS m^{-1} with SAR 4)	5.0	9.7	2.5	7.5	12.5	25.0

Adapted by Mashali (1989) based on data reported by Rady (1990)

Crops now grown are mostly forage, cereals, and vegetables. In the Delta, saline waters of EC 2.5–4 dS m^{-1} has been used successfully to grow vegetables under greenhouse conditions. In the New Valley (oases Siwa, Bahariya, Farafra, Dakhla, and Kharga), there is potential to irrigate about 60×10^3 ha utilizing groundwater (EC ranging from 0.5 to 6.0 dS m^{-1}), of which 17×10^3 ha is already under cultivation. Siwa Oasis has the largest naturally flowing springs in the New Valley. Siwa once contained a 1,000 springs, of salinity ranging from ECw 2 to 4 dS m^{-1}, which were used successfully to irrigate olive and date-palm orchards, with some scattered forage areas. At present, 3,600 ha is irrigated from about 1,200 wells. Of these 1,000 wells are hand dug to depths of 20–25 m (EC ranging from 3.5 to 5.0 dS m^{-1} and in some locations as much as 10 dS m^{-1}), and the remaining 200 wells were drilled deep (70–130 m) with EC 2.5–3.0 dS m^{-1} and SAR values varying from 5 to 20 (mmoles $l^{-1})^{0.5}$. Presently about 235 MCM $year^{-1}$ is being used successfully to irrigate olive and date-palm orchards, alfalfa, cereals, and wood trees, of which 60 MCM is sourced from continuing flowing springs. Due to over-irrigation without appropriate drainage facilities, seepage as well as runoff to low-lying land has increased as well as salinity and waterlogging have developed in some lands of the oasis. To reduce drainage water volumes and minimize water pollution and safely disposal of unusable final drainage water, new strategies are being developed and implemented by the government authorities in Siwa Oasis (similar problems exist in Dakhla Oasis). These include the use of natural flowing springs to irrigate winter crops such as cereals and forage; use of saline water over 5 dS m^{-1} to irrigate salt-tolerant crops like barley, vetches, rhodes grass, and sugar beet; use of drainage water for the production of windbreak and growing wood trees; use of drainage water for stabilization of sand dunes; reuse of drainage water (average EC 6.0 dS m^{-1} and SAR of 10–15 (mmoles $l^{-1})^{0.5}$ after blending with good quality water (recently drilled deep well of salinity EC 0.4 dS m^{-1} with SAR of 5 (mmoles $l^{-1})^{0.5}$); or alternating the drainage water with good water.

37.2 Conclusions

It is concluded that Egypt requires to boost agriculture production to meet the food demand of the existing as well as fast-growing population. This can be achieved through an integrated approach including soil and water conservation, use of alternate water resources (saline, drainage, treated sewage), amending soil sodicity using gypsum, reducing water salinity through blending with fresh water and using the concept of LF and LR, and the selection of appropriate irrigation system based on soil types and the nature of crop.

References

Abdel-Aal RM (1971) Effect of groundwater and parent material on different soil characteristics in the north east Nile Delta. Ph.D. thesis, Cairo University, Egypt

Amer MH, Ridder NA (1988) Land drainage in Egypt, Drainage Research Institute, technical report, project FAO, UNDP, EGY/80/21, Cairo, Egypt

Eid HM (1999) Estimation of water consumptive use for Egyptian crops. In: Proceedings of third conference of on-farm irrigation and agro-climatology, 25–27 Jan 1999, pp 764–779

El-Mowlhi N, Abd El-Hafez SA (1999) Water requirements and field irrigation for 50 years in Egypt. Soils, Water and Environmental Research Institute, Agriculture Research Center, Ministry of Agriculture, Egypt pp 1–10

El-Nahal RM, Abd El-Aal AA, Abd El-Wahed AA (1977) Soil studies on the Nile Delta. Egy J Soil Sci 17(1):55–65

FAO (1985) Water quality for agriculture. FAO, Rome, pp 71–76

Khater AA, Al-Sharif A, Abdellah AA (1997) Potential effect of groundwater table on some pedogenic formations in the soils of Kafr El-Sheikh Governorate, Egypt

Mashali AM (1985) Amelioration and development of deteriorated soils-Egypt. Technical report, Project FAO/UNDP EGY/79/020, Cairo, Egypt, 1984–1985

Mashali AM (1989) Salinization as a major process of soil degradation in the near East. In: 10th session of the regional commission on land and water use in the near East, Amman, Jordan, 10–14 Dec 1989

Mass EV, Hoffman GJ (1977) Crop salt tolerance-current assessment. ASCE J Irrig Drain Div 103(IR2):115–134

Rady AHM (1990) Water, soil and crop management relating to the use of saline water. In: Proceedings expert consultation on water soil and crop management relating to the use of saline water, Oct 1989. AGL/MISC/16/90, FAO, Rome

Ragab R (1983) The effect of sprinkler intensity and energy of falling drops on soil surface sealing. Soil Sci 136:117–123

Ragab R, Feyen J, Hillel D (1984) Simulating two-dimensional infiltration into sand from a trickle line source using the matric flux potential concept. Soil Sci 137:120–127

Ramadan SA, Ibrahim MM, El-Leithi AA, Gaheen SA (1989) Evaluation of some drainage water of some drain tributaries in Dakahlia governorate, North-East of Nile Delta, Egy J Agric Res Tanta Univ 15(4):746–756

Rhoades JD (1972) Quality of water for irrigation. Soil Sci 113:277–284

Zein FI, Omar EH, Zein MA, El-Kady FA (1997) Effect of some water quality and rhizobium inoculation on soybean. Menofiya J Agric Res 22(5):1441–1445

Zein FI, Omar EH, Abd-Allah MAA, El-Yamani MS (2000) Influence of irrigation water quality on some crops, soils and water table and their content of heavy metal, at the Northern Nile Delta-Egypt. Egy J Soil Sci 4(1):31–42

Part III
New Trends in Land Degradation and Desertification

Chapter 38
New Trends in Land Degradation and Desertification Research and the Role of the Association DesertNet International in Sharing Knowledge and Promoting Sustainable Land Management

Pandi Zdruli

Abstract Efforts of the international and national organisations to halt or reverse land degradation have produced mixed results. Problems are much accentuated in the drylands due to their natural fragility coupled with human-induced pressures and further exacerbated when land degradation is combined with naturally occurring drought. The recent terminology adopted by the United Nations Convention to Combat Desertification (UNCCD) involves desertification, land degradation and drought (DLDD). However, confusion still exists to distinct between them, and this hampers actions from decision-makers as often maps and databases do not make a clear distinction between the potential risks or status of degradation. Soil information could remedy these shortcomings, but soil information must be supported by new surveys to update obsolete soil data. The recommendations of the First Scientific Conference of the Committee on Science and Technology (CST) of the UNCCD held at COP9 in Buenos Aires, Argentina, in September 2009 suggest that desertification research and mitigation should be based on ten priority areas, and the trend is to support sustainable land management rather than focusing only on combating land degradation. The positive outcomes of this paradigm shift emphasise the role of soil as a nonrenewable resource and endorse a biophysical and socio-economic ecosystem-based approach for assessment and monitoring. The Association DesertNet International (DNI) formed on the grounds of the former European DesertNet is aiming to translate scientific knowledge for improved land management of the drylands. As a non-governmental scientific entity, DNI is open to all of them who have interest in land degradation/desertification research.

P. Zdruli (✉)
Land and Water Resources Management Department, International Centre for Advanced Mediterranean Agronomic Studies, Mediterranean Agronomic Institute of Bari, Via Ceglie 9, 70010 Valenzano, Bari, Italy
e-mail: pandi@iamb.it

S.A. Shahid et al. (eds.), *Developments in Soil Classification, Land Use Planning and Policy Implications: Innovative Thinking of Soil Inventory for Land Use Planning and Management of Land Resources*, DOI 10.1007/978-94-007-5332-7_38,

Keywords Association DesertNet International • European DesertNet • Dryland Development Paradigm • Human-environment systems • Soil and vegetation degradation • UNCCD

38.1 Introduction

Land degradation and desertification are some of the most pressing challenges for today's societies due to their impacts on both environmental quality and development process. In addition, there are other derived consequences that affect the livelihoods of billions of people with direct impacts on the well-being of entire communities, including also the devastating effects on ecosystem's stability, functions, goods and services, loss of biodiversity and a number of other related negative severances (Zdruli et al. 2010). By far, the extent of degradation in the drylands covering one-third of the planet and hosting around two billion people has reached critical levels (Adeel et al. 2009).

Both desertification and drought globally cause great losses every year to agricultural production and contribute thus to food insecurity, famine and poverty which can give rise to social, economic and political tensions that can cause conflicts and further impoverishment (UNCCD 2008). Worldwide losses of land degradation in terms of lost productivity equal some US$65 billion annually (Adeel et al. 2009). However, the terminology used to describe resource degradation creates often confusion, especially for the decision-makers, since no clear indication is made about the differences between soil degradation, vegetation degradation, land degradation and desertification. The latter is seldom considered as the final product of degradation and includes all of the above forms but is used mainly to characterise dryland degradation. Problems are further exacerbated when land degradation, mostly a human-induced process, is combined with naturally occurring drought. It is for these reasons that the most recent terminology adopted by the UNCCD involves altogether three terms such as desertification, land degradation and drought. This represents a major shift for the UNCCD giving it a global mandate (as of 2009, 193 countries have ratified the UNCCD). This represents the largest number ratifications among the Rio Conventions and bringing in closer to similar UN Conventions like the Biological Diversity (CBD) and the UN Framework Convention on Climate Change (UNFCCC). However, for the UNCCD, the major focus will be still placed on drylands and particularly in Africa.

The scientific community has invested more than half a century of research into land degradation and desertification, and much is now known compared to when, in the 1950s, Auberville (1949) first coined the term "desertification", interestingly enough not in the drylands but in the tropical forests of Africa with precipitation rates that are far from what conforms to present-day definition. He wrote: "These are real deserts that are being born today, under our eyes, in the regions where the annual rainfall is from 700 to 1,500 mm" (Ce sont de vrais déserts que naissent aujourd'hui, sous nos yeux, dans des pays où il tombe cependant annuellement de

700 à plus de 1,500 mm de pluies) (p. 332). Yet, this situation sounds very pertinent! Auberville noted that the influence of human activities, that is, cultivation, deforestation and accelerated erosion contributed to the process of transformation of tropical forests into savannas and finally into desert-like environments.

Yet today, "we need to know where DLDD is happening, how severe it is, how much harm it does to affected people and ecosystems and the economic costs that it incurs" (the opening speech of Mr. Luc Gnacadja, Executive Secretary of the UNCCD at the First UNCCD CST Scientific Conference, Buenos Aires, Argentina, 22–24 September 2009). These remarks summarise the strongest criticisms that the UNCCD has received, as many stakeholders in the desertification-affected areas around the world could question its more than a decade achievements. A number of assessment technologies have been proposed, involving the use of remote sensing technology combined with ground truthing (Roder and Hill 2009); however, they still remain problematic. Safriel (2007) proposes assessing desertification, especially on a global scale, by involving three stages: (a) generating numerical data based on ground observations and measurements, (b) transforming the numerical data to map units, and (c) extracting statistics by subjecting map units to various analyses.

The GEF-UNEP-FAO (Global Environment Facility-United Nations Environment Programme-Food and Agricultural Organisation of the United Nations)-sponsored LADA project (Land Degradation Assessment in Drylands) was engaged too in the development of standard methods to assess global land degradation (GLADA), and results were reported by Bai et al. (2008). Lambin et al. (2009) proposed an eight-step integrated methodology to assess desertification by involving both biophysical and socio-economic indicators. In fact, such methodology that includes a combination of remote-sensing techniques with local measurements could be applied also to other areas and not necessarily only to drylands. The methodology strives to link theoretical thinking with practical application of human-environment system assessments to establish present status, predict trends and propose policy interventions.

The issue of economic costs of land degradation and desertification is also difficult to address. A number of global assessments have pointed out very different results. Wiebe (2003) estimates the economic effect of soil erosion globally only at 0.05% per year of the total production value. Other authors admit similar values but point out that off-site effects are much higher in economic terms. Controversially, studies in the mid-1990s predicted higher figures, reaching as much as 10% of the value of agricultural production each year according to a joint study of UNEP, United Nations Development Programme (UNDP) and FAO (Pimentel et al. 1995). These estimates reintroduce the crucial question of data availability and their quality. Especially soil data from a number of developing countries are obsolete or even missing, requiring thus the launching of new soil survey programmes. Many shortcomings in land degradation/desertification assessments are also derived from the lack of cause-effect relationships between severity of degradation and agricultural productivity, as elaborated for instance by Nachtergaele (2003). It is also for these reasons that the second UNCCD CST Scientific Conference planned for 2012 will address the theme "Economic assessment of desertification, sustainable land management and resilience of arid, semiarid and dry subhumid areas".

Results "on the ground" over the last decade are not yet convincing many local stakeholders that progress has been made (Zdruli et al. 2009), despite numerous excellent examples of sustainable natural resources management worldwide (i.e. Liniger and Critchley 2007). Recent trends ask for a paradigm shift in support of sustainable land management rather than simply focusing on combating land degradation (Thomas 2008). Climate change will continue to dominate the environmental agenda, and its effects will impact also land degradation-affected areas that will experience additional adversities, especially the drylands (Alcamo et al. 2007; IPCC 2007; NOAA 2011). While recognising the needs for further mitigation actions to alleviate climate change effects, adaptation to the new climatic conditions will be the final unavoidable choice as the history of natural evolution has shown.

There are also a number of other important issues that require attention. They include the links between science and policymaking, as they appear to be weak and the information flow among many stakeholders involved in the combat against land degradation and desertification that is slowly moving.

38.2 The Quest for New Paradigm Shifts

Reynolds et al. (2007) analyse desertification from the perspective of the so-called Dryland Development Paradigm (DDP), which is based on five coupled human-environment (H-E) principles that assess both biophysical aspects of desertification as well as human management and socio-economic components, all incorporated in an integrated ecosystem assessment approach (Fig. 38.1). Drylands are largely backward areas due to their harsh climatic conditions, poor soil fertility and extreme

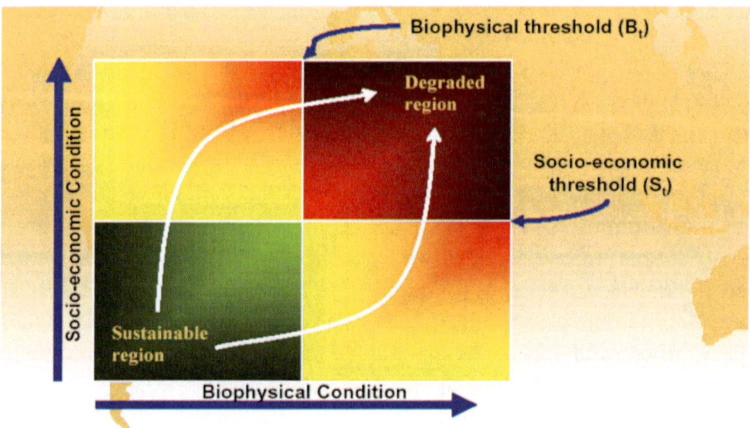

Fig. 38.1 Interaction of biophysical and socio-economic thresholds and their impact on resource base degradation: (a) degradation is a combined process of human and environmental drivers, (b) slow variables are critical determinants of system dynamics and (c) thresholds are crucial and may change over time (Source: Stafford Smith and Reynolds 2002)

scarcity of water coupled with excessive distances from global markets and from centres of political power. In very few cases like the western desert parts of the USA most typically in the Las Vegas area or in some parts of the Arabian Peninsula (Abu Dhabi and Dubai), drylands and deserts are major hubs of business and development (Stafford Smith 2008).

Within the DDP approach, principle P1 states that "human-environment systems are coupled, dynamic and co-adapting" and this is reflected with the closer dependency of most dryland livelihoods on the state of the environment. Principle P2 uses "fast" and "slow" variables to evaluate the impacts of various indicators on desertification trends that are critical to determine the H-E system dynamics. Examples of fast variables for rangeland soils for instance could be soil moisture content, while slower ones are infiltration rate and water-holding capacity. In the case of rangeland pest dynamics, fast thresholds are linked with insect migration (e.g. locust infections), and slower ones relate to foliage type and distribution of shrubs. Similar examples could be given for socio-economic indicators. Emphasis though should be placed on "slow" variables (i.e. soil physical properties) as they control permanent changes of state; instead, "fast" variables (i.e. water availability) reflect quick variations and are not reliable to make long-term predictions. Following these observations, it is very important to establish threshold levels as early warning systems beyond which degradation is not reversible. Prevention is cheaper than cure!

Principle P3 identifies thresholds in key slow variables that define the state of H-E systems. This could be linked for instance with the capacity to invest in recovering from undesirable levels of soil fertility, but this could be hampered by the costs of intervention that increase non-linearly with increasing land degradation. Principle P4 indicates that coupled H-E systems are hierarchical, nested and networked; in other words, the dryland H-E systems must be managed at the appropriate scale since linkages among scales from local to international level are often weak and require additional institutional assistance. The last P5 principle deals with a very important issue or the local environmental knowledge (LEK) that is the key to adaptation and well functioning of H-E systems. But LEK must be combined also with the development and implementation of appropriate modern science-based programmes that should be supplemental to indigenous local knowledge.

Safriel (2009) uses also the term Desertification Paradigm and describes this as a combination of demographic, social, economic and policy processes that lead to pressures on naturally and inherently low productivity of drylands (Safriel and Adeel 2008). These human-induced pressures ultimately result in soil erosion and/or salinisation, as major effects of desertification that are clearly expressed in the eventual reduction of productivity below the natural or potential levels. Such pressures bring to further dwindle of livelihood conditions manifested by increased poverty, migration or even famine. The conclusion is that the world's poor becomes poorer. Safriel links Desertification Paradigm with the Environmental Security Narrative, which addresses particularly the demographic indicators such as population growth, a rate that in the case of the drylands was 18.5% during the last decade of the twentieth century or 11% higher than the global population growth during that period (Safriel 2009).

To face the above-described problems, the inhabitants of these harsh environments have developed a number of ingenious techniques (described by Reynolds et al. (2007) as Principle 5 of the DDP) that are well documented in the literature (Reij and Waters-Bayer 2001; Kapur and Akca 2004; Reij 2005; Liniger and Critchley 2007, WOCAT database www.wocat.net). Safriel (2009) describes also the Counter Paradigm or the "Dryland Livelihood Counter Paradigm" (Fig. 38.2) when a combination of indigenous technologies of agriculture production and soil and water conservation are combined with new opportunities for development such as ecotourism, solar energy production and aquaculture. These are very important aspects that require attention since time has come that dryland's production and development opportunities should be better explored beyond the traditional agricultural, farming or pastoral domains.

38.2.1 Analyses of Two Important Biophysical Components of Resource Base Degradation

Other important features in the recent terminology in land degradation and desertification research include a clear distinction between soil degradation and vegetation degradation as described in the White Paper of Working Group 1 of the Dryland Science for Development consortium (DSD 2009). In this context, degradation of the natural environment involves a decline or change in the quantitative and qualitative characteristics of two major types of ecosystem variables: vegetation and soils and characteristics of these variables are described according to their current "states". Vegetation degradation could be assessed using quantitative and qualitative state variables. The first are described by examples like the proportion of land covered by the vegetation canopy and its biomass density. Examples of qualitative state variables include plant communities e.g. the Sahel is composed of grassland and savannah (typical grasses include *Schoenefeldia gracilis* and *Aristida stipoides*) interspersed with numerous woody plants and shrubs (the dominant ones being acacias, e.g. *Acacia tortilis* and *Acacia senegal*). Degradation can involve a change in species composition or plant functional types that are less productive than the ones they replaced or, in some instances, total production may be unchanged but is due to undesirable or less palatable species (Huenneke et al. 2002).

Vegetation degradation is often caused by human activities like overgrazing through the concentration of too many animals in an area, either permanently or seasonally (e.g. around a borehole used by nomadic herds) that can deplete valuable perennial grasses and replace them by less nutritious annual plants. This practice reduces vegetation density, compact the soil through trampling by livestock and degrades vegetation facilitating thus the process of soil erosion. Deforestation and woodland degradation is another form of human-induced vegetation degradation. Wooded areas are cleared, or deforested, for various reasons, such as expansion of cultivated areas, to obtain fuel wood or charcoal, lopped for fodder or overbrowsed by animals. This reduces the density of trees and the amount of biomass and carbon each tree contains.

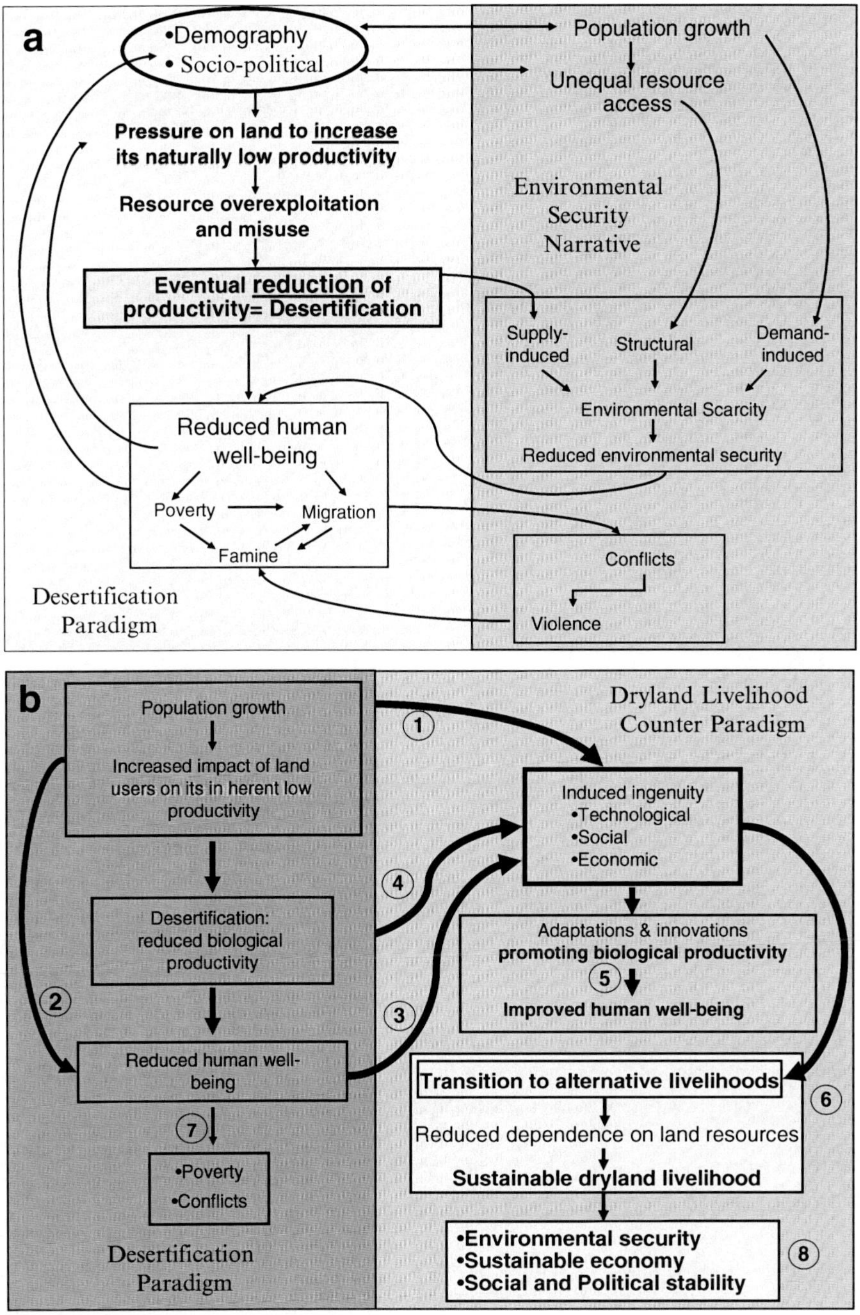

Fig. 38.2 Three views of desertification: (**a**) Desertification in the "Desertification Paradigm" (*left*) is exacerbated by the "Environmental Security Narrative"; (**b**) the "Desertification Paradigm" can be replaced by its counter paradigm, the "Dryland Livelihood Paradigm" in which the downward spiral leading to desertification is disrupted and dryland livelihoods, including alternative ones, leading to a sustainable land use in the drylands (Source: Safriel 2009. Used by permission by the author)

Fig. 38.3 Erosion sediment deposition in a reservoir in the region of Kairouan in Tunisia

A variety of state variables may be required to adequately account for the various forms of soil degradation. The European Union (EU) Thematic Strategy for Soil Protection (European Commission 2006) identifies erosion, decreasing organic matter content, compaction, salinisation, landslides, contamination, sealing and biodiversity decline as the major threats to European soils. These threats, however, are real in many other soils outside the EU. Attempts have also been made in the EU to quantify in economic terms the total costs of soil degradation that just for erosion, organic matter decline, salinisation, landslides and contamination, on the basis of available data, would be up to 38 billion annually for EU25 (Montanarella 2010). Similar problems have been reported also for the Mediterranean countries (Zdruli et al. 2007). Many of the above processes prevail in most dryland soils in the forms described below:

38.2.1.1 Water Erosion

Water erosion changes soil fertility when it is washed away by rain due to the lack of protection of vegetation cover. Erosion can also change the micro-topography of the land if it continues unhindered as water flows may concentrate in small channels or rills. In time, these rills can develop into recognisable gullies, which in some cases grow into deep canyons. Side effects are sedimentation of water reservoirs (Fig. 38.3), destruction of infrastructure and increased flooding intensity associated with extensive property losses. Often, off-side economic losses are higher than on-site ones.

38.2.1.2 Wind Erosion

Wind erosion changes soil quality when wind blows away its finer components, such as silt, clay and organic matter, leaving behind inert sand, gravel and other coarse particles. It can also change the character of surface cover, leading to the formation of sandy areas and sand dune environments. Mobilisation of sand dunes and sand encroachments can overwhelm nearby cropland and human settlements.

38.2.1.3 Soil Compaction

Soil compaction changes the permeability of the soil surface. Consequently, run-off increases lead to erosion, less water entering into the soil for use by plants, which makes it difficult for plants to germinate and establish roots. The depth of compaction can vary considerably and depends on the type of agricultural machinery used.

38.2.1.4 Salinisation and Waterlogging

The salinisation and waterlogging affect the physical and chemical properties of poorly drained irrigated cropland, consequently reducing plant growth and yields. Simple waterlogging can be followed by salinisation or the formation of the Solonchak soils as described by the International Union of Soil Sciences (IUSS) Working Group on World Reference Base for Soil Resources as excess water evaporates from the surface, leaving behind dissolved salts mainly sodium chloride and sodium sulphate and other minerals either near or on the surface. In alkalinisation or sodification (Solonetz soils, IUSS Working Group WRB 2006), clay particles are enriched in exchangeable sodium and/or magnesium ions. These are the most problematic soils of the world. Altogether saline, sodic and gypsum constraint soils cover an estimated area of 6% of the world soils, and the majority of them are concentrated in the drylands (Fischer et al. 2000).

38.2.1.5 Reduction of Soil Organic Matter and Soil Biodiversity Losses

The reduction of organic matter and loss of biodiversity ultimately reduce soil fertility and water-holding capacity and have direct consequences on crop productivity and other soil/ecosystem functions and services. Aridic and Xeric soils naturally exhibit limited to very low fertility and organic matter content; consequently losses that may appear small by midlatitude standards often have drastic, sometimes irreversible, consequences in the drylands. The Aridic and Xeric terms are used in the United States Department of Agriculture (USDA) Soil Taxonomy to classify soil moisture regimes (SMR). Aridic soils are dry in all parts for more than half of the cumulative days per year when soil temperature at a depth of 50 cm from the surface is above 5°C, and moist in some or all parts for less than 90 consecutive days when the soil temperature at a depth of 50 cm is above 8°C. They typically occur in arid areas and a few in semiarid climates. Xeric soils are typical for Mediterranean environments where winters are moist and cool and summers warm and dry (Soil Survey Staff 2006).

38.2.2 *Climate Variations, Ecosystem Goods and Services and the Drylands*

Drylands are particularly vulnerable to climate variability, and even slight changes in rainfall could lead to major shifts in productivity (Reynolds et al. 2004) and could provide misleading information about the real causes of biomass productivity changes. The case of the Sahel droughts, for example, has been thoroughly described in literature (Tucker et al. 1994; Grainger 1990; Safriel 2009; Udelhoven and Hill 2009), and they point at the fact that links between drought and desertification are not straightforward. The Sahel droughts of the period 1968–1974 and then those of 1983–1985 were combined with overexploitation of land resources and the explosion of population increase (Safriel 2009) that were the main cause for adopting in 1977 the first Plan of Action to Combat Desertification (PACD) and in 1992 the Agenda 21. But, with very little action on the ground, a "greening of the Sahel" followed the period between 1985 and 2000 as the drought period came to the end (Safriel and Adeel 2005) questioning therefore the fact if human-induced desertification was the cause of biomass reduction or the naturally occurring drought. It was also for these reasons that the Millennium Ecosystem Assessment (MEA 2005) modified the definition of desertification by introducing the word "persistent" reduction of biological productivity. On the contrary, soil degradation reflects longer past and present severe disturbances of the ecosystems. Finally, climate variability and the degree of degradation have many implications as assessments are always related to the initial baseline information to which comparisons have to be made. Moreover, if one considers also the anticipated climate change impacts, these analyses become even more complicated.

Another important issue that has received wide attention is the subject matter of ecosystem goods and services as described by the Millennium Ecosystem Assessment. They include four groups identified by the following services: provisioning services (food, water, wood, fishing, hunting, genetic resources, etc.), regulating services (carbon sequestration, air quality, climate, floods, diseases), cultural services (aesthetic, spiritual, religious, educational and recreation) and supporting services (soil, nutrient supply, crop growth). The latter include also services such as the net primary production, habitat for biodiversity, soil formation and the cycling of water and nutrients. A study published almost a decade and half ago in 1997 (Costanza et al. 1997) valued biosphere and ecosystem services at some US$33 trillion per year. That is worth nearly double the world's GDP! In more practical terms, in order to assess for instance the value of ecosystem goods and services provided by forests and vegetation cover, one has to account not only the value of timber and fodder but also other derived benefits such as carbon sequestration, clean purified water, soil preservation, biodiversity sheltering, crop pollination, recreation values and may be even more features.

In the case of land degradation and desertification, it is typically the reduction of *supporting services* that ultimately lead to the persistent decrease in the ability of the system to provide provisioning and regulatory services. However, all of them have direct impacts on the human well-being since involve a range of services that are essential for life such as health, nutrition, income, basic materials, good social relations and finally environmental security.

38.2.3 Establishing Effective Non-profit Institutions to Accelerate Knowledge Dissemination: From the European DesertNet (EDN) to Association Desertnet International (DNI)

The need for accelerating information flow in land degradation and desertification research has become a pressing priority. The UNCCD is particularly attentive to such initiatives, as it has realised that a large number of its stakeholders often complain for not having easy access to research findings, and for worst, decision-makers are not even aware of the stage the research has reached. To fill these gaps, at least at the European level, a handful of concerned scientists decided to join forces and establish an entity called European DesertNet (EDN).

The first ideas started in 2001 based on discussions within Desert*Net Germany for the creation of a regional scientific network. In May–June 2005, at the third session of the Committee for the Review of the Implementation of the Convention (CRIC3) of the UNCCD, such discussions were further expanded between Desert*Net Germany and Le Comitè Scientifique Français de la Dèsertification (CSFD) or the French Scientific Committee on Desertification and the Belgian Expert Group on Desertification. The three bodies agreed to write a Declaration on Desertification that emphasised mostly the needs for further applied research on desertification within and outside Europe. In September 2005, the declaration was finalised, and within a few months, more than 80 European scientists signed it. In October 2006, the European DesertNet was officially launched in Bonn, Germany, in a meeting attended by 38 scientists from 8 European countries and from representatives from the UNCCD/CST, European Commission (DG Research and DG Joint Research Centre) and UNESCO's Man and Biosphere Programme (UNESCO/MAB) as well as representatives of European ministries of environment. The meeting concluded with the establishment of the goals and the structure of EDN. In January 2007 in Brussels, the first working meeting was held that established (1) Working Groups, (2) the launching of a regular electronic newsletter and (3) the launching of an Internet webpage. A second working meeting was held in Madrid, Spain, in the occasion of UNCCD COP8 where the legal status and funding mechanisms of the organisation were discussed. During the period 2007–2008, EDN became very active and was present at various COPs, CST and CRIC meetings of the UNCCD. A very important event was the organisation of a desertification-hearing meeting with members of the European Parliament in Brussels in May 2008.

As EDN was becoming more international in membership, a new phase appeared ahead, and the need was felt to transform it from a European entity into an international one in the form of an association. In March 2008, five international organisations, that is, International Centre for Agricultural Research in Dry Areas (ICARDA), International Crop Research Institute for Semiarid Tropics (ICRISAT), European Commission Joint Research Centre – Institute for Environment and Sustainability (EC-JRC-IES), United Nations University – Institute for Water Environment and Health (UN-INWEH) and the EDN formed a consortium called Dryland Science for Development (DSD) that was in charge of organising the first scientific-style UNCCD CST Conference at COP9 in Buenos Aires, Argentina, in September 2009. The author of this chapter was a member of DSD Working Group 1 "Integrated methods for monitoring and assessment of desertification/land degradation processes and drivers" and is presently member of the Steering Committee of the DNI. This was a great achievement for EDN that opened the way for transforming it in an international association. This was officially accomplished at Illkirch/Strasbourg court in France in 30 November 2009 that finally legalised the creation of the Association DesertNet International (DNI).

38.2.3.1 The DNI Approach

Networking has been identified as the best tool for knowledge dissemination. As a multidisciplinary network, DesertNet International provides a platform for scientifically based discussions and exchange of ideas (http://www.european-desertnet.org/). DNI also addresses knowledge gaps and identifies research areas that are relevant to land degradation and desertification. As a think tank, DNI communicates with stakeholders and policymakers by identifying issues and priorities of importance for them. The network responds to demands for assessment and information needs and translates knowledge into laymen's terms to improve public awareness of desertification, its costs and implications even beyond the drylands. DNI cooperates with the EC Research Directorate, the EC DG Joint Research Centre, ministries, development agencies and UN bodies (specifically UNCCD). DesertNet International has no government affiliation, and it is independent of any form of lobby or private interests. The organisational structures of the association are composed of the General Assembly, the Steering Committee, the Advisory Board, the Users Board and the Board Intern. These structures have become operational after the 1st General Assembly held in Rome in September 2010. Members of the association are persons as opposed to institutions, and the criterion for membership is individually based. Scientists of all disciplines with expertise in dryland research are invited to join. Students may also join but do not have the right to vote and are not allowed membership on the Association's Steering Committee or the Advisory Board.

DNI activities are organised in five Working Groups (WG) that include (1) Science-Policy Interface, (2) Dryland Observation System, (3) Economic Drivers and Social-Economic Consequences, (4) Training and Capacity Building and (5) Working Group on Structures. Some of these WGs are more active, for instance, the Science Policy Interface in Spring 2007 received an official ministerial request from Germany to assess the UNCCD 10-year strategic plan and to analyse the framework to enhance the implementation of the convention for the period 2008–2018. At COP8 of the UNCCD in Madrid, this working group also organised a side event entitled "Desertification: Science, Policy and the Web of Stakeholders" in cooperation with the OASIS consortium (http://www.icarda.org/INRMsite/Oasis_Subweb/partner.htm).

The Working Group on Dryland Observation System was established in January 2007, and it addresses specific demands for instruments to observe and analyse processes of desertification as controlled by climate change and land use changes. In September 2007, this Working Group carried out its first international and interdisciplinary workshop on Observation of Desertification in Hamburg, Germany. The meeting formulated a declaration that stressed the needs to assess, monitor and improve the understanding of the present state and trends in dryland ecosystems and their services as tools for supporting actions leading to sustainable land use and desertification mitigation. It also proposed to identify a suitable and focused set of harmonised variables and protocols for use in remote sensing and ground observations in support of the UNCCD and other UN Conventions and to promote synergies between them.

Other DNI WGs are in preparation stage. They include environmental management, food security, livestock and modelling.

DesertNet International is currently discussing its role in the Second CST Scientific Conference, which is planned for 2012 and will focus on the economics of desertification and thus specifically address DNI WG "economics" objectives. The WG on Training and Education has addressed the need of a science-based Manual of Desertification that could be used by a variety of stakeholders and relevant national and international organisations, including administrators in developing countries, NGOs and private entities operating in the dry areas.

As of 28 September 2011, DNI has 259 members from 45 countries worldwide. Many of them were former members of European DesertNet. Membership at DNI is open to all qualified scientists that have interest in the topics of land degradation and desertification and endorse a multidisciplinary approach. DNI is looking forwards to collaborating with international organisations, programmes and agencies in need of scientific information or advice and is prepared to put its knowledge and understanding to the service of all stakeholders involved with combating desertification and creating sustainable livelihoods in drylands through sound scientific cooperation.

38.3 Conclusions

Combating land degradation and desertification proved to be not an easy task as may be originally thought. Even after more than a decade of numerous meetings and international events, most typically the Conference of Parties (COPs) of the UNCCD, results on the ground show little progress, and UNCCD itself appears to be at the crossroads. This is due to various reasons ranging from lack of financial resources to shortcomings in science-based policy advice. One thing is sure: resource base degradation has both implications for the environment as well as for the development process, and land degradation is both cause and consequence of poverty: people will never spare the last tree for the sake of landscape loveliness if they feel cold inside their home.

The issue of land and soils as natural capitals should be at the centre of the political debates as they are not inexhaustible resources in front of so many competing interests, where agriculture is often the looser. However, no nation could afford to degrade its land and abandon it, and the time has come to shift to a new paradigm from exploiting the land to renewing the land (Adeel et al. 2009). Land degradation has no political borders and is affecting in various forms both the poor and the rich nations; hence, global assessment and global agreements should be able to address various levels of decision-making as important livelihood concerns are at stake.

The UNCCD CST COP9 1st Scientific Conference identified ten priority areas (UNCCD 2009) for monitoring and assessing land degradation and desertification and for enhancing sustainable land management. The *first* priority is the establishment of scientific networks that are able to develop science-based frameworks for assessing and monitoring desertification, land degradation and drought (DLDD). *Second* is the development of Integrated Assessment Models (IAMs) that collate biophysical knowledge with social, policy, economic and institutional data that facilitate participatory and trans-disciplinary monitoring and assessment and enable decision-makers to understand priorities and trade-offs. *Third* are the tools relevant to different decision-making levels that should be relevant to each level but compatible between levels to allow for upscaling and downscaling. *Fourth* deals the importance of sustainable land management (SLM) as assessment should not address only DLDD but most importantly the recovery of land and of its functions through best SLM practices. The best way to describe this is to paraphrase Mahatma Gandhi: "To forget how to tend the soil is to forget ourselves." *Fifth* relates to synergies deriving by improved SLM that increases carbon sequestration in the soil mitigating thus climate change impacts. SLM increases also crop yields, farm income, improves soil biodiversity and acts as a common denominator for other UN Conventions such as CBD and the UNFCCC. *Sixth* is cost-benefit analyses that can help decision-makers to make the right decisions at the appropriate time to combat DLDD based on evaluation of ecosystem services that include both monetary and nonmonetary benefits to the society. *Seventh* is the knowledge management platform needed to integrate the know-how to combat DLDD with effective dissemination of both success stories and failures in natural resources use and management. The *eighth* pillar relates to cross-sector capacity building that should be able to tackle the great complexity of DLDD and SLM paradigms that must be endorsed by governmental policy agendas. The *ninth* priority deals with the strengthening of scientific capacities and the establishment of an IPCC style, independent, international and interdisciplinary scientific mechanism able to provide policy-based advice to the UNCCD based on proved scientific records. The last but not least, the *tenth* priority is the establishment of a DLDD/SLM monitoring system that must be harmonised and coordinated with other similar systems already operational at UNFCCC and CDB.

Acknowledgement The author would like to thank Dr. Mariam Akhtar-Schuster, Chair of the DesertNet International at the Biocentre Klein Flottbek and Botanical Garden, University of Hamburg in Germany for her kind support in describing both European DesertNet and the DesertNet International Association.

References

Adeel Z, Dent DL, Dobie P, Mersmann C, Niamir-Fuller M, Quatrini S, Sokona Y (2009) Revitalizing the UNCCD. United Nations University-International Network on Water, Environment and Health (UNU-INWEH). Hamilton Ontario, Canada. http://www.inweh.unu.edu/drylands/docs/Revitalizing_the_UNCCD-English.pdf. Accessed 2 Feb 2010

Alcamo J, Florke M, Marker M (2007) Future long-term changes in global water resources driven by socio-economic and climatic changes. Hydrol Sci J 52:247–275

Auberville A (1949) Climats, Forêts et Désertification de l'Afrique Tropicale. Société d'Editions Géographiques, Maritimes et Coloniales, Paris

Bai ZG, Dent DL, Olsson L, Schaepman ME (2008) Proxy global assessment of land degradation. Soil Use Manag 24:223–234

Costanza R, d'Arge R, de Groot R, Farber S, Grasso M, Hannon B, Limburg K, Naeem S, O'Neill RV, Paruelo J, Raskin RG, Sutton P, van den Belt M (1997) The value of the world's ecosystem services and natural capital. Nature 387:253–260

Dryland Science for Development (2009) White paper of DSD working group 1. http://dsd-consortium.jrc.ec.europa.eu/documents/WG1_White-Paper_Draft-2_20090818.pdf

European Commission (2006) Thematic strategy for soil protection, COM (2006), p 231

Fischer G, van Velthuizen H, Nachtergaele F, Medow S (2000) Global agro-ecological zones (Global-AEZ). Web page available on-line on http://www.fao.org/ag/AGL/agll/gaez. Food and Agriculture Organization of the United Nations (FAO) and International Institute for Applied Systems Analysis (IASA), Rome/Laxenburg. Accessed 29 Jan 2010

Grainger A (1990) The threatening desert: controlling desertification. Earthscan, London

Huenneke LF, Anderson JP, Remmenga M, Schlesinger WH (2002) Desertification alters patterns of aboveground net primary production in Chihuahuan ecosystems. Glob Change Biol 8:247–264

IPCC (2007) Climate change 2007. The physical science basis. In: Solomon S, Qin D, Manning M, Chen Z, Marquis M, Averyt KB, Tignor M, Miller HL (eds) Contribution of working group 1 to the fourth assessment report of the intergovernmental panel on climate change. Cambridge University Press, Cambridge/New York, p 996. http://www.ipcc.ch/ipccreports/ar4-wg1.htm

IUSS Working Group WRB (2006) World reference base for soil resources 2006. World soil resources reports no 103. FAO, Rome

Kapur S, Akca E (2004) Indigenous environmentally friendly technologies. Encycl Soil Sci. doi:10.1081/E-ESS 120006648, Marcel Dekker

Lambin EF, Geist H, Reynolds JF, Mark D, Stafford Smith M (2009) Coupled human-environment system approaches to desertification: linking people to pixels. In: Roder A, Hill J (eds) Recent advances in remote sensing and geoinformation processing for land degradation assessment, ISPRS book series. CRC Press/Balkema, Leiden

Liniger HP, Critchley W (eds) (2007) Where the land is greener: case studies and analysis of SWC worldwide. CTA and University of Berne, Berne

MEA (2005) Ecosystems and human well-being: synthesis. Island Press, Washington, DC

Montanarella L (2010) Moving ahead from assessments to actions: could we win the struggle with soil degradation in Europe? Chapter 2. In: Zdruli P, Pagliai M, Kapur S, Faz Cano A (eds) Land degradation and desertification: assessment, mitigation and remediation. Springer, Berlin/Heidelberg/New York, pp 15–23. doi:10.1007/978-90-481-8657-0_2

Nachtergaele F (2003) Land degradation assessment in drylands: the LADA project. In: Montanarella L, Jones RJA (eds) Land degradation, EUR 20688 EN. Office for Official Publications of the European Communities, Luxembourg, p 324

NOAA (2011) Human-caused climate change a major factor in more frequent Mediterranean droughts. http://www.noaanews.noaa.gov/stories2011/20111027_drought.html. Last accessed 07 Nov 2011

Pimentel D, Harvey C, Resosudarmo P, Siclair K, Kurz D, McNair M, Crist S, Shpritz L, Fitton L, Saffouri R, Blair R (1995) Environmental and economic costs of soil erosion and conservation benefits. Science 267:1117–1123

Reij C (2005) Silent transformation: processes of innovation in African agriculture. In: Zdruli P, Trisorio Liuzzi G (eds) Promoting participatory management of the land system to enhance soil conservation, MEDCOASTLAND publications 3. IAM, Bari, pp 119–132

Reij C, Waters-Bayer A (2001) Farmer innovation in Africa: a source of inspiration for agricultural development. Earthscan, London

Reynolds JF, Kemp PR, Ogle K, Fernández RJ (2004) Modifying the 'pulse-reserve' paradigm for deserts of North America: precipitation pulses, soil water, and plant responses. Oecologia 141:194–210

Reynolds JF, StaffordSmith M, Lambin E, Turner B, Mortimore M, Batterbury S, Downing T, Dowlatabadi H, Fernandez R, Herrick J, Huber-Sannwald E, Jiang H, Leemans R, Lynam T, Maestre F, Ayarza M, Walker B (2007) Global desertification: building a science for dryland development. Science 316:847–851

Roder A, Hill J (eds) (2009) Recent advances in remote sensing and geoinformation processing for land degradation assessment, ISPRS book series. CRC Press/Balkema, Leiden, p 400

Safriel UN (2007) The assessment of global trends in land degradation. In: Sivakumar MVK, Ndiang'ui N (eds) Climate and land degradation. Springer, Berlin, pp 1–38

Safriel UN (2009) Status of desertification in the Mediterranean region. In: Rubio JL, Safriel UN, Daussa R, Blum WEH, Pedrazzini F (eds) Water scarcity, land degradation and desertification in the Mediterranean region, NATO science for peace and security series C: environmental security. Springer Science + Bussines Media BV, Dordrecht, pp 33–73

Safriel UN, Adeel Z (2005) Dryland systems. In: Hassan R, Scholes R, Ash N (eds) Ecosystems and human well being: current state and trends. Island Press, Washington, DC, pp 623–662

Safriel UN, Adeel Z (2008) Development paths of drylands – is sustainability achievable? Sustain Sci J 3(1):117–123

Soil Survey Staff (2006) Keys to soil taxonomy, 10th edn. USDA-Natural Resources Conservation Service, Washington, DC

Stafford Smith M (2008) The 'desert syndrome'- causally-linked factors that characterise outback Australia. Rangeland J 30:3–14

Stafford Smith M, Reynolds JF (2002) Desertification: a new paradigm for an old problem. In: Reynolds JF, Stafford Smith M (eds) Global desertification: do humans cause deserts? Dahlem workshop report 88. Dahlem University Press, Berlin, pp 403–424

Thomas R (2008) Critical review: 10th anniversary review: addressing land degradation and climate change in dryland agroecosystems through sustainable land management. J Environ Monit 10:595–603. doi:10.1039/b801649f

Tucker CJ, Newcomb WW, Dregne HE (1994) AVHRR data sets for determination of desert spatial extent. Int J Rem Sens 15:3547–3565

Udelhoven T, Hill J (2009) Change detection in Syria's rangelands using long-term AVHRR data (1982–2004). In: Roder A, Hill J (eds) Recent advances in remote sensing and geoinformation processing for land degradation assessment, ISPRS book series. CRC Press/Balkema, Leiden

UNCCD (2008) UN launches international year of deserts and desertification. Original text published in 1 January 2006 and updated in 2008: http://www.unccd.int/IYDD/index.htm

UNCCD (2009) Land matters: enhancing synergies among the Rio Conventions on land use and sustainable land management. Science brief series no 1. http://www.unccd.int/knowledge/docs/Science%20Briefs%20Series%20No.1.pdf. Accessed on 4 Feb 2010

Wiebe K (2003) Land quality, agricultural productivity and food security. In: Keith W (ed) Land quality, agricultural productivity and food security: biophysical processes and economic choices at local, regional and global levels. Edward Elgar, Northampton

Zdruli P, Lacirignola C, Lamaddalena N, TrisorioLiuzzi G (2007) The EU-funded MEDCOASTLAND thematic network and its findings in combating land degradation in the Mediterranean region. In: Sivakumar MVK, Ndiang'ui N (eds) Climate and land degradation. WMO and UNCCD/Springer, Berlin/Heidelberg, pp 422–434

Zdruli P, Biancalani R, Nachtergaele F (2009) Results of the second worldwide user survey of LADA project. In: First UNCCD CST scientific conference on Understanding Desertification and Land Degradation Trends: bio-physical and socio-economic monitoring and assessment of desertification and land degradation to support decision making in land and water management. Buenos Aires, Argentina, 22–24 Sept 2009

Zdruli P, Pagliai M, Kapur S, Faz Cano A (2010) What we know about the saga of land degradation and how to deal with it. Chapter 1. In: Zdruli P, Pagliai M, Kapur S, Faz Cano A (eds) Land degradation and desertification: assessment, mitigation and remediation. Springer, Berlin/Heidelberg/New York, pp 3–15. doi:10.100/978-90-481-8657-0_1

Chapter 39
Land-Use Planning for Controlling Land Degradation in Kuwait

Fikry I. Khalaf, Jasem Al-Awadhi, and Raafat F. Misak

Abstract Several land degradation indicators have been recognized in Kuwait; these are soil loss by wind and/or water; soil crusting, sealing, and compaction; soil contamination by oil; soil salinization; deterioration of vegetation cover and its biodiversity; and hydrological degradation. Aeolian processes, as manifested by soil deflation, drifting sand, and migrating dunes, may be considered as one of the primary causes of soil degradation in Kuwait. The present status of land degradation as reflected by the severity of degradation of vegetation cover, soil erosion/deflation, and hydrological drought has been assessed based on visual comparison between data recorded in the early 1980s and recent field surveys (early 2010). It has been concluded that the northwestern and southern parts of Kuwait are severely degraded. Anthropogenic activities (off-road vehicle traffic, excessive grazing, camping, and quarrying) as well as military operations (Gulf Wars and present defense structures) are the main causes of land degradation in Kuwait. However, aeolian processes, nature of surface sediments (soil type), and climatic conditions play a significant role. It was also recognized that the sandy soils (Torripsamments) that cover most of the southern part of Kuwait are the most vulnerable to aeolian processes, particularly soil loss by deflation. Hence, mitigation measures for the maintenance of Kuwait desert ecosystem are recommended.

F.I. Khalaf (✉) • J. Al-Awadhi
Earth and Environmental Science Department, Faculty of Science,
Kuwait University, Kuwait, Kuwait
e-mail: fikry_khalaf@hotmail.com

R.F. Misak
Environment and Urban Development Division, Coastal and Air Pollution Department,
Kuwait Institute for Scientific Research (KISR), P. O. Box 2488513109 Safat, Kuwait
e-mail: rmisak@kisr.edu.kw

S.A. Shahid et al. (eds.), *Developments in Soil Classification, Land Use Planning and Policy Implications: Innovative Thinking of Soil Inventory for Land Use Planning and Management of Land Resources*, DOI 10.1007/978-94-007-5332-7_39,

Keywords Anthropogenic activities • Kuwait • Land degradation • Land-use planning • Military operations

39.1 Introduction

Land degradation is believed to be one of the most severe and widespread global environmental problems (Dregne et al. 1991; Dregne 1992; Oldeman et al. 1992; UNEP 1992; UNCED 1992; Maingnet 1994; Lal and Stewart 1994; Johnson and Lewis 1995; Middleton and Thomas 1997; Lal et al. 1997; Reynolds and Smith 2002; Al-Awadhi et al. 2005a). The term land degradation, as defined by UNEP (1992), is a process, which lowers the current and/or the potential capability of land to produce goods or services (quantitatively and/or qualitatively).

Dry lands are particularly susceptible to degradation, and although they cover about 41% of Earth's land surface (Safriel and Adeel 2005), estimates of the extent and severity of degradation vary greatly (Lepers et al. 2005). There are estimates that 70% of the world's dry lands are affected and that at least one third of the present deserts are man-made (UNCED 1993; Le Houérou 1996; Warren et al. 1996; Ravi et al. 2010). According to Kassas (1995) and Agnew and Warren (1996), it is particularly the semiarid regions of the world that are most susceptible. At the same time, these regions, both rich and poor, are experiencing some of the highest population growth rates worldwide (Warren et al. 1996).

Land degradation stems from an imbalance between the fragile environment and human economic activity, among many other factors. In other words, land degradation in ecologically vulnerable region is the outcome of irrational unsustainable land use and lack of proper resource management. Rational and effective measures for controlling and rehabilitation of degraded lands should be based on a thorough understanding of the causes and pattern of land degradation.

In recent decades, Kuwait has suffered severe land degradation, as noted by several authors (e.g., Khalaf 1989; Omar 1991; Zaman 1997; Brown 2003; Brown and Porembski 1997, 1998, 2000; Howle 1998; Misak et al. 1999; Shahid et al. 1998, 1999, 2003; Al-Dousari et al. 2000; Al-Awadhi et al. 2001; Omar et al. 2001, 2005). The main causes of this problem can be attributed principally to overgrazing, but also to recreational activities such as off-road driving and camping, as well as industrial practices, particularly quarrying (Khalaf 1989). In addition, serious damage was inflicted on the natural environment during the recent Iraqi occupation and the hostilities associated with the liberation of the country during the Gulf War (Shahid et al. 1999; Al-Awadhi 2001). The study of Misak et al. (1999) indicated that at least 76% of the desert ecosystem of Kuwait suffers different degrees of land degradation, 44% moderately, and 32% being severely degraded.

Although globally, the world main concern of land degradation is due to a decrease in land capabilities and productivity, the main concern in Kuwait may be attributed to the associated deterioration in the desert ecosystem particularly environmental quality and biodiversity. Land capability for grazing activities has high cultural importance; however, it is economically insignificant.

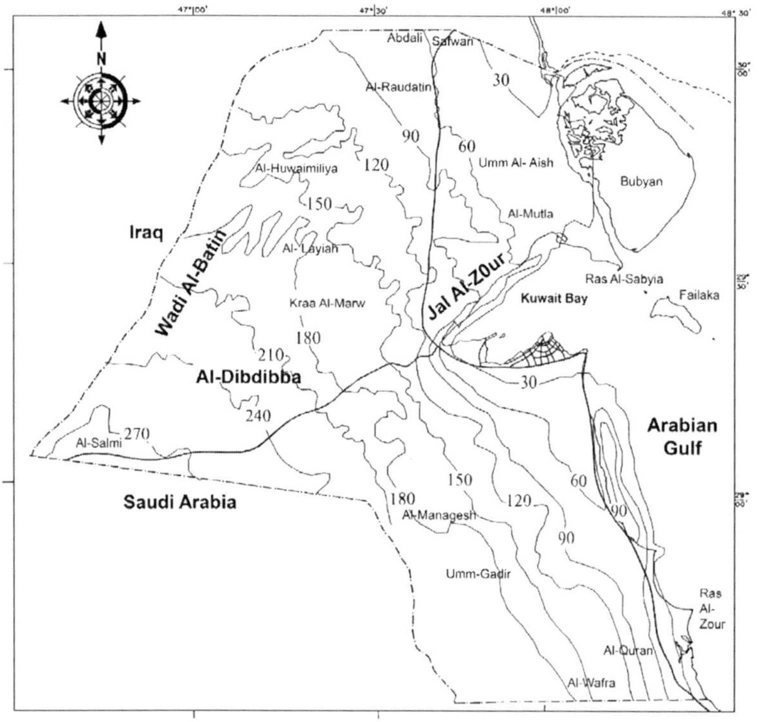

Fig. 39.1 Location map of Kuwait

This chapter provides a review of published work on land degradation of Kuwait. Also, the present status of land degradation as reflected by the severity of degradation of vegetation cover, soil erosion/deflation, and hydrological drought are assessed based on visual comparison between records and observation by the authors in early 1980s of the last century and their recent field surveys during the period of January–March 2010. An assessment of the vulnerability of the various soil types in the northwestern and southern areas of Kuwait to aeolian processes, particularly soil loss by deflation, is discussed. Causes and consequences of land degradation are also discussed and mitigation measures to alleviate the severity of land degradation in Kuwait are recommended.

39.1.1 Natural Environment of Kuwait

The State of Kuwait is situated at the northwestern corner of the Arabian Gulf, and it covers an area of 17,818 km^2 (Fig. 39.1). Summer is very hot, especially in July and August, with mean temperatures of 37.4°C and maximum mean temperatures of 45°C. On a monthly basis, potential evapotranspiration far exceeds precipitation throughout the year. Rainfall is scarce and irregular; it varies from 28 mm (1963–1964) to 260 mm (1975–1976), and the total average of rainfall is approximately 115 mm $year^{-1}$, with an evaporation rate of 16.6 mm day^{-1}.

Wind direction in Kuwait is fairly consistent; about 75% of the winds blow from the west, northwest, and north. The average monthly wind speed ranges from 8.7 knots (in January and December) to 11.2 knots (in June and July). Sand and dust storms are frequent during the summer period.

The surface of Kuwait is carved from the Mio-Pleistocene calcretic formation during early Holocene pluvial periods. The carved depressions and water courses are filled by the fluvial and alluvial sediments. By the advent of aridity, aeolian processes started to act on these fluvial sediments as the present-day aeolian landforms were developed. These landforms can be grouped into erosional and depositional landforms. Erosional landforms are represented by (a) calcretic and gypcretic rock exposures, which occur as flat terrain and small humps and some are yardangs; (b) desert pavement represented by pebbly lag, granule lag, and mixed lag; and (c) granule ripples. Depositional landforms include (a) sand dunes, (b) nabkhas and sand drifts, (c) sand sheets, and (d) anthropogenic-related sand accumulations. Sabkhas and tidal flats are the most dominant landforms in the coastal zone.

The open rangeland is flat to slightly undulating desert plain sloping gently toward the northeast from about 280 m above sea level near Wadi Al-Batin in the extreme southwestern corner of the country toward the Arabian Gulf coast in the east. It is broken by occasional ridges, low hills, wadies (drainage lines), depressions, and sand dunes. Two prominent ridges dissect the northwestern part of the rangeland, namely, Al-Layiah and Kraa Al Marw, which are made up of gravelly gypcretic calcrete. Less prominent discontinuous smaller ridges are parallel to these ridges. They are separated by relatively flat desert plains covered by desert pavement (pebbly lag). Isolated calcretic hills are frequent in the southwestern area. Permanent or even intermittent streams are lacking in Kuwait. Dense internal drainage systems occur in the northern area especially in the Rawdatain and Umm Al-Aish areas where they converge into a fairly large depression. Small depressions (playas) are encountered in the north, central, and western parts. Wadi Al-Batin, the largest wadi in the country, marks the northwestern border with Iraq. It reaches a width of 8–11 km and a depth of about 50 m below the level of the adjacent plateau.

Due to the harshness of the climate, pedogenesis is minimal, leaving the sandy or gravelly parent material little altered by the soil-forming processes. In most southern areas, Torripsamments are prevalent, whereas Petrogypsids, Calcigypsids, and Haplocalcids predominate in many northern and western parts of the country (Kuwait Institute for Scientific Research 1999; Omar et al. 2000) (Fig. 39.2).

According to Halwagy and Halwagy (1974) and Halwagy et al. (1982), three distinct vegetation communities are present, namely, (a) a dwarf shrub community of Rhanterium epapposum dominated the southern half of the country; (b) Cyperus conglomeratus which replaces R. epapposum as the key perennial species, probably as a result of overgrazing; and (c) the chenopod Haloxylon salicornicum community that is prevalent in the northern Kuwait.

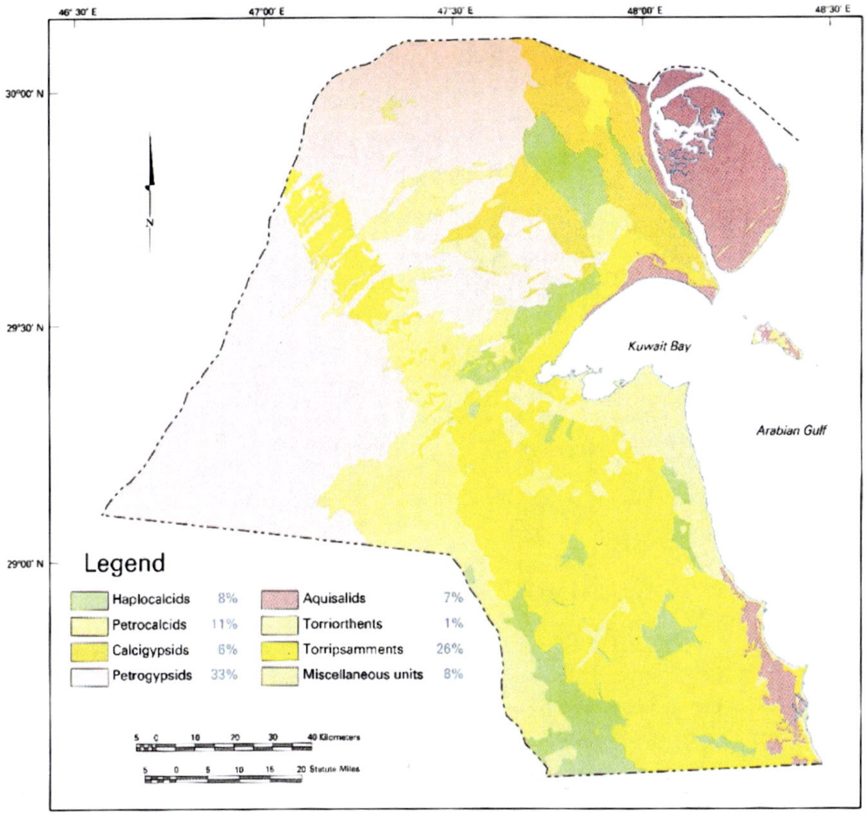

Fig. 39.2 Generalized great group soil map of Kuwait

39.1.2 Land Use

Nineteen land-use types are recognized within Kuwait territory (Fig. 39.3). These can be grouped into six main land-use types, namely, open rangeland, oil fields, protected areas, agricultural areas, military areas, and urban areas. Open rangeland, which is the subject of the present study, constitutes about 75% of Kuwait's terrestrial area.

39.2 Manifestation of Land Degradation

Land degradation processes are prevailing in the majority of the terrestrial environment of Kuwait in general and in the open rangeland in particular. This is mainly manifested by the destruction of the vegetation cover and soil erosion by aeolian

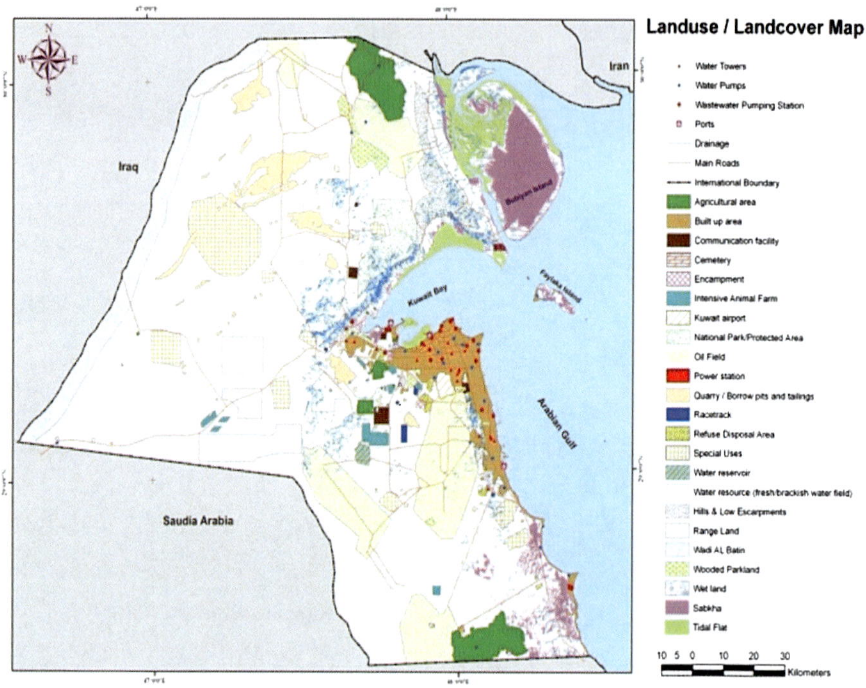

Fig. 39.3 Current land-use map of Kuwait

and fluvial processes. The following is a description of the observed changes in the vegetation cover, surface sediments/soil, and the pattern of surface hydrology.

39.2.1 Deterioration of the Vegetation Cover

A comparison between the densities of the vegetation cover (mainly perennial desert shrubs) recorded during the early 1980s to that observed in February 2010, revealed a state of severe degradation. The first author of this chapter prepared a sedimento-morphic map for the State of Kuwait during early 1980s (Khalaf et al. 1984). This map indicates that most of the western and southern parts of Kuwait were covered by rugged vegetated sand sheets (RVSS), which were developed by the coalescence of nabkhas (Fig. 39.4a). These areas were characterized by dense to moderate vegetation cover.

At present, living desert shrubs (Haloxylon salicornicum) were rarely observed within these areas. The RVSS have been replaced by thin sand sheets, mostly covered by granule lag. The densely populated shrubs, that at one time use to stabilize the RVSS, have been almost totally weathered or died, with remnants of these shrubs represented by dead or dying stems and roots still

Fig. 39.4 Rugged vegetated sand sheets in 1982 (**a**), eroded rugged vegetated sand sheets (2010), note remnants of eroded nabkhas with residues of dried up Haloxylon roots and stems (**b**), granule ripples (2010) (**c**), well-developed nabkhas at Ras Al Subbiya (2010) (**d**)

marking the original occurrences of the once existed nabkhas (Fig. 39.4b). However, in most areas the only existing nabkhas have totally been disappeared and the bedrock is exposed.

The depositional nabkhas which flourished during early 1980s have been extensively eroded. The present erosional phase of nabkhas started after the destruction of the vegetation cover. Where sand mounds that formed the nabkhas sediments have been deflated, and these have been replaced by granule residuals mostly in the form of granule ripples (Fig. 39.4c).

Some of the more intact Haloxylon stands were encountered within the Ras Al Subbiya area. Also, the vegetation cover in the demilitarized zone along the northern and northwestern border area between Kuwait and Iraq is relatively dense, where grazing and recreational pressures are considerably forbidden. In these areas well-developed H. salicornicum nabkhas are flourishing (Fig. 39.4d). Brown and Porembski (1997) broadly recognized four distinct stages of desertification within the Haloxylon community. Stages 1 through 3 involve a gradual decrease of Haloxylon cover and loss of sand, while severe degradation occurred in the fourth stage. The latter was characterized by the near total loss of Haloxylon cover and severe reduction in vegetation productivity. The present observation contradicts the conclusion of Wang et al. (2006), who stated that nabkha formation is a good indicator of wind erosion and land degradation.

39.2.2 Soil Erosion

Two processes are responsible for soil erosion in Kuwait, namely, aeolian processes (wind erosion), which is the most predominant, and fluvial processes (water erosion). Active aeolian processes in Kuwait are attributed to a combination of the nature and fragility of the desert ecosystem, prevalent climatic conditions, and the uncontrolled human activities. In particular, the following conditions are accelerating soil erosion: (a) the location of Kuwait downwind of the high deflation area of the Mesopotamian flood plain (which forms a source for regional sand supply), (b) the higher proneness of both unconsolidated and consolidated deposits to wind and water erosion (this ultimately forms as the source for local sand supply), and (c) the near absence of effective biological soil stabilization (that is usually provided by deep-rooted shrubs and trees) (Khalaf and Al-Hashash 1983; Khalaf and Al-Ajmi 1993; Al-Awadhi and Misak 2000).

The terrestrial environment of Kuwait especially its central part (about 60% of total area) is an open theater for extremely active aeolian processes, where significant active sand bodies (dunes and sand sheets) dominate. Application of a sand transport model developed by Al-Awadhi and Al-Awadhi (2009) showed that the calculated average monthly rate of sand transport during summer is about 5,900 kg m^{-1} toward the southeast. They recognized that the highest sand transport rate occurs within the Al Huwaimiliya sand dune corridor (Fig. 39.5).

Wind erosion (soil drifting) is very common in the areas of fragile sandy soils which cover more than 50% of the terrestrial environment of Kuwait. Recent field measurements indicate that the 2007/2008 dry season (less than 30 mm rainfall) resulted in severe soil losses through wind erosion. Over an extended period of time, winds removed 10–15 cm thick of top soils (around 1,000 m^3 ha^{-1}) (Fig. 39.4c).

In the early 1980s, the desert sediment approached equilibrium, that is, areas of deposition were nearly equal to erosional areas (Khalaf and Al-Ajmi 1993). At present, in 2010, the distribution pattern of the depositional and erosional landforms has been drastically changed. The sediment (deposition) budget has deviated from a balanced budget to a budget dominated by erosion. With the exception of sand dunes, most of the aeolian depositional landforms have been altered to erosional landforms; for example, the previously identified RVSS has been altered to granule lag or desert pavement. Relatively thick sand sheets that used to dominate the southern part of the country have been transformed to thin smooth sand sheets covered with a thin veneer of granule lag. In the early 1980s, these thick sand sheets consisted of a well-developed sandy soil, particularly in areas where the dense community of perennial shrub Cyperus conglomeratus acted as a stabilizing agent and enhanced the depositional aeolian processes.

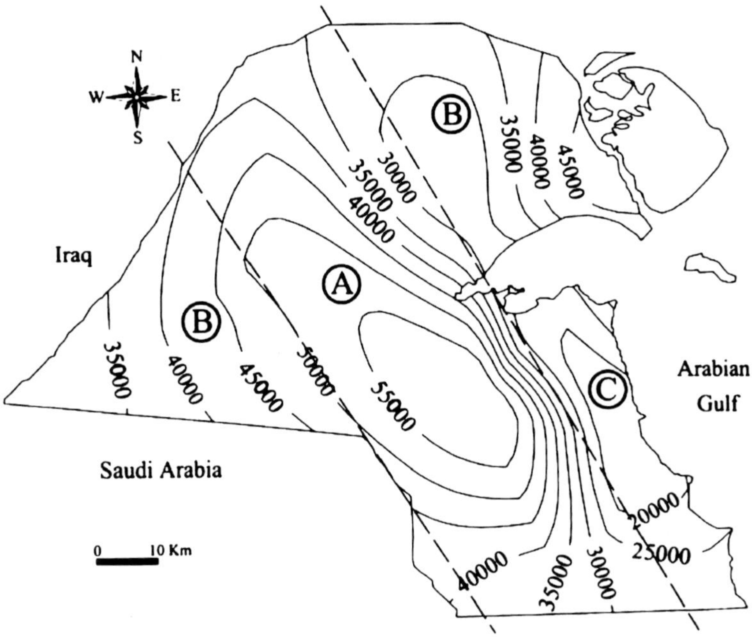

Fig. 39.5 Spatial variation of sand transport rate (kg m^{-1}) in Kuwait. Letters indicate zones of sand transport rate: (A) severe, (B) moderate, and (C) slight

The country is crossed from NW to SE by a deflational corridor (Al Huwaimiliya belt). It is bounded by elevated grounds and dissected transversely by elevated calcretic and gypcretic gravelly ridges. The low-lying areas between these ridges are desert plain mostly covered by pebbly lag. In the early 1980s, abundant small-size barchan dunes (about 3 m high) were scattered within these low-lying areas. Field observation in February 2010 revealed the occurrence of two types of dunes, namely, (a) the Al Huwaimiliya small barchan dunes that were recorded in 1984 and (b) large barchan dunes (about 10–12 m in height). The latter type was not observed in 1984 and has most probably developed since the end of the Gulf War in 1990. There exists a need to investigate the causes and mechanism of the development of these dunes.

Although rainfall in Kuwait is scarce, in some rainy seasons, intensive showers are experienced for short duration (usually in excess of 20 mm day^{-1}). This may result in the occurrence of erosive surface runoff. Misak et al. (2001) identified three degrees of water erosion in Kuwait based on the depth of gullies and rills, namely, strong, moderate, and slight. Strong water erosion creates gullies ranging in depth from 50 to 150 cm and rills of 20 to 50 m interspacing. Accelerated water runoff in Kuwait usually prevails in areas where natural drainages are degraded by grazing and other human activities.

39.2.3 *Vulnerability of Soil Classes to Degradation*

In Kuwait, eight soil types were mapped by Kuwait Institute for Scientific Research (1999) at great group level using the USDA soil taxonomy. These are Haplocalcids (8%), Petrocalcids (11%), Haplogypsids (0.5%), Calcigypsids (6%), Petrogypsids (33%), Aquisalids (7%), Torriorthents (1%), and Torripsamments (27%), in addition to a miscellaneous unit (7%). These soil types exhibit different vulnerability to degradation and drought conditions.

It is very difficult to categorize soil types based on their vulnerability to degradation; however, three categories are tentatively identified. These are high, moderate, and low vulnerable. Torriorthents (1%) and Torripsamments (27%) have high vulnerability to degradation, while Aquisalids (7%) have medium vulnerability. The remaining have low vulnerability.

39.2.4 *Abundance of Bedrock Exposures*

A comparison between the present occurrence of exposed calcretic and gypcretic bedrock (October 2010) with that of the early 1980s revealed a significant increase in the locations of the exposed bedrock, particularly in the northwestern and southern areas. Such relative abundance of the exposed bedrocks may be attributed to the severe deflation/erosion of the loose surface sediments. The increase in the bedrock exposures indicates the erosion of the loss sediments/soil that previously covered the bedrock of these areas. Accurate assessment of the increase of bedrock exposures requires a detailed study for the temporal change detection using sequential satellite images.

The exposed bedrock occurs as almost flat to low humps of calcrete and gypcretic calcrete in the western and southern parts of Kuwait. Most of these humps are oriented parallel to the prevailing wind direction, that is, NW-SE, with some having a yardang morphology.

39.2.5 *Enhancement of Sand Encroachment*

Most of the developed areas and defense facilities that are located within the main passage of the wind corridor (classified as high to very high encroachment zones) are subject to extensive and intensive sand encroachment. Al-Awadhi and Misak (2000) identified 13 mobile sand bodies in Kuwait (Fig. 39.6). These facilities include installations of oil fields (pipelines, well heads, etc.), installations of groundwater fields, air bases, military encampments, animal production farms, roads and highways, electrical transmission stations, and water storage facilities. The total sand removed from such installations, including motorways, oil wells and gathering centers, air bases, and electric power stations, is estimated at about 4.39×10^6 m^3. The estimated cost for removal of the wind deposited material is a staggering figure of nearly Kuwaiti Dinar (KD) 1,680,000 $year^{-1}$ (Al-Awadhi and Misak 2000). Some

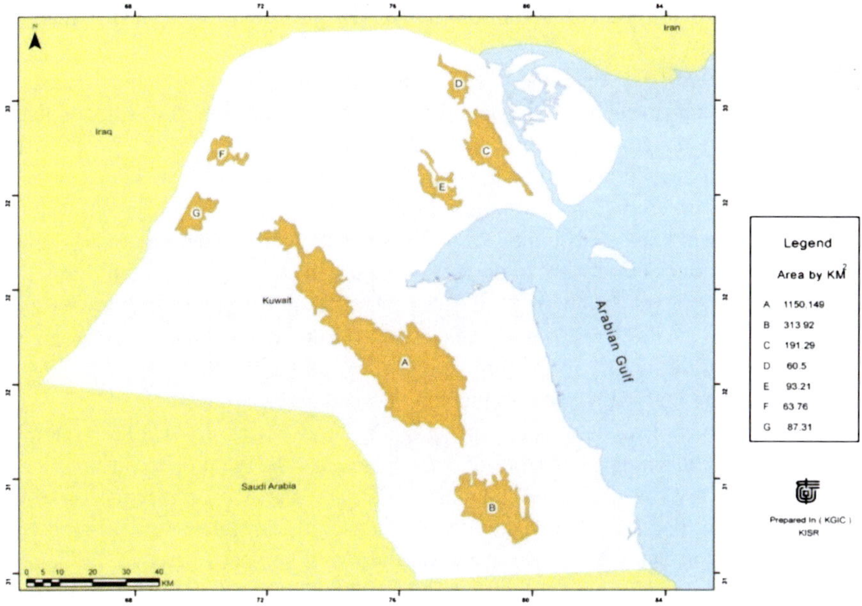

Fig. 39.6 Mobile sand bodies in Kuwait

development sectors spend millions of KD for sand clearance, for example, "KD 6.2 millions" for KOC in the period 2003–2008. Currently, Kuwait lacks a sustainable action plan (SAP) for managing the hazards of sand encroachment.

The 2007/2008 drought resulted in massive wind erosion, severe sand encroachment, and extensive soil loss even in protected areas. Also, occurrence of severe dust storms (for days and sometimes week) was recorded. During this period road accidents significantly increased on the main highways.

The recent field survey (January 2010) in the central part of the country (Al Qurain-Wafra) indicates that northern farms of Wafra agricultural area, Wafra-Ras Azour oil flow lines, and a water storage facility are being severely encroached by shifting sands. These facilities are located at the extreme downwind side of the natural wind corridor (Al Huwaimiliya-Wafra). The source of shifting sands is the huge undulated sandy plain stretching from the upwind side of the attacked facilities. The thickness of accumulated sand ranges between 50 and 120 cm.

39.2.6 Disruption of Surface Hydrological Conditions

Recent field observations (2009/2010) indicate that a number of land uses have negative impact on recharge conditions of vital shallow groundwater reservoirs. The natural flow of runoff water of several catchment areas is disrupted by bund walls, roads, and highways. Blocking of runoff water during floods is the most common mechanism of hydrological disruption in several parts in Kuwait, specially its northern and western fringes.

In the 2009/2010 wet season, huge amounts of runoff water were blocked against bund walls (berms) which act as dykes (1.5–2 m in height, 2–3 m in width, and tens of km in length). A good example is the disruption of the drainage system within Rawdatain-Umm El Eish catchment, where at least 40% of runoff water does not reach to Ar-Rawdatain-Umm El Eish depressions (discharging areas). Also, a bund wall running for about 75 km and at about 2 km to the south of the Kuwait northern border with Iraq currently blocks a network of drainage systems and disrupts the runoff drain to the nearby depressions in southern Iraq. Since most of the drainage systems in Kuwait (particularly at the north) are running in an east to west direction, most of the roads and highways that are trending in a north-south direction (Al Mutla-Safwan and Ras Al Subbiya-Abdali highways) are also blocking normal surface runoff.

Bund walls and roads crossing drainage systems act as dams and accumulate surface runoff during rainy seasons at their upstream side. Field observation after the last rainfall (in November 2009–January 2010) revealed the occurrence of several extensive water ponds within shallow depressions at the upstream side of bund walls and highways at many localities in the northern and southern parts of the country. Some of these ponds extend for about 1 km^2 and the quantity of accumulated runoff water was estimated at 200,000 m^3. Assuming that such ponds deposit about 5 cm thick of outwashed material, this may result in the deposition of about 50,000 m^3 of fine-grained sediments. The deposited mud dries up and desiccated cracks developed.

Blocking of runoff at the upstream of the drainage systems by the bund walls and roads and highways is responsible for hydrological drought at their downstream. This may result in significant drop of soil moisture and consequently deflation of wadi fill at downstream of the bund wall is accelerated. In addition, off-road traffic and grazing along the dry cracked mud flats result in their fragmentation and disaggregation and present a local source of dust storms during active winds.

39.3 Causes of Land Degradation

Although anthropogenic and military activities are the main causes of land degradation, aeolian processes, nature of surface sediments (soil type), and climatic conditions have played a significant role.

39.3.1 Anthropogenic Activities

A leading cause of land degradation is the influence of the human population, which has led to land misuse and overexploitation of desert resources. Increase in population growth rate (3.8%) and welfare and the socioeconomic transformation in

Table 39.1 Summary of the impacts of human activities on land degradation

Activity	Immediate impact	Long-term impact
Excessive grazing	Depletion of biomass, forage loss, soil trampling, and sediment disturbance	Increase in dust and sand storms and in drifting sands, increase in rate of water erosion as a result of disappearance of the vegetation cover
Off-road vehicles and camping	Soil compaction, decrease in permeability, and destruction of vegetation cover	Increase in rate of wind and water erosion, development of sand drifts and active sand sheets, loss of soil fertility
Gravel quarrying	Breaking of the armor layer of pebbles and gravel, exposing finer sediments to wind erosion, disruption and rupture of surface and near-surface sediments, destruction of vegetation cover	Increase in dust and sand in downwind areas, disturbance of pattern of surface runoff, and loss of running water in pits and quarries

general have led to a significant pressure on the available natural resources, leading to land degradation. Anthropogenic activities include excessive grazing (Omar et al. 1999), off-road vehicle traffic (Al-Dousari et al. 2000; Al-Awadhi 2001), spring camping (Al-Sudairawi and Misak 1999), and overexploitation of sand and gravel (Al-Awadhi 2001) (Table 39.1).

39.3.1.1 Overgrazing

Several studies reported that overgrazing in arid regions could change plant community composition and reduce biodiversity (Shaltout et al. 1996; Li and Jiang 1997; Friedel et al. 2003; El-Keblawy 2003; Li et al. 2005; Tefera et al. 2007; El-Keblawy et al. 2009). Camel grazing affects over 90% of land on the Arabian Peninsula, of which 44% is severely or very severely degraded (Ferguson et al. 1998). Excessive grazing by camels is recognized as the single greatest threat to the inland desert ecology of the Gulf states (Hellyer et al. 2001; Gallacher and Hill 2006); Ghazanfar (2004) suggested that overgrazing has increased the proportion of unpalatable shrubs and dwarf shrubs. Brown and Porembski (1998) suggested that the current dwarf shrub ecology in northern Kuwait probably arose through tens or hundreds of years of excessive grazing. In Saudi Arabia and Kuwait, livestock enclosures were reported to show a rapid recovery in plant biomass (Zaman 1997; Barth 1999), and similar observations have been made in the United Arab Emirates (Khan 1980, 1981; Oatham et al. 1995). A field survey in January 2010 revealed that camel and sheep grazing was observed over the entire desert of Kuwait.

39.3.1.2 Off-Road Vehicles Traffic

Off-road driving has long been recognized as a major detrimental factor, causing widespread damage to the vegetation and producing tracks on the soil surface (Adams et al. 1982; Brown and Schoknecht 2001). A number of studies have attempted to quantify the damage caused by off-road vehicles, particularly in the Mojave Desert in eastern California and southern Nevada (Wilshire and Nakata 1976; Webb 1982; Watts 1998). Apart from the obvious damage to shrubs, vegetation is also affected due to soil compaction in tracks (Adams et al. 1982; Wilhelm and Mielke 1988; Bainbridge and Virginia 1990; Rundel and Gibson 1996; Lovich and Bainbridge 1999).

In Kuwait, there has been a dramatic increase in off-road driving activity during the past decades. Vehicle tracks on the desert surface, often devoid of vegetation, are a common sight throughout many parts of the country. Off-road vehicle traffic is also responsible for soil compaction. At least 65% of Kuwait soils are affected by some degrees of compaction, which in turn inhibits the infiltration capacity of soils by 40–100% (Al-Awadhi et al. 2005b). Al-Dousari et al. (2000) reported that about 9% of the western area of Kuwait is assessed as highly compacted. The combined effect of the off-road vehicles traffic and the trampling caused by large herds of livestock may be responsible for the severe deterioration of the desert vegetation (Brown and Porembski 1997, 1998).

39.3.1.3 Camping and Livestock Enclosures

One of the social traditions in Kuwait is camping in desert areas for about 2 weeks during spring time. In the early 1960s, camping activities were relatively limited and restricted to few locations. Also, livestock enclosures (mostly for sheep) were few and mostly scattered in the northern areas. During a survey conducted in the winter of 2010, it was observed that all desert areas are now crowded by camps, mostly permanent, and are in very close proximity. Also, animal enclosures for camels and sheep are now counted by hundreds. Within these locations, vegetation cover has totally been destroyed and the soils have been severely compacted.

39.3.1.4 Quarrying

Gravel and gatch pits are extensively scattered over the desert of Kuwait. The extensive urban development over the past five decades created a great demand for gravel aggregate for housing projects and road construction. Gravel deposits are located within Al Dibdibba alluvial sheet in northern Kuwait, particularly within Al-Layiah and Kraa Al Marw ridges. Gravel quarrying started in early 1960s and continued till the 1980s. During this period extensive quarrying operations took place without any planning or regulation. This resulted in the occurrence of abundant open gravel pits that have drastically changed the natural landforms and eliminated the native vegetation cover over vast areas. Most of these gravel pits are located within the active wind corridor and, therefore, have significantly contributed to the abundance of drifting sands.

In addition, the overexploitation of the near-surface solidified calcareous and gypsiferous sands (gatch) for road construction and as fill material has resulted in the occurrence of hundreds of large pits over the desert of Kuwait. This is responsible for destruction of the vegetation cover, changes in the micro-topography of the desert areas, disrupting the surface water runoff system, and an increase of dust and shifting sand in downwind areas.

39.3.2 Military Activity

39.3.2.1 The Gulf War

The Iraqi invasion of Kuwait on 2 August 1990 and the resulting Gulf War activities have left many scars on the fragile but balanced desert ecosystem of Kuwait (Holden 1991; Karrar et al. 1991; Al-Ajmi et al. 1994). The direct effect of these activities has been the mechanical removal of the native plants and extensive and severe disturbance of the soil (surface cover). The heavy bombing and trench digging during the war made the Kuwaiti desert even more vulnerable to intense and frequent sandstorms. In addition, maneuvers and off-road vehicle use caused severe compaction or loss of topsoil. The most severe aspect of Kuwait's environmental crisis was the burning of 727 oil wells and the gushing wells which made rivers of crude oil that ran into the low lands forming oil pools and lakes (Al-Besharah 1992). The presence of SO_2 in such a massive quantity at the time of the oil burning is thought to have led to the formation of acid rain which would have a damaging effect on vegetation entering the food chain (Al-Hassan 1992; Al-Houty et al. 1993).

It was estimated by El-Baz (1992) that about 30.6% of the land area of Kuwait was adversely affected by the Gulf War.

39.3.2.2 Postliberation Activities

After liberation, the first activity to be initiated was the removal of mines and ordinance from the Kuwaiti desert. This, in addition to the digging of numerous deep ditches for the detonation of collected ordinance, has resulted in the physical scraping of the topsoil and the associated biomass over much of Kuwait. This alternation has many detrimental effects, let alone the unknown impacts of the detonation process itself. Such operations have resulted in an extensive and severe rupture of the surface and near-surface sediments.

After the liberation of Kuwait, the following defense measures were implemented:

(a) The construction of a trench with a depth and width of 5 m along the entire northern and northwestern borders of Kuwait
(b) Installation of a chain link fence along the northern and northwestern borders to delineate a buffer zone of about 5 km wide

The trench acted as a backstop, trapping all migrating sands from southern Iraq. The buffer zone is considered as a protected area and, therefore, encourages the growth and flourishing of desert shrubs which also acted as a windbreak trapping migrating sands from the northern and northwestern directions. These measures have halted the sand supply from outside Kuwait, from southern Iraq. This is responsible from the enhancement of wind action on the loose surface sediments in Kuwait.

Several bund walls that extend for tens of kilometers; have been constructed mostly in an east-west direction. In addition, most of the military sites are fenced by similar bund walls. These are playing a significant role in the pattern and mode of both aeolian and fluvial sediment transport. They disrupt the continuity of sediment migration due to their interference with the near-surface aerodynamic regime. They act as windbreaks, accelerating soil erosion on the upwind side. Detailed study is recommended for thorough assessment of the impacts of these bund walls on soil stability, particularly in the northern areas of Kuwait.

39.3.3 Climatological Factors

Climatological factors influence the land degradation processes through the following phenomena:

1. The scarcity and irregularity of rainfall. During the period from 1957 to 1998, total seasonal rainfall fluctuated between 23.1 mm (1963–1964) and 260.2 mm (1975–1976) compared with the average seasonal rainfall of around 115 mm.
2. The prevalence of drought periods (years with a remarkably low rainfall compared with the average). In 1962, for example, there was 27.2 mm of rain, in 1964 some 19.9 mm, in 1973 a total of 34.8 mm, and in 1987–1989 between 67.5 and 71 mm. During these drought periods, the vegetation cover deteriorated and the soil temperature increased. Consequently, desiccation processes cause shrinkage and cracking and the problem of sand encroachment.
3. The occurrence of intensive rainfall (65–105 mm in a single storm) and the consequent accumulation of very large amounts of dominantly sand and silt outwash material (Al-Sudairawi et al. 1999) have subsequently contributed to the sand supply.
4. The prevalence of strong northwesterly winds (maximum 30 m s^{-1}) during the dry season (May–September). This phenomenon encourages sand transport and soil erosion (Al-Awadhi and Misak 2000).

39.4 Conclusions and Recommendations

The present review and recent field surveys revealed that the biotic and abiotic components of the desert ecosystem of Kuwait have witnessed progressive severe and extensive degradation. This is manifested by the drastic deterioration of the vegetation cover, severe erosion and compaction of the soil, significant changes in micro-topography and imbalance in the surface sediment budget, and disturbance of the surface water runoff system which accelerate hydrological drought. The causes and driving forces for land degradation include uncontrolled irrational anthropogenic activities, aggressive military operations, and climatologic events. It is believed that both anthropogenic activities and military operations are the main causes of land degradation.

Mitigation of the adverse impacts of the aforementioned activities, rehabilitation of the degraded components of the desert ecosystem, and preservation of Kuwait's natural desert heritage require the implementation of a national action plan. The following are recommended measures:

Assessment of the present practice of grazing within a socioeconomic framework. The results of such an assessment should be used for the design of appropriate rules and regulations to maintain optimum soil conditions and allow maximum sustainable grazing capacity in Kuwait. The number of animals to be allowed for grazing should consider the present carrying capacity of Kuwait desert ecosystem.

Rational control of entertainment camping practices and off-road vehicles traffic. Suggested control measures should be based on assessment of the present situation and the carrying capacity of Kuwait desert ecosystem.

Reconditioning of the natural topography through filling quarries and pits and even removal of unnecessary bund walls and any man-made obstructions in order to reestablish the original aerodynamic conditions and therefore maintenance of the natural sand migration conditions while avoiding the imbalance in the surface sediment budget.

Reestablish the natural shrub vegetation through regeneration, restoration, and establishment of enclosures in selected areas for a certain period of time. However, it should be noted that recovery of the vegetation is presumably dependent on the extent of degradation when remediating measures are invoked, as well as climatic factors. In this regard use should be made of previous relevant studies including Charley and West (1975), Thalen (1979), Krueger (1990), Omar et al. (1990), Vetaas (1992), Le Houérou (1995, 1996), Tongway and Ludwig (1996), Ludwig and Tongway (1996), Zaman (1997), Brown and Porembski (1997, 1998), Shachak et al. (1998), and Brown and Al-Mazrooei (2001, 2003).

Sand encroachment problems resulted from the erection of development projects in the pathway of migrating sands; therefore, planning for these projects should consider the nature of aeolian sediment transport regime. Avoid as much as possible the stopping of creeping sands, with the use of passive control measures. If active control measures are required, it should be preceded by a thorough impact assessment.

Emphasis should be placed on policies, regulations and legislation, as well as national public education and awareness programs.

References

Adams JA, Endo AS, Stolzy LH, Rowlands PG, Johnson HB (1982) Controlled experiments on soil compaction produced by off-road vehicles in the Mojave Desert, California. J Appl Ecol 19:167–175

Agnew C, Warren A (1996) A framework for tackling drought and land degradation. J Arid Environ 33:309–320

Al-Ajmi D, Misak R, Al-Dousari A, Al-Enezi A (1994) Impact of the Iraqi war machinery and ground fortifications on the surface sediments and aeolian processes in Kuwait. In: Proceedings of the international conference on the effect of the Iraqi aggression on the State of Kuwait, Kuwait, vol 3, 2–6 Apr 1994, pp 229–248

Al-Awadhi JM (2001) Impact of gravel quarrying on the desert environment of Kuwait. Environ Geol 41(3–4):365–371

Al-Awadhi JM, Al-Awadhi A (2009) Modeling the aeolian sand transport for the desert of Kuwait: constraints by field observations. J Arid Environ 73:987–995

Al-Awadhi JM, Misak RF (2000) Field assessment of aeolian sand processes and sand control measures in Kuwait. Kuwait J Sci Eng 27(1):156–176

Al-Awadhi JM, Al-Dousari A, Al-Enezi A (2001) Barchan dunes in northern Kuwait. Arab Gulf J Sci Res 18(1):32–40

Al-Awadhi JM, Al-Helal A, Al-Enezi A (2005a) Sand drift potential in the desert of Kuwait. J Arid Environ 63:425–438

Al-Awadhi JM, Omar SAS, Misak RF (2005b) Land degradation indicators in Kuwait. Land Deg Dev 16(2):163–176

Al-Besharah J (1992) The Kuwait oil fires and oil lakes – facts and numbers. In: Proceeding of the international symposium on the environmental and health impact of Kuwait oil fires, University of Birmingham, pp 12–15

Al-Dousari AM, Misak R, Shahid SA (2000) Soil compaction and sealing in AL-Salmi area, Western Kuwait. Land Deg Dev 11:401–418

Al-Hassan JM (1992) Iraqi invasion of Kuwait, an environmental catastrophe. Fahad Al-Marzooq Press, Kuwait

Al-Houty W, Abdal M, Zaman S (1993) Preliminary assessment of the Gulf War on Kuwait desert ecosystem. J Environ Sci Health 28(8):1705–1726

Al-Sudairawi M, Misak R (1999) Challenges and problems confronting the sustainable development in the desert of Kuwait. In: Proceedings of the 9th international conference on environmental protection is a must, Alexandria, Egypt

Al-Sudairawi M, Misak R, Kwarteng A, Al-Awadhi JM, Al-Dousari A, Gharib I (1999) Study of sand control at Kuwait Oil Company (KOC) operational areas of southeast, west and north Kuwait. Kuwait Institute for Scientific Research, report no KISR5550, Kuwait

Bainbridge DA, Virginia RA (1990) Restoration in the Sonoran desert of California. Restor Manag Notes 8:3–13

Barth HJ (1999) Desertification in the Eastern Province of Saudi Arabia. J Arid Environ 43:399–410

Brown G (2003) Factors maintaining plant diversity in degraded areas of northern Kuwait. J Arid Environ 54:183–194

Brown G, Al-Mazrooei S (2001) Germination ecology of Haloxylon salicornicum from Kuwait. Botanische JahrbucKher FuKr Systematik, Pflanzengeschichle und Pflanzengeographie 123:235–247

Brown G, Al-Mazrooei S (2003) Rapid vegetation regeneration in a seriously degraded Rhanterium epapposum community in northern Kuwait after 4 years of protection. J Environ Manag 68:387–395

Brown G, Porembski S (1997) The maintenance of species diversity by miniature dunes in a sand-depleted Haloxylon salicornicum community in Kuwait. J Arid Environ 37:461–473

Brown G, Porembski S (1998) Flora and vegetational aspects of miniature dunes in a sand-depleted Haloxylon salicornicum community in the Kuwait desert. Flora 193:133–140

Brown G, Porembski S (2000) Phytogenic hillocks and blow-outs as 'safe sites' for plants in an oil-contaminated area of northern Kuwait. Environ Conserv 27:242–249
Brown G, Schoknecht N (2001) Off-road vehicles and vegetation patterning in a degraded desert ecosystem in Kuwait. J Arid Environ 49:413–427
Charley JL, West NE (1975) Plant-induced soil chemical patterns in some shrub-dominated semi-desert ecosystems of Utah. J Ecol 63:945–964
Dregne HE (ed) (1992) Degradation and restoration of arid lands. Texas Technical University, Lubbock
Dregne HE, Kassas M, Rozanov B (1991) A new assessment of the world status of desertification. Desertification Control Bull 20:6–19
El-Baz F (1992) Preliminary observations of environmental damage due to Gulf War. Nat Resour Forum 16(1):71–75
El-Keblawy A (2003) Effect of protection from grazing on species diversity, abundance and productivity in two regions of Abu-Dhabi Emirate, UAE. In: Alsharhan AS, Wood WW, Goudie AS, Fowler A, Abdellatif EM (eds) Desertification in the third millennium. Swets & Zeitlinger Publisher, Lisse, pp 217–226
El-Keblawy A, Ksiksi T, El-Alqamy H (2009) Camel grazing affects species diversity and community structure in the deserts of the UAE. J Arid Environ 73:347–354
Ferguson M, McCann I, Manners G (1998) Less water, more grazing. ICARDA Caravan 8:9–11
Friedel MH, Sparrow AD, Kinloch JE, Tongway DJ (2003) Degradation and recovery processes in arid grazing lands of central Australia. Part 2: vegetation. J Arid Environ 55:327–348
Gallacher DJ, Hill JP (2006) Effects of camel grazing on the ecology of small perennial plants in Dubai (UAE) inland desert. J Arid Environ 66:738–750
Ghazanfar SA (2004) Biology of the central desert of Oman. Turk J Bot 28:65–71
Halwagy R, Halwagy M (1974) Ecological studies on the desert of Kuwait. II. The vegetation. J Univ Kuwait (Sci) 1:87–95
Halwagy R, Moustafa AF, Kamel SM (1982) On the ecology of the desert vegetation in Kuwait. J Arid Environ 5:95–107
Hellyer P, Al-Abed I, Vine P (2001) United Arab Emirates: a new perspective. Trident Press, London
Holden C (1991) Kuwait's unjust deserts: damage to its desert. Science 251:1175–1181
Howle S (1998) The long term environmental consequences of the Gulf War in northeastern Kuwait. MSc thesis, University of Massachusetts, Boston, MA, USA
Johnson DI, Lewis LA (1995) Land degradation: creation and destruction. Blackwell, Oxford
Karrar G, Batanouny KH, Mian MA (1991) A rapid assessment of the impact of the Iraqi-Kuwait conflict on terrestrial ecosystem, part II. The State of Kuwait. Report prepared for International Fund for Agricultural Development, Rome, Italy
Kassas M (1995) Desertification: a general review. J Arid Environ 30:115–128
Khalaf FI (1989) Desertification and aeolian processes in the Kuwait desert. J Arid Environ 16:125–145
Khalaf FI, Al-Ajmi D (1993) Aeolian processes and sand encroachment problems in Kuwait. Geomorphology 6:111–134
Khalaf FI, Al-Hashash M (1983) Aeolian sedimentation in the northwestern part of the Arabian Gulf. J Arid Environ 6:319–332
Khalaf FI, Gharib IM, Al-Hashash M (1984) Types and characteristics of the recent surface deposits of Kuwait. Arab Gulf J Arid Environ 7:9–33
Khan MIR (1980) Al bujair nursery in the western region. Emirates Nat Hist Group Bull 10:16–18
Khan MIR (1981) Al babha plantation. Emirates Nat Hist Group Bull 13:17–19
Krueger WC (1990) Animal range relations on arid and semiarid rangelands. In: Halwagy R, Taha FK, Omar SA (eds) Advances in range management in arid lands, 201–205. Kegan Paul International, London/New York, p 234
Kuwait Institute for Scientific Research (1999) Soil survey for the state of Kuwait, vol 1. Executive summary. AACM International, Adelaide, Australia, CD-ROM

Lal R, Stewart BA (eds) (1994) Land degradation, vol 11, Advances in soil science. Springer, New York

Lal R, Blum WEH, Valentin C, Stewart BA (eds) (1997) Methods for assessment of land degradation. CRC, Boca Raton

Le Houérou HN (1995) Eco-climatic and biogeographic comparison between the rangelands of the iso-climatic Mediterranean arid zone of North Africa and the Near East. In: Omar SAS, Razzaque MA, Alsdirawi F (eds) Range management in arid zones, 25–40. Kegan Paul International, London/New York, p 397

Le Houérou HN (1996) Climate change, drought and desertification. J Arid Environ 34:133–185

Lepers E, Lambin EF, Janetos AC, DeFries R, Achard F, Ramankutty N (2005) A synthesis of information on rapid land-cover change for the period 1981–2000. BioScience 55:115–124

Li JH, Jiang P (1997) The effects of different rotational grazing intensities on the soil, grassland and sheep productions in the northern Tianshan in China. Archivos de Zootecnia 46:301–310

Li JH, Li ZQ, Ren JZ (2005) Effect of grazing intensity on clonal morphological plasticity and biomass allocation patterns of Artemisia frigida and Potentilla acaulis in the Inner Mongolia steppe. N Z J Agric Res 48:57–61

Lovich JE, Bainbridge D (1999) Anthropogenic degradation of the southern California desert ecosystem and prospects for natural recovery and restoration. Environ Manag 24:309–326

Ludwig JA, Tongway DJ (1996) Rehabilitation of semiarid landscapes in Australia. II. Restoring vegetation patches. Restor Ecol 4:398–406

Maingnet M (ed) (1994) Desertification: natural background and human mismanagement. Springer, Berlin

Middleton N, Thomas DSG (1997) World Atlas of desertification, 2nd edn. Published for UNEP. Edward Arnold, London, p 182

Misak R, Al-Awadhi JM, Al-Sudairawi M (1999) Assessment and controlling land degradation in Kuwaiti desert ecosystem. In: Proceedings of the conference on the impact of environmental pollution on the development in the Gulf Region, Kuwait, 15–17 Mar 1999

Misak RF, Al-Awadhi JM, Al-Sudairawi M (2001) Assessing and controlling land degradation in Kuwaiti desert ecosystem. In: (Al-Sarawi M, Massoud M (eds). The Impact of Environmental Pollution on Development in the Gulf Region, Vol. II Environmental Public Authority, State of Kuwait, p209–223

Oatham MP, Nicholls MK, Swingland IR (1995) Manipulation of vegetation communities on the Abu Dhabi rangelands. I. The effects of irrigation and release from longterm grazing. Biodivers Conserv 4:696–709

Oldeman LR, Hakkeling RTA, Sombroek WG (1992) World map of the status of human-induced soil degradation: an explanatory note. ISRIC, Wageningen

Omar SAS (1991) Dynamics of range plants following 10 years of protection in arid rangelands of Kuwait. J Arid Environ 21:99–111

Omar SAS, Taha FK, Nassef A (1990) Ecological monitoring of vegetation in Kuwait's rangelands. In: Halwagy R, Taha FK, Omar SAS (eds) Advances in range management in arid lands. Kegan Paul International, London/New York, pp 57–71

Omar SAS, Shahid SA, Misak R (1999) Assessing damage magnitude and recovery of the terrestrial eco-system. Follow up of natural and induced desert recovery. Kuwait Institute for Scientific Research, Kuwait

Omar SAS, Briskey EJ, Misak R, Asem A (2000) The Gulf war impact on the terrestrial environment of Kuwait. In: Proceedings of international conference addressing environmental consequences of war: legal, economic and scientific perspectives, Washington, DC, USA

Omar SAS, Misak R, King P, Shahid SA, Abo-Rizq H, Grealish G, Roy W (2001) Mapping the vegetation of Kuwait through reconnaissance soil survey. J Arid Environ 48:341–355

Omar SAS, Bhat NR, Shahid SA, Assem A (2005) Land and vegetation degradation in war-affected areas in the Sabah Al-Ahmad nature reserve of Kuwait: a case study of Umm Ar Rimam. J Arid Environ 62:475–490

Ravi S, Breshears D, Huxman TE, D'Odorico P (2010) Land degradation in drylands: interactions among hydrologic-aeolian erosion and vegetation dynamics. Geomorphology 116:236–245

Reynolds JF, Smith S (2002) Do humans create deserts? In: Global desertification: do humans create deserts? Dahlem University Press, Berlin, pp 1–22

Rundel PW, Gibson AC (1996) Ecological communities and processes in a Mojave Desert ecosystem: Rock Valley, Nevada. Cambridge University Press, Cambridge, p 369

Safriel U, Adeel Z (2005) Dryland systems. Ecosystems and human well-being: current state and trends. Island Press, Washington, DC, p 948

Shachak M, Sachs M, Moshe I (1998) Ecosystem management of desertified shrublands in Israel. Ecosystems 1:475–483

Shahid SA, Omar SAS, Grealish G, King P, El-Gawad M, Al-Mesabahi A (1998) Salinization as an early warning of land degradation in Kuwait. Probl Desert Dev 5:8–12

Shahid SA, Omar SAS, Al-Ghawas S (1999) Indicators of desertification in Kuwait and their possible management. Desert Control Bull 34:61–66

Shahid SA, Omar SAS, Misak R, Abo-Rizq H (2003) Land resource stresses and degradation in the arid environment of Kuwait: an overview. In: Alsharhan AS, Wood WW, Goudie AS, Fowler A, Abdellatif EM (eds) Desertification in the third millennium. Swets & Zeitlinger Publisher, Lisse, pp 351–360

Shaltout KH, El Halawany EF, El Kady HF (1996) Consequences of protection from grazing on diversity and abundance of the coastal lowland vegetation in eastern Saudi Arabia. Biodivers Conserv 5:27–36

Tefera S, Snyman HA, Smit GN (2007) Rangeland dynamics of southern Ethiopia: (2) Assessment of woody vegetation structure in relation to land use and distance from water in semi-arid Borana rangelands. J Environ Manag 85:443–452

Thalen DC (1979) Ecology and utilization of desert shrub rangelands in Iraq. Junk, The Hague, p 430

Tongway DJ, Ludwig JA (1996) Rehabilitation of semiarid landscapes in Australia. I. Restoring productive soil patches. Restor Ecol 4:388–397

UNCED (1992) Managing fragile ecosystems: combating desertification and drought. Report no Agenda 21, Chapter 12: United Nations conference on environment and development

UNCED (1993) Agenda 21: earth summit – the United Nations programme of action from Rio: managing fragile ecosystems: combating desertification and drought. United Nations conference on environment and development, rio de Jenario, p 294

UNEP (1992) Status of desertification and implementation of United Nations plan of action to combat desertification, pp 1–15

Vetaas OR (1992) Micro-site effects of trees and shrubs in dry savannas. J Veg Sci 3:337–344

Wang X, Wang T, Dong Z, Liu X, Qian G (2006) Nebkha development and its significance to wind erosion and land degradation in semi-arid northern China. J Arid Environ 65:129–141

Warren A, Sud YC, Rozanov B (1996) The future of deserts. J Arid Environ 32:75–89

Watts SE (1998) Short-term influence of tank tracks on vegetation and microphytic crusts in shrub steppe habitat. Environ Manag 22:611–616

Webb RH (1982) Off-road motorcycle effects on a desert soil. Environ Conserv 9:197–208

Wilhelm WW, Mielke LN (1988) Winter wheat growth in artificially compacted soil. Canad J Soil Sci 68:527–535

Wilshire HG, Nakata JK (1976) Off-road vehicle effects on California's Mojave Desert. Calif Geol 29:123–132

Zaman S (1997) Effects of rainfall and grazing on vegetation yield and cover of two arid rangelands in Kuwait. Environ Conserv 24:344–350

Chapter 40
Methodological Approach to Estimate In-Site Costs of Desertification When Empirical Data Are Not Available

Heitor Matallo Junior

Abstract This chapter addresses the issue of economic losses due to soil erosion and discusses some well-known methodologies used since the 1990s by different authors in the USA and Europe. Issues related to the reliability of data are analyzed, and the consistency of the basic assumptions in the literature is questioned. A new approach is used considering the lack of empirical and reliable data in order to determine the economic losses of land degradation. The Universal Soil Loss Equation is applied to estimate the economic losses in 11 Latin American countries. Estimations of the GNP losses are also presented. Finally, it is pointed out that the importance of the economic losses is not proportional to the political attention governments have been giving to the problem.

Keywords Desertification • Economic losses • Economics of desertification • Environmental economics • Soil erosion

40.1 Introduction

The word "desertification" was mentioned for the first time by the French researcher Louis Lavauden in 1927 and was made popular by Andre Aubreville in the 1940s after one decade of hard experiences related to land degradation in the American Great Plains, caused mainly by deforestation, intensive overexploitation of soils, and 8 years of drought that affected the region in the years 1929–1936.

H. Matallo Junior (✉)
UNCCD Secretariat Bonn, Bonn, Germany
e-mail: hmatallojr@yahoo.com.br

S.A. Shahid et al. (eds.), *Developments in Soil Classification, Land Use Planning and Policy Implications: Innovative Thinking of Soil Inventory for Land Use Planning and Management of Land Resources*, DOI 10.1007/978-94-007-5332-7_40,

More than 80 years after Aubreville, the United Nations has declared 2006 the International Year of Deserts and Desertification. In 2008, the United Nations Convention to Combat Desertification (UNCCD) celebrated its 14th anniversary, and the UN Conference on Desertification, held in Nairobi in 1977 in Kenya, celebrated its 32nd anniversary. Still in 2008, the Brundtland Report, the most important report on sustainable development ever made, completed 21 years of existence with many references on the risks of desertification.

The figures mentioned above have no mathematical harmony or beauty but reflect a set of initiatives at international level that shows desertification as a problem that has been affecting drylands for a time long enough to justify that effective measures should already be taken by countries and international community to tackle the problem. But what we can see nowadays is a lack of effective and efficient policies and programs to combat desertification in the affected countries worldwide.

This is due, in part, to the lack of consensus on the meaning of the concept of desertification among the politicians, decision-makers, researchers, and academics and, particularly, on the ways to transform the concept in concrete policies and programs. The concept of desertification has its referents (in real terms) in the manifold social and environmental aspects such as water, soils, forests, climate, social, and economic structure and the very complex cultural traditions. By its complexity and transdisciplinary aspect, the concept of "desertification" goes beyond the scientific and institutional disciplines, and for this reason, it is not so easy to understand its meaning in terms of the different fields of research and also to fit it into the academy.

In general terms, the scientific knowledge is characterized as a disciplinary and sectorial enterprise with a long tradition, and the academy and other research institutions reflect this aspect. It means that the scientific problems in general, and those related to environment in particular, respond to the way science has been structured and cover a wide range of disciplines such as biology, botany, soil science, and climatology. In the same way the administrative and political structures in countries respond to the social organization needs organized in terms of sectors such as agriculture, education, and health. In this way, social demands are managed as sectorial demands against the administrative structure in countries. For this reason, it is not so easy to "locate" desertification as a problem in the institutional structure of countries. The UNCCD Focal Points are located generally in the environmental divisions, but they could be in the agricultural, social, or other divisions as well. Desertification as a concept and as a concrete problem requires a distinctive and transdisciplinary approach and requires a new paradigm and a new pattern of action and a very high level of coordination.

This is an important aspect to be considered and could explain in part the unbalance among the three Rio Conventions (UNFCCC, CBD, and UNCCD) in terms of implementation, technical and financial resources allocated, and public attention.

Another aspect that should be taken in consideration is referred to the economy of drylands. Desertification is a phenomenon that affects the arid and semiarid regions, and these dry ecosystems have a particular dynamics in terms of economic and ecological response vis-à-vis to the humid areas. In real terms it means that drylands have a slower ecological and economic response capability when compared to the humid areas due to a limited availability of water and the occurrence of recurrent droughts. This is a conditioning factor for economic competitiveness in these areas and a conditional factor for investments when compared to the humid ecosystems. It means that if one dollar is invested in agriculture in both areas (dry and humid), the profit obtained from each one is different, and probably the profit obtained from the humid areas could be around 20 or 30% higher.

The environmental vulnerabilities in drylands represent an economic risk for the productive process and at the same time an environmental risk that can affect the costs of production and the environmental sustainability. The environmental dryland vulnerabilities as drought and water scarcity represent economic externalities, which sometimes cannot be incorporated to the costs of production in order to maintain the competitiveness of the agricultural products in the markets. This is the root of the unsustainable environmental and economic processes. The final cost of this unsustainable cycle is land degradation and desertification.

40.2 The Problem of Data Availability

When it comes to data availability on land degradation and desertification at global level, the issue of "reliability" has to be discussed. As mentioned by Sivakumar and Ndiang'ui (2007), the five global assessments on land degradation carried out between 1997 and 2003 present data in different formats and scales, leading to some conclusions, which are far from reality. The author says that the figures related to land degradation have been "overly exaggerated." The rationale behind is that "the bigger is the problem, the bigger is the attention and consequently, the bigger is the assignment of financial resources." But this seems to be not true in the case of land degradation and desertification. Actually, the opposite effect is happening because there is no good basis for decision-makers to believe in the "big numbers" presented. This situation is mainly due to the lack of reliable data and, more important, the lack of a methodology to estimate land degradation and its economic losses.

What is presented in this chapter is a methodology to estimate economic losses due to soil erosion when empirical data is not available. The estimations are based on a research made by the author in Latin America and on some available estimation.

40.3 The Assessment of Economic Losses of Soil Erosion in Drylands

In March 2003, the OECD organized a meeting on "soil erosion and biodiversity indicators" in order to get information on the policy-relevant indicators that could track the current state and trends in soil erosion and soil biodiversity related to agriculture at global level, in particular in Europe, and also the current approaches for modeling the economic valuation of soil erosion. Some papers were presented in this meeting, and some sort of data and methodological discussions came out as the most recent overview on the economic issue of soil erosion even though the papers were not referred specifically to drylands.

The main conclusion contained in the studies prepared for the meeting was that soil erosion should not be of much concern in developed countries, particularly in the USA and Europe. According to some data presented by Crosson (2003), the estimated costs of in-farm soil erosion in the USA are around US$ 100 million annually (US$ 0.60 ha^{-1}). The author mentions other alternative assessments, including the one offered by Pimentel et al. (1995) that has assumed an economic loss around US$ 25 billion due to soil erosion. According to Crosson (2003), Pimentel et al. (1995) do not show any good evidence for their estimations, and their figures cannot be accepted. In the same line, the author mentions some data regarding the situation in China and Indonesia and concludes that for these countries, soil erosion does not represent major concern even when some research shows a decline in topsoil depth. Maybe Crosson (2003) had made the mistake as Pimentel et al. (1995) did regarding the lack of evidence.

When the problems of soils erosion come to drylands, the methodologies and data are even less accurate, and we have to rely on the studies conducted almost 30 years ago by Harold Dregne, who has designed the methodology to assess the costs of land degradation in drylands during the 1980s (Dregne and Chou 1992); Crosson (2003) recognizes also that it is the only one referred to desertification. The outcomes presented by Dregne related to the amount of degraded areas, its intensity, and the further estimation of costs have been used by many institutions for more than 25 years and have been taken as "quasi-official" by many institutions, including the UNEP assessment of the costs of land degradation (Crosson (2003) accepted Dregne's methodology but has made new calculations).

But looking carefully to the foundations of such methodology and data, it seems that it is not accurate enough to be credible. According to Mr. Dregne, the data and the estimations lie on a very weak source of data. This is the author's view on his own sources of data:

> The information base upon which the estimates in this report were made is poor. Anecdotal accounts, research reports, travelers' descriptions, personal opinions, and local experience provided most of the evidence for the various estimates. Some data were available for Australia and the United States. Both of these countries have conducted comprehensive assessments of land degradation on irrigated, rainfed farming, and rangelands. For the country data, it is impossible to estimate the error in the numbers of hectares in each degradation class because there are no accepted values against which to make comparisons. To our knowledge, no one except the senior author has ever attempted a global assessment, and very few have published national assessments. (Dregne and Chou 1992)

Regarding the economic losses, Mr. Dregne considers two components: (a) the costs of losses in rainfed and irrigated agriculture and also rangelands and (b) the costs of restoration in the three mentioned categories. For each category Mr. Dregne gives the following figures based on the US and Australian experience:

(a) Costs of land degradation – economic losses:

- Irrigated land – US$ 250.00 ha^{-1} $year^{-1}$
- Rainfed cropland – US$ 38.00 ha^{-1} $year^{-1}$
- Rangeland – US$ 7.00 ha^{-1} $year^{-1}$

(b) Costs of rehabilitation:

- Irrigated areas – US$ 2,000.00 ha^{-1}
- Rainfed cropland – US$ 400.00 ha^{-1}
- Rangeland – US$ 40.00 ha^{-1}

It is clear that the figures presented above are linked with the US economy, and the values estimated to the losses and restoration should be adapted for the economies in developing countries. As pointed out by Dregne,

> At the global scale, it is difficult to select a single figure for the cost of degraded irrigated land, for example, because the cash equivalent value of the crop, whether it is wheat or sorghum or corn, varies greatly from country to country. Subsidies, price controls, and foreign exchange rates, among other factors influence price. Despite the variations, one figure was used as the amount of income foregone on irrigated, rainfed, and rangeland when the degradation was at least moderate in severity. The number used represents, approximately, a 40% loss in productivity. A 40% loss means that the actual yield was 40% less than it would have been in the absence of any degradation. For irrigated land, that represents a $250 (U.S.) per hectare per year reduction in income, $38 on rainfed cropland and $7 on rangeland. The numbers represent our estimates, based upon a relatively small amount of data, most of it from the United States and Australia. (Dregne and Chou 1992)

In the text quoted, there is no clear mention or indication about the methodology or sources used to come up with the figures related to the economic losses according to different land uses. It seems that the only reason to accept those figures at that time was due to the lack of other alternative research and reliable data and also because of the political support given by UNEP to the mentioned study.

It is worth to mention that at the time Mr. Dregne came up with his assessment, the different land uses in drylands were roughly covering rangelands in 88%, rainfed crops in 9%, and irrigated crop production in 3%.

It means that for each 100 ha of agricultural land, it can be assumed that 88 ha was referred to rangelands, 9 ha for rainfed crops, and only 3 ha for irrigated crops. Considering the situation above mentioned and the value of economic losses established by Dregne, it can be assumed that the economic losses for each 100 ha in affected drylands were the following:

88 ha	×	7.00 US$	=	616.00 US$
9 ha	×	38.00 US$	=	342.00 US$
3 ha	×	250.00 US$	=	750.00 US$
Total (100 ha)			=	1,708.00 US$
Average loss			=	17.08 US$ ha^{-1} $year^{-1}$

It has to be clear that Mr. Dregne has not made the above-mentioned estimation related to the average of losses per hectare and according to the different land uses. He has only mentioned in general terms the amount of land used for different purposes. But the logical conclusion based on the Dregne's assessment leads us to the mentioned figures, even considering his warning that the data applies to US and Australian economy only.

As we know, in most developing countries, the dryland's economy is not well integrated to international markets or even national markets, and the economic value of soil losses and restoration would be possibly smaller than those related to developed countries.

If this is the case, we should consider a "k factor" for adjusting the figures for drylands in developing countries. Based on the existing experience in terms of the costs of production and the prices for some agricultural inputs, we can estimate a "k factor" as around at least 20% less than the prices of the same commodities or agricultural inputs in developed countries (Matallo and Vasconcelos 1999). Considering the same situation proposed by Dregne but now applied to drylands in developing countries, it can be concluded that the average of the economic losses could be around 13.6 US\$ ha^{-1} year^{-1} as shown below:

88 ha	×	5.60 US\$	=	492.80 US\$
9 ha	×	30.40 US\$	=	273.60 US\$
3 ha	×	200.00 US\$	=	600.00 US\$
Total (100 ha)			=	1,366.40 US\$
Average loss			=	13.60 US\$ ha^{-1} year^{-1} (k factor applied)

However, almost 30 years after the estimations made by Mr. Dregne, the situation is quite different for both developing and developed countries.

According to the ICID (http://www.icid.org/index.html0), the average of irrigated land vis-à-vis the arable and permanent crop areas for each continent has been improved and can be seen in Table 40.1. It should be noticed that the data were taken in general terms and are not specifically referred to drylands.

Table 40.1 shows that the amount of irrigated area in the world is 6 times higher than during the 1980s when Mr. Dregne came up with his analysis, and if the trends in land use changes kept in 2009 the same patterns as in the 1980s, the economic losses due to irrigation could be, at least, 6 times higher.

Following the trends presented in Table 40.1, it could be considered for a particular region as Latin America that the average of irrigated area would have grown from 3 to 10 ha for each 100 ha in 25 years, which would imply an increase of 300%. But the situation is not simple like that. Irrigation is something special in drylands because it depends on the quality of soil and, most important, the availability of water, which is a limitation by definition. Data available for some countries (Chile, Brazil, and Argentina mainly) shows that irrigated area has increased around 100% in the last 25 years. It means that it can be assumed that irrigated area grew from 3 to 6 ha for each 100 ha.

Since we do not have the data regarding rangelands, we can assume that the irrigated area has grown over the previous rainfed agriculture and that the expansion of rainfed

Table 40.1 Total geographical, arable, permanent cropped, and irrigated area in the continents

Continent	Total geographical area (million ha)	Arable and permanent crop area (APC) (million ha)	Irrigated area (million ha)	(%) of irrigated area to APC
America	3,795.50	377.77	41.8	11.0
Asia	3,002.25	556.18	195.5	35.0
Europe	2,172.01	292.58	26.6	9.0
Africa	2,199.30	176.96	13.5	7.0
Oceania	801.17	51.97	2.9	5.0
World	11,970.23	1,455.57	280.3	19.0

Sources: ICID-http://www.icid.org/imp-data.pdf

agriculture (with the same growth rate) was made over rangelands with the same proportion. These assumptions lead us to the following estimations for each 100 ha:

82 ha (rangelands)	×	5.40 US$	=	442.80 US$
12 ha (rainfed agriculture)	×	30.40 US$	=	364.80 US$
6 ha (irrigated land)	×	200.00 US$	=	1,200.00 US$
Total (100 ha)			=	2,007.60 US$
Average loss			=	20.07 US$ ha^{-1} $year^{-1}$

The new figures express mainly the development of irrigated agriculture in developing countries. However, it seems that these numbers are extremely high (Crosson 2003).

It is quite clear that the economic losses resulting from land degradation cannot be estimated easily. However, it is absolutely crucial for sustainable development and the fight against desertification to have at least a general idea on how much money land degradation represents.

Considering the lack of consensus on the methodology to establish the economic losses of soil erosion (as assumed by Dregne or Crosson), it could be suggested to consider the economic losses due to soil degradation in drylands as US$ 10.00 ha^{-1} $year^{-1}$. This means a bit more than 50% of the average estimation emerged from Dregne's methodology. This assumption is reasonable and acceptable for general estimations, particularly in the absence of a more detailed and acceptable methodology and empirical data.

40.4 Desertification in Latin America

As mentioned before, the source of data and information on desertification in the world is very limited. Many countries do not have reliable data on the extension of land degradation or the population affected, and many others do not present official documents and figures on the extent of desertification. It means that we do not have

Table 40.2 Total area, population, and areas in process of desertification

Country	Total area (ha)	Total population	Areas in process of desertification (ha)	Total population in areas in process of desertification
Argentina	279,181,000	36,223,947	195,426,700	108,671,841
Brazil	851,420,490	169,799,170	66,554,300	15,748,769
Colombia	114,174,800	44,000,000	19,351,000	20,900,000
Costa Rica (data from 2003)	5,106,000	4,089,609	51,654	–
Ecuador	25,637,000	12,156,608	7,060,437	1,000,000
El Salvador	2,104,079	6,329,091	363,000	650,414
Mexico	195,924,800	104,213,503	58,689,150	–
Panamá (data from 2003)	7,551,700	2,839,117	1,876,920	662,236
Paraguay	40,675,200	5,163,198	1,000,000	–
Dominican Republic	4,769,300	8,562,541	3,290,817	5,908,153
Venezuela	91,645,500	23,232,553	9,883,100	6,119,112
Total	1,635,811,369 ha 16,358,113 km^2	419,809,337	363,547,078 ha 3,635,470 km^2	52,055,868

precise information that allows us to have a general and coherent view on land degradation in the world. In this context LAC region is not an exception.

The Facilitation Unit of the UNCCD and the Argentinean National Focal Point, in its capacity as coordinator of the Technical Regional Programme on Benchmarks and Indicators, have conducted a research among countries in order to get information on the status of desertification in the region. The questionnaire was elaborated and applied in the framework of the TPN1 Benchmarks and Indicators and was sent to all LAC countries. We mention only the countries that have answered the questionnaire. The main results can be seen in the Table 40.2.

As it can be seen, the total degraded area in its different levels in the mentioned countries is of 3,635,470 km^2, 22% of the total area of the same countries. The affected population in these countries (exception of Mexico, Costa Rica, and Paraguay) is of 52 million or 12.4%.

Considering the mentioned information, we cannot establish in detail the different levels of degradation or the economic impact of land degradation on countries and their population, but we consider that an economic evaluation of desertification is crucial for policy elaboration process on land degradation and poverty reduction. With this idea in mind, and taking into consideration the "economic exercise" made for the dry regions of Brazil (Matallo and Vasconcelos 1999), it is possible to develop some hypothesis for obtaining an estimation of the costs of desertification in the above-mentioned countries.

As known, soil erosion is a natural phenomenon even in areas with no human activity. But in the areas under agricultural activities, particularly on the areas under intensive and inadequate use of soils, the erosion is intensified and leads to changes in landscape with impacts on other natural resources as water and forests.

Table 40.3 Qualitative and quantitative risk of erosion

Erosion rate	Losses (t ha^{-1} $year^{-1}$)
Very high	>20
High	10–20
Moderate	5–10
Low	2–5
Very low	0–2

The Universal Soil Loss Equation (USLE) is a quantitative and empirical model for the prediction of soil losses during a period of time and under specific circumstances such as precipitation, soil texture, and the land use system. This formula predicts the physical soil erosion, and even considering its limitations, it can be extremely useful for estimating the economic losses of land degradation in a situation of "lack of research and empirical information" and in offering decision-makers an approximate dimension of the desertification. Our hypothesis is based on the fact that erosion is probably the major problem for the maintenance of sustainability of land use and management and that the erosion rate can be different for different types of soils or management systems and different cultivation practices.

The risk of erosion can be expressed qualitatively as "very high, high, moderate, low, and very low" or quantitatively as "tons per hectare per year (t ha^{-1} $year^{-1}$)" (Table 40.3). The technical literature agrees on the following general figures for soil losses.

The types of soils or productive systems are not considered in these figures, and for this reason, they are considered as theoretical values. Generally speaking, the concrete situations are much more complex than that. Using these figures, economical losses can be estimated from soil erosion and from water degradation, since soil erosion impacts watersheds and dams through sedimentation. It means that water reservoirs have been affected in their capacity of water storage and there are other possible hydrologic cycle disturbances.

In order to estimate the financial cost of soil erosion in LAC region, we assume that the affected areas mentioned by countries in the table above have a moderate level of degradation of 7.5 t ha^{-1} $year^{-1}$ (that is a very modest estimation). This means that the soil losses for the entire region are 357,247,078 ha × 7.5 t ha^{-1} $year^{-1}$ that is equal 2,726,603,148 t of soils per year (2.7 billion of t $year^{-1}$).

The cost estimation for different types of agricultural practices as irrigated crops or rainfed crops and grazing was discussed in the previous section, and for our purposes and considering that we do not know how is the composition of land uses in agriculture in the affected areas in terms of rainfed or irrigated agriculture or grazing, it can be assumed as an average loss of US$ 10.00 ha^{-1} as mentioned before. Considering this amount, the losses are of more than 27 billion US$ per year.

Table 40.4 shows the total losses and its relationship with the national growth product (GNP) for the mentioned countries in 2004. The most impressive case is Argentina, where the losses caused by desertification represent more than 9% of the GNP.

At this point we should consider another aspect of land degradation and its economic impacts, that our estimation is annual but desertification is a process in time, and for this reason, we must consider the data for a certain period of time. For estimation purposes, we assume the hypothesis that desertification has been harming

Table 40.4 Total losses and their relationship with the national growth product (GNP)

Country	GNP (2004) (million US$)	Costs of soils and water losses (2005) (million US$)	Losses/ GNP (%)
Argentina	153,014	14,730.3	9.00
Brazil	603,973	5,016.5	0.60
Colombia	97,718	1,458.6	1.00
Costa Rica	18,496	3.9	0.02
Ecuador	30,282	532.2	1.70
El Salvador	15,824	27.4	0.10
México	676,497	4,423.7	0.60
Panamá	13,733	141.5	0.01
Paraguay	7,343	75.4	0.01
Dominican Republic	18,673	248.0	0.01
Venezuela	110,104	741.6	0.01
Total	11,745,657	27,399.1	

Source of GNP: World Bank

countries in the last 12 years (again a very modest assumption), since the approval of the convention in 1994.

During the last 12 years, the average economic growth was around 3% annually, and this is the figure we suppose is the annual increment of the losses due to desertification. The calculations show that the accumulative economic losses represent more than US 150 billion dollars for the 11 countries considered. It means that the deficit per capita is more than US$ 3,500.00 and it is higher than the per capita income regional average. This means a real impoverishment of the population.

40.5 Conclusion

An overview on the desertification process in the mentioned 11 Latin American countries, based on the existing information and the economic assumptions we have made, allows us to recognize the catastrophic dimensions of the problem in the region. Even considering that the existing methodology for economic assessment shall be improved and consolidated in order to offer a better and reliable data, the estimation of losses due to soil erosion represents a huge economic loss that affects millions of people and contributes to poverty and social vulnerability. At the same time the problem has received no proportional attention from the authorities and from the international community. It means that there is an unbalance in terms of attention and financial resources received from the international community, particularly if compared with the other Rio Conventions. The approach offered here for assessing economic impacts of soil erosion is a reasonable alternative that certainly could be improved if some additional data were available. In the other hand, this is crucial for the future of sustainable land management and to overcome the economic and environmental limits land users have been facing in drylands.

References

Crosson P (2003) The economics of soil erosion and maintaining soil biodiversity, Conference paper, Rome

Dregne HE, Chou NT (1992) Global desertification dimensions and costs. In: Degradation and restoration of arid lands. Texas Tech. University, Lubbock

Matallo H, Vasconcelos R (1999) Estimativa de Perdas Econômicas provocadas pelo processo de desertificação na Região do Semi-Arido do NE, in Matallo H, Schenkel C. Desertificação, Unesco, Brasília, Brasil, pp 27–37

Pimentel D, Harvey C, Resosudarmo P, Sinclair K, Kurz D, McNair M, Crist S, Shpritz L, Fitton L, Saffouri R, Blair R (1995) Environmental and economic costs of soil erosion and conservation benefits. Science 267:1117–1123

Sivakumar MVK, Ndiang'ui N (2007) Climate and land degradation. Springer, New York, pp 1–38

Chapter 41
Managing the Hazards of Drought and Shifting Sands in Dry Lands: The Case Study of Kuwait

Raafat F. Misak, Fikry I. Khalaf, and Samira A.S. Omar

Abstract Dry lands cover about 41% of the global terrestrial areas. These are characterized by low average annual rainfall and large variations. Drought is a serious natural hazard in Kuwait and its adjacent countries. During the last four decades, Kuwait experienced a number of dry seasons with rainfall below average (<110 mm year^{-1}). During 2007–2008 and 2008–2009 dry seasons, total rainfall of 35 and 65 mm was recorded, respectively. The consequences of the drought seasons were the massive soil losses (750–1,000 m^3 ha^{-1} in the west Managish area in July 2008); severe sand encroachment even in areas protected for decades, e.g., KISR experimental station at Kabd; relatively longer period of sand and dust storms (May–September 2008 and July–August 2009); and depletion of soil moisture and dryness of natural vegetation.

Sustainable land-use planning in Kuwait is the first defensive step to mitigate the consequences of drought and to reduce land degradation. In the past 15 years, significant changes in land use were observed in Kuwait. Some of these changes have positive and others have negative ecological and environmental impacts. Establishment of the buffer zone (15 km wide and >200 km long) between Iraq and

R.F. Misak (✉)
Environment and Urban Development Division, Coastal and Air Pollution Department, Kuwait Institute for Scientific Research (KISR), P. O. Box 24885, Safat 13109 Kuwait, Kuwait
e-mail: rmisak@kisr.edu.kw

F.I. Khalaf
Earth and Environmental Science Department, Faculty of Science, Kuwait University, Kuwait, Kuwait
e-mail: fikry_khalaf@hotmail.com

S.A.S. Omar
Food Resources and Marine Sciences Division, Kuwait Institute for Scientific Research, P. O. Box 24885, Safat 13109 Kuwait, Kuwait
e-mail: somar@kisr.edu.kw

S.A. Shahid et al. (eds.), *Developments in Soil Classification, Land Use Planning and Policy Implications: Innovative Thinking of Soil Inventory for Land Use Planning and Management of Land Resources*, DOI 10.1007/978-94-007-5332-7_41,

Kuwait in 1993–1994 enhanced the vegetation cover and improved biodiversity and soil conditions, while border trenches (3 m deep, 5 m wide, and hundreds of kilometers long) and the construction of bund walls (2–3 m high, 3–5 m wide, and hundreds of kilometers long) have negatively affected surface water and natural vegetation. It is visualized that in Kuwait sustainable measures to mitigate the consequences of drought are not well adopted. Based on the vast KISR experience in managing dry lands, four programs are proposed to manage the hazards of drought in Kuwait. These are watershed management and restoration, mitigating hydrological drought, managing the hazards of shifting sands, and setting up sustainable land-use plans. The main objective of this study was to adopt integrated approach to mitigate drought in Kuwait. To achieve the objective, intensive fieldwork including experiments and surveys accompanied by analyses and interpretation of remote sensing data were carried out and reported in this chapter.

Keywords Hydrological drought • Kuwait • Rainwater harvesting • Shifting sands • Sustainable land-use plans

41.1 Introduction

Kuwait covers 17,818 km^2 area, of which 85% is covered by terrestrial environment, where 19 land uses are identified. Of these uses, the rangeland grazing constitutes about 75% (KISR 1999). The soil survey of Kuwait (KISR 1999) mapped eight major soil great groups in the terrestrial environment of Kuwait, of which the Petrogypsids constitute 33% (KISR 1999). A vegetation map of Kuwait was published by Omar et al. (2001).

The map was prepared through field surveys and integrating soil and vegetation information in a geographic information system (GIS) environment. The vegetation map presents eight vegetation units: Haloxyletum (22.7%), Rhanterium (2.1%), Cypertum (26.9%), Stipagrostietum (39.3%), Zygophylletum (0.3%), Centropodietum (1%), Panicetum (0.7%), and Halophyletum (1.9%). The study (Omar et al. 2001) also revealed intensive land degradation and retrogression of shrub by species in particular Rhanterium epapposum.

The rainfall in Kuwait is scanty and irregular; the rainy season extends between October and April. The average annual rainfall is about 110 mm. The rainfall fluctuates annually ranging from a reported high of 351.7 mm at Al Ahmadi in the south during 1971–1972 season to the lows of only 20.1 mm at Umm El Eish, in the north, during 1963–1964 season (Omar et al. 2001). During the last four decades, Kuwait experienced a number of dry seasons (2007–2008, 2008–2009) during which rainfall was below the average level of 110 mm. The prevailing wind direction is from northwest and average speed is about 4 m s^{-1}.

During dry seasons of 2007–2009, there was limitless supply of drifting sands in the desert of Kuwait, especially in areas of wind corridors (Huwaimiliyah-Wafra and Umm Qasr-Ras Al-Sabiyah). Massive soil losses (750–1,000 m^3 ha^{-1}) in the west Managish area in July 2008 and severe sand encroachment even in areas protected for decades, e.g., KISR experimental station at Kabd, were the consequences of the dry seasons. Rainfall deficiency resulted in rapid depletion of soil moisture and degradation of vegetation cover. Wide areas experienced unusual sand and dust storms (Civil Aviation of Kuwait April 2009). The absence of drought preparedness plan and adequate early warning systems in Kuwait has exacerbated the impacts of droughts in the last 2 years.

Five-year (2003–2008) protection of highly degraded terrains, e.g., Al Liyah area (north of Al Jahra City), resulted in soil stabilization and recovery of great number of native plants and animals. Use of eco-friendly mulching materials (eco-mat and plant residues) in desert sandy soils resulted in immediate soil stabilization and soil improvement. In these mulched soils, soil moisture in the upper 60-cm depth was 2.5 times greater than the moisture in untreated soils (Misak et al. 2007).

A national action plan to control land degradation and to mitigate the drought effects currently does not exist in Kuwait, and hence, sustainable land-use planning is lacking. Activities like grazing and military exercises are allowed in highly vulnerable areas, such as Al Edairah in northwestern parts of the country. Water harvesting programs for large catchment areas such as Wadi Al Batin (western part of the country) and Jal Az Zour (northeastern part of Kuwait) are not designed. However, measures to manage flash floods are applied in several areas, e.g., Jahra City (west of Kuwait City) and Shuaiba Industrial Area (south of Kuwait City).

In order to address certain issues in the deserts of Kuwait, a study was started with the objective to formulate a management plan to mitigate the impacts of drought in the deserts of Kuwait.

41.1.1 Drought Events and Consequences

During the past four decades, Kuwait has experienced a number of dry seasons. During these dry seasons, rainfall was below the annual average level (110 mm $year^{-1}$). A meteorological record of dry seasons shows total rainfall as 28.1 mm (1963–1964), 39.7 mm (1972–1973), 31.6 mm (1988–1989), 28.3 mm (1993–1994), 35 mm (2007–2008), and 65 mm (2008–2009). In addition to short dry seasons, Kuwait experienced drought periods that lasted for more than three seasons, e.g., 1962–1967 (28.1–87.7-mm rainfall) and 1987–1990 (31.6–84-mm rainfall). Figure 41.1 presents rainfall data from 1957 to 2005.

Field observations during the dry seasons of 2007–2008 and 2008–2009 revealed that landforms and local ecosystems have differently responded to drought. Eolian forms (barchans, falling dunes, nabkhas, and sand sheets), hydrographic basins (wadis and depressions), and wetlands (coastal and inland sabkhas) are sensitive, in varying degrees to drought (Table 41.1).

In relative sense, native plant species exhibit three degrees of susceptibility to drought: extremely high, high, and medium. The *Stipagrostis plumosa* is most

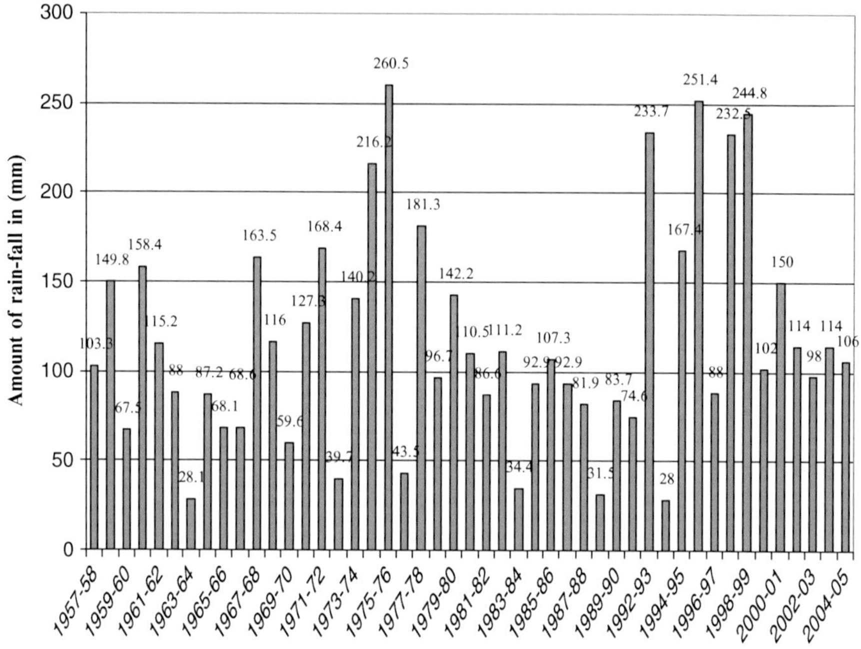

Fig. 41.1 Rainfall in Kuwait 1957–2005 (Al-Dousari et al. 2007)

Table 41.1 Responses of several ecosystems to drought in Kuwait

Landform/ecosystem	Response to drought	Remarks
Eolian	Depletion of stored moisture	Eolian landforms cover about 50% of the surface of the desert of Kuwait
	Disappearance of annuals	
	High rates of soil losses (by wind)	
	Degradation of perennial vegetation (through wind blasting and exposing roots)	
	Deterioration of wildlife	
Wadis and depressions	Drop of water table of wadi fill aquifers	
	Vegetation degradation	
	Increase of soil dryness	
	Increase of soil losses by wind and water (in coming wet seasons)	
Sabkha	Concentration of surface salts	Coastal and inland sabkhas cover about 8% of the surface of Kuwait
	Increase groundwater salinity	
	Deterioration of natural vegetation, especially annuals growing in sandy patches within the sabkhas	
	Erosion of fine particles from topsoil	

susceptible to drought, while *Haloxylon salicornicum* and *Cyperus conglomeratus* are highly susceptible. On the other side, *Rhanterium epapposum* has medium susceptibility to drought conditions (due to its ability to become dormant during summer and low-rainfall conditions).

In the terrestrial environment of Kuwait, drought indicators are diversified and are differentiated into physical and biological. Physical indicators are sand drifting, formation of active sandy sheets and prevalence of dust and sand storms, development of salt crusts, mud cracking, and others, while biological indicators are represented by root exposure and removal and dryness of vegetation.

41.2 Methodology

To achieve the objective of the present study, a number of activities were accomplished. Kuwait was geographically classified into four sectors: NE (2,200 km^2), SE (4,627 km^2), SW (4,750 km^2), and NW (3,981 km^2). The soils, vegetation cover, landforms, and land-use types of different sectors were generally reviewed. These include soil (KISR 1999), vegetation (Omar et al. 2001), and land-use/land cover maps of Kuwait (KISR 1999).

Latest information on the soils, vegetation, land use, land degradation types, and hydrologic disruption was obtained through intensive field surveys (May 2008–May 2010) and measurements. To complete this task, ten pilot sites representing the prevailing conditions and land-use types were selected (Fig. 41.2).

Interpretations of Landsat Thematic Mapper (TM) band 2, 4, and 7 color composite (March 1995); Landsat 7 Enhanced Thematic Mapper (ETM+) band 2, 4, and 7 color composite mosaicked images recorded 31 January and 26 March 2000; Landsat Thematic Mapper (TM) of 2004; and aerial photos of 2003 were made. These images were used in the delineation of main landforms, areas of hydrologic disruption, extent of catchment areas, and areas affected by shifting sands. Ground truth for remote sensing information was conducted in more than 50 ground stations distributed in different parts of the terrestrial environment of Kuwait. The ESRI Arc Map GIS & Arc Info GIS is used to map 11 surface hydrologic units. A field criterion for the assessment of the long-term impacts of the main land-use types in the terrestrial environment of Kuwait was established. The criterion indicates the vertical extent of damage into the soils. Based on this criterion, land uses are differentiated into three categories, extremely destructive (damage extends 1.5–5 m), destructive (damage extends 50–75 cm), and nondestructive (no observable damage).

Assessment of the sand encroachment problem in Kabd area (southwest of Kuwait city) was made using field surveys and remote sensing information. The locations of encroached facilities were recorded using GPS equipment (Global Positioning System). The current measures and practices of mobile sand control were assessed based on their efficiency and durability.

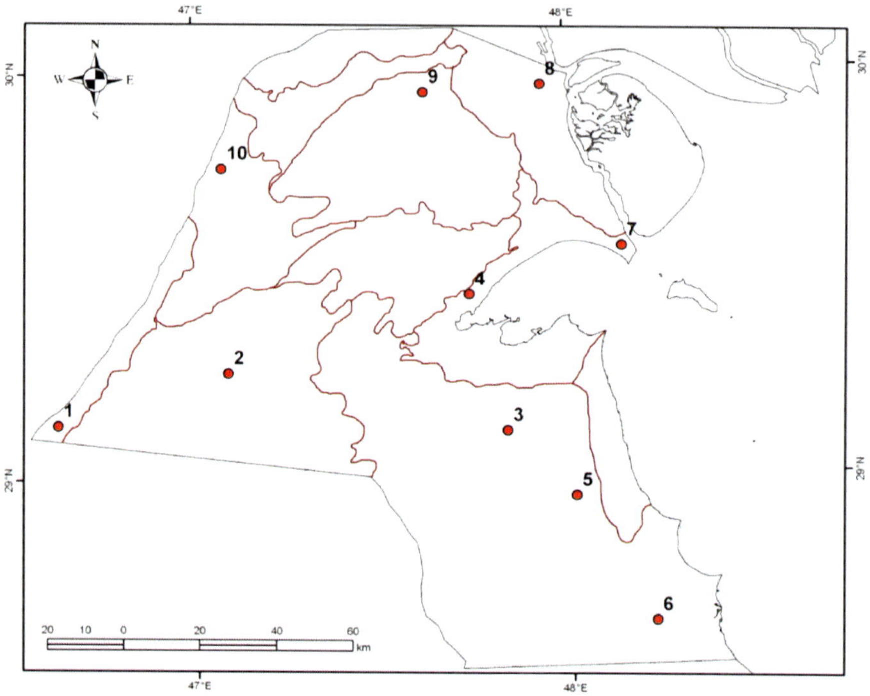

Fig. 41.2 Map showing the locations of pilot sites

A number of field tests and experiments were designed and implemented to control mobile sands, combating land degradation and managing micro watershed areas. These tests include gravel paving of active sandy sheet (Managish pilot site); development of checkerboard system on active sandy sheets (Managish pilot site); protection of a desert facility through erection of fences made up of plant residues, about 1 m high (Managish pilot site); mulching of sandy soils using ecomat, plant residues, and gravel (Al Liyah, Shuaiba, and Subahiyah sites); trapping and then biological stabilization of drift sands using vetiver (Al Managish pilot site); shattering thick soil crusts to enhance soil infiltration capacity and to encourage the growth of natural vegetation (Al Liyah and Subahiyah sites); and water harvesting techniques (mainly earth dykes and terraces) (Liyah, Shuaiba, and Subahiyah sites).

41.3 Results and Discussion

To achieve successful results in mitigating the effects of drought requires detailed understanding of local and regional environmental conditions, potential of natural resources, and past and current land-use and socioeconomic conditions. In addition, it is essential to assess and monitor drought conditions (duration, frequency, and severity) and drought impacts on short-, medium-, and long-term basis. Based on

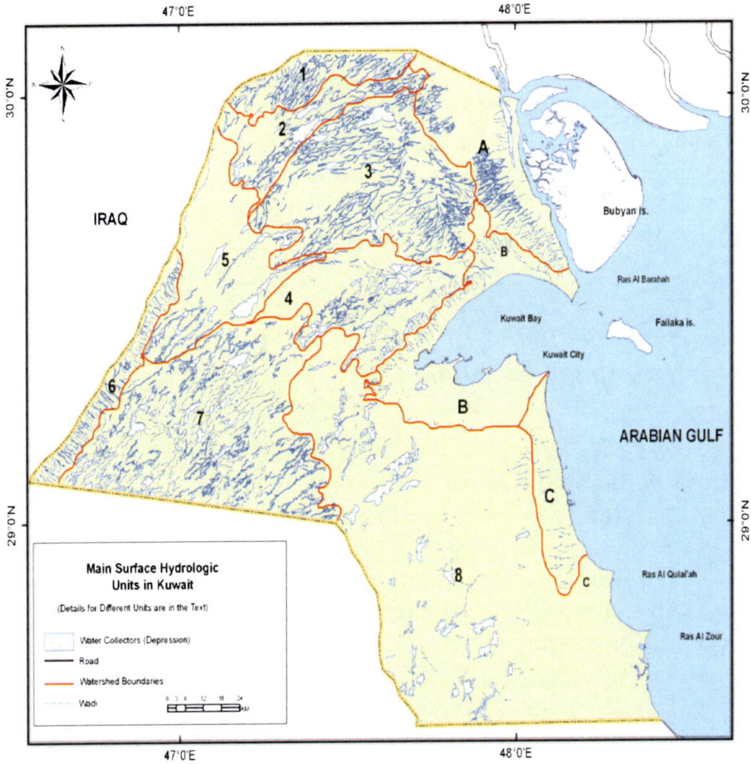

Fig. 41.3 Main surface hydrologic units in Kuwait

the long experience of KISR in managing dry lands and restoration of degraded terrains, the following programs are proposed for mitigating the drought in Kuwait: (1) watershed management, (2) mitigating hydrological drought, (3) managing the hazards of shifting sands, and (4) setting up of sustainable land-use plans

41.3.1 Watershed Management

Based on the nature and location of discharging sites, hydrographic basins in Kuwait are distinguished into the following two main categories, described below.

41.3.1.1 Exterior Basins

This set of basins discharges into surrounding water bodies, i.e., Khor As Sabiyah, Kuwait Bay, and Arabian Gulf. These basins are geographically classified into three subsets. These occur from north to south (Fig. 41.3): Khor As Sabiyah (A), Kuwait Bay (B), and Arabian Gulf (C).

41.3.1.2 Interior Basins

This set of basins discharges into inland hollows (playas) and plains, e.g., Rawdatain-Umm El Eish collectors (locally called khabari). These basins are differentiated into several units which exhibit wide variations in size, features, and land use. These basins include the following (Fig. 41.3): Ritqa (1), Abdaly (2), Rawdatain-Umm El Eish (3), Al Liyah-Umm Al Rimmam (4), Umm Ruwaysat (5), Wadi Al Batin (6), Dibidibah (7), and Kabd-Wafra (8).

41.3.2 Assessment and Controlling of Surface Hydrologic Disruption

In a previous study, Ud Din et al. (2007) mapped and evaluated the Rawdatain catchment area (northeastern part of Kuwait). According to this study, the total precipitation of Rawdatain basin is estimated at 334 million cubic meters (MCM) for the year 2003. More recent studies (Misak 2009) indicate that huge amount of runoff water is blocked before reaching to Rawdatain basin. This factor should be considered in evaluating the recharge conditions of this basin.

Recent field investigations (2009–2010 season) reveal that almost all of the hydrologic units (including Rawdatain catchment) are subjected to hydrologic disruption. This disruption is represented by the following forms.

41.3.2.1 Degradation of Ground Cover (Soils and Vegetation)

This form of degradation is caused by various human activities specially overgrazing and camping and recreation. It negatively influences the local hydrologic conditions (mainly soil infiltration capacity). Al-Dousari et al. (2000) assessed the impact of ground cover degradation on local hydrologic conditions in a part of hydrologic unit no. 7 (Dibidibah). According to this study, the infiltration capacity of the degraded soil has decreased by 18.46–91.96% in comparison with nondegraded soils in the same unit. In Ahmadi-Al Daher area (hydrologic unit C), the infiltration rate of degraded soils has decreased by 61.45% in comparison to non-degraded soils of the same unit (Al-Awadhi et al. 2005).

41.3.2.2 Modification/Blockage

Modification/blockage of the natural flow of runoff water and in turn disruption of recharge conditions of shallow aquifers have also been observed, such as roads of Mutlaa-Abdaly (about 75 km) and bund walls such as those of Sabah Al Ahmad Natural Reserve (about 117-km length) and Wadi Al Batin that block and modify runoff water.

The areas, land use, vegetation and soil types, mechanism of hydrologic disruption, and proposed mitigation measures for the different surface hydrologic units are presented in Table 41.2.

Table 41.2 Field data on surface hydrologic units

Hydrologic unit	Map symbol	Area (km^2)	Land use	Vegetation types (Omar et al. 2001)	Soil types (KISR 1999)	Mechanisms of surface hydrologic disruption	Recommended mitigation measures
Khor As Sabiyah	A	1,010	R, C	Hs, Cc, Re, Sp, H	Gc, Ch, Sa	Dg	Md
Kuwait Bay	B	1,097	U	H, Hs, Sp	Sa, To	Dg	
Arabian Gulf	C	439	U, R, I	Cc, H	Sa, To	Dg	
Ritqa	1	473	P	Hs	Gp	Dg, Ms	
Abdaly	2	613	A, R, Me	Hs, Sp	Gp	Dg, Ms	
Rawdatain-Umm El Eish	3	1,886	R, O	Hs, Cc, Sp	Gc, Ch	Dg, Ms, Mg	Md, Re, Br
Al Liyah-Umm Al Rimmam	4	1,122	R	Sp, Cc	Gp, Ch, Cp	Dg, Ms	Md
Umm Ruwaysat	5	1,209	R, Me	Hs	Gp	Dg	
Wadi Al Batin	6	402	P	Hs, Sp	Gp	Ms, Mg	Md
Dibidibah	7	2,608	R, C	Sp	Gp	Dg	Md, Re, Br
Kabd-Wafra	8	5,559	R, O, A, C	Cc, Sp	Ts, Sa, Ch	Dg	

Land use
Rangelands grazing (R), camping and recreation (C), military exercises (Me), oil extraction (O), protected areas (P), agricultural (A), urban (U), industrial (I)
Vegetation type:
Haloxylon salicornicum (Hs), *Cyperus conglomerates* (Cc), *Stipagrostis plumose* (Sp), *Panicum turgidum* (Pt), *Rhanterium epapposum* (Re), *Halophyletum* (*Nitraria retusa, Tamarix aucheriana, and others*) (H)
Soil types
Haplocalcids (Ch), Petrocalcids (Cp), Calcigypsids (Gc), Petrogypsids (Gp), Aquisalids (Sa), Torriorthents (To), Torripsamments (Ts)
Mechanism of surface hydrologic disruption:
Degradation of groundcover (Dg), modification/blockage of the surface runoff (Ms), disruption of groundwater recharge conditions (Mg)
Recommended mitigation measures
Maintenance of drainage: establishing open cuts in the bund walls or culverts under roads to allow the free movement of runoff water to basins (Md); revegetation of native plants (Re); breaking soil crusts to enhance infiltration rates (Br)

41.3.3 Floods in Kuwait

In Kuwait floods occur during heavy rainstorms with rainfall amounting 30–40 mm in one storm (Al-Dousari et al. 2007). Historic and recorded flood events in Kuwait took place in the following dates: 27 December 1934, 30 November 1954, 2 February 1993 (40 mm within 6–8 h), 11 November 1997 (105 mm within 3–4 h), January 2004, January 2007, April 2008, and November 2009.

41.3.4 Water Harvesting (A Case Study)

In a previous study, Kwarteng et al. (2000) mapped large depressions and playas, paleo-drainage patterns, and catchment areas in northeastern part of Kuwait using aerial photographs, Landsat Thematic Mapper (TM), and digital elevation model (DEM) datasets. They stated that under specific conditions in Kuwait, large volume of rain accumulates in depressions and recharges shallow groundwater aquifers.

Runoff water is developed in the majority of watershed areas during intensive rainfall. These watersheds include Jal Az Zour (northeastern part of Kuwait), Al Ahmadi (south of Kuwait City), and Wadi Al Batin (western part of Kuwait). In general, considerable amount of rainwater is often wasted due to inappropriate land-use types.

To minimize the impact of drought in Kuwait, local water collectors need to capture rainfall water and store it as much as possible. Even under low- and variable-rainfall conditions, efficient soil moisture management before, during, and after rainy seasons is a good approach for improving local water supplies for vegetation use. In addition, runoff water should be collected and stored. In 2006, water harvesting systems were established in Shuaiba Industrial Area (Al-Dousari et al. 2007). The system consists of check dykes (1.5–2 m high, 2–3 m wide) in the main course of an active wadi (Fig. 41.4).

Our long experience in Kuwait indicates that there is reasonable potential for water harvesting in Jal Az Zour, Wadi Al Batin, and Ahmadi ridge during good rainy seasons (when total annual rainfall exceeds 120 mm and to be distributed in a number of heavy showers each at least 20 mm). Water harvesting systems have numerous benefits. In addition to collection and storage of rainwater, they minimize the hazards of flash flooding (if properly designed).

41.3.5 Mitigation of Hydrological Drought

Although climate is a primary contributor to hydrological drought, other factors such as changes in land use, e.g., clearance of vegetation; construction of bund walls (1.5–2 m high, 3–5 m wide, and 10 km long); and digging of trenches (2–3 m deep, 3–5 m wide, and 10 km long) negatively affect the local hydrological

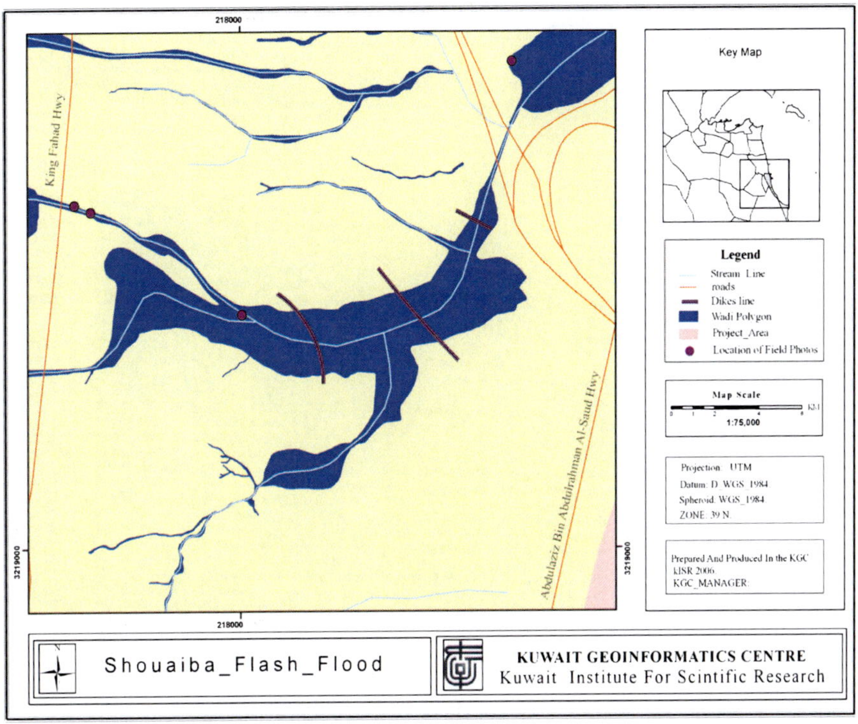

Fig. 41.4 Location of check dykes in the course of main Wadi at Shuaiba Industrial Area (Al-Dousari et al. 2007)

characteristics and the mechanisms of shallow groundwater recharge. In some cases, bund walls act as dams or check dykes, which block the surface runoff and prevent recharge at downstream portions. This is the case at Umm El Eish and Rawdatain basins (northeastern parts of Kuwait) and to some extent Wadi Al Batin (western parts of Kuwait).

41.3.5.1 Forms of Hydrological Drought

Recent field observations (2009–2010 season) indicate that a number of land uses have negative impact on recharge conditions of vital shallow groundwater reservoirs. The flow of runoff water of several catchment areas is disrupted. In some cases such as the case of Rawdatain-Umm El Eish catchment, at least 40% of runoff water does not reach to discharging areas. Figure 41.5 shows the main catchment area of Rawdatain-Umm El Eish basin. In the case of Umm El Eish and Rawdatain basins, hydrological drought is mainly represented by deficiency of runoff water in the downstream portions as a direct result of land-use changes. Runoff water is blocked in the upstream portions by solid bund wall (2 m high). This bund wall is

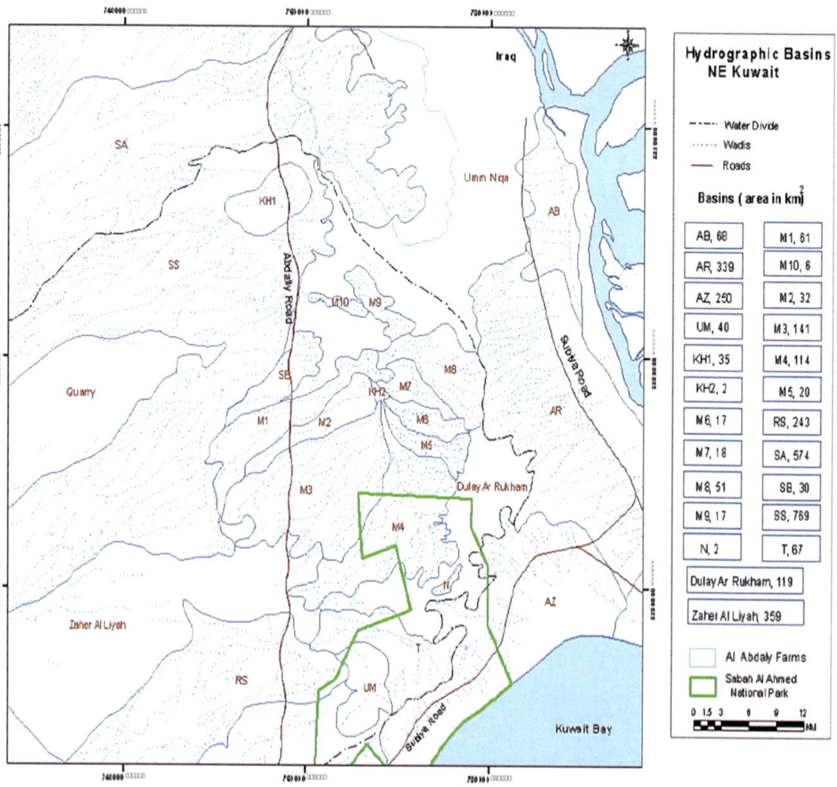

Fig. 41.5 Rawdatain-Umm El Eish catchment area (Misak et al. 2008)

constructed at the northern side of Sabah Al Ahmad Natural Reserve (Fig. 41.6). This results in acute shortage of runoff in the downstream portions and the consequent depletion of soil moisture, i.e., hydrological drought.

41.3.5.2 Restoration of Hydrologic Drought (A Case Study)

To mitigate hydrologic drought at the southern part of Umm El Eish catchment area (case study), it is proposed to (1) map the catchment area using Arc GIS 9.2; (2) locate the bund walls which cut the catchment area using GPS; (3) assess the impact of bund walls on surface water conditions, i.e., runoff and soil conditions; (4) compare between the soil conditions (moisture and texture) and vegetation (types, density, biomass, and biodiversity) in two sites, one in the upstream of the bund wall and the second in downstream side; (5) identify sites where runoff of main water courses is blocked by the bund walls; (6) design an action plan to modify the bund walls through establishing culverts (cuts) to allow the free flow of surface runoff to its discharging basin (Umm El Eish water collector) (culverts in bund

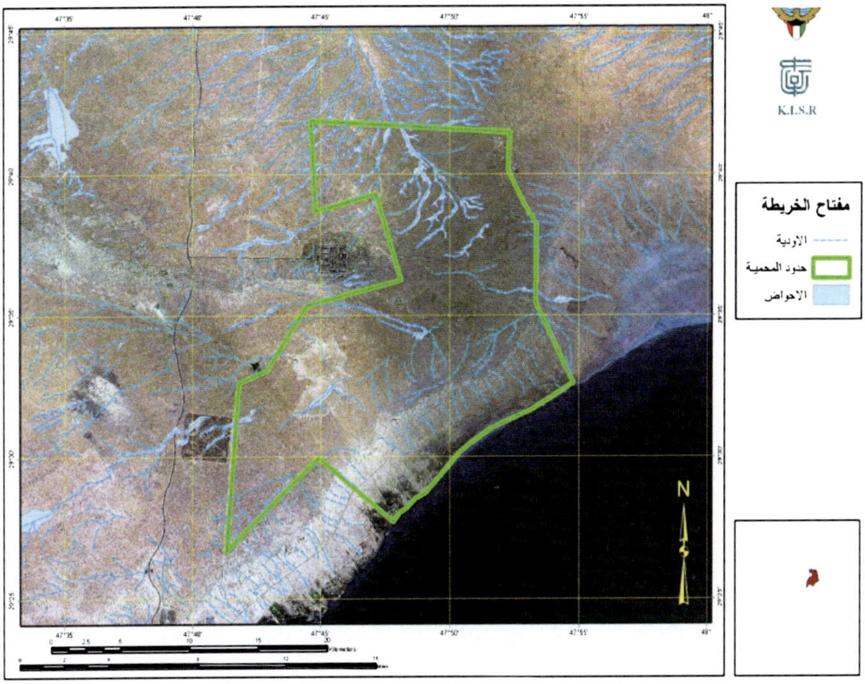

Fig. 41.6 An image showing Sabah Al Ahmad Natural Reserve (fenced area). Note the dissection of the drainage basins by the outer fence (bund wall)

walls were successively applied in Shuaiba Industrial Area by the Ministry of Public Works in 2009); (7) establish the terms of reference (TOR) for bund walls maintenance (specifications and location of cuts); and (8) estimate the cost of bund walls maintenance.

41.3.6 Managing the Hazards of Shifting Sands (Different Scenarios)

Al-Awadhi and Misak (2000) mapped 14 desert facilities which are subjected to sand encroachment. Most facilities are centrally located in the desert of Kuwait (Al Huwaimiliyah-Wafra natural wind pass). The measures used in sand control include bund walls (sand ridges), porous fences, and greenbelts.

Al-Awadhi et al. (2003) mentioned that very severe desertification prevails in Kuwait due to increasing formations of new active sandy bodies, deterioration of many areas of natural vegetation cover to less than 10%, and limited water resources for large-scale forage production. Recent mapping of areas affected by shifting sands using Arc GIS 9.2 indicates that Al Atraf-Kabd area

Table 41.3 Statistics on sand encroachment problem in Kuwait

Facility/ infrastructure	Total number	Number of affected facilities	Percentage affected (%)	Sustainable action plans for mobile sand control
Agricultural areas (farms)	3	2	66.6	Not available
Animal production	2	2	100	Not available
Desert highways	7	6	85	Not available
Air bases	2	2	100	Not available
Military camps	8	4	50	Not available
Oil fields	14	14	100	Available (not implemented)
New cities	3	3	100	Available (not implemented)

(about 1,655 km^2) is severely encroached by active sands. This area is located 50 km to the south of Kuwait City, where at least 11 strategic facilities including airbases, military camps, and highways are threatened by shifting sands (Misak 2009).

As stated by Misak and Alhajraf (2005), development sectors of Kuwait suffer serious problems caused by sand encroachment (Table 41.3). In the agricultural sector, the majority of farms and animal production facilities are affected by sand encroachment. Wafra and Sulaibiyah farms, as well as Wafra and Kabd animal production facilities, are examples of severely encroached facilities in the agricultural sector. In the public works sector, 85% of the highways are influenced by shifting sands. Wafra-Mina Abdullah, Kabd, Abdaly, and Salmi roads are highly impacted by shifting sands. In the defense sector, 100% of the airbases and 50% of the military camps are suffering from serious sand encroachment problems. Ali Al-Salem and Ahmad Al Jaber Air Bases are severely affected by shifting sands. In the oil sector, 100% of the oil fields are affected by shifting sands of different degrees. Managish, Umm Gudair oil fields (west Kuwait), and Wafra oil field (south) are examples of highly affected fields. In the housing sector, 100% of the new cities are affected by sand encroachment. Wafra, Ras Al-Sabiyah, and Erfjan new cities are additional examples where sand encroachments exist (Misak and Alhajraf 2005).

It has been observed that the development sectors (oil, defense, public works, agriculture, etc.) deal with the sand encroachment problem in their affected sites, but without coordination. Their approach for controlling drift sands depends in almost, if not all cases, on the periodical removal of the accumulated sand, which is dropped in open desert areas. Dropping of drift sands in open desert areas causes many problems as they can be re-shifted and attack new areas. Establishing green-belts of drought-resistant plants in open desert areas to minimize sand encroachment

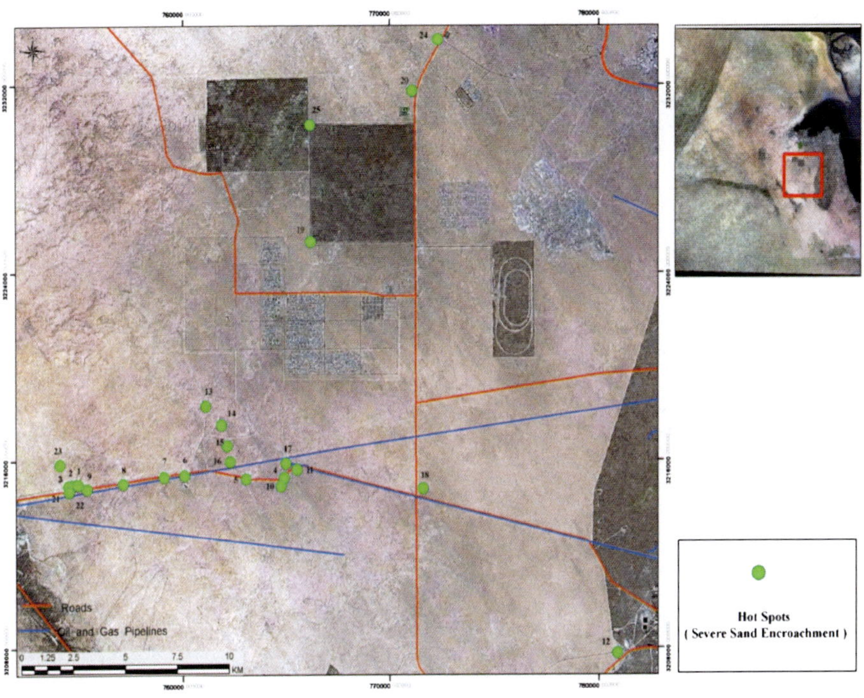

Fig. 41.7 Image showing facilities influenced by shifting sands in 25 locations

problems has not yet been considered in future plans; however, afforestation is focused in the urban areas.

41.3.6.1 Magnitude of Sand Encroachment in South Kabd Area (Case Study)

South Kabd area (30×40 km) covers 1,200 km^2 (about 7% of Kuwait). The concerned area is located in the heart of the main natural wind pass extending between Al Huwaimiliyah at the northwest and Al Wafra at the southeast. This area and nearby areas are severely attacked by shifting sands. All the facilities including flow lines, water wells, farms, roads, fences, and encampments are encroached by drifting sands. In May–July 2008, the concerned area, as well as other areas in Kuwait, experienced severe sand encroachment problems and terrible dust storms. In south Kabd area, a number of facilities are severely attacked by sands (Fig. 41.7). Several stretches of the existing roads, e.g., West Burgan-Ahmad Al Jaber Air Base, Abdaliyah-Managish, and Kabd main road (Road 604), were severely encroached by sands. Information on the location and nature of sand-encroached facilities is shown on Table 41.4.

The concerned area is characterized by heavy human pressure and intensive land uses including oil exploration and exploitation, agriculture and animal production,

Table 41.4 Field information on 25 sand-encroached facilities in the south Kabd area

Serial No.	GPS	Locality	Type of affected facility	Remarks
1	E: 0754883 N:3214988	Managish (close to gate 2)	Flow lines (KOC)	Huge amount of shifting sands
2	E: 0754658 N:3214961	Managish (close to gate 2)	Flow lines (KOC)	
3	E: 0754496 N:3214938	Managish (close to gate 2)	Flow lines (KOC)	
4	E: 0764925 N:3215261	Managish-Abdaliyah	Managish main road	
5	E: 0763057 N:3215218	Managish-Abdaliyah	Managish main road	
6	E: 0760092 N:3215374	Managish-Abdaliyah	Managish main road	
7	E: 0759097 N:3215304	Managish-Abdaliyah	Managish main road	
8	E: 0757123 N:3215020	Managish-Abdaliyah	Managish main road	
9	E: 0755374 N:3214804	Managish-Abdaliyah	Managish main road	
10	E: 0764760 N:3214908	Managish-Abdaliyah	Water well	
11	E: 0765544 N:3215624	Managish-Abdaliyah	Security facility	
12	E: 0780894 N:3207775	West Burgan	Burgan (GC2)-Ahmad Al Jaber road	
13	E: 0761097 N:3218318	North Abdaliyah	Flow line	
14	E: 0761866 N:3217519	Abdaliyah	Water well	
15	E: 0762146 N:3216629	Abdaliyah	Desert camp	
16	E: 0762275 N:3215947	Abdaliyah	Water well	
17	E: 0764993 N:3215877	Abdaliyah	Pump house	Severe encroachment
18	E: 0771605 N:3214798	East Abdaliyah	Road	
19	E: 0766182 N:3225332	Kabd	Kabd transmission station (NE corner)	
20	E: 0771092 N:3231748	Kabd	Farm	Very severe encroachment
21	E: 0754536 N:3214692	Managish	Gate no. 2	
22	E: 0754962 N:3214996	Managish	Oil facility	
23	E: 0754091 N:3215845	Managish	Bund wall	
24	E: 0772311 N:3233931	Kabd	Gasoline station	New facility
25	E: 0766159 N:3230295	Kabd	Experimental farm	

military activities, recreation and camping, and grazing. Controlling wind erosion and sand encroachment problems in the area will definitely minimize the problem in areas located at the downwind side, e.g., Wafra and southwest Burgan oil field.

41.3.6.2 Proposed Action Plan for Managing the Hazards of Shifting Sands

To manage the hazards of shifting sands in the terrestrial environment of Kuwait, three scenarios are proposed (Misak et al. 2009).

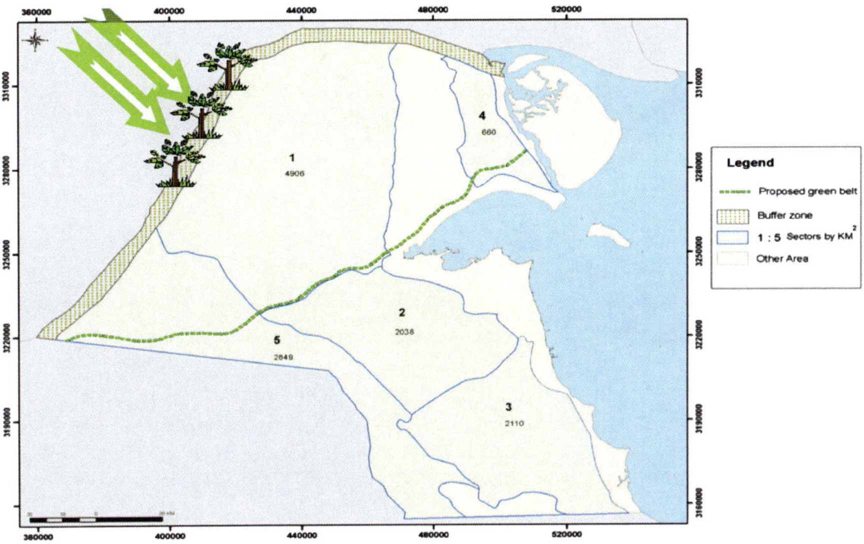

Fig. 41.8 First scenario, frontal greenbelt: Huwaimiliyah, ten-row greenbelt (25-km length and 187,500 trees). *Arrows* indicate the prevailing NW winds

First Scenario

The first scenario proposed by Misak (2009) includes the following measures:

- Establishment of two greenbelts (each has at least 10 rows of *Prosopis juliflora, Ziziphus spina-christi, and Tamarix aphylla* trees). The first belt (25-km length) is at Al Huwaimiliyah area (Fig. 41.8). The other belt (130-km length) is at Ras Al-Sabiyah-Al Salmi (Fig. 41.9). The selection of sites for two belts is based on information gained through field survey and remote sensing imagery interpretation of the area.
- Stabilization of active sandy bodies between the two belts as well as some active bodies in the south, using environmentally friendly materials: ecomat, coir, plant residues, etc.

Second Scenario

The second scenario consists of the following main measures:

- Afforestation: Plantation of *Prosopis juliflora, Ziziphus spina-christi, and Tamarix aphylla* trees along Huwaimiliyah-Al Wafra corridor (Fig. 41.10). The areas proposed for afforestation cover about 615 km^2. Treated sewage water has the potential to be used to irrigate these plantations for at least 1 year.
- Conservation/revegetation of native shrubs such as *Haloxylon salicornicum* and *Rhanterium epapposum* which are very effective in trapping shifting sands.

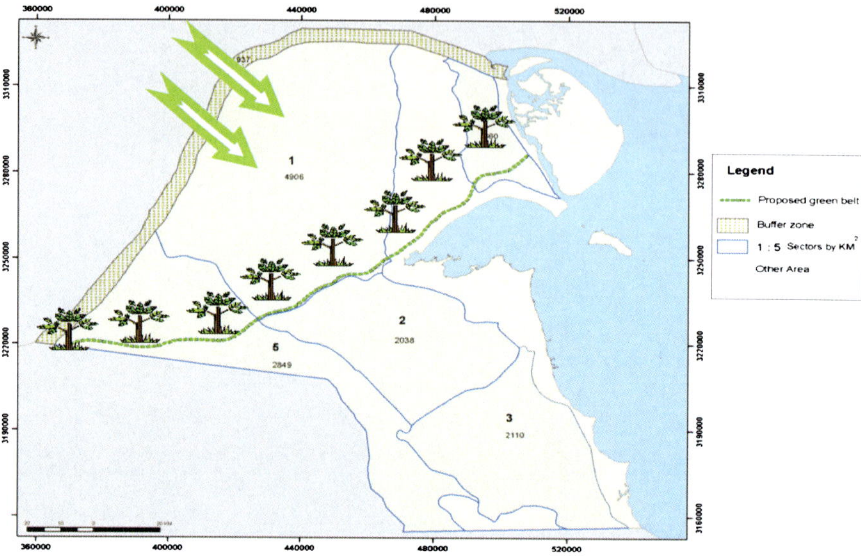

Fig. 41.9 First scenario, second belt: Ras Al-Sabiyah-Al Salmi, 10-row greenbelt (130-km length and 975,500 trees). *Arrows* indicate the prevailing NW winds

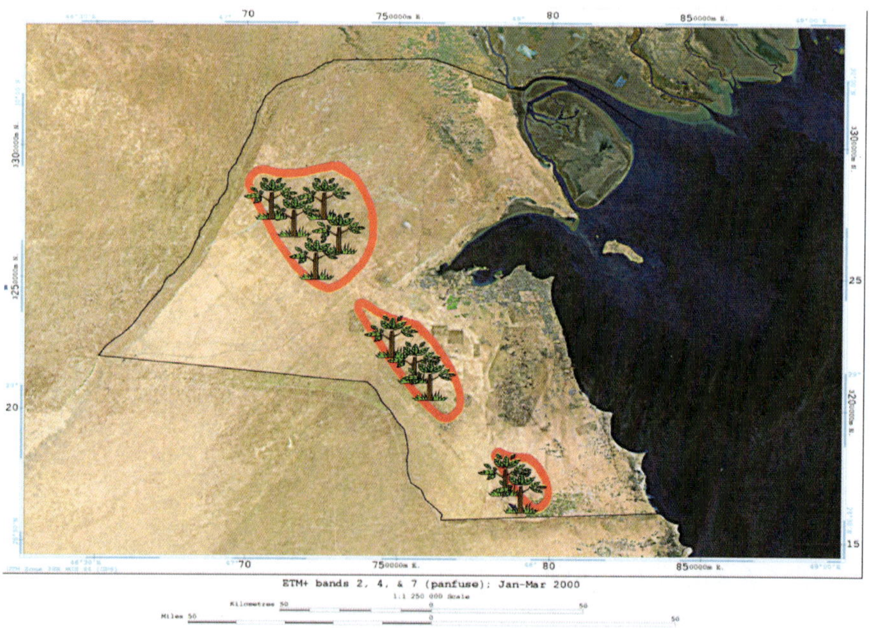

Fig. 41.10 An image showing areas proposed for afforestation (second scenario)

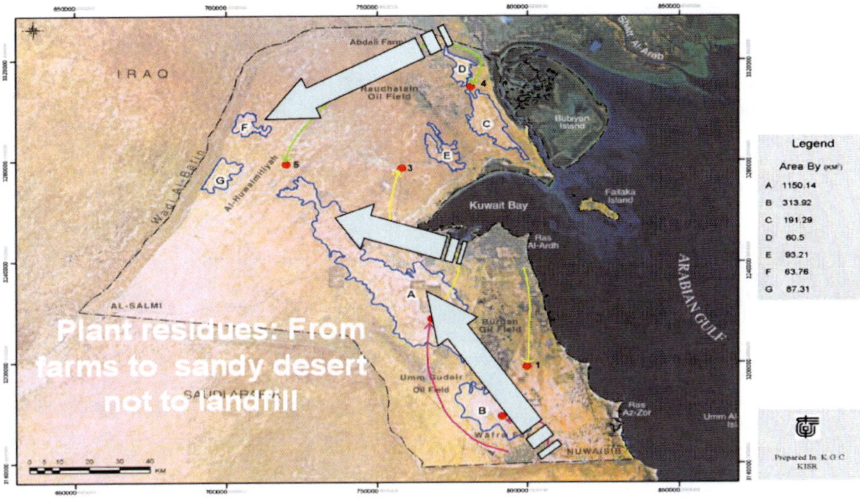

Fig. 41.11 An image showing the locations of main sources of greenery residues

Table 41.5 Second scenario: areas proposed for afforestation

Site	Area (km²)	Remarks
Huwaimiliyah	151	Sand dunes and sand sheet complex
Patches in Al Atraf-Kabd	150	Sand sheets and dunes
North Wafra	314	Sand sheets

- Stabilization of active sandy bodies along segments of Huwaimiliyah-Al Wafra corridor using greenery residues of Abdaly, Kabd, Wafra, and urban areas in Kuwait (recycling of greenery residues). Figure 41.11 shows the locations of main sources of greenery residues. Areas proposed for afforestation are shown in Table 41.5.

Third Scenario

The third scenario consists of the following main measures:

- Establishments of greenbelts in northern and central parts of Kuwait (Fig. 41.12).
- Revegetation of native plants such as *Rhanterium epapposum* in the southern parts of Kuwait.
- Immediate stabilization of active sandy bodies and local sources of dust using greenery residues and mulching sheets, e.g., ecomat and other environmentally friendly mulching materials.

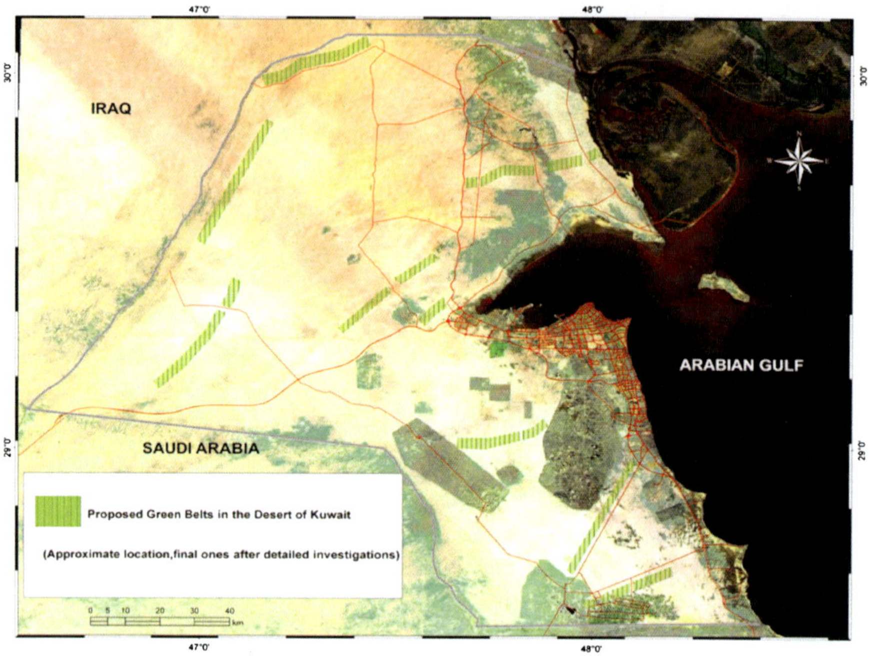

Fig. 41.12 An image showing the proposed greenbelts for the third scenario

The cost, efficiency, and environmental sustainability of the proposed scenarios will be assessed by a multidisciplinary team of experts. For the final design of greenbelt, wind tunnel experiments will be conducted, and the most cost-effective and environmentally sound scenario will be selected.

41.4 Setting Up Guidelines for Sustainable Land-Use Plans

Under hyperarid environmental conditions of Kuwait, sustainable land-use planning is the first defensive line in managing the drought and desertification. During the last 15 years, significant changes in land use were observed in Kuwait. Some of these changes have positive environmental impacts, while others have negative. Establishment of the buffer zone (15 km wide and more than 200 km long) between Iraq and Kuwait in 1993–1994 resulted in the enhancement of ecological conditions, while digging of border trench (2–3 m deep, 3–5 m wide, and about 200 km long) and construction of long bund walls (about 2–3 m high) have negative impact on soil, surface hydrologic conditions, and natural vegetation. Update information on the current land use suggests corrective measures are significantly required for managing drought. Figure 41.13 shows the current land-use types in Kuwait. Table 41.6 presents information on specific land-use types in the terrestrial environment of Kuwait.

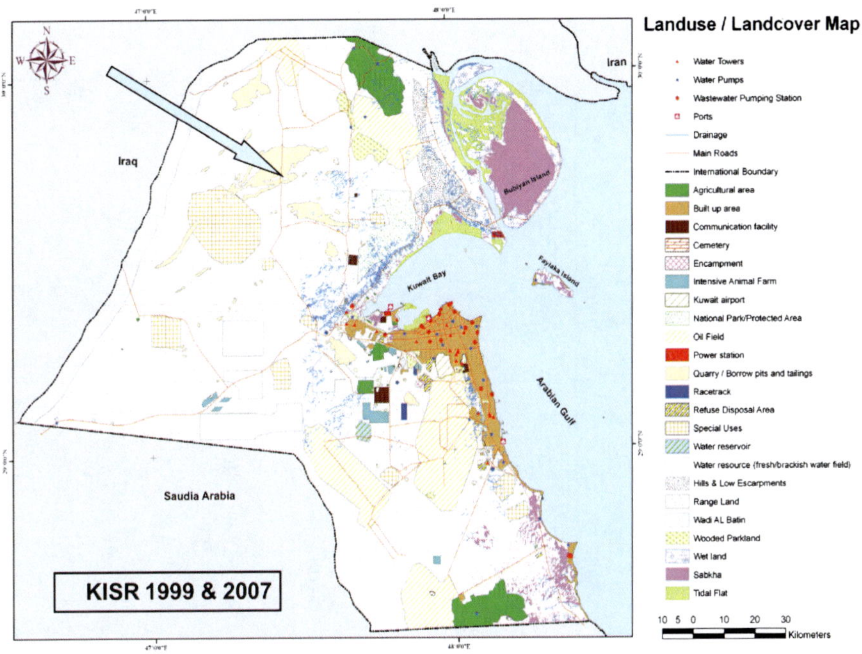

Fig. 41.13 Land-use/land cover map, *arrow* indicates prevailing wind direction (KISR 2006)

Generally, the current land-use types are differentiated into the following three categories.

41.4.1 Extremely Destructive

This category includes current military exercises and life shooting in the northwestern parts of Kuwait, establishment of bund walls and trenches in wide areas of the country, and the development of great number of gatch pits in Kuwait Oil Company operational areas (for clean soil quarrying). In KOC oil fields, some 83 gatch pits are found (KOC 2010). Damage caused by these types of land uses extends between 1.5 and 5 m from ground surface.

41.4.2 Destructive

This category includes rangeland grazing (close to 75% of total land-use types) and camping and recreation in scattered areas in Kuwait (5–10% of Kuwait). Damage caused by these types of land uses is restricted to the most upper part of soils (50–75 cm thick).

Table 41.6 Examples of land-use types and their impacts

Degree of impact	Impacts	Location(s)	Duration/year	Land-use type
Extremely destructive	Soil losses by wind, soil mining, soil compaction and sealing, surface deformation, hydrological disruption, vegetation degradation, loss of biodiversity, and deterioration of wildlife habitats	Al Edairah (northwest)	About 9 months/year	Military exercises and shooting ranges
Destructive	Soil compaction and sealing, vegetation degradation, and loss of biodiversity	Open desert areas	Ongoing	Rangeland grazing
Destructive	Vegetation degradation, soil compaction and sealing, loss of biodiversity, and deterioration of wildlife habitats	Open desert areas	October–April	Camping and recreation
Extremely destructive	Soil losses by wind, soil mining, soil compaction and sealing, surface deformation, hydrological disruption, vegetation degradation, loss of biodiversity, and deterioration of wildlife habitats	Al Huwamiliyah (northwest)	Ongoing	Sand quarrying
Nondestructive	Soil stabilization/enhancement of soil properties, vegetation cover, wildlife, and biological diversity	Al Salmi-Umm Qasr buffer zone	Ongoing	Land protection/ conservation

41.4.3 Nondestructive

This category includes protected and restricted areas in Kuwait which constitute 12–15% of the country. Natural recovery of ecosystems occurs under protection conditions. In this category, damage to ground surface is not noticeable (with the exception of war-related damage).

41.5 Proposed Sustainable Land-Use Plans (The Case of Rangeland Grazing)

Rangeland grazing represents about 75% of land use (KISR 1999). Based on the study of Omar et al. (2001), some 925,000 sheep, 112,000 goats, and 23,000 camels are the grazing animals in the rangelands of Kuwait. Currently, grazing is allowed in the western half of the country, while it is prohibited in the eastern side. We suggest the current map of grazing areas (PAAFR 1988) should be considered for updating. According to personal communication (Omar, 13 September 2009, Personal communication), it was apparent that the map was prepared without considering range management aspects (type of vegetation cover, animal numbers and distribution, seasonal variation, and intensity of grazing).

For sustainable rangeland grazing, it is suggested to develop a new map considering the shortcomings of the current map. From the authors' point of view, the shortcomings are discussed in the following part:

- Dividing the State of Kuwait into two parts, western (where grazing is allowed) and eastern (where grazing is prohibited), is not scientifically and environmentally justified.
- The western half of Kuwait is cut by a NW-SE natural hollow (about 25-km average width and 145-km length, approximate area is about 3,625 km^2). The hollow starts in the Iraqi territories and acts as an active wind corridor for the prevailing NW winds. Grazing in the wind corridors (as indicated in the current map) is a destructive land-use type.
- More than 70% of the soils of the mentioned corridor are Torripsamments (sandy soils), which have very high vulnerability to wind erosion relative to developed soils (Haplocalcids, Haplogypsids, Petrocalcids, Petrogypsids, etc.) especially during dry years. Grazing in this corridor accelerates soil deflation processes causing sand and dust storms especially during summer and spring seasons.
- The natural vegetation represented by Haloxylon salicornicum and Stipagrostis plumosa in the corridor is severely to completely degraded. Vegetation cover is less than 5%. Dry relics of natural vegetation are common in the natural corridor. Thus, it is essential to restore vegetation in the corridor.
- The eastern half of Kuwait is influenced by another natural wind corridor. This corridor extends for about 50 km between Umm Qasr and Ras Al-Sabiyah (north-eastern part of Kuwait). The corridor width ranges between 10 km (at its middle

Table 41.7 History of land-use of Wadi Al Batin

Period	Land use and activities	Remarks
Pre-Iraqi invasion of Kuwait (before August 1990)	Open area for livestock grazing	Borders with Iraq and Saudi Arabia were opened (no barriers)
Invasion and occupation of Kuwait (August 1990–February 1991)	Military activities *Iraqi troops*: establishment of mine fields, development of oil trenches, and construction of bunkers and weapon pits *Coalition forces*: crossing Wadi Al Batin during liberation of Kuwait (desert storm)	Severe ecological deterioration
March 1991–July 1993	Demining and EOD (Explosive Ordnance Demolition), refilling and ground leveling of bunkers	Degradation of soils and vegetation cover
1993–1995	*UN*: border demarcation and buffer zone establishment *Kuwait*: establishment of security system, border trench, electric fence, eastern trench, and bund walls	Positive ecological and environmental changes

part) and about 2 km (at its extreme southern part). Grazing in this corridor enhances sand movement. Shifting sand is a hazard for the new Ras Al-Sabiyah settlement (population 0.60–0.55 million).

- For range management, grazing in both wind corridors should be prohibited at least for 5–10 years.

41.5.1 Proposed Corrective Measures (The Case of Wadi Al Batin)

Wadi Al Batin constitutes a huge catchment area and is located at the western part of Kuwait. It extends in northeast direction from Al Salmi area (extreme southwestern part of Kuwait) for about 100 km. This wadi consists of a main channel of about 10-km width and eastern and western cliffs. The western cliff is located in Iraq. The Kuwaiti-Iraqi border runs in the main channel of the wadi.

During 1993–1995, a system of security measures was established by Kuwait along Iraq-Kuwait border. This system consists of two trenches. One of these trenches was dug along the borderline. The width and depth of this trench average 5 and 3 m, respectively, and the length is 212 km. The second trench is located about 5 km to the east of the border trench. The history of land use of Wadi Al Batin is shown in Table 41.7.

41.5.1.1 Consequences of the Current Land Use of Wadi Al Batin

- Hydrological disruption leading by time to ecological deterioration: dryness of plant species and deterioration of wildlife (Misak 2009)
- Loss of runoff water through discharging into trenches (border and eastern trenches)
- Disturbance of the mechanism of local groundwater recharge from the wadi banks to the main channel, where a shallow wadi fill aquifer exists

41.5.1.2 Proposed Action Plan for the Restoration of Wadi Al Batin

As stated by Alenezil et al. (2010), oil-polluted soils of Wadi Al Batin (about 125,000 m^3) should be excavated. Clean soils (fresh eolian sands) should replace oil-polluted soils. The excavated heavy polluted soils could be used for road paving for off-road tracks to minimize land degradation.

To restore hydrological disruption and to recover the long-term impact of destructive land use of Wadi Al Batin, the following are proposed:

- Refilling the trenches (border and eastern trenches) to maintain surface hydrologic and recharge conditions
- Leveling north-south bund walls stretching parallel to Wadi Al Batin to avoid blocking of runoff flowing to the main channel of the wadi
- Breaking soil crusting, sealing, and compaction in the main water course of the wadi to enhance the infiltration of rainfall into the soil (mechanical breaking and plantation)

41.6 Conclusions and Recommendations

To accomplish the objective, managing the hazards of drought and shifting sands in Kuwait, intensive field work accompanied by analyses and interpretation of satellite images was carried out. We are proposing four programs, the watershed management, mitigating hydrological drought, managing the hazards of shifting sands, and setting up sustainable land-use plans.

Based on the findings of the present study, the following recommendations are made. It is recommended to review the current land-use plans considering number of important issues, such as exploring new site for military exercises as an alternative to Al Edairah highly sensitive area (northwestern part of Kuwait). This area has been used since two decades for exercises and shooting ranges. Currently Al Edairah area (about 400 km^2) acts as a main source of sands and dust in Kuwait. Protecting the area for about 5 years is likely to enhance soil properties and vegetation potentials and in turn increase soil stability. Managing livestock grazing in highly fragile areas such as the two natural corridors of drift sand, i.e., Al Huwaimiliyah-Wafra (northwestern part of Kuwait) and Umm Qasr-Ras Al-Sabiyah (northeastern part of

Kuwait). Controlling grazing and camping in these corridors for at least 5 years is a sustainable land-use approach. Establishment of an action plan for watershed management including restoration of hydrologic disruption and water harvesting. First priority should be given to Rawdatain and Umm El Eish basins (northeastern part of Kuwait) and Wadi Al Batin (western part of Kuwait). Second priority to be given to Jal Az Zour (northeastern part of Kuwait) and Ahmadi (south of Kuwait City). Guidelines for this action plan are available in the section of watershed management and mitigation of hydrologic drought of this chapter. Establishment of an integrated action plan for managing the hazards of shifting sands considering the proposed three scenarios of mobile sand control. Development of a drought preparedness plan for the State of Kuwait including the establishment of a drought monitor consisting of advanced remote sensing systems and ground observatory stations.

References

Al-Awadhi JM, Misak RF (2000) Field assessment of aeolian sand processes and sand control measures in Kuwait. Kuwait J Sci Eng 27(1):156–176

Al-Awadhi JM, Misak R, Omar SAS (2003) Causes and consequences of desertification in Kuwait. A case study of land degradation. Bull Eng Geol Environ 62:107–115

Al-Awadhi JM, Omar SAS, Misak RF (2005) Land degradation indicators in Kuwait. Land Deg Dev 16:163–176

Al-Dousari AM, Misak RF, Shahid SA (2000) Soil compaction and sealing in Al Salmi area, western Kuwait. Land Deg Dev 11:401–418

Al-Dousari AM, Misak RF, Al Gamily H, Neelamani N (2007) Integrated system for flood management in Shuaiba area and its vicinities. KISR EC055C final report, KISR 8910

Alenezil A, R Misak, Al-Abdullah H (2010) Assessment and mapping of heavy oil contamination Southwest of Kuwait. Int J Soil Sediment Contam (Paper under review)

Civil Aviation of Kuwait (2009) Meteorological data and information

KISR (1999) Soil survey for the State of Kuwait. AACM International, Adelaide. ISBN 095770 030x

KISR (2006) Land-use/land cover map, Geoinformatics center

KOC (Kuwait Oil Company) (2010) SEED project, KNFP Oil Lake/Soil remediation forum, pp 22–25, Mar 2010, SEED

Kwarteng A, Viswanathan M, Al Senafy M, Rashid T (2000) Formation of groundwater in northern Kuwait. J Arid Environ 46:137–155

Misak RF (2009) Managing land degradation and hazards of shifting sands in the terrestrial environment of Kuwait (A sustainable management plan: 2010–2025). In: Abstracts, conference on land degradation in dry environments, Kuwait University, 8–14 Mar 2009

Misak RF, Alhajraf S (2005) Kuwaiti experience in managing land degradation with emphasis on mobile sand control. In: Gulf conference on environment and sustainability, 3–5 Dec 2005, Kuwait

Misak RF, Al-Dousari A, Alhajraf S (2007) Combating land degradation using eco-friendly materials. Desertification control in the arid region 12–15 May 2007, Kuwait Institute for Scientific Research

Misak RF, Omar SAS, Al Gamily H (2008) Ecohydrological features of Sabah Al Ahmad Nature Reserve (in the book entitled Physical Features and Natural Resources of Sabah Al Ahmad Nature Reserve, in Arabic)

Misak RF, Al Sudairawi M, Al-Dousari A, Al Gamilly H (2009) Long-term national program for managing the hazards of shifting sands in the terrestrial environment of Kuwait. Proposal, EUD, KISR

Omar SAS, Misak RF, King P, Shahid SA, Abo Rizq H, Grealish G, Roy W (2001) Mapping the vegetation of Kuwait through reconnaissance soil survey. J Arid Environ 48:341–355

PAAFR (Public Authority for Agriculture and Fish Resources) (1988) Reports on livestock grazing and rules of grazing

Ud Din S, Al-Dousari A, Al Ghadban A (2007) Sustainable fresh water resources management in northern Kuwait-A remote sensing view from Rawdatain basin. Int J Appl Earth Obs Geoinf 9:21–31

Chapter 42
Determining Degraded Soils of Southern Kazakhstan Through Assessing Stability of Soil Aggregates

Akmal Karimov, A. Noble, R. Kurbantoev, and N. Salieva

Abstract Degradation of irrigated soils is often the consequence of changes in physical and chemical properties. Such changes are reflected in the loss of soil structure, dispersion of soil particles, compaction, reduction in permeability, and hydraulic conductivity. These processes are widespread on the irrigated soils in the midstream of Syrdarya and Zarafshan river basins of Central Asia. The Arys-Turkestan Canal (ATC) command area in the southern Kazakhstan is a typical example of degraded irrigated soils. Cotton crops cultivated on these soils have suboptimal aboveground growth and weak root systems that drastically reduce yields. Determining the degradation level of the irrigated soils is important in assisting the farmers to select appropriate soil management strategies and reduce the risk of accelerating degradation. This study focuses on identifying the degree of degradation by determining the stability of soil aggregates. Water-stable aggregates were analyzed of topsoil from the ATC zone under four contrasting management practices: undisturbed virgin soil, productive irrigated soil, low-productive degraded soil, and salt-affected abandoned soil. The results of these studies clearly show a trend in a reduction in the quantity of the stable macroaggregates on the degraded soils. Among a range of different aggregate stability indices, the stable macroaggregate index (SMAI) was found to be the most suitable to differentiate highly degraded soils. The SMAI values of less than 10% are found to be indicative

A. Karimov (✉)
International Water Management Institute (IWMI),
Central Asia Regional Office, Tashkent, Uzbekistan
e-mail: a.karimov@cgiar.org

A. Noble
International Water Management Institute (IWMI),
National Agriculture and Forestry Research Institute, Ban Nongviengkham,
Xaythany District Vientiane Lao PDR

R. Kurbantoev • N. Salieva
The Soil Science Institute, The Academy of Science
of Uzbekistan, Tashkent, Uzbekistan

S.A. Shahid et al. (eds.), *Developments in Soil Classification, Land Use Planning and Policy Implications: Innovative Thinking of Soil Inventory for Land Use Planning and Management of Land Resources*, DOI 10.1007/978-94-007-5332-7_42,

of soils having poor physical properties. Sensitivity analyses found that the SMAI is most sensitive to the changes in the following soil properties: quantity of particles <1-mm size, electrical conductivity of the soil, organic matter, and gypsum content.

Keywords Central Asia • Degraded soil • Soil structure • Southern Kazakhstan • Stability of soil aggregates

42.1 Introduction

Degradation of soils is often the consequence of changes in physical and chemical properties of soil associated with the conventional irrigation practices. Such changes are reflected in the loss of soil structure, dispersion of soil particles, compaction, and reduction in permeability (McNeal et al. 1968; Suarez et al. 1984; Lebron et al. 1994; Kjaergaard et al. 2004). These negative processes are widespread on irrigated soils of Central Asia in the midstream of Syrdarya and Zarafshan river basins (Bodruhina 1973; Vyshpolsky et al. 2008). In southern Kazakhstan, the area of irrigated soil having poor soil structure and reduced permeability exceeds 70,000 ha (Vyshpolsky et al. 2008). On these soils, crops invariably have poor aboveground growth and weak root systems that drastically reduce yields. Farmers often abandon land where the productivity of soils has declined to a level where growing crops is no longer economically viable.

The Arys-Turkestan Canal (ATC) command zone located in the southern Kazakhstan typically represents irrigated areas in the region that are afflicted by poor soil physical and chemical properties. The topsoil is highly compacted with permeability at 0.1–0.3 meters per day (m day^{-1}). The organic matter (OM), gypsum, and total soluble salts contents are very low in the soil profile at 7–10, 0, and 1.0–1.4 g kg^{-1}, respectively. Soil aggregates lose their stability during irrigation events, disperse, and form compacted massive structures during the drying phase. Tillage of the soil causes the formation of massive clods directly affecting the growth of cotton crops resulting in yields of 1.4–1.6 t ha^{-1} and eventually land abandonment when yields decline to uneconomic levels. The focus of this study is on the stability of aggregates of degraded soils in the ATC command zone with the following objectives: (1) to assess aggregate-size stability distribution of a range of soils of the ATC zone, (2) to identify a suitable soil aggregate stability index for delineating degraded soils with massive structures, and (3) to determine the sensitivity of the soil aggregate stability to changes in physical and chemical properties of soil.

Different methods have been proposed to determine soil aggregate stability (Kemper 1966; Kemper and Rosenau 1986). The methods of Pavlov and Savvinov (Kachinski 1965, 1970; Dj Bairov et al. 2004) are applied under local conditions to analyze the stability of soil aggregates. In this study, the method used to determine soil aggregate-size distribution and stability was that of Kemper and Rosenau (1986). Several studies have used capillary-wetted and slaked pretreatments (Elliot 1986; Cambardella and Elliot 1993; Six et al. 2000a; Marquez et al. 2004). The studies of Marquez et al. (2004) have demonstrated that the use of a subsequent slaking treatment following the standard capillary-wetted pretreatment provides the means to accurately determine the amount of stable and unstable macroaggregates. This study applies the combined use of the capillary-wetted and the slaked pretreatments to determine changes in aggregate-size distributions on a range of the soils of the ATC command zone.

Different indices were applied to quantify soil aggregate stability. The mean weight diameter (MWD) has often been used to assess the effect of different management practices on the soil structure (Angers and Mehuys 1993). The stable aggregate index (SAI) and the stable macroaggregate index (SMAI) were proposed and tested for studying soil stability based on aggregate resistance to slaking (Marquez et al. 2004). Further, the normalized stability index (NSI) was proposed and tested to characterize soil stability, eliminate confounding effects of pretreatment and antecedent water content, and correct effects of differences in sand size distribution among soils (Six et al. 2000a). In this study, these indices have been evaluated to delineate degraded massive structure soils of the ATC command zone.

Sensitivity analyses were undertaken to determine the relationship between low stability of soil aggregates and physical and chemical properties of soil. It has been concluded that this relationship between soil attributes and aggregate stability is complex (Remezov 1957; Six et al. 2000a; Dontsova and Norton 2002; Levy and Mamedov 2002; Kjaergaard et al. 2004; Yilmaz et al. 2005). Under conditions where exchangeable sodium percentage (ESP) is <15 and the electrical conductivity (EC) is low, other factors may affect the dispersion; the most important includes clay mineralogy, the surface charge, pH, OM, and the presence of Fe- and Al-oxides. Boix-Fayos et al. (2001) found a positive correlation between water stability of microaggregates and clay content, whereas the stability of macroaggregates depended on the OM content only when the OM content was $>5\%$. When the OM was $<5\%$, aggregate stability was strongly affected by the $CaCO_3$ content. Keren and Ben-Hur (2003) indicated that $CaCO_3$ acting as a cementing agent decreased aggregate slaking of a Chromoxerert-sand mixture. Levy et al. (2003, 2005) studied the combined effects of salinity, sodicity, wetting rate, and soil texture on aggregate stability and found that the wetting rate has no effect on aggregate slaking of soils with low clay content. As clay content increases, there is a significant interaction among the variables affecting aggregate stability. Tedeshi and Dell'Aquila (2005) used aggregate stability as an index to estimate the degradation of the physical properties of soil caused by irrigation with saline water and found that aggregate stability reduces as ESP increases. Ruiz-Vera and Wu (2006) found that aggregate stability is affected by clay mineralogy, clay content, and pre-wetting rate. In summary, soil aggregate stability responds to a combination of physical and chemical properties of soil. Although there has been considerable interest in this area, limited studies have included degraded massive structure soils with high $[Mg^{2+}]$ and low $[Na^{+}]$ contents on the cation exchange capacity.

42.2 Materials and Methods

42.2.1 *Research Site*

The study was carried out at the Eski Ikan farm located in the ATC command zone in southern Kazakhstan. Mean annual rainfall in the study area (150 mm) predominantly occurs from October to May. The relief is flat, with slopes of 0.1–0.2 degrees. The dominating soil of the area is classified as silt loam light serozem of low salinity. The organic matter content in the topsoil is less than 1%, and pH varies from 8.1 to 8.4. The soil is characterized by poor physical properties, and it is highly dispersive during irrigation and when dry forms a massive structure. The bulk density of the top 15-cm soil layer is 1.45–1.50 g cm^{-3}, and permeability is 0.1–0.3 m day^{-1}. Conventional deep plowing on these soils results in the formation of large impermeable clods. Cotton yields do not exceed 1.4–1.6 t ha^{-1} which is below the average of 2–2.2 t ha^{-1}. Groundwater levels are 1.5 m below the surface in spring and decline to 2.5 m in autumn.

42.2.2 *Soil Sampling*

Soil samples were collected from the 0- to 5-cm depth interval in the fall of 2007. Four sites were selected to undertake this assessment. The first site sampled (Ikan A) was virgin land surrounded by irrigated land that had not been under agronomic production and had been used for grazing livestock. The soil was a silt loam and nonsaline. At the second site (Ikan B), the irrigated soil did not show evidence of degradation. Stands of cotton were well established, and the yield of raw cotton was in the range of 2.2–2.6 t ha^{-1}. The soil was classified as silt loam to loam by texture. On the third site (Ikan C), the soil was saline, and the farmer had abandoned the plot due to low yields of cotton and resulting low income from this area. Soil was silt to heavy loam by texture. On the fourth site (Ikan D), the irrigated soil showed clear evidence of massive structures. The soil was classified as silt to silt loam by texture. The soil was highly compacted along irrigated furrows with deep surface cracks. Soil aggregates were dispersive and lose their stability during irrigation with consequent compaction and decreased permeability when dry. Cotton plants growing in the plot were poor, with yields of raw cotton not exceeding 1.5 t ha^{-1}.

42.2.3 *Physicochemical Analyses*

Particle size was measured using the pipette method (Dj Bairov et al. 2004). The soil pH, EC, and composition of water soluble ions were determined in a 1:5 soil to water extraction. Sodium cations were measured by flame emission spectrophotometer, and the other major cations and anions were measured using standard titration procedures

(Arinushkina 1970). Exchangeable K^+, Na^+, Ca^{2+}, and Mg^{2+} were determined using the Pfeffer method (0.1 M NH_4Cl+70% ethanol) (Arinushkina 1970; Dj Bairov et al. 2004). The soil effective cation exchange capacity (ECEC) was calculated by summation of exchangeable Ca^{2+}, Mg^{2+}, Na^+, and K^+. The gypsum content was measured using 0.2N HCl solution (Arinushkina 1970). The Tyurin method, based on oxidation by $K_2Cr_2O_7$, was chosen for measuring OM content (Arinushkina 1970).

Sodium adsorption ratio (SAR) was calculated using Eq. 42.1:

$$\mathrm{SAR} = \frac{\mathrm{Na}^{+}}{[(\mathrm{Ca}^{2+} + \mathrm{Mg}^{2+})/2]^{0.5}} \tag{42.1}$$

where Na^+, Ca^{2+}, and Mg^{2+} are concentrations of cations in meq L^{-1}.

Exchangeable sodium percentage (ESP) and exchangeable magnesium percentage (EMgP) were calculated using Eqs. 42.2 and 42.3:

$$\mathrm{ESP} = \left(\frac{\mathrm{Na}^{+}}{\mathrm{CEC}}\right) \times 100 \tag{42.2}$$

$$\mathrm{EMgP} = \left(\frac{\mathrm{Mg}^{+}}{\mathrm{CEC}}\right) \times 100 \tag{42.3}$$

where Na^+ and Mg^{2+} are concentrations of cations in the cation exchange capacity (CEC) expressed in $cmol_c\,kg^{-1}$.

42.2.4 Quantification of Soil Aggregates

Aggregate analysis was performed on disturbed soil samples sieved through 8-mm mesh screens after air-drying. Two 100-g subsamples of air-dried soil were used to analyze the aggregate-size stability distribution. Two pretreatments are applied before wet sieving: air-drying followed by rapid immersion in water (slaked) and air-drying plus capillary rewetting (capillary wetted) (Six et al. 1998; Marquez et al. 2004). The capillary-wetted pretreatment applied before wet sieving is air-drying of the soil samples plus capillary rewetting above the field capacity by 5%. Both subsamples were stored overnight in a refrigerator at 4°C before wet sieving. Aggregates were physically separated into four aggregate-size fractions: (1) large macroaggregates >2,000 mm in diameter, (2) small macroaggregates between 250 and 2,000 mm in diameter, (3) microaggregates between 53 and 250 mm in diameter, and (4) mineral fraction <53 mm in diameter.

After wet sieving, all the fractions were oven-dried at 70°C, except the large and small macroaggregates obtained by the capillary-wetted pretreatment. These macroaggregates were air-dried and used for separation of large and small stable aggregates. Sand corrections were performed by subtracting the total sand content of each size fraction from the amount of sample retained on each size fraction. The total sand content of each aggregate-size fraction was determined by weighing the material that was retained on the sieve with a 53-mm screen upon dispersal of the aggregates with sodium hexametaphosphate (5 g L^{-1}). The method is described in detail by Marquez et al. (2004).

42.2.5 *Estimation of Stable and Unstable Aggregates*

Indices used for evaluating soil aggregate stability were as follows:

1. Mean weight diameter (MWD) (van Bavel 1949):

$$\text{MWD} = \sum_{i=1}^{n} (x_i y_i) \tag{42.4}$$

   where x_i = mean diameter of each size fraction (mm) i, and y_i are the proportion of the total sample weight occurring in the size fraction i. This index was calculated separately for slaked (MWD_s) and capillary-wetted (MWD_{cp}) pretreatments.
2. Geometric mean diameter (GMD) (Mazurak 1950):

$$\text{GMD} = exp\left(\frac{\sum_{i=1}^{n}(y_i \text{Ln}(x_i))}{\sum_{i=1}^{n}(y_i)}\right) \tag{42.5}$$

3. Water-stable aggregates (WSA) (Kemper 1966; USDA 1998):

$$\text{WSA}_{(\%\ \text{of soil}>250\ \text{mm})} = \frac{\text{weight of dry aggregates}}{\text{weight of dry soil} - \text{sand}} \times 100 \tag{42.6}$$

4. Normalized stability index (NSI) (Six et al. 2000b):

$$\text{NSI} = 1 - \left[\frac{\text{DL}}{\text{DL}_{\max}}\right], \text{and} \tag{42.7}$$

$$\text{DL} = \frac{1}{n}\sum_{i=1}^{n}\left[(n+1) - i\right]\text{DLS}_i$$

$$\text{DLS}_i = \frac{\{[(P_{io} - S_{io}) - (P_i - S_i)] + |(P_{io} - S_{io}) - (P_i - S_i)|\}}{[2(P_{io} - S_{io})]}$$

$$\text{DL}_{\max} = \frac{1}{n}\sum_{i=1}^{n}\left[(n+1) - i\right]\text{DLS}_{i\max}$$

$$\text{DLS}_{i\max} = \frac{[(P_{io} - P_p) + |P_{io} - P_p|]}{[2(P_{io} - S_{io})]}$$

where DL is the soil disruption level, n the number of aggregate-size classes, DL_{max} the maximum disruption level, DLS_i the disruption level for each size class i, P_{io} the proportion of total sample weight in size class i before disruption (i.e., rewetted), P_i the proportion of total sample weight in size class i after disruption (i.e., slaked), S_{io} and S_i the proportions of sand with size i in aggregates of size i before and after disruption, DLS_{imax} the maximum disruption, and P_p the primary sand particle content with the same size as the aggregate-size class after complete disruption of the whole soil. Based on weight losses, $i=1$ for the smallest size class.

5. Stable aggregates (SAI) and stable macroaggregates (SMAI) indices (Marquez et al. 2004):

$$\text{SAI} = \frac{\left(\sum_{j=1}^{n}\left[(n+1)-j\right]S_j\right)}{\left(\sum_{j=1}^{n}\left[(n+1)-j\right]T_j\right)} \tag{42.8}$$

$$\text{SMaI} = \frac{\left(n\sum_{j=1}^{m}\left[(m+1)-j\right]S_j\right)}{\left(m\sum_{j=1}^{n}\left[(n+1)-j\right]T_j\right)} \tag{42.9}$$

where $j=1$ for the largest size class, m is the total number of size classes larger than 250 mm, S_j the amount of stable aggregates in fraction j, and T_j the total amount of aggregates in fraction j upon capillary wetting. Each index was calculated separately for slaked and capillary-wetted pretreatments.

42.2.6 Sensitivity Analysis for the Stability of Soil Aggregates

Relationship between quantity of soil water-stable aggregates and chemical and physical properties of soil was described by a linear relation:

$$\text{SSI} = \sum_{i} \alpha_i P_i + \mu \tag{42.10}$$

where SSI = calculative value of a soil aggregate stability index which is most suitable among those tested for specifying the degraded soils, a the design variable, i = soil physical or chemical property, and P_i the value of the soil property i.

Soil physical and chemical properties applied to determine the relationship with the soil stability index were as follows: OM content, gypsum, particles less than 1-μm size, EC of the soil solution, $CaCO_3$, $MgCO_3$, HCO_3^-, and exchangeable Mg^{2+} and Na^+.

Values of design variables were found using GAMS 2.25 software (Brooke et al. 1992) with the objective function

$$\text{MinZ} = \sum_{1}^{n}\left[\left(SSI_a - SSI_c\right)\left(SSI_a - SSI_c\right)\right] \tag{42.11}$$

where SSI_a is the actual value of the soil aggregate stability index obtained from the soil laboratory studies.

A sensitivity analysis was carried out to determine key factors affecting soil aggregate stability. In a single-factor sensitivity analysis, i.e., for each output value, only one input was changed from its base case value.

42.3 Results and Discussion

42.3.1 Physical and Chemical Properties of the Degraded Soils of the ATC Command Zone

Analysis of soil samples collected from the four different sites in the ATC command zone in the fall of 2007 indicated high variations in the different soil properties (Table 42.1). The virgin soil (Ikan A) was classified as a silt loam with the OM content varying from 23 to 32 g kg^{-1} of dry soil in the topsoil. The quantity of particles that were in the size class of <1-mm size was in the range of 9.4–14.7%. Among anions HCO_3^- was dominant followed by SO_4^{2-}, and soluble Ca^{2+} was dominant among cations followed by Mg^{2+}. The pH of the soil solution was at 7.4–7.7. Exchangeable Ca^{2+} occupied 60–70% of the exchange complex, with Mg^{2+} 19–29% and Na^+ less than 2% of the ECEC. The soil was nonsaline. However, a very low sodium adsorption ratio (SAR) and EC of the soil solution were indicative that this soil may become unstable and susceptible to dispersion under the influence of external factors, such as irrigation.

Productive irrigated soil (Ikan B) was classified as a silt loam with low organic matter content in the range of 7.1–8.2 g kg^{-1} of dry soil. The quantity of the particles <1-mm size was in the range of 13–14.2%. Among the anions, SO_4^{2-} was dominant followed by HCO_3^-, while soluble Ca^{2+} followed by Mg^{2+} were the two most dominant cations. The pH of the soil solution was 8.0–8.2. Exchangeable Ca^{2+} dominated the ECEC (56–60%), with Mg^{2+} ranging from 34 to 38% and Na^+ <4%. The soil was nonsaline. Higher SAR and EC of the irrigated soil indicated that this soil was more stable than the virgin soil. Increases in the content of the exchangeable Mg^{2+} on the ECEC are another difference of the irrigated soils as compared to the virgin soils. Accumulation of carbonates of Ca^{2+} ($CaCO_3$) and Mg^{2+} ($MgCO_3$) is noted after irrigating the virgin soil. Cotton plants were well developed in spite of the high soil pH and the high content of the exchangeable Mg^{2+}.

Soil samples collected from the abandoned site (Ikan C) showed evidence of saline patches. The soils on these patches were classified as silt loam with organic matter content from 17 to 23 g kg^{-1} of dry soil. The quantity of the particles with <1-mm size was found to be much higher than the virgin and the productive irrigated soil, at 19.8–22.8%. Among anions, SO_4^{2-} was dominating followed by Cl^-, and soluble Ca^{2+} among cations followed by Mg^{2+}. The pH of the soil solution was at 7.65–7.8. Exchangeable Ca^{2+} dominated the ECEC (43–50%) with Mg^{2+} and Na^+ occupying

Table 42.1 Selected physical and chemical properties of the studied soils of the ATC zone collected (in the fall of 2007) from the 0- to 5-cm topsoil interval

		Soil			
		Virgin soil	Irrigated silt loam	Abandoned silt loam	Irrigated silt loam
Soil property	Unit	(Ikan A)	(Ikan B)	(Ikan C)	(Ikan D)
Particle size group					
Sand	g kg^{-1}	250 (±119)	342 (±58)	158 (±75)	237 (±60)[a]
Silt	g kg^{-1}	526 (±109)	523 (±64)	591 (±76)	552 (±78)
Clay	g kg^{-1}	144 (±28)	135 (±16)	251 (±20)	212 (±47)
Particles less than 1-mm size	g kg^{-1}	118 (±23)	135 (±5)	212 (±12)	212 (±2)
OM	g kg^{-1}	25.5 (±3.4)	7.5 (±0.3)	19.6 (±1.6)	10.2 (±0.4)
Gypsum	g kg^{-1}	3.8 (±0.8)	0	5.6 (±1.1)	0
pH		7.6 (±0.13)	8.1 (±0.05)	7.7 (±0.04)	8.1 (±0.05)
HCO_3^-	$mmol_c L^{-1}$	6.5 (±0.67)	4.33 (±0.52)	4.2 (±0.16)	6.0 (±0.16)
Cl^-	$mmol_c L^{-1}$	0.8 (0)	2.47 (±1.03)	16.1 (±12.1)	1.0 (0)
SO_4^{2-}	$mmol_c L^{-1}$	6.23 (±0.78)	14.03 (±5.74)	50.77 (±16.21)	4.3 (±0.86)
Ca^{2+}	$mmol_c L^{-1}$	6.75 (±0.43)	9.0 (±2.83)	33.43 (±15.56)	5.0 (±1.08)
Mg^{2+}	$mmol_c L^{-1}$	2.5 (±0.5)	7.4 (±1.78)	17.13 (±6.87)	3.67 (±0.24)
Na^+	$mmol_c L^{-1}$	4.33 (±1.3)	4.2 (±1.96)	13.63 (±6.41)	2.03 (±0.21)
EC	dS m^{-1}	0.71(±0.2)	2.12 (±0.62)	6.98 (±2.75)	1.12 (±0.1)
SAR	$(mmol\ L^{-1})^{0.5}$	0.1 (±0.03)	1.4 (±0.46)	2.7 (±0.97)	1.4 (±0.46)
$CaCO_3$	g kg^{-1}	169 (±16)	233 (±2)	265 (±3)	237 (±1)
$MgCO_3$	g kg^{-1}	18 (±1)	29 (±1)	26 (±1)	29 (±1)
Exchangeable cations					
$[Ca^{2+}]$	cmolc kg^{-1}	6.3 (±0.6)	5.0 (±0.2)	5.9 (±0.1)	6.5 (±0.1)
$[Mg^{2+}]$	cmolc kg^{-1}	2.2 (±0.4)	3.0 (±0.2)	4.5 (±0.2)	4.0 (±0.1)
$[K^+]$	cmolc kg^{-1}	0.6 (±0.1)	0.3 (±0.03)	0.7 (±0.1)	0.6 (±0.02)
$[Na^+]$	cmolc kg^{-1}	0.1 (±0.03)	0.2 (±0.05)	1.5 (±0.3)	0.2 (±0.02)
ECEC	cmolc kg^{-1}	9.19 (±0.7)	8.6 (±0.4)	12.6 (±0.4)	11.2 (±0.1)
ESP	(%)	1.3 (±0.5)	2.3 (±0.8)	3.2 (±1.5)	1.7 (±0.4)
EMgP	(%)	23.6 (±3.8)	35.4 (±1.6)	43.6 (±4.1)	35.7 (±1.9)

[a]Values within parentheses are standard deviation of the mean

38–48 and 1–5%, respectively. High soil EC (>6 dS m^{-1}) and relatively low SAR (<3 $(mmol\ L^{-1})^{0.5}$) indicated that this soil was moderately saline and non-sodic. Further, increasing $CaCO_3$ but not $MgCO_3$ is noted on these saline soils. The soil was abandoned due to the low yield of crops affected by salinity of a moderate level and high content of Cl^- exceeding 0.1% of dry soil.

The survey of the degraded irrigated spots (Ikan D) with poor standings of cotton plants showed poor soil structure, compaction, and low permeability. The soil had a massive structure evident up to 50-cm depth. The soil samples had low organic matter content at 10.2 g kg^{-1}. The quantity of the particles <1-mm size was very high, in the range of 20.9–21.4%.

Among anions, HCO_3^- was dominant followed by SO_4^{2-} and soluble Ca^{2+} among cations followed by Mg^{2+}. The pH of the soil solution was 8–8.2. Exchangeable Ca^{2+} was occupied 56–60%, Mg^{2+} 33–38%, and Na^+ = 1–2% of the ECEC. These data show an increase in the share of exchangeable Mg^{2+} of almost twice that in the degraded soils as compared to the virgin soil. The yield of cotton on this degraded soil was 1.4–1.6 t ha^{-1}. Farmers often abandon these lands due to the high cost of their preparation and low income from crop production. The study focused on the soil with the described soil properties at the site Ikan D.

The data presented in Table 42.1 indicates that the soil with the massive structure (Ikan D) is characterized by high level of particles with 1-mm size, higher than 20% of the dry soil; low EC of soil/water suspension; low content of OM at 10 g kg^{-1}; negligible content of gypsum; dominated by HCO_3^- among anions; and high content of Mg^{2+} in the cation exchange capacity at 35%. Vyshpolsky et al. (2008) made an attempt to link the degradation level of the irrigated soils of the ATC zone with the level of Mg^{2+} on the cation exchange complex. Since the data presented above indicate that the relation is a more complex function, a stability of aggregates in the studied soils was analyzed first, and then an attempt was made to correlate aggregate stability with the physical and chemical properties of soil.

42.3.2 *Aggregate-Size Stability Distribution of Soil Aggregates*

Table 42.2 presents the aggregate-size stability distribution of soil aggregates among different-size fractions as determined after wet sieving. The quantity of stable aggregates was taken as equal to the amount of aggregates after the slaked pretreatment for large macroaggregates, after subsequent slaked for small macroaggregates, and after capillary-wetted pretreatment for microaggregates. The aggregate-size stability distribution shows significant differences in the distribution of the stable large macroaggregates (>2,000 mm) within the soils sampled. The amount of stable large macroaggregates (>2,000 mm) followed the descending order: virgin (Ikan A) silt loam nonsaline soil (7.3%)>silt loam saline abandoned (Ikan C) soil (6.3%)>silt loam nonsaline (Ikan B) irrigated soil (3.6%)>silt loam highly dispersive nonsaline (Ikan D) irrigated soil (1.6%).

Slightly different trends were observed in the distribution of stable small macroaggregates. The amount of small stable macroaggregates observed over the studied soil sampling sites was of the following order: the silt loam low-saline irrigated (Ikan B) soil (8.0%), the highly saline abandoned (Ikan C) soil (6.9%), the virgin (Ikan A) soil (5.2%), and the silt loam highly dispersive (Ikan D) soil (2.1%). The low percent of the stable large macroaggregates on the irrigated soils (Ikan B) as compared to the small stable macroaggregates indicates that the stable large macroaggregates are affected by irrigation practices.

The total amount of the stable macroaggregates (>250 mm), which is the sum of the stable aggregates of size from 250 to 2,000 mm and >2,000 mm, followed the order: the silt loam highly saline abandoned (Ikan C) soil (13.2%)>the silt loam

Table 42.2 The aggregate-size stability distribution for three different soil conditions

		Water pretreatments		Aggregate-size stability distribution		
Size fraction	Slaked	Capillary wetted	Subsequent slaked	Stable	Unstable	Gains
mm	(%)	(%)	(%)	(%)	(%)	(%)
Ikan A						
>2,000	7.3 (12.2)	11.7 (14.1)	3.1 (5.2)	7.3 (2.6)	4.4 (2.3)	
250–2,000	13.6 (22.0)	23.3 (27.0)	5.2 (8.6)	5.2 (1.3)	18.0 (4.4)	8.3 (3.8)
53–250	21.6 (35.0)	16.8 (28.3)		16.8 (0.4)		4.8 (1.5)
<53	29.4 (29.4)	28.5 (28.5)		28.5 (6.0)		2.5 (2.5)
Total	71.9 (98.6)	80.2 (97.9)		TS = 57.7 *T* = 80.2	TU = 22.5	TG = 15.6
Ikan B						
>2,000	3.6 (5.8)	5.4 (6.8)	1.7 (2.7)	3.6 (1.6)	1.8 (0.6)	
250–2,000	15.6 (25.7)	27.2 (33.2)	8.0 (13.1)	8.0 (0.6)	19.2 (1.2)	7.7 (0.4)
53–250	22.0 (36.3)	17.1 (28.3)		17.1 (0.1)		4.9 (0.7)
<53	30.0 (30.0)	30.2 (30.2)		30.2 (1.9)		1.0 (1.0)
Total	71.2 (97.8)	79.8 (98.5)		TS = 58.8 *T* = 79.8	TU = 21.0	TG = 13.6
Ikan C						
>2,000	6.3 (9.5)[a]	9.4 (11.4)	3.1 (4.7)	6.3 (0.6)[b]	3.2 (1.3)	
250–2,000	17.6 (26.4)	27.3 (31.0)	6.9 (10.3)	6.9 (1.8)	20.5 (3.7)	10.8 (3.2)
53–250	22.6 (36.2)	17.0 (27.3)		17.0 (0.1)		5.5 (0.5)
<53	25.2 (25.2)	23.6 (23.6)		23.6 (0.4)		1.6 (0.9)
Total	71.6 (97.3)	77.4 (93.2)		TS = 53.7 *T* = 77.4	TU = 23.7	TG = 17.9
Ikan D						
>2,000	1.6 (2.5)	2.8 (3.6)	0.9 (1.4)	1.6 (0.3)	1.2 (0.8)	
250–2,000	3.3 (5.2)	5.7 (7.7)	2.1 (3.3)	2.1 (0.3)	3.6 (0.5)	1.2 (0.5)
53–250	27.4 (43.2)	29.7 (47.3)		29.7 (1.3)		0 (0.1)
<53	42.6 (42.6)	41.6 (41.4)		41.6 (1.4)		1.0 (0.8)
Total	74.9 (92.7)	79.7 (100)		TS = 74.9 *T* = 79.7	TU = 4.8	TG = 2.2

[a]Values within parentheses are grams of dry weight of soil without sand correction for columns related to water pretreatments

[b]Values within parentheses are standard deviation of mean for columns related to aggregate-size stability distribution

virgin (Ikan A) soil (12.5%) > the silt loam irrigated soil under low-saline (Ikan B) conditions (11.6%) > the silt loam irrigated highly dispersive (Ikan D) soil (3.7%). These data indicate that the quantity of macroaggregates was much less in the case of the degraded soils (Ikan D) than in the other studied soil conditions. The quantity of unstable macroaggregates was calculated by subtracting the amount of stable macroaggregates produced by the slaking treatment from the total amount of macroaggregates obtained from the capillary-wetted treatment. The calculations suggest that the total amount of unstable macroaggregates (>250 mm) followed the order: the silt loam highly saline abandoned soil (Ikan C) (23.7%) > the silt loam virgin soil (Ikan A)

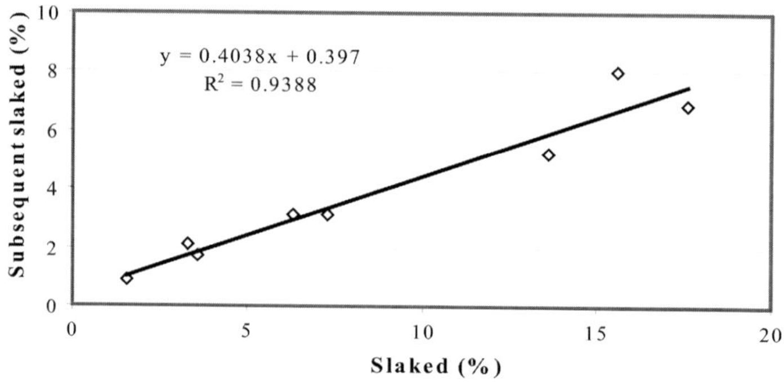

Fig. 42.1 Relationship between the percentage of large macroaggregates >2,000 mm quantified by slaked pretreatment and stable large macroaggregates >2,000 mm quantified by the subsequent slaking pretreatment. Values are expressed as percentages of soil which is dry and on a sand-free basis

(22.5%) > the silt loam irrigated soil under low-saline conditions (Ikan B) (21.0%) > silt loam irrigated highly dispersive soil (Ikan D) (4.8%). These data indicate that 33% of the large macroaggregates determined by capillary-wetted pretreatment was unstable under the silt loam low-saline (Ikan B) conditions, 34% of large macroaggregates was unstable under highly saline soil conditions (Ikan C), 38% was unstable under the virgin silt loam soil conditions (Ikan A), and 43% was unstable under the silt loam highly dispersive (Ikan D) irrigated soil conditions.

Table 42.2 presents the aggregate-size stability distribution for three different soil conditions. Values are data from 2007 to 2008 expressed as percentages of dry weight of soil and on a sand-free basis ±0.1 in each size fraction. TS is the total percentage of stable aggregates, and TU is the total percentage of unstable aggregates. TG is the total gain in aggregates from other fractions, and *T* is total percentage of soil aggregates where *T* = TS + TU.

Different trends were observed with respect to the behavior of microaggregates. The quantity of microaggregates (53–2,500 mm) in ascending order was as follows: virgin soil (Ikan A) (16.8%), abandoned silt loam soil (Ikan C) (17%), silt loam irrigated soil (Ikan B) (17.1%), and silt loam degraded irrigated soil (Ikan D) (29.7%).

The highest propensity for microaggregates and minimum macroaggregates was found on degraded silt loam highly dispersive irrigated (Ikan D) soils. In contrast, silt loam, moderately saline, abandoned soil (Ikan C) had more macroaggregates and less microaggregate than irrigated (Ikan B) soil. In spite of variations in the amount of stable macro or microaggregates in the studied soils, the total quantity of aggregates was in the range of 77.4–80.2% in all the studied soil samples.

The results suggest that highly degraded soil with a massive structure (Ikan D) was characterized by a low amount of stable macroaggregates less than 10% of dry weight of soil; high quantity of stable microaggregates, almost 30% of dry weight of soil; and high quantity of mineral fraction exceeding 40% of dry weight of soil.

The regression analysis between the quantity of large aggregates that withstood the subsequent slaking was highly correlated ($r^2 = 0.9388$) to the amount of large macroaggregates that survived the slaking pretreatment for the studied field sites (Fig. 42.1).

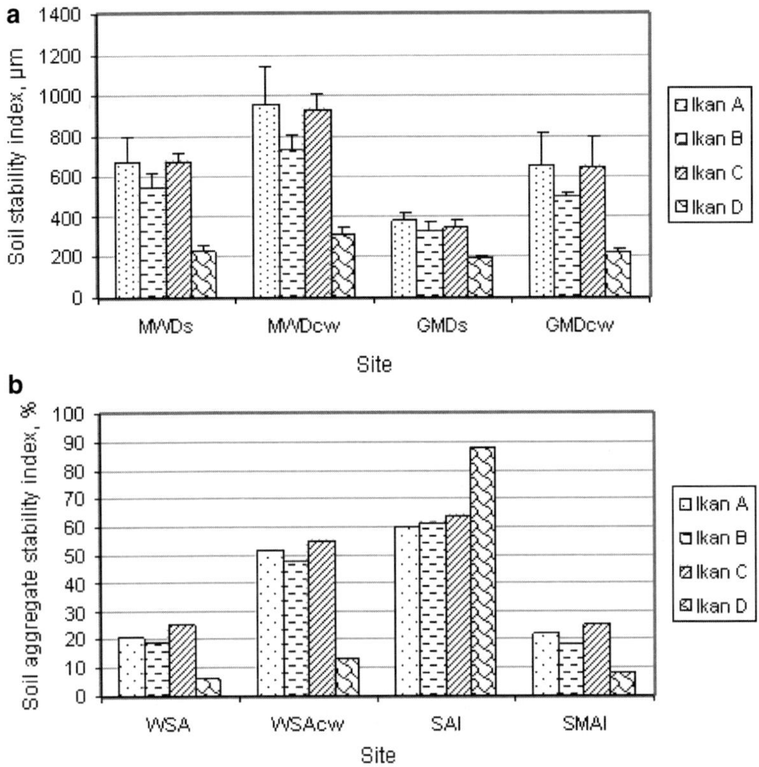

Fig. 42.2 Mean weight diameter of slaked (MWDs) and capillary-wetted (MWDcw), geometric water diameter of slaked (GMDs) and capillary-wetted (GMDcw), water-stable aggregates of slaked (WSAs) and capillary-wetted (WSAcw), and stable aggregate index (SAI) and stable macroaggregate index (SMAI) of soils under different uses of the ATC command zone. *Error bars* represent the standard deviations of means

This relation could be further evaluated for the degraded soils of the ATC command zone to simplify the aggregate stability analysis for practical needs.

Figure 42.2 contains treatment values for different soil aggregate stability indices. Mean weight diameter determined after slaked (MWDs) and capillary-wetted (MWDcw) pretreatments showed clear trends across the studied soils. A large reduction in the MWD on the degraded soils (Ikan D) was observed upon slaking and capillary-wetted pretreatments. The value of MWDs and MWDcw in descending order was as follows: virgin soil (Ikan A) > abandoned silt loam soil under moderate-saline conditions (Ikan C) > silt loam irrigated (Ikan B) > silt loam irrigated under nonsaline conditions (Ikan D). The indices of geometric mean diameter (GMDs, GMDcw) and water-stable aggregates (WSAs, WSAcw) showed similar trends for slaked and capillary-wetted pretreatments.

The values of the indices for the virgin silt loam (Ikan A), the irrigated productive silt loam (Ikan B), and the abandoned silt loam (Ikan C) soils were close to one another. All applied indices indicated the same tendency to decline with respect to the amount of water-stable aggregates at location Ikan D, having highly dispersive gypsum-free soils. Significant differences were observed for the quantity of the stable aggregate index (SAI) between the massive structure soil and the other studied soils. The high value of the SAI for degraded soil (Ikan D) indicates that along with a reduction of macroaggregates, there is an accumulation of microaggregates and mineral fraction in the degraded soils.

Calculation of the indices allowed the quantification of differences in the stability of aggregates between the studied soils (Table 42.3). Highly dispersive soil (Ikan D) had a mean weight diameter (MWD) in the range of 202–256 mm and 280–355 mm after slaking and capillary-wetted pretreatments, respectively. These values are 2–3 times less than the other studied soils. The degraded soil (Ikan D) had a geometric mean diameter (GMD) in the range of 179–201 mm and 200–243 mm after slaking and capillary-wetted pretreatments, respectively. These values are 1.5–4 times less than in the other studied soils. The percentage of water-stable aggregates was less than 8 and 15% on highly dispersive soil (Ikan D) after slaking and capillary-wetted pretreatments, respectively. This index for the other soils was in the range of 15–38% and 41–60%, after slaking and capillary-wetted pretreatments, respectively. The stable aggregate index (SAI) was very high, from 86 to 93% for the dispersive soils against 55–79% for the other soils. The stable macroaggregate index (SMAI) was from 7 to 10% for the dispersive soils against 16–35% for the other soils.

The values of the indices presented in Table 42.3 for the irrigated heavy loam soils (Ikan D) could be applied for specifying the degraded massive structure of soils. In the next section, an attempt was made to correlate the poor stability of soil aggregates represented by the SMAI values with the physical and chemical properties of soil.

42.3.3 Causes of Low Stability of Soil Aggregates

Changes in chemical and physical properties of soil were analyzed as potential factors affecting the stability of soil aggregates. A linear relationship was found between the quantity of soil stable macroaggregates (SMAI) and the physical and chemical properties of the studied soils:

$$\begin{aligned} \mathrm{SMAI_c} &= 0.231 - 0.009 \times \mathrm{p001} + 0.021 \times \mathrm{EC} + 0.026 \times \mathrm{OM} \\ &\quad + 0.044 \times \mathrm{Gypsum} \end{aligned} \tag{42.12}$$

where the quantity of the particles less than 1-mm size is in % of dry soil, EC in dS m^{-1}, and OM and gypsum contents are in g kg^{-1}.

The correlation coefficient (r^2) between the calculated and actual values of the stable macroaggregate index was 0.95, and the mean square error of the correlation coefficient was 0.081.

The dependence obtained indicates that the SMAI was inversely dependent on the quantity of particles with less than 1-mm size. The quantity of the stable

Table 42.3 Values of the soil stability indices for different soil degradation levels in the ATC command zone

Soil stability indices	Unit	Soil Virgin soil (Ikan A)	Irrigated silt loam (Ikan B)	Abandoned silt loam (Ikan C)	Irrigated silt loam (Ikan D)
Mean weight diameter (after slaking pretreatment (MWDs))	mm	483–783	472–645	608–730	201–256
After capillary-wetted pretreatment (MWDcw)	mm	732–1,224	665–830	849–1,004	280–355
Geometric mean diameter: after slaking pretreatment (GMDs)	mm	317–413	282–383	294–382	179–201
After capillary-wetted pretreatment (GMDcw)	mm	472–767	465–520	503–858	200–243
Water-stable aggregates: after slaked pretreatment (WSAs)	(%)	15–26	17–22	17–38	5–7
After capillary-wetted pretreatment (WSAcw)	(%)	41–60	46–49	51–60	11–15
Stable aggregate index (SAI)	(%)	55–66	60–63	51–79	86–93
Stable macroaggregate index (SMAI)	(%)	16–27	15–22	19–35	7–10

Table 42.4 Single-factor sensitivity analysis of the stability of macroaggregates and physical and chemical properties of the soils of the ATC command zone

	Value			SMAI		
Soil property	Initial	Minimum	Maximum	Initial	Minimum	Maximum
Particles 1-mm size	11.8	9.40	22.80	0.221	0.244	0.127
EC (dS m^{-1})	0.71	0.55	10.85	0.221	0.210	0.253
OM (%)	2.55	0.71	3.13	0.221	0.158	0.223
Gypsum (%)	0.38	0.00	0.68	0.221	0.194	0.225

macroaggregates was directly dependent on the EC of the soil and the OM and gypsum contents. The results of the studies did not show an obvious direct effect of the other soil chemical properties including exchangeable Mg^{2+} on the stability of the soil aggregates. However, this aspect requires further analysis.

Single-factor sensitivity analyses were applied using Eq. 42.12 to find out changes in soil macroaggregates stability (SMAI) as affected by changes in the measured soil parameters. Initial values of the soil parameters applied to calculate the SMAI represent the virgin soil (Table 42.4). Minimum and maximum values are levels of the parameters found at one of the other studied soil sites.

The data presented in Table 42.4 show that SMAI was most sensitive to the accumulation of the particles of 1-mm size, followed by a reduction of the OM content and loss of gypsum from the soil profile. These data indicate that irrigation of virgin soil in the ATC zone was followed by increasing the mineral fraction and that leaching the organic matter and gypsum from the soil profile has led to the formation of highly dispersive soils of the ATC zone.

Increasing the quantity of particles with less than 1-mm size is the consequence of the disruption of unstable macroaggregates during irrigation events. However, the chemical properties of the soil may affect the quantity of particles with less than 1-mm size and, that way, on the stability of aggregates of the studied degraded soils. Therefore, Eqs. 42.10 and 42.11 were applied to find a relationship between the quantity of the particles with less than 1-mm size and the chemical properties of the soils. These analyses showed a good linear relationship between the quantity of the particles with less than 1-mm size and the chemical properties of the studied soils:

$$\mathrm{P001_c} = 0.026 + 0.041 \times [\mathrm{Mg}^{2+}] + 0.009 \times \mathrm{SAR} - 0.001 \times \mathrm{SO_4}^{2-} \quad (42.13)$$

Correlation coefficient (r^2) between the calculated and actual quantity of particles with less than 1-mm size was 0.85, and the mean square error of the correlation coefficient was 0.07.

Single-factor analysis using Eq. 42.13 showed that the quantity of particles with less than 1-mm size was the most sensitive to the accumulation of Mg^{2+} on the exchange complex that contributes to increasing the concentration of SO_4^{2-} in the soil solution that retards dispersion of the soil aggregates.

The data presented in Table 42.5 indicates that the quantity of particles with less than 1-mm size is the most sensitive to accumulation of Mg^{2+} on the exchange complex. The increasing quantity of the particles follows increasing exchangeable Mg^{2+}, which is opposite to the quantity of the OM. The relation between SAR of

Table 42.5 Single-factor sensitivity analysis of the quantity of particles with less than 1-mm size and chemical properties of the soils of the ATC command zone

	Value			$P001_c^a$, (%)		
Soil property	Initial	Minimum	Maximum	Initial	Minimum	Maximum
Mg^{2+} (cmolc kg^{-1})	2.32	1.58	6.51	11.0	9.0	31.3
SO_4^{2-} (mmolc L^{-1})	17.30	3.30	71.90	11.0	12.2	5.9
SAR (mmol $L^{-1})^{0.5}$	0.11	0.06	3.40	11.0	10.9	11.2

[a]Means the calculative quantity of particles with less than 1-mm size, % indicates % of dry soil, initial means initial for the virgin soil, and minimum and maximum mean minimum and maximum levels of the soil parameters

the soil solution and the quantity of the particles with less than 1-mm size was minor for the studied soils.

42.4 Conclusions

The study found that highly degraded irrigated soils with massive structures and subject to dispersion can be identified by determining water-stable aggregates in the topsoil. The SMAI was found to be most suitable to analyze the casual factors contributing to the problem. The SMAI values of less than 10% were found to be indicative of soil that is highly dispersive. A linear relationship was found between the quantity of SMAI, the quantity of particles with less than 1-mm size, EC, OM, and gypsum content in the topsoil. Irrigation of nonsaline virgin soil having low EC at 1 dS m^{-1}, followed by the leaching OM and gypsum from the soil profile and accumulation of particles with less than 1-mm size, was found to be associated with the formation of massive structures under conditions where there was no perched water table. Accumulation of particles with less than 1-mm size was associated with increasing Mg^{2+} on the exchange complex and lowering the concentration of SO_4^{2-} in the soil solution. Findings of this study could be used to determine alternative technologies to ameliorate degraded soils with massive structures. Alternative technologies could be water-saving technologies to reduce slaking and disruption effect of irrigation water on soil aggregates, temporary raising of the water table in early summer and subirrigation, groundwater use in conjunction with canal water and increasing EC of the soil solution, application of amendments such as phosphogypsum, and increasing the gypsum content in the topsoil.

References

Angers DA, Mehuys G (1993) Aggregate stability to water. In: Carter MR (ed) Soil sampling and methods of analysis, Canadian society and soil science. Lewis Publ, Boca Raton, pp 651–657

Arinushkina EV (1970) Handbook for chemical analysis of soil (in Russian). Publishing House of the Moscow State University, Moscow, p 489

Bodruhina AG (1973) Saline soil leaching in northern part of Yavan valley. In: Proceedings of the Tadjik Research Institute of Soil Science Dushanbe 16:61–74

Boix-Fayos CA, Calvo-Cases AC, Imeson AC, Soriano-Soto MD (2001) Influence of soil properties on the aggregation of some Mediterranean soils and the use of aggregate size and stability as land degradation indicators. Catena 44:47–67

Brooke A, Kendrick D, Meeraus A (1992) GAMS: a user's guide: release 2.25, The Scientific Press series. Boyd and Fraser Publishing Co, Danvers

Cambardella CA, Elliot ET (1993) Methods for physical separation and characterization of soil organic matter fractions. Geoderma 56:449–457

Dj Bairov A, Tashkuziev MM, Kurbantaev R, Karimberdiev AA, Sattarov DS (2004) Manual for soil chemical and agro-physical analysis for soil monitoring (in Russian). State Committee of Uzbekistan on Land Resources, Tashkent, p 259

Dontsova KM, Norton LD (2002) Clay dispersion, infiltration and erosion as influenced by exchangeable Ca and Mg. Soil Sci 167:184–193

Elliot ET (1986) Aggregate structure and carbon, nitrogen, and phosphorus in native and cultivated soils. Soil Sci Soc Am J 50:627–633

Kachinski NA (1965) Soil physics, vol 1. High School, Moscow (In Russian)

Kachinski NA (1970) Soil physics, vol 2. High School, Moscow (In Russian)

Kemper WD (1966) Aggregate stability of soils from western United States and Canada. USDA-ARS, New Mexico

Kemper WD, Rosenau RC (1986) Aggregate stability and size distribution. In: Klute A (ed) Methods of soil analysis. Part 1, 2nd edn, Agronomy monograph 9. ASA, Madison, pp 425–442

Keren R, Ben-Hur M (2003) Interaction effects of clay swelling and dispersion and $CaCO_3$ content on saturated hydraulic conductivity. Aust J Soil Res 41:979–989

Kjaergaard C, de Jonge LW, Moldrup P, Schjønning P (2004) Water-dispersible colloids: effects of measurement method, clay content, initial soil matric potential, and wetting rate. Vadoze Zone J 3:403–412

Lebron I, Suarez DC, Alberto F (1994) Stability of a calcareous saline-sodic soil during reclamation. Soil Sci Soc Am J 58:1753–1762

Levy GJ, Mamedov AI (2002) High-energy-moisture-characteristic aggregate stability as a predictor for seal formation. Soil Sci Soc Am J 66:1603–1609

Levy GJ, Mamedov AI, Goldstein D (2003) Sodicity and water quality effects on slaking of aggregates from semi-arid soils. Soil Sci 168:552–562

Levy GJ, Goldstein D, Mamedov AI (2005) Saturated hydraulic conductivity of semiarid soils: combined effects of salinity, sodicity, and rate of wetting. Soil Sci Soc Am J 69:653–662

Marquez CO, Garcia VJ, Cambardella CA, Schultz RC, Isenhart TM (2004) Aggregate-size stability distribution and soil stability. Soil Sci Soc Am J 68:725–735

Mazurak AP (1950) Effect of gaseous phase on water-stable synthetic aggregates. Can J Soil Sci 70:33–42

McNeal BL, Layfield DA, Norwell WA, Roades JD (1968) Factors influencing hydraulic conductivity of soils in the presence of mixed-salt solutions. Soil Sci Soc Am Proc 32:187–190

Remezov NP (1957) Soil colloids and soil absorption complex (in Russian). State Publishing House of Agricultural Literature, Moscow, p 223

Ruiz-Vera, Victor M, Laosheng Wu (2006) Influence of sodicity, clay mineralogy, prewetting rate, and their interaction on aggregate stability. Soil Sci Soc Am J 70:1825–1833

Six J, Elliott ET, Paustian K, Doran J (1998) Aggregation and soil organic matter accumulation in cultivated and native grassland soils. Soil Sci Soc Am J 62:1367–1377

Six J, Paustian K, Elliot ET, Combrink C (2000a) Soil structure and organic matter: I Distribution of aggregate-size classes and aggregate-associated carbon. Soil Sci Soc Am J 64:681–689

Six J, Elliot ET, Paustin K (2000b) Soil structure and soil organic matter: II. A normalized stability index and the effect of minerology. Soil Sci Soc Am J 64:1042–1049

Suarez DL, Rhoades JD, Lavado R, Grieve CM (1984) Effect of pH on saturated hydraulic conductivity and soil dispersion. Soil Sci Soc Am J 48:50–55

Tedeshi A, Dell'Aquila R (2005) Effects of irrigation with saline waters, at different concentrations, on soil physical and chemical characteristics. Agric Water Manag 77:308–322

USDA (1998) Soil quality test kid guide. USDA-ARS, Washington, DC

Van Bavel C (1949) Mean weight diameter of soil aggregates as a statistical index of aggregation. Soil Sci Soc Am Proc 14:20–23

Vyshpolsky F, Qadir M, Karimov A, Mukhamedjanov K, Bekbaev U, Paroda R, Aw-Hassan A, Karajeh F (2008) Enhancing the productivity of high-magnesium soil and water resources through the application of phosphogypsum in Central Asia. Land Deg Dev 19:45–56

Yilmaz K, Celik I, Kapur S, Ryan J (2005) Clay minerals, Ca/Mg ratio and Fe-Al-oxides in relation to structural stability, hydraulic conductivity and soil erosion in Southeastern Turkey. Turk J Agric For 29:29–37

Part IV
Modeling of Soil and Groundwater Contamination

Chapter 43
Application of a Screening Model to Evaluate Pesticide Contamination in Soil and Groundwater for Sustainable Agriculture in Oman

Basma Nasser Al-Shidhani, Mushtaque Ahmed, Salem Al-Jabri, and Farid Talukder

Abstract In the recent years, agricultural activities in Oman have been increased significantly that has increased the use of pesticides. Consequently, soil and groundwater quality is likely to deteriorate. This chapter presents prediction of pesticide movement through the unsaturated zone to groundwater. This will help the decision makers in assessing the likelihood of soil and groundwater contamination in Oman under various land use practices. A model (PESTAN) based on the analytical solution of 1-D advective-dispersive-reactive transport equation is used to estimate the vertical migration of the dissolved organic solutes as well as for conducting initial screening assessment of the potential for contamination of soil and groundwater. PESTAN calculates the movement of organic chemicals with a linear isotherm, first-order degradation, and hydrodynamic dispersion. Input data include the following: soil, chemical, and management practice parameters. Data on Omani soil, climatic, and irrigation practices as well as chemical parameters of most common pesticides used are required for modeling. Reliable assumptions are made to compensate for missing data. The main objective was to model the top ten pesticides used in Oman for 5 years to calculate the time each particular pesticide would take to reach the water table that is estimated to be at 10-m underground surface. Three scenarios were developed, after selecting three variables which mainly contribute to pesticides' fate: recharge, soil texture, and application rate of pesticides. For each variable, maximum and minimum values were modeled to be compared with a base run describing the recommended conditions. The results suggest that the more the recharge and application rates, the faster the pesticides reach groundwater. In coarse-textured soil, the pesticides penetrate faster through the soil profile. Overall, the simulations demonstrated that none of the ten pesticides is reaching the water table

B.N. Al-Shidhani (✉) • M. Ahmed • S. Al-Jabri • F. Talukder
College of Agricultural and Marine Sciences, Sultan Qaboos University,
P.O. Box 34, Al-Khod 123 Muscat, Oman
e-mail: basmaalshidhani@gmail.com

S.A. Shahid et al. (eds.), *Developments in Soil Classification, Land Use Planning and Policy Implications: Innovative Thinking of Soil Inventory for Land Use Planning and Management of Land Resources*, DOI 10.1007/978-94-007-5332-7_43,

to a depth of 10 m within 5 years by assuming a single application, where application frequency is concerned. Reducing irrigation amount and increasing the organic matter content of the soil will help reduce groundwater contamination.

Keywords Contamination • Oman • PESTAN • Pesticides • Recharge estimation

43.1 Introduction

Oman being an arid country due to drought conditions and rainfall scarcity depends mainly on groundwater as the main source of water. This situation has prompted the government to improve the standard of water management, conservation, and augmentation in Oman. Protecting groundwater is a crucial obligation on the part of the government and the population. Its cleanup process is costly and time consuming. With regard to groundwater contamination by agricultural activities, Chilton et al. (1995) found that intensive agriculture could produce a serious deterioration in groundwater quality. Animal waste, inorganic fertilizers, irrigation return flows, and pesticides are four major sources of agricultural pollutants.

A pesticide can be defined as any substance or mixture of substances intended for preventing, destroying, repelling, or mitigating any pest. Pesticides are considered as diffuse sources of groundwater contamination. Even minute concentrations can have serious consequences regarding the potability of groundwater. Analytical test of water in the USA revealed at least 17 pesticides have been detected in groundwater in 23 states as a result of agricultural practice in 1986 (Garner et al. 1986). A local study on pesticide use and groundwater quality in Al-Batinah region of Oman was conducted by MSc student in soil and water management showed that over 22 water supply wells contained no pesticide traces (Al-Masehli 2003). However, it is essential to monitor and assess the likelihood of groundwater contamination by pesticides or any other chemicals to ensure that agricultural practices are not affecting groundwater quality.

The main objective of this study is to apply a screening model for evaluating pesticide movement within the vadose (unsaturated) zone, where most pesticide concentrations are held. The risk of pesticides leaching to the water table is valid especially in areas where the water table is shallow and irrigation is excessive.

Scientists have employed modeling techniques to understand the flow and transport dynamics in aquifers (Bear 1972; Anderson and Woessner 1992). Unsaturated zone modeling has also been widely used. To meet the objective of this study, a model based on unsaturated flow is used.

43.1.1 Model Selection

Before developing a model to solve the problem, it is important to understand what best suits the need of targeted study. There are many packages available in the market; some are offered free by well-known organizations like the United States Environmental Protection Agency (USEPA).

In this study, it was rational to select a 1-dimensional model for a couple of reasons; on top of them, the main concern is focusing on developing an initial assessment of the potential for contamination of soil and groundwater. There are other models besides PESTAN which could be useful such as Hydrus1D or Scigrow2.3. However, the lack of data constrained their use and has led to go for a less sophisticated model. Comparing the two models with PESTAN, Hydrus1D is a data-demanding model, whereas Scigrow2.3 is very simplistic. PESTAN on the other hand allows compensating the missing data with reliable assumptions.

43.2 Methodology

43.2.1 PESTAN Summary

PESTAN (pesticide analytical) model is a computer code for estimating the transport of organic solutes through soil to groundwater. The model is based on a closed-form analytical solution of the advective-dispersive-reactive transport equation. The model was developed to be used for initial screening assessments to evaluate the potential for groundwater contamination by pesticides. PESTAN has been tested under field and laboratory conditions. Although the model is based on a simple analytical solution, it may be useful in making preliminary or risk assessments as long as the user is fully aware of its assumptions and limitations.

43.2.1.1 Assumptions and Limitations

1. PESTAN conceptualization assumes the leachate concentration equals the maximum possible concentration.
2. The slug enters the soil at the velocity of the pore water, which is the ratio of the recharge rate to the pore-water content.
3. Steady-state flow conditions are assumed in the code.
4. Homogeneous soil conditions are assumed in the model.
5. Linear isotherms describe the partitioning of the pollutant between the liquid and soil phases. Local or instantaneous equilibrium between these phases is assumed.
6. First-order degradation of the pollutant is assumed.
7. The water content of the soil is related to the hydraulic conductivity as described by Campbell (1974):

$$\frac{K}{K_{\text{sat}}} = \left(\frac{\theta}{\theta_{\text{sat}}}\right)^{2b+3} \tag{43.1}$$

where K = hydraulic conductivity, θ = volumetric water content, K_{sat} = hydraulic conductivity of soil at saturated water content, and θ_{sat}, b = characteristic curve coefficient for soil. This relationship assumes steady-state conditions for the flow.

8. The model does not account for nonaqueous phase liquids or any flow conditions derived from variable density.

43.2.1.2 Solute Transport Theory in PESTAN

A brief discussion of the mathematical development and important aspects of the model is presented below.

The vertical transport of a pollutant dissolved in water through the soil can be described by the following equation:

$$\frac{\partial C}{\partial t} = D\frac{\partial^2 C}{\partial X^2} - v\frac{\partial C}{\partial X} - \frac{\rho_b}{\theta}\frac{\partial S}{\partial t} - k_l C \tag{43.2}$$

where C=liquid-phase pollutant concentration (mass of pollutant in water/volume of water) (M/L3), t=time (T), x=distance along the flow path (L), D=dispersion coefficient (L2/T), v=interstitial or pore-water velocity (L/T), ρ_b=bulk density (M/L3), q=volumetric water content (volume of pore water/total volume) (L3/L3), S=solid-phase concentration (mass of pollutant in soil/mass of soil) (M/M), and K_l=first-order decay coefficient in liquid phase (T^{-1}).

The term ∂S, t is the rate of loss of solute from liquid phase to solid phase due to sorption. Under the assumption of linear, instantaneous sorption, ∂S, t can be evaluated as follows:

$$\frac{\partial S}{\partial t} = k_d \frac{\partial C}{\partial t} \tag{43.3}$$

where k_d=linear Freundlich sorption coefficient.

Slug thickness can be calculated by the following equation:

$$x_o = \frac{M_a \exp(-k_s t_r)}{S(\theta + k_d \rho_d)} \tag{43.4}$$

where x_o=slug thickness in (L), M_a=total pollutant mass applied per unit area (M/L2), k_s=solid-phase decay coefficient (T^{-1}), t_r=time lapse between application and recharge (T), and S=solubility of pollutant in water (M/L3).

43.2.1.3 Conceptual Framework

From the conceptual sketch (Fig. 43.1), it is clear that our zone of interest in this study is the vadose zone. It represents the medium where pesticide reactions that contribute in soil and groundwater contamination take place. The input parameters in the system include the amount of recharge, pesticide used and its application rate, and the soil texture in which this process takes place. The output describes the

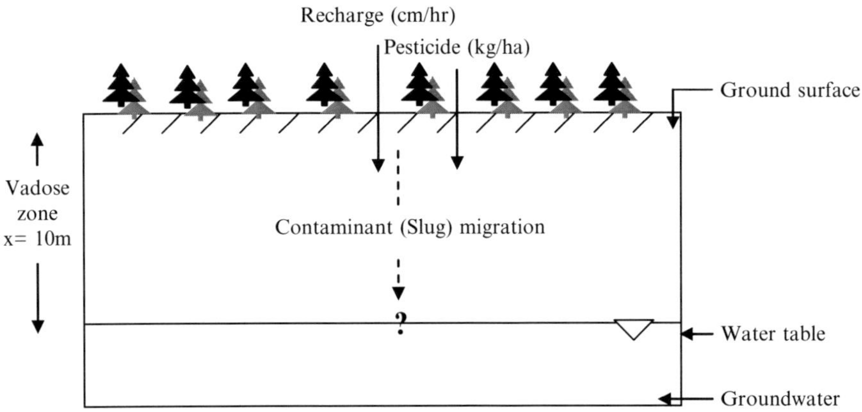

Fig. 43.1 Schematic drawing of the study problem

length of the slug that represents the pesticide fate, its velocity and concentration in a particular depth and time. The main interest is the concentration and depth at which the pesticide reaches. The following are the ten pesticides most commonly used in Oman: abamectin, carbendazim, chlorpyrifos, cypermethrin, decis, dimethoate, glyphosate, malathion, mancozeb, and metalaxyl. They were modeled in PESTAN for over 5 years after a single application of recommended dose to see whether they will reach the water table that is 10m below ground surface within this period of time.

Pesticide fate is a complicated issue, because not all of pesticide's concentration ends up in killing the targeted pest. Some of the concentrations may volatilize in air or might get absorbed by the crop and topsoil; as well they can reach surface water or groundwater through soil profile. Fate processes can be separated into three major types: adsorption, which binds pesticides; transfer processes, which move pesticides; and degradation processes, which break pesticides down. Nature of pesticide, soil type, climatic factors, and handling practices can promote or prevent each process. An understanding of the fate processes can help ensure that pesticide applications are not only effective but also environmentally safe. Through modeling by trying different scenarios, understanding the fate process will become clear and manageable.

43.2.1.4 Modeled Scenarios

Guidelines followed in organizing three different scenarios are presented below:

1. By looking at PESTAN model sheet, it is clear that inputs required by PESTAN are divided into five main categories:

   (a) Reliable assumed values
   (b) Simulation parameters

(c) Pesticide-related parameters
(d) Soil-related parameters
(e) Management practices

2. From PESTAN model sheet, some parameters will be distinguished as variable parameters that can be controlled, and the rest will be constant parameters. Variable parameters are:

   (a) Recharge (maximum, recommended, minimum)
   (b) Soil texture (loamy sand, sandy loam, loam)
   (c) Application rate (maximum, recommended, minimum)

3. For each pesticide, the attempted scenarios will be run according to the maximum and minimum values for the selected variables except for soil texture in which there is no maximum or minimum; instead, three soil groups will be simulated. Along there will be a base run describing the optimum or recommended condition for comparison purpose; consequently, the number of simulations will be 90 for all ten pesticides.
4. In order to understand how the selected variables play a role in controlling a pesticide's fate, the following statements are considered:

   (a) First scenario: How does recharge affect the pesticide movement and concentration in soil and groundwater?
   (b) Second scenario: How does application rate affect the pesticide movement and concentration in soil and groundwater?
   (c) Third scenario: How does soil texture affect the pesticide movement and concentration in soil and groundwater?

5. By taking each problem or variable at a time leaving everything else constant, the effect of a particular parameter on pesticide's fate will be clear.

In the next few paragraphs, some of the important parameters are explained.
Recharge Estimation
I: irrigation, *R*: rainfall, ET: evapotranspiration, and LF: leaching fraction

$$\text{(43.5) Recharge} = I + R - \text{ET}(I = \text{ET} \times \text{LF}) \rightarrow \text{Recharge} = \text{ET} \times \text{LF} + R - \text{ET}$$

Assuming constant R over the year = 100 mm year^{-1} = 0.274 mm day^{-1} (averaged value)
Summer: ET = 7.6 mm day^{-1}
Winter: ET = 3.3 mm day^{-1}
For this study, the simulations will be run for only summer:

$$\boxed{\text{Recharge in summer} = 7.6 \times \text{LF} + 0.0274 - 7.6}$$

LF values for:
Maximum recharge LF = 150% ET
Minimum recharge LF = 100% ET
Recommended recharge LF = 120% ET
Therefore,
Maximum recharge = 7.6 × 1.5 + 0.274–7.6 = 4.074 mm day^{-1} = 0.0170 cm h^{-1}

Minimum recharge = 7.6 × 1 + 0.274 – 7.6 = 0.274 mm day^{-1} = 0.0011 cm h^{-1}
Recommended = 7.6 × 1.2 + 0.274 – 7.6 = 1.794 mm day^{-1} = 0.0075 cm h^{-1}

Water Table Depth

Minimum X-value to be 0 cm referring to the ground surface level and the maximum X-value to be 1,000 cm as the water table depth below ground surface which is quite reasonable for the Batinah area which is the most important agricultural area of Oman.

Simulation Parameters

This part depends on the modeler preferences; in this study, the model was run at:

- A maximum of 5 years = 1,825 days.
- The output time interval could be narrowed or expanded, but a maximum of ten intervals this software can present.
- Leachate breakthrough curve presents the concentration of leachate in the soil at a specified depth and time.
- Leachate total mass flux graph presents the mass flux passing through the specified depth and time.
- Soil concentration profile presents the concentration profile in the soil, along the soil column, at the specified time. The specified depth in this case is 10 cm, and time is set at the end of simulation 1,825 days.

Pesticide-Related Parameters

Once the pesticide is selected for simulation, the relevant parameters can be found from different pesticide databases from the web (www.syngenta.com.au; http://npic.orst.edu/ppdmove.htm). Table 43.1 provides all information about pesticide properties for all ten pesticides used for this study.

Soil-Related Parameters

Soil texture is an important parameter. Once the soil texture is known, the rest of the information regarding this parameter can be found in Table 43.2. PESTAN manual index provides all necessary data for many soil textures, but this study is focused only on three types of soils that Table 43.3 summarizes.

Management Practices

For this section, the number of applications can be set as one. In reality, farmers are applying a lot more than that, but for the sake of simplicity (as the interest is in the

Table 43.1 Pesticide properties database

Common name	Pesticide movement rating	Soil half-life $t_{0.5}$ (days)	Water solubility (mg L^{-1})	Organic carbon partition coefficient Koc (ml g^{-1})	Sorption constant K_d=Koc × foc ml g^{-1} (cc g^{-1})	Decay rate K=0.693/t 0.5 (1 h^{-1})	EC active ingredient (g L^{-1})	Recommended application rate (kg ha^{-1})	Maximum application rate (kg ha^{-1}) 1.5 times recommended	Minimum application rate (kg ha^{-1}) 0.5 times of recommended
Abamectin	Very low	28	5	5,000	50.000	0.001031	18	0.021	0.0315	0.0105
Carbendazim	Moderate	120	8	400	4.000	0.000241	500	0.275	0.4125	0.1375
Chlorpyrifos	Very low	30	0.400	6,070	60.700	0.000963	239	1.116	1.6740	0.5580
Cypermethrin	Extremely low	30	0.004	100,00	1,000.000	0.000963	100	0.050	0.0750	0.0250
Decis	–	14	0.002	4.6	0.046	0.00206	24	0.215	0.3225	0.1075
Dimethoate	Moderate	7	39,800	20	0.200	0.004125	400	0.340	0.5100	0.1700
Glyphosate	Extremely low	47	900,000	24,000	240.000	0.000614	450	3.240	4.8600	1.6200
Malathion	Extremely low	1	130	1,800	18.000	0.028875	958	1.340	2.0100	0.6700
Mancozeb	Low	70	6	2,000	20.000	0.000413	1,485	3.375	5.0625	1.6875
Metalaxyl	Very high	70	8,400	50	0.500	0.000413	480	0.240	0.3600	0.1200

Assume solid-phase decay rate = liquid-phase decay rate

Assumed dispersion coefficient be constant = 0.336 cm^2 h^{-1} (Al-Masehli 2003)

For most Omani, soil organic content does not exceed 1% so the organic carbon content = 0.01 ml g^{-1}

EC: active ingredient is important in calculating the actual application rate of the constituent pesticide excluding the inert ingredients from the whole product

Table 43.2 Soil properties database

Soil type	Bulk density g cm^{-3}	Saturated water content	Characteristic curve coefficient	Saturated hydraulic conductivity cm h^{-1}
Loamy sand	1.55	0.410	4.38	56.28
Sandy loam	1.47	0.435	4.90	12.48
Loam	1.35	0.451	5.39	2.50

Table 43.3 List of input parameter values

Summary of the constant values to be used:
Recharge:
Maximum = 0.0170 cm h^{-1}
Minimum = 0.0011 cm h^{-1}
Recommended = 0.0075 cm h^{-1}
Water table:
Maximum = 1,000 cm
Simulation time:
Maximum = 1,825 day (5 years)
Dispersion coefficient = 0.336 cm^2 h^{-1}
Organic carbon content (foc) = 0.01 ml g^{-1}
Number of applications = 1 application

Table 43.4 Pesticide application rates

Pesticides	Active ingredient (AI) g L^{-1}	Application rate (AR) ml ha^{-1}	Application rate in kg ha^{-1}
Abamectin	18	1,200	0.0216
Carbendazim	500	550	0.2750
Chlorpyrifos	239	4,671	1.1100
Cypermethrin	100	500	0.0500
Decis	24	8,958.3	0.2150
Dimethoate	400	850	0.3400
Glyphosate	450	7,200	3.2400
Malathion	958	1,400	1.3400
Mancozeb	750 g kg^{-1}	4.5 kg ha^{-1}	3.3750
Metalaxyl	480	500	0.2400

relative potential for groundwater contamination by the ten pesticides), we opted to model only one application.

The application rates data for each pesticide are calculated using pesticide labels provided by an Australian website (Syngenta) where most pesticides in Oman come from. Each label provides the application rates in different units. PESTAN operates in kg ha^{-1}; therefore, Table 43.4 shows all calculations used to get the ultimate application rates. These application rates can be considered as recommended rates.

The starting time here is meant by time of application prior to recharge. It is usually a range of few days; 2 days is a convenient starting time. Table 43.2 lists some other input parameters to be used in the simulation process.

43.3 Results and Discussion

In the first scenario (where recharge is the variable), results suggest that three out of ten pesticides are reaching the water table (Fig. 43.2). They are the following: decis, dimethoate, and metalaxyl. However, considering their extremely low concentrations, the possibility of them reaching the water table in detectable concentrations within 5 years is almost nil. The breakthrough curves (Fig. 43.3) show total concentrations of pesticides at 10-cm depth at different times.

In the second scenario (where application rate is the variable), results suggest that two out of ten pesticides have the potential to reach the water table (Fig. 43.4) and they are decis and dimethoate. They also reach the water table with extremely low concentrations. Realistically speaking, we can safely assume that nothing reaches the water table within 5 years after a single application at the recommended rate.

In the third scenario (where soil texture is the variable), results suggest that two out of ten pesticides are reaching the water table (Fig. 43.5). They also reach the water table with undetectable concentrations, which means nothing reaches the water table within 5 years.

Generally results show that the more the recharge and application rates, the faster the pesticides reach groundwater. In coarse-textured soil, the pesticides penetrate faster through the soil profile. The effect of increasing the organic matter delays the pesticide arrival to the groundwater as shown by simulating Decis (Fig. 43.6).

43.4 Conclusions and Recommendations

Overall, the simulations demonstrated that all ten pesticides are not reaching the water table within 5 years after a single application. However, dimethoate, decis, and metalaxyl show a faster movement than the others. As such, farmers should be extra cautious when dealing with these pesticides especially since the farmers apply more than one application in a rate higher than the recommended dose. For management of pesticide contamination problem, we can reduce irrigation amount and increase the organic matter content of the soil.

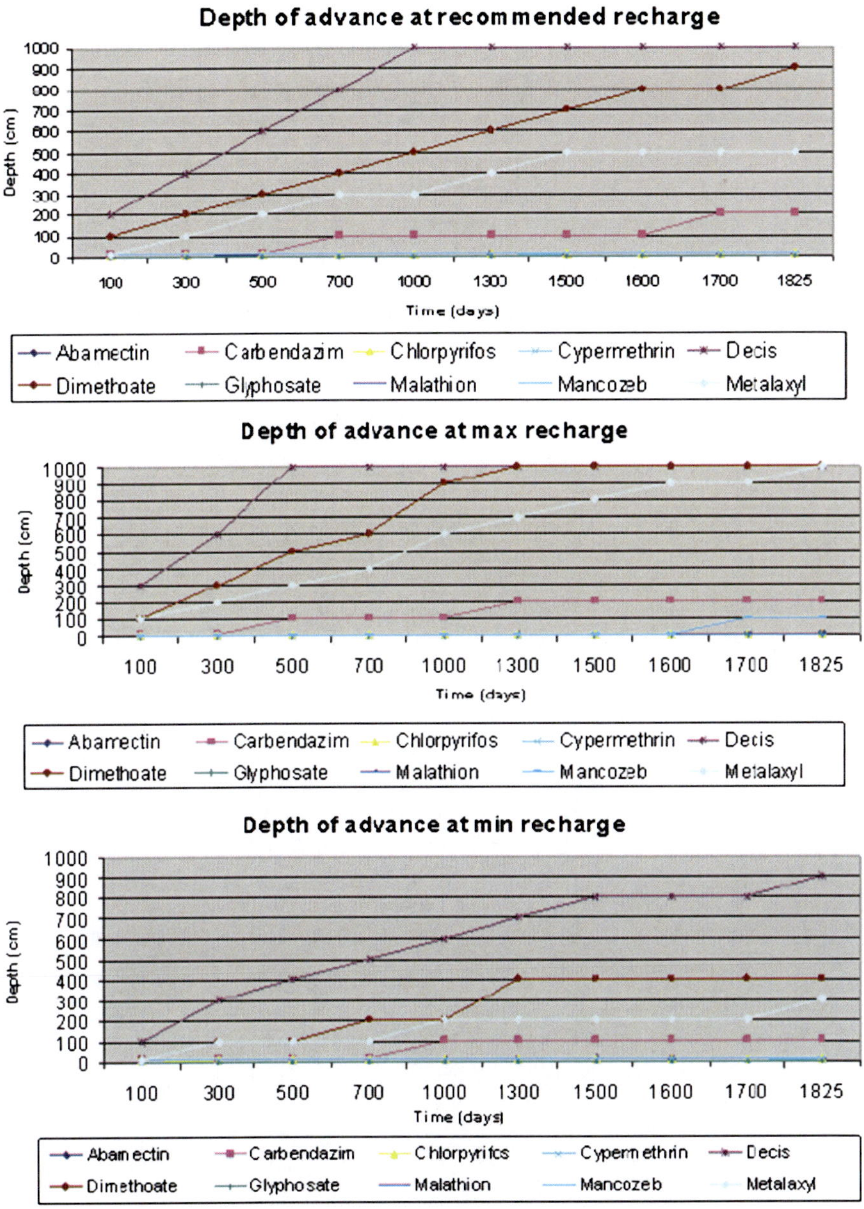

Fig. 43.2 Depth of pesticides advance at three different recharge rates

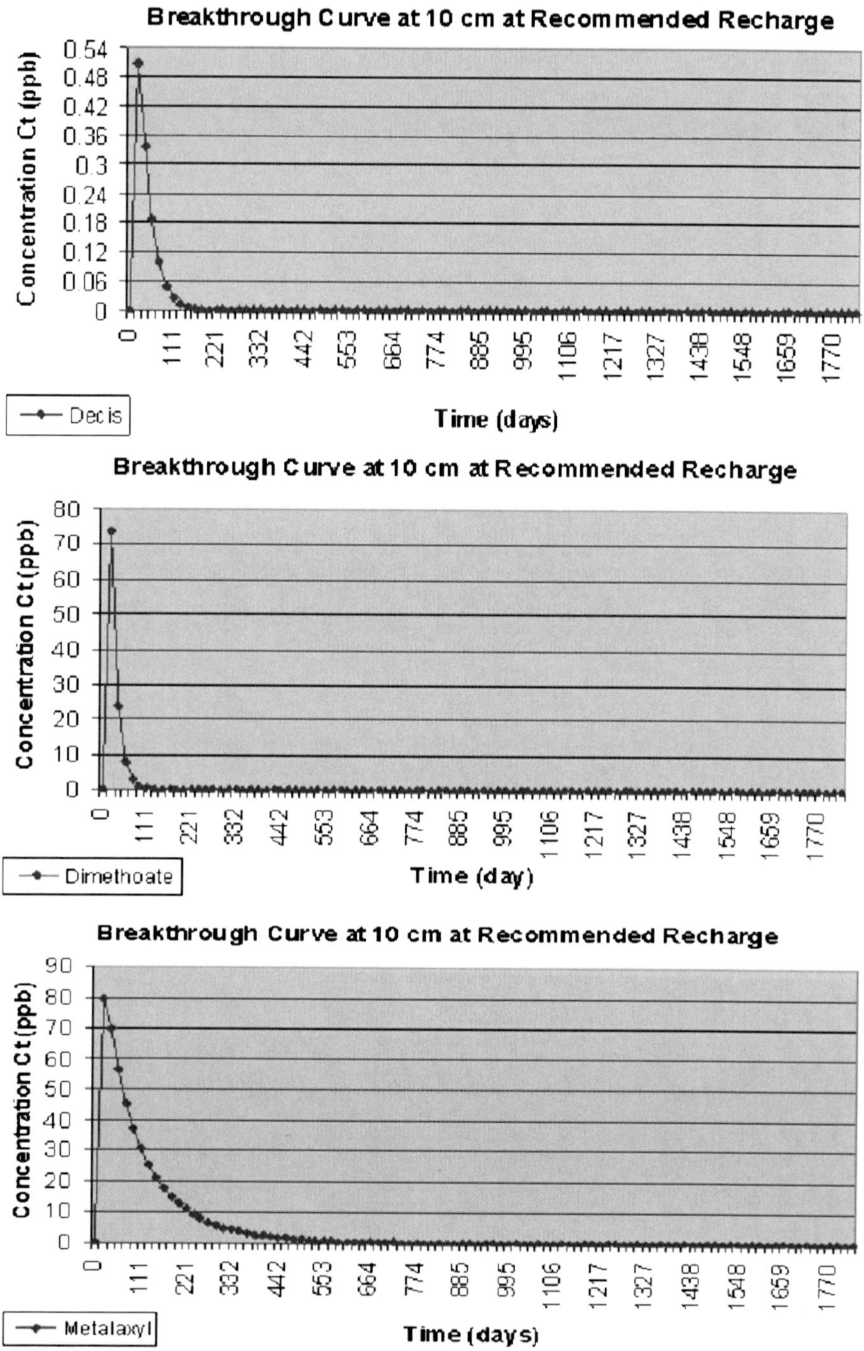

Fig. 43.3 Breakthrough curve at 10 cm at different recharge rates

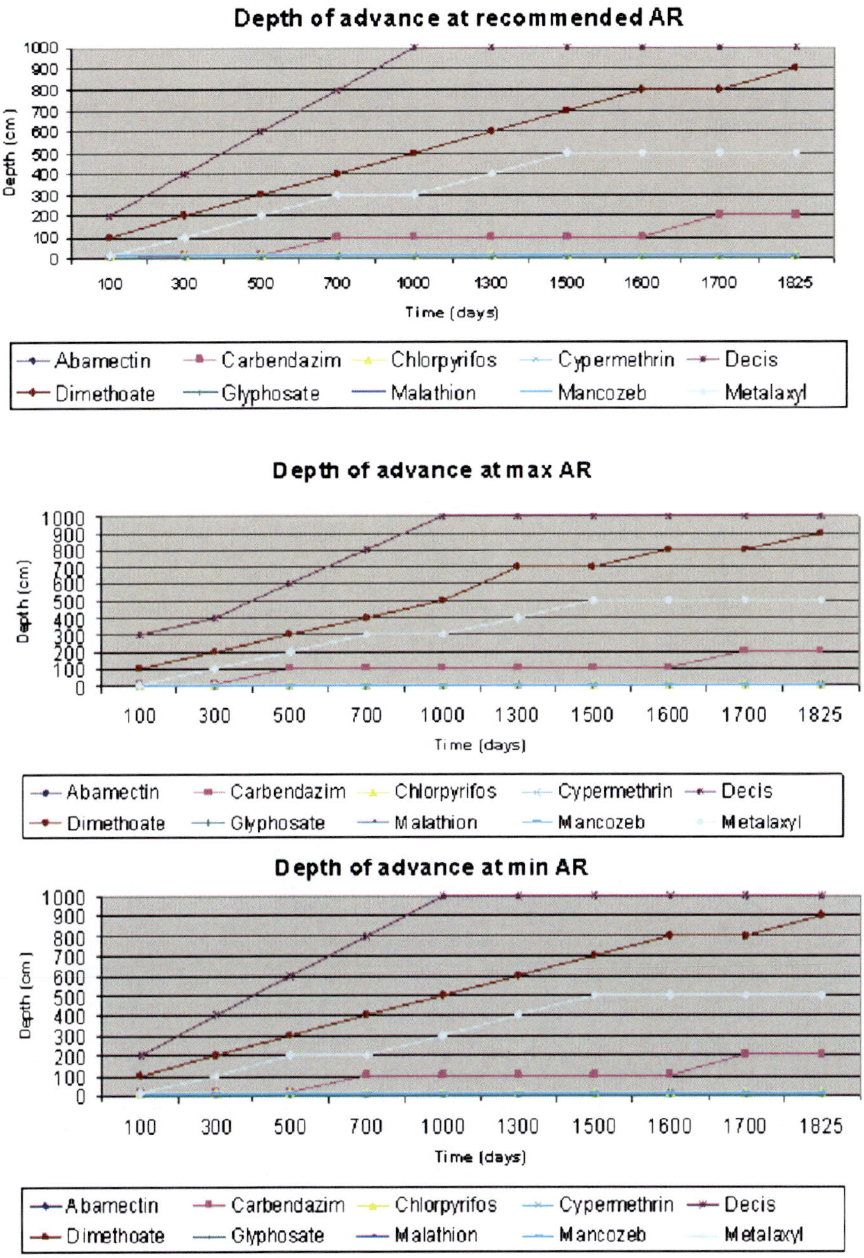

Fig. 43.4 Depth of advance at different application rates

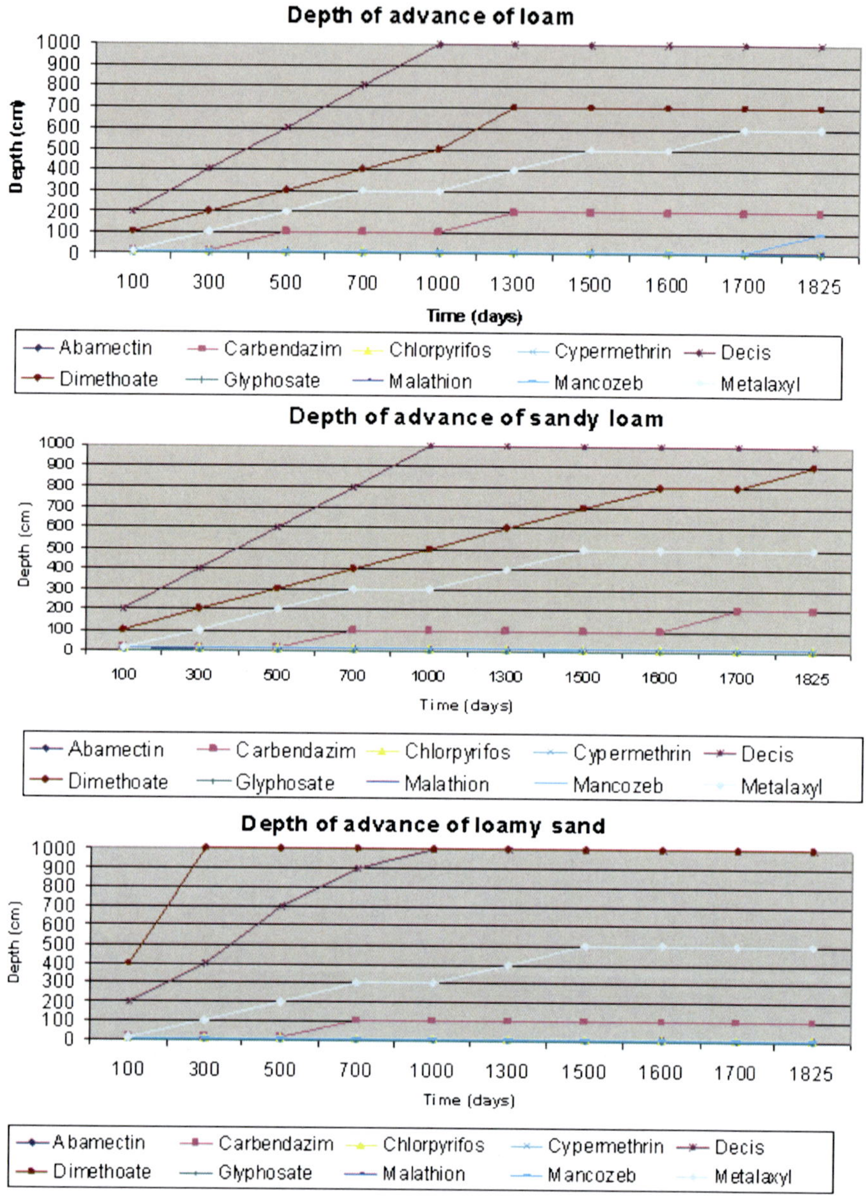

Fig. 43.5 Depth of advance of pesticides in different soils

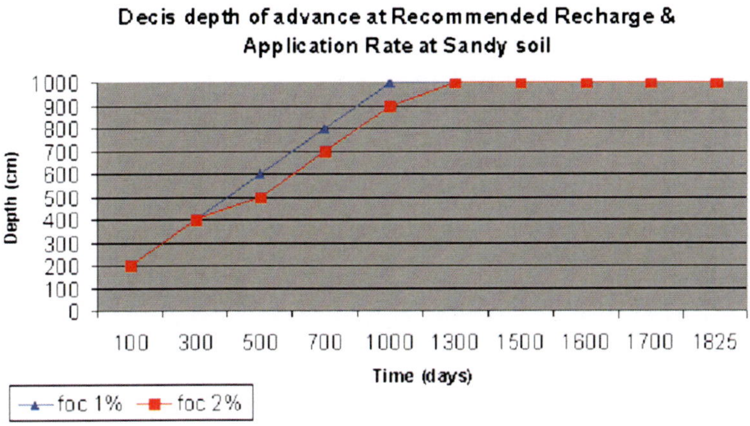

Fig. 43.6 Depth of advance (decis) at recommended recharge and application rate in sandy soil with different organic matter content

References

Al-Masehli M (2003) Pesticide use and groundwater quality in Al-Batinah region of Oman. Unpublished MS thesis, Sultan Qaboos University, Sultanate of Oman

Anderson MP, Woessner WW (1992) Applied groundwater modeling. Academic, San Diego

Bear J (1972) Dynamics of fluids in porous media. American Elsevier, New York

Campbell GS (1974) A simple method for determining unsaturated conductivity from moisture retention data. Soil Sci 117(6):331–314

Chilton PJ, Jegat HJ, Stuart ME (1995) Groundwater and agriculture: the interrelationship. British Geological Survey Technical Report WD/95/26

Garner WY, Honeycutt RC, Nigg HN (1986) Evaluation of pesticides in groundwater. American Chemical Society, Washington, DC, New York

PESTAN manual electronic reference: http://www.epa.gov/ada/csmos/models/pestan.html

Pesticide database electronic reference: http://npic.orst.edu/ppdmove.htm

Pesticide labels electronic reference: www.syngenta.com.au

Chapter 44
Modelling of Soil Contamination and its Remediation by In-Situ Solvent Flushing

Anwar A. Khan

Abstract The consequence of industrialization and other development activities have led to serious degradation of land and water quality. Its remediation has become an important necessity in this scenario. To overcome this issue, it is essential to understand the nature of contaminant transport and their adsorption and desorption substance characteristics. In this study, both organic and inorganic contaminants have been considered, and their movement through the soil together with adsorption and desorption characteristics as well as chemical/biochemical degradation has been investigated. The transport has been described using a set of equations numerically solved with different boundary conditions to represent several scenarios experienced in several cases in the arid environment. The modelling has been confined to one dimension assuming the soil is homogeneous. Contaminant transport may be visualised in many ways; however, in this study, two specific cases are considered. First case is the movement of leachate from soil waste dumping to ground, where toxic organic and inorganic chemicals were also dumped. The second case is the large-scale spillage of some toxic chemicals/oil/chlorinated solvents/non-chlorinated solvents on the soil and their subsequent movement through the soil. There can be several variations of these scenarios. In this chapter, variation with respect to rainfall in the area followed by dry seasons is considered. There is no single technique universally applicable for all contaminated sites. This chapter presents an analysis of the site restoration techniques that may be employed in a variety of contaminated site cleanup programmes.

Keywords Contamination • Effluent • Environmental degradation • Modelling • Soil flushing

---

A.A. Khan (✉)
Department of Drinking Water, Govt. of Uttarakhand, Dehradun 248001, India
e-mail: envanwar@yahoo.com

S.A. Shahid et al. (eds.), *Developments in Soil Classification, Land Use Planning and Policy Implications: Innovative Thinking of Soil Inventory for Land Use Planning and Management of Land Resources*, DOI 10.1007/978-94-007-5332-7_44,

44.1 Introduction

Rapid industrialization and associated developmental activities have taken major position in environmental degradation in the present scenarios. This has polluted the environment in many ways. The hazardous pollutants which have adverse effects on humans as well as the environment include organics like perchloroethylene (PERC), trichloroethylene (TCE), benzene, toluene, xylene, phenol, pyrene, acrolein, chloroform, vinyl chloride and their derivatives, and inorganic metals like chromium, zinc, iron, copper, lead, mercury, cyanide, nickel, cadmium, arsenic and their compounds. The major causes for contaminated sites are underground tanks, accidental spills, landfills, surface impoundments, land application of wastes and pesticides, septic tanks, acid mine drainage and the disposal of radioactive waste. These pollutants percolate through the soil and contaminate the underground water, often used for drinking purposes. Moreover, if pollutants contaminate the aquifer, it is likely that they would be carried to places far away from the source. There are several instances of such uncontrolled disposal causing severe problems for sub-surface water quality.

Many sites have been rendered useless due to the presence of hazardous wastes. The distillery waste disposal on land and toxic organic chemicals from dye and other manufacturing industries are some of the most visible examples of such contamination. In most cases after the initial dumping, further disposals are stopped; however, the rainfall moves the pollutants and causes extensive damage to the areas (Bou-Zeid and El-Fadel 2004). In addition, the accidental discharge of chemicals on land takes place with increasing frequency and has caused several problems in some areas. The need has arisen in recent times to remedy waste sites having contaminated soil to bring them in use. The sites can be cleaned by methods classified in five broad categories: isolation, immobilisation, toxicity reduction, physical separation and extraction (Cokca 2003).

This chapter presents "in-situ soil flushing" method classified under the extraction category. In this method, an extracting fluid is passed over the contaminated soil to mobilise the contaminants through solubilisation and desorption, which is then pumped out using extraction wells, and the contaminated fluid is treated before recycling. This remediation technology is less costly as it requires lower energy and less labour. There is no need to dig soil and transport for treatment, as required in an ex-situ method. Moreover, the extracting fluids are easily available. It is difficult to remediate less permeable soils by this method. The success of remediation depends on the site characteristics, levels of contamination and the extent of remediation desired (Mulligan et al. 2001).

The type of soil texture (particle size distribution) can affect the extent of contamination at a site. Fine-textured soils are highly contaminated than coarse-textured soils due to higher surface area. The nature of soil particle (mineral type) to which a metal is adsorbed determines the remediation technologies, e.g. soil washing (Dzombak et al. 1994). The amount of dissolved minerals, pH and redox potential of the soil water depends on the soil moisture content which also influence the degree of contamination. Soil structure can affect contaminant mobility by limiting the degree of contact between groundwater and contaminants (Evanko and Dzombak 1997).

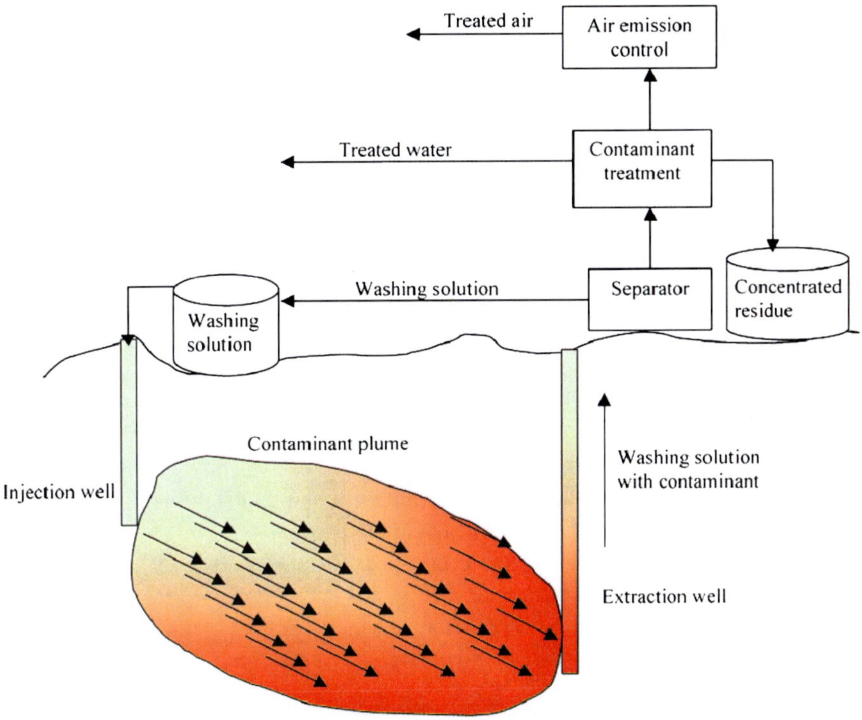

Fig. 44.1 In-situ soil flushing process

Soil porosity, bulk density and permeability can influence remediation process. Less permeable soils are not amenable to leaching or flushing methods of cleaning. Differences in soil properties can lead to inconsistent cleaning. The soils with high organic matter content can restrict desorption of the contaminants (Grasso 1993).

44.1.1 The Process of In-Situ Soil Flushing

In this method, the contaminants are mobilised by leaching with a suitable solvent. An aqueous extracting solution is injected into or sprayed onto the contaminated area to mobilise the contaminants usually by solubilisation. The solution can be applied by various methods, such as surface flooding, using sprinklers, leaching, vertical or horizontal injection wells and basin infiltration or trench infiltration systems (USEPA 1996). After the extracting solvent is being contacted with the contaminant, the resultant solutions are collected using extraction wells or pumps. At times certain additives (acids/bases, chelating agents, oxidising/reducing agents and surfactants/co-solvents) may be added to the extracting solvent to improve pollutant extraction. The schematic diagram of in-situ soil flushing process is shown in Fig. 44.1.

Apart from the extraction of the desired contaminants, it is important that the extracting fluid should have low toxicity and low volatility under ambient conditions. Under such conditions, it is easier to recover for reuse (Grasso 1993). The "tailing" resulting from slow desorption of pollutants from soil particles, dissolution of solid-phase contaminants, and release of contaminants from the fine pores in the soil matrix usually represents the practical limit for remediation.

The fate and transport of a contaminant in soil and groundwater depend significantly on the chemical form and speciation of the metal (Allen and Torres 1991). Many types of reactions (acid/base, precipitation/dissolution, oxidation/reduction, sorption or ion exchange) may occur which can influence the mobility of contaminants. In addition, the precipitation, sorption, and ion exchange reactions can retard the movement of contaminants. Groundwater flow characteristics also influence the transport of contaminants.

In this investigation, appropriate differential equations have been developed to model the flow of contaminants through the soil matrix. They have been numerically solved using the finite difference method. A computer simulation in MATLAB has been prepared. The objective of the simulation was to develop experimentally verified models which can then be used during field studies for a variety of purposes. In this chapter, theoretical modelling has been done to predict the contaminant transport during washing by rainwater, which has the potential to enhance the remediation process.

44.2 Theoretical Modelling of Contaminant Transport Through Soil

The equation governing the transport of contaminant in groundwater is a statement of the law of conservation of mass. The basic transport equation can be written, considering advection and dispersion for a reactive contaminant, as follows:

$$\left(D_x \frac{\partial^2 C}{\partial x^2} + D_y \frac{\partial^2 C}{\partial y^2} + D_z \frac{\partial^2 C}{\partial z^2}\right) - \left(v_x \frac{\partial C}{\partial x} + v_y \frac{\partial C}{\partial y} + v_z \frac{\partial C}{\partial z}\right) + r_x \left(C, t\right) + \frac{\partial C}{\partial t} \tag{44.1}$$

where D_x, D_y and D_z = hydrodynamic dispersion coefficients in space.

v_x, v_y and v_z = are the components of the advective transport.

The assumptions to simplify the equation:

1. The porous medium is homogenous and isotropic.
2. The contaminant is ideal with constant density and viscosity.
3. Saturated flow is considered with constant seepage velocity.
4. The fluid is incompressible.

In such a case, the equation reduces to the one-dimensional form for nonreactive contaminant as follows:

$$D_x \frac{\partial^2 C}{\partial x^2} - v_x \frac{\partial C}{\partial x} = \frac{\partial C}{\partial t} \tag{44.2}$$

Analytical solutions of Eq. (44.2) exist for some simple cases of boundary conditions (Augustijn et al. 1994; Bedient et al. 1994; Charbeneau 2000). There may be several special cases considered for the transport of contaminant through soil depending on the nature of the contaminant and the physical phenomenon of washout condition. The following special cases with special boundary conditions are described below.

44.2.1 Advection-Dispersion-Sorption Transport

Soil desorption is the most important process for physical removal of many inorganic or organic contaminants. The adsorption of the solute retards its movement and slows down its transport to aquifer. The mass balance equation in this case can be written as follows:

$$D_x \frac{\partial^2 C}{\partial x^2} - v_x \frac{\partial C}{\partial x} = (\rho_b / \eta) \frac{\partial S}{\partial t} = \frac{\partial C}{\partial t} \tag{44.3}$$

The concept of adsorption isotherm is used to relate the amount of contaminant adsorbed by the solids S to the concentration in solution C. Using the Freundlich's isotherm,

$$S = K_d C^b \tag{44.4}$$

If b =1, the isotherm is linear. Therefore, the following transformation can be made in the mass balance equation using

$$\frac{\partial S}{\partial C} = K_d$$

$$D_x \frac{\partial^2 C}{\partial x^2} - v_x \frac{\partial C}{\partial x} = (1 + \frac{\rho_b K_d}{\eta}) \frac{\partial C}{\partial t} \tag{44.5}$$

where $(1 + \rho_b \frac{K_d}{\eta}) = \mathrm{R}$, also called the retardation factor, which has the effect of slowing down the movement of the adsorbed species relative to the advective transport of the contaminant with the flow of water. It is a useful term to mathematically analyse the problems having linear isotherms with fast, reversible adsorption.

44.2.2 Advection-Dispersion-Sorption-Decay Transport

In several contaminants, significant changes occur due to decay by chemical reaction with soil matrix or among the different contaminants or microbial degradation.

The latter is possible through simple decomposition or by the biological oxidation process with the oxygen diffusing into the soil or produced locally by a secondary reaction.

In case of chemical decay, the rate expression of simple first-order kinetics has been used, whereas in case of biodegradation, the rate has been described by Monod kinetics. In one dimension, the mass transport equation for first-order decay by chemical reaction may be written as follows:

$$\frac{D_x}{R}\frac{\partial^2 C}{\partial x^2} - \frac{v_x}{R}\frac{\partial C}{\partial x} - \frac{kC}{R} = \frac{\partial C}{\partial t} \tag{44.6}$$

For Monod kinetics, the rate equation is given by $r = \dfrac{\mu_m CX}{K_s + C}$

As a result, the mass balance equation takes the form

$$\frac{D_x}{R}\frac{\partial^2 C}{\partial x^2} - \frac{v_x}{R}\frac{\partial C}{\partial x} - \frac{\mu_m CX}{K_s + C} = \frac{\partial C}{\partial t} \tag{44.7}$$

44.3 Results and Discussion

The model equations described above for different cases under consideration were solved by numerical technique using finite difference method. The simulations were carried out using different boundary conditions required to define the particular condition.

44.3.1 Advection-Dispersion-Sorption for Continuous Source

In case the contaminants do not undergo chemical or biochemical decay but retardation occurs due to adsorption process, in such cases, the dispersion effect due to the flow velocity becomes an important factor. In this case, a simulation was carried out where a liquid contaminant flows through a soil-column depth of 20 cm and source concentration ($C_\circ = 20$ mg l^{-1}).

Curve 1 in Fig. 44.2 shows the outlet concentration profile for advection-dispersion transport without and adsorption having dispersion coefficient $Dx = 0.01$ day^{-1}, $v = 3.19 \times 10^{-5}$ m s^{-1} and initial concentration as zero. It is a typical S-shaped breakthrough curve. However, if the background concentration is greater than the source concentration, as in the case of curve 2, relative washout occurs initially giving a desorption-type breakthrough curve. For both these cases, D_x and "v" remain unchanged. The effects of adsorption can be seen in curve 3. For this, curve 2 was taken retardation factor "R". The curve shows the extent of retardation in the form of appearance of the contaminant as a very slow rise in concentration as compared to curve 1. This implies that the contaminant was retained by the soil and slowing

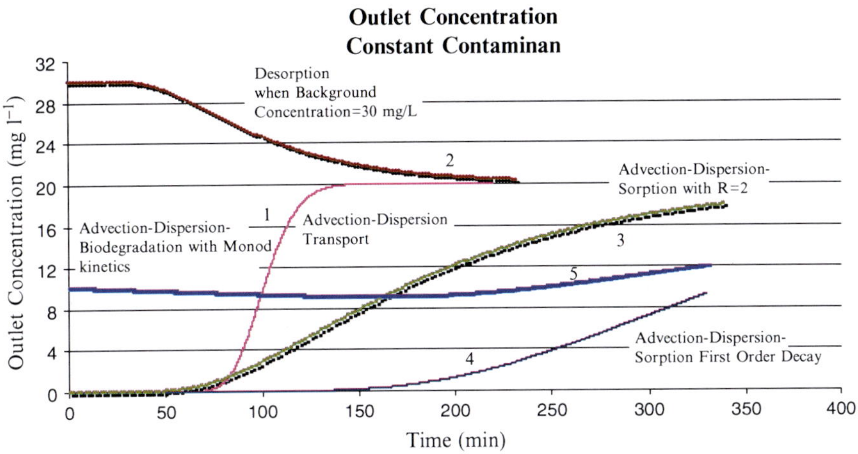

Fig. 44.2 Effluent concentration for transport due to a continuous source

down its transport. This has significant importance in terms of the self-cleaning of the soil occurring due to rainfall as described later.

The effect of the chemical/biochemical decay is shown in Fig. 44.2 as curve 4. The first-order decay constant has been taken as 1.6×10^{-6} s^{-1}. As a result, the contaminant is consumed as it is transported through the length of the column. The curve 5 depicts the concentrations of profile, where, in addition to adsorption, biodegradation of the contaminant is also considered. The maximum utilisation rate has been assumed to be 0.374 S^{-1}, and the half-saturation constant as 73.6 mg l^{-1}. In curve 5, the microbial concentration is assumed to be uniform across the column at 0.05 mg l^{-1}.

44.3.2 Advection-Dispersion-Sorption Transport of Inorganics in Soils for Discontinuous Source

There are several cases where a particular site gets contaminant disposal over the land for a limited period with further movement being carried out by the rainfall occurring during seasonal rain for about 15 days. This is followed by a dry period and then again a period of rain.

The simulation results in terms of concentration profile with time at different depths of the soil are shown in Fig. 44.3. The results show a continuous rise in the concentration for a long time after the contaminant flow stopped and then a very sharp drop in concentration followed by a static level for a period of about 4–5 months. This was following by another sharp decrease due to further rainfall after a stable period of no rain and no contaminant discharge. However, the concentration at a depth of 30 m or more again rises in due to the accumulation of contaminants

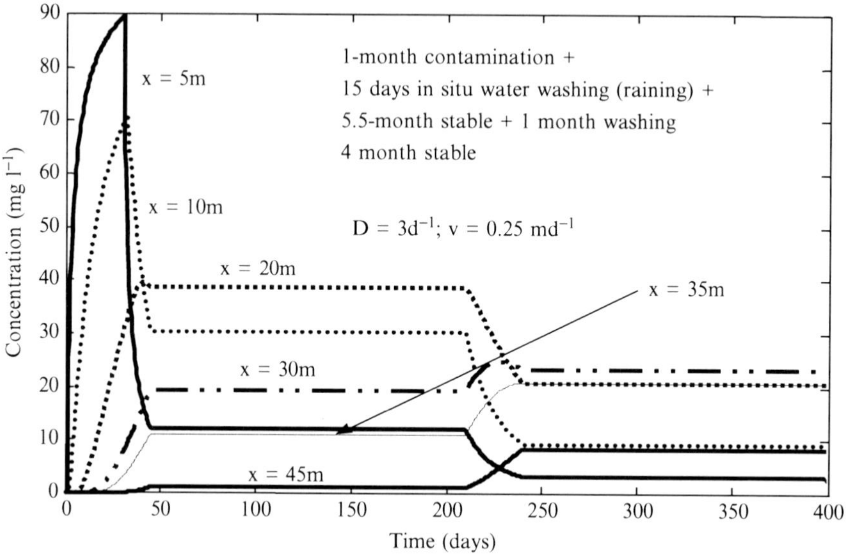

Fig. 44.3 Concentration profile with time at various depths of soil (discontinuous source)

by rainwater washout from upper layers. A most interesting observation is the very low concentration attained at a depth of 45 m and its very slow but steady rise in the yearly cycle. In the absence of any chemical or biological degradation, this accumulation will ultimately result in the contaminant reaching the underground water table and contaminants being transported by the underground aquifer.

44.3.3 *Advection-Dispersion-Sorption Transport with Biological Decay of Organics in Soils for Discontinuous Source*

Similar simulations were carried out for the spillage of organic compounds as shown in Figs. 44.4 and 44.5. The organic contaminant concentration in the spillage was taken as 10 mg l^{-1}. In this case, in addition to the adsorption, biodegradation has also been considered. The topsoil shows concentration build-up is similar to that of inorganic contaminants during the early periods, but the long-term concentration values are much lower due to the biodegradation. Further, the extent of increase of the concentration levels at a depth of 30 m and more is much less pronounced due to biodegradation. It was found that at the end of year, i.e. an entire cycle of source and washout due to rainfall, starting with a contaminant source of 10 mg l^{-1}, a final concentration of about 0.2 mg l^{-1} reaches the 45-m depth, where the groundwater table is assumed to exist. Figure 44.4 shows the case where it has been assumed that rainfall is distributed in two phases at an interval of about 5 months. In Fig. 44.5, similar analyses have been carried for the same source strength with the rainfall occurring

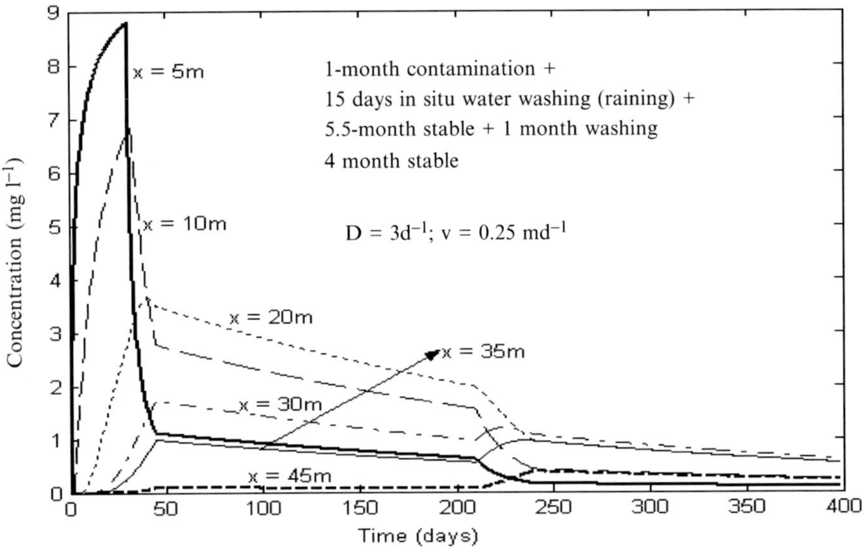

Fig. 44.4 Advection-dispersion-sorption and biological decay transport of organics in soils for a discontinuous source

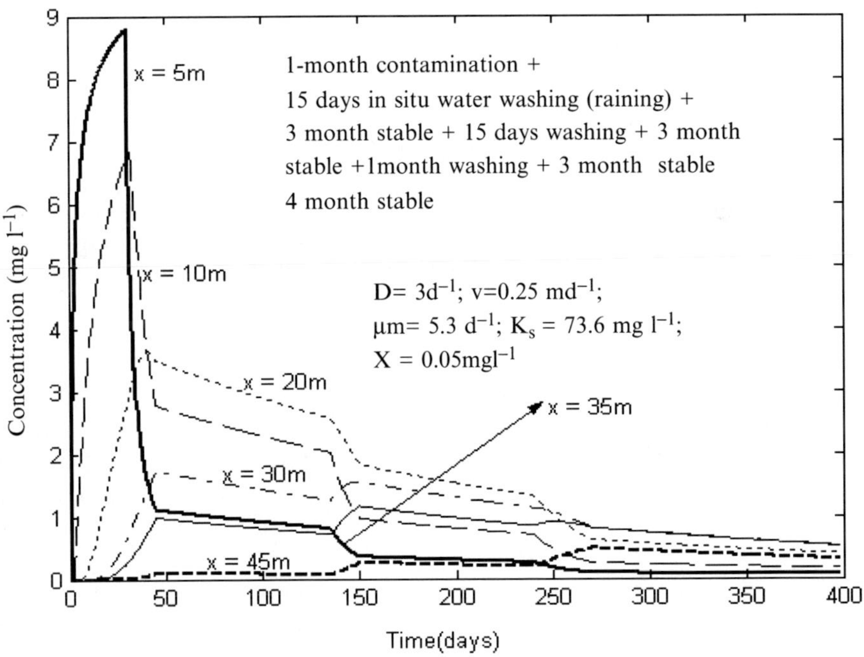

Fig. 44.5 Advection-dispersion-sorption and decay transport of organic contaminant in soils for discontinuous sources

in three stages. The results show that there is continuous decrease in the concentration of the contaminant up to a depth of about 20 m, but at the depth of 35 m, marginal increase in concentration occurs due to carry over from higher layer by washout. However, the concentration levels reaching the groundwater aquifer level are quite low to the extent of less than 0.01 mg l^{-1}. This shows the importance of the biological degradation in terms of contaminant depletion and its impact on the soil degradation.

44.4 Conclusions

In this chapter, the theoretical analysis/mathematical modelling was done to predict the behaviour of organic and inorganic contaminants in waste sites and how their presence can affect the groundwater. It is important to contain the contaminants when their transport can reach any underground aquifer and carry them further far away areas and thus, increase their spread. For this purpose, various methods of land redemption are available, and one can choose the method depending on the site and contaminant characteristics. The in-situ soil remediation seems to be more technical and economical viable option for remediation of soil contaminated with inorganic and organic chemicals, as addition of water can provide substantial desorption of contaminants from soil particles. The mathematical analysis of the soil pollution and its washout by rainfall shows interesting results for the inorganic and organic contaminants. In case of inorganic chemicals, the absence of chemical decay makes the adsorption phenomena as the major contaminant removal mechanism. But gradual washout due to rainfall ultimately carries all the contaminant to the underground water table. In case of biodegradation, the chemical depletion becomes a major factor in the transport of pollutant. The concentration reaching the groundwater table becomes quite low, and this can be further reduced by the application of biological agents so designed to enhance the biodegradation. The information provided in this chapter should be useful in screening technologies early in the remedy evaluation and selection process.

The mathematical model presented in this chapter needs to be verified for its applicability and accuracy in actual soil systems. For this purpose, experiments needed to be carried out in laboratory to try and match the results obtained with the simulations. These experiments would also be helpful in better understanding the theoretical concepts regarding contaminant and water flow through the soil described above. Values of parameters—seepage velocity (v_x), dispersion coefficient (D_x) and the retardation factor (R)—have been taken from empirical correlations or typical values found in the literature. Thus, the authors are only able to approximate the actual results. However, in real life situations also, the values of these parameters are generally approximated. In spite of this, modelling is useful because these approximations are not far off from the actual results and one is able to predict the contaminant flow with sufficient accuracy for the desired purpose.

Nomenclature

Symbol	
A	= cross-sectional area (m^2)
C	= concentration (mg l^{-1})
D_x	= dispersion coefficient in the flow direction (m^2 s^{-1})
h	= depth (m)
Kd	= distribution coefficient
Ks	= half-saturation constant
S	= mass of chemical constituent adsorbed per unit mass of the solid aquifer material (mg gm^{-1})
t	= time
v_x	= average seepage velocity in X-direction (m s^{-1})
x	= distance (m)
Greek Letters	
η	= porosity
ρ_b	= bulk density of porous media (gm cm^{-3})
μ_m	= maximum utilisation rate

References

Allen JP, Torres IG (1991) Recent developments in separation science, vol V. CRC Press, West Palm Beach

Augustijn DCM, Jessup RE, Rao PSC, Wood AL (1994) Remediation of contaminated soils by solvent flushing. J Environ Eng 120(1):42–57

Bedient PB, Rifai HS, Newwell CJ (1994) Groundwater contamination: transport and remediation. Prentic Hall PTR, Englewood Cliffs

Bou-Zeid E, El-Fadel M (2004) Parametric sensitivity analysis of leachate transport simulations at landfills. Waste Manag 24:681–689

Charbeneau RJ (2000) Groundwater hydraulics and pollutant transport. Prentice Hall, Upper Saddle River

Cokca E (2003) A computer program for the analysis of 1-D contaminant migration through a soil layer. Environ Model Software 18:147–153

Dzombak DA, Rosetti PK, Evanko CR, DeLisio RF (1994) Treatment of fine particles in soil washing processes. In: Proceedings of the specialty conference on Innovative Solutions for Contaminated Site Management. Water Environment Federation, Alexandria, pp 473–484

Evanko CR, Dzombak DA (1997) Remediation of metals-contaminated soils and groundwater. GWRTAC, Pittsburgh

Grasso D (1993) Hazardous waste site remediation-source control. Lewis Publishers, Boca Raton

Mulligan CN, Yong RN, Gibbs BF (2001) Remediation technologies for metal-contaminated soils and groundwater: an evaluation. Eng Geol 60:193–207

USEPA (1996) Recent developments for in situ treatment of metals contaminated soils. US Environmental Protection Agency, Office of Solid Waste and Emergency Response, Washington, DC

Part V
Innovations in Research, Development, Education and Extension

Chapter 45
Sharing Information to Accelerate Implementation of Reclamation and Improvement of Degraded Lands: WASWAC Experience

Samran Sombatpanit

Abstract Degraded lands are lands that have become inferior in quality, essentially in terms of crop productivity or in providing environmental services. Previously, it was the matter of land conservation and improvement that occupied researchers' mind – but since 20–30 years ago, the interest had shifted to the point that people pay more interest to the causes of degradation and how to correct them. It thus seems to be a decent way to implement any program concerning soil or land. For the research part, there have been many projects and programs that do research on land degradation today, the executors of which are universities, government agencies, research institutions, etc. From doing research, there are a lot of results that are produced each year. The next step is to present the research findings that researchers go to attend conference at various venues or publish in peer-reviewed scientific journals. There are several series of meetings, that is, ICLD, ISCO, COMLAND, ESSC, WASWAC, IECA, SWCS, and many other national societies of soil and water conservation. For the implementation part, the noted work to combat land degradation commenced in USA by the agency called Soil Conservation Service (SCS), which later became Natural Resources Conservation Service (NRCS). This agency is a part of the US Department of Agriculture (USDA). Many visitors from abroad go to learn soil conservation techniques and take back to apply in their own countries – often with certain degrees of modification.

Keywords China • COMLAND • Degraded lands • Reclamation • WASWAC

S. Sombatpanit (✉)
World Association of Soil and Water Conservation (WASWAC),
67/141, Amonphant 9, Senanikom 1, Lat Phrao, Bangkok 10230, Thailand
e-mail: sombatpanit@yahoo.com; samran_sombatpanit@yahoo.com; sombatpanits@gmail.com

S.A. Shahid et al. (eds.), *Developments in Soil Classification, Land Use Planning and Policy Implications: Innovative Thinking of Soil Inventory for Land Use Planning and Management of Land Resources*, DOI 10.1007/978-94-007-5332-7_45,

45.1 Introduction

Degraded lands are lands that have become inferior in quality, essentially in terms of crop productivity or in providing environmental services. The degradation may be due to either natural or human-induced causes, for example, acid soil, saline and sodic soils, mine tailing, waterlogged land, eroded land, and land in dry environments, to cite just a few, and the way to correct each type of degraded land is quite different from one to another (Fig. 45.1).

To make use of degraded lands profitably, one needs to study about the causes and how they can be restored to a suitable fertility status. Two main components needed for this to achieve are research and extension. Previously, it was the matter of land conservation and improvement that occupied researchers' mind – but since 20–30 years ago, the interest had shifted to the point that people pay more interest to the causes of degradation and how to correct them. It thus seems to be a decent way to implement any program concerning soil or land.

For the research part, there have been many projects and programs that do research on land degradation today, the executors of which are universities, government agencies, research institutions, etc. From doing research, there are a lot of results that are produced each year. The next step is to present the research findings

Fig. 45.1 Soil erosion by various agents is a basic factor that causes land degradation (All photos from WOCAT)

that researchers go to attend conference at various venues or publish in peer-reviewed scientific journals. There is one series of meetings, that is, International Conference on Land Degradation (ICLD), that meets every 2–3 years in different parts of the world. Also, there is a series of meetings called ISCO (International Soil Conservation Organization) that meets every 2–3 years at various places too. An organization that is dealing strongly with land degradation is COMLAND (Commission on Land Degradation and Desertification), which is a part of the International Geographical Union. The organization has coordinated several COMLAND meetings in many parts of the world and has produced many excellent publications. Apart from that, there are several other meetings that feature land degradation as an important theme, such as those organized by the ESSC (European Society for Soil Conservation), WASWAC (World Association of Soil and Water Conservation), IECA (International Erosion Control Association), SWCS (Soil and Water Conservation Society), and many other national societies of soil and water conservation.

For the implementation part, the noted work to combat land degradation commenced in USA by the agency called Soil Conservation Service (SCS), which later changed its name to Natural Resources Conservation Service (NRCS). This agency is a part of the US Department of Agriculture (USDA). Many visitors from abroad go to learn soil conservation techniques and take back to apply in their own countries – often with certain degrees of modification.

However, there has been a big gap in the success in technology research and application in developed and developing countries. Research being done in cooperation between researchers in developing and developed countries during past 30 years has brought about several findings, one very important of which would be the participatory approach, which is almost the *only* way to assure the success of technology application in the farmers' fields. At the same time, the technologies involved in mitigating land degradation have been refined so much and can be classified into several categories – with different amounts of cost that suit the land of different terrains and people of differing wealth levels.

Since 1983, the World Association of Soil and Water Conservation (WASWAC) was born out of the need to have a forum for soil and water conservationists in various parts of the world to work together. For two decades, the association was operating from USA, a first world country, but it proved to be less than suitable at a later stage. So it moved out to locate in the People's Republic of China since April 2003, where it can handle and create many activities that serve members from both north and south countries very well. With good management, it is hoped that WASWAC will function properly in bonding people who take care of the health of the land, with a result that most of the world's population will lead their lives normally and far away from hunger.

This chapter describes the activities that WASWAC has been operating during the past decade, which from a solitary task of producing quarterly newsletter, now the association is able to handle almost one dozen activities, as well as function as a forum where people living in different environments can contact and work together through the new technology – Internet – which has become available just in time to enhance the association to expand its sphere and make itself more useful to people living anywhere in the world.

45.2 WASWAC and Its History

The WASWAC is an international nongovernment organization that was established 28 years ago with the objective of promoting the wise use of the earth's land resources (Moldenhauer et al. 2009). In doing this, it acts as a forum for people from all over the world, who are interested in soil and water conservation (Figs. 45.2 and 45.3); it provides a network, through which we can make contact with people with similar interests, problems, and projects to our own from any country or region in the world; and it provides information – mainly through the quarterly newsletter and monthly Hot News – about matters of mutual interest. WASWAC also sponsors and cosponsors conferences, seminars, and workshops every now and then and synthesized and compiled them into proceedings or books. Over the last 28 years, it has been responsible for producing more than 20 books, which have been made available to members at reasonable prices. WASWAC books have had a very big influence on the way soil and water conservation is now perceived and approached in many countries, especially the small booklet "Land Husbandry – a Framework for Soil and Water Conservation"; it was one of WASWAC's first publications, and it has been widely distributed and given guidance to many soil and water conservationists in many parts of the world.

The WASWAC is run by a hardworking council, whose members give their time to the association for free. The 30 council members come from different countries all over the world. As you can see, WASWAC council members are very international group representing a large number of regions, countries, interests, and experiences. Although the members of the council live in different parts of the world, they are in contact practically every day by e-mail, sometime several times a day; they conduct nearly all the businesses and meetings on the e-mail.

At present, WASWAC has approximately 1,000 members from about 70 countries, and the membership numbers are growing rapidly. Apart from that, it has around 4,000 guest members. WASWAC also has a number of vice presidents who represent different countries. It is the responsibility of the VPs to represent WASWAC at their country or region, to advise the president on what is happening, and to organize all kinds of activities.

The WASWAC works very closely with national soil and water conservation societies and soil science societies. WASWAC has also developed strong links with other related organizations like the International Union of Soil Sciences (IUSS), the International Erosion Control Association (IECA), the International Soil Conservation Organization (ISCO), the European Society for Soil Conservation (ESSC), and the European Confederation of Soil Science Societies (ECSSS).

45.3 The SWCS and the Move to China

When WASWAC was established in 1983, it was sponsored and helped by the Soil and Water Conservation Society of America (SWCSA, now the Soil and Water Conservation Society – SWCS). They provided a secretariat and ran the affairs from

Fig. 45.2 Structural measures (**a**), vegetative measures (**b**), agronomic measures (**c**), and management measures for soil and water conservation (**d**) (Both photos from WOCAT)

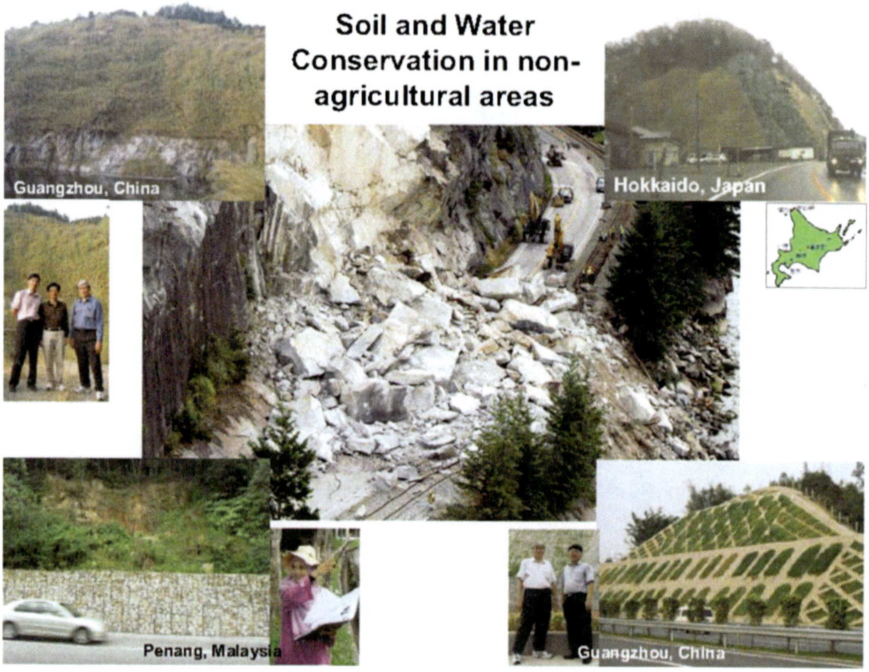

Fig. 45.3 Soil and water conservation in nonagricultural areas

their office in Ankeny, Iowa. They looked after the administration; they kept the records, handled the finances, and published the books and newsletters. This arrangement worked well for many years, but recently there were some troubles. The main problem was that WASWAC had to pay for the services that it received, and, being in America, it was expensive – in fact, WASWAC was beginning to run at a loss and started using up its small savings.

WASWAC looked at several options, and then, in August 2002, it received an offer from China. The Department of Soil and Water Conservation of the Ministry of Water Resources in Beijing offered to provide the facilities for the secretariat. They offered very generous terms. WASWAC had signed a memorandum of understanding (MoU) with the Ministry of Water Resources; under this MoU, the Chinese took over responsibilities for the administration of the association from April 1, 2003. In this way, the WASWAC has now canceled the charity status in America and is in the process of registering with the authority in China as a nonprofit organization.

45.3.1 Vice Presidents

Vice presidents are appointed by the council, and the activities of members in the country can be organized through some new or existing organization, such as national soil and water conservation society.

It is like doing a development project, which is more or less a process for establishing some kind of movement that is sustainable for a long time. At this point, WASWAC is rather certain to be in a right direction, since the membership has climbed up to more than 1,000 from approximately 70 countries. It is certain that there is much need in having a central organization. But the important point is that this kind of organization needs personnel who are willing to work for global soil and water conservationists worldwide, normally without any compensation; those who belong to such category should therefore step forward and join hands with the administration in managing the WASWAC now and in the future.

45.3.2 Products and Services of WASWAC

The WASWAC has fulfilled its commitments to share knowledge gathered from all over the world and provide services to those who need. Examples are the following: WASWAC Newsletter in ten languages (English, Spanish, French, Chinese, Portuguese, Bahasa, Russian, Vietnamese, Arabic, Thai) (quarterly); monthly Hot News; Journal and Proceedings of WASWAC; The Land Journal; special publications once a year; Websites, operated from Guangzhou, Beijing, and Tokyo (http://waswac.soil.gd.cn, www.waswac.org); photo Websites, operated from Bangkok (http://community.webshots.com/user/WASWAC and http://community.webshots.com/user/WASWAC1); forum for discussion, for example, law and policy, climate change, monitoring and evaluation (M&E), and conservation agriculture (including no-till); contacts among professionals in various countries; coordination among various WASWAC chapters; and coordinating the LANDCON series of meetings, with the use of Guidelines for Successful Meetings developed by WASWAC in 2009. WASWAC supports LANDCON meetings worldwide with "cooperation package," presenting the Norman Hudson Memorial Award annually, plus other awards. Following are the details of some of the main products and services:

45.3.2.1 The Newsletter, Bulletin, and Hot News

For over 20 years, the WASWAC has produced a newsletter every 3 months. For a long time, this came out as a small printed publication, usually of only four pages. During the time of President David Sanders, we made a big effort to improve the newsletter by making it longer and to contain more articles that would be of interest and value to the members. From 2002 the council had put more pages and divided into several sections that may serve most members in a satisfactory way. A breakthrough came when we were able to produce the electronic newsletter and send out by e-mail, the idea being strongly supported by David Sanders' predecessor, Hans Hurni. One previous problem was the difficulty in getting members to contribute with articles. This has now changed; we are now receiving a wide range of articles, book reviews, and other items from all over the world. The title of this publication has changed to WASWAC Bulletin in 2010 and is available every 4 months.

The bulletin now concentrates more on writings that show the outcome from technical meetings and other activities like project implementation.

45.3.2.2 The WASWAC Hot News

All upcoming news pieces (e.g., conferences, jobs, funds, exhibitions, study tours, publications) are accommodated in monthly WASWAC Hot News that we send directly to every member. The WASWAC Hot News was born from the separation of certain urgent news items from the WASWAC Newsletter and became available for the first time in 2006 and has become a main source of information that members depend on for enriching their knowledge and database and enhancing international involvement.

45.3.2.3 The Journal and Proceedings

WASWAC has since 2006 started the Journal and Proceedings to publish peer-reviewed articles and non-peer-reviewed articles, respectively. Though these cannot be published as hard copies, they are posted on the Website instead. The WASWAC Website is going to be used more efficiently for many purposes. The Journal and Proceedings have temporarily stopped in 2010 and will resume in the printed and online form in 2012.

45.3.2.4 The Special Publication

Under this program, WASWAC is able to publish special publication to give members for free once a year (Fig. 45.4). An issue is selected that may attract interest from majority of members in the form of long articles, which can be used as additional reading for students. So far the titles of WASWAC special publications are as follows:

No. 1 (2003): The USLE Story by John Laflen and W.C. Moldenhauer
No. 2 (2004): Carbon Trading, Agriculture and Poverty by Mike Robbins
No. 3 (2008): No-Till Farming Systems by Tom Goddard et al.
No. 4 (2009): Soil and Water Assessment Tool (SWAT) by Jeff Arnold et al.
No. 5 (2011): Conservation Agriculture in Southeast Asia by SANREM team of editors
No. 6 (2011): Vegetable Agroforestry Systems in Indonesia, Philippines and Vietnam by TMPEGS team of editors
No. 7 (2011): LANDCON e-LIBRARY: Our Gift to the Earth (a set of DVDs) by Samran Sombatpanit, Karika Kunta, and Winai Wangpimool
No. 8 or 9 (2013): Conservation Agriculture – A Text Book by international group of editors

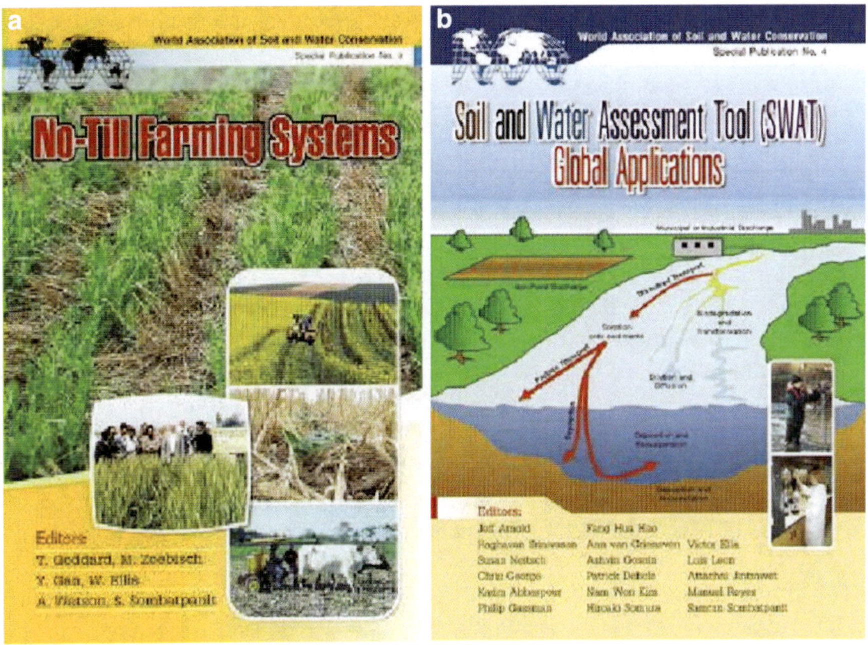

Fig. 45.4 WASWAC special publications no. 3 (**a**) and 4 (**b**)

45.3.2.5 The Norman Hudson Memorial Award

WASWAC started to give an award called the Norman Hudson Memorial Award to a deserving person since 2004. This is the highest honor bestowed on an individual by the association, one person a year. It is given for distinguished service in recognition of international accomplishments in soil and water conservation. The award is named for Dr. Norman Hudson, whose exemplary professional career was devoted to the cause of global soil and water conservation. Persons receiving it were Calvin Rose (Australia 2004), Rolf Derpsch (Paraguay 2005), John Greenfield (New Zealand 2006), Hans Hurni (Switzerland 2007), Michael Stocking (United Kingdom 2008), Rattan Lal (USA 2009), and J. S. Bali (India 2010).

45.3.2.6 The Vision and Mission Statement and Slogan of WASWAC

The council is very keen that the WASWAC should not only grow in numbers of members in various categories but that it should also become more responsive to the needs of members. In other words, association wants to become more useful to everyone. For this reason, discussions have been made with the members in 2003 and have acquired the "Vision and Mission Statement," as is shown below, which is intended to be followed as much as possible.

Vision A world in which all soil and water resources are used in a productive, sustainable, and ecologically sound manner.

Mission To promote worldwide the application of wise soil and water management practices that will improve and safeguard the quality of land and water resources so that they continue to meet the needs of agriculture, society, and nature.

In 2009, the acronym WASWC was changed to WASWAC during deliberation with all stakeholders. The change was desired to attract more members by making the acronym easily pronounceable. The membership enhancement is necessary for such an association.

The WASWAC Slogan The WASWAC slogan "Conserving soil and water worldwide – join WASWAC" was made in 2004.

It is understood that the vision, mission, and slogan may be amended once the need arises.

45.3.2.7 Guidelines for Successful Meetings and LANDCON Series of Meetings (See http://waswac.soil.gd.cn/consti-decentra.html)

This is one of the most recent acquisitions of WASWAC. The association had deliberated about it in 2008 and it is in effect now. This was done to assist/support people/organizations/institutes, etc. who organize meetings, so they will successfully obtain good results and it is a way to help academics and professionals to know which meetings they should go to attend, so not to waste their time and money. There were eight meetings in 2009 and nine meetings in 2010. Any organizer who thinks the event will pass at least 7 out of 10 points of our guidelines will be able to register as a LANDCON meeting with us and obtain a specific number, for example, LANDCON 1004 for the INTERPRAEVENT meeting that took place in April 2010 in Taiwan. Table 45.1 shows the list of LANDCON meetings from the year 2009 onward.

45.3.2.8 LANDCON e-LIBRARY: Our Gift to the Earth

This is a project of WASWAC to collect 10,000 papers/documents to facilitate research and implementation of land, soil, and water degradation, rehabilitation, management, care, conservation, improvement, and preservation of available works for posterity, as well as to improve the visibility of works, authors, institutions, and publishers to a wider audience. The project is to be completed in 2011 (Fig. 45.5), and the DVD set will be available from 2012 onward.

The idea in producing this work came from the fact that, in the world, almost everyone uses computer and thus has an access to digital technology and should benefit from the publications that have been prepared during the past 2–3 decades. Many of such works were published in various publications and posted on the Website. However, not everyone who is working in soil and water fields have access to the Internet, especially the fast one that can be used for downloading works from

Table 45.1 LANDCON series of meetings

No.	Meeting topic	Place	Time
0902	Conference of the International Erosion Control Association (IECA): Environmental Connection 09:	Reno, Nevada, USA	February 9–12, 2009
0903	Conference on Land Degradation in Dry Environments	Kuwait City, Kuwait	March 8–14, 2009
0905a	Global Change, Challenge for Soil Management: From Degradation – Through Soil and Water Conservation – To Sustainable Soil Management	Tara Mt., Serbia	May 26–30, 2009
0905b	Conference on "The Sustainable Development of Water Resources," organized by Hydrotechnics Faculty of "Politehnica" University of Timisoara, Romania	Timisoara, Romania	May 27–28, 2009
0906a	The 10th International Meeting on Soil with Mediterranean Type of Climate	Beirut, Lebanon	June 22–26, 2009
0906b	Conference on "Challenges and Opportunities of Bioindustrial Watershed Development for the Prosperity of the Farming Community," organized by the Soil Conservation Society of India	Bengaluru, India	June 25–27, 2009
0909	International Conference on Desertification in Memory of Professor John B. Thornes	Murcia, Spain	September 16–18, 2009
0911	International Seminar on Upland for Food Security, Faculty of Agriculture, University of Soedirman, Purwokerto, Central Java. Contact Budi Prakoso.	Purwokerto, Central Java, Indonesia	November 7–8, 2009
1004a	INTERPRAEVENT Conference in the Pacific Rim.	Taipei, Taiwan	April 26–30, 2010
1004b	6th National Seminar on Watershed Management and 4th National Seminar on Soil Erosion and Sediment	Noor, Iran	April 28–29, 2010
1005	First Conference on Soil and Roots Engineering Relationships	Ardabil Province, Iran	May 24–26, 2010
1009	FAME 2010, an annual event organized by Green World Conferences	Berlin, Germany	September 13–14, 2010
1010	First WASWC Council Meeting and Int. Conf. on Combating Land Degradation in Agricultural Areas	Xian, Shaanxi, China	October 11–15, 2010
1011	International Conference on Cooling the Earth	Pantnagar, India	November 15–17, 2010
1012a	International Geographical Union (IGU) Commission Seminar on Land Use, Biodiversity and Climate Change	Guwahati, Assam, India	December 11–13, 2010

(continued)

Table 45.1 (continued)

No.	Meeting topic	Place	Time
1012b	ASADWA Arid and Semi-arid Development through Water Augmentation, Optimizing water resources in a world of water shortage	Valparaiso, Chile	December 13–17, 2010
1101	SWAT SEA II Workshop and Conference	Ho Chi Minh City, Vietnam	January 4–8, 2011
1103	BioWise 2011, an annual event organized by Green World Conferences	Kuala Lumpur, Malaysia	Mar 14–15, 2011
1107a	2nd Conference on Conservation Agriculture in Southeast Asia	Phnom Penh, Cambodia	July 4–7, 2011
1107b	SWAT Regional Conference and Training Workshop	Guangzhou, China	July 24–30, 2011
1112	1st International Symposium on Cashew Nut	Madurai, Tamil Nadu, India	December 9–12, 2011
1305	2nd WASWAC World Conference and Council Meeting	Pattaya, Thailand	May 14–18, 2013

Fig. 45.5 Special publication no. 7 to be completed in 2011

millions of Websites available these days. Many people put their trust on the Internet and are not convinced that putting works in the plastic discs called CD or DVD is necessary – perhaps one may think that while collecting works to burn into a CD or DVD, all the works will have been old and many new works will become available. Anyhow, after deliberating with the WASWAC stakeholders, the association got good comments and support from almost 100 members in doing this project. One prominent member, ESSC President Prof. Jose L. Rubio of University of Valencia in Spain, when asked if he liked the e-LIBRARY project, said:

> My reaction is YES. It looks like a very interesting and worthwhile initiative. More and more we will be using electronic formats, which is also a way to avoid losing useful information. Thanks for your efforts. – José Luis Rubio, Spain

After new year 2010, WASWAC started to collect materials for this LANDCON e-LIBRARY. At this time, a subtitle of "Our Gift to the Earth" is added to it. This subtitle came after the earthquake hit Haiti so hard that killed over 200,000 people. The idea was that humans have lived and enjoyed their lives on earth for a long time and rather carelessly and, as a result, have caused many scars to appear on its surface, including solid, liquid, and gaseous parts like what is happening now all over the world. There is every reason for us to do something now to repair those unfavorable human footprints.

Accordingly, what scientists have in possession now in the digital forms of technical papers, documents, presentations, films, and videos should be provided for making the collection of such works possible, to be used wider to help heal such scars. Besides this, the good thing of sending your works out as requested would be an efficient way of preserving them for a long, long time – long after we are gone, I assure you. However, what WASWAC will give to the earth this time seems very little when compared with what we have got from, or done to, it.

Rationale. WASWAC is going to produce *LANDCON e-LIBRARY: Our Gift to the Earth*, which will comprise many thousands of papers and other documents in the various fields of soil and water sciences. These works have been presented at various meetings during the past three decades, and we want to make them available to people who may need them for doing their work, both research and implementation, as well as to preserve the works to posterity before they are forgotten and lost. All the works will be put in a set of four DVDs. This will form a wealth of information that will be useful to you and your colleagues for many years.

45.3.2.9 Partnership with Some Research and Development Programs

WASWAC has been working with members from developed and developing countries with a role to enhance better and quicker participation in certain projects/programs. In 1992 WASWAC President Hans Hurni initiated a long-term program, "World Overview of Conservation Approaches and Technologies (WOCAT)," based in Berne, Switzerland, and had a landmark WOCAT global overview book "where the land is greener" published in 2006. During the 1990s WASWAC has supported Jim Cheatle's "Organic Matter Management Network" based in Nairobi, Kenya. WASWAC is also closely allied with Reseau Erosion, a project of Vice President Eric Roose, based in Montpellier, France, and operating mainly in Africa. WASWAC is allied to ISCO and cooperates with planning and conducting its biennial conferences. WASWAC is requested and very willing to cosponsor conferences, symposia, and workshops it feels will further its philosophy and objectives. The most recent partnership occurred when WASWAC helped organize a Soil and Water Assessment Tool (SWAT) regional conference and training workshop in Guangzhou, China, in July 2011; it was agreed that WASWAC from then on considers SWAT as its associate program and will help spread the use of it whenever and wherever possible, and a great synergy can be expected. At this time discussion is going on with few more programs about future cooperation.

45.3.2.10 Some Things That WASWAC Is Not

The WASWAC often receives requests that cannot be met. The WASWAC is primarily a professional association of soil and water conservationists. It is not a funding agency. Unfortunately, some members do not understand and send requests for funding or for help in setting up projects. Of course, association cannot help with these requests other than putting the people involved in touch with some agency that may be able to help.

References

Arnold J, Srinivasan R, Neitsch S, George C, Abbaspour K, Gassman P, Hao FH, Van Griensven A, Gosain A, Debels P, Kim NW, Somura H, Ella V, Leon L, Jintrawet A, Reyes M, Sombatpanit S (eds) (2009) Soil and Water Assessment Tool (SWAT) global applications, World Association of Soil and Water Conservation Special Publication No 4. World Association of Soil and Water Conservation, Beijing, p 417

Goddard T, Zoebisch M, Gan Y, Ellis W, Watson A, Sombatpanit S (eds) (2008) No-till farming systems, World Association of Soil and Water Conservation special publication no 3. World Association of Soil and Water Conservation, Beijing, p 540

Laflen JM, Moldenhauer WC (2003) Pioneering soil erosion prediction. The USLE story, World Association of Soil and Water Conservation special publication no 1. WASWC, Beijing, p 54

Moldenhauer W, Sanders DW, Sombatpanit S (2009) WASWC: Its history and operation. In: Jeff Arnold et al. (eds) Soil and Water Assessment Tool (SWAT): global applications, World Association of Soil and Water Conservation. ISBN 978-974-613-722-5, pp 410–415

Robbins M (2004) Carbon trading, agriculture and poverty, World Association of Soil and Water Conservation special publication no 2. World Association of Soil and Water Conservation, Beijing, p 48

WASWAC (2009) Guidelines for successful meetings. The World Association of Soil and Water Conservation. http://waswac.soil.gd.cn/consti-decentra.html

Chapter 46
Innovations in Soil Chemical Analyses: New ECs and Total Salts Relationship for Abu Dhabi Emirate Soils

Shabbir A. Shahid, Mahmoud A. Abdelfattah, and Henda Mahmoudi

Abstract Soil analysis is important to understand composition of soils for many reasons including developing soil management options for sustainable agricultural activities. There exists a relationship between total soluble salts (meq l^{-1}) and ECs (dS m^{-1}) developed in 1954 and published in the USDA Handbook 60. In the present study, an attempt has been made to correlate total soluble salts and ECs values from a range of soils from Abu Dhabi Emirate to test the validity of USDA relationship to local soils. The USDA ratios (TSS/ECs) of 10 and 16 exist for ECs 1 and 200 dS m^{-1}, respectively ($R^2=0.9577$). Whereas TSS/ECs ratio from Abu Dhabi Emirate soils was found to be 10 and 11.38 for ECs 1 and 200 dS m^{-1}, respectively, and for ECs 500 dS m^{-1} a ratio of 12 was found ($R^2=0.9711$), the USDA does not present such a ratio for ECs more than 200 dS m^{-1}. The present study has rejected the hypothesis that same relationship exists between TSS/ECs on soils of Abu Dhabi Emirate as that of USDA and, therefore, this relationship cannot be used for Abu Dhabi Emirate conditions. This chapter presents a new ratio for Abu Dhabi soils to avoid misinterpretation of soil analytical data for quality assurance (QA) purposes and to formulate soil management options. The newly developed relationship was

S.A. Shahid (✉)
International Center for Biosaline Agriculture, P.O. Box 14660, Dubai, UAE
e-mail: s.shahid@biosaline.org.ae

M.A. Abdelfattah
Soil and Land Use Management, Environment Agency-Abu Dhabi,
P.O. Box 45553, Abu Dhabi, UAE

Soils and Water Sciences Department, Fayoum University, Fayoum, Egypt
e-mail: mabdelfattah@ead.ae; mahmoudalia@yahoo.com

H. Mahmoudi
International Center for Biosaline Agriculture,
P.O. Box 14660, Dubai, United Arab Emirates
e-mail: hmj@biosaline.org.ae

S.A. Shahid et al. (eds.), *Developments in Soil Classification, Land Use Planning and Policy Implications: Innovative Thinking of Soil Inventory for Land Use Planning and Management of Land Resources*, DOI 10.1007/978-94-007-5332-7_46,

validated by measuring sodium adsorption ratio (SAR) by various methods, that is, using Na derived by difference between TSS and Ca+Mg and through measured Na and Ca+Mg values from Abu Dhabi Soil Survey. The SAR calculated by using Na derived from newly developed relationship fits close to the SAR values from actually measured Na and Ca+Mg contents; these SAR values deviate to those measured from USSL and USDA relationships. In the light of present finding, it is recommended that other regions where soil and environmental conditions are similar to Abu Dhabi Emirate, a similar relationship most suited to their local conditions should be developed or the results to be correlated with that established from Abu Dhabi soils for validation.

Keywords Abu Dhabi Emirate • ECs • Handbook 60 • Salt concentration • UAE

46.1 Introduction

Irrigated agriculture has faced the challenge of sustaining its productivity for centuries. One of the major threats to irrigated agriculture productivity is soil salinity, which is developed either through high water table, capillary rise, and subsequent evaporation or through using alternative irrigation water sources where the quality of irrigation water is often low (variable salinity) and there is variable ionic composition in these solutions. Salinity can cause various salt stresses such as physiological drought (salt-induced drought), potential nutrient and element toxicities or management problems, nutrient imbalances and induced deficiencies, and inhibition of soil water and oxygen due to soil structure breakdown.

Salinity is a measure of the concentration of all the soluble salts in soil or water. It is expressed as deciSiemens per meter (dS m^{-1}) or milliSiemens per centimeter (mS cm^{-1}). Understanding soil salinity levels and the ionic composition across irrigated fields and landscapes is essential for better management. Salinity may exhibit considerable spatial and in depth variability (i.e., salt levels vary from one location to another and in depth) across the agricultural fields because of water movement, infiltration rate, runoff, and evapotranspiration patterns (Shahid and Rehman 2011; Shahid et al. 2010). Among other soil parameters, salinity measurement is one of the simplest and least expensive tools.

The measurement of electrical conductivity (EC) is an indirect way of salinity diagnosis. This can be accomplished by using routine EC meter and/or modern equipment (EM38). The EC meters are not difficult to use and eliminate the guesswork. There are different makes and models of salinity meter to suit a range of budgets and uses. If brackish or saline water is used for irrigation, then an EC meter is a valuable asset.

The choice of equipment/procedure depends upon the purpose of salinity determination, size of the area being evaluated, the depth of soil to be assessed, the number and frequency of measurements needed, the accuracy required, and the availability of manpower. For many reasons, soil scientists still believe that electrical conductivity of soil saturation extract (ECs) is the most common technique for assessing soil salinity and other potential hazards. This is due to the fact that the amount of water that a soil holds at saturation (saturation percentage) is related to a number of soil parameters, such as texture, surface area, clay content, and cation-exchange capacity. The lower soil-water ratio (1:1 or 1:2) makes extraction easier, but cautioned, as less related to field moisture condition than the saturated paste.

Soil sodicity is measured either as sodium adsorption ratio (SAR) or through measurement of exchangeable sodium percentage (ESP) using SAR values or exchangeable Na and cation-exchange capacity values using standard calculation procedures. The SAR requires soluble sodium to be determined in the soil extract from saturated soil paste; this can be achieved by analyzing the extract for Na using flame photometer or atomic absorption spectrophotometer (AAS), or soluble Na can be determined by difference, in samples where soluble K is either absent or present in negligible quantities. The latter requires the ECs as well as total soluble salt (meq l^{-1}) values to find soluble Na by difference. This is usually determined using USDA (Richards 1954) graph (Fig. 4 page 12 Handbook 60), which may suit to US local conditions, but may or may not be suitable for Abu Dhabi Emirate soils.

In this chapter, a new relationship between TSS/ECs is developed that differs from earlier ones developed by USDA (Richards 1954). This has been done through testing a hypothesis.

46.2 Hypothesis to Be Tested

The present study uses same approach as used by the United States Salinity Laboratory Staff (Richards 1954) to develop ECs and salt concentration relationship (Fig. 46.1). The hypothesis to test is that same relationship exists between ECs and total salts on soils of Abu Dhabi Emirate as that of USDA (Richards 1954) and, therefore, relationship as developed earlier (Richards 1954) can be used confidently in the soils of Abu Dhabi Emirate. If this hypothesis is supported, then a separate relationship for Abu Dhabi soils will not always be required and QA on ECs and total soluble salt concentration can be made confidently using the existing USDA relationship (Richards 1954). If however the relationship on ECs and salt concentration on Abu Dhabi Emirate soils does not match with that of USDA (Richards 1954), then QA of ECs and salt concentration is to be made using the developed relationship and recommendations made for sodicity appraisals and salinity management in Abu Dhabi Emirate.

46.3 Study Area

Abu Dhabi is the largest of the seven emirates of United Arab Emirates, accounting for 87% of the country's area. The emirate lies between latitude 22° 29′ N and 24° 53′ N and longitude 56° 10′ E and 51° 37′ E. Abu Dhabi Emirate is subjected to a tropical desert climate with typically high temperatures and low rainfall.

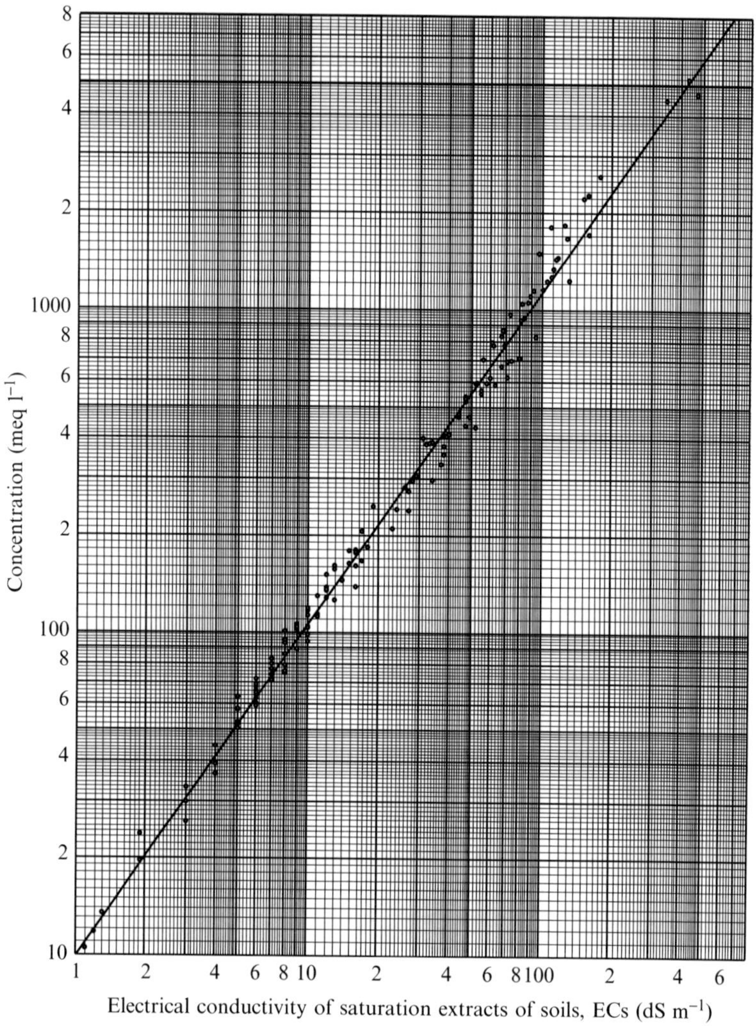

Fig. 46.1 Relationship between electrical conductivity of soil saturation extract (dS m^{-1}) and total salts concentration (meq l^{-1}) from Abu Dhabi Emirate soils

The climatic conditions are arid with extremely harsh and dry summers, when temperatures regularly exceed 45°C, and mild to warm winters with very little, sporadic rainfall, with two distinct seasons: a prolonged dry summer of very high summer temperatures between April and November and a winter of mild to warm temperatures with little rainfall between December and March. The soil temperature regime of most of Abu Dhabi can be regarded as hyperthermic (the mean annual soil temperature is 22°C or higher, and the difference between the mean summer and mean winter soil temperature is more than 6°C either at a depth of 50 cm from the soil surface or at a densic, lithic, or paralithic contact, whichever is shallower) as per Soil Survey Staff (2010). The landscape of the emirate is diverse, ranging from level coastal plains and sabkha to undulating desert sandplain; extensive areas of linear, transverse, and barchan sand dunes; and a single mountainous rocky outcrop, Jabal Hafeet.

46.3.1 Soils of Abu Dhabi Emirate

Recent Soil Survey of Abu Dhabi Emirate (EAD 2009) revealed variety of soils. The objective of this chapter is not to detail these but to give a general overview of the soils. The reader is referred to EAD (2009) and Shahid et al. (2013) for further details on the soil description. At the great group level, the soils are Haplocalcids (0.4%), Petrocalcids (0.04%), Calcigypsids (0.24%), Petrogypsids (2.88%), Aquisalids (3.07%), Haplosalids (6.53%), Torriorthents (0.59%), and Torripsamments (81.09%), whereas rock outcrops account for 0.57% only and the miscellaneous area 0.97% (urban, military, private areas, refilled, dumps, rubbish tips, quarries). This explains that sandy soils (Torripsamments) are the dominant soils in Abu Dhabi Emirate. They occur in the form of sand sheet and various types of sand dunes. These native sandy soils are poor in physical and chemical properties and in inherent soil fertility and, therefore, require significant efforts for economical agriculture.

46.4 ECs and Total Soluble Salts

The ECs values are related to total salt concentration in the liquid phase (solution chemistry) when expressed in meq l^{-1}. In the year 1954, United States Salinity Laboratory Staff published a historical book on Diagnosis and Improvement of Saline and Alkali Soils as Agriculture Handbook 60 (Richards 1954). Later much work has been done on salinity aspects worldwide; however, soil scientists around the world still believe this book to be a foundation to most of the salinity-related work. Among others, this book presents relationships between EC (millimhos cm^{-1} or new unit dS m^{-1}) and single salts ($CaSO_4$, Na_2SO_4, $NaHCO_3$, NaCl, $MgSO_4$, $CaCl_2$, and $MgCl_2$) and related salt concentration in meq l^{-1}. However, soil is a complex natural system consisting of a number of salts from various sources including parent material on which soil developed through a combination of five soil-forming factors (climate, parent material, living organisms, topography, and time). Therefore, an ECs from soil extract represents all salts, and their contribution to EC appears as weighted average conductivities of all salts that the soil is containing. The USDA (Richards 1954) in the historical book published such a relationship as Fig. 4 on page 12 of the Agriculture Handbook No 60. This relationship was developed on soils from widely separated areas in Western United States, which shows higher concentration of salts with respect to EC compared with those curves developed on single salt concentrations.

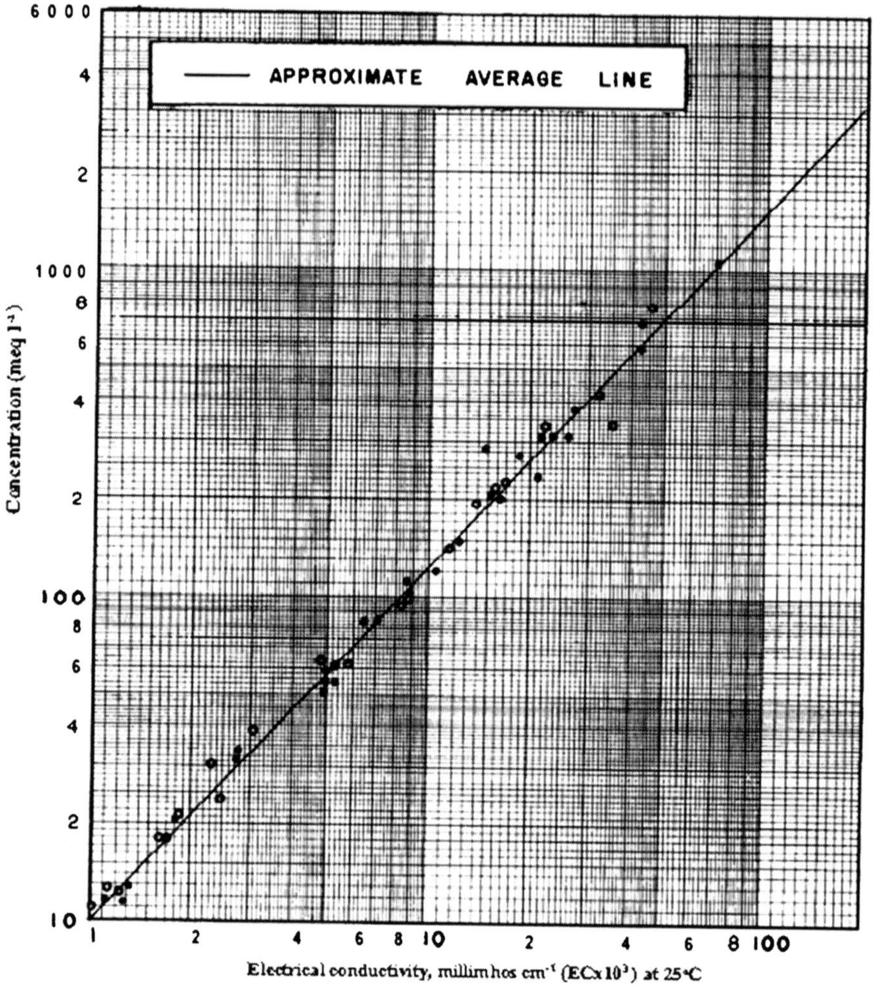

Fig. 46.2 Concentration of saturation extracts of soils in milliequivalents per liter as related to electrical conductivity (cf. Fig. 4, page 12, USDA Handbook 60)

This USDA relationship (Figs. 46.2 and 46.3) has been used intensively around the world by soil scientists and other professionals working on soil salinity issues to assess the quality of the data (ECs, anion and cation concentrations) in saturation extracts for further predictions of sodium adsorption ratio (SAR) which is then used for sodicity assessment (exchangeable sodium percentage – ESP), and to prepare salinity management plans on high ECs level soils.

Soils are heterogeneous, and great difference in soil characteristics occurs within short distances; this has been proved worldwide and recently in Abu Dhabi Emirate (EAD 2009), where a uniform-looking desert landscape presented 62 different soil types at the family level of USDA Soil Taxonomy hierarchy (Soil Survey Staff 1999, 2010).

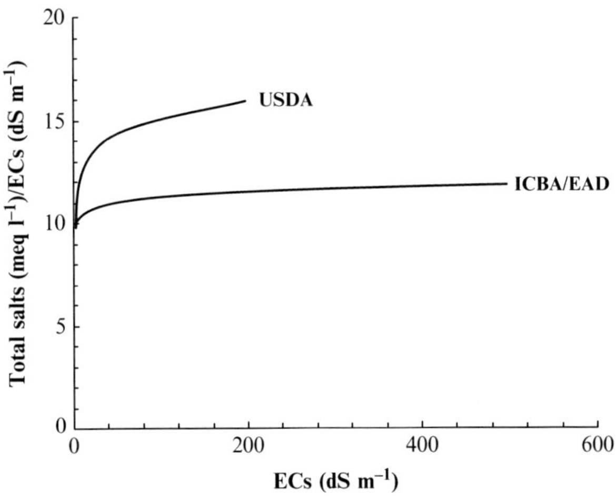

Fig. 46.3 Comparative representation of relationship between ECs and total soluble salts from *USDA* (Richards 1954) curve versus soils of Abu Dhabi Emirate (*ICBA/EAD* curve)

These desert soils are different to those in Western America, and therefore, it is hypothesized that different relationship may exist between total soluble salt concentrations and ECs in Abu Dhabi Emirate soils.

46.4.1 Importance of ECs in Crop Yields

The plant physiologists determine salinity tolerance of plants using the ECs of the root zone and crop yields. Plants can tolerate salinity up to certain levels without a measurable loss in yield (this is called threshold level). At salinity levels greater than the threshold, crop yield reduces linearly as salinity increases. Using the salinity values in a salinity/yield model developed by Maas and Hoffman (1977), predictions of expected yield loss can be made. Typically, plant growth is suppressed when a threshold value of root zone salinity is exceeded. Maas and Hoffman expressed salt tolerance of many plants by this relationship: $\mathrm{Yr} = 100 - s(\mathrm{ECs} - t)$, where Yr = percentage of the yield of plants grown in saline conditions relative to that obtained on nonsaline conditions, t = threshold salinity level where yield decrease begin, and s = percent yield loss per unit increase of ECs (dS m^{-1}) in excess of t. The ECs is the electrical conductivity of soil saturation extract. In this model, it is assumed that crops respond primarily to the osmotic potential of soil solution, and specific ion effects are of secondary importance. Therefore, soil salinity assessment is essential for plant selection growing in salinized fields and to take necessary management actions for better yields. Salinity monitoring helps understand the root zone salinity levels, whether below or above threshold level of crop in the field. The latter will require extra water to be applied based on the leaching fraction to maintain the root zone salinity below threshold salinity.

Soil salinity is a widespread limitation to agricultural production and a feature of land degradation in semiarid and arid region soils throughout the world. The accumulation of soluble salts in the soil profile curtails crop growth by increasing the osmotic potential of the soil solution and inducing specific ion toxicity or nutrient imbalances. Soil structure is affected by salinity and sodicity. A number of approaches have been devised to characterize soil salinity. Most conventional methods employ aqueous or direct extraction of the soil solution and subsequent analysis of salt concentrations. The standard way of establishing soil salinity is through the measurement of electrical conductivity of the soil saturation extract (ECs).

46.4.2 Why Total Salt/ECs Relationship Is Required

There are two main reasons to develop total salt/ECs relationship:

1. Many soil analytical laboratories carry detailed solution chemistry analysis and measure cations and anions in soil saturation extract. Major cations in arid and semiarid region soils are Na^+, K^+, Ca^{2+}, and Mg^{2+}, and anions are CO_3^{2-}, HCO_3^-, Cl^-, and SO_4^{2-}; however, heavily fertilized soils do contain other cations (NH_4^+) and anions (NO_3^-, PO_4^{3-}). In the desert soils of Abu Dhabi Emirate, other ions were mainly undetectable, or their amounts were insignificant compared with major anions and cations. Once the laboratory analytical data is available, the next step is to conduct quality assurance to develop confidence on data quality.

   Soil saturation extracts collected from saturated soil pastes are electrically neutral, and, therefore, the total charges on the cations and anions reported in the analysis should be equal. These total cations and anions have relationship with associated ECs values of the saturation extracts. Therefore, the relationship between total salts and ECs is essential to confirm the quality of the analytical data. This QA test is usually performed in many laboratories by determining total salts (meq l^{-1}) against respective ECs (dS m^{-1}) values using already established relationship by USDA (Richards 1954). Any difference between the reported values and the USDA relationship is taken as guidelines to predict the quality of analytical data (Shahid and Ahmad 2004).
2. In many developing countries, the analytical laboratories lack sufficient modern analytical facilities to determine total cations and anions separately. These laboratories use routine titration procedures (Ca and Mg), and some of the parameters are obtained through difference between total salts as predicted from USDA total salt/ECs relationship (Richards 1954). The common ions determined through difference are Na^+ and SO_4^{2-}. Na is obtained through the following relationship, where K is considered as insignificant quantities in arid and semiarid region soils, for example, the absence of K-containing minerals (mica and microcline feldspar) in Abu Dhabi Emirate soils. This compromise is made in many laboratories to make the analyses affordable, rapid, and cost effective (Bresler et al. 1982).

$$\text{Na} = \left[(\text{Total soluble salts}) - (\text{Ca} + \text{Mg})\right]$$

Na is then used in the following relationship to determine sodicity hazards (sodium adsorption ratio – SAR), which is then used to calculate exchangeable sodium percentage (ESP):

$$\text{SAR} = \text{Na} / \left[(\text{Ca} + \text{Mg})/2\right]^{0.5}$$

$$\text{ESP} = \left[100\,(-0.0126 + 0.01475\text{SAR})/1 + (-0.0126 + 0.01475\text{SAR})\right]$$

where Na and Ca+Mg concentrations are in milliequivalents per liter and SAR is expressed as $(\text{mmol l}^{-1})^{0.5}$.

Firstly, by determining Na through difference, any quantity of K is added into Na, which is overestimated; secondly, the total salts obtained from USDA curve developed from Western American soils may or may not be representative to the study area in the country. Such a practice leads to overestimate sodicity hazard in waters or in saturation extract of soils leading to wrong prediction and management options. Therefore, for each area, total salt/ECs relationship is essential to be developed for correct land management decisions.

46.5 Materials and Methods

For a wider coverage and representation of emirate soils and landscapes, soil samples were collected from the entire emirate (excluding offshore islands) and from different landscape positions, and their GPS coordinates (latitude and longitude) recorded. The present study is part of the major Soil Survey of Abu Dhabi Emirate undertaken between 2006 and 2009 (EAD 2009). The Soil Survey of Abu Dhabi Emirate was jointly conducted by Environment Agency-Abu Dhabi (EAD) and International Center for Biosaline Agriculture (ICBA) through an international contractor, GRM International Pty. Ltd. The survey mapped 62 soil types in the entire emirate at the soil family level of the USDA-NRCS soil hierarchy (Soil Survey Staff 1999, 2010). A total of 22,000 sites have been investigated to a 200-cm depth. Additional 500 sites were studied through typical soil profiles of representative soil families and soil samples collected for physical, chemical, engineering, and mineralogical investigations. The detailed description of soil families (site and morphological description and associated soil analyses) mapped in Abu Dhabi Emirate is reported in Volume I of the soil survey reports (EAD 2009). Reader is referred to these reports for further reading of Abu Dhabi soils. The objective of this study is not to repeat what has been reported in detailed reports. However, the soil analytical results required for this study have been extracted and further analyzed.

To establish the chemical characteristics of the soils, the representative soil samples are subjected to standard routine chemical determinations. To obtain these results, the soil samples are used to prepare standard saturated soil pastes to collect saturation extract under vacuum. The saturation extract is analyzed for the measurement of electrical conductivity, (ECs) for salinity establishment, and for anions (CO_3^{2-}, HCO_3^-, Cl^-, SO_4^{2-}, NO_3^-, PO_4^{3-}) and cations (Na^+, K^+, Ca^{2+}, Mg^{2+}).

Cation and anion contents provide information about the inorganic constituents of the soil solution that impact the soil's suitability for plant growth and many physical and chemical properties. Analyses of soil extracts were carried out using the standard USDA procedures (Burt 2004). The saturation extract was obtained from saturated soil paste through vacuum extraction and filtration and analyzed for soluble Ca^{2+}, Mg^{2+}, Na^+, and K^+ using inductively coupled plasma-atomic emission spectrometry (ICP-AES). Soluble HCO_3^- were determined by titration with 0.011N H_2SO_4 in the presence of methyl orange indicator and CO_3^{2-} by titration with 0.011N sulfuric acid in the presence of phenolphthalein indicator and soluble Cl^-, NO_3^-, PO_4^{3-}, and SO_4^{2-} by ion chromatography.

The analyses were performed at the Central Laboratory Unit (CLU) of the United Arab Emirates University and the Central Analytical Laboratory (CAL) of the Dubai-based International Center for Biosaline Agriculture. The CLU is an accredited laboratory under ISO/IEC 17025:2005. The ionic composition in the saturation extract is expressed on equivalent weight basis rather than on their actual weight basis. Equivalent is a large value and is not expected in soil solutions, and, therefore, a small value as milliequivalent (meq) is used per liter basis. Having measured the cations and anions, according to the above rule, theoretically, the anions must equal cations when expressed on a meq l^{-1} basis and, therefore, is one basis for quality assessment of the chemical soil data through ion balance assessment. Many laboratories accept up to 5% ion balance error (IBE) between total cations and anions as shown by Shahid and Ahmad (2004).

46.5.1 *Correlation Curves*

Correlation between total soluble salts (meq l^{-1}) and ECs (dS m^{-1}) from the soils of Abu Dhabi Emirate was developed by using SigmaPlot graphing software from Systat. The log-log graph (two-dimensional graph of numerical data that uses logarithmic scales on both the horizontal and vertical axes) was used for plotting ECs and total soluble salt values (Fig. 46.1), and simple scatter regression was used to develop total salt/ECs correlation.

46.5.2 *Total Salt and ECs Relationship of USDA (Richards 1954) and Soils of Abu Dhabi Emirate*

Recent publication (Burt 2004) reported the use of ECs (EC saturation extract) to estimate the total cation or anion concentration (meq l^{-1}) of the solution (Richards 1954) as total cations = 10 × ECs (mmhos cm^{-1}) and total anions = 10 × ECs (mmhos cm^{-1}),

Table 46.1 The ECs, associated total salts, and total salts/ECs relationship

USDA Richards (1954)			Soils of Abu Dhabi Emirate EAD (2009)		
ECs (dS m^{-1})	Associated salt concentration (meq l^{-1})	Total salts/ ECs	ECs (dS m^{-1})	Associated salt concentration (meq l^{-1})	Total salts/ ECs
1	10.0	10.00	1	10	10.00
2	21.0	10.50	2	20	10.00
3	32.7	10.90	3	30	10.00
5	56.0	11.20	5	52	10.40
8	96.0	12.00	8	83	10.38
10	123.0	12.30	10	105	10.50
15	190.0	12.67	15	150	10.00
20	262.0	13.10	20	215	10.75
25	330.0	13.20	25	270	10.80
30	405.0	13.50	30	330	11.00
35	480.0	13.71	35	390	11.14
40	560.0	14.00	40	450	11.25
45	635.0	14.11	45	490	10.89
50	710.0	14.20	50	550	11.00
60	870.0	14.50	60	670	11.17
70	1,025.0	14.64	70	790	11.29
80	1,200.0	15.00	80	900	11.25
90	1,370.0	15.22	90	1,000	11.11
100	1,520.0	15.20	100	1,120	11.20
150	2,350.0	15.67	150	1,720	11.47
200	3,200.0	16.00	200	2,275	11.38
			250	2,900	11.60
			300	3,450	11.50
			350	4,050	11.57
			400	4,700	11.75
			450	5,200	11.56
			500	6,000	12.00

where ECs is at 25°C. We believe this relationship is for general prediction of total salts; however, different relationship exists between ECs and total soluble salts at lower end, that is, ECs 1 dS m^{-1} gives TSS 10 meq l^{-1} (1:10), whereas at the upper end, ECs 200 dS m^{-1} gives TSS 3,200 meq l^{-1} (1:16) (Richards 1954).

Using the total salt concentration (meq l^{-1}) and ECs (dS m^{-1}) relationship from United State Salinity Laboratory Staff (Richards 1954) (Fig. 46.2), and similar results from the soils of Abu Dhabi Emirate (Table 46.1), correlation lines were developed. Table 46.1 clearly illustrates that total salt/ECs relationship from USDA curve (Richards 1954) ranges between 10 at 1 dS m^{-1} and 16 at 200 dS m^{-1}, which has also been shown in Figs. 46.3 and 46.4. The ICBA/EAD curve shows better correlation between ECs and total soluble salts ($R^2=0.9711$) compared with that of USDA for same parameters ($R^2=0.9577$). No values reported above 200 dS m^{-1} ECs values in USDA log-log graph. The total salt/ECs relationship ranges between 10 at 1 dS m^{-1} and 12 at 500 dS m^{-1} in the soils of Abu Dhabi Emirate (Figs. 46.3 and 46.4).

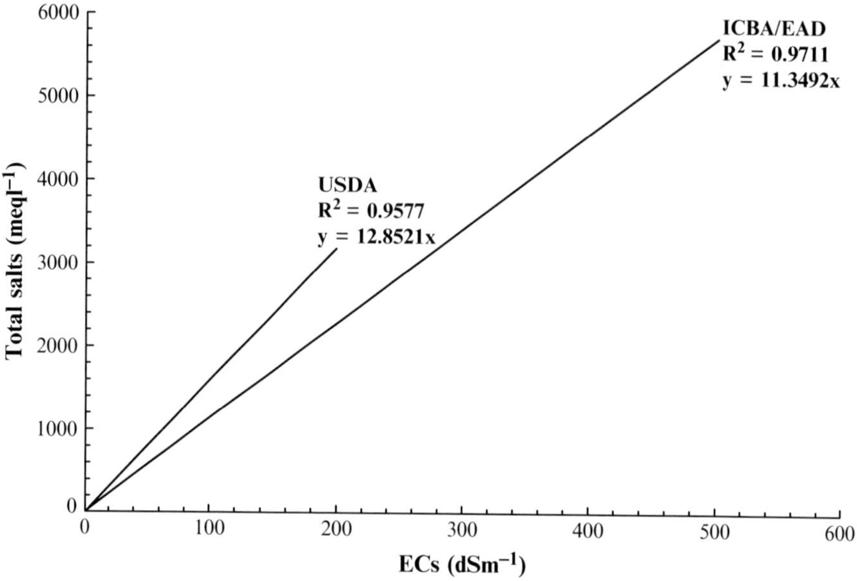

Fig. 46.4 Average lines showing relationship between ECs and total salts from *USDA* (Richards 1954 line) versus soils of Abu Dhabi Emirate (*ICBA/EAD* line)

The nonmatching of both curves confirms that USDA curve as such cannot be used in Abu Dhabi Emirate soils for predictions such as total salts from ECs values and for further quality assurance of soil data or to determine soluble sodium through difference. The USDA curve may be site specific, and relevant to original site from where it was constructed, and could be used where such soils and similar composition of salts may be existing elsewhere in the world.

The soils of Abu Dhabi Emirate present negligible quantities of K in the soil saturation extracts; this can be explained by the absence of K-bearing minerals (mica and microcline feldspar – $KAlSi_3O_8$) in the Abu Dhabi soils; however, minor to major quantities of plagioclase group containing both albite ($NaAlSi_3O_8$) and anorthite ($CaAl_2Si_2O_8$) were detected in most of Abu Dhabi soils (EAD 2009). This also supports that the curve developed in the present study can be confidently used to determine soluble Na by difference to calculate SAR values and consequently the ESP values.

46.5.3 Validation of New ECs and TSS Relationship (ICBA-EAD) Through Measuring SAR

In order to validate the newly developed relationship between ECs and TSS on Abu Dhabi soils, we made an attempt to calculate SAR by following ways (Table 46.2).

Table 46.2 Measurements of SAR in different soils by various methods[a]

Horizon	Depth	ECs	Ca	Mg	Na	K	SAR			
							$(mmol\ l^{-1})^{0.5}$			
							[b]EAD	[c]USSL	[d]USDA	[e]ICBA/EAD
	cm	$dS\ m^{-1}$	$meq\ l^{-1}$							
Ck2	60–110	5	25	5	23	0.6	5.90	7.23	5.16	5.68
Ckz	45–60	10	46	42	306	4	23.60	30.08	22.71	24.18
Bky2	80–130	10	25	3	84	1	22.40	25.92	19.24	20.58
Bkyz	15–50	51	50	38	480	9	72.40	95.28	63.62	69.65
2Ckz	90–135	55	47	37	480	6	74.10	98.91	71.91	73.45
2Ckyz1	80–110	101	66	103	990	14	107.70	155.67	91.49	104.00
Ckz1	8–45	101	25	47	1,063	22	177.20	254.67	156.33	175.50

[a,b]EAD: We used actual values of Na and Ca + Mg from EAD (2009) to calculate SAR
[c]USSL: We determined TSS against ECs (Figure 4 page 12 Handbook 60), and Na was obtained by difference [Na = TSS – (Ca + Mg)]. Ca + Mg were taken from EAD (2009)
[d]USDA: We determined TSS against ECs by using factor of 10 (TSS = ECs × 10), and Na was obtained by difference [Na = TSS – (Ca + Mg)]. Ca + Mg were taken from EAD (2009)
[e]ICBA-EAD: We determined TSS against ECs using ICBA-EAD curve (Fig. 46.1), and Na was obtained by difference [Na = TSS – (Ca + Mg)]. Ca + Mg were taken from EAD (2009)

Table 46.2 illustrates SAR values from four methods of calculation. The comparison shows that the SAR values calculated by using newly developed relationship between ECs and TSS (ICBA-EAD curve) give close values as to EAD, confirming negligible quantities of potassium in Abu Dhabi Emirate soils.

The SAR values determined by USSL method are higher relative to those measured by EAD and through using ICBA-EAD newly developed curve; however SAR values calculated by USDA relationship are lower. This implies that a factor of 10 (TSS = ECs × 10) as given by USDA cannot be used on all ranges of ECs; this factor works well at lower ECs level. From above it is concluded that ICBA-EAD curve can be confidently used to obtain TSS against ECs and to calculate Na by difference. This will avoid analysis of Na in the laboratory, saving precious time and resources, and also will help determine quality of laboratory results of solution chemistry. Notably, this curve is applicable to soils where potassium is either absent or present in insignificant amount.

46.6 Conclusions

The present study on relationship between total salts and ECs from the soils of Abu Dhabi Emirate and its comparison with standard USDA relationship has rejected the hypothesis that same relationship exists between ECs and total salts on soils of Abu Dhabi Emirates as that of USDA and, therefore, relationship as developed earlier by USDA cannot be used for Abu Dhabi Emirate conditions. Therefore, a separate relationship developed through present study is recommended to avoid misinterpretation of soil analytical data for QA purposes and soil management. In the light of present study, it is recommended that other regions, where soil and environmental conditions similar to Abu Dhabi are occurring, develop similar relationship most suited to their local conditions.

References

Bresler E, McNeal BL, Charter DL (1982) Saline and sodic soils. Principles-dynamics-modelling, Advanced series in agricultural sciences 10. Springer, Berlin/Heidelberg/New York, pp 1–236

Burt R (ed) (2004) Soil survey laboratory methods manual. SSIR No 42, Ver 4.0. US Govt Printing Office. Washington, DC

EAD (2009) Soil survey of Abu Dhabi Emirate-Extensive survey. Environment Agency-Abu Dhabi, United Arab Emirates, Volume I, p xx + 506

Maas EV, Hoffman GJ (1977) Crop salt tolerance-current assessment. J Irrig Drain Div Am Soc Civ Eng 103:115–134

Richards LA (ed) (1954) Diagnosis and improvement of saline and alkali soils. USDA Handbook No 60. Washington, DC

Shahid SA, Ahmad S (2004) Guidelines for laboratory quality controls and quality assessment of soil analytical data. ERWDA Soils Bulletin No 1, p viii + 30. Published by Environment Agency Abu Dhabi, United Arab Emirates. (ISBN 9948-408-25-x). Available on line (www.ead.ae)

Shahid SA, Rehman K (2011) Chapter 2: Soil salinity development, classification, assessment and management in irrigated agriculture. In: Mohammed Passarakli (ed) Handbook of plant and crop stress, 3rd edn. The University of Arizona, Tucson, AZ 85721. Published by CRC Press, pp 23–39

Shahid SA, Abdefattah MA, Omar SAS, Harahsheh H, Othman Y, Mahmoudi H (2010) Mapping and monitoring of soil salinization-remote sensing, GIS, modeling, electromagnetic induction and conventional methods-case studies. In: Mushtaque Ahmad, Salim Ali Al-Rawahy (eds) Proceedings of the international conference on soils and groundwater salinization in Arid countries. Published by Sultan Qaboos University, Muscat Sultanate of Oman, pp 59–97

Shahid SA, Abdelfattah MA, Othman Y, Kumar A, Taha FK, Kelley JA, Willson MA (2013) Chapter 1: Innovative thinking for sustainable use of terrestrial resources in Abu Dhabi Emirate through scientific soil inventory and policy development. In: Shahid SA, Taha FK, Abdelfattah MA (eds). Developments in soil classification, land use planning and policy implications-innovative thinking of resource inventory for sustainable use and management of land resources, pp 3–50

Soil Survey Staff (1999) Soil taxonomy: a basic system of soil classification for making and interpreting soil surveys, 2nd edn. USDA-Natural Resources Conservation Service, Agriculture Handbook Number 436. US Govt Printing Office, Washington, DC

Soil Survey Staff (2010) Keys to soil taxonomy, 11th edn. US Govt Printing Office, Washington, DC

Chapter 47
Quality Assurance Standards: USDA Perspective of the Extensive Soil Survey of Abu Dhabi Emirate

John A. Kelley, Michael A. Wilson, Mahmoud A. Abdelfattah, and Shabbir A. Shahid

Abstract Soil scientists from USDA-Natural Resources Conservation Service were invited by the International Center for Biosaline Agriculture (ICBA) and the Environment Agency-Abu Dhabi (EAD) to participate in a quality assurance review of the extensive soil survey of Abu Dhabi. The review was of interest to USDA not only due to the application of US soil survey mapping and classification standards but the emphasis on developing soils within the emirate into useful and productive agricultural areas. Meetings were conducted with scientists of ICBA and EAD and the soil survey management team of GRM International (Australian mapping contractor) to review various aspects of soil mapping, soil survey documentation, and day-to-day operational procedures. Field visits included examination of representative polygons of preselected map units with on-site investigation by backhoe and hand dug pits traversing a widely diverse set of landscapes and landforms across

J.A. Kelley (✉)
United States Department of Agriculture-Natural Resources Conservation Service, Raleigh, NC, USA
e-mail: bettmark.john@gmail.com

M.A. Wilson
United States Department of Agriculture-Natural Resources Conservation Service, Lincoln, NE, USA
e-mail: mike.wilson@lin.usda.gov

M.A. Abdelfattah
Soil and Land Use Management, Environment Agency-Abu Dhabi, P.O. Box 45553, Abu Dhabi, UAE

Soils and Water Sciences Department, Faculty of Agriculture, Fayoum University, Fayoum, Egypt
e-mail: mabdelfattah@ead.ae; mahmoudalia@yahoo.com

S.A. Shahid
International Center for Biosaline Agriculture, P.O. Box 14660, Dubai, UAE
e-mail: s.shahid@biosaline.org.ae

S.A. Shahid et al. (eds.), *Developments in Soil Classification, Land Use Planning and Policy Implications: Innovative Thinking of Soil Inventory for Land Use Planning and Management of Land Resources*, DOI 10.1007/978-94-007-5332-7_47,

several regions of the Abu Dhabi Emirate. Processes for soil pedon examination, recording of soil properties, classification of soils, mapping procedures, and determination of map unit composition were evaluated. Additional methods used within the USA were presented for consideration when conducting future soil surveys. Laboratory procedures used in the analysis of soil samples were based primarily on USDA-NRCS methods, and derivations unique to the Abu Dhabi soil survey were reviewed. In support of ongoing cooperative efforts between the agencies of USDA, ICBA, and EAD, soil samples were collected for detailed analysis in an effort to develop new methodology for identification and quantification of anhydrite, as well as elucidate the mechanism and controlling factors for the formation of this mineral in soils of the region. Field and laboratory methods, standards, and procedures implemented by the project team (both ICBA/EAD and GRM) with their high level of technical skills, knowledge, and experience have ultimately yielded an array of high-quality soil survey products greatly enhancing the wise use of one of the region's most precious natural resources—its soil.

Keywords Abu Dhabi Emirate • Map unit • Quality assurance • Soil mapping • USDA-NRCS

47.1 Introduction

Soil survey quality assurance is a process of providing oversight and review to ensure that soil survey products meet the stated objectives outlined in their guiding documents, such as "terms of reference and scope of work" for the soil survey of Abu Dhabi Emirate (ICBA-EAD 2005). The extensive soil survey of Abu Dhabi Emirate was completed using fourth-order survey standards of USDA with some modifications to fit Abu Dhabi conditions. A total of 22,000 observations over an area of 5.5×10^6 ha were completed. The survey covered the entire emirate but excluded offshore islands, restricted areas such as military and urban, and the previously surveyed coastal land area. More information can be seen in Chap. 1 of this book (Shahid et al. 2013).

Assistance was requested from USDA-Natural Resources Conservation Service (NRCS) by the Dubai-based International Center for Biosaline Agriculture (ICBA) and the Environment Agency-Abu Dhabi (EAD) for a quality assurance (QA) review of the ongoing extensive soil survey of Abu Dhabi. Two soil scientists (John Kelley and Michael Wilson) conducted a QA review of extensive soil survey of Abu Dhabi Emirate (Fig. 47.1a) from November 7 to 20, 2008.

Phase I is an extensive soil survey of the entire emirate conducted at a scale of 1:100,000 (EAD 2009a), while phase II is an intensive survey of 400,000 ha of the lands with the greatest agricultural potential conducted at a scale of 1:25,000 (EAD 2009b). The review is of interest to USDA not only due to the application of the US

Fig. 47.1 (**a**) USDA staff visit to UAE University laboratory; (**b**) USDA staff meeting with high-level project management; (**c**) USDA staff interaction with project staff; (**d**) witnessing sand spear—an innovative equipment to sample loose sandy samples; (**e**) USDA staff met soil mineralogist at UAE University; (**f**) sampling anhydrite profiles in the coastal sabkha

soil survey mapping and classification standards (Soil Survey Division Staff 1993; Schoeneberger et al. 2002; Soil Survey Staff 1999, 2006) but also the emphasis on developing soils within this country into useful and productive agricultural areas. The effort to use soil survey to identify these areas, to determine the limitations for agricultural use, and then to identify methodologies to overcome these limitations has worldwide implications for land-use management.

47.2 Review Process

47.2.1 Initial Familiarization

In the initial stage of the review, meetings with scientists of ICBA and EAD and the soil survey management team of GRM International (Australian mapping contractor) were conducted to familiarize the review team with the soil survey project terms of reference (TOR) and scope of work (ICBA-EAD 2005). Additional time was spent examining various aspects of the mapping, soil survey documentation, and day-to-day operational procedures used by field staff and the management team (Fig. 47.1b).

47.2.2 Field Visits

The review team was provided ample opportunity to examine office, laboratory, and field mapping protocols. Field visits included examination of representative polygons of preselected map units with on-site investigation by backhoe and hand dug pits. A widely diverse set of landscapes and landforms were traversed across several regions of the Abu Dhabi Emirate. The process for soil pedon examination, recording of soil properties, and classification of soils was demonstrated by the survey staff. Required information needed in pedon descriptions and concepts regarding use of horizon suffixes, uniformity of descriptions, and cementation classes were also discussed and ideas for their usage exchanged (Fig. 47.1c).

At the initial field site, a new and unique sampling procedure (use of the sand spear) was demonstrated that will prove to be very beneficial in examining/sampling similar soils in the USA (Fig. 47.1d). A typical soil survey field office was visited, and mapping and documentation procedures used by the soil survey crew leaders were demonstrated. Noted was the excellent working relationship (rapport and informational exchange) between the management and field mapping personnel of GRM International. Interaction of this nature is critical to the success of any survey project.

47.2.3 Databases

Database management of field information was initially based on the USDA pedon data entry system and later modified to meet the unique needs of the Abu Dhabi survey. Pedon descriptions were scanned and entered into a system that appeared very user friendly. The system utilized facilitated data entry and examination of completed pedon descriptions. Data output was formatted to allow pedon descriptions to have a universal format. A sufficient system of quality assurance of these data is in place from training of the field soil surveyors, review of the collected data by the senior soil scientist, and QA of the computerized data compared to field sheets. The soil information system is called "Abu Dhabi Soil Information System (ADSIS)," and it is now available online at www.adsis.ae.

47.2.4 Mapping Procedures and Determination of Map Unit Composition

Remote sensing of Landsat imagery facilitated the creation of map units. The process of initial delineation of mapping units or polygons was routinely completed in the office, followed by on-site transects of 12–15 points to discern polygon boundary placement. GRM has established a standard for minimal field information required. The procedures used for determining the composition of mapping units and the development of map unit names were reviewed. A variety of methods used within the USA were discussed, and the advantages and disadvantages are identified. Methods and analysis used by the survey staff, although somewhat modified from the USDA procedures, are well within the principles established in the USDA procedural guide.

All pedon descriptions examined were accurate and well documented. During site investigations, soil survey project mapping crew leader and members were eager to interact and discuss taxonomic criteria as well as field procedures with the USDA representatives in order to validate record collection/documentation according to USDA standards. At several of the sites, they enthusiastically demonstrated their field procedures for determining chemical and physical soil properties (e.g., electrical conductivity, pH). The knowledge and skill level by the field staff of complex landscape/soil relationships was very apparent.

47.2.5 Laboratory Data

Laboratory procedures used in the analysis of samples were based on USDA-NRCS methods (Burt 2004). Deviations from current USDA methodologies are generally procedural and would not impact the overall quality of the data or the placement of pedons in the soil classification system. One deviation noted that may need to be further explored and tested was the analysis of 33-kPa water retention (water retained at field capacity) using sieved, <2-mm soil material rather than natural soil aggregates as is common in the NRCS laboratory.

In addition, the validation and cross-checking of laboratory data for quality was discussed. This process is ongoing via interlaboratory comparison of data with the USDA-NRCS laboratory and a laboratory at the University of Western Australia. Also, the mineralogy data was examined. There were concerns regarding the methodology for determining the semiquantitative mineral classes from the results of x-ray diffraction analysis.

47.3 Summary

Over the extent of the visit, the excellent spirit of cooperation and technical expertise shared among the representatives of the ICBA/EAD and the staff of GRM became very apparent. The ICBA/EAD and GRM International teams exhibited a sincere interest in making changes or implementing recommendations identified by the USDA representatives. Detailed discussions were conducted throughout the visit (most in an informational context). Suggestions were directed toward immediate improvements in the soil survey, ways to best utilize the completed soil survey products (descriptions, maps, laboratory data, etc.), and the identification of future needs.

The dedication of project management and project staff (both ICBA/EAD and GRM) with their high level of technical skills, knowledge, and experience has ultimately yield high-quality soil survey products greatly enhancing the wise use of one of the regions' most precious natural resources.

47.4 USDA Evaluation

The following observations were offered as part of the USDA QA review:

- GRM International is meeting and/or exceeding USDA standard field mapping procedures and data collection protocols for an order 4 soil survey based on our review of data in the office and field.
- The examination of typical pedons revealed that properties described fit within the range of taxonomic criteria ensuring correct classification of soils using the USDA Soil Taxonomy system.
- The system of soil correlation currently being used by GRM is consistent with USDA standards and provides for an accurate accounting of the various soil components commonly found in mapping units. It was suggested that documentation of naming conventions for mapping units and a summary explanation of map unit composition be provided in the final soil survey report. An initial review draft of this documentation was provided by GRM personnel and reviewed and approved by the USDA staff.
- The database appears to be designed for future use in soil survey maintenance and updates and capable of storing of additional data when the survey is complete.
- The laboratory procedures appear to be established, well designed, and sufficiently tested (Fig. 47.1e). The overall completed set of laboratory data should be cross-checked for potential analytical problems (Shahid and Ahmad 2004) as some inconsistencies (revealing possible analytical errors) were apparent in some older

data. This task appears to be ongoing with interlaboratory comparisons as well as examination of internal data consistency. These consistency checks should include evaluation of sample dispersion during particle-size analysis by use of the 1,500-kPa water retention/clay ratio, as well as use of TPL (total pretreatment loss) and LAT (loss on acid treatment) analyses to validate other extraction/quantification methods for salts, gypsum, and carbonates.

- The laboratory procedures of TPL and LAT should be better documented in the soil survey. They are cited as based on USDA procedures, but these methods are not currently used within the NRCS laboratory. The TPL procedure is used only (and infrequently) within USDA as a pretreatment method for particle-size analysis and was not designed within our laboratory to be quantitative. This is not to say that the method is not reliable, quantitative, or useful but appears to the reviewers as a more fully developed method than has been available in the past.
- Reporting conventions should be modified regarding reporting of laboratory data. Modifications should specifically target lower reported values, including an understanding and reporting of analytical detection limits and establishing "trace" and "nondetectable" thresholds. Elimination of "zero" values is necessary in the reported data. Also, blank values and dashes in the data tables should be defined. Within the USDA system, "tr" is used for measurable values below quantitative limits (e.g., trace), "--" for analyses performed but no analyte detected, and a blank is used when the analysis was not performed.
- The method to determine the semiquantitative class for reporting soil minerals by x-ray diffraction should be documented in the soil survey. This documentation should include the initial mineral quantification procedure used (e.g., relative mineral percentage based on peak height above background) as well as specific class placement. The class placement may be unique to the survey based on reasonable distribution of results.
- The reporting of Atterberg limits within the survey laboratory data differs from standard USDA protocols (Soil Survey Staff 2009) that are based on American Society for Testing and Materials' standards (Method ASTM D 4318). Within this ASTM method, liquid limit and plastic limit are reported as percent water content of the soil in the defined states, while plastic index is the range of water content where the soil behaves plastically and is defined as the difference in water content between the two states. For samples where either or both the liquid limit or plastic limit cannot be measured, the plastic index is reported as "nonplastic."

47.5 Continued Cooperation Between USDA and ICBA/EAD

The cooperative efforts between the agencies of USDA, ICBA, and EAD were initiated over a year ago with questions of the presence and quantification of the mineral anhydrite in soils of coastal and inland sabkhas. Cooperation has been further facilitated by our review of the use of USDA soil survey standards and soil taxonomy for the ongoing soil survey of the emirate of Abu Dhabi. Use of the USDA standards in this desert environment will help us to further test the applicability of soil survey mapping protocols around the world. The fact that Abu Dhabi Emirate has invested in a soil survey and is interested in protecting and preserving soils for agriculture production is very commendable. The continued cooperation between the ICBA, EAD, and USDA will be valuable for all parties.

The most immediate avenue for cooperation has been the sampling of selected soils in Abu Dhabi. A sequence of pedons along a coastal sabkha from the tidal influenced soils to inland areas were evaluated and sampled to develop the methodology for identification and quantification of anhydrite, as well as elucidate the mechanism and controlling factors for the formation of this mineral in soils of the region (Fig. 47.1f). Collected samples were shipped to the USA and analyzed by the NRCS Soil Survey Laboratory.

A quantitative method for anhydrite has been developed using the difference between the routine water-acetone procedure of NRCS to quantify gypsum and anhydrite together with a weight loss procedure for quantification of gypsum alone (Wilson et al. 2012). This study found an increasing amount of anhydrite in soils furthest from the coast. These results should help refine the maps of the Abu Dhabi coastal area.

Future efforts in cooperation can result in developing land-use interpretations for the survey of the emirate. Once developed, the application will prove useful to surveys both in the USA and in other arid areas of the world.

47.6 Future Development and Use of Soil Survey in Abu Dhabi

The information (maps, field and laboratory data) collected during the soil survey will prove to be a very valuable resource tool for land management and development. It is vital that a well-trained and experienced staff be maintained to interpret and help utilize the spatial and tabular data. For example, in the USA, four to six experienced soil scientists per state are employed on a full-time basis in order to provide technical soil services. The provided services are directed toward specific as well as general user needs and include interpretation of soil data, education, and limited training. Natural resource education for the public must begin at a young age to ensure an appropriate respect for soil.

In addition to the resource soil scientists, a cadre of over 400 soil scientists is located in 144 permanent soil survey offices across the USA. These offices are assigned responsibility for major land resource areas and cross provincial (state and county) boundaries. Each office is responsible for the maintenance and update of the published soil surveys.

The US soil survey information is available on the Internet at http://websoilsurey.nrcs.usda.gov/app/HomePage.htm. This site has greatly expanded the accessibility of soil information by the public. It has fostered greater use of natural resource information as well as increased contact with users. This contact with the public is important to ensure the appropriate use and interpretation of the soil survey information.

The mapping of the soils is only the beginning in terms of natural resource management and protecting this valuable resource. Soil scientists must be available to provide assistance to agricultural persons and others in understanding utility and limitations of the maps and provide a range of assistance on topics such as suitability of soils for irrigation, nutrient management, land reclamation, or erosion and salinity control. Soils are one part of the ecosystem that controls the sustainability of life. Desertification is an extremely important problem to this region that must be addressed by the understanding of the soil. How we manage soils and what tools are appropriate to use affect the long-term sustainability of this resource.

The utilization of the survey is not limited to agriculture. The soil survey is the primary tool to prevent loss of agricultural lands by urbanization. The importance of soils is also related to urban land use and the environment, such as construction of homes and buildings, wind or water erosion, and groundwater protection. Increasingly, there is a greater problem with contamination of urban soils by pathogens and other contaminants. Many of the diseases related to human and animal health are soilborne problems, often a deficiency or excess of certain trace elements. Food security and safety and the global carbon cycle are also important issues that are addressed by this science.

Thus, once the soil survey is made available to the public, a great demand for the time and expertise of soil scientists can be expected. The level of expertise of these scientists must be maintained in order to ensure an in-depth knowledge of soils and soil survey, but also an understanding of how the survey process is conducted and the benefits and limitations of the product. Soil scientists are increasingly asked to help develop public policy and regulations related to land-use management. Maintaining a well-trained cadre of soil scientists is critical for full utilization of the soil information.

47.7 Conclusions and Recommendations

It is gratifying that the soil survey standards of USDA-NRCS are being used and successfully applied in the soil survey of the Abu Dhabi Emirate (EAD 2009a, b). Universal application of procedural guides such as those provided in the USDA National Soil Survey Handbook and Soil Survey Manual ensure uniformity and consistency in soil survey products worldwide. The use of soil taxonomy as an international classification system for the mapping and interpretation of soils was an original intent of the designers of the system. Using this system has proven to be both challenging and rewarding to the Abu Dhabi survey team. The research and contributions by EAD soil scientist Dr. Mahmoud Abdelfattah and ICBA soil scientist Dr. Shabbir A Shahid in the area of anhydrite mineralogy are acknowledged and greatly appreciated.

The field survey staff, the survey management team, and the sponsoring administrators are to be commended for conducting an outstanding soil survey program. Their expertise in soil science and management has resulted in the development of an excellent soil survey product that will prove to be of great value to the people of Abu Dhabi for years to come.

References

Burt R (ed) (2004) Soil survey laboratory methods manual. United States Department of Agriculture, Natural Resources Conservation Service, Soil survey investigations report no 42, version 4.0

EAD (2009a) Soil survey of Abu Dhabi Emirate. Extensive survey, vol I. Environment Agency, Abu Dhabi, United Arab Emirates, p xx+506

EAD (2009b) Soil survey of Abu Dhabi Emirate. (Intensive survey), vol III. Environment Agency, Abu Dhabi, United Arab Emirates, p xviii+435

ICBA-EAD (2005) Terms of reference and scope of work. Soil survey of Abu Dhabi Emirate. Jointly prepared by International Center for Biosaline Agriculture and Environment Agency Abu Dhabi, p 107

Schoeneberger PJ, Wysocki DA, Benham EC, Broderson WD (eds) (2002) Field book for describing and sampling soils, version 2.0. Natural Resources Conservation Service, National Soil Survey Center, Lincoln, NE

Shahid SA, Ahmad S (2004) Guidelines for laboratory quality controls and quality assessment of soil analytical data. ERWDA Soil Bulletin No 1, p viii+30

Shahid SA, Abdelfattah MA, Othman Y, Kumar A, Taha FK, Kelley JA, Wilson MA (2013) Chapter 1: Innovative thinking for sustainable use of terrestrial resources in Abu Dhabi Emirate through scientific soil inventory and policy development. In: Shahid SA, Taha FK, Abdelfattah MA (eds). Developments in soil classification, land use planning and policy implications-innovative thinking of resource inventory for sustainable use and management of land resources, pp 3–50

Soil Survey Division Staff (1993) Soil survey manual. Soil Conservation Service. US Department of Agriculture Handbook 18

Soil Survey Staff (1999) Soil taxonomy-a basic system of soil classification for making and interpreting soil surveys, 2nd edn, Agriculture handbook no. 436. United States Department of Agriculture – Natural Resources Conservation Service, Washington, DC, p 869

Soil Survey Staff (2006) Keys to soil taxonomy, 10th edn. US Government Printing Office, Washington, DC

Soil Survey Staff (2009) Soil survey field and laboratory methods manual. In: Burt R (ed) Soil survey investigations report no 51, version 1.0. US Department of Agriculture, Natural Resources Conservation Service

Wilson MA, Shahid SA, Abdelfattah MA, Kelley JA, Thomas JE (2013) Chapter 8: Anhydrite formation on the coastal sabkha of Abu Dhabi, United Arab Emirates. In: Shahid SA, Taha FK, Abdelfattah MA (eds). Developments in soil classification, land use planning and policy implications-innovative thinking of resource inventory for sustainable use and management of land resources, pp 175–201

Chapter 48
The Role of Mycorrhiza in the Reclamation of Degraded Lands in Arid Environments

Ghazi N. Al-Karaki

Abstract Land disturbance and degradation is recognized as one of the most important environmental problems worldwide caused by many factors like human activities and adverse climatic factors like that occurred in most parts of the arid and semiarid regions (e.g., the Middle East and Arabian Peninsula). In the arid environments, land degradation is mainly caused by wind erosion and salinization with loss of productive surface soil and loss of vegetation as primary indicators. In these regions, soil conservation and rehabilitation of degraded lands are essential for sustainable agriculture and improvement of dry land ecosystem. Revegetation is one of the most effective means to control soil degradation and to rehabilitate degraded lands. However, in arid environments, low rainfall, harsh climatic conditions, and frequent droughts are major limitations for natural rehabilitation. There is a general consensus that biotechnology can be a valuable tool to mitigate water scarcity and to improve quality of degraded lands. Microbial technology, e.g., use of mycorrhizal fungi, has been considered a valuable tool in the rehabilitation of disturbed and degraded lands. Mycorrhizal fungi play a crucial role in enhancing plant growth and survival through enhancing plant nutrient uptake, water relations, ecosystem establishment, plant diversity, and productivity of plants. Mycorrhiza also protects plants against root pathogens and abiotic stresses such as drought and salinity and improves soil structure by enhancing soil aggregation and water-holding capacity. This chapter provides an insight into how mycorrhizal fungi might play a role in reclamation and revegetation of degraded lands in arid regions.

Keywords Arbuscular mycorrhiza • Degradation • Disturbed lands • Rehabilitation • Revegetation

---

G.N. Al-Karaki (✉)
Faculty of Agriculture, Jordan University of Science and Technology,
P.O. Box 3030, Irbid, Jordan
e-mail: gkaraki@just.edu.jo

S.A. Shahid et al. (eds.), *Developments in Soil Classification, Land Use Planning and Policy Implications: Innovative Thinking of Soil Inventory for Land Use Planning and Management of Land Resources*, DOI 10.1007/978-94-007-5332-7_48,

48.1 Introduction

Land disturbance and degradation is currently recognized as one of the most important environmental problems worldwide especially in arid and semiarid regions due to various factors including climatic variations and human activities (UNCCD 2009). The major human activities which cause land degradation include urbanization, nonregulated overgrazing, cultivation (overexploitation of land and water resources), deforestation, mining disturbances, and road construction (Skujins and Allen 1986; Lambin et al. 2001). The perturbations associated with land degradation include denudation, erosion, drought, nutrient deficiencies, increased heavy metal concentration, salinization, diminution of belowground microbiota with associated decrease in nutrient cycling, and ecosystem energy budget changes.

The primary visible evidence of land degradation is the loss of productive soil and vegetation, and the effects have been most pronounced in the arid and semiarid regions like those found in most parts of the Arabian Peninsula and the Middle East regions. The conservation and rehabilitation of degraded lands are essential parts of sustainable agricultural development and ecosystem improvement. In order to reclaim degraded lands, revegetation is considered very important practice (Al-Karaki 2011a; Scherr and Yadav 1996). However, the low availability of soil moisture due to scarce and irregular precipitation and frequent drought is a major obstacle to the successful revegetation of lands existing in the arid and semiarid regions. Urbanization and continuing population growth, however, are seriously threatening the sustainable use of natural resources in these regions. Recently, efforts are made by the governments and research centers to identify means of tackling threats to existing natural resources that have been brought upon by the rapidly rising levels of water consumption. Overexploitation and recycling have depleted the groundwater resources to an alarming extent and increased water salinity (Jaradat 2005; Rimawi and Al-Ansari 1997). The areas where groundwater is used for farming, landscaping, or afforestation are showing the signs of severe soil salinity in many parts of arid regions (e.g., the Arabian Peninsula). Millions of trees planted during the last decade are dying off or being damaged as a result of drought and/or salinity stress. Dust storms and land erosion are becoming increasingly numerous where there is increase in soil salinity (Wiede 2005).

Long-term success in establishing plant growth under arid or semiarid conditions can only be seen in the context of soil improvement that takes into consideration the close monitoring of soil alkalinity, salinity, and nutrition deficiency and the use of appropriate water supply for sustainable agricultural development (Al-Karaki 2011a; Wiede 2005). Many soil amendments have been considered in the Arabian Peninsula region to replenish soil organic matter. A disadvantage of some of these commercial soil amendment products is that they function only for a few years before they decompose or dissolve (Wiede 2005).

There is a general consensus that biotechnology can be used as a valuable tool to mitigate water scarcity and improve land quality. Microbial technology, e.g., use of mycorrhizal fungi, has been considered a valuable tool in the rehabilitation of disturbed and degraded lands. Mycorrhizal fungi are extremely important to improve soil health and in turn to the health of plants, which might in turn make a critical contribution in reducing the degradation of lands through enhancing revegetation programs. This chapter provides an insight into how mycorrhizal fungi might play a role in reclamation and revegetation of degraded lands in arid regions.

48.2 What Are the Mycorrhizal Fungi?

Mycorrhizae are beneficial symbiotic fungi that form an association on the roots of most of the world's plants (Marschner 1995). Mycorrhizal fungi are found in most natural terrestrial ecosystems including arid and saline environments (Brundrett et al. 1996). The major two most common associations are the arbuscular endomycorrhizae (AM) and the ectomycorrhizae. The ectomycorrhizal fungi are important in about 3% of plant species which are mostly forestry trees, whereas AM fungi are the most widely spread and are associated with the roots of over 80% of the terrestrial plant species (Smith and Read 2008) including halophytes (Khan 2003; Khan and Belik 1995). Therefore, in this chapter discussions will be restricted to AM fungi.

Bidirectional movement of nutrients characterizes the fungus plant symbioses, where carbon flows to the fungus and inorganic nutrients move to the plant, thereby, providing a critical linkage between the plant root and soil (Finlay 2008; Gupta et al. 2000). Mycorrhizal plants, in comparison with nonmycorrhizal plants, have greater nutrient and water uptake capacity from soil because AM fungi develop an extensive network of external fungal filaments (hyphae) that act as an extension of the root absorbing area (Fig. 48.1).

AM fungi associations have several advantages for their hosts, including increased growth and yield and reproductive success (Al-Karaki 2000; Al-Karaki and Hammad 2001), nutrient acquisition (Al-Karaki 2006; Stanley et al. 1993), reducing fertilizer requirements (Gemma et al. 1997; Al-Karaki et al. 2007), providing protection against some root pathogens (Azcón-Aguilar et al. 2002; Govindaraju et al. 2005; Whipps 2004), improving water relations (Subramanian et al. 1997; Al-Karaki et al. 2004), improving soil structure through glomalin excreted by fungal hyphae (Bearden and Petersen 2000), and greatly improving tolerance to drought and salinity stresses (Al-Karaki et al. 2001; Barea and Jeffries 1995; Kaya et al. 2009; Yano-Melo et al. 2003; Tian et al. 2004). Knowledge of the relationship between plants and the fungi is important for successful utilization of AM fungi under particular conditions.

48.3 Mycorrhizal Fungi in Degraded Soils

Mycorrhizal associations are the most widespread symbiosis between plants and microorganisms (Marschner 1995). AM fungi are found in most natural terrestrial ecosystems including dry environments which are characterized by harsh and

Fig. 48.1 Mycorrhizal plants (*right*) with extensive network of hyphae extending from roots and nonmycorrhizal plants (*left*)

unfavorable conditions for growing plants (Al-Karaki 2011a; Brundrett et al. 1996). In the past decades, natural environments were full of mycorrhizae. Healthy soil contains billions of beneficial microorganisms which play a role in nutrition and nutrient recycling. However, over time, and due to application of chemicals (e.g., fertilizers, pesticides), desertification, erosion, drought, soil compaction, loss of organic matter, and other degradation variables, these symbiotic fungi have become less prevalent in soils, and the continual tilling of land for crops has greatly reduced the benefits of these fungi in many lands (Mozafar et al. 2000; Oehl et al. 2003).

In a mycorrhizal association with plants, AM fungus is considered an obligate symbiont, in which fungal response is affected by any disturbance affecting the plant (Skujins and Allen 1986; Dubey and Fulekar 2011). Many studies have indicated that interacting changes in soils, climate, and land management result in soil degradation, which in turn results in reduction of plant production (Dodd 2000), in addition to changes in the diversity and numbers of microbiota (Allen and MacMahon 1985). Many cultural practices used in cultivation of soils in arid regions have been reported to reduce AM fungal densities and alter their diversity resulting in great reduction of their benefits in these lands. For example, high rates of inorganic fertilizer (especially P) and fungicide applications inhibit AM fungi persistence (Amaya-Carpio et al. 2009). Irrigation with slightly saline water can lead to salinization and reduce AM fungi colonization of plant roots. Overgrazing might seriously reduce the presence and persistence of AM fungi in soils of arid lands, although the effect is probably indirect. High grazing intensity can reduce AM fungi populations due to soil erosion (Powell 1980).

48.4 Contribution of Mycorrhizal Fungi to Revegetate Degraded Lands

Revegetation is one of the most effective means for controlling soil degradation and for the reclamation of abandoned lands in arid and semiarid areas. The low availability of soil moisture due to scarce and irregular precipitation and frequent drought is a major obstacle to the successful revegetation of these areas (Albaladejo 1990). Accelerated soil erosion is also among the most pressing of environmental problems, resulting in degradation of ecosystem function, decreased productivity, and agricultural sustainability (Albaladejo 1990; Powell 1980; Kumar et al. 2010). Moreover, salinization is found in all eco-zones around the world, which is considered as a significant factor limiting agricultural productivity (Al-Khaliel 2010).

The fundamental importance of the mycorrhizal association in restoration and to improve revegetation of disturbed lands has been well recognized. The destruction of mycorrhizal fungal network in soil system is the vital event of soil disturbance and degradation, and its reintroduction is an essential approach of habitat restoration (Quoreshi 2008). Mycorrhizal symbiosis confers numerous benefits to host plants including improved plant growth and mineral nutrition and tolerance to diseases and stresses such as drought, salinity, temperature fluctuation, and metal toxicity (Al-Karaki 2000; Al-Karaki et al. 2004; Gemma et al. 1997; Borowics 2001). Mycorrhizal plants, in comparison with nonmycorrhizal plants, have greater nutrient and water uptake from soil because AM fungi develop an extensive network of external fungal filaments (hyphae) that act as an extension of the root absorbing area. Furthermore, AM fungi may play a role in the formation of stable soil aggregates, building up a macroporous structure of soil that allows penetration of water and air and prevents erosion (Jeffries et al. 2003). All of these beneficial effects on plant health and soil fitness mean that AM fungi may play an important role in the reclamation of degraded ecosystem.

48.5 Nutrient Cycling and Mycorrhizae

Conventional land management practices often include tillage, and significant inputs of fertilizers (nutrients), herbicides, and pesticides which can have negative impacts on the number of mycorrhizal species present and can, in effect, marginalize mycorrhizal and microbial functioning (Gosling et al. 2006). Preservation of soil microbiology in less intensively managed sites can contribute to self-regulation of fundamental ecosystem processes, particularly nutrient recycling, without need for further nutrient inputs. Enhancing the availability of P and N as well as many other nutrients to host plants is considered the most important function of mycorrhizae (Smith and Read 2008), and nutrient availability in sustainable systems is often dependent on mycorrhizal activity. Phosphorus is a major nutrient required by plants, although in soil it is usually present in very low concentrations as soluble P (Smith and Read 2008). AM fungi have shown to have a positive influence on the nutrient status of plants (Abbaspour et al. 2006; Al-Karaki 2000, 2006; Sannazzaro et al. 2007) by enhancing nutrient uptake facilitated by the extensive hyphae of the fungus which allow them to explore more soil volume than the nonmycorrhizal plants. Closure of major nutrient cycles, such as N cycling, is one of the most important factors in ecologically sustainable systems, as it lowers the amount of N leached out or lost in gaseous form. It is estimated that external hyphae deliver up to 80% of a plant's P requirements (Matamoros et al. 1999). AM fungi can effectively react with relatively insoluble forms of soil inorganic P, such as rock phosphate and iron phosphates (Smith and Read 2008).

48.6 Effects of Mycorrhizal Fungi on Soil Structure

The stability of soil macroaggregates is highly dependent on the growth and decomposition of roots and mycorrhizal hyphae. The AM fungi appear to be the most important mediator of soil aggregation (Rilling et al. 2000). In AM associations, the external hyphae provide a direct physical link between the host plant and soil resource. External hyphae of AM fungi can bind the small soil particles into microaggregates by producing a glycoprotein (glomalin) which alone can account for 30–60% of C in undisturbed soils (Treseder and Allen 2000), and the resultant entanglement of microaggregates creates macroaggregates that finally lead to improved structure and aggregation stability in soils (Fig. 48.2) with a wide range of texture, e.g., sandy, loamy, and clayey soils (Bearden and Petersen 2002). Thus, AM fungi are considered as key determinants of soil quality.

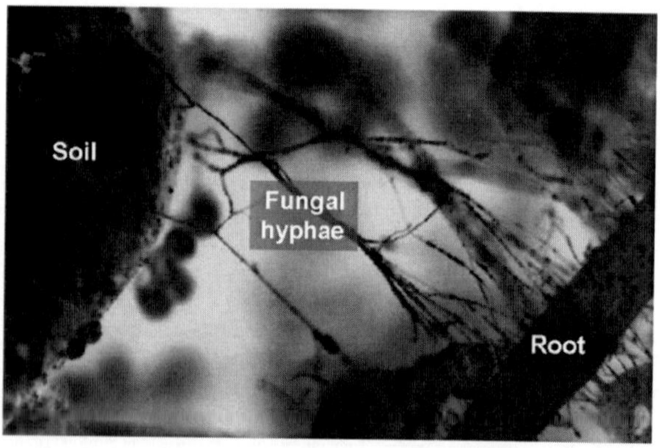

Fig. 48.2 Mycorrhizal fungi filaments (*hyphae*) which enhance soil aggregation and play an important role in soil structure development

48.7 Mycorrhizal Role in Tolerance to Drought Stress

A major feature of lands in arid zones is the low precipitation. In addition, they are characterized by widespread diurnal mean values of temperature, diversity of soils, variability in climatic regimes, and extreme patchiness of soils characterized by clumping of vegetation (Skujins and Allen 1986). It is apparent that the functioning of AM fungi in arid communities is based on several interacting physiological mechanisms to overcome the extreme variability of environmental conditions (Allen 1984).

Nutrient uptake and plant productivity in arid and semiarid areas are greatly limited by drought. However, AM fungi have been shown to increase water uptake and to increase drought tolerance of several plant species (Al-Karaki et al. 2004; Safir and Nelsen 1985). Increased water uptake by mycorrhizal plants has been attributed to many factors like increased P uptake and stomatal responses (altering stomatal control) (Augé 2001). Transport of water to plants by fungal hyphae has also been suggested (Allen 1984, 2007).

48.8 Role of Arbuscular Mycorrhiza in the Alleviation of Salinity

One of the most widespread agricultural problems in arid and semiarid regions is soil salinity which restricts plant growth and biomass production. Salinization of soil is a major concern where saline water is used in irrigation. Improving plant tolerance to saline stress is very significant in agriculture. Mycorrhizal fungi have been reported to improve the growth of many plant species under a variety of stresses such as salinity (Al-Karaki 2000; Al-Karaki et al. 2004; Al-Karaki and Hammad 2001). Several mechanisms have been proposed to explain the role of AM fungi in protecting plants against salt stress which results from a combination of nutritional, biochemical, and physiological effects. These include increased nutrient uptake (Al-Karaki 2006, 2000), accumulation of osmoregulators (Kaya et al. 2009; Sharifi et al. 2007), increase in photosynthetic rate and water use efficiency, changes in plant hormones (Danneberg et al. 1992), increased leaf gas exchange and photosynthetic rate (Ruiz-Lozano et al. 1996), direct hyphal water uptake from the soil and transfer to the host plant (Ruiz-Lozano and Azcon 1995), enhanced activity of enzymes involved in antioxidant defense (Ruiz-Lozano et al. 1996), enhanced water uptake through improved hydraulic conductivity and increasing leaf conductance and photosynthetic activity (Dell'Amico et al. 2002), osmotic adjustment (Al-Garni 2006) and changes in cell-wall elasticity (Augé et al. 1987), and cell membrane stability (Kaya et al. 2009).

Several researchers have reported that inoculation with AM fungi increased plant growth (shoot and root) and decreased yield losses under saline conditions. These enhancement effects have been reported for many plant species, including bell pepper (Kaya et al. 2009), tomato (Al-Karaki 2006; Al-Karaki et al. 2001),

Table 48.1 Enhancement effects (percentage change[a]) on growth (shoot, root) and fruit yields due to inoculation with AM fungi (AM) against no inoculation (nonAM)

Salinity level	Host species	Shoot yield	Root yield %	Fruit yield	Reference
2.15 dS m^{-1}	Bell pepper	11	3	4	Kaya et al. (2009)
7.15		19	7	24	
12.5		40	15	54	
1.7 dS m^{-1}	Tomato	26	42	29	Al-Karaki (2006)
4.4		106	217	60	
7.1		59	133	–[b]	Al-Karaki et al. (2001)
50 mM NaCl	Soybean	58	52	–	Sharifi et al. (2007)
100		70	82	–	
150		17	67	–	
200		10	20	–	
2 dS m^{-1}	Onion	1,620	–	–	Cantrell and Linderman (2001)
4		1,533	–	–	
8		750	–	–	
12		700	–	–	
0 mM NaCl	Phragmites australis	5	15	–	Al-Garni (2006)
50		10	17	–	
100		55	35	–	
150		81	103	–	
200		195	173	–	
250		334	254	–	
300		312	271	–	

[a]Growth yield (GY) change $= \text{GY}_{\text{AM}} - \text{GY}_{\text{nonAM}} \times 100 / \text{GY}_{\text{nonAM}}$

[b]Not reported

soybean (Sharifi et al. 2007), onion (Cantrell and Linderman 2001), and Phragmites australis (Al-Garni 2006), with most results indicating that inoculation with AM fungi greatly enhanced the growth (shoot and root) and fruit yield under different salinity levels (Table 48.1).

48.9 Strategies for the Management of Mycorrhiza in Degraded Soil

In past decades, natural environments were full of mycorrhiza that associated with roots of most plants living naturally. However, over time, due to heavy chemical application, desertification, soil erosion, drought, soil compaction, loss of organic matter, and other degradation factors, these symbiotic fungi have become less prevalent in soils, and their benefits for crops might be reduced completely.

A number of scientists have suggested that reclamation of degraded lands could be accelerated by the inoculation of mycorrhizal fungi or through the manipulation of their indigenous populations (Reeves et al. 1979; Allen 1991; Kumar et al. 2010). The application of AM inoculums, however, would seem to be true where the density of mycorrhizal propagules is very low (Miller and Jastrow 1992). The propagules of AM fungi are generally concentrated in the uppermost few centimeters of soil and reach their maximum concentration in the rhizosphere (Bellgard 1993). When soil is disturbed or is partially removed, a decrease in the number of mycorrhizal propagules occurs (Miller 1979; Reeves et al. 1979).

In general, plants from mature ecosystems require the presence of mycorrhizae for their development (obligatory mycotrophs) (Janos 1980; Brundrett 1991). Thus, if disturbance causes a loss of mycorrhizal propagules, recovery of the degraded areas is only possible if these propagules are reintroduced into soils.

In recent studies conducted under natural dry climates and poor soils (e.g., Bahrain), in which a commercial AM fungal inoculum has been tested on the marigold (*Tagetes erecta* L.) seedling quality and survival under drought (Al-Karaki 2011b) and on establishment of turfgrass water and water use efficiency (Al-Karaki et al. 2007). Results of these studies have indicated that inoculation with AM fungi was effective in enhancing early growth and development of tested plant species. For example, Al-Karaki et al. (2007) found that inoculation of grass lawn plots with mycorrhizal inoculum has resulted in faster establishment of lawns by twofolds over the uninoculated plots (Fig. 48.3). Moreover, the inoculation of the ornamental

Fig. 48.3 The effect of AM fungi inoculation on development of grass lawn and land coverage after 7 weeks under arid land conditions (Bahrain)

Fig. 48.4 The effective commercial AM fungi increased early establishment and development of marigold at nursery

plant (marigold) seeds with AM fungi inoculum has resulted in faster growth and establishment of seedlings in comparison to uninoculated controls (Fig. 48.4; Al-Karaki 2011b).

48.10 Conclusions

It is apparent that mycorrhizal fungi are essential components of both agricultural and native vegetation communities. Inoculation techniques for the rehabilitation of degraded lands may be used, especially with improved inocula production methods development. Rehabilitation of disturbed sites should include techniques designed to stimulate reformation of AM fungi symbioses. However, in view of the increasingly high costs of fertilizers and the negative impacts of continued fertilization on land degradation and environment pollution (e.g., underground water), studies should be conducted to increase yields of arid area crops by utilizing mycorrhizal fungi.

References

Abbaspour H, Fallahyan F, Fahimi H, Afshari H (2006) Response of Pistacia vera L. in salt tolerance to inoculation with arbuscular mycorrhizal fungi under salt stress. Acta Hortic 726:383–389

Albaladejo J (1990) Impact of the degradation processes on soil quality in arid mediterranean environment. In: Rubio JL, Rickson J (eds) Strategies to combat desertification in Mediterranean Europe. Commission of the European Communities, Luxembourg, pp 193–215

Al-Garni SMS (2006) Increasing NaCl – salt tolerance of a halophytic plant Phragmites australis by mycorrhizal symbiosis. Am-Euras J Agric Environ Sci 1:119–126

Al-Karaki GN (2000) Growth of mycorrhizal tomato and mineral acquisition under salt stress. Mycorrhiza 10:51–54

Al-Karaki GN (2006) Nursery inoculation of tomato with arbuscular mycorrhizal fungi and subsequent performance under irrigation with saline water. Sci Hortic 109:1–7

Al-Karaki GN (2011a) Mycorrhizal fungi role in reducing the impacts of environmental climate change in arid regions. In: Solh M, Saxena M (ed) Food security and climate change in dry areas: Proceedings of the International Conference 1–4 February 2010, Amman, Jordan. International Center for Agricultural Research in the Dry Areas, Aleppo, Syria, pp 304–313

Al-Karaki GN (2011b) Effect of commercial mycorrhizae on seedling quality of marigold grown in nursery and subsequent survivability under drought. The first international symposium on microbial horticulture, Alnarp, Sweden, 15–19 May 2011

Al-Karaki GN, Hammad R (2001) Mycorrhizal influence on fruit yield and mineral contents of tomato grown under salt stress. J Plant Nutr 24:1311–1323

Al-Karaki GN, Hammad R, Rusan M (2001) Response of salt sensitive and tolerant tomato cultivars inoculated with mycorrhizal fungi to salt stress. Mycorrhiza 11:43–47

Al-Karaki GN, McMichael B, Zak J (2004) Field response of wheat to arbuscular mycorrhizal fungi and drought stress. Mycorrhiza 14:263–269

Al-Karaki GN, Othman Y, Al-Ajmi A (2007) Effects of mycorrhizal fungi inoculation on landscape turf establishment under Arabian Gulf region conditions. Arab Gulf J Sci Res 25(3):147–152

Al-Khaliel AS (2010) Effect of salinity stress on mycorrhizal association and growth response of peanut infected by Glomus mosseae. Plant Soil Environ 56:318–324

Allen EB (1984) The role of mycorrhiza in mined land diversity. In: Proceedings of the third biennial Symposium surface mine reclamation on the great plains, Montana, 19–21 March 1984, pp 273–295

Allen MF (1991) The ecology of mycorrhizae. Cambridge University Press, Cambridge

Allen MF (2007) Mycorrhizal fungi: highways for water and nutrients in arid soils. Vadose Zone J 6:291–297

Allen MF, MacMahon JA (1985) Impact of disturbance on cold desert fungi: comparative dispersion patterns. Pedobiologia 28:215–224

Amaya-Carpio L, Davies FT Jr, Fox T, He C (2009) Arbuscular mycorrhizal fungi and organic fertilizer influence on photosynthesis, root phosphatase activity, nutrition, and growth of Ipomoea carnea ssp. Fistulosa. Photosynthetica 47:1–10

Augé RM (2001) Water relations, drought and vesicular-arbuscular mycorrhizal symbiosis. Mycorrhiza 11:3–42

Augé RM, Schekel KA, Wample RL (1987) Leaf water and carbohydrate status of VA mycorrhizal rose exposed to drought stress. Plant Soil 99:291–302

Azcón-Aguilar C, Jaizme-Vega MC, Calvet C (2002) The contribution of arbuscular mycorrhizal fungi to the control of soil borne plant pathogens. In: Gianinazzi S, Schuepp H, Haselwandter K, Barea JM (eds) Mycorrhizal technology in agriculture. From genes to bioproducts. ALS Birkhauser Verlag, Basel, pp 187–197

Barea JM, Jeffries R (1995) Arbuscular mycorrhizas sustainable soil plant system. In: Varma A, Hock B (eds) Mycorrhiza structure, function, molecular biology and biotechnology. Springer, Heidelberg, pp 521–560

Bearden BN, Petersen L (2000) Influence of arbuscular mycorrhizal fungi on soil structure and aggregate stability of a vertisol. Plant Soil 218:173–183

Bellgard SE (1993) The topsoil as the major store of propagules of vesicular-arbuscular mycorrhizal fungi in southeast Australian sandstone soils. Mycorrhiza 3:19–24

Borowics VA (2001) Do arbuscular mycorrhizal fungi alter plant pathogen relations? Ecology 82:3057–3068

Brundrett M (1991) Mycorrhizas in natural ecosystems. In: Macfayden A, Begon M, Fitter AH (eds) Advances in ecological research. Academic, Town, pp 171–313

Brundrett M, Bougher N, Dell B, Grave T, Malajczuk N (1996) Working with mycorrhizas in forestry and agriculture. Australian Centre for International Agricultural Research Monograph 32, Canberra, p 374

Cantrell IC, Linderman RG (2001) Preinoculation of lettuce and onion with VA mycorrhizal fungi reduces deleterious effects of soil salinity. Plant Soil 233:269–281

Danneberg G, Latus C, Zimmer W, Hundeshagen B, Schneider-Poetsch HJ, Bothe H (1992) Influence of vesicular-arbuscular mycorrhiza on phytohormone balances in maize (Zea mays L.). J Plant Physiol 141:33–39

Dell'Amico J, Torrecillas A, Rodriguez P, Morte A, Sanchez-Blanco M (2002) Responses of tomato plants associated with the arbuscular mycorrhizal fungus Glomus clarum during drought and recovery. J Agric Sci 138:387–393

Dodd JC (2000) The role of arbuscular mycorrhizal fungi in agro-and natural ecosystems. Outlook Agric 29:63–70

Dubey KK, Fulekar MH (2011) Mycorrhizosphere development and management: the role of nutrients, micro-organisms and bio-chemical activities. Agric Biol J North Am 2(2):315–324

Finlay RD (2008) Ecological aspects of mycorrhizal symbiosis: with special emphasis on the functional diversity of interactions involving the extraradical mycelium. J Exp Bot 59:1115–1126

Gemma JN, Koske RE, Roberts EM, Jackson N, De Antonis KM (1997) Mycorrhizal fungi enhance drought resistance in creeping bentgrass. J Turfgrass Sci 73:15–29

Gosling P, Hodge A, Goodlas G, Berding GD (2006) Arbuscular mycorrhizal fungi and organic farming. Agric Ecosys Environ 113:17–35

Govindaraju M, Pfeffer PE, Jin HR, Abu Baker J, Douds DD, Allen JWB, Bucking H, Lammers PJ, Shchar H (2005) Nitrogen transfer in the arbuscular mycorrhizal symbiosis. Nature 435:819–823

Gupta V, Satyanarayana T, Garg S (2000) General aspects of mycorrhiza. In: Mukherji KG, Chamola BP, Singh J (eds) Mycorrhizal biology. Kluwer, Dordrecht, pp 27–44

Janos DP (1980) Mycorrhizae influence tropical succession. Biotropica 12:56–64

Jaradat AA (2005) Saline agriculture in the Arabian Peninsula: management of marginal lands and saline water resources. J Food Agric Environ 3(2):302–306

Jeffries P, Gianinazzi S, Perotto S, Turnau K, Barea JM (2003) The contribution of arbuscular mycorrhizal fungi in sustainable maintenance of plant health and soil fertility. Biol Fert Soils 37:1–16

Kaya C, Ashraf M, Sonmez O, Aydemir S, Tuna AL, Cullu MA (2009) The influence of arbuscular mycorrhizal colonisation on key growth parameters and fruit yield of pepper plants grown at high salinity. Scientia Hort 121:1–6

Khan AG (2003) Mycotrophy and its significance in wetland ecology and wetland management. In: Wong MH (ed) Proceedings of the Croucher Foundation Study Institute: wetland ecosystems in Asia – function and management, Hong Kong, pp 11–15

Khan AG, Belik M (1995) Occurrence and ecological significance of mycorrhizal symbioses in aquatic plants. In: Verma A, Hork B (eds) Mycorrhiza: structure, function, molecular biology and biotechnology. Springer, Heidelberg, pp 627–666

Kumar A, Raghuwanshi R, Upadhyay RS (2010) Arbuscular mycorrhizal technology in reclamation and revegetation of coal mine spoils under various revegetation models. Engineering 2:683–689

Lambin EF, Turner BL, Geist HJ, Agbola SB, Angelsen A, Bruce JW, Coomes OT, Dirzog R, Fischer G, Folke C, George PS, Homewood K, Imbernonl J, Leemansm R, Lin X, Morano EF, Mortimore M, Ramakrishnan PS, Richards JF, Skåness H, Steffent W, Stone GD, Svedin U, Veldkamp TA, Vogel C, Xuy J (2001) The causes of land-use and land-cover change: moving beyond the myths. Glob Environ Change 11(4):261–269

Marschner H (1995) Mineral nutrition of higher plants, 2nd edn. Academic, New York, p 889

Matamoros MA, Baird LM, Escuredo PR, Dalton DA, Mixhin FR, Iturbe-Ormaetxe I, Rubio MC, Moran JF, Gordon AJ, Becana M (1999) Stress-induced legume root nodule senescence. Physiological, biochemical and structural alterations. Plant Physiol 121:97–111

Miller RM, Jastrow JO (1992) The application of VA mycorrhizae to ecosystem restoration and rec1amation. In: Allen MF (ed) Mycorrhizal functioning, an integrative plant-fungal process. Chapman & Hall, New York, pp 438–467

Mozafar A, Anken T, Ruh R, Frossard E (2000) Tillage intensity, mycorrhizal and nonmycorrhizal fungi, and nutrient concentrations in maize, wheat, and canola. Agron J 92:1117–1124

Oehl F, Sieverding E, Ineichen K, Mäder P, Boller T, Wiemken A (2003) Impact of land use intensity on the species diversity of arbuscular mycorrhizal fungi in agroecosystems of central Europe. Appl Environ Microbiol 69:2816–2824

Powell CL (1980) Mycorrhizal infectivity of eroded soils. Soil Biol Biochem 12:247–250

Quoreshi AM (2008) The use of mycorrhizal biotechnology in restoration of disturbed ecosystem. In: Siddiqui ZA et al (eds) Mycorrhizae: sustainable agriculture and forestry. Springer Science+Business Media B.V., pp 303–320

Reeves FB, Wagner D, Moorman T, Kiel J (1979) The role of endomycorrhizae in revegetation practices in the semi-arid west. I. A comparison of incidence of mycorrhizae in severely disturbed vs. natural environments. Am J Bot 66:6–13

Rilling MC, Hernandez GY, Newton PCD (2000) Arbuscular mycorrhizae respond to elevated atmospheric CO_2 after long-term exposure: evidence from a CO_2 spring in New Zealand supports the resource balance model. Ecol Lett 3:475–478

Rimawi O, Al-Ansari NA (1997) Groundwater degradation in the northeastern part of Mafraq area, Jordan. Freshwater contamination (Proceedings of Rabat Symposium S4, April–May 1997). IAHS Publ no 243

Ruiz-Lozano JM, Azcon RM (1995) Effects of arbuscular mycorrhizal Glomus species on drought tolerance: physiological and nutritional plant responses. Appl Environ Microbiol 61:456–460

Ruiz-Lozano JM, Azcon R, Gomez M (1996) Alleviation of salt stress by arbuscular mycorrhizal Glomus species in Lactuca sativa plants. Physiol Plant 98:767–772

Safir GR, Nelsen CE (1985) VA mycorrhizas: plant and fungal water relations. In Molina R (ed) Proceedings of the sixth North American conference on mycorrhizae, Forest Research Laboratory, Corvallis, OR, pp 161–164

Sannazzaro AI, Echeverria M, Edgardo AO, Ruiz AO, Menandez A (2007) Modulation of polyamine balance in Lotus glaber by salinity and arbuscular mycorrhiza. Plant Physiol Biochem 45:39–46

Scherr SJ, Yadav S (1996) Land degradation in the developing world: implications for food, agriculture, and the environment to 2020. Food, agriculture, and the environment discussion paper 14. International Food Policy Research Institute, Washington, DC

Sharifi M, Ghorbanli M, Ebrahimzadeh H (2007) Improved growth of salinity stressed soybean after inoculation with salt pre-treated mycorrhizal fungi. J Plant Physiol 164:1144–1151

Skujins J, Allen MF (1986) Use of mycorrhizae for land rehabilitation. MIRCEN J 2:161–176

Smith SE, Read DJ (2008) Mycorrhizal symbiosis, 3rd edn. Academic Press, London

Stanley MR, Koide RT, Shumway DL (1993) Mycorrhizal symbiosis increases growth, reproduction and recruitment of Abutilon theophrasti Medic in the field. Oecologia 94:30–35

Subramanian KS, Charest C, Dwyer LM, Hamilton RI (1997) Effects of arbuscular mycorrhizae on leaf water potential, sugar content and P content during drought and recovery of maize. Can J Bot 75:1582–1591

Tian CY, Feng G, Li XL, Zhang FS (2004) Different effects of arbuscular mycorrhizal fungal isolates from saline or non-saline on salinity tolerance of plants. Appl Soil Ecol 26:143–148

Treseder KK, Allen MF (2000) Mycorrhizal fungi have a potential role in soil carbon storage under elevated CO_2 and nitrogen deposition. New Phytol 147:189–200

UNCCD (2009) Report on Land Day, <www.unccd.int/publicinfo/landday/menu.php>

Whipps JM (2004) Prospects and limitations for mycorrhiza in biocontrol of root pathogens. Can J Bot 82:1198–1227

Wiede J (2005) Saline soils: a landscaping challenge. http://www.csbe.org/water_conserving_landscapes/publications/saline_soils/saline_soils_main3.htm. On 16 March 2010

Yano-Melo AM, Saggin OJ, Maia LC (2003) Tolerance of mycorrhized banana plantlets to saline stress. Agric Ecosys Environ 95:343–348

Chapter 49
Preliminary Interpretation of Environmental Isotope Data in the Ain El Atti Area (Tafilalet)

Mohamed Aoubouazza, Willibad Stichler, and Piotr Maloszewski

Abstract Present study on the sustainable use of saline land and saline groundwater for agricultural production has been conducted in the pre-Saharian area of Ain El Atti through the application of environmental isotopes supported by the hydrochemistry. In the study area, a network of 20 water points has been the focus of the isotope analysis ($\delta^{18}O$, $\delta^{2}H$, ^{3}H and ^{14}C) and physical chemistry. The samples were collected once every 3 months from artesian groundwater of "the Infracenomanian" (4), the Turonian (4), the Senonian (1) and the Quaternary aquifer (5) and from the precipitation of the years 2001, 2002 and 2003. The results show that (i) the stable isotope from the Infracenomanian is very poor and they are without tritium, confirming the fact that this aquifer is confined and it is not evaporated. Its strong salinity is due to the dissolution and the lixiviation of the geological formation; (ii) the Turonian, the Senonian and the Quaternary aquifers are not confined, and their stable isotope contents more or less important as the tritium, signifying that they receive recent recharge. The first one is affected by the artesian well and it is not evaporated and it has high salinity. The second and the third one are not affected by the artesian well, but the influence of the precipitation and the flood is clear. Their groundwater is not evaporated and their salinity is moderate; (iii) however, the Ziz surface water isotopic elements are rich, signifying an actual recharge. Its water is highly evaporated and its salinity is variable.

Keywords Agricultural production • Ain El Atti area • Environmental isotope data • Saline groundwater • Saline land

---

M. Aoubouazza (✉)
Service d'Ecologie, de Biodiversité et de la Conservation des Sols,
Centre de la Recherche Forestière, BP. 763, Agdal-Rabat, Morocco
e-mail: abouzzayna@gmail.com

W. Stichler • P. Maloszewski
GSF-Institute of Hydrology, D-85764, Neuherberg b, Munich, Germany

S.A. Shahid et al. (eds.), *Developments in Soil Classification, Land Use Planning and Policy Implications: Innovative Thinking of Soil Inventory for Land Use Planning and Management of Land Resources*, DOI 10.1007/978-94-007-5332-7_49,

49.1 Introduction

The groundwater quality, its availability and its variations in intensive farming are still a concern for many users and researchers. The age and origin of these aquifers is an additional difficulty for solving scientific problems and sensitive issues of hydrogeology including current trends, development and management of water resources. Over the past 10 years, the use of isotope techniques is of economic interest with regard to the conditions of overexploitation of groundwater and identification of areas likely to provide the groundwater quality and quantity. These isotopic techniques relate to methods for determining the isotopic data subject, the residence time of groundwater and the forecast arrival of pollutants to the aquifer horizons. The resolution of these problems by hydrogeological investigations can improve the effectiveness of their implementation and accuracy of the estimated recoverable reserves of groundwater. They also increased measures to protect these reserves against pollution and overexploitation.

49.1.1 Geography, Hydrology and Hydrogeology

The region of Ain El Atti is located immediately upstream of the plain Tafilalt, about 60 km south of the city of Errachidia and 10 km north of Erfoud, in the southeastern Morocco (Fig. 49.1).

The study area has arid climate with average annual rainfall of 120 mm. Annual rainfall is variable and irregular, in the order of 129 mm in the northern zone (Errachidia) and 50 mm in the southern zone (Erfoud). Autumn and spring are the rainy seasons. July is the hottest month with 37.6°C as monthly mean maximum

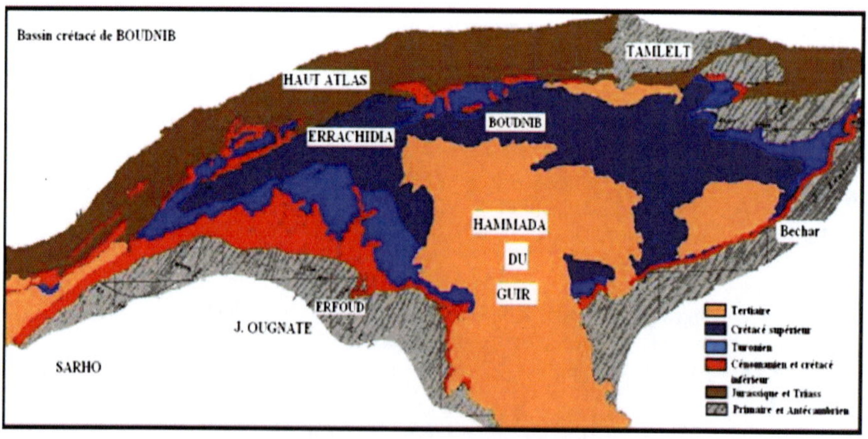

Fig. 49.1 Location of Errachidia-Boudenib basin

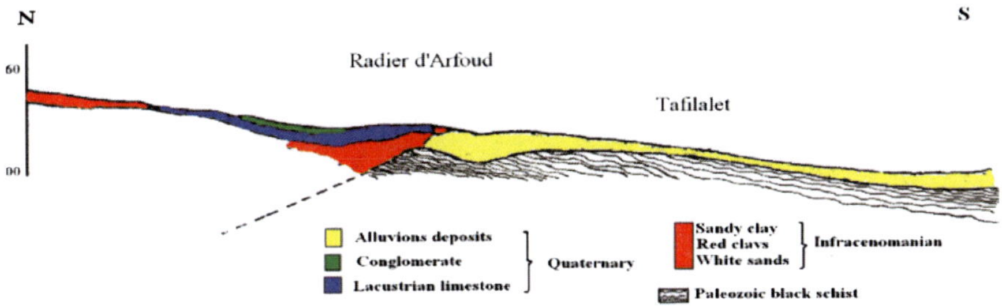

Fig. 49.2 Interpretative geological section of the Errachidia basin

temperature. January is the coldest with −0.8°C as monthly mean minimum temperature. The dry season extends to often months. Evaporation is very high (4,266 mm per year) with extreme values in January (121.8 mm) and July (668.3 mm) resulting in very low recharge rate of 6–3 mm during six cycle periods (1980–1986) and 44–14 mm during only one cycle period (1979–1980). From a hydrogeological (Fig. 49.2) point of view, this area consists of a multilayer aquifer system consisting of four main aquifer levels more or less communicating them through layers of semipermeable clastic lithology (Margat 1954).

49.1.2 The Quaternary Alluvial Deposits

The Quaternary alluvial deposits are present in the south part of the basin; they are reduced to 100 m width along the river. The recharge is mainly assured by floodwaters in the river and rainfall; thickness ranges from 15 to 20 m; permeability from 10^{-3} to 3×10^{-2} m s^{-1} with sand, gravel and alluvium; transmissivity ranges from 3×10^{-3} to 9×10^{-5} m^2 s^{-1} for gravel and sand horizons; the storage coefficient is 10^{-3} to 10^{-4}; mean discharge is 0.6×10^5 m^3 day^{-1}; free flow is 50–100 m^3 day^{-1}; hydraulic gradient is 1.7–3‰; and porosity is 5.5 to 2.5‰.

49.1.2.1 The Senonian

This Senonian aquifer is located in the north part of Errachidia basin. It is a detritic layer with some sandstone, sand, red marl, gypsum, sodium chloride and limestone. The transmissivity is of the order of 5×10^{-4} m^2 s^{-1}. It is mainly used as traditional wells with bandwidths from 2 to 10 L s^{-1} as a discharge. The deepest wells (100–130 m) do not exceed 20 L s^{-1} as a discharge. Water quality is highly variable and the dry residue varies between 0.5 and 10 g L^{-1}. The general flow direction is from north to south, with areas of preferential flow in the same direction as the Turonian. These lines converge towards the sources from Meski, Aoufous and Tarda, which indicates that the Senonian is drained by these sources.

49.1.2.2 The Turonian

The Turonian aquifer consists mainly of fractured dolomitic limestone to a depth of 100 m; transmissivity is 10^{-3} to 2×10^{-5} m^2 s^{-1}. Its recharge is done by drainage upstream and upper aquifers. They are natural outlets and are represented by the Meski-Tarda and Tifounassine. In the Ziz Valley, in addition to traditional wells, the aquifer is exploited by pumping station group, with rates varying between 60 and 100 L s^{-1}. Apart from some wells with salinity increasing with depth, the dry residue varies between 1 and 2 g L^{-1}. The general sense of flow is from north to south. These natural outlets are presented by many springs, which contribute to agricultural development.

49.1.2.3 The Infracenomanian

This artesian aquifer is made of clays and stone and gypsum sands. It is the Paleozoic bedrock as a substratum. Its recharge is done from the north border of the basin and flood influents. This aquifer is 400–600 m below the surface at the upstream of catchment area (near the dam) and close to 0 m at the upstream part of Tafilalt plain; thickness is 350 m (argillaceous liking and gypseous sands). The transmissivities are low and vary between 2×10^{-4} and 2.2×10^{-3} m^2 s^{-1} (Directorate of Research, Planning and Water 1989). The storage coefficient is 10^{-3} to 10^{-4}; the discharge is 0.5×10^{-5} m^3 day^{-1}. Water quality varies greatly from one sector to another: good quality at the west of the study sector (dry residue varies from 0.8 to 1.6 g L^{-1}), brackish at Aoufous (2–3 g L^{-1}) and highly mineralized between Douira and foundation raft Erfoud (the residue reached 14.4 g L^{-1}).

49.2 Materials and Methods

Starting from the 20th of May 1997, 15 monitoring campaigns and sampling were conducted every 3 months in the study area. This monitoring includes in situ measurements of hydrodynamic (piezometric level, depth of wells, etc.) and physicochemical properties (temperature, pH, salinity indirectly using electrical conductivity, TDS and alkalinity) for a sampler of groundwater by a piezometric probe, a conductivity meter and pH meter (WTW LF-320/SA). The chemical analysis (cations and anions) was carried out in the laboratories of INRA-Rabat and the AGR-Rabat (Directorate of Management Water). The isotopic analysis was performed in the laboratory of Radiation and Isotope Application Division, PINSTECH, Islamabad, from 1997 to 2000, and Hydrology Laboratory in Vienna, Austria, during 2000 and 2004. Moreover, empty bottles (500 and 50 ml) were used to collect rainwater.

49.2.1 Isotope Analysis

The range of $\delta^{18}O$ [−9.98 to −0.81‰] and of δD [−69.2 to −12.3‰] is large; it indicates that these aquifers are highly depleted (Sajjad 1997). The diagrams of (Figs. 49.3

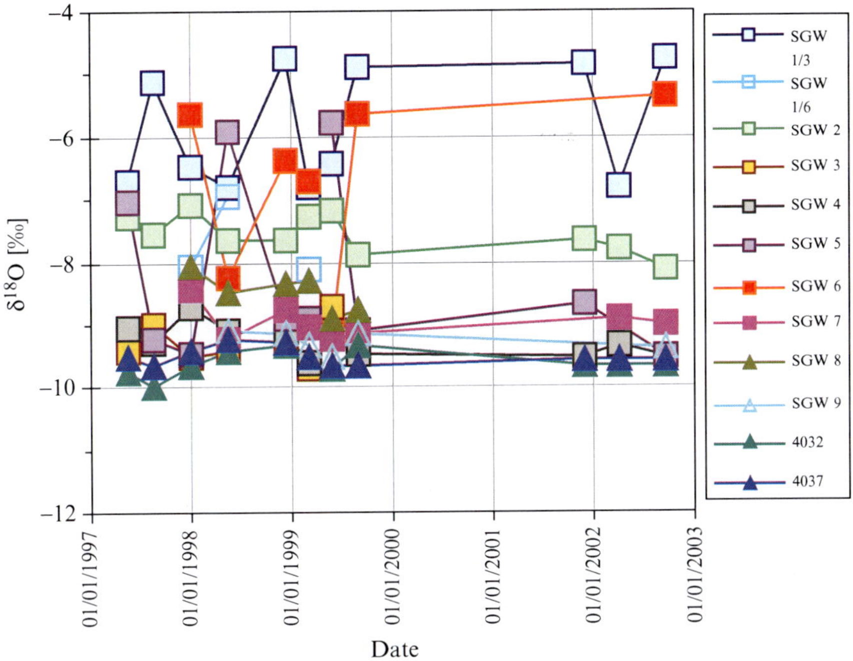

Fig. 49.3 $\delta^{18}O$ versus date of sampling

and 49.4) $\delta^{18}O$ versus the date of sampling showed that water exchange occurs between the Infracenomanian and the Turonian aquifers. However, this exchange did not happen with the rest from other aquifers (Senonian, Quaternary). The diagram δD–$\delta^{18}O$ (Fig. 49.5) shows that there are at least four sources of recharge (Stumpp et al. 2009).

The first group of water points capturing Infracenomanian "4032/57", "4037/57", "SGW3" and "SGW4" "located on the GMWL" is represented by two artesian wells at 100 m depth and they have almost constant discharge of 15 and 10 L s^{-1}, respectively. They have $\delta^{18}O$ as mean value [−9.52‰]. The mean δD is [−66.19‰] and d-excess is [−9.95‰]. These wells are characterized as zero [0] UT indicating that these recharges are very old. It can be inferred that recharge of the aquifer is very old and the water has undergone no evaporation, d-excess [−9.95‰]. The absence of tritium [0] UT confirms that this aquifer is confined.

The source of recharge is distant and is at high altitude as indicated by the values of depleted $\delta^{18}O$–EC (Fig. 49.6). Figure 49.6 indicates that high salinity (EC = 14.4 dS m^{-1}) is attributed to leaching from the aquifer. Artesian well number (4037/57) compared with that of the experimental station indicates that it is richer in stable isotopes and tritium, but it is less salty (EC = 11.6 dS m^{-1}). In the south sector, groundwaters (SGW1/6) have the same isotopic character but are evaporated. Increasing salinity (EC = 22.4 dS m^{-1}) indicates a layer in isolated pockets.

The second group of items captured the Cenomanian-Turonian "SGW7 and SGW9". This point group is also low in stable isotopes, $\delta^{18}O$ [−9.54 to −8.45‰], and is located slightly below the first. Low levels of tritium from [0.40] UT, indicating

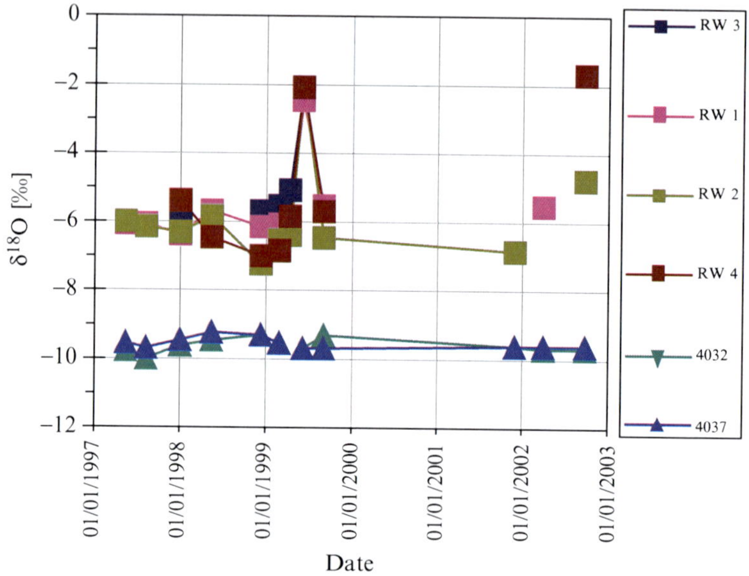

Fig. 49.4 δ^{18}O versus date of sampling

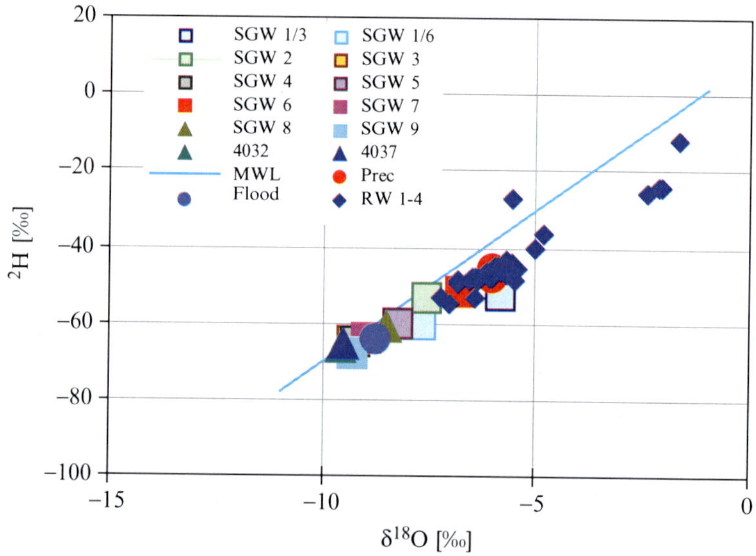

Fig. 49.5 δD–δ^{18}O [‰]–[‰]

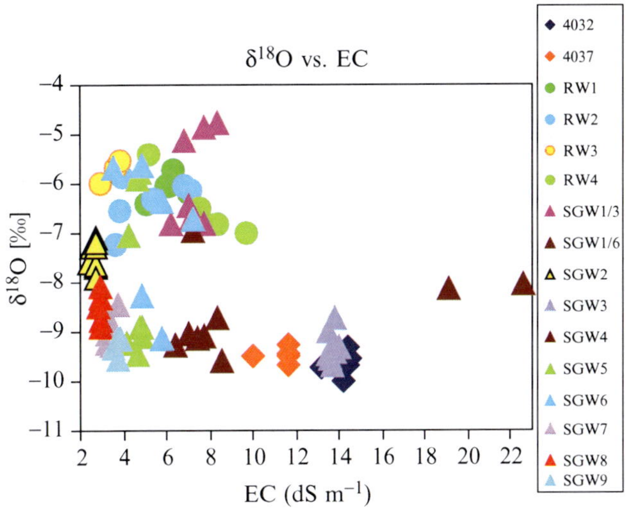

Fig. 49.6 $\delta^{18}O$–EC [‰]–[dS m^{-1}]

an old recharge, and values of d-excess [6.84‰] indicate that the water has undergone no evaporation. This aquifer influenced by Infracenomanian has a high salinity (EC = 3.6 dS m^{-1}).

The third group of items captured the Senonian "SGW5 and SGW8". This point group is relatively richer in stable isotopes, $\delta^{18}O$ [−8.89 to −8.08‰] and δD [−63.34 to −59.29‰], and away from the first group. These waters contain significant concentrations of tritium [0.92–2.65] UT, indicating a recent recharge. In addition, values of d-excess [6.91‰] indicate that the water of this aquifer is not evaporated. This aquifer is not affected by the artesian aquifer, but the effect of rainfall and floodwaters is clear. The salinity is moderate (EC = 2.6 dS m^{-1}).

The fourth group of items captured the Quaternary "SGW2 and SGW6". This group of points is very rich in stable isotope $\delta^{18}O$ [−7.89 to −5.65‰], and δD is [−63.34 to −47.13‰] and located near the water surface of Ziz wadi. The levels of tritium are of the order of waters of precipitation, [6.3–8.35‰] UT, indicating recent recharge. The d-excess of [7.5‰] indicates that the water of this aquifer is not evaporated. The groundwater quality is average (EC = 2.5 dS m^{-1}).

49.2.2 Surface water

The composition of isotopic elements of the group of points of surface water RW_{1-4}, is between $\delta^{18}O$ [−7.01 to −2.08‰] and δD [−53.71 to −12.3‰]. The very high content of tritium [9.43] UT (Fig. 49.7) confirms that water is actual. This group of water points is extremely evaporated, with d-excess [0] and a variable salinity (EC: 1 to 9.7 dS m^{-1}).

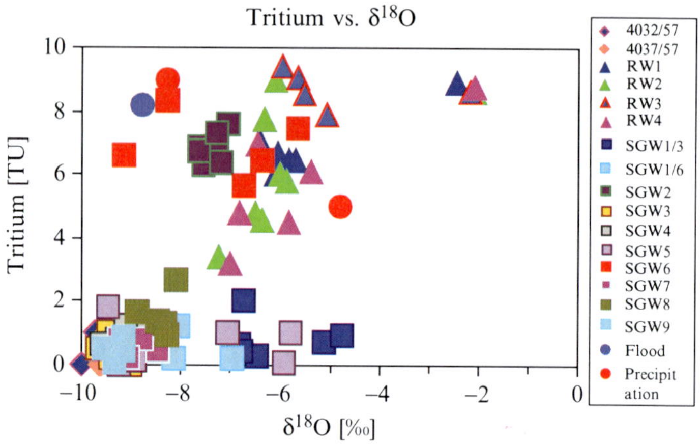

Fig. 49.7 Tr–$\delta^{18}O$ [UT]–[‰]

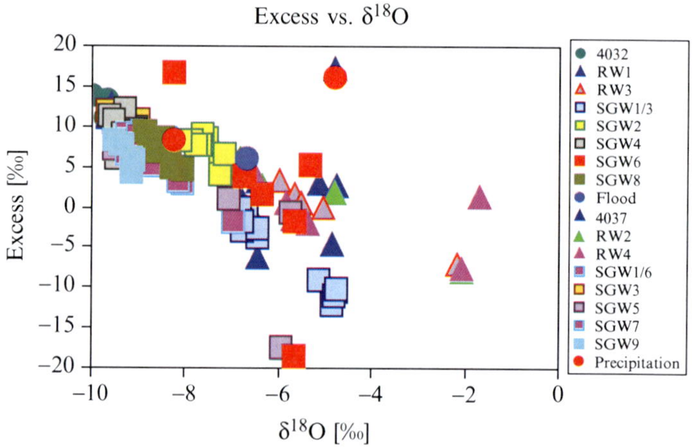

Fig. 49.8 d-Excess–$\delta^{18}O$ [UT][‰]–[‰]

49.3 Conclusions

The area of Ain El Atti consists of a multilayer aquifer system consisting of four main aquifer levels more or less communicating them through layers of semipermeable clastic lithology. The waters are generally highly depleted. The Infracenomanian is low and stable isotopes do not contain tritium, confirming that this aquifer is confined. The source of recharge is distant and located at high altitude as indicated by the values of depleted δD and $\delta^{18}O$. On the other hand, the aquifers of the Turonian, Senonian and Quaternary receive recharge from the recent flood of the river Ziz and local rainfall. Furthermore, isotopic analysis shows that water exchange occurs mainly between the aquifer and the Infracenomanian, the Turonian and in any case with other Senonian aquifers and overlying Quaternary, indicating the presence of a geological barrier. The quick response and instantaneous of these aquifers (Fig. 49.8) to dam inflows and precipitation (Maloszewski 1994; Maloszewski 1996; Maloszewski and Zuber 1996) is for a feed "piston flow effect" through training Turonian basin Ziz. The isotopic composition of rainwater is well correlated with the Global Meteoric Water Line ($\delta D = 8\ \delta^{18}O + 10$). The relationship δD–8 $\delta^{18}O$ samples of surface water of the Ziz wadi coincides with the evaporative line ($\delta D = 5.74$–12.55). The enrichment of wadi water ($\delta D = -43.88$‰ and $\delta^{18}O = -5.48$‰) as a result of evaporation is evident. Increasing salinity is attributed to the dissolution and leaching of geological aquifers. Variations in isotopic $\delta^{18}O$ and tritium versus chlorides do not appear to show a clear compositions effect of evaporation. Increasing the concentration of chloride is related to the leaching of materials (sodium chloride) and/or mixing waters with the brine.

References

DRPE (1989) Establissement et mise au point de modèles de gestion des nappes du bassin de Goulmima, Errachidia

Maloszewski P (1994) Mathematical modelling of tracer experiments in fissured aquifers. Freiburger Schriften zur Hydrogeologie, vol 2, Freiburg

Maloszewski P (1996) Program flow PC version 2-3-September. GSF-Institut fuer Hydrologie D-85764 Neuherberg B Muenchen, Fed Rep Germany

Maloszewski P, Zuber A (1996) Lumped parameter models for the interpretation of environmental tracer data. IAEA – TECDOC – 910

Margat J (1954) Mémoire explicatif de la carte hydrogéologique de la plaine du Tafilalet au 1/50.000. Notes et Mém Serv Géol Maroc, no 150 bis

Sajjad I (1997) Report for the inter-regional project on sustainable utilization of salt-affected wasteland and saline groundwater for plant production- assistance to counterparts in initiating isotope hydrology work. Chief Scientific Officer. Head, Radiation & Isotope Application Division. PINSTECH, Islamabad, Pakistan

Stumpp C, Stichler W, Maloszewski P (2009) Application of environmental isotope $\delta^{18}O$ to study water flow in unsaturated soils planted with different crops: case study of weighable lysimeter from the research field in Neuherberg, Germany. J Hydrol 368:68–78

Chapter 50
Use of Lombrica to Enhance Soil Fertility for Crop Production in Madagascar

Razafindravoniarisoa Euphrasie

Abstract Eighty percent of Malagasy population are peasants and practise agriculture. Continuing population growth increases food demand, and to achieve food demand, there is a need to find ways for rational use and management of soil resources for better agriculture production. This requires more soils to be brought into cultivation; however, the country lacks sufficient cultivable soils to meet food demand. The Malagasy government strongly supports agriculture development and encourages the farmers to find new soils for cultivation and crop diversification. In order to bring more soils into cultivation and to improve crop productivity, farmers need more fertilizers to offset nutrient requirement of crops and to improve soil fertility status. As per local experience, we believe that the use of 'lombricompost' fertilizer is essential for Malagasy farmers. Owing to the high price of chemical fertilizers and its impact on soils, we found the 'lombricompost' a beneficial fertilizer with affordable price and easy to produce locally. Besides the exploitation of 'lombricus worm', it generates income and contributes enormously to the amelioration and protection of the environment. In this chapter, the benefits of using lombrica to farmer's daily work, daily life and environment are discussed.

Keywords Crop production • Lombrica • Madagascar • Soil fertility • Soil resources

---

R. Euphrasie (✉)
Environmental Education and Cultural Center,
17 bis, rue Commandant Escande Porte 9 Place Kabary, Antsiranana 201, Madagascar
e-mail: zafivoniarisoa@yahoo.fr

S.A. Shahid et al. (eds.), *Developments in Soil Classification, Land Use Planning and Policy Implications: Innovative Thinking of Soil Inventory for Land Use Planning and Management of Land Resources*, DOI 10.1007/978-94-007-5332-7_50,

50.1 Introduction

Madagascar is a large island situating between the Indian Ocean and Mozambique Channel. Eighty percent of its population is engaged in agriculture. The main source of food for Malagasy population is from local agriculture. Continuing population growth requires more food to be produced locally. There are many constraints in meeting the present and future food demand, such as low income (poverty), no access to modern agriculture farming and, the most alarming is population growth. According to National Institute of Statistics, currently the population of Malagasy is about 20 millions. To meet the food requirements, it is necessary to find more soils where cultivation can be made. However, there are some constraints, such as most of the island constitutes land with various reliefs; this limits finding sufficient 'cultivable' soils to further agriculture activities. It is, therefore, essential that the government must take urgent action for land use planning and examine in detail the policy implication (African Development Fund Report 1994; Randrianarisoa and Minten 2001).

The government of Malagasy encourages the peasants to exploit new soils for agriculture production to meet local food demand. The government has set the policy that after 5 years of cultivation, the relevant peasant becomes the owner of the land. However, the parcel should be in the arid or semiarid area. The newly selected land should have the capacity to produce crops and, therefore, should not require heavy investments that peasants cannot afford. However, if these suitable soils are not properly managed, these can be degraded; therefore, these soils require good management for their sustainable use.

In order to have optimum crop yield, it is necessary to determine the nutrient availability in soils through analysing soil samples in a standard laboratory, and the difference between the crops requirements should be compensated through adding chemical or organic fertilizers. The misuse of chemical fertilizers usually leads to the contamination of the soil and ground water and hence degrades the resource quality. An alternate to chemical fertilizers is the use of either organic fertilizers or the compost (Rosen and Bierman 2005). These are not only supplementing plant nutrients but also improving soil's physical health through improving soil structure that enhances soil's nutrient and water-holding capacities. It is sensible to choose the right source of materials to improve soil fertility that provide sufficient nutrients required by the plants, improve soil structure and at the same time save the environment. In order to achieve this, we have used lombrica in composting process and to improve soil fertility for better crop production.

50.1.1 Composting and Beneficial Uses

Compost is a valuable soil amendment. The use of compost is an effective way to improve plant growth. Composting is an aerobic biological process that converts organic waste into a stable organic product that can be used on-site or transported off-site for use. Composting reduces the volume of water and kills pathogens while preserving more of the nutrients for use by crops (Rosen and Bierman 2005). The composted material improves soil fertility, tilth (tilled earth), and water-hold capacity.

Compost can be used for bioremediation of soil and pollution prevention (USEPA 1997a), reduce erosion and nutrient runoff (USEPA 1997b), alleviate soil compaction, help control disease and pest infestation in plants and animals (USEPA 1997c) and help wetland and habitat restoration (USEPA 1997d). These beneficial uses of compost can increase healthy plant production, help save money, reduce the use of chemical fertilizers and conserve natural resources. In the poultry industry, composting also has become a cost-effective method of mortality management. It destroys disease organisms and creates a nutrient-rich product that can be used or sold.

In nature's laboratory, there are a number of organisms (micro and macro) that have the ability to convert organic waste into valuable resources containing plant nutrients and organic matter, which are critical for maintaining soil productivity. Microorganisms and earthworms are important biological organisms helping nature to maintain nutrient flows from one system to another and also minimize environmental degradation. The earthworm population is about 8–10 times higher in uncultivated area. This clearly indicates that earthworm population decreases with soil degradation and thus can be used as a sensitive indicator of soil degradation (Wani 2002).

Vermicomposting is a simple biotechnological process of composting, in which certain species of earthworms are used to enhance the process of waste conversion and produce a better end product. Vermicomposting differs from composting in several ways (Gandhi et al. 1997). It is a mesophilic process, utilizing microorganisms and earthworms that are active at 10–32°C (not ambient temperature but temperature within the pile of moist organic material). The process is faster than composting; because the material passes through the earthworm gut, a significant but not yet fully understood transformation takes place, whereby the resulting earthworm castings (worm manure) are rich in microbial activity and plant growth regulators and fortified with pest repellence attributes as well. In short, earthworms, through a type of biological alchemy, are capable of transforming garbage into 'gold' (Vermi Co. 2001; Tara Crescent 2003).

50.2 Materials and Methods

50.2.1 Study Area

In order to complete this study, we have selected TATA centre (Tanora Andrin'ny Tontolo Ambanivohitra). The TATA centre is specialized in exploiting the use of lombrica worms. It also provides the material to peasants (individual or collective) in collaboration with the BIMTT (Birao Ifandraisan'ny Mpampiofana eo amin'ny

Table 50.1 Characteristics of epiges

Number	Kind	Length (cm)	Aspect	Activities (role)
01	'Menabota' Fisepia Foetida	3–3.5	Scarlet red, but back is stressed. It lives in the domestic garbage and excrement of cow or horse	To digest
02	'Menalava' (hybrid) with menabota and njila fisaka	4.0	Colour is red. It lives in domestic garbage and excrement of cow and horse	To decompose
03	Njila fisaka	20–23	Colour is red. It lives in domestic garbage and excrement of cow and horse	To digest

Tontolon'ny Tantsaha). This centre is situated in the Fokontany of Antanetibe, district of Ambohimanambola, region of Analamanga. The centre has started working on lombrica in 1997, and initially it has used only 100 lombrica. Prior to starting this work, the team decided to visit the centre to meet the relevant scientists and to be aware of the lombrica technology in producing biofertilizers/composting.

In addition to visiting the centre, in parallel we have visited five farmers who were using the lombrica technology at their farms. The objective of this visit was to get farmers' views about the impact of the lombrica fertilizer technology on their farms. The team worked hand in hand with the scientists at the centre for 1 month. This way we were able to develop collaboration with the scientists and the peasants in the area. The work at the centre was also useful in learning more information about the lombrica production technology and the neighbourhood of the centre.

50.2.2 The Lombrica

Lombrica is useful in the fertilizer production. There are two types of lombrica: the one living inside the soil is called 'endoges', and the one living outside or on the soil is called 'epiges' (Table 50.1; Fig. 50.1). The latter is used to make compost or fertilizer. The lombrica has specific characteristics and nature. It transforms atmospheric CO_2 for plants; adds 10–90 tonnes per hectare (t ha^{-1}) good quality humus including nutrients such as nitrogen (N), phosphorous (P), calcium (Ca) and magnesium (Mg); strengthens the plants; improves soil fertility and structure; and protects environment from contamination/pollution (Lombricompost 2011).

Moreover, the residue of epiges contains group of 'streptomycetes' and 'actinomycetes' kinds of antibiotics that protect the plants against diseases.

The 'lombrica' needs food to survive, and it uses 'precompost' material. The precompost is composed of domestic garbage including residues of unused vegetables, fruit, fish bones and dead leaves, which is mixed with the excrements of cow

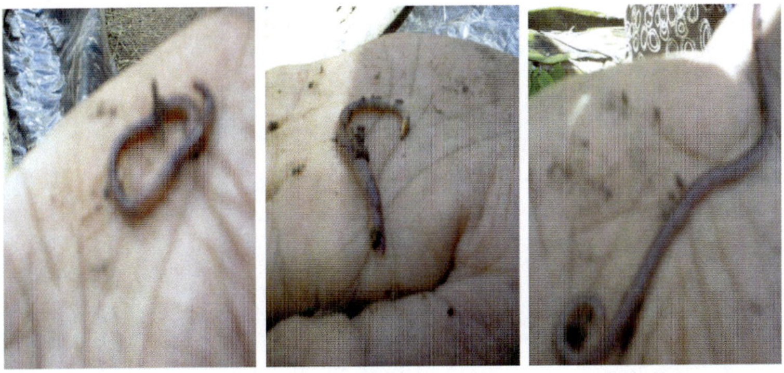

Menabota Menalava Njila fisaka

Fig. 50.1 Three types of epiges 'lombrica'

Fig. 50.2 A structure for composting with lombrica

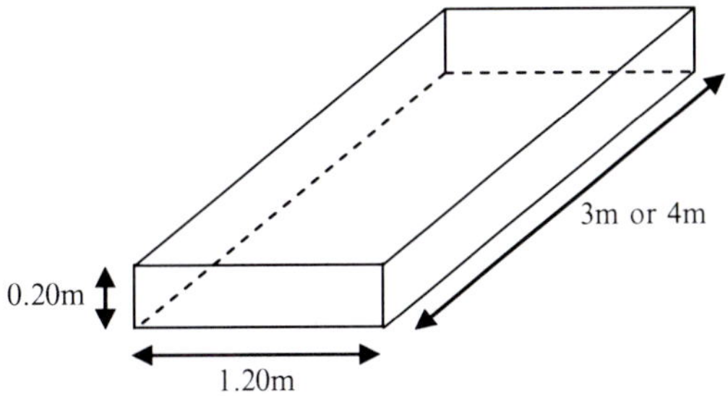

or horse (dung). This precompost material should be left to ferment for at least 10 days before lombrica is introduced. The lombrica should be added in the precompost material in quantities such as 10–30% adult lombrica and 70–75% medium and little lombrica.

In order to achieve composting through using lombrica, a concrete structure (Fig. 50.2) or solid base soil is to be prepared. Its dimension should be at least 3–4 m × 1.2 m × 0.2 m = (0.72–0.96 m^3). This is an ideal place where composting can be archived; its size can be increased slightly, keeping in mind to give a conducive environment for composting the material.

For the size of 0.73–0.96-m^3 structure (Fig. 50.2), about 1,500–3,000 lombrica can be accommodated. The rate of composting depends on the number of lombrica, the nature of the precompost material, conditions during composting (moisture, temperature etc.) and the nature of food for lombrica. The optimum temperature where lombrica can survive and work perfectly is around 26°C; it cannot survive very high or very low temperatures. So to maintain this temperature, the structure has to be modified using a kind of roof to protect them. For reproduction, they need a conducive environment. The structure should be constructed in a way that there are six pits: two for the precompost and four for the 'lombrica'.

The present study was completed using an integrated participatory strategy including socio-economic, biological and ecological approaches. The data was collected through reviewing already published literature, interviewing the farmers and synthesizing the information.

50.3 Results and Discussion

50.3.1 Fertilizer Production

The first product after composting and with the action of lombrica is the composted material which can be called as biofertilizer, organic fertilizer or compost. The final product is like humus rich in N, P, Ca and Mg. We have observed that the TATA centre weekly produces approximately 400 kg biofertilizer or compost material. The composition of the lombrica compost depends on the need of the soil.

50.3.2 Lombrica Multiplication and Generating Income

The reproduction of lombrica is faster. They increase in number over a period of time. Lombrica lays one or two eggs per week. In the third week, every egg gives 2–20 little lombrica; however, experience has shown that only four can reach to the adult age.

In order to produce beneficial fertilizer, the compost must be separated from the lombrica. TATA center sells biofertilizer to the farming communities in different localities of Madagascar: Mananjary (Côte Est de Madagascar, Ambohidratrimo, Ambatofotsy), and to those farmers living close by the centre. The cost of 1 kg of lombrica fertilizer (Fig. 50.3) is ariary (Ar) 1,000 (0.8 US$).

Every month, the centre earns Ar 16,000 (865 US$). In addition to lombrica fertilizer, the centre also sells lombrica to the farmers to allow them local production of fertilizer at their farms. The price of 1 kg of lombrica is about Ar 200,000 (for the lombrica only); however, the cost of 1 kg of lombrica with compost is Ar 40,000. Table 50.2 shows how the centre earns money by selling the lombrica and the lombrica fertilizer. So the exploitation of lombrica involves income-generating activities.

50.3.3 Plant Beneficial Impacts

Many farmers are already using lombrica fertilizer for various purposes (Fig. 50.4). The plant requirement for lombrica is given in Table 50.3. They are of the opinion that the fertilizer is very beneficial for their crops. They have found that with the use of lombrica fertilizer, their plants grow very well, and there is no negative impact on the soil characteristics. The government instructions are to cultivate many kinds of

Fig. 50.3 Lombrica fertilizer (**a**) and tree growth in soil where lombrica was applied (**b**)

Table 50.2 Earning of centre per month

	Product (kg)	Price per kg in [a]Ar	Total price per week in Ar	Total price per month in Ar
Fertilizer	400	1,000	400,000	1,600,000
Lombrica only	4	200,000	800,000	3,200,000
			Total earning	4,800,000

[a]1 US$ = 1,850 ariary; it changes based on the exchange rate

crops including vegetables, fruit trees and rice. In any case, these farmers have control about use of lombrica fertilizer.

If the farmers follow the instructions of the extension staff carefully, they can gain high benefits through enhanced crop production. It has been found that each year the yield of tomatoes is increasing. At his new parcel of small plot, the farmer produced tomatoes in year 2009 (950 kg), 2010 (1,000 kg) and 2011 (1,025 kg).

50.3.4 Environmental Protection

The lombrica transforms the organic matter to mineral matter (mineralization). During this process, some nutrients are released which are taken up by plants, and the organic matter (humus) improves the soil structure and therefore provides conducive environment for plant growth. The recommended dose is 140 t of residue per hectare of land and about 1,000 t of lombrica fertilizer to be ploughed per hectare

Fig. 50.4 The lombrica compost is widely used in plant production

Table 50.3 Plants' requirement for lombrica fertilizer

Plants	Quantity of lombrica fertilizer
Pepiniere	¼ fertilizer, ¾ soil
Fruit trees	1.3 kg $foot^{-2}$
Garden (vegetable) and rice	50 g to 250 g m^{-2}
Other culture	50 g to 250 g m^{-2}
Fertilizer in liquid	1 kg of lombrica for 9 l of water

per year or about 10–90 t of humus to be applied per hectare. The addition improves the agronomic quality of soil and it improves soil fertility. The addition of lombricompost has similar effect on soil fertility improvement. This way can avoid using chemical fertilizers which most of the time causes soil and ground water contamination, for example, nitrate pollution of ground water is very common.

The exploitation of lombrica ameliorates the soil and protects the environment. The lombrica decomposes the garbage and waste, converts into mineral matter and releases nutrients beneficial for plant growth. Moreover, the lombrica fertilizer has many positive impacts for plant growth and in improving quality of arable soils. In Malagasy the farmers have used the lombrica fertilizer in fruit trees, vegetables, rice, manioc, spud, corn, beans and flowers and found that the growth was significantly improved. In order to have maximum benefits, it is essential that the farmers follow the instructions carefully. They should select the waste or garbage based on what nutrients are deficient in soil and what garbage contains these nutrients. It is also important to analyse soil samples from representative sites to know the fertility status of the soils where crops are to be grown.

Regular meetings of the farmers to share the information through discussion about the performance of lombricompost and earning from their uses are essential. This way the success stories can be shared with farmers who are not using this product. This helps to understand the impact of the use of lombricompost on the socio-economic conditions of the farmers. In order to achieve this, it is important that the farmers are properly trained on the production process of lombricompost and its proper use to get more benefits.

50.4 Conclusions

It is concluded that in order to increase local food production, it is essential to find new soils which have the potential for agriculture activities. These soils are to be properly managed and fertility status improved by using lombricompost to avoid soil and ground water contamination from chemical fertilizers. This way the farmers will be highly benefitted through soil health improvement which consequently increases plant growth and farmers' income. It is recommended that the farmers should get training on the production of lombricompost and proper use for sustainable agriculture production.

References

African Development Fund Report (1994) Project completion report. Tsiribihina irrigated agricultural dDevelopment project. Mananbolo Phase I, Madagascar, p 91

Gandhi M, Sangwan V, Kapoor KK, Dilbaghi N (1997) Composting of household wastes with and without earthworms. Environ Ecol 15(2):432–434

Lombricompost (2011) Il faut nourrir le sol pour nourrir la plante, p 12. www.bimtt.mg/DocWeb/LombriCompost.pdf. Last accessed 17 Nov 2011

Randrianarisoa JC, Minten B (2001) Agricultural production, agricultural land and rural poverty in Madagascar. Cornell University, Ithaca, p 46
Rosen CJ, Bierman PM (2005) Using manure and compost as nutrient sources for fruit and vegetable crops. Publication of the Department of Soil, Water, and Climate University of Minnesota. Publication No M1192, University of Minnesota, USA. http://www.extension.umn.edu/distribution/horticulture/M1192.html. Last accessed 17 Nov 2011
Tara Crescent (2003) Vermicomposting. Development Alternatives (DA) Sustainable Livelihoods. http://www.dainet.org/livelihoods/default.htm
USEPA (1997a) Innovative uses of compost: bioremediation and pollution prevention, USEPA publication no. EPA530-F-97-042. USEPA, Washington, DC, p 6
USEPA (1997b) Innovative uses of compost: erosion control, turf remediation, and landscaping, USEPA publication no. EPA530-F-97-043. USEPA, Washington, DC, p 8
USEPA (1997c) Innovative uses of compost: disease control for plants and animals, USEPA publication no. EPA530-F-97-044. USEPA, Washington, DC, p 4
USEPA (1997d) Innovative uses of compost: reforestation, wetlands restoration, and habitat revitalization, USEPA publication no. EPA530-F-97-046. USEPA, Washington, DC, p 4
Vermi Co (2001) Vermicomposting technology for waste management and agriculture: an executive summary. http://www.vermico.com/summary.htm.Vermi Co, PO Box 2334, Grants Pass, OR 97528, USA
Wani SP (2002) Improving the livelihoods: new partnerships for win-win solutions for natural resource management. Paper submitted in the 2nd international agronomy congress held at New Delhi, India during 26–30 Nov 2002

Acronyms and Abbreviations

AACM	Australian Agricultural Consultant Management
AAS	Atomic Absorption Spectrophotometer
ADSIS	Abu Dhabi Soil Information System
ADUPC	Abu Dhabi Urban Planning Council
AM	Arbuscular Mycorrhiza
ANN	Artificial Neural Network
API	Arial Photo Interpretation
ASS	Acid Sulfate Soils
BMPs	Best Management Practices
CLRs	Common Land Resources
COMLAND	Commission on Land Degradation and Desertification
COP	Conference of Parties
CPRs	Common Property Resources
CSIRO	Commonwealth Scientific and Industrial Research Organization
DBMS	Data Base Management System
DDP	Dryland Development Paradigm
DEM	Digital Elevation Model
DNI	DesertNet International
DSM	Digital Soil Mapping
DTM	Digital Terrain Model
EAD	Environment Agency-Abu Dhabi
EC	Electrical Conductivity
ESP	Exchangeable Sodium Percentage
FAO	Food and Agriculture Organization
GIS	Geographic Information System
GPS	Global Positioning System
GST	Global Scan Technologies
ICARDA	International Center for Agricultural Research in Dry Areas
ICBA	International Center for Biosaline Agriculture

S.A. Shahid et al. (eds.), *Developments in Soil Classification, Land Use Planning and Policy Implications: Innovative Thinking of Soil Inventory for Land Use Planning and Management of Land Resources*, DOI 10.1007/978-94-007-5332-7,

ICP	Inductively Coupled Plasma
IECA	International Erosion Control Association
IRS	Indian Remote Sensing
ISCO	International Soil Conservation Organization
IWMI	International Water Management Institute
KISR	Kuwait Institute for Scientific Research
KOC	Kuwait Oil Company
IUSS	International Union of Soil Sciences
LADA	Land Degradation Assessment in Drylands
LDD	Land Development Department
LIDAR	Light Detection and Ranging
MOEW	Ministry of Environment and Water
NDVI	Normalized Difference Vegetation Index
NRCS	Natural Resources Conservation Service
PAFAFR	Public Authority for Agriculture and Fish Resources
RS	Remote Sensing
SAR	Sodium Adsorption Ratio
SCS	Soil Conservation Service
SIS	Soil Information System
SLDPM	Spatial Land Degradation Process Model
SRTM	Shuttle Radar Topography Mission
TGA	Thermal Gravimetric Analyzer
TM	Thematic Mapper
UAE	United Arab Emirates
UNCCD	United Nations Convention to Combat Desertification
UNDP	United Nations Development Program
UNEP	United Nations Environmental Program
UNFCCC	United Nations Framework Convention on Climate Change
USDA	United States Department of Agriculture
USEPA	United States Environmental Protection Agency
USGS	United States Geological Survey
USLE	Universal Soil Loss Equation
WASWAC	World Association of Soil and Water Conservation
WRB	World Reference Base for Soil Resources
XRD	X Ray Diffraction

Author Index

S.A. Shahid et al. (eds.), *Developments in Soil Classification, Land Use Planning and Policy Implications: Innovative Thinking of Soil Inventory for Land Use Planning and Management of Land Resources*, DOI 10.1007/978-94-007-5332-7,

Subject Index

S.A. Shahid et al. (eds.), *Developments in Soil Classification, Land Use Planning and Policy Implications: Innovative Thinking of Soil Inventory for Land Use Planning and Management of Land Resources*, DOI 10.1007/978-94-007-5332-7,

MIX
Papier aus verantwortungsvollen Quellen
Paper from responsible sources
FSC® C105338

Printed by Books on Demand, Germany